Signal and Linear System Analysis

Signal and Linear System Analysis

Gordon E. Carlson
University of Missouri—Rolla

Houghton Mifflin Company ■ Boston ■ Toronto
Dallas ■ Geneva, Illinois ■ Palo Alto ■ Princeton, New Jersey

Cover image: Robert Delauney, *Rythme sans fin*. Courtesy of the galerie Louis Carré & Cie, Paris.

Sponsoring Editor: Rodger H. Klas
Development Editor: Maria A. Morelli
Project Editor: Maria A. Morelli
Assistant Design Manager: Patricia Mahtani
Senior Production Coordinator: Renée LeVerrier
Manufacturing Coordinator: Holly Schuster
Marketing Manager: Michael Ginley

Printed in the U.S.A.

Library of Congress Catalog Number: 91-71997

ISBN: 0-395-51538-6

ABCDEFGHIJ-D-954321

Contents

5 Time-Domain Analysis of Continuous-Time Systems   188

6 Spectral Analysis of Continuous-Time Systems   203

7 Analysis of Continuous-Time Systems Using the Laplace Transform   245

8 Continuous-Time Filters   306

Preface

Signal and Linear System Analysis is designed for the signals and systems analysis course that engineering students take commonly in their junior year. A standard component of electrical engineering curricula, these analysis concepts are also an integral part of many engineering disciplines.

Although the text is written primarily for the two-semester format, it is easily adapted to the one-semester course featured in some curricula. Both continuous-time and discrete-time signals and systems are covered.

It is assumed that the student has completed courses in college calculus and differential equations, as well as a course in college physics. To appreciate fully some of the examples and applications, many of which are set in the context of electrical engineering, an introductory course in electric circuits that includes the methods of AC circuit analysis is desirable. Nonetheless, without this background, the student is still able to understand the basic analysis tools presented.

General Format

The text is divided into three parts. Part I introduces students to the preliminary concepts of the course. Part II covers analysis of continuous-time signals and systems while Part III covers that of discrete-time signals and systems. Chapters are organized conceptually; each chapter is devoted to a single concept in signal or system analysis (such as the time-function representation of signals, the time-domain analysis of systems, and so on) rather than to individual analysis tools. In addition, the sequence of topics in Part II is parallel to the progression of those in Part III—an organization that offers the instructor flexibility in structuring the course and the student pedagogical benefits.

Pedagogy

The material for this text was developed through many years of teaching signal and linear system analysis to electrical engineering students at the University of Missouri—Rolla. As a result of this experience, I have structured the text to permit the student to see the basis of signal and linear system analysis procedures and to enable him or her to work a relatively wide range of problems in linear systems. The pedagogical features and benefits are listed below:

- **Continuous- and discrete-time analyses treated as separate but similar concepts**   I have chosen to develop the discussions of continuous-time analysis and discrete-time analysis in separate parts to allow the student to learn these distinct concepts with less confusion. Consequently, the parallel progression of chapter topics in each part allows the similarities between the two concepts to be explored effectively. Continuous-time analysis is covered first because knowledge of continuous-time spectra uniquely enhances the understanding of discrete-time signal spectra and discrete Fourier transform (DFT) characteristics, such as aliasing and spectrum leakage.

- **Chapters arranged conceptually**   Arranging chapters conceptually enables the student to perceive the structure of signal and linear system analysis more readily. Furthermore, the student is able to concentrate on well-defined topics and understand their rationale.

- **Foundation for further studies**   The signal and linear system analysis tools developed in this text form the foundation for further studies of systems such as communication systems and control systems. Included are introductory discussions of the characteristics and design of electrical filters, which exemplify use of the analysis tools and provide a useful background for subsequent study of signal processing systems.

- **Early coverage of convolution**   Convolution is explained early in the text to allow the student to master this useful mathematical operation before it needs to be applied. Also, by introducing convolution relatively early, we are able to show its usefulness in signal representation and spectral manipulation *in addition* to system analysis, where it has a major application.

- **Signal analysis precedes system analysis**   Analysis of signals including their spectra precedes that of systems. By doing so, the discussion of signal and signal spectrum concepts is more focused.

Pedagogical apparatus

- **Definitions**   Key definitions are concisely defined and easily identified.

- **Examples**   Approximately 200 well-chosen and thoroughly worked out examples are drawn mainly from an electrical engineering context.

- **Summaries**   A summary at the end of each chapter reinforces the chapter's objectives and summarizes its concepts.

- **Problems**   Nearly 500 end-of-chapter problems drill the material appropriately and allow the student to apply the techniques learned.

- **Illustrations**   Of the nearly 700 illustrations, numerous depictions of signals and spectra are provided.

- **Appendixes**   For convenient reference, appendixes include matrix properties and operations, mathematical tables, transform tables, and a computer program for FFT and IFFT computation.

Organization and Coverage

Part I encompasses Chapters 1 and 2, which introduce preliminary concepts. Chapter 1 discusses basic definitions, general characteristics of signals and systems, and mathematical modeling techniques for both continuous-time and discrete-time signals and systems. Chapter 2 explains the convolution operation for both continuous time and discrete time.

Chapters 3 through 9 are organized in Part II, "Continuous-Time Signals and Systems." Signal time-function representation is covered in Chapter 3. In Chapter 4, characteristics of signal spectra are presented, and for motivation, the signal spectrum concept is initiated with a discussion of sinusoidal signals. Then, the complex exponential Fourier series, the reduction of this series to a trigonometric Fourier series, and the Fourier transform are introduced to provide tools useful in determining signal spectra.

Chapters 5, 6, and 7 present single-input, single-output system analysis. The techniques discussed in these three chapters are time-domain analysis, spectral analysis, and analysis using the Laplace transform, respectively. The first two techniques involve the previously defined convolution operation and spectral concepts. After considering the reasons for using the third technique, the Laplace transform, I develop its discussion and apply the technique to system analysis.

To introduce continuous-time filtering concepts in Chapter 8, I employ the signal and system analysis techniques developed in the previous chapters. Both ideal and practical filters are considered. This material exemplifies well the application of the analysis tools and provides a useful background for later signal processing studies.

Chapter 9 introduces the student to state-space concepts and analysis for continuous-time systems, although it is not designed to explore this topic in detail. Some solution techniques for state equations are presented.

Part III, "Discrete-Time Signals and Systems," encompasses Chapters 10 through 17. Except for Chapter 17, each chapter in Part III has a counterpart in Part II. Because the presentation of Chapters 10 through 16 relies on many of the concepts defined in Part II, redefining concepts such as spectrum and frequency response is not required.

Chapter 17 presents the discrete Fourier transform (DFT) and a fast Fourier transform (FFT) algorithm that is used to compute the DFT rapidly. Introducing these topics in a textbook at this level is important because the DFT and its FFT implementation are often the tools used to perform spectral analysis and system implementation. The DFT is developed in considerable detail with many signal and spectrum illustrations used to highlight signal truncation effects (spectrum leakage); spectrum aliasing effects; the periodic-extension nature of the signal being transformed; the relation between the length of the signal being transformed and the spectrum sample spacing, and so forth. Because the DFT is often used to compute the spectrum samples of a continuous-time signal, the development of the DFT is initiated with the Fourier transform of a continuous-time signal. In this way, the student can understand the nature of the computed spectrum samples.

Solutions Manual

A solutions manual provides detailed solutions to all the problems in the book. It is obtained by contacting Houghton Mifflin Company.

Acknowledgments

The development of a text such as this one requires considerable effort; to produce this effort the author cannot work in a vacuum. I thank my colleagues in the electrical engineering department at the University of Missouri—Rolla for their encouragement and suggestions. In particular, encouragement and support from the department chairman, Dr. Walter J. Gajda, has been extremely helpful. I am grateful to my students for their patience as I tested new methods of presenting signal and linear system analysis. The refinement of these methods resulted in the approach featured in this text.

The following reviewers provided me with many suggestions that have greatly improved the clarity of my presentation. They have my sincere thanks for their conscientious effort and their meaningful and helpful comments.

Anders H. Andersen, *University of Kentucky*
Er-Wei Bai, *University of Iowa*
Jerome R. Breitenbach, *California Polytechnic State University*
Leon W. Couch, *University of Florida*
Dean T. Davis, *Ohio State University*
John A. Fleming, *Texas A&M University*
Gary E. Ford, *University of California, Davis*
Jeffrey W. Gluck, *University of Illinois, Chicago*
John C. Kieffer, *University of Minnesota*
Hua Lee, *University of California, Santa Barbara*
Weiping Li, *Lehigh University, Pennsylvania*
Russell M. Merserseau, *Georgia Institute of Technology*
Thomas W. Moore, *Drexel University, Pennsylvania*
Steven C. Nardone, *Southeastern Massachusetts University*
William D. O'Neill, *University of Illinois, Chicago*
Kevin J. Parker, *University of Rochester, New York*
Karen L. Payton, *Southeastern Massachusetts University*
William A. Porter, *University of Alabama, Huntsville*
Vasant B. Rao, *University of Illinois, Urbana-Champaign*
James R. Rowland, *University of Kansas*
Michael Rudko, *Union College, New York*
John J. Shynk, *University of California, Santa Barbara*
Marvin Siegal, *Michigan State University*
Malur K. Sundareshan, *University of Arizona*
Barry Van Veen, *University of Wisconsin, Madison*
Pravin Varaiya, *University of California, Berkeley*

John F. Wager, *Oregon State University*
John V. Wait, *University of Arizona*
Rao Yarlagadda, *Oklahoma State University*

In addition, the staff at Houghton Mifflin has been very supportive and helpful throughout the production of this text. In particular, I wish to thank Raymond Deveaux, who provided the impetus and expertly guided me through the writing and review stages. Also, I thank Rodger Klas, who succeeded Ray as sponsoring editor and gave me continued support; and Maria Morelli, who skillfully guided me through the production phase.

An important part of developing a manuscript is the typing of that manuscript. Special thanks are due to Tonda Davis, who single-handedly and expertly typed the manuscript and its revisions. She spent many hours at the computer terminal typing the text and producing the many illustrations.

Finally, I want to thank my wife, Kathy, for her patience and support during the writing of this text.

G.E.C.

Using This Text in One-Semester and Two-Semester Courses

Because of its flexible organization, this text can be used in two-semester courses that present continuous-time (ct) and discrete-time (dt) signal and system analysis either sequentially or integrated. It can also be conveniently adapted for use in a one-semester course.

The structure of this text readily lends itself for use in the two-semester course in which continuous-time analysis is treated first semester and discrete-time analysis, second semester. As mentioned previously, each topic in Part II has a counterpart in Part III, and the sequence of topics in both parts is parallel; therefore, instructors who wish to integrate the presentation of continuous-time and discrete-time analysis topics in the two-semester course can choose chapters alternately from Part II and Part III.

In the one-semester course, one possibility is to cover Chapters 1 through 7 and Chapters 10 through 14, with some decrease in the detail of coverage. Once again, it is possible to alternate chapters from Part II and Part III when integrating the presentation of the two concepts.

For the convenience of users who are alternating chapters in Part II and Part III, tabs were designed on the right-hand pages to indicate the parts of the book.

On the following pages, the topic sequences suggested for one- and two-semester courses are perhaps the most obvious, although they are not the only ones possible. Through proper selection of material and depth of coverage, different sets of topics can be emphasized.

Two-Semester Course: Sequential Presentation of ct and dt Analyses

Semester One

Chapter 1	Signals and Systems
Chapter 2	Convolution
Chapter 3	ct Signals
Chapter 4	Spectra of ct Signals
Chapter 5	Time-Domain Analysis (ct systems)
Chapter 6	Spectral Analysis (ct systems)
Chapter 7	Laplace Transform
Chapter 8	ct Filters

Semester Two

Chapter 10	dt Signals
Chapter 11	Spectra of dt Signals
Chapter 12	Time-Domain Analysis (dt systems)
Chapter 13	Spectral Analysis (dt systems)
Chapter 14	z-Transform
Chapter 15	dt Filters
Chapter 17	The DFT

Two-Semester Course: Integrated Presentation of ct and dt Analyses

Semester One

Chapter 1	Signals and Systems
Chapter 2	Convolution
Chapter 3	ct Signals
Chapter 10	dt Signals
Chapter 4	Spectra of ct Signals
Chapter 11	Spectra of dt Signals
Chapter 5	Time-Domain Analysis (ct systems)
Chapter 12	Time-Domain Analysis (dt systems)
Chapter 6	Spectral Analysis (ct systems)
Chapter 13	Spectral Analysis (dt systems)

Semester Two

Chapter 7	Laplace Transform
Chapter 14	z-Transform
Chapter 8	ct Filters
Chapter 15	dt Filters
Chapter 17	The DFT

Note: In regard to the two-semester format, if coverage of state-space concepts is desired, then Chapter 9 can replace Chapter 8 while Chapter 16 can replace Section 15.5.

One-Semester Course: Sequential Presentation of ct and dt Analyses

Sequence		*Coverage*
Chapter 1	Signals and Systems	
Chapter 2	Convolution	
Chapter 3	ct Signals	*Omit Sec. 3.6.*
Chapter 4	Spectra of ct Signals	*Omit Fourier transform derivation and power-density spectrum.*
Chapter 5	Time-Domain Analysis (ct systems)	*Omit systems with nonzero initial conditions.*
Chapter 6	Spectral Analysis (ct systems)	*Omit Sec. 6.4 and 6.5.*
Chapter 7	Laplace Transform	*Omit Sec. 7.5.*
Chapter 10	dt Signals	
Chapter 11	Spectra of dt Signals	*Omit energy-density spectrum and power-density spectrum.*
Chapter 12	Time-Domain Analysis (dt systems)	*Omit systems with nonzero initial conditions and Sec. 12.3.*
Chapter 13	Spectral Analysis (dt systems)	
Chapter 14	z-Transform	

Note: In the one-semester course, omitting the generalized Fourier series and the Fourier transform derivation means that the Fourier series and Fourier transform are merely defined rather than additionally supported by their derivations. The instructor will need to include motivational material from the derivation sections when presenting solely the definitions.

See the next page for suggestions for the integrated presentation of ct and dt analyses in a one-semester course.

One-Semester Course: Integrated Presentation of ct and dt Analyses

Sequence		*Coverage*
Chapter 1	Signals and Systems	
Chapter 2	Convolution	
Chapter 3	ct Signals	*Omit Sec. 3.6.*
Chapter 10	dt Signals	
Chapter 4	Spectra of ct Signals	*Omit Fourier transform derivation and power-density spectrum.*
Chapter 11	Spectra of dt Signals	*Omit energy-density spectrum and power-density spectrum.*
Chapter 5	Time-Domain Analysis (ct systems)	*Omit systems with nonzero initial conditions.*
Chapter 12	Time-Domain Analysis (dt systems)	*Omit systems with nonzero initial conditions and Sec. 12.3.*
Chapter 6	Spectral Analysis (ct systems)	*Omit Sec. 6.4 and 6.5.*
Chapter 13	Spectral Analysis (dt systems)	
Chapter 7	Laplace Transform	*Omit Sec. 7.5.*
Chapter 14	z-Transform	

Signal and Linear System Analysis

I

Preliminary Concepts

1 Signal and System Characteristics and Models

This text is devoted to the development of techniques that are useful in analyzing signals and systems. The definitions; general concepts; and characteristics for signals, systems, and their mathematical models presented in this chapter provide the necessary background and overview of signal and system concepts and characteristics needed as a framework for the analysis techniques developed in subsequent chapters.

1.1 Definitions and General Concepts

At the outset, we must indicate the meaning of the terms *signal* and *system*. Also, we must define and discuss the concept of a mathematical model of a signal or a system because such models are used in signal and system analysis. We will also address the distinction between continuous-time and discrete-time signals and systems in this section.

Signals and Systems

The terms *signal* and *system* are used broadly in reference to problems in diverse fields such as engineering, science, economics, health care, and politics, to name a few. Three examples serve to illustrate an application of signals and systems—that of analysis—with which this text is concerned. As one example, an electrical engineer may be analyzing how an electric circuit operates. In this case, the electrical circuit is the system; the resistors, capacitors, inductors, transistors, and so forth that make up the circuit are the system components; and the voltages and currents in the circuit are the signals. As a second example, a mechanical engineer may be analyzing the attachment of a wheel to an automobile; that is, the wheel suspension. Here, the wheel suspension is the system; the wheel, tire, spring, shock absorber, and so forth are the system components; and the position and velocity of the various system components are the signals. Finally, as a third (nonengineering) example, consider the government economist who is analyzing the economic performance of the United States. In this context, the economic system is made up of system components, such as the available labor force, the available plant capacity, and the available capital. Signals include money supply, prime interest rate, government deficit, and gross national product.

Signals that enter a system from some external source are referred to as *input signals*. These include voltage sources and current sources for the electric circuit, the vertical displacement of the road surface for the wheel suspension, and the money supply, prime interest rate, and government deficit for the economic system. Signals produced by the system in its response to the input signals are called the *output signals*. They are some or all of the voltages and currents in the electric circuit, the vertical wheel displacement for the wheel suspension, and the gross national product for the economic system. Signals that occur within a system and therefore are neither input nor output signals are called *internal signals*.

Specifically, a system is an interacting group of physical objects or physical conditions that are called system components. The system responds to one or more input quantities, called input signals, to produce one or more output quantities, called output signals. Quantities present within the system are called internal signals. The quantities, or signals, are functions of an independent variable such as time, distance, or force.

Most systems are quite complex. They may contain system components and signals of different types. For example, an audio amplifier system contains microphones that convert acoustic signals to electrical signals, amplifiers that condition and amplify the electrical signals, and speakers that convert electrical signals to acoustic signals. Because this text is primarily for electrical engineering students, we will focus on electrical systems when exemplifying the analysis techniques developed; however, it should be remembered that these techniques are equally applicable to other types of systems.

Although signals and system components may be invariant with respect to parameters of interest, such as time or location, most signals are not invariant. In particular, time-varying signals are very common and will be considered throughout this text. System components may also vary with time, but only very slowly in the case of many systems. Thus, most often they can be considered as constant over the period for which the system is being considered.

Mathematical Models

The ability to analyze signals and systems enables us to determine signal characteristics and/or system performance. A characteristic of a signal might be its time waveform; the performance of a system might be given by the output signal power as a function of the input signal power. Many times the signal and system analysis must be done before the signals or systems are actually constructed and available for testing. It is desirable that the signal and system analysis be performed by mathematical means so that quantitative characteristic or performance results can be obtained. Thus, mathematical models, which are mathematical equations representing signals and systems, must to be generated. For example, the signal model for a household voltage is given by the equation

$$v(t) = 115\sqrt{2}\cos[2\pi(60)t] \qquad -\infty < t < \infty \qquad (1.1)$$

As an example of a system model, a resistor can be modeled by the equation

$$v(t) = Ri(t) \qquad (1.2)$$

where $v(t)$ is the voltage across the resistor, and $i(t)$ is the current through the resistor.

Mathematical models permit quantitative design of signals or systems to satisfy given requirements. Design is accomplished by varying signal or system parameter values and determining the effects on system characteristics or performance so that the best choice of parameters can be made. In many cases, signals are given and the analysis or design is for a system. For example, an engineer may want to design a filter for a radio signal to remove unwanted portions of input signals, such as atmospheric noise. In other cases, signals may be changed to provide optimum overall signal and system performance. In radio systems, for example, the input signal is changed by amplitude modulation or frequency modulation so that it can be efficiently radiated. Where signals may be changed, the engineer uses system analyses to determine the relative merits of various signal types and the optimum choice of signal type for a specific case.

It is not practical to produce an exact mathematical model for a physical signal or system—too many variable parameters are involved. For example, the household voltage model of eq. (1.1) is not an exact mathematical model for several reasons. First, if you look carefully at the physical household voltage signal, you will see that it is not exactly sinusoidal in shape and that it also contains some small variations due to noise. Moreover, it has not existed since $t = -\infty$, nor will it continue to exist until $t = +\infty$. Likewise, the constant-resistance mathematical model for a resistor given by eq. (1.2) is not an exact model because the resistance R of the resistor is actually a function of its temperature, which depends on the atmospheric temperature, the current through the resistor, and heat-transfer characteristics. The resistance also changes slowly with time, and the random motion of electrons in a resistor generates a noise voltage across its terminals. These factors and many others would all need to be included in a detailed mathematical model of a resistor.

The objective is to make the mathematical model as simple as possible for ease of analysis but still retain those parameters that significantly affect characteristics or performance being measured. Therefore, we use simplified mathematical models such as eqs. (1.1) and (1.2) in signal and system analysis: they will produce reliable results if the signal and system parameters that were ignored in generating the mathematical models produce negligible effects on signal characterizations and performance measures that we use.

Many times, all the characteristics of the input signal are not known for a system of interest. Instead, only the general class of possible input signals for the system in question is known. In a communication system, for example, we do not know the specific signal to be sent, for there would be no need for us to send it if it were known. In this case, system analysis and design can sometimes be done on the basis of such general signal characteristics as maximum signal values, signal bandwidths, or maximum signal power. In other cases, probabilistic signal and system models and analyses must be used to determine average system performance.

Signal and system models serve as examples in this text, but the development of models for specific physical signals and system components is not considered. Our intent is to concentrate on the important first step of mastering analysis techniques. Although signal- and system-modeling equations will be referred to simply as signals or systems throughout the remainder of the text, the reader is cautioned to bear in mind that the equations are not the signals and systems, but rather models of them.

Continuous-Time and Discrete-Time Signals and Systems

One of the ways by which signals and systems may be categorized is according to whether they are *continuous time* or *discrete time*. We use separate but very similar analysis tools for these two categories of signals and systems. Following Part I on preliminary concepts, the text is divided into two parts: Parts II centers on developing analysis tools for continuous-time signals and systems; Part III, on doing the same for discrete-time signals and systems.

A continuous-time signal is denoted by a function of time such as $x(t)$. As shown in Figure 1.1a, the waveform of a continuous-time signal does not have to be a continuous function of time. It merely needs to be defined for all points

Figure 1.1
Example of
Continuous-time and
Discrete-time Signals

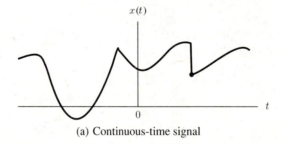

(a) Continuous-time signal

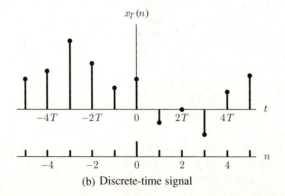

(b) Discrete-time signal

in time. At the discontinuity shown in the example, the defined waveform value is indicated by a dot.[†]

A continuous-time system accepts only continuous-time input signals. It produces continuous-time internal and output signals.

Definition _____

A continuous-time signal has a value defined for each point in time, and a continuous-time system operates on and produces continuous-time signals.

As illustrated in Figure 1.1b, a discrete-time signal is defined only at discrete points in time. The signal can have any value at each of the discrete points in time. A discrete-time system accepts only discrete-time signals and produces discrete-time internal and output signals.

Definition _____

A discrete-time signal has a value defined only at discrete points in time and a discrete-time system operates on and produces discrete-time signals.

The discrete-time signal shown in Figure 1.1b is a sequence of values defined at points in time that are equally spaced by T seconds. In most cases, the sequence values are called *samples* for simplicity, and the space between them is called the *sample spacing*. (Although it is not necessary for discrete-time signal samples to be equally spaced in time, unequally spaced samples are more difficult to analyze. Such samples are not considered in this text.) When the sample spacing is equal, it is convenient to express the sequence of values as a function of the signed integer n. In the function $x(n)$, for example, n is an integer indicating the sample number as counted from a chosen time origin. Note that $n = 0$ corresponds to the time origin and that negative values of n correspond to negative time. We refer to the function of n as a *sequence of samples*, or *sequence* for short. The complete specification of a discrete-time signal consists of its defining sequence, $x(n)$, and the sample spacing, T. Thus, we use the notations $x_T(n)$ and $[x(n), \quad T]$ for a discrete-time signal—the first to represent the signal in general and the second to represent it specifically by its particular sequence and sample spacing.

If the time separation between the first nonzero sample and the last nonzero sample of a discrete-time signal is finite, then its corresponding sequence consists of a finite number of nonzero samples. It is sometimes convenient to

_____

[†]In most cases, it is not critical whether the defined waveform value is its upper or lower value, or some value in between, at a discontinuity. Therefore, after we have completed this section, we will omit the dot notation except when it is necessary or enhances waveform interpretation.

represent a short sequence by listing its values in order from the first nonzero sample to the last nonzero sample using the notation

$$x(n) = \{x_1, x_2, x_3, x_4, x_5\} \qquad n_1 = N \qquad (1.3)$$

where x_i is the ith sample value in the list and the integer value N specified for the index n_1 indicates the value of n corresponding to the first sample listed. The value of n_1 is not written if it is zero. That is, a sequence with no value specified for n_1 is a sequence in which the first nonzero value occurs at $n = 0$. Two examples of the sequence notation given by eq. (1.3) are shown in Figure 1.2.

A sequence $x(n)$ represents different discrete-time signals $x_T(n)$ when T is different, the difference being a scale factor along the time axis. Thus, a sequence can be thought of as a time-normalized version of the corresponding discrete-time signal. The time-normalized discrete-time signal (that is, sequence) is often used in analysis of discrete-time signals and systems for simplicity and generality. Appropriate scaling of results to reflect the actual sample spacing is required to obtain results for specific cases.

Figure 1.2
Examples Using
Finite-Sequence
Notation

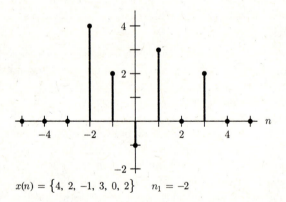

$$x(n) = \{4, 2, -1, 3, 0, 2\} \qquad n_1 = -2$$

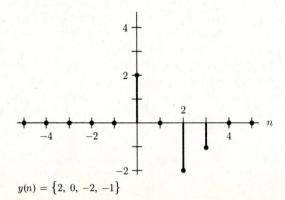

$$y(n) = \{2, 0, -2, -1\}$$

When sketching pictures of discrete-time signals, or sequences corresponding to discrete-time signals, we indicate the values by dots. Lines are drawn from the dots to the time axis to indicate the discrete-time values more clearly.

A signal may take on only a set of values that is countable. Two examples are signal values limited to $\pm 3i^2$, where the integer i is one of the counting numbers, and signal values limited to $-0.5, 0, 0.5$, and 1. We refer to a signal of this type as a *quantized signal*. Examples of a continuous-time and a discrete-time quantized signal are shown in Figure 1.3. A signal that is both discrete-time and quantized is referred to as a *digital signal*. Digital signals are a special case of discrete-time signals. They are not considered separately in this text.

We have specified time as the independent variable for signals and systems and the concepts of continuous-time and discrete-time for purposes of categorization. The same concepts would apply for other independent variables (time, distance, force, temperature, number of items, and the like). Thus, the concepts of continuous-time and discrete-time signals and systems and their analysis are considerably more general than their names indicate and have much wider applicability. For example, assume that you are given the height $h(x)$ of the road surface above sea level from your home to your friend's home as a function of the horizontal distance x along the road. You want to determine the slope $s(x)$

Figure 1.3
Examples of
Quantized Signals

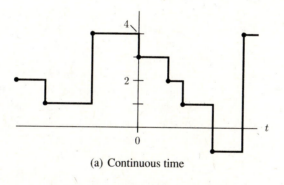

(a) Continuous time

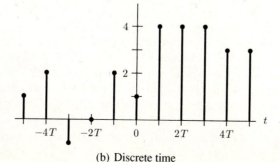

(b) Discrete time

of this road as a function of x. In signal and system terminology, you will compute the continuous-distance output signal $s(x)$ from the continuous-distance input signal $h(x)$ by using the continuous-distance system characterized by the equation $s(x) = dh(x)/dx$.

Throughout this text, we present signal and signal analysis concepts prior to analysis concepts for systems since signal and signal characteristics are required for system analysis. For this reason, we will describe in the next section general signal characteristics before we discuss general system characteristics. In Part II, we discuss continuous-time signals prior to continuous-time systems. We follow the same procedure in Part III for discrete-time signals and systems. This organization provides focused discussions of signals and signal analysis concepts and of systems and system analysis concepts.

1.2 General Signal Characteristics

Continuous-time and discrete-time signals can be further categorized in terms of various characteristics they possess. In this section, we will examine several frequently used signal characteristics.

Deterministic and Random Signals

A *deterministic signal* behaves in a fixed known way with respect to time. Thus, as illustrated in Figure 1.4, a deterministic signal can be modeled by a known function of time, t, for continuous-time signals, or a function of a sample number, n, for discrete-time signals. For simplicity, if no range of values for t or n is shown, then the function is valid for all values.

On the other hand, a *random signal* takes on one of several possible values at each time for which a signal value is defined. Because there is a probability of occurrence associated with each of the resulting time functions or sequence of signal values that can occur, random signals require probabilistic models; hence, such signals will not be considered in this text.

Piecewise-Defined and Simply Defined Signals

The deterministic signals shown in Figure 1.4 are represented by a single equation for all time. Some very useful signals cannot be represented by a single equation for all time but can be represented by a set of equations, with each equation valid over a defined time span. Examples of these *piecewise-defined signals*, as they are called, are shown in Figure 1.5. We will refer to signals represented by a single equation for all time as *simply defined signals* when we need to distinguish them from piecewise-defined signals.

A few signals that can be defined by a single equation are more conveniently analyzed if a piecewise definition is employed. A continuous-time example of such a signal is $x(t) = e^{-|t|}$, $-\infty \le t \le \infty$. The single equation model of

Figure 1.4
Examples of
Deterministic Signals

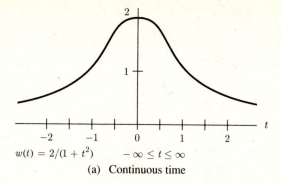

$w(t) = 2/(1 + t^2)$ $-\infty \leq t \leq \infty$

(a) Continuous time

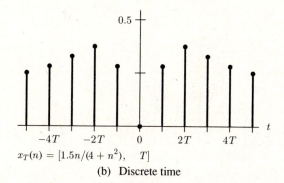

$x_T(n) = [1.5n/(4 + n^2),\quad T]$

(b) Discrete time

this signal is a compact method for describing it. However, when operations such as integration are to be performed on the signal, then the signal must be broken into two parts; hence, the piecewise-defined model

$$x(t) = \begin{cases} e^t & t \leq 0 \\ e^{-t} & t > 0 \end{cases}$$

is more convenient.

Periodic and Aperiodic Signals

A *periodic signal* has a waveform that repeats over and over, with the time between repeats defined as the period of the signal. We refer to a signal which is not periodic as an *aperiodic signal*.

Definition _____

The continuous-time signal $x(t)$ is periodic if and only if

$$x(t + T_0) = x(t) \qquad \text{for all } t \tag{1.4}$$

where T_0 is positive and is the period of the signal measured in units of time.

Figure 1.5
Examples of
Piecewise-Defined
Signals

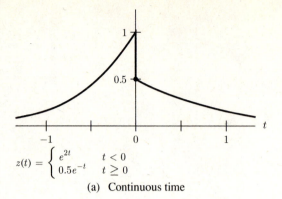

$$z(t) = \begin{cases} e^{2t} & t < 0 \\ 0.5e^{-t} & t \geq 0 \end{cases}$$

(a) Continuous time

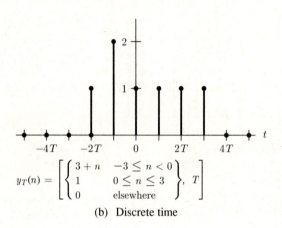

$$y_T(n) = \left[\begin{cases} 3 + n & -3 \leq n < 0 \\ 1 & 0 \leq n \leq 3 \\ 0 & \text{elsewhere} \end{cases} \right\}, \; T \right]$$

(b) Discrete time

Definition

The discrete-time signal $x_T(n)$ is periodic if and only if

$$x_T(n + N_0) = x_T(n) \qquad \text{for all } n \tag{1.5}$$

where N_0 is positive and is the period of the signal measured in units of number of sample spacings.

The smallest value of T_0 or N_0 for which the definitions hold is called the *fundamental period* of the periodic signal.

In the general case, a periodic continuous-time signal can be expressed as

$$x(t) = \sum_{i=-\infty}^{\infty} x_p(t - iT_0) \tag{1.6}$$

where, for an arbitrary time t_1,

$$x_p(t) = \begin{cases} x(t) & t_1 \le t < t_1 + T_0 \\ 0 & \text{elsewhere} \end{cases} \tag{1.7}$$

For a discrete-time signal,

$$y_T(n) = \left[\sum_{i=-\infty}^{\infty} y_p(n - iN_0), \quad T \right] \tag{1.8}$$

where, for an arbitrary sample index n_1,

$$y_p(n) = \begin{cases} y(n) & n_1 \le n < n_1 + N_0 \\ 0 & \text{elsewhere} \end{cases} \tag{1.9}$$

Examples of such periodic signals are shown in Figure 1.6. Note that three dots are shown at each end of the portion of the periodic signal illustrated. This notation signifies that the signal is periodic; therefore, it repeats over the range of all t or n.

Figure 1.6
Examples of Periodic
Signals

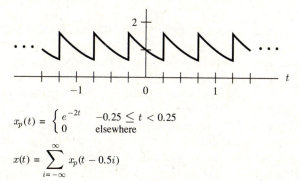

$$x_p(t) = \begin{cases} e^{-2t} & -0.25 \le t < 0.25 \\ 0 & \text{elsewhere} \end{cases}$$

$$x(t) = \sum_{i=-\infty}^{\infty} x_p(t - 0.5i)$$

(a) Continuous time

$$y_p(n) = \begin{cases} -n & -2 \le n < 0 \\ n & 0 \le n < 2 \\ 0 & \text{elsewhere} \end{cases}$$

$$y_T(n) = \left[\sum_{i=-\infty}^{\infty} y_p(n - 4i), \quad 0.1 \text{ s} \right]$$

(b) Discrete time

A few periodic signals can be modeled simply by single equations that are valid for all time. Important examples are the continuous-time sinusoidal signal,

$$x(t) = A \cos\left(\frac{2\pi t}{T_0} - \theta\right)$$

and the discrete-time sinusoidal signal,

$$z_T(n) = \left[C \sin\left(\frac{2\pi n}{N_0} - \beta\right), \quad T\right]$$

These examples will be discussed in more detail in Chapters 3 and 10.

The sum of N periodic continuous-time signals is not necessarily periodic. It is periodic with period T_0 if and only if the condition

$$\frac{T_0}{T_{0i}} = n_i \qquad 1 \le i \le N \tag{1.10}$$

is satisfied, where T_{0i} is the period of the ith signal in the sum and n_i is an integer. Simultaneous solution of eq. (1.8) for the first signal in the sum paired with each additional signal in the sum gives

$$\frac{T_{01}}{T_{0i}} = \frac{n_i}{n_1} \qquad 2 \le i \le n \tag{1.11}$$

Thus, an equivalent necessary and sufficient condition for a sum of periodic continuous-time signals to be periodic is that all the ratios of the period of the first signal in the sum to the period of another signal in the sum be rational; that is, ratios of integers.

The following steps can be taken to determine the period of the sum signal, if it is periodic. First convert each ratio from eq. (1.11) to a ratio of integers. If this conversion is not possible, the sum signal is not periodic. Then factor the greatest common divisor (g.c.d.) from the numerator and denominator of each individual ratio. The least common multiple (l.c.m.) of the denominators of the resulting ratios is the value of n_1. The period of the sum signal is computed as

$$T_0 = T_{01} n_1 \tag{1.12}$$

The preceding development is also applicable to the sum of discrete-time periodic signals if T_0 is replaced by N_0 and each T_{0i} is replaced by N_{0i}. Note that a sum of discrete-time periodic signals is always periodic because the period ratios N_{01}/N_{0i} are always rational.

Example 1.1

Determine whether the continuous-time signals $x(t)$, $y(t)$, and the sequence $z(n)$ are periodic. Compute the period for any that are periodic. For $x(t)$,

$$x(t) = x_1(t) + x_2(t) + x_3(t)$$

where $x_1(t)$, $x_2(t)$, and $x_3(t)$, have periods of 8/3, 1.26, and $\sqrt{2}$ s respectively.

For $y(t)$,

$$y(t) = y_1(t) + y_2(t) + y_3(t)$$

where $y_1(t)$, $y_2(t)$, and $y_3(t)$, have periods of 1.08, 3.6, and 2.025 s, respectively.
For $z(n)$,

$$z(n) = z_1(n) + z_2(n)$$

where $z_1(n)$ and $z_2(n)$ have periods of 90 and 54, respectively.

Solution
For $x(t)$

$$\frac{T_{01}}{T_{02}} = \frac{(8/3)}{1.26} = \frac{8}{3.78} = \frac{800}{378} = \frac{400}{189} \qquad \text{rational}$$

$$\frac{T_{01}}{T_{03}} = \frac{(8/3)}{\sqrt{2}} = \frac{8}{3\sqrt{2}} \qquad \text{not rational}$$

The signal $x(t)$ is not periodic since T_{01}/T_{03} is not rational.
For $y(t)$,

$$\frac{T_{01}}{T_{02}} = \frac{1.08}{3.6} = \frac{108}{360} = \frac{3}{10} = \frac{3}{(2)(5)} \qquad \text{rational}$$

$$\frac{T_{01}}{T_{03}} = \frac{1.08}{2.025} = \frac{1080}{2025} = \frac{8}{15} = \frac{8}{(3)(5)} \qquad \text{rational}$$

The signal $y(t)$ is periodic since the ratios are rational.

$$n_1 = \text{l.c.m. of denominators} = (2)(3)(5) = 30$$

$$T_0 = T_{01}n_1 = (1.08)(30) = 32.4 \text{ s}$$

The sequence $z(n)$ is periodic since $z_1(n)$ and $z_2(n)$ are periodic.

$$\frac{N_{01}}{N_{02}} = \frac{90}{54} = \frac{5}{3} \qquad \text{rational}$$

$$n_1 = \text{l.c.m. of denominators} = 3$$

$$N_0 = N_{01}n_1 = (90)(3) = 270$$

It should be noted that no physical signals are actually periodic since they all begin at some time and/or cease to exist at some later time. However, as we indicated earlier for the household voltage model given by eq. (1.1), in many cases a repeating signal is on for such a long period of time with respect to the period of analysis and the time constants of the systems being considered that it can be modeled as a periodic signal.

1.3 System Representations and Models

System representations and system mathematical models that are useful in continuous-time and discrete-time system analysis are discussed in this section. The representations provide visual illustrations of system component intercon-nections and, as indicated earlier, the mathematical models permit analysis of system characteristics and performance.

System Representations

Systems may be represented by diagrams that feature symbols representing various components in the system. This convention is often used for continuous-time electrical network systems such as the one represented in Figure 1.7 and referred to as a *circuit diagram*.

For systems in general, we often find it more convenient to use *block-diagram* system representations, such as those illustrated in Figure 1.8. In block dia-grams, operations on signals are shown as boxes, with the operations labeled inside. Lines connecting the various boxes are drawn to indicate paths of sig-nal flow, with arrowheads at box entries indicating signal-flow direction. The particular signal present in a path is indicated beside or at the end of the path. For simplicity, we will refer to input signals and output signals as inputs and outputs throughout this text.

The label O_c inside a box in a continuous-time system representation means that the output of the box, $y(t)$, can be computed from the input to the box, $x(t)$, by

$$y(t) = O_c[x(t)] \tag{1.13}$$

where O_c is a continuous-time operation, such as a derivative, an integral, or a multiplication by a constant. The subscript d is used for discrete-time operations in discrete-time systems. Discrete-time operations include multiplication by a constant or a delay of one or more sample spacings.

We use circles rather than boxes for two common operations: the addi-tion/subtraction operation and the multiplication operation. This convention serves to highlight these operations and simplify the block diagram. The sym-

Figure 1.7
Example of an Electrical Network System Representation

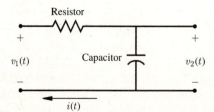

Figure 1.8
Examples of System
Block Diagrams

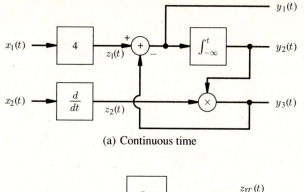

(a) Continuous time

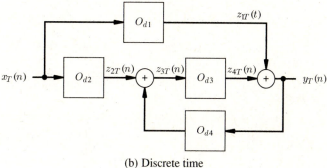

(b) Discrete time

bol placed in the circle is $\times$ for multiplication and $+$ for addition/subtraction. Plus signs on both input lines of an adder/subtractor indicates that addition is being performed. A plus on one input line and a minus on the other indicates that subtraction is being performed. Frequently, the signs on the input lines are included only if subtraction is being performed; that is, input lines without signs indicate addition.

The discrete-time system of Figure 1.8b has only a single input and a single output. Such systems, referred to as *single-input, single-output systems*, are the type we will consider throughout most of this text. Since multiple-input, multiple-output systems are interconnections of many single-input, single-output subsystems, the analysis tools developed for the latter are useful for these more complex systems, too; however, a structured approach employing vector signals and matrix operations is most useful in their analysis. We present one such approach in Chapters 9 and 16.

System Mathematical Models

Mathematical models for systems consist of equations that relate signals of interest. These equations can be obtained from system block diagrams. For example, we can see from the system block diagram of Figure 1.8a that

$$y_1(t) = z_1(t) - y_3(t)$$

$$= 4x_1(t) - z_2(t)y_2(t)$$

$$= 4x_1(t) - \frac{dx_2(t)}{dt} \int_{-\infty}^{t} y_1(\tau)\, d\tau \qquad (1.14)$$

is the integro-differential equation that relates the output signal $y_1(t)$ of the continuous-time system to its input signals when $y_2(t) = 0$ at $t = -\infty$. From Figure 1.8b, we see that

$$y_T(n) = z_{1T}(n) + z_{4T}(n)$$

$$= O_{d1}\left[x_T(n)\right] + O_{d3}\left[z_{3T}(n)\right]$$

$$= O_{d1}\left[x_T(n)\right] + O_{d3}\left\{O_{d2}\left[x_T(n)\right] + O_{d4}\left[y_T(n)\right]\right\} \qquad (1.15)$$

is the equation that relates the output signal of the discrete-time system to its input signal.

Two useful operations for modeling components of discrete-time systems are multiplication by the constant C giving

$$y_T(n) = O_d\left[x_T(n)\right] = Cx_T(n) \qquad (1.16)$$

and the delay of the input signal sequence by one sample (T seconds) giving

$$y_T(n) = O_d\left[x_T(n)\right] = x_T(n-1) \qquad (1.17)$$

The delay operation is implemented in a discrete-time system by a storage register that holds its input value for an interval of time equal to one sample-spacing before releasing it as an output. Thus, it is an energy-storage element. If the four operations in eq. (1.15) are defined to be

$$O_{d1} = O_{d3} = a \qquad (1.18)$$

$$O_{d2} = c \qquad (1.19)$$

$$O_{d4} = \text{one sample delay} \qquad (1.20)$$

then eq. (1.15) becomes

$$y_T(n) = ax_T(n) + a\left[cx_T(n) + y_T(n-1)\right]$$

$$= ay_T(n-1) + (a + ac)x_T(n)$$

$$\equiv ay_T(n-1) + bx_T(n) \qquad (1.21)$$

where $b = a + ac$.

A discrete-time system equation of this type; that is, one containing both current and past sample values of input and output signals is referred to as a *difference equation*. The solution for $y_T(n)$ for $n \geq n_0$ when the input signal is $x_T(n)$ can be found from eq. (1.21) by repeated substitution. We begin by substituting

$$y_T(n-1) = ay_T(n-2) + bx_T(n-1) \qquad (1.22)$$

in eq. (1.21) to give

$$y_T(n) = a^2 y_T(n-2) + b a x_T(n-1) + b x_T(n) \tag{1.23}$$

We then substitute

$$y_T(n-2) = a y_T(n-3) + b x_T(n-2) \tag{1.24}$$

in (1.23) to obtain

$$y_T(n) = a^3 y_T(n-3) + b a^2 x_T(n-2) + b a x_T(n-1) + b x_T(n) \tag{1.25}$$

Continued substitution finally gives the equation

$$y_T(n) = a^{(n-n_0+1)} y_T(n_0-1) + b \sum_{i=n_0}^{n} a^{(n-i)} x_T(i) \qquad n \geq n_0 \tag{1.26}$$

In eq. (1.26), $x_T(i)$ are the discrete-time input signal values for $i \geq n_0$ and $y_T(n_0 - 1)$ is a constant value established by the initial energy stored in the discrete-time system. This constant value must be specified.

Equation (1.26) can also be obtained using classical techniques for solving difference equations that are very similar to classical techniques for solving differential equations. In this text, however, we will focus on alternative techniques that are often much simpler, besides being useful in the analysis of many systems. Examples using classical solution techniques will be shown only for continuous-time systems in the remainder of this chapter.

System equations can also be developed from system-component diagrams if the mathematical models for the system components are specified. In Examples 1.2 and 1.3 for electric circuits, the equations that model the components are $v(t) = Ri(t)$ for a resistor and $i(t) = \frac{1}{L} \int v(t)\, dt$ for an inductor. The values of R and L are placed next to the component.

Example 1.2

The electric circuit shown in Figure 1.9 is constructed in the laboratory to demonstrate various circuit-theory concepts. Find the equation model that relates the output signal, current $y(t)$, to the input signal, voltage $x(t)$.

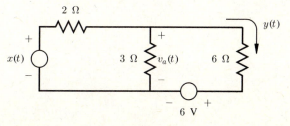

Figure 1.9 Electric Circuit for Example 1.2

Solution

$$\frac{v_a(t) - x(t)}{2} + \frac{v_a(t)}{3} + \frac{v_a(t) - 6}{6} = 0$$

$$6v_a(t) - 3x(t) - 6 = 0$$

But,

$$v_a(t) = 6y(t) + 6$$

Substituting gives

$$36y(t) = 3x(t) - 30$$

$$y(t) = \left(\frac{1}{12}\right) x(t) - \frac{5}{6}$$

Example 1.3

The electric circuit in Figure 1.10 can be used to produce an output signal $y(t)$ that does not contain the dc component present in the input signal $x(t)$. Find the equation that relates the output signal to the input signal for $t \geq t_0$ when the energy stored in the inductor at $t = t_0$ is such that $i(t_0) = I_0$.

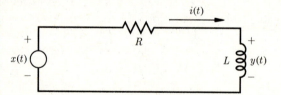

Figure 1.10 Electric Circuit for Example 1.3

Solution

$$x(t) = Ri(t) + y(t)$$

But

$$i(t) = \frac{1}{L} \int_{t_0}^{t} y(\tau) \, d\tau + I_0 \qquad t \geq t_0$$

Substituting gives

$$x(t) = y(t) + RI_0 + \frac{R}{L} \int_{t_0}^{t} y(\tau) \, d\tau \qquad t \geq t_0 \qquad (1.27)$$

Figure 1.11
Block Diagram
Representing the
Electric Circuit of
Example 1.3 for
$t \geq t_0$

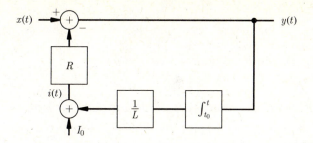

An alternative equation relating $y(t)$ to $x(t)$ for the circuit in Example 1.3 can be developed by differentiating eq. (1.27) to give

$$\frac{dx(t)}{dt} = \frac{dy(t)}{dt} + \frac{R}{L}y(t) \tag{1.28}$$

and finding the boundary condition needed to solve eq. (1.28) for $t \geq t_0$ (that is, the value of $y(t)$ at $t = t_0$) by solving the integral equation for $y(t_0)$ to yield

$$y(t_0) = x(t_0) - RI_0 \tag{1.29}$$

A boundary condition in this case is specified at the initial time considered, $t = t_o$, and is referred to as an *initial condition*. Classical solution techniques for differential equations can then be used to solve for $y(t)$ for $t \geq t_0$ with a given $x(t)$.

We can easily draw the block diagram representing the electric circuit of Example 1.3 for $t \geq t_0$ by rewriting the eq. (1.27) as

$$y(t) = x(t) - R\left[I_0 + \frac{1}{L}\int_{t_0}^{t} y(\tau)\,d\tau\right]$$

and then using this equation to define the block diagram shown in Figure 1.11.

1.4 General System Characteristics

Now that we have illustrated useful system representations and mathematical models, we can turn our attention to defining system characteristics and categories of systems and relating them to mathematical model and physical system properties. Characteristics will be defined only for single-input, single-output systems for simplicity; however, they also apply to multiple-input, multiple-output systems since, as previously mentioned, such systems can be represented as interconnections of single-input, single-output systems. As an aid to discussing the characteristics, we use

$$x(t) = x_1(t) \quad \text{yields} \quad y(t) = y_1(t)$$

or $\tag{1.30}$

$$x_T(n) = x_{3T}(n) \quad \text{yields} \quad y_T(n) = y_{3T}(n)$$

to indicate that the system output signals $y_1(t)$ and $y_{3T}(n)$ result from the system input signals $x_1(t)$ and $x_{3T}(n)$ for continuous-time and discrete-time systems, respectively. Also, $y_1(t)$ and $y_{3T}(n)$ are sometimes referred to as the *responses* of the systems to the inputs $x_1(t)$ and $x_{3T}(n)$.

Distributed-Parameter and Lumped-Parameter Systems

In the most general sense, all physical systems contain distributed parameters because of the physical size of the components. For example, the resistance of a resistor is distributed throughout its volume. Such *distributed-parameter systems* must be modeled with partial differential equations if they are continuous-time systems, or with partial difference equations if they are discrete-time systems. Fortunately, the action of a system component behaves as if it all were occurring at a point, given that the size of the component is small with respect to the wavelength of the highest frequency present in the signals associated with it. If this is true for all components in a system, then the system is said to be a *lumped-parameter system*. Such a system can be modeled with ordinary differential or difference equations.

In some systems, however, we find some components that are large with respect to the signal wavelength and others that are small. An example is a 60-Hz electric power system. The wavelength of the signal in this case is approximately 3100 miles. Obviously, the electrical system inside a building can be treated as a lumped-parameter system—but the same is not true for the long-distance transmission lines. In this case we usually divide the analysis problem into one part for the building and one part for the transmission lines, with partial differential equations required only for the latter part.

Since many systems can be treated as lumped-parameter systems, they are the only type we will consider in this text. Therefore, the differential and difference equations we will employ are all ordinary differential and difference equations.

Static and Dynamic Systems

Systems can be categorized as static or dynamic. A *static system* has no memory. Physically, it contains no energy-storage elements. The equation relating its output signal to its input signal contains no derivatives, integrals, or signal delays.

Definition _____

A static system is a system with an output signal that at any specific time depends on the value of the input signal at only that time.

_____

The circuit of Example 1.2 is a static system. Note that it indeed has no energy-storage elements such as inductors or capacitors and that its output is expressed in terms of its input, with no derivatives, integrals, or signal delays.

A *dynamic system* has one or more energy storage-elements. If the input–output equation is a differential equation for a continuous-time system, or a difference equation for a discrete-time system, then the system is dynamic. Continuous-time systems with signal delays are also dynamic.

Definition _____

A dynamic system is a system with an output signal that at any specified time depends on the value of the input signal at both the specific time and at other times.

The circuit of Example 1.3 is a dynamic continuous-time system. The block diagram of Figure 1.8b defines a dynamic discrete-time system when the operations are defined as in eqs. (1.18), (1.19), and (1.20).

The output signal of a static system at time t or sample number n is the solution of its input–output equation and depends on values of the input signal and any internal signal sources in the system at time t or sample number n only. An example of an internal signal source is the 6-V voltage source in Example 1.2. On the other hand, the output of a dynamic system after a specified time t_0 or sample number n_0 is the solution for $t \geq t_0$ or $n \geq n_0$ of its input–output differential or difference equation. This solution depends on the input signal and any internal signal sources for $t \geq t_0$ or $n \geq n_0$, and on the energies stored at $t = t_0$ or $n = n_0$ in the energy-storage components of the system. These stored energies define the initial conditions used in solving the differential or difference equation.

Causal Systems

Frequently, the output of a system at a given time depends on only the present and previous inputs to the system. A system that possesses this characteristic is called a *causal system*.

Definition _____

A causal system is one for which the value of the output signal at a specified time depends only on the values of the input signal that occur at times preceding or equal to the specified time.

Note that static systems are always causal systems. The continuous-time systems shown in Figure 1.8 and in Examples 1.2 and 1.3 are causal systems. We can see from their input–output equations that any integrals are only over preceding signal data. The discrete-time system shown in Figure 1.8 is also causal when the operations are those specified in eqs. (1.18), (1.19), and (1.20). If we change the operation O_{d1} to be a one-unit advance rather than a constant, then the input–output equation for this system is

$$y_T(n) = a y_T(n-1) + a c x_T(n) + x_T(n+1) \qquad (1.31)$$

and the system is noncausal since its output at a given time depends on the input at a later time. Physical systems that produce output signals from input signals as the input signals arrive are always causal since the system cannot know future values of the input signal when it constructs the output signal.

System Order

Differential equations and difference equations of different orders possess different and known solution characteristics. Therefore, it is convenient to characterize a system by the order of its corresponding differential or difference equation.

Definition

The order of a single-input, single-output system for which the output is related to the input by a differential equation or difference equation is the order of the differential or difference equation.

The order of a continuous-time system corresponds to the highest derivative of the output signal that appears in the input–output differential equation. The order of a discrete-time system corresponds to the largest number of units of delay of the output signal sequence appearing in the input–output difference equation. The system order is equal to the number of energy-storage elements in the system when the system is reduced to a configuration with the fewest energy-storage elements that produces the same input–output equation; for example, parallel capacitors reduced to one capacitor.

Example 1.4

The input–output equations for two systems are:

$$\frac{d^3 y(t)}{dt^3} + 3 \frac{dy(t)}{dt} + 2y(t) = \frac{dx(t)}{dt} + x(t)$$

and

$$y_T(n) = y_T(n-2) - 4 y_T(n-1) + x_T(n) - x_T(n-1)$$

Find the order of the systems.

Solution

The continuous-time system is a third-order system since the highest order derivative of $y(t)$ is three. The discrete-time system is a second-order system since the largest number of units of delay of the output appearing in the equation is two.

Linear Systems

Before stating the definition for a *linear system*, we must define three additional terms: *zero-initial-energy response*, *zero-input response*, and *decomposable system response*.

Definition

The zero-initial-energy response of a system is the system response to an input signal when all initial stored energies are zero.

The zero-initial-energy response is denoted as $x_{oe}(t)$ and $x_{oeT}(n)$ for continuous-time and discrete-time systems, respectively. In Chapters 9 and 16, we refer to it as the *zero-state response* in discussing state-space techniques.

Definition

The zero-input response of a system is the system response to initial stored energies when the input signal is zero.

The zero-input response is denoted by $x_{oi}(t)$ and $x_{oiT}(n)$ for continuous-time and discrete-time systems, respectively.

Definition

A decomposable system response can be written as the sum of the system's zero-initial-energy response and zero-input response.

A system with a decomposable response is said to possess the *decomposition property*. The discrete-time system of Figure 1.8b, with operations specified by eqs. (1.18), (1.19), and (1.20) possesses the decomposition property, as is apparent from its system response given by eq. (1.26).

Now, let us consider the following definition for a linear system:

Definition _____

A system is linear if and only if it satisfies the following three properties:

Property I.   Its response is decomposable.
Property II.   Superposition holds for its zero-initial-energy response.
Property III. Superposition holds for its zero-input response.

_____

For a continuous-time system, the superposition property holds for the zero-initial-energy response if and only if the following is true:
If

$$x(t) = x_1(t) \quad \text{yields} \quad y_{oe}(t) = y_1(t) \tag{1.32}$$

and

$$x(t) = x_2(t) \quad \text{yields} \quad y_{oe}(t) = y_2(t) \tag{1.33}$$

then

$$x(t) = Ax_1(t) + Bx_2(t)$$

yields   (1.34)

$$y_{oe}(t) = Ay_1(t) + By_2(t)$$

for all $x_1(t)$, $x_2(t)$, A, and B. That is, if the input is a weighted sum of two signals, then the output is the same weighted sum of the outputs resulting from each of the signals acting alone.

Likewise, for a continuous-time system, the superposition property holds for the zero-input response if and only if

$$y_{oi}(t) = Ay_a(t) + By_b(t) \tag{1.35}$$

when the initial stored energies are A times those which produce the zero-input response $y_a(t)$ and B times those producing the zero-input response $y_b(t)$ for all possible initial energy values and all values of A and B. The preceding definitions apply to discrete-time systems if the t's are changed to n's.

For a static system, the third property of the linear system definition is meaningless because such a system does not have any initial stored energy. It follows, then, that its total response is equal to its zero-initial-energy response. Thus, Property I is satisfied for a static system if its zero-input response equals zero for all time.

In the following examples, we will determine whether some of the systems previously illustrated are linear.

_____   _____

Example 1.5

Determine whether a single-resistor system with input $i(t)$ and output $v(t)$ and modeled by eq. (1.2) is linear.

Solution

In this static system, Property I holds since

$$\text{Zero-input response} = v_{oi} = 0$$

Property II holds since

$$i(t) = i_1(t) \quad \text{yields} \quad v_{oe}(t) = v_1(t) = Ri_1(t)$$

$$i(t) = i_2(t) \quad \text{yields} \quad v_{oe}(t) = v_2(t) = Ri_2(t)$$

and

$$i(t) = Ai_1(t) + Bi_2(t)$$

yields

$$v_{oe}(t) = R[Ai_1(t) + Bi_2(t)] = Av_1(t) + Bv_2(t)$$

Therefore, the system is linear.

In the preceding example, the single-resistor system would be nonlinear if the resistance were a function of the current value, $R = r[i(t)]$. In that case,

$$i(t) = Ai_1(t) + Bi_2(t)$$

would yield

$$
\begin{aligned}
v_{oe}(t) &= r\,[i(t)]\,i(t) \\
&= r\,[Ai_1(t) + Bi_2(t)]\,[Ai_1(t) + Bi_2(t)] \\
&= Ar\,[Ai_1(t) + Bi_2(t)]\,i_1(t) + Br\,[Ai_1(t) + Bi_2(t)]\,i_2(t) \\
&\neq Ar\,[i_1(t)]\,i_1(t) + Br\,[i_2(t)]\,i_2(t) = Av_1(t) + Av_2(t)
\end{aligned}
$$

Hence, Property II would not hold.

Example 1.6

Determine whether the system (circuit) of Example 1.2 is linear.

Solution

In this static system, Property III need not be checked. Property I does not hold since

$$\text{Zero-input response} = y_{oi}(t) = -5/6 \neq 0$$

Property II does not hold for all A and B since

$$x(t) = x_1(t) \quad \text{yields} \quad y_{oe}(t) = y_1(t) = \left(\frac{1}{12}\right) x_1(t) - \frac{5}{6}$$

$$x(t) = x_2(t) \quad \text{yields} \quad y_{oe}(t) = y_2(t) = \left(\frac{1}{12}\right) x_2(t) - \frac{5}{6}$$

and

$$x(t) = Ax_1(t) + Bx_2(t)$$

yields

$$y_{oe}(t) = \left(\frac{1}{12}\right)[Ax_1(t) + Bx_2(t)] - \frac{5}{6}$$

$$= Ay_1(t) + By_2(t) - \left(\frac{5}{6}\right)(1 - A - B)$$

$$\neq Ay_1(t) + By_2(t)$$

except in the special case when $A + B = 1$. Therefore, the system is nonlinear. Note that Property II did not need to be checked when the system failed the Property I check. It was done here just to illustrate the method of checking Property II. The feature that accounts for the system's nonlinearity is the constant term in the system input–output equation that results from the voltage source in the circuit.

Example 1.7

Determine if the discrete-time system of Figure 1.8b is linear when the operations are specified by eqs. (1.18), (1.19), and (1.20) with $a = 2$ and $c = 0.5$.

Solution

The input–output equation for the system is given by eq. (1.21) as

$$y_T(n) = 2y_T(n - 1) + 3x_T(n)$$

The solution for the discrete-time output signal for $n \geq 0$ is given by eq. (1.26) as

$$y_T(n) = 2^{n+1}y_T(-1) + 3\sum_{i=0}^{n} 2^{n-i}x_T(i) \qquad n \geq 0$$

The system is dynamic and causal since it depends only on present and past inputs.

Property I holds since

$$\text{Zero-input response} = y_{oiT}(n) = 2^{n+1}y_T(-1) \qquad n \geq 0$$

and

$$\text{Zero-initial-energy response} = y_{oeT}(n) = 3\sum_{i=0}^{n} 2^{n-i}x_T(i) \qquad n \geq 0$$

This can be easily verified by repeated substitution as we did to obtain eq. (1.26); thus,

$$y_T(n) = y_{oeT}(n) + y_{oiT}(n) \qquad n \geq 0$$

Property II holds since

$$x_T(n) = x_{1T}(n) \quad \text{yields} \quad y_{oeT}(n) = y_{1T}(n) = 3\sum_{i=0}^{n} 2^{n-i} x_{1T}(i) \qquad n \geq 0$$

$$x_T(n) = x_{2T}(n) \quad \text{yields} \quad y_{oeT}(n) = y_{2T}(n) = 3\sum_{i=0}^{n} 2^{n-i} x_{2T}(i) \qquad n \geq 0$$

and

$$x_T(n) = Ax_{1T}(n) + Bx_{2T}(n)$$

yields

$$y_{oeT}(n) = 3\sum_{i=0}^{n} 2^{n-i} \left[Ax_{1T}(i) + Bx_{2T}(i) \right]$$

$$= A3\sum_{i=0}^{n} 2^{n-i} x_{1T}(i) + B3\sum_{i=0}^{n} 2^{n-i} x_{2T}(i)$$

$$= Ay_{1T}(n) + By_{2T}(n) \qquad n \geq 0$$

Property III holds since

$$y_T(-1) = Y_1 \quad \text{yields} \quad y_{oiT}(n) = y_{1T}(n) = 2^{n+1} Y_1 \qquad n \geq 0$$

$$y_T(-1) = Y_2 \quad \text{yields} \quad y_{oiT}(n) = y_{2T}(n) = 2^{n+1} Y_2 \qquad n \geq 0$$

and

$$y_T(-1) = AY_1 + BY_2$$

yields

$$y_{oiT}(n) = 2^{n+1} \left(AY_1 + BY_2 \right)$$

$$= A2^{n+1} Y_1 + B2^{n+1} Y_2$$

$$= Ay_{1T}(n) + By_{2T}(n) \qquad n \geq 0$$

Therefore, the system is linear.

The preceding examples illustrate several properties of physical systems and their equations that are implied by the linearity of the system. These properties are stated here without formal proof. We see from the examples that equations for linear systems cannot contain terms that do not depend on the input or output, or terms that are nonlinear functions of input and/or the output. That is, the system equation must be a linear equation without a constant term. This means that a linear physical system cannot contain any multipliers, internal sources, or components with parameters that depend on the values of signals in the system. Thus, linear system components are all linear and each has a mathematical model that is a linear equation without a constant term.

Example 1.8

Determine whether the electric circuit of Example 1.3 is linear.

Solution

From Example 1.3 the equation for the circuit is

$$x(t) = y(t) + RI_0 + \frac{R}{L} \int_{t_0}^{t} y(\tau)\, d\tau \qquad t \geq t_0$$

Property I holds since

$$-RI_0 = y_{oi}(t) + \frac{R}{L} \int_{t_0}^{t} y_{oi}(\tau)\, d\tau \qquad t \geq t_0$$

and

$$x(t) = y_{oe}(t) + \frac{R}{L} \int_{t_0}^{t} y_{oe}(\tau)\, d\tau \qquad t \geq t_0$$

which gives, by equation addition,

$$x(t) = [y_{oi}(t) + y_{oe}(t)] + RI_0 + \frac{R}{L} \int_{t_0}^{t} [y_{oi}(\tau) + y_{oe}(\tau)]\, d\tau \qquad t \geq t_0$$

and thus

$$y(t) = y_{oi}(t) + y_{oe}(t)$$

Property II holds since the equations relating the zero-initial-energy response to the inputs $x(t) = x_1(t)$ and $x(t) = x_2(t)$ are

$$x_1(t) = y_1(t) + \frac{R}{L} \int_{t_0}^{t} y_1(\tau)\, d\tau \qquad t \geq t_0$$

and

$$x_2(t) = y_2(t) + \frac{R}{L} \int_{t_0}^{t} y_2(\tau)\, d\tau \qquad t \geq t_0$$

which gives

$$[Ax_1(t) + Bx_2(t)] = [Ay_1(t) + By_2(t)] + \frac{R}{L} \int_{t_0}^{t} [Ay_1(\tau) + By_2(\tau)]\, d\tau \qquad t \geq t_0$$

when the equations are multiplied by A and B and added. Thus

$$y(t) = Ay_1(t) + By_2(t) \quad \text{when} \quad x(t) = Ax_1(t) + Bx_2(t)$$

Property III can be shown to hold by using the procedure used for Property II.

Since the circuit of Example 1.3 satisfies all three properties, it is a linear system.

Note that we were able to determine system linearity for the circuit of Example 1.3 without finding the expression for the output signal. This is possible only when the input signal and initial condition enter the system equation in a direct and additive fashion.

Time-Invariant Systems

In broad terms, a system that does not change with time is a time-invariant system. A more careful definition of the time-invariant characteristics follows.

Definition

If both the input of a system and the time at which the internal energies in the system are specified are shifted in time by the same increment and the resulting output does not change shape but shifts in time by the same time increment, then the system is said to be time-invariant.

This definition can be expressed mathematically for continuous-time and discrete-time systems as follows:

If

$$x(t) = x_1(t) \quad \text{yields} \quad y(t) = y_1(t) \tag{1.36}$$

then

$$x(t) = x_1(t - \tau) \quad \text{yields} \quad y(t) = y_1(t - \tau) \tag{1.37}$$

and if

$$x_T(n) = x_{1T}(n) \quad \text{yields} \quad y_T(n) = y_{1T}(n) \tag{1.38}$$

then

$$x_T(n) = x_{1T}(n - m) \quad \text{yields} \quad y_T(n) = y_{1T}(n - m) \tag{1.39}$$

where it is understood that the time at which the internal stored energies are specified is also shifted by τ or m. Figure 1.12 illustrates example input and output signals for time-invariant systems.

Algebraic, differential, and difference equations that are mathematical models for time-invariant systems all have constant coefficients for the signal terms in them. These constant coefficients are produced by physical systems with components that do not vary with time.

Figure 1.12
Example of
Time-Invariant
System Responses

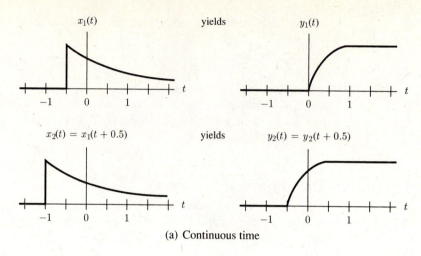

(a) Continuous time

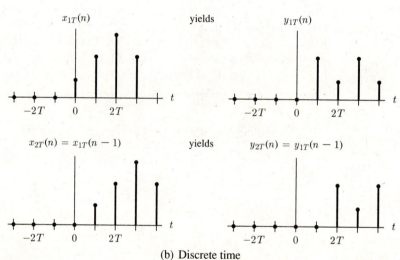

(b) Discrete time

Example 1.9

A continuous-time system is modeled by the equation $y(t) = tx(t) + 4$ and a discrete-time system is modeled by $y_T(n) = x_T^2(n)$. Are these systems time invariant?

Solution

For the continuous-time system,

$$x(t) = x_1(t)$$

yields

$$y(t) = y_1(t) = tx_1(t) + 4$$

and

$$x(t) = x_1(t - \tau)$$

yields

$$y(t) = tx_1(t - \tau) + 4 \neq (t - \tau)x_1(t - \tau) + 4 = y_1(t - \tau)$$

The continuous-system is not time-invariant.

For the discrete-time system

$$x_T(n) = x_{1T}(n)$$

yields

$$y_T(n) = y_{1T}(n) = x_{1T}^2(n)$$

and

$$x_T(n) = x_{1T}(n - m)$$

yields

$$y_T(n) = x_{1T}^2(n - m) = y_{1T}(n - m)$$

The discrete-time system is time-invariant.

The systems considered in this text are restricted to linear time-invariant systems unless otherwise indicated. Fortunately, many practical systems can be adequately modeled as linear time-invariant systems; hence, the tools developed are quite powerful.

System Stability

A *bounded signal* has an amplitude that remains finite. We want a physical system to have a bounded output for any bounded input so that its output does not grow unreasonably large. We say that such a system possesses *bounded-input, bounded-output (BIBO) stability*.

Definition

A system possesses bounded-input, bounded-output (BIBO) stability if every bounded input signal produces a bounded output signal.

We will discuss system characteristics that produce BIBO stability in Chapters 7 and 14. Note that the system in Figure 1.8b, with operators defined by eqs. (1.18), (1.19), and (1.20), is unstable if $a > 1$. We see from eq. (1.26) that, when $a > 1$, then the system output grows without bound as n increases for many bounded input signals, such as $x_T(n) = 1$, $n \geq 0$, or for any bounded input if the initial energy stored is nonzero.

Example of Classical Solution for System Output

As a final step in the discussion of system models and characteristics, let us consider an example of a general solution for the output of a continuous-time system. We could use a similar example for a discrete-time system employing classical solution techniques for difference equations. However, as indicated earlier, a continuous-time example is used because it is assumed that the reader is familiar with classical solution techniques for differential equations.

Example 1.10

The circuit shown in Figure 1.13 is a simple form of a filter circuit designed to reduce high-frequency additive interference that is present with the desired lower frequency message signal at the input. Find the output signal, voltage $y(t)$, for $t \geq t_0$ for the input signal, voltage $x(t)$, when the energy stored in the capacitor at $t = t_0$ is such that $y(t_0) = Y_0$.

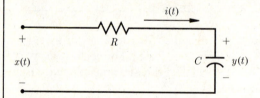

Figure 1.13 Electric Circuit for Example 1.10

Solution

We first find the input–output equation by using the circuit equations

$$x(t) = Ri(t) + y(t) \quad \text{and} \quad i(t) = C\frac{dy(t)}{dt}$$

to produce

$$RC\frac{dy(t)}{dt} + y(t) = x(t)$$

Next, we solve the differential equation for $t \geq t_0$, subject to the boundary condition $y(t_0) = Y_0$. The homogeneous equation for the system is

$$RC\frac{dy(t)}{dt} + y(t) = 0$$

From classical differential-equation theory, we know that the homogeneous solution is of the form $y(t) = Ae^{pt}$. Substituting this solution in the homogeneous equation gives

$$RCApe^{pt} + Ae^{pt} = 0$$

$$A(RCp + 1)e^{pt} = 0$$

In order that the equation be satisfied for all t, $p = -1/RC$. Therefore,

$$y(t) = Ae^{-t/RC}$$

The next step required is to find the particular solution of the differential equation for $t \geq t_0$ so that it can be added to the homogeneous solution to determine the total solution. Actually, the total solution for a general input can be found by the method of variation of parameters. Using this method, we assume that the arbitrary constant in the homogeneous solution is a function of time $A(t)$ and solve for it. Thus, we define $y(t)$ as

$$y(t) = A(t)e^{-t/RC}$$

Substituting this expression for $y(t)$ in the differential equation gives

$$RC\left[\frac{dA(t)}{dt}e^{-t/RC} - \frac{1}{RC}A(t)e^{-t/RC}\right] + A(t)e^{-t/RC} = x(t)$$

Canceling terms and rearranging the equation gives

$$\frac{dA(t)}{dt} = x(t)\left[\frac{1}{RC}e^{t/RC}\right]$$

Integrating the equation to solve for $A(t)$ for $t \geq t_0$ yields

$$A(t) = \int_{t_0}^{t} x(\tau)\left[\frac{1}{RC}e^{\tau/RC}\right] d\tau + A(t_0) \qquad t \geq t_0$$

Then, substituting this back in the defined expression for $y(t)$ gives

$$y(t) = e^{-t/RC}\int_{t_0}^{t} x(\tau)\left[\frac{1}{RC}e^{\tau/RC}\right] d\tau + A(t_0)e^{-t/RC} \qquad t \geq t_0$$

At the initial solution time of $t = t_0$, the output is

$$y(t_0) = 0 + A(t_0)e^{-t_0/RC}$$

which can be solved for $A(t_0)$ to yield

$$A(t_0) = y(t_0)e^{t_0/RC} = Y_0 e^{t_0/RC}$$

Therefore, the output signal is

$$y(t) = \int_{t_0}^{t} x(\tau)\left[\frac{1}{RC}e^{-(t-\tau)/RC}\right] d\tau + Y_0 e^{-(t-t_0)/RC} \qquad t \geq t_0 \qquad (1.40)$$

The output of the system in Example 1.10 as expressed by eq. (1.40) is clearly the sum of the zero-initial-energy response (first term) and the zero-input response (second term). Thus, the system satisfies Property I for a linear system. We also see that when

$$x(t) = Ax_1(t) + Bx_2(t) \quad \text{and} \quad Y_0 = 0$$

then

$$y(t) = \int_{t_0}^{t} [Ax_1(t) + Bx_2(t)] \left[\frac{1}{RC} e^{-(t-\tau)/RC} \right] d\tau$$

$$= A \int_{t_0}^{t} x_1(t) \left[\frac{1}{RC} e^{-(t-\tau)/RC} \right] d\tau$$

$$+ B \int_{t_0}^{t} x_2(t) \left[\frac{1}{RC} e^{-(t-\tau)/RC} \right] d\tau \qquad t \geq t_0$$

so superposition holds for the zero-initial-energy response; therefore, the system satisfies Property II for a linear system. Likewise, when

$$Y_0 = AY_{01} + BY_{02} \quad \text{and} \quad x(t) = 0$$

then

$$y(t) = [AY_{01} + BY_{02}] e^{-(t-t_0)/RC}$$

$$= AY_{01} e^{-(t-t_0)/RC} + BY_{02} e^{-(t-t_0)/RC}$$

so superposition holds for the zero-input response; therefore, the system satisfies Property III for a linear system. The simple filter circuit of Example 1.10 is a linear system since all three properties are satisfied. (In the Problem section at the end of the chapter, you are asked to show that this system is also time-invariant [see Problem 1.20].)

Now, let us assume that $t_0 = -\infty$ in Example 1.10 so that the solution for the system output can be found for any desired time. Let us also assume that $y(-\infty) = y(t_0) = Y_0 = 0$, so that the system has no energy stored initially. In addition, for convenience, we will define the function $u(t)$ such that $u(t) = 0$ for $t < 0$ and $u(t) = 1$ for $t \geq 0$.

Using the preceding assumptions and definition, the output signal can be written as

$$y(t) = \int_{-\infty}^{\infty} x(\tau) \left[\frac{1}{RC} e^{-(t-\tau)/RC} u(t-\tau) \right] d\tau \qquad (1.41)$$

since $u(t - \tau) = 0$ for $\tau > t$. Further, if we define the function $h(t)$ as

$$h(t) = \frac{1}{RC} e^{-t/RC} u(t) \qquad (1.42)$$

then the output signal for the circuit of Example 1.10 can be written as

$$y(t) = \int_{-\infty}^{\infty} x(\tau)h(t - \tau)\, d\tau \qquad (1.43)$$

We will see in Chapter 5 that this particular form of the equation for the output of the RC circuit of Example 1.10, obtained when the initial energy stored is zero, applies more generally to any linear, time-invariant, continuous-time system with zero initial conditions. The only difference is in the form of the function $h(t)$. Thus, the integral operation of eq. (1.43) on the two functions $x(t)$ and $h(t)$ is sufficiently important to be given a special name. It is referred to as the *convolution operation for continuous-time functions*. This integral operation also occurs in other system analyses for functions of frequency, as well as for functions of time.

A similar summation operation on two discrete-time sequences is called the *convolution operation for discrete-time sequences*. The solution of the difference equation that models a linear, time-invariant, discrete-time system with zero-initial conditions results in such a convolution operation, as we will see in Chapter 12.

Since the convolution operation is so important for system analyses, we will discuss its evaluation in detail in the next chapter.

1.5 Summary

The terms *signal*, *system*, and *system component* represent important concepts associated with a physical device or phenomenon. For example, an electric circuit is a system with system components consisting of resistors and capacitors and signals consisting of voltages and currents. Signal, system, and system-component representations possess general characteristics that help to describe them. Mathematical models, which are useful in their analysis, can be generated.

The categorization of signals and systems as continuous time or discrete time is very useful in that each category employs separate but very similar analysis tools. The remainder of this text is divided into two parts—one for each category.

All signals can be categorized as deterministic signals or random signals. In this text, we will consider only deterministic signals. Signals can be further categorized as either periodic or aperiodic signals and as either piecewise-defined or simply defined signals.

System characteristics of particular importance are linearity, time-invariance, causality, and stability. Lumped-parameter systems are the only type we will consider, with an emphasis on linear time-invariant systems. A wide variety of systems possess these characteristics approximately, and a number of mathematical tools are effective in their analysis.

In the last example of Chapter 1, the analysis of a continuous-time system consisting of a simple RC filter circuit illustrates several system concepts and indicates one use for the important mathematical operation called convolution.

Problems

1.1 Sketch the continuous-time signals defined by the following equations; also indicate which of these signals are piecewise defined:

a. $w(t) = (2 + t^2)/(1 + t^2)$ $-\infty \le t \le \infty$

b. $x(t) = 0.5 |t|$ $-\infty \le t \le \infty$

c. $y(t) = \begin{cases} 2 + t & -2 \le t < 0 \\ 2 - t & 0 \le t < 2 \\ 0 & \text{elsewhere} \end{cases}$

d. $z(t) = \begin{cases} 4 & 1 \le t \le 2 \\ 0 & \text{elsewhere} \end{cases}$

1.2 Sketch the continuous-time signals defined by the following equations:

a. $x(t) = \begin{cases} t - 1 & 1 \le t < 2 \\ 2 & 2 \le t < 5 \\ 0 & \text{elsewhere} \end{cases}$

b. $x(t) = e^{-2|t|}$ $-\infty \le t \le \infty$

c. $x(t) = \dfrac{t - 1}{0.25t^2 + 0.5}$ $-\infty \le t \le \infty$

1.3 Sketch the discrete-time signals defined by the following equations; also indicate which of these signals are piecewise defined:

a. $z_T(n) = \left[1 - \exp\left[-0.1n^2 \right], \quad 2 \text{ s} \right]$
$$-\infty \le n \le \infty$$

b. $x_T(n) = \left[\left\{ \begin{matrix} 2e^{-0.5n} & 0 \le n \le 5 \\ 0 & \text{elsewhere} \end{matrix} \right\}, \quad 0.1 \text{ s} \right]$

c. $w_T(n) = \left[4 - |n|, \quad 3 \text{ ms} \right]$ $-\infty \le n \le \infty$

d. $y_T(n) = \left[\left\{ \begin{matrix} 1/(1 + n) & n \ge 0 \\ 1/(1 - n) & n < 0 \end{matrix} \right\}, \quad 0.01 \text{ s} \right]$

1.4 Sketch the discrete-time signal sequences defined by the following equations:

a. $x(n) = e^{-|0.25n|}$ $-\infty \le n \le \infty$

b. $x(n) = \begin{cases} n + 1 & n \le 2 \\ 6 & 3 \le n \le 5 \\ 8 - n & n \ge 6 \end{cases}$

c. $x(n) = (n - 1)/(n^2 + 1)$ $-\infty \le n \le \infty$

1.5 Write the equations for the piecewise-defined signals shown in Figure 1.14.

a.

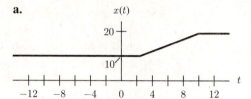

b.

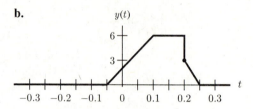

c.

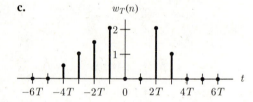

d.

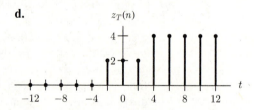

Figure 1.14

1.6 Write the equations for the piecewise-defined signals and sequences shown in Figure 1.15.

a.

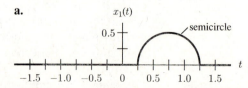

Figure 1.15

b.

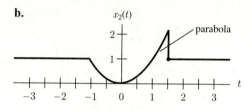

parabola

c.

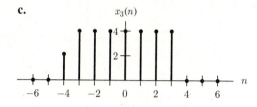

d.

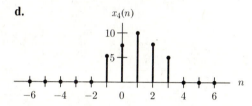

Figure 1.15 *(continued)*

1.7 Determine whether the following signals and sequences are periodic, and compute the period for any that are periodic:

a. $x(t) = e^{-j5t}$ $-\infty \le t \le \infty$

b. $y(n) = 4 + \cos\left(0.25\pi n^2\right)$ $-\infty \le n \le \infty$

c. $w(t) = \begin{cases} 3\sin t & -10 \le t \le 10 \\ 0 & \text{elsewhere} \end{cases}$

d. $z_T(n) = \left[3^{\cos(0.1\pi n)}, \quad 10 \text{ s}\right]$ $-\infty \le n \le \infty$

1.8 Sketch the following periodic signals for three periods:

a. $x(t) = \sum\limits_{i=-\infty}^{\infty} x_p(t - 5i)$

$x_p(t) = \begin{cases} t+1 & -1 \le t < 0 \\ 1 & 0 \le t < 1 \\ 0 & \text{elsewhere} \end{cases}$

b. $y(t) = \sum\limits_{i=-\infty}^{\infty} y_p(t - 0.1i)$

$y_p(t) = \begin{cases} t & 0.05 \le t < 0.15 \\ 0 & \text{elsewhere} \end{cases}$

c. $z_T(n) = \left[\sum\limits_{i=-\infty}^{\infty} z_p(n - 6i), \quad T\right]$

$z_p(n) = \begin{cases} 3e^{-0.25n} & 0 \le n < 6 \\ 0 & \text{elsewhere} \end{cases}$

d. $w_T(n) = \left[\sum\limits_{i=-\infty}^{\infty} w_p(n - 3i), \quad 0.2\right]$

$w_p(n) = \begin{cases} 3, 4, \text{ and } 2 & \text{for } n = -1, 0, \text{ and} \\ & 1, \text{ respectively} \\ 0 & \text{elsewhere} \end{cases}$

1.9 Determine whether the following signals and sequences are periodic, and compute the period for any that are periodic:

a. $x(t) = x_1(t) + x_2(t)$

$x_1(t)$ has period $T_{01} = 2\pi$ s

$x_2(t)$ has period $T_{02} = 1.5\pi$ s

b. $w(t) = w_1(t) + w_2(t) + w_3(t) + w_4(t)$

$w_1(t)$ has period $T_{01} = 0.21$ s

$w_2(t)$ has period $T_{02} = 1.275$ s

$w_3(t)$ has period $T_{03} = 2.55$ s

$w_4(t)$ has period $T_{04} = 4.95$ s

c. $z(t) = z_1(t) + z_2(t)$

$z_1(t)$ has period $T_{01} = 3\pi$ s

$z_2(t)$ has period $T_{02} = 5$ s

d. $y(n) = y_1(n) + y_2(n) + y_3(n)$

$y_1(n)$ has period $N_{01} = 10$

$y_2(n)$ has period $N_{02} = 15$

$y_3(n)$ has period $N_{03} = 36$

1.10 Write the input–output equation for the electric circuit shown in Figure 1.16 on page 40. The input is the voltage $x(t)$ and the output is the current $y(t)$.

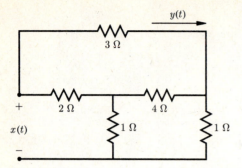

Figure 1.16

1.11 Write the input–output equation for the electric circuit shown in Figure 1.17. The voltages $x(t)$ and $y(t)$ are the input and the output, respectively.

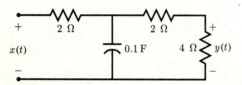

Figure 1.17

1.12 The block diagrams for feedback-control systems using various types of performance compensation are shown in Figure 1.18. As a first step in the analysis of these systems, write the input–output equations for them.

a.

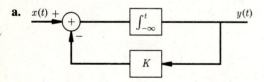

b.

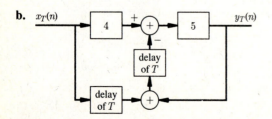

Figure 1.18

c.

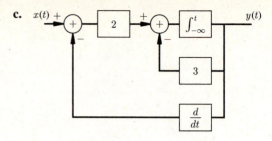

d.

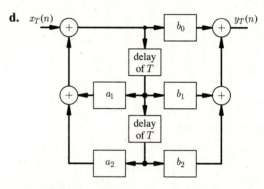

Figure 1.18 *(continued)*

1.13 Draw block-diagram representations for the systems characterized by the following input–output equations, where $x(t)$ or $x_T(n)$ is the input and $y(t)$ or $y_T(n)$ is the output:

a. $\dfrac{d^2 y(t)}{dt^2} + 3\dfrac{dy(t)}{dt} + 2y(t) = x(t) + \dfrac{dx(t)}{dt}$

b. $\dfrac{dy(t)}{dt} + 4y(t) + 0.5 \displaystyle\int_{-\infty}^{t} y(\tau)\, d\tau = x(t)$

c. $y_T(n) - 0.5 y_T(n-1) - 0.2 y_T(n-2) =$
$\quad x_T(n-1) + x_T(n-2)$

d. $y_T(n) - 0.75 y_T(n-3) = x_T(n) - 2 x_T(n-2)$

1.14 Draw block-diagram representations for the systems modeled by the following sets of equations, where $x(t)$ and $x_T(n)$ are the input signals and $y(t)$ and $y_T(n)$ are the output signals:

a. $\dfrac{dy(t)}{dt} + y(t) - az(t) + \displaystyle\int_{-\infty}^{t} z(\tau)\, d\tau = x(t)$

$\displaystyle\int_{-\infty}^{t} y(\tau)\, d\tau + bz(t) = c\dfrac{dx(t)}{dt}$

b. $y_T(n) = y_T(n-1) + Az_T(n-1) +$

$$Bx_T(n) + Cx_T(n-1)$$

$$z_T(n) = Dz_T(n-2) - Ey_T(n-1) + x_T(n)$$

1.15 Determine whether the systems having the following input–output equations are static or dynamic. If a system is dynamic, what is its order? State the reasons for your answers.

a. $y(t) = 3x^2(t) + \dfrac{dx(t)}{dt}$

b. $\dfrac{d^2y(t)}{dt^2} = y(t) + 4tx(t) - \dfrac{dx(t)}{dt}$

c. $y(n) = 0.5x(n) + 3$

d. $y(t) - 2x(t)\dfrac{dy(t)}{dt} = \dfrac{d^2x(t)}{dt^2}$

e. $y(n) = y(n-1) + 2x(n) - 3x(n-1) + 2x(n-2)$

f. $y(n) - \dfrac{y(n-2)}{n} = 2x(n)$

1.16 Show whether Property II for system linearity holds for each of the systems of Problem 1.15.

1.17 Determine whether the systems of Problem 1.15 are time-invariant or time-varying. State the reasons for your answers.

1.18 Show whether the systems with the following outputs are linear. The input is $x(t)$ or $x_T(n)$, the output is $y(t)$ or $y_T(n)$, and the initial energy stored is E_0.

a. $y(t) = E_0 x(t) + 4 \displaystyle\int_{t_0}^{2t} x(\tau)\, d\tau \qquad t \geq t_0$

b. $y_T(n) = 3E_0 + 0.5[x_T(n) + x_T(n-1)]$

$$n \geq n_0$$

c. $y(t) = tE_0 + \displaystyle\int_{t_0}^{t} x^2(\tau)h(t-\tau)\, d\tau \qquad t \geq t_0$

d. $y_T(n) = E_0 + \dfrac{x_T(n) - x_T(n+2)}{1 + x_T(n-1)} \qquad n \geq n_0$

1.19 Determine whether the systems of Problem 1.18 are causal. State the reasons for your answers.

1.20 Show that the simple RC filter system of Example 1.10 is time-invariant.

2 Convolution

Convolution is a mathematical operation applied to two functions of the same independent variable, producing a third function of that independent variable. The third function is referred to as the convolution of the other two functions. We refer to convolution of continuous-variable functions as *continuous convolution* and convolution of discrete-variable functions as *discrete convolution*.

In the discussion following Example 1.10 of the previous chapter, we indicated that the output of a simple RC circuit with zero initial energy stored in the capacitor is the continuous convolution of the input signal, as a function of time, and a time function dependent on the resistance and capacitance values. We further indicated that this convolution relationship between the output and the input applies to a broad class of continuous-time and discrete-time systems. Convolution also proves to be useful in some system analyses that use functions of the independent-variable frequency, as we will see in Chapters 4 and 11.

At this point, we will consider the mathematical characteristics of convolution and techniques for its evaluation so as to avoid interrupting the discussion of convolution applications in later chapters.

2.1 Continuous Convolution

Let us first consider continuous convolution of two functions of the independent variable α.

Definition _____

Continuous convolution of the two functions $f_1(\alpha)$ and $f_2(\alpha)$ of the independent variable α is the function $f_3(\alpha)$ given by the integral

$$f_3(\alpha) = \int_{-\infty}^{\infty} f_1(\beta) f_2(\alpha - \beta)\, d\beta \equiv f_1(\alpha) * f_2(\alpha) \qquad (2.1)$$

where β is a dummy variable used in the integration.

_____

We often use the short notation defined on the right side of eq. (2.1) to indicate continuous convolution of the two functions $f_1(\alpha)$ and $f_2(\alpha)$. Equation (2.1) is called the *convolution integral*.

Continuous-Convolution Evaluation

It is helpful to present examples of the evaluation of continuous convolution before indicating some of its properties. Our first example considers two functions, each defined by a single equation for all values of the independent variable (that is, simply defined functions). The convolution integral is conceptually very easy to evaluate because it consists of a single direct evaluation of the integral and produces a simply defined function. However, as for any integral, the actual ease with which the evaluation can be performed depends on the form of the integrand; hence, the evaluation may be quite difficult for some relatively simple functions.

Example 2.1

Find the convolution of the two continuous-variable functions

$$f(x) = e^{-x^2} \quad \text{and} \quad g(x) = 3x^2, \text{ for all } x.$$

Solution

$$h(x) = f(x) * g(x) = \int_{-\infty}^{\infty} e^{-y^2} \left[3(x-y)^2\right] dy$$

$$= 3x^2 \int_{-\infty}^{\infty} e^{-y^2} dy - 6x \int_{-\infty}^{\infty} ye^{-y^2} dy + 3 \int_{-\infty}^{\infty} y^2 e^{-y^2} dy$$

$$= 3x^2\sqrt{\pi} - 0 + 1.5\sqrt{\pi}$$

$$= 5.318x^2 + 2.659 \qquad \text{for all } x$$

Next, we will consider the convolution of a piecewise-defined function and a simply defined function. In this case the integral must be evaluated in two or more parts. However, the resulting convolution is a simply defined function; hence, the integral needs only to be evaluated once to yield a result that is valid for all values of the independent variable.

Example 2.2

Find the convolution of the two continuous-time functions

$$x(t) = 3\cos 2t \qquad \text{for all } t$$

and

$$y(t) = e^{-|t|} = \begin{cases} e^t & t < 0 \\ e^{-t} & t \geq 0 \end{cases}$$

Solution

$$z(t) \equiv x(t) * y(t) = \int_{-\infty}^{\infty} x(\tau)y(t-\tau)\, d\tau \qquad \text{for all } t$$

Now

$$x(\tau) = 3\cos 2\tau \qquad \text{for all } \tau$$

and
$$y(t - \tau) = \begin{cases} e^{(t-\tau)} & t - \tau < 0 & \rightarrow & \tau > t \\ e^{(\tau-t)} & t - \tau \geq 0 & \rightarrow & \tau \leq t \end{cases}$$

Therefore,

$$z(t) = \int_{-\infty}^{t} \{3\cos 2\tau\} \{e^{(\tau-t)}\} \, d\tau$$

$$+ \int_{t}^{\infty} \{3\cos 2\tau\} \{e^{(t-\tau)}\} \, d\tau$$

$$= 3e^{-t} \int_{-\infty}^{t} e^{\tau} \cos 2\tau \, d\tau + 3e^{t} \int_{t}^{\infty} e^{-\tau} \cos 2\tau \, d\tau$$

$$= 3e^{-t} \left[\frac{e^{\tau} (\cos 2\tau + 2 \sin 2\tau)}{5} \right]_{-\infty}^{t}$$

$$+ 3e^{t} \left[\frac{e^{-\tau} (-\cos 2\tau + 2 \sin 2\tau)}{5} \right]_{t}^{\infty}$$

$$= \frac{6}{5} \cos 2t \qquad \text{for all } t$$

Example 2.3

Two continuous-frequency functions are

$$x(f) = \frac{1}{1 + f^2} \qquad \text{for all } f$$

and

$$y(f) = \begin{cases} 0 & f < 0 \\ 1 & f \geq 0 \end{cases}$$

Find their convolution.

Solution

$$w(f) \equiv x(f) * y(f) = \int_{-\infty}^{\infty} x(\beta) y(f - \beta) \, d\beta \qquad \text{for all } f$$

Now

$$x(\beta) = \frac{1}{1 + \beta^2} \qquad \text{for all } \beta$$

and

$$y(f - \beta) = \begin{cases} 0 & f - \beta < 0 & \rightarrow & \beta > f \\ 1 & f - \beta \geq 0 & \rightarrow & \beta \leq f \end{cases}$$

Therefore,

$$w(f) = \int_{-\infty}^{f} \left[\frac{1}{1+\beta^2}\right][1]\,d\beta + \int_{f}^{\infty}\left[\frac{1}{1+\beta^2}\right][0]\,d\beta$$

$$= \int_{-\infty}^{f} \frac{d\beta}{1+\beta^2} = \left[\tan^{-1}\beta\right]_{-\infty}^{f}$$

$$= \tan^{-1} f + \frac{\pi}{2} \qquad \text{for all } f$$

In the last two examples, we carefully expressed the two functions that were multiplied to form the integrand before substituting them in the integral. Including this step takes a little more time when we construct the solution to a convolution evaluation. This is time well spent, however, since the step is effective in reducing errors in function substitution. It is also very helpful in determining the correct intervals into which the integration is to be divided; thus, in establishing the correct limits for the individual integrals. It is highly recommended that this step always be included to reduce errors.

Convolution of two functions that are piecewise defined produces a convolution integral that must be evaluated in two or more parts. Since the resulting convolution is also a piecewise-defined function, the convolution integral must be evaluated as many times as there are pieces in the resulting function to find the mathematical expressions for those pieces. For the evaluation of the convolution of two piecewise-defined functions, it is helpful to express carefully the functions being multiplied to form the integrand and to sketch these functions in positions corresponding to each of the integral evaluations that must be made. Since there are a number of tasks to be done in evaluating the convolution of two piecewise-defined functions, it is best to organize the evaluation in some well-defined steps. The steps in evaluating the convolution defined by eq. (2.1) are:

Step 1 Write equations and draw sketches for $f_1(\alpha)$ and $f_2(\alpha)$.

Step 2 Write equations and draw sketches for $f_1(\beta)$ and $f_2(\alpha - \beta)$.

Step 3 Perform integrations to find $f_3(\alpha)$ for all α.

The following example illustrates the use of the evaluation steps.

Example 2.4

Find the convolution of the two continuous-time functions

$$f(t) = e^{-|t|} \qquad \text{for all } t$$

and

$$g(t) \begin{cases} = 0 & t < 1 \\ = e^{-2t} & t \geq 1 \end{cases}$$

Solution

$$h(t) \equiv f(t) * g(t) = \int_{-\infty}^{\infty} f(\tau)g(t - \tau)\,d\tau$$

Step 1   Write the equations and draw sketches for $f(t)$ and $g(t)$, as shown in Figure 2.1.

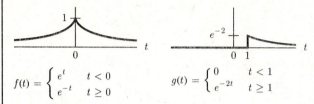

$$f(t) = \begin{cases} e^t & t < 0 \\ e^{-t} & t \geq 0 \end{cases} \qquad\qquad g(t) = \begin{cases} 0 & t < 1 \\ e^{-2t} & t \geq 1 \end{cases}$$

Figure 2.1 Step 1 for Example 2.4

Step 2   Write equations and draw sketches for $f(\tau)$ and $g(t - \tau)$, as shown in Figure 2.2.

$$f(\tau) = \begin{cases} e^\tau & \tau < 0 \\ e^{-\tau} & \tau \geq 0 \end{cases} \qquad g(t - \tau) = \begin{cases} 0 & t - \tau < 1 \rightarrow \tau > t - 1 \\ e^{-2(t-\tau)} & t - \tau \geq 1 \rightarrow \tau \leq t - 1 \end{cases}$$

Figure 2.2 Step 2 for Example 2.4

Step 3   Perform integrations to find $h(t)$ for all values of t.

Case I

$$t - 1 < 0 \quad \rightarrow \quad t < 1$$

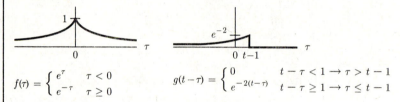

Range of τ for which integrand is nonzero

$f(\tau)$
$g(t - \tau)$

Figure 2.3 Function Sketch for Step 3, Case I

$$h(t) = \int_{-\infty}^{t-1} \left[e^\tau \right]\left[e^{-2(t-\tau)} \right]\,d\tau + \int_{t-1}^{\infty} [f(\tau)][0]\,d\tau$$

$$= e^{-2t} \int_{-\infty}^{t-1} e^{3\tau}\, d\tau = e^{-2t} \left[\frac{e^{3\tau}}{3} \right]_{-\infty}^{t-1}$$

$$= \frac{1}{3} e^{(t-3)} \qquad t < 1$$

Case II

$$t - 1 \geq 0 \quad \rightarrow \quad t \geq 1$$

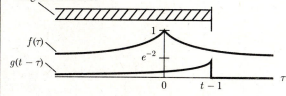

Range of τ for which integrand is nonzero

$f(\tau)$

$g(t - \tau)$

Figure 2.4 Function Sketch for Step 3, Case II

$$h(t) = \int_{-\infty}^{0} \left[e^{\tau} \right]\left[e^{-2(t-\tau)} \right] d\tau + \int_{0}^{t-1} \left[e^{-\tau} \right]\left[e^{-2(t-\tau)} \right] d\tau + \int_{t-1}^{\infty} \left[e^{-\tau} \right][0]\, d\tau$$

$$= e^{-2t} \left[\int_{-\infty}^{0} e^{3\tau} d\tau + \int_{0}^{t-1} e^{\tau} d\tau \right]$$

$$= e^{-2t} \left\{ \left[\frac{e^{3\tau}}{3} \right]_{-\infty}^{0} + \left[e^{\tau} \right]_{0}^{t-1} \right\}$$

$$= e^{-(t+1)} - \frac{2}{3} e^{-2t} \qquad t \geq 1$$

In summary,

$$h(t) = \begin{cases} \frac{1}{3} e^{(t-3)} & t < 1 \\ e^{-(t+1)} - \frac{2}{3} e^{-2t} & t \geq 1 \end{cases}$$

The solution is illustrated by the sketch in Figure 2.5.

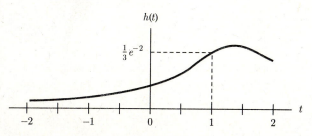

Figure 2.5 Convolution Result for Example 2.4

Note that $g(t-\tau)$ as a function of τ has the same form as $g(t)$ as a function of t, except that the former is reversed and shifted. The reversal occurs because τ enters the argument of $g(t-\tau)$ with a negative sign; the shift occurs because of the presence in the argument of t, which is an arbitrary constant when $g(t-\tau)$ is considered as a function of τ.

Step 1 is included in the solution to aid us in expressing the correct mathematical formulas for the functions being multiplied in the integrand and in making the correct sketches to evaluate the correct limits on the integrals. The solution in Example 2.4 is intended to be complete; thus, the given functions are again specified in Step 1.

Since $h(t)$ in Example 2.4 is evaluated for all values of t, it is best to consider these values of t from smallest to largest, as indeed we did. That is, t was varied from $-\infty$ to ∞ so that $g(t-\tau)$ was shifted from left to right. We encounter a new case to evaluate when a boundary of a piecewise-defined portion of $g(t-\tau)$ passes a boundary of a piecewise-defined portion of $f(\tau)$, as $g(t-\tau)$ is shifted to the right.

A second example will serve to reinforce the preceding evaluation concepts for the convolution of two piecewise-defined functions.

Example 2.5

Find the convolution of the two continuous-variable functions shown in Figure 2.6.

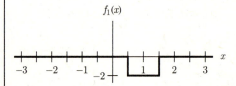

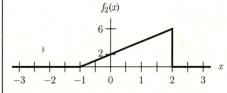

Figure 2.6 Functions of the Variable x for Example 2.5

Solution

$$f_3(x) \equiv f_1(x) * f_2(x) = \int_{-\infty}^{\infty} f_1(y)f_2(x-y)\,dy$$

Step 1   Write equations and draw sketches for $f_1(x)$ and $f_2(x)$, as shown in Figure 2.7.

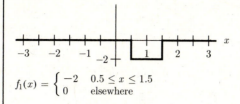

$$f_1(x) = \begin{cases} -2 & 0.5 \le x \le 1.5 \\ 0 & \text{elsewhere} \end{cases}$$

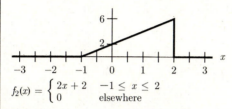

$$f_2(x) = \begin{cases} 2x + 2 & -1 \le x \le 2 \\ 0 & \text{elsewhere} \end{cases}$$

Figure 2.7 Step 1 for Example 2.5

Step 2   Write equations and draw sketches for $f_1(y)$ and $f_2(x - y)$, as shown in Figure 2.8.

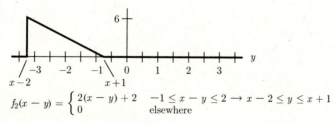

$$f_1(y) = \begin{cases} -2 & 0.5 \le y \le 1.5 \\ 0 & \text{elsewhere} \end{cases}$$

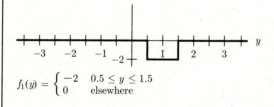

$$f_2(x - y) = \begin{cases} 2(x - y) + 2 & -1 \le x - y \le 2 \to x - 2 \le y \le x + 1 \\ 0 & \text{elsewhere} \end{cases}$$

Figure 2.8 Step 2 for Example 2.4

Step 3   Perform integrations to find $f_3(x)$ for all values of x.

Case I

$$x + 1 < 0.5 \quad \to \quad x < -0.5$$

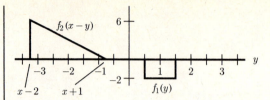

Figure 2.9 Function Sketch for Step 3, Case I

Since the product in the integrand is zero for all y,

$$f_3(x) = \int_{-\infty}^{\infty} f_1(y) f_2(x - y)\, dy = 0$$

Case II

$$0.5 \le x + 1 < 1.5 \quad \rightarrow \quad -0.5 \le x < 0.5$$

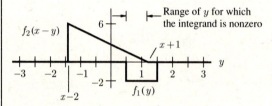

Figure 2.10 Function Sketch for Step 3, Case II

$$f_3(x) = \int_{-\infty}^{0.5} [0][f_2(x - y)]\, dy$$

$$+ \int_{0.5}^{x+1} [-2][2(x - y) + 2]\, dy$$

$$+ \int_{x+1}^{\infty} [f_1(y)][0]\, dy$$

$$= 4 \int_{0.5}^{x+1} [y - (x + 1)]\, dy$$

$$= 4 \left[\frac{y^2}{2} - (x + 1)y \right]_{0.5}^{x+1}$$

$$= -2x^2 - 2x - 0.5 \qquad -0.5 \le x < 0.5$$

Case III

$$x + 1 \ge 1.5 \quad \rightarrow \quad x \ge 0.5$$

and $\qquad\qquad\qquad\qquad\qquad\qquad\qquad\qquad\qquad\qquad$ $0.5 \le x < 2.5$

$$x - 2 < 0.5 \quad \rightarrow \quad x < 2.5$$

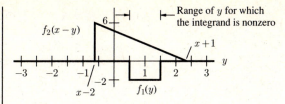

Figure 2.11 Function Sketch for Step 3, Case III

$$f_3(x) = \int_{-\infty}^{0.5} [0][f_2(x-y)]\, dy$$

$$+ \int_{0.5}^{1.5} [-2][2(x-y)+2]\, dy$$

$$+ \int_{1.5}^{\infty} [0][f_2(x-y)]\, dy$$

$$f_3(x) = 4 \left[\frac{y^2}{2} - (x+1)y \right]_{0.5}^{1.5}$$

$$= -4x \qquad 0.5 \le x < 2.5$$

Case IV

$$0.5 \le x - 2 < 1.5 \quad \rightarrow \quad 2.5 \le x \le 3.5$$

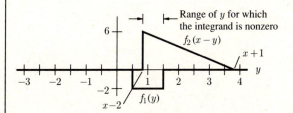

Figure 2.12 Function Sketch for Step 3, Case IV

$$f_3(x) = \int_{-\infty}^{x-2} [f_1(y)][0]\, dy$$

$$+ \int_{x-2}^{1.5} [-2][2(x-y)+2]\, dy$$

$$+ \int_{1.5}^{\infty} [0][f_2(x-y)]\, dy$$

$$= 4 \left[\frac{y^2}{2} - (x+1)y \right]_{x-2}^{1.5}$$

$$= 2x^2 - 2x - 17.5 \qquad 2.5 \le x < 3.5$$

Case V

$$x - 2 \geq 1.5 \quad \rightarrow \quad x \geq 3.5$$

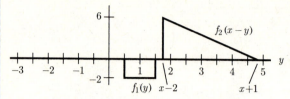

Figure 2.13 Function Sketch for Step 3, Case V

Since the product in the integrand is zero for all y,

$$f_3(x) = \int_{-\infty}^{\infty} f_1(y) f_2(x - y) = 0$$

In summary,

$$f_3(x) = \begin{cases} -2x^2 - 2x - 0.5 & -0.5 \leq x < 0.5 \\ -4x & 0.5 \leq x < 2.5 \\ 2x^2 - 2x - 17.5 & 2.5 \leq x < 3.5 \\ 0 & \text{elsewhere} \end{cases}$$

which is plotted in Figure 2.14.

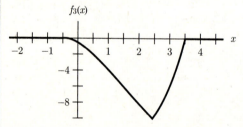

Figure 2.14 Convolution Result for Example 2.5

In learning how to evaluate convolution integrals, students should follow the detailed structured approach illustrated in the solutions to Examples 2.4 and 2.5. Once again, we emphasize that such an approach will serve to significantly reduce errors in working solutions of this type.

Continuous-Convolution Properties

Continuous convolution is *commutative*; that is

$$f(x) * g(x) = g(x) * f(x) \tag{2.2}$$

The commutative property is easily shown as follows; when written out in detail, the left side of eq. (2.2) is

$$f(x) * g(x) = \int_{-\infty}^{\infty} f(y)g(x - y)\, dy \qquad (2.3)$$

If we perform a change of variables by letting $x - y = \alpha$, then $y = x - \alpha$, $d\alpha = -dy$, $\alpha \to \infty$ as $y \to -\infty$, and $\alpha \to -\infty$ as $y \to \infty$. Therefore,

$$f(x) * g(x) = \int_{\infty}^{-\infty} f(x - \alpha)g(\alpha)[-d\alpha]$$

$$= \int_{-\infty}^{\infty} g(\alpha)f(x - \alpha)\, d\alpha$$

$$= g(x) * f(x) \qquad (2.4)$$

Considering this property can be helpful when performing convolution evaluation because the integration may be simpler for $g(x) * f(x)$ than for $f(x) * g(x)$.

Example 2.6

Repeat Example 2.3 by computing $y(f) * x(f)$ rather than $x(f) * y(f)$. Which computation is preferred?

Solution

$$w(f) = y(f) * x(f) = \int_{-\infty}^{\infty} y(\beta)x(f - \beta)\, d\beta \qquad \text{for all } f$$

Now

$$y(\beta) = \begin{cases} 0 & \beta < 0 \\ 1 & \beta \geq 0 \end{cases}$$

and

$$x(f - \beta) = \frac{1}{1 + (f - \beta)^2} = \frac{1}{1 + f^2 - 2f\beta + \beta^2} \qquad \text{for all } \beta$$

Therefore,

$$w(f) = \int_{-\infty}^{0} [0]\left[\frac{1}{1 + f^2 - 2f\beta + \beta^2}\right] d\beta$$

$$+ \int_{0}^{\infty} [1]\left[\frac{1}{1 + f^2 - 2f\beta + \beta^2}\right] d\beta$$

$$= \int_{0}^{\infty} \frac{d\beta}{1 + f^2 - 2f\beta + \beta^2}$$

$$= \left[\frac{2}{\sqrt{4\left(1+f^2\right)-4f^2}} \tan^{-1} \left[\frac{2\beta - 2f}{\sqrt{4\left(1+f^2\right)-4f^2}} \right] \right]_0^\infty$$

$$= \frac{\pi}{2} - \tan^{-1}(-f) = \tan^{-1} f + \frac{\pi}{2} \qquad \text{for all } f$$

This result is the same as that obtained in Example 2.3, but in this case the integrand and, thus, the integral evaluation is more complicated. It is preferable, then, to compute $w(f) = x(f) * y(f)$ rather than $w(f) = y(f) * x(f)$.

It is generally preferable to choose the simplest function to be the second function when performing convolution for the reason that the single independent-variable argument must be replaced by a difference argument for the second function.

Continuous convolution is *associative*; that is,

$$f(x) * [g(x) * h(x)] = [f(x) * g(x)] * h(x) \qquad (2.5)$$

To show associativity, the left side of eq. (2.5) is written out to give

$$f(x) * [g(x) * h(x)] = f(x) * \left[\int_{-\infty}^\infty g(\alpha)h(x-\alpha)\,d\alpha \right]$$

$$= \int_{-\infty}^\infty f(\beta) \left[\int_{-\infty}^\infty g(\alpha)h(x-\beta-\alpha)\,d\alpha \right] d\beta \quad (2.6)$$

A change of variables is performed by letting $\alpha = y - \beta$, which also yields $d\alpha = dy$, $y \to \infty$ as $\alpha \to \infty$, and $y \to -\infty$ as $\alpha \to -\infty$. Therefore, eq. (2.6) becomes

$$f(x) * [g(x) * h(x)] = \int_{-\infty}^\infty \int_{-\infty}^\infty f(\beta)g(y-\beta)h(x-y)\,dy\,d\beta \qquad (2.7)$$

Interchanging the order of integration, which is permissible except for some classes of functions of little interest in signal and system analysis, gives the desired result

$$f(x) * [g(x) * h(x)] = \int_{-\infty}^\infty \left[\int_{-\infty}^\infty f(\beta)g(y-\beta)\,d\beta \right] h(x-y)\,dy$$

$$= \int_{-\infty}^\infty [f(y) * g(y)]\, h(x-y)\,dy$$

$$= [f(x) * g(x)] * h(x) \qquad (2.8)$$

Since continuous convolution is associative, there is no need to indicate which convolution is to be performed first; hence, the square brackets can be removed in eq. (2.5). Since continuous convolution is also commutative, then

$$f(x) * g(x) * h(x) = f(x) * h(x) * g(x) = \cdots = h(x) * g(x) * f(x) \qquad (2.9)$$

That is, convolution of several functions can be performed in any order. Some orders may produce integrals that are easier to evaluate than others. The particular order that is best depends on the specific functions considered.

Continuous convolution is *distributive*; that is

$$f(x) * [g(x) + h(x)] = f(x) * g(x) + f(x) * h(x) \qquad (2.10)$$

It is easy to show that eq. (2.10) is true; therefore, we leave this proof as a problem for the reader. The distributive property is of interest because it is convenient to decompose a function into the sum of two simpler functions, perform two convolutions of the simpler functions with another function, and add the results than to perform one convolution on the original function and the other.

One property of the continuous convolution of two functions that can be noted from the examples previously shown is that the location of the left edge of the nonzero portion of the continuous convolution of two functions equals the sum of the locations of the left edges of the nonzero portions of the two functions. The same is true for the right edges and defines a second property.

These two properties produce a third property; that is, the width of the nonzero extent (the interval of time between the first and last nonzero values) of the continuous convolution of two functions equals the sum of the widths of the nonzero extents of the two functions.

We will now illustrate the three properties by means of an example.

Example 2.7

The two continuous-time functions $x(t)$ and $y(t)$ are shown in Figure 2.15. Where will the right and left edges of the convolution of $x(t)$ and $y(t)$ be located? What will be the width of the nonzero extent of the convolution?

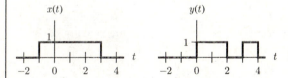

Figure 2.15 Functions for Example 2.7

Solution

$$\text{Right edge of } x(t) = RE_x = 3$$

$$\text{Right edge of } y(t) = RE_y = 4$$

$$\text{Left edge of } x(t) = LE_x = -1$$

$$\text{Left edge of } y(t) = LE_y = 0$$

$$\text{Width of } x(t) = W_x = RE_x - LE_x = 3 - (-1) = 4$$

Width of $y(t) = W_y = RE_y - LE_y = 4 - 0 = 4$

$z(t) \equiv x(t) * y(t)$

Right edge of $z(t) = RE_z = RE_x + RE_y = 3 + 4 = 7$

Left edge of $z(t) = LE_z = LE_x + LE_y = (-1) + (0) = -1$

Width of $z(t) = W_z = W_x + W_y = 4 + 4 = 8$

The result of the convolution of $x(t)$ and $y(t)$ is shown in Figure 2.16. One can easily verify that it is correct. Note that the right-edge and left-edge locations and the nonzero extent of $z(t)$ match the preceding computations.

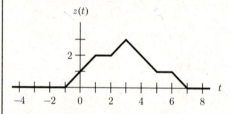

Figure 2.16 Convolution of Functions in Figure 2.15

The three edge and width properties we have illustrated provide indicators of the location of a continuous-convolution result before it is computed. We can use them as partial checks of continuous-convolution results.

2.2 Discrete Convolution

We will now consider discrete-variable functions of the independent variable α, the samples of which have a sample spacing of A.

Definition

The discrete convolution of the two discrete-variable functions, $x_A(n)$ and $y_A(n)$, is the discrete-variable function $z_A(n)$ given by the summation

$$z_A(n) = \sum_{m=-\infty}^{\infty} x_A(m)y_A(n-m) = x_A(n) * y_A(n)$$

$$= \sum_{m=-\infty}^{\infty} y_A(m)x_A(n-m) = y_A(n) * x_A(n) \qquad (2.11)$$

where m is a dummy index used in the summation.

Equation (2.11) is called the *convolution sum*.

Note that eq. (2.11) is of the same form as the continuous convolution defined by eq. (2.1), except that the integral in eq. (2.1) is replaced with a summation and the functions are discrete-variable functions rather than continuous-variable functions.

Discrete convolution has meaning for discrete-variable functions only when the two functions used in the convolution have the same sample spacing. A discrete-variable function with the same sample spacing will result. For example, if we want to compute the convolution of two discrete-time functions, then both must have the same sample spacing T. The convolution produces a discrete-time function with sample spacing T. We actually perform the convolution of two discrete-variable functions, eq. (2.11), by computing

$$z(n) = \sum_{m=-\infty}^{\infty} x(m)y(n-m) = x(n) * y(n)$$

$$= \sum_{m=-\infty}^{\infty} y(m)x(n-m) = y(n) * x(n) \qquad (2.12)$$

which is the discrete convolution of the sequences corresponding to the discrete-variable functions, keeping in mind that the discrete-variable functions corresponding to the sequences have a sample spacing of A. Thus, this chapter concentrates on the discrete convolution of sequences rather than discrete convolution of discrete-variable functions.

We often use the short notation defined on the right of eqs. (2.11) and (2.12) to indicate discrete convolution of the discrete-variable functions $x_A(n)$ and $y_A(n)$, respectively, without writing out the convolution summation. The notation is the same as that used for continuous convolution. The type of convolution intended is determined by the type argument of the functions—either continuous variable or sample index.

Discrete-Convolution Evaluation

The following examples serve to illustrate techniques that can be used to evaluate discrete convolution. The first example uses two sequences, each defined by a single equation for all values of n; that is, simply-defined sequences. In fact, the two sequences have sample values that are taken at one-unit intervals from the continuous-variable functions used in Example 2.1.

Example 2.8

Find the convolution of the two sequences $f(n) = e^{-n^2}$ and $g(n) = 3n^2$, for all n.

Solution

$$h(n) \equiv f(n) * g(n) = \sum_{m=-\infty}^{\infty} f(m)g(n-m)$$

$$= \sum_{m=-\infty}^{\infty} e^{-m^2} \left[3(n-m)^2 \right]$$

$$= 3n^2 \sum_{m=-\infty}^{\infty} e^{-m^2} - 6n \sum_{m=-\infty}^{\infty} me^{-m^2}$$

$$+ 3 \sum_{m=-\infty}^{\infty} m^2 e^{-m^2}$$

$$= An^2 + B \qquad \text{for all } n$$

where the constants A and B are

$$A = 3 \sum_{m=-\infty}^{\infty} e^{-m^2} = 5.318 \quad \text{and} \quad B = 3 \sum_{m=-\infty}^{\infty} m^2 e^{-m^2} = 2.654$$

In Example 2.8, we found the closed-form solution with constants accurate to three decimal places by simply summing several terms in the two summations. This worked because the two series being summed converge quite rapidly. Such rapid convergence will not always be the case, however, so that it may be difficult at times to evaluate the required summations simply. Note that the convolution of two simply defined sequences produces a simply defined sequence. This result parallels the result we saw for continuous convolution.

If one of the sequences to be convolved is piecewise defined and the other is simply defined, then we must evaluate the summation in two or more parts. However, the resulting convolution is a simply defined sequence, again paralleling the continuous-convolution result.

Example 2.9

Find the convolution of the two sequences

$$x(n) = e^{-n^2} \qquad \text{for all } n$$

and

$$y(n) = \begin{cases} e^n & n < 0 \\ e^{-n} & n \geq 0 \end{cases}$$

Solution

$$z(n) \equiv x(n) * y(n) = \sum_{m=-\infty}^{\infty} x(m)y(n-m)$$

Now

$$x(m) = e^{-m^2} \qquad \text{for all } m$$

and

$$y(n-m) = \begin{cases} e^{n-m} & n-m < 0 \quad \rightarrow \quad m > n \\ e^{-(n-m)} & n-m \geq 0 \quad \rightarrow \quad m \leq n \end{cases}$$

Therefore,

$$z(n) = \sum_{m=-\infty}^{n} e^{-m^2} e^{-(n-m)} + \sum_{m=n+1}^{\infty} e^{-m^2} e^{(n-m)}$$

$$= e^{-n} \sum_{m=-\infty}^{n} e^{(m-m^2)} + e^{n} \sum_{m=n+1}^{\infty} e^{-(m+m^2)} \qquad \text{for all } n$$

The sequence that results from the convolution performed in Example 2.9 is indeed simply defined. However, since the limits on the summations required are functions of n, it cannot be written as a simple closed-form function of n. We can determine a finite number of samples by computing $z(n)$ for several values of n. However, two infinite sums are required and can only be evaluated simply if they converge rapidly.

The two examples considered show that the need to compute infinite sums creates a difficulty in obtaining solutions for discrete convolutions. The infinite sums do not occur if sequences have zero values prior to some starting time. We see that if

$$x(n) = 0 \qquad n < k_1$$

and

$$y(n) = 0 \qquad n < k_2$$

(2.13)

then

$$x(m) = 0 \qquad m < k_1$$

and

$$y(n-m) = 0 \qquad n - m < k_2 \quad \rightarrow \quad m > n - k_2$$

(2.14)

Therefore,

$$z(n) \equiv x(n) * y(n) = \begin{cases} \sum_{m=k_1}^{n-k_2} x(m)y(n-m) & n \geq k_1 + k_2 \\ 0 & n < k_1 + k_2 \end{cases} \qquad (2.15)$$

which is a finite sum, not an infinite sum.

The second part of eq. (2.15) follows from the upper limit of the summation being less than the lower limit of the summation, which says that there is no value of m for which the product of $x(m)$ and $y(n-m)$ is nonzero. That is, the nonzero portions of $x(m)$ and $y(n-m)$ do not overlap. Further examples of the evaluation of discrete convolution will be restricted to functions that are zero before some starting time so that results can be left in other than summation form. Such functions are often encountered in system analysis. They are piecewise-defined, and the convolution of two of them will be a piecewise-defined function, as it was for continuous convolution.

Example 2.10

Find the convolution of the two sequences

$$f(n) = \begin{cases} 0 & n < -5 \\ (1/2)^n & n \geq -5 \end{cases}$$

and

$$g(n) = \begin{cases} 0 & n < 3 \\ (1/3)^n & n \geq 3 \end{cases}$$

Solution

$$h(n) \equiv f(n) * g(n) = \sum_{m=k_1}^{n-k_2} f(m)g(n-m)$$

where $k_1 = -5$ and $k_2 = 3$.
Since

$$f(m) = \begin{cases} 0 & m < -5 \\ (1/2)^m & m \geq -5 \end{cases}$$

$$g(n-m) = \begin{cases} 0 & n-m < 3 \quad \rightarrow \quad m > n-3 \\ (1/3)^{n-m} & n-m > 3 \quad \rightarrow \quad m < n-3 \end{cases}$$

then

$$h(n) = \begin{cases} 0 & n < -2 \\ \sum_{m=-5}^{n-3} (1/2)^m (1/3)^{n-m} & n \geq -2 \end{cases}$$

Letting $i = m + 5$, then for $n \geq -2$,

$$h(n) = \sum_{i=0}^{n-8} (1/2)^{i-5} (1/3)^{n-i+5}$$

$$= (2/3)^5 (1/3)^n \sum_{i=0}^{n-8} (3/2)^i$$

$$= (2/3)^5 (1/3)^n \left[\frac{1 - (3/2)^{n-7}}{1 - 3/2} \right]$$

$$= -2 (2/3)^5 \left[(1/3)^n - (2/3)^7 (1/2)^n \right]$$

and the solution can be written as

$$h(n) = \begin{cases} 0 & n < -2 \\ -64 (1/3)^{n+5} + 2 (2/3)^{12} (1/2)^n & n \geq -2 \end{cases}$$

For many functions that are zero before some starting time, it is not possible to arrive simply at a closed-form expression for a resulting convolution sequence. Consequently, we often compute the values of the first k nonzero terms in the convolution sequence. That is, $z(n)$ is computed for $(k_1 + k_2) \leq n < (k_1 + k_2 + k)$ by using eq. (2.15). The convolution sequence is then known for $n \leq (k_1 + k_2 + k - 1)$, since it is known to be zero for $n < (k_1 + k_2)$.

The convolution sum for $n \geq (k_1 + k_2)$ shown in eq. (2.15) can be written as

$$z(n) = x(k_1) y(n - k_1) + x(k_1 + 1) y[n - (k_1 + 1)]$$

$$+ \cdots + x(n - k_2) y(k_2) \tag{2.16}$$

Thus, its evaluation for each value of n is the sum of the products of a set of $x(i)$ and $y(j)$ values for which the sum of the arguments of x and y is n (that is, $i + j = n$). In the summation, i begins at $i = k_1$ and increases, and j decreases and ends at $j = k_2$.

The evaluation of the first k terms of the convolution sequence is particularly useful in two instances: first, if the resulting convolution approaches a constant rapidly enough with n so that its values can be inferred after some period of time and, second, if the two sequences being convolved cease to exist (become zero) for values of n greater than some value. In the second case, the convolution can be determined for all n, since it becomes zero for values of n greater than some value. Convolution computation techniques for these cases are illustrated in the following two examples.

Example 2.11

Find the sequence values for $n \leq 5$ for the convolution of the two sequences

$$\alpha(n) = \begin{cases} 0 & n < 4 \\ 1 & n \geq 4 \end{cases}$$

and

$$\beta(n) = \begin{cases} 0 & n < -2 \\ (1/2)^n & n \geq -2 \end{cases}$$

Solution

$$\gamma(n) \equiv \alpha(n) * \beta(n) = \sum_{m=k_1}^{n-k_2} \alpha(m)\beta(n-m) \qquad n \geq k_1 + k_2$$

where $k_1 = 4$ and $k_2 = -2$. Therefore,

$$\gamma(n) = 0 \qquad n < k_1 + k_2 = 2$$

$$\gamma(2) = \alpha(4)\beta(-2) = (1)(1/2)^{-2} = 4$$

$$\gamma(3) = \alpha(4)\beta(-1) + \alpha(5)\beta(-2)$$

$$= (1)(1/2)^{-1} + (1)(1/2)^{-2} = 6$$

$$\gamma(4) = \alpha(4)\beta(0) + \alpha(5)\beta(-1) + \alpha(6)\beta(-2)$$

$$= (1)(1/2)^0 + (1)(1/2)^{-1} + (1)(1/2)^{-2} = 7$$

$$\gamma(5) = \alpha(4)\beta(1) + \alpha(5)\beta(0) + \alpha(6)\beta(-1) + \alpha(7)\beta(-2)$$

$$= (1)(1/2) + (1)(1/2)^0 + (1)(1/2)^{-1} + (1)(1/2)^{-2} = 7.5$$

We can see that the convolution sequence approaches

$$\gamma(\infty) = 4 + 2 + \sum_{i=0}^{\infty} (1/2)^i = 4 + 2 + 2 = 8$$

quite rapidly—the next four terms are $\gamma(6) = 7.75$, $\gamma(7) = 7.875$, $\gamma(8) = 7.9375$, and $\gamma(9) = 7.96875$.

Example 2.12

Find the convolution of the two sequences

$$x(n) = \begin{cases} n+1 & 0 \leq n < 2 \\ 5-n & 2 \leq n < 5 \\ 0 & \text{elsewhere} \end{cases}$$

and

$$y(n) = \begin{cases} -\frac{n}{2} & 2 \leq n < 5 \\ 0 & \text{elsewhere} \end{cases}$$

Solution

$$z(n) \equiv x(n) * y(n) = \sum_{m=k_1}^{n-k_2} x(m)y(n-m) \qquad n \geq k_1 + k_2$$

where $k_1 = 0$ and $k_2 = 2$. Therefore,

$$z(n) = 0 \qquad n < k_1 + k_2 = 2$$

$$z(2) = x(0)y(2) = (0+1)(-2/2) = -1$$

$$z(3) = x(0)y(3) + x(1)y(2)$$
$$= (0+1)(-3/2) + (1+1)(-2/2) = -3.5$$

$$z(4) = x(0)y(4) + x(1)y(3) + x(2)y(2)$$
$$= (0+1)(-4/2) + (1+1)(-3/2) + (5-2)(-2/2) = -8$$

$$z(5) = x(0)y(5) + x(1)y(4) + x(2)y(3) + x(3)y(2)$$

Since $y(5) = 0$,

$$z(5) = x(1)y(4) + x(2)y(3) + x(3)y(2)$$
$$= (1+1)(-4/2) + (5-2)(-3/2) + (5-3)(-2/2) = -10.5$$

For $z(6)$,

$$z(6) = x(2)y(4) + x(3)y(3) + x(4)y(2)$$
$$= (5-2)(-4/2) + (5-3)(-3/2) + (5-4)(-2/2) = -10$$

where we have not written the first two terms of the summation since $y(6) = y(5) = 0$. For $z(7)$,

$$z(7) = x(3)y(4) + x(4)y(3) + x(5)y(2)$$

Since $x(5) = 0$,

$$z(7) = x(3)y(4) + x(4)y(3)$$
$$= (5-3)(-4/2) + (5-4)(-3/2) = -5.5$$

For $z(8)$,

$$z(8) = x(4)y(4) = (5-4)(-4/2) = -2$$

where we have not written the first four terms and last two terms of the summation, since $y(8) = y(7) = y(6) = y(5) = 0$ and $x(5) = x(6) = 0$. Obviously, $z(n) = 0$ for $n \geq 9$ for the reason that all the nonzero values for $y(j)$ in the summation correspond to $j \le 4$, meaning that they are multiplied by $x(i)$ values corresponding to $i = n - j \ge 9 - 4 = 5$, and these values are all zero.

The preceding sequences and their resulting convolution are illustrated in Figure 2.17.

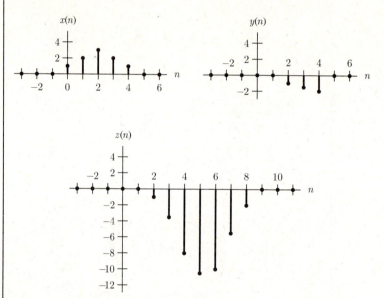

Figure 2.17 Sequences for Example 2.12

Discrete-Convolution Properties

We can show that discrete convolution is commutative, associative, and distributive. The reader will be asked to provide proof details in the end-of-chapter problems. The commutative and associative properties permit discrete convolution of several sequences to be performed sequentially and in different orders. As for continuous convolution, evaluation may be easier for some choices of order.

The discrete convolution of two sequences has the same type of edge properties as the continuous convolution of two continuous-variable functions. That is, the location of the first nonzero sequence value (left edge) of the discrete convolution of two sequences equals the sum of the locations of the first nonzero sequence values (left edges) of the two sequences. The same is true for the last nonzero sequence values (right edges). It is obvious that these properties are true for the discrete convolution in Example 2.12, where $LE_x + LE_y = 0 + 2 = 2 = LE_z$ and $RE_x + RE_y = 4 + 4 = 8 = RE_z$. Performing these computations before proceeding with the convolution of two finite length sequences indicates which values must be computed for the convolution.

Example 2.13

Compute the convolution of the two sequences shown in Figure 2.18.

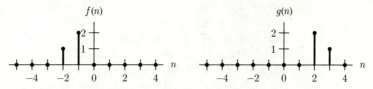

Figure 2.18 Sequences for Example 2.13

Solution

$$h(n) \equiv f(n) * g(n) = \sum_{m=k_1}^{n-k_2} f(m)g(n - m) \qquad n \geq k_1 + k_2$$

where

$$k_1 = -2 \quad and \quad k_2 = 2$$

$$LE_h = LE_f + LE_g = -2 + 2 = 0$$

$$RE_h = RE_f + RE_g = -1 + 3 = 2$$

Therefore,

$$h(n) = 0 \qquad n < 0$$

$$h(n) = 0 \qquad n > 2$$

$$h(0) = f(-2)g(2) = (1)(2) = 2$$

$$h(1) = f(-2)g(3) + f(-1)g(2) = (1)(1) + (2)(2) = 5$$

$$h(2) = f(-1)g(3) = (2)(1) = 2$$

We found earlier that the width of the nonzero extent of the continuous convolution of two functions is the sum of the widths of the nonzero extents of the two functions. A similar property that holds for discrete convolution is that the number of samples in the nonzero extent of a discrete convolution of two sequences is one less than the sum of the number of samples in the nonzero extents of the two sequences. This is readily apparent from Examples 2.12 and 2.13. We now demonstrate that it is true in general.

The number of sequence values in the nonzero extent of $f(n)$ is $N_f = RE_f - LE_f + 1$ and the number of sequence values in the nonzero extent of $g(n)$ is $N_g = RE_g - LE_g + 1$. Likewise, the number of sequence values in the nonzero extent of $h(n) = f(n) * g(n)$ is $N_h = RE_h - LE_h + 1$. Since $RE_h = RE_f + RE_g$, and $LE_h = LE_f + LE_g$, then $N_h = RE_f + RE_g - LE_f - LE_g + 1 = N_f + N_g - 1$, which is the desired result.

2.3 Summary

Convolution is an important operation in determining output signals and signal-frequency characteristics. The focus of this chapter was on the mathematical characteristics of convolution and on techniques for its evaluation so as to provide the necessary background for discussion of convolution applications to system analysis in subsequent chapters.

The convolution of continuous-variable functions is called continuous convolution; that of discrete-variable sequences is called discrete convolution. Continuous convolution involves the evaluation of an integral, called the convolution integral, whereas discrete convolution involves the evaluation of an infinite sum, called the convolution sum. A structured approach is required to perform these evaluations without error. We have illustrated this approach with a number of examples using different types of signals. Properties of the location and width of the nonzero portion of a convolution are useful in determining some characteristics of the convolution of two functions before performing the evaluation.

Problems

2.1 Find the convolution of $f(x) = e^{-4x^2}$ and $g(x) = 2\cos 2x$.

2.2 Compute the convolution of the two continuous-frequency functions

$$x(f) = \begin{cases} 1 & |f| \leq 4 \\ 0 & |f| > 4 \end{cases}$$

and

$$y(f) = 3\sin(0.5\pi f) \qquad \text{for all } f$$

2.3 Find and sketch $h(t) = f(t) * g(t)$ when

$$f(t) = \begin{cases} 1+t & -1 \leq t < 0 \\ 1-t & 0 \leq t < 1 \\ 0 & \text{elsewhere} \end{cases}$$

$$g(t) = e^{-|t|} \qquad \text{for all } t$$

2.4 Find and sketch the convolution of $f(x)$ with itself when $f(x)$ is the function shown in Figure 2.19.

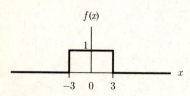

$f(x)$

$-3 \quad 0 \quad 3$

Figure 2.19

2.5 Find the convolution of the two continuous-time functions given and sketch the result.

$$x(t) = \begin{cases} 1 & t \geq 3 \\ 0 & t < 3 \end{cases}$$

$$h(t) = \begin{cases} e^{-2t} & t \geq 0 \\ 0 & t < 0 \end{cases}$$

2.6 Repeat Problem 2.5 with $x(t)$ replaced by

$$x(t) = \begin{cases} 1 & 3 \leq t \leq 4 \\ 0 & \text{elsewhere} \end{cases}$$

2.7 Compute the convolution of the two continuous-time functions given and sketch the result.

$$f(t) = \begin{cases} 1 - e^t & t < 0 \\ 0 & t \geq 0 \end{cases}$$

$$g(t) = \begin{cases} 2 & 1 \leq t \leq 3 \\ 0 & \text{elsewhere} \end{cases}$$

2.8 Find and sketch the convolution of the functions shown in Figure 2.20.

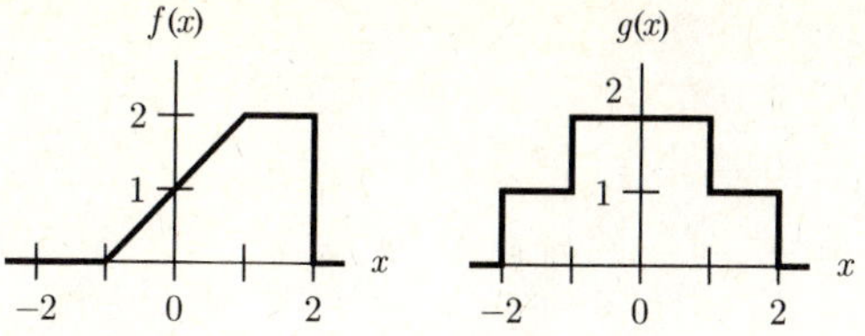

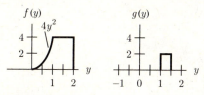

Figure 2.20

2.9 Repeat Problem 2.8 for the functions in Figure 2.21.

Figure 2.21

2.10 Repeat Problem 2.8 for the functions in Figure 2.22.

2.11 Repeat Problem 2.8 for the functions in Figure 2.23.

2.12 Without performing the convolutions, find the left- and right-edge locations and the widths of the functions that result from the convolutions of Problems 2.5 through 2.7.

2.13 Without performing the convolutions, find the left- and right-edge locations and the widths of the functions that result from the convolutions of Problems 2.8 through 2.11.

2.14 Show that the distributive property holds for continuous convolution.

2.15 Find the closed-form expression for the sequence $h(n) = f(n) * g(n)$, with constants accurate to two decimal places, when

$$f(n) = \frac{1}{1 + n^4} \quad \text{and} \quad g(n) = n^2, \text{ for all } n.$$

2.16 For Example 2.9, write and use a simple computer program to evaluate values of $z(n)$ for $-10 \leq n \leq 10$ that are accurate to two decimal places.

2.17 Find the first six nonzero values for $w_T(n) = x_T(n) * y_T(n)$ when

$$x_T(n) = \left[\left\{ \begin{matrix} 2e^{-0.5n} & n \geq 0 \\ 0 & n < 0 \end{matrix} \right\}, \quad 2 \text{ s} \right]$$

and

$$y_T(n) = \left[\left\{ \begin{matrix} 1 & 2 \leq n \leq 4 \\ 0 & \text{elswhere} \end{matrix} \right\}, \quad 2 \text{ s} \right]$$

Sketch the resulting signal.

2.18 Find and sketch the convolution of $x(n)$ with itself when

$$x(n) = \left[\begin{matrix} 1 & -2 \leq n \leq 2 \\ 0 & \text{elsewhere} \end{matrix} \right.$$

2.19 Find the values of $z(n) = x(n) * y(n)$ for $-6 \leq n \leq 6$ and sketch them when

$$x(n) = \left\{ \begin{matrix} \frac{1}{n^2} & n \leq -1 \\ 0 & n > -1 \end{matrix} \right.$$

and

$$y(n) = \left\{ \begin{matrix} n - 0.5 & 1 \leq n \leq 3 \\ 0 & \text{elsewhere} \end{matrix} \right.$$

2.20 Find and sketch the convolution of the two sequences shown in Figure 2.24.

Figure 2.22

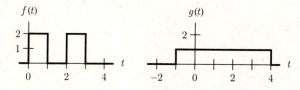

Figure 2.23

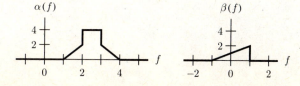

Figure 2.24

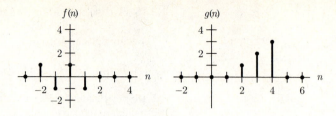

Figure 2.25

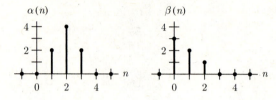

2.21 Repeat Problem 2.20 for the sequences

$$x(n) = \{3, 2, 0, 0, 2\} \qquad n_1 = 3$$

and

$$y(n) = \{1, -2, 1, 1\} \qquad n_1 = -1$$

2.22 Repeat Problem 2.20 for the sequences shown in Figure 2.25.

2.23 Repeat Problem 2.20 for the sequences

$$f(n) = \{2, 2, 3, 3, 2, 2\}$$

and

$$y(n) = \{1, 2, 3, 2\} \qquad n_1 = 2$$

2.24 Without performing the convolutions, find the locations of the first nonzero sequence value (left edge) and the last nonzero sequence value (right edge), as well as the number of sequence values in the nonzero extent of the sequences that result from the convolutions in Problems 2.17, 2.18, and 2.19.

2.25 Repeat Problem 2.24 for the convolution sequences that result from Problems 2.20, 2.21, 2.22, and 2.23.

2.26 Show that discrete convolution is commutative.

2.27 Show that discrete convolution is associative.

2.28 Show that discrete convolution is distributive.

II

Continuous-Time Signals and Systems

3 Continuous-Time Signals

In Chapter 1, we indicated that a deterministic continuous-time signal is defined for all values of time and that it can be modeled by either a single mathematical expression for all time (a simply-defined signal) or a set of mathematical expressions, with each valid over a defined portion of the time axis (a piecewise-defined signal). In this chapter, we will consider the definitions and properties of several simple mathematical expressions that are useful by themselves and in combinations for modeling continuous-time signals in continuous-time signal and system analysis. Recall that signals may be functions of time, frequency, distance, temperature, and the like. Here, we consider them as functions of time.

In addition to discussing useful signal models, we will define the concept of signal energy and signal power. We also describe one method for representing a continuous-time signal over a time interval by a sum of signals. The resulting sum of signals is referred to as a *generalized Fourier series* when the signals summed are chosen to satisfy certain general characteristics. These characteristics, along with the nature of the convergence of the series as more terms are added, will also be described.

3.1 Sinusoidal and Complex-Exponential Signals

As previously mentioned, no physical signals are periodic; all are turned on or begin at some time and are turned off or cease to exist at some later time. However, the continuous-time sinusoidal signal

$$x(t) = A \cos \left[\frac{2\pi t}{T_0} + \theta \right] = A \sin \left[\frac{2\pi t}{T_0} + \theta + \frac{\pi}{2} \right] \tag{3.1}$$

is periodic with period T_0, and is quite useful in signal and system analysis. For example, when analyzing an electric power system we use the sinusoidal signal to represent signal waveforms which, in this system, are approximately sinusoidal in shape and extend for a long period of time. The sinusoidal signal is also useful in providing background and insight into the concepts of signal-frequency content and bandwidth and of frequency response to be discussed in later chapters.

We can write the sinusoidal signal of eq. (3.1) as

$$x(t) = A \cos \left(2\pi f_0 t + \theta \right) \tag{3.2}$$

The parameter A in eq. (3.2) is called the *signal amplitude* and has units consistent with the signal type (for example, a voltage has units of volts). The parameter $f_0 = 1/T_0$ is called the *signal frequency* and has units of hertz (abbreviated Hz). A signal with a frequency of f_0 Hz repeats f_0 times in one second. We say that it contains f_0 cycles per second. The parameter θ, expressed in units of radians, is called the *phase angle* or *phase* of the signal. It is the angular difference between the argument of the specified cosine signal and the argument of the reference cosine signal $x_r(t) = A\cos(2\pi f_0 t)$.

The phase angle of a sinusoidal signal is proportional to the time shift of the sinusoidal signal with respect to the reference sinusoidal signal. We show this by rewriting eq. (3.2) as

$$x(t) = A\cos\left\{2\pi f_0\left[t - \left(-\frac{\theta}{2\pi f_0}\right)\right]\right\} = A\cos\left\{2\pi f_0(t - t_d)\right\} \qquad (3.3)$$

from which we see that the time delay of $x(t)$ with respect to the reference signal $x_r(t) = A\cos(2\pi f_0 t)$ is

$$t_d = -\left(\frac{1}{2\pi f_0}\right)\theta \qquad (3.4)$$

The constant of proportionality is negative, which means that a negative phase angle corresponds to a time delay and a positive phase angle corresponds to a time advance. Also, the absolute value of a sinusoidal signal's phase angle is directly proportional to its frequency for a given time shift. These facts are relevant in later considerations of system effects.

The preceding definitions and quantities for a sinusoidal signal that is referenced to a cosine reference signal are shown in Figure 3.1a. We can also choose the reference sinusoidal signal to be the sine signal. The same definitions and quantities apply and are illustrated in Figure 3.1b. Note that, for a given reference signal, the three parameters A, f_0, and θ completely specify the sinusoidal signal. In our discussion, we will restrict the reference signal to a cosine, so that the three parameters will be sufficient to specify completely the sinusoidal signal.

Euler's theorem states that the complex exponential $e^{j\beta}$, can be written

$$e^{j\beta} = \cos\beta + j\sin\beta \qquad (3.5)$$

Thus, we can write the cosine signal of eq. (3.3) as

$$x(t) = Re\left[Ae^{j(2\pi f_0 t + \theta)}\right] = Re\left[x_p(t)\right] \qquad (3.6)$$

where, as shown in Figure 3.2, $x_p(t)$ represents a phasor of constant amplitude A that rotates at a uniform angular rate of $2\pi f_0 = \omega_0$ rad per second. Figure 3.2 shows that the rotating phasor $x_p(t)$ is periodic with period $1/f_0 = T_0$, since $x_p(t)$ repeats at multiples of the time interval T_0.

We write the rotating phasor as

$$x_p(t) = \left[Ae^{j\theta}\right]e^{j2\pi f_0 t} \equiv \mathbf{X}e^{j2\pi f_0 t} \qquad (3.7)$$

where $\mathbf{X}$ is a complex number, the value of which is the value of the rotating phasor at $t = 0$. Since the magnitude of $\mathbf{X}$ is the amplitude of $x(t)$, and the angle of $\mathbf{X}$ is the phase of $x(t)$, then $\mathbf{X}$ specifies the signal $x(t)$, except for its frequency. It is referred to as the *phasor representation of $x(t)$*.

Figure 3.1
Sinusoidal Signals

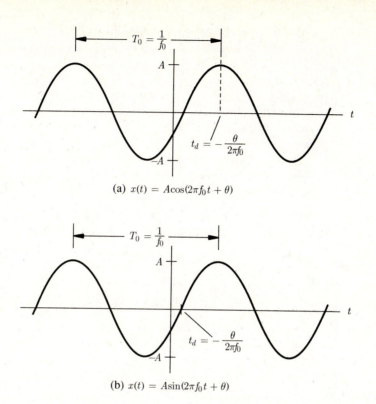

(a) $x(t) = A\cos(2\pi f_0 t + \theta)$

(b) $x(t) = A\sin(2\pi f_0 t + \theta)$

The phasor representation of $x(t)$ is convenient for analyzing a linear time-invariant system with input signals that are all sinusoids of the same known frequency. Each signal in the system is a sinusoid of the known frequency, and all that is needed to characterize the signal is its amplitude and phase. Since the rotating phasors corresponding to all system signals rotate at the same angular rate in this case, they maintain the same relative position with respect to each other at all times. Thus, we can analyze signal characteristics

Figure 3.2
Phasor Definitions

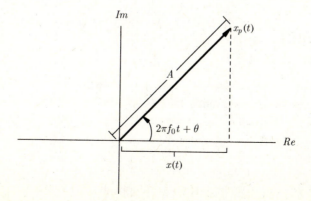

(amplitudes and phases) at any time instant. With the choice of $t = 0$, the signal characteristics are determined by using the phasor representation for the signals, algebraic equations, and complex arithmetic. It is assumed that the reader has already completed a study of steady-state sinusoidal circuit analysis in which this method was developed and used; hence, no further details will be presented here. However, as a review, we include the following example of the use of the phasor method in adding two sinusoidal signals of the same frequency.

Example 3.1

A single household circuit supplies electrical power to a fan motor and a lamp. They are connected in parallel and the electric currents supplied to them are

$$i_m(t) = 6.5 \cos\left[2\pi(60)t - \pi/6\right]$$

and

$$i_\ell(t) = 2 \cos\left[2\pi(60)t + \pi/40\right]$$

Find the total current supplied to the circuit.

Solution

$$\mathbf{I}_m = 6.5e^{-j\pi/6} = 6.5\underline{/-\pi/6} = 5.629 - j3.250$$

$$\mathbf{I}_\ell = 2e^{+j\pi/40} = 2\underline{/+\pi/40} = 1.994 + j0.157$$

$$\mathbf{I}_t = \mathbf{I}_m + \mathbf{I}_\ell = 7.623 - j3.093$$

$$= 8.227\underline{/-0.385} = 8.227e^{-j0.385}$$

where we have included the polar notation $A\underline{/\theta} = Ae^{j\theta}$ for review and all angles are expressed in radians (rad). Therefore,

$$i_t(t) = 8.227 \cos[2\pi(60)t - 0.385]$$

In the preceding example, we used the polar notation commonly used to represent a complex number; that is,

$$\mathbf{X} = 3e^{j\pi/4} \equiv 3\underline{/\pi/4} \tag{3.8}$$

where the angle has units of radians. When complex quantities are to be expressed in general, rather than as specific numerical values, it is convenient to use a slightly different definition for the polar-form notation:

$$\mathbf{X} = |\mathbf{X}|\underline{/\mathbf{X}} = |\mathbf{X}|e^{j\underline{/\mathbf{X}}} \tag{3.9}$$

where $|\mathbf{X}|$ is read "magnitude of $\mathbf{X}$" and $\underline{/\mathbf{X}}$ is read "angle associated with $\mathbf{X}$". For the specific complex number in eq. (3.8), $|\mathbf{X}| = 3$ and $\underline{/\mathbf{X}} = \pi/4$.

The polar-form notation defined in eq. (3.9) allows magnitudes and angles of different complex quantities to be easily distinguished (for example, $|\mathbf{X}|$ and $|\mathbf{Y}|$; $\underline{/\mathbf{X}}$ and $\underline{/\mathbf{Y}}$).

We see from eq. (3.5) that

$$\frac{\alpha}{2}e^{j\beta} + \frac{\alpha}{2}e^{-j\beta} = \frac{\alpha}{2}[\cos\beta + j\sin\beta + \cos\beta - j\sin\beta] = \alpha\cos\beta \qquad (3.10)$$

Thus, the cosine signal of eq. (3.2), which is a real function of time, can be alternatively written as the sum of two complex functions of time, which are complex conjugates. That is,

$$x(t) = \frac{A}{2}e^{j(2\pi f_0 t + \theta)} + \frac{A}{2}e^{-j(2\pi f_0 t + \theta)} \equiv x_{p2}(t) + x_{p2}^*(t) \qquad (3.11)$$

where the asterisk superscript indicates complex conjugate. While this appears to be a more complicated representation of the cosine function, we often find that it is mathematically convenient when considering signal properties and analyzing system performance.

Equation (3.11) shows that a sinusoidal signal can be represented as the sum of two counterrotating phasors, as illustrated in Figure 3.3. It can also be rewritten as

$$x(t) = \left[\frac{A}{2}e^{j\theta}\right]e^{j2\pi f_0 t} + \left[\frac{A}{2}e^{-j\theta}\right]e^{-j2\pi f_0 t} \qquad (3.12)$$

With the equation in this form, we see that the phasor rotating in the positive direction has a complex value at $t = 0$, the magnitude of which is one-half the amplitude of $x(t)$ and the angle of which is the phase of $x(t)$. Thus, this phasor specifies the signal $x(t)$, except for its frequency.

Figure 3.3
Counterrotating Phasor Representation of a Cosine Function

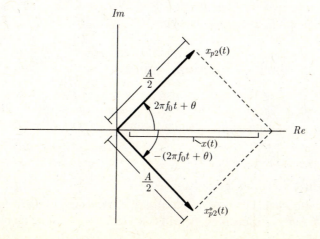

Example 3.2

Write $x(t) = 6\cos[2\pi(100)t - 0.45]$ as a sum of counterrotating phasors with complex magnitudes.

Solution

$$x(t) = \left[3e^{-j0.45}\right]e^{j2\pi(100)t} + \left[3e^{j0.45}\right]e^{-j2\pi(100)t}$$

$$= [2.701 - j1.305]e^{j2\pi(100)t} + [2.701 + j1.305]e^{-j2\pi(100)t}$$

where the complex magnitudes of the rotating phasors have been expressed in two forms to highlight their complex and conjugate values.

3.2 Exponential Signals

The transfer of energy from energy-storage components to energy-using components in a continuous-time, linear, time-invariant system often produces signals that decrease in amplitude with time in an approximately exponential manner. Thus, a useful continuous-time signal is

$$x(t) = \begin{cases} Ae^{-\alpha t} & t \geq t_1 \\ 0 & t < t_1 \end{cases} \tag{3.13}$$

where $\alpha > 0$. If there is more than one energy-storage component in the system, then the energy may oscillate back and forth between them during the time that it is being transferred to energy-using components. This gives signals that are approximately exponentially decreasing sinusoids (referred to as *damped sinusoids*), expressed by

$$x(t) = \begin{cases} Ae^{-\alpha t}\cos[2\pi f_0 t + \theta] & t \geq t_1 \\ 0 & t < t_1 \end{cases} \tag{3.14}$$

where $\alpha > 0$. Examples of these two types of signal models are shown in Figure 3.4. Signals for which $\alpha < 0$ can occur in unstable systems.

3.3 Singularity-Function Signals

A group of functions, referred to as *singularity functions*, are useful as signal models. They can all be derived by successive differentiation or integration of the basic singularity function, which is known as the *unit impulse function*, or *delta function*.

Unit Impulse Function

Before indicating the physical signals that might be effectively modeled by a unit impulse function, we will consider how this function is defined and discuss

Figure 3.4
Examples of
Exponential Signals

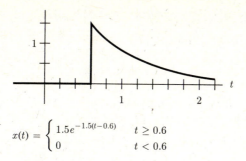

$$x(t) = \begin{cases} 1.5e^{-1.5(t-0.6)} & t \geq 0.6 \\ 0 & t < 0.6 \end{cases}$$

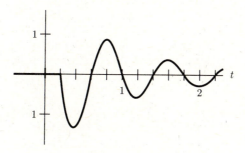

$$x(t) = \begin{cases} 2e^{-t} \cos\left[2\pi(1.25)t\right] & t \geq 0.2 \\ 0 & t < 0.2 \end{cases}$$

its properties. Actually, the unit impulse function is probably most useful in continuous-time system analysis in an abstract signal sense; that is, as a signal not actually used, but which generates a system response providing fundamental information about the system characteristics.

The unit impulse function, $\delta(t)$, is not a mathematical function in the usual sense of defining a value of the dependent variable for each value of the independent variable. Instead, it is a generalized function defined by the integral

$$\int_{-\infty}^{\infty} x(t + t_0)\delta(t) \, dt = x(t_0) \tag{3.15}$$

where the integral obeys the formal properties of integrals, t_0 is an arbitrary constant time, and $x(t + t_0)$ is any function that is continuous at $t = 0$. That is, $x(\alpha)$ is continuous at $\alpha = t_0$, so that $x(t_0)$ is defined. By using the change of variables $t = \tau - t_0$, we can write eq. (3.15) as

$$\int_{-\infty}^{\infty} x(\tau) \, \delta(\tau - t_0) \, d\tau = x(t_0) \tag{3.16}$$

where $\delta(\tau - t_0)$ is the unit impulse function shifted t_0 units to the right and $x(\tau)$ is continuous at $\tau = t_0$. We refer to eq. (3.16) as the *sifting integral*, since it sifts out the value of the function $x(t)$ at $t = t_0$. Equation 3.16 can be used as the definition of the unit impulse function rather than eq. (3.15).

Impulse-Function Properties and Notation

We can multiply eq. (3.15) by the constant A to obtain

$$\int_{-\infty}^{\infty} x(t + t_0)[A\delta(t)] \ dt = Ax(t_0) \tag{3.17}$$

Thus, the function $A\delta(t)$ is a more general (nonunit) impulse function. We refer to it as an impulse function of strength A. Equation (3.16) can also be multiplied by the constant A to give

$$\int_{-\infty}^{\infty} x(\tau)[A\delta(\tau - t_0)] \ d\tau = Ax(t_0) \tag{3.18}$$

which is useful in the following discussion of impulse-function properties. Assuming that the concept of the impulse function is new to most readers, we will demonstrate that the properties are valid in considerable detail.

Property I $A\delta(-t) = A\delta(t)$ \hfill (3.19)

To demonstrate the validity of Property I, we first define $y(t + t_0) = x(-t + t_0)$, where $x(-t + t_0)$ is any function that is continuous at $t = 0$. Thus, $y(t + t_0)$ is continuous at $t = 0$ and $y(t_0) = x(t_0)$. Using this result in eq. (3.17) gives

$$\int_{-\infty}^{\infty} y(\tau + t_0)[A\delta(\tau)] \ d\tau = Ay(t_0) = Ax(t_0) \tag{3.20}$$

Therefore, since $y(\tau + t_0) = x(-\tau + t_0)$, then

$$\int_{-\infty}^{\infty} x(-\tau + t_0)[A\delta(\tau)] \ d\tau = Ax(t_0) \tag{3.21}$$

We make the change of variable $\tau = -t$ to give

$$\int_{-\infty}^{\infty} x(t + t_0)[A\delta(-t)] \ dt = Ax(t_0) \tag{3.22}$$

Since $x(t + t_0)$ is any function continuous at $t = 0$, then eq. (3.22) has the form of eq. (3.17), which defines the function $[A\delta(-t)]$ in the integrand as an impulse function of strength A. Therefore, $A\delta(-t) = A\delta(t)$, which is the desired property.

Property II $\int_{-\infty}^{\infty} A\delta(t) \ dt = A$ \hfill (3.23)

That is, the area under an impulse function equals its strength. Property II is easily shown by setting $x(t + t_0) = 1$ for all t in eq. (3.17) to give

$$\int_{-\infty}^{\infty} A\delta(t) \ dt = Ax(t_0) = A \tag{3.24}$$

Property III $A\delta(t) = 0$ for $t \neq 0$

We can show that this property is valid by setting

$$x(t + t_0) = \begin{cases} 1 & |t| \leq \tau \\ 0 & \text{elsewhere} \end{cases}$$

where τ can be chosen to be arbitrarily small. With this choice of $x(t + t_0)$, eq. (3.17) becomes

$$\int_{-\tau}^{\tau} A\delta(t)\, dt = Ax(t_0) = A \tag{3.25}$$

Since this integral equals the same value as the integral in eq. (3.24), regardless of the value of τ, it is apparent that all of the contribution to the integral comes at $t = 0$; hence, $A\delta(t) = 0$ for $t \neq 0$.

The preceding properties support the intuitive description of an impulse function of strength A as a pulse with area A and zero width, located at $t = 0$. Since the area is finite and the width is zero, the amplitude of the pulse is infinite. Thus, as shown in Figure 3.5, we use an arrow located at $t = t_0$ to

Figure 3.5

Examples of Shifted Impulse Functions

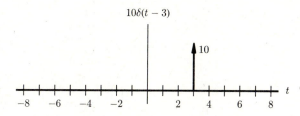

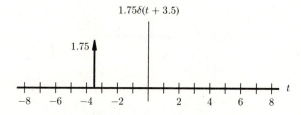

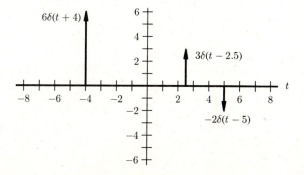

designate an impulse function located t_0 units to the right of the origin; that is, $A\delta(t - t_0)$. Since the area of such a shifted impulse function is concentrated at $t = t_0$, the shifted impulse function is referred to as an impulse function or impulse located at $t = t_0$ for simplicity.

We indicate the strength of an impulse located at $t = t_0$ by writing its value next to the arrow or by making the length of the arrow proportional to the impulse strength, as is also shown in Figure 3.5. Making the arrow length proportional to the strength makes the relative strength of impulses at different locations apparent at a glance. However, it is important to remember when doing this that the length of the arrow is *not* proportional to the amplitude of the impulse, since the amplitude is infinitely large.

Property IV $A\delta(t - t_0) + B\delta(t - t_0) = (A + B)\delta(t - t_0)$ (3.26)

This property states that the sum of two impulses at the same location is an impulse with a strength equal to the sum of the strengths of the impulses summed. The demonstration of the validity of this property is easily accomplished using eq. (3.18), reasoning in the manner following eq. (3.22). This task is left for the reader in the end-of-chapter problems.

Property V $[y(t)][A\delta(t - t_0)] = Ay(t_0)\delta(t - t_0)$ (3.27)

if $y(t)$, is continuous at $t = t_0$.

That is, the product of a function $y(t)$, continuous at $t = t_0$, and an impulse of strength A, located at $t = t_0$, is an impulse of strength $Ay(t_0)$, located at $t = t_0$. We can demonstrate the validity of this property by letting $x(t) = z(t)y(t)$, where $y(t)$ is the function just defined and $z(t)$ is any function continuous at $t = t_0$. Then, from eq. (3.18),

$$[Ay(t_0)] z(t_0) = Az(t_0) y(t_0) = \int_{-\infty}^{\infty} [z(t)y(t)][A\delta(t - t_0)]\ dt$$

$$= \int_{-\infty}^{\infty} z(t)\{[y(t)][A\delta(t - t_0)]\}\ dt \qquad (3.28)$$

But, since $z(t)$ is any function continuous at $t = t_0$, then eq. (3.28) is of the form of eq. (3.18), which defines the function $\{[y(t)][A\delta(t - t_0)]\}$ in the integrand as an impulse of strength $Ay(t_0)$, located at $t = t_0$. Therefore, $[y(t)][A\delta(t - t_0)] = Ay(t_0)\delta(t - t_0)$, which is the desired relation.

Example 3.3

Find and plot the signals $y_1(t) = x_1(t) + x_3(t)$, $y_2(t) = x_1(t) + x_4(t)$, $y_3(t) = x_1(t) + z(t)$, $y_4(t) = x_1(t)z(t)$, $y_5(t) = x_2(t)z(t)$, and $y_6(t) = x_3(t)z(t)$ when $x_1(t) = 3\delta(t - 0.5)$, $x_2(t) = -0.5\delta(t - 2)$, $x_3(t) = 4\delta(t + 1)$, $x_4(t) = -1.5\delta(t - 0.5)$, and

$$z(t) = \begin{cases} 3t & 0 \le t < 2 \\ -3t + 12 & 2 \le t < 4 \\ 0 & \text{elsewhere} \end{cases}$$

Solution

$$y_1(t) = 3\delta(t - 0.5) + 4\delta(t + 1)$$

$$y_2(t) = 3\delta(t - 0.5) - 1.5\delta(t - 0.5) = 1.5\delta(t - 0.5)$$

$$y_3(t) = 3\delta(t - 0.5) + z(t)$$

$$y_4(t) = [z(0.5)][3\delta(t - 0.5)] = (1.5)(3)\delta(t - 0.5) = 4.5\delta(t - 0.5)$$

$$y_5(t) = [z(2)][-0.5\delta(t - 2)] = (6)(-0.5)\delta(t - 2) = -3\delta(t - 2)$$

$$y_6(t) = [z(-3)][4\delta(t + 3)] = [0][4]\,\delta(t + 3) = 0$$

These signals are shown in Figure 3.6.

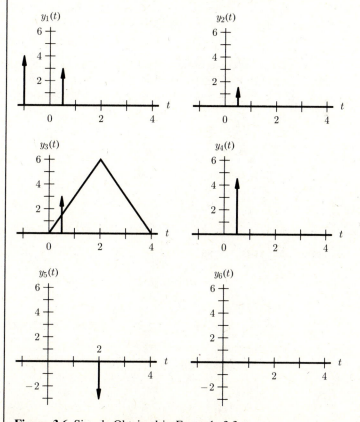

Figure 3.6 Signals Obtained in Example 3.3

Note that since $y_3(t)$ in Example 3.3 is the sum of an impulse function and an ordinary function, it cannot be plotted as a single function. Each component of the sum is thus plotted separately. The vertical scale not only gives the value for the ordinary function $z(t)$ but also gives the strength of the impulse (area under the impulse). The relative height of the two components has no meaning.

Property VI $x(t) * A\delta(t - t_0) = Ax(t - t_0)$ (3.29)

That is, the convolution of any function and an impulse is the original function multiplied by the strength of the impulse and shifted by an amount equal to the shift of the impulse from the origin. This property is easily shown for continuous functions, in which case,

$$x(t) * A\delta(t - t_0) = \int_{-\infty}^{\infty} x(\lambda)\{A\delta[(t - \lambda) - t_0]\}\, d\lambda$$

$$= A \int_{-\infty}^{\infty} x(\lambda)\delta[-(t - \lambda - t_0)]\, d\lambda$$

$$= A \int_{-\infty}^{\infty} x(\lambda)\delta[\lambda - (t - t_0)]\, d\lambda$$

$$= Ax(t - t_0) \qquad\qquad\qquad (3.30)$$

where the second step follows, since $\delta(t) = \delta(-t)$, and the fourth step follows from the sifting integral, eq. (3.16). We can see that the property would also hold if $x(t)$ had discontinuities where it was undefined, since eq. (3.30) holds for all points of $x(t)$ except the undefined discontinuous points.

Property VI also holds when $x(t)$ is an impulse. That is,

$$B\delta(t - t_1) * A\delta(t - t_0) = AB\delta[(t - t_1) - t_0] = AB\delta[t - (t_1 + t_0)] \quad (3.31)$$

We can show that eq. (3.31) is valid by showing that

$$\int_{-\infty}^{\infty} x(t)[B\delta(t - t_1) * A\delta(t - t_0)]\, dt = ABx(t_1 + t_0) \qquad (3.32)$$

because it follows, by eq. (3.18), that $[B\delta(t - t_1) * A\delta(t - t_0)]$ in the integrand is defined as the impulse $AB\delta[t - (t_1 + t_0)]$. Therefore, we write

$$\int_{-\infty}^{\infty} x(t)[B\delta(t - t_1) * A\delta(t - t_0)]\, dt$$

$$= \int_{-\infty}^{\infty} x(t)\left[\int_{-\infty}^{\infty} B\delta(\tau - t_1)\, A\delta[(t - \tau) - t_0]\, d\tau\right] dt$$

$$= AB \int_{-\infty}^{\infty} \delta(\tau - t_1)\left[\int_{-\infty}^{\infty} x(t)\delta[t - (\tau + t_0)]\, dt\right] d\tau$$

$$= AB \int_{-\infty}^{\infty} x(\tau + t_0)\delta(\tau - t_1)\, d\tau$$

$$= AB\delta(t_1 + t_0) \qquad\qquad\qquad (3.33)$$

where the first step uses the definition of the convolution integral, the second step interchanges the order of integration, and the third and fourth step use eq. (3.18).

Example 3.4

Consider the signals $x(t) = 1.5\delta(t + 1.5) + 1.5\delta(t - 1.5)$, $y(t) = 2\delta(t - 0.5) + 4\delta(t - 1)$, and

$$z(t) = \begin{cases} t & 0 \le t < 2 \\ 0 & \text{elsewhere} \end{cases}$$

Plot $x(t)$, $y(t)$, $z(t)$, $x(t) * y(t)$, and $x(t) * z(t)$.

Solution

The signals are shown in Figure 3.7.

$$
\begin{aligned}
x(t) * y(t) &= [1.5\delta(t + 1.5) + 1.5\delta(t - 1.5)] * [2\delta(t - 0.5) + 4\delta(t - 1)] \\
&= 1.5\delta(t + 1.5) * 2\delta(t - 0.5) + 1.5\delta(t + 1.5) * 4\delta(t - 1) \\
&\quad + 1.5\delta(t - 1.5) * 2\delta(t - 0.5) + 1.5\delta(t - 1.5) * 4\delta(t - 1) \\
&= 3\delta(t + 1) + 6\delta(t + 0.5) + 3\delta(t - 2) + 6\delta(t - 2.5)
\end{aligned}
$$

and

$$
\begin{aligned}
x(t) * z(t) &= [1.5\delta(t + 1.5) + 1.5\delta(t - 1.5)] * z(t) \\
&= 1.5z(t + 1.5) + 1.5z(t - 1.5) \\
&= \begin{cases} 1.5(t + 1.5) & -1.5 \le t < 0.5 \\ 1.5(t - 1.5) & 1.5 \le t < 3.5 \\ 0 & \text{elsewhere} \end{cases}
\end{aligned}
$$

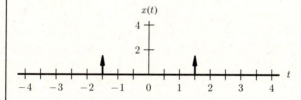

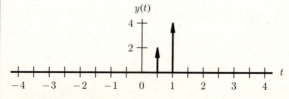

Figure 3.7 Signals for Example 3.4

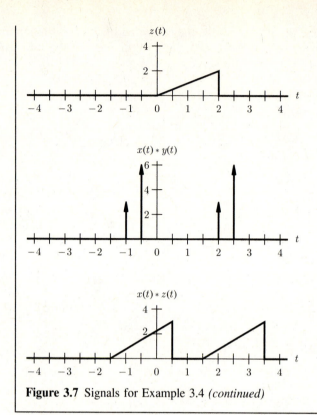

Figure 3.7 Signals for Example 3.4 *(continued)*

We can obtain an intuitive feel for the nature of the impulse function of strength A by constructing a rectangular pulse of width τ, and height A/τ that is centered at the origin. If τ is allowed to approach zero, as illustrated in Figure 3.8, then the width of the pulse approaches zero and the height of the pulse approaches infinity, but its area remains constant at the value A. Thus, the pulse's properties when τ approaches zero are like those of an impulse function of strength A. In some cases, we can use this approximation of an impulse function to find the response of a system to an impulse input signal. The procedure is first to find the response of the system to an input signal, which is a rectangular pulse of the form defined in Figure 3.8, and then to determine the limit of the response as τ approaches zero. Actually, the shape of the pulse considered does not need to be rectangular. We can use any shape as long as it has an area that remains constant as its width is decreased. Examples are

$$x(t) = \frac{1}{2\tau} \exp(-|t|/\tau) \qquad x(t) = \frac{\sin(t/\tau)}{\pi t}$$

and

$$x(t) = \begin{cases} \frac{1}{\tau}\left(1 - \frac{|t|}{\tau}\right) & |t| < \tau \\ 0 & \text{elsewhere} \end{cases}$$

Figure 3.8
Rectangular Pulses
of Area A

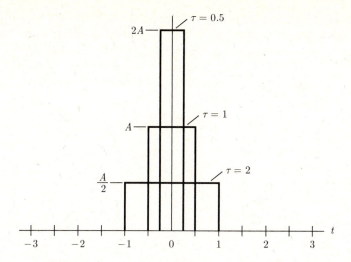

The preceding intuitive description of the properties of an impulse function indicates that the impulse function can be used as an approximate signal model for a pulse of any shape when the pulse is very narrow and the impulse-function model is given a strength equal to the area under the narrow pulse. This is true even though the impulse function is not a function in the usual sense and cannot actually be generated by a physical system. An example of a signal for which an impulse would be a reasonable model, if small-scale signal features were not of interest, would be the pulse of current that flows into a capacitor when it is suddenly connected to a battery with low internal resistance by low resistance wires. Another example would be the narrow pulses used to transmit digital data.

Integrals of Impulse Functions

The integral and multiple integrals of the impulse functions are also referred to as singularity functions. The integral of the impulse function $A\delta(t)$ gives

$$\int_{-\infty}^{t} A\delta(\tau)\, d\tau = \begin{cases} A & t > 0 \\ 0 & t < 0 \end{cases} \tag{3.34}$$

since the integral gives the area under $\delta(t)$ from $-\infty$ to t and the area under $\delta(t)$ is A and is all concentrated at $t = 0$. The value at $t = 0$ is not specified. We can choose it to be any finite value, with the values A and $A/2$ being frequent choices, but in most cases no problems are encountered by leaving it unspecified. The function defined by eq. (3.34) is shown in Figure 3.9, where it is called a *step function* and given the notation $Au(t)$. The function

$$u(t) = \begin{cases} 1 & t > 0 \\ 0 & t < 0 \end{cases} \tag{3.35}$$

also shown in Figure 3.9, is called the *unit step function*. It occurs when $A = 1$, and thus corresponds to the integral of the unit impulse function. Therefore, we can interpret the unit impulse function as the derivative of the unit step function. To permit representations of step functions that are shifted in time and/or are reversed in time, the general step function is written as

$$Au(at + b) = \begin{cases} A & at + b > 0 \\ 0 & at + b < 0 \end{cases} \tag{3.36}$$

where $a \neq 0$.

An example of the use of the step function as a signal model is the representation of the voltage across a resistor that is suddenly connected to a constant voltage source, as illustrated in Figure 3.10. In the physical circuit being modeled, switch closing is not actually instantaneous. Stray capacitance and inductance in the circuit prevent the voltage from changing instantaneously. However, the time required for the voltage to change from 0 to A across the resistor is small. Thus, $Au(t - t_0)$ is a good model for this voltage, except when the time scale is sufficiently expanded to permit observation of the effects we have just described.

The step function is also useful for providing a simple notation for signals that begin or cease at some given time. Examples are shown in Figure 3.11. The last signal shown is a rectangular pulse; it is of considerable importance, since

Figure 3.9
Step Functions

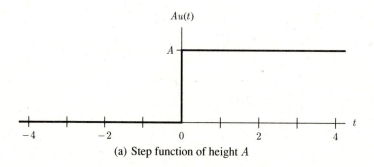

(a) Step function of height A

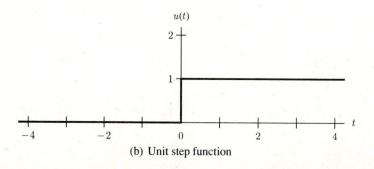

(b) Unit step function

Figure 3.10
Step-Signal Model
for Electric Circuit

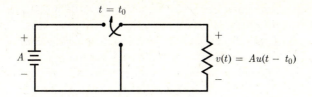

Figure 3.11
Use of the Unit Step
Function in Signal
Representation

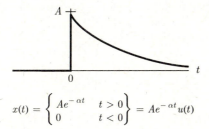

$$x(t) = \begin{cases} Ae^{-\alpha t} & t > 0 \\ 0 & t < 0 \end{cases} = Ae^{-\alpha t}u(t)$$

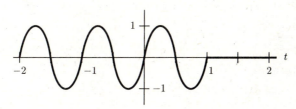

$$y(t) = \begin{cases} \sin(2\pi t) & t < 1 \\ 0 & t > 1 \end{cases} = \sin(2\pi t)u(-t + 1)$$

$$z(t) = \begin{cases} 1 & |t| < \tau/2 \\ 0 & |t| > \tau/2 \end{cases} = u(t + \tau/2) - u(t - \tau/2)$$

$$= \Pi\left(\frac{t}{\tau}\right)$$

nearly rectangular pulses often occur in such practical systems as computer and radar systems. Consequently, the rectangular pulse signal is given its own notation and is sometimes referred to as a *unit amplitude pulse of width* τ. It also can be shifted along the time axis and multiplied by another signal to turn the other signal on and off at specified times. For example,

$$x(t) = \begin{Bmatrix} Ae^{-\alpha t} & 3 < t < 5 \\ 0 & \text{elsewhere} \end{Bmatrix} = Ae^{-\alpha t}\Pi\left(\frac{t-4}{2}\right) \qquad (3.37)$$

If the impulse function $A\delta(t)$ is integrated twice, the result is

$$\int_{-\infty}^{t}\int_{-\infty}^{\alpha} A\delta(\tau)\,d\tau\,d\alpha = A\int_{-\infty}^{t} u(\alpha)\,d\alpha$$

$$= \begin{Bmatrix} At & t > 0 \\ 0 & t < 0 \end{Bmatrix} = Atu(t) \equiv Ar(t) \qquad (3.38)$$

where $r(t)$ is the integral of a unit step function. We call $r(t)$ the *unit ramp function*, since it has the appearance of a ramp with unit slope. $Ar(t)$ is a ramp function with slope A. In general, we write the ramp function as

$$Ar(at+b) = A(at+b)u(at+b) = \begin{cases} A(at+b) & at+b > 0 \\ 0 & at+b < 0 \end{cases} \qquad (3.39)$$

where $a \neq 0$. This general ramp function starts at $t = -b/a$ and has a slope of Aa. Examples are shown in Figure 3.12.

Any piecewise-defined signal that is made up of straight-line segments can be represented by a compact expression consisting of a sum of step and ramp functions. We will see in Chapter 7 that this compact representation makes it easy to determine the Laplace transform of such a signal. First of all, the Laplace transform of the signal can be computed term by term and the results added; second, the Laplace transforms of time-shifted steps and ramps are simple.

The following three examples serve to illustrate the step- and ramp-sum representation of a signal. The first example merely illustrates the signal corresponding to the sum of step and ramp functions. The second and third examples

Figure 3.12
Example Ramp
Functions

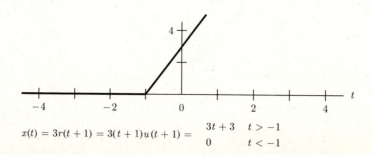

$$x(t) = 3r(t+1) = 3(t+1)u(t+1) = \begin{array}{ll} 3t+3 & t > -1 \\ 0 & t < -1 \end{array}$$

Figure 3.12
Example Ramp
Functions
(continued)

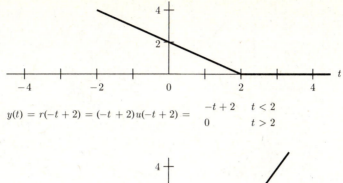

$$y(t) = r(-t + 2) = (-t + 2)u(-t + 2) = \begin{cases} -t + 2 & t < 2 \\ 0 & t > 2 \end{cases}$$

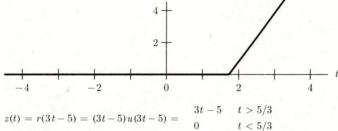

$$z(t) = r(3t - 5) = (3t - 5)u(3t - 5) = \begin{cases} 3t - 5 & t > 5/3 \\ 0 & t < 5/3 \end{cases}$$

illustrate the generation of the sum of step- and ramp-function representations for given piecewise-defined functions.

Example 3.5

Plot the two signals

$$x(t) = u(-t + 2) + r(t + 1) - r(t - 1)$$

and

$$y(t) = 3u(t + 3) - r(t + 2) + 2r(t) - 2u(t - 2) - r(t - 3) - 2u(t - 4)$$

Solution

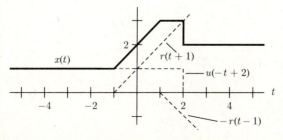

Figure 3.13 Signal Plots for Example 3.5

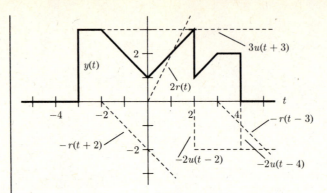

Figure 3.13 Signal Plots for Example 3.5 *(continued)*

Note that we produce the plots in Example 3.5 by plotting the individual components and graphically adding them. When obtaining the sum of step and ramp functions that represent a piecewise-defined signal, it is convenient to work from the left on a plot of the signal and add terms as needed at each signal-segment boundary. This procedure is illustrated in Examples 3.6 and 3.7.

Example 3.6

Find a sum of step and ramp functions that represents the signal shown in Figure 3.14.

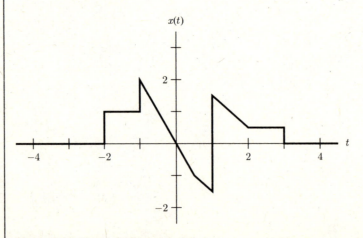

Figure 3.14 Signal for Example 3.6

Solution

1. Starting from the left, the first thing we encounter is a step of height +1 beginning at $t = -2$. Therefore, the first term is $+u(t + 2)$.
2. A second step of height +1 is encountered at $t = -1$. Therefore, the second term is $+u(t + 1)$.
3. The slope also changes from 0 to -2 at $t = -1$. A ramp with slope -2 must be added at this point to change the slope. Therefore, the third term is $-2r(t + 1)$.
4. The slope changes to -1 at $t = 0.5$. Therefore, a ramp with slope +1 must be introduced at this point to reduce the negative slope. This gives the fourth term, $+r(t - 0.5)$.
5. A step of +3 units is encountered at $t = 1$, after which the slope remains -1. Therefore, the fifth term required is $+3u(t - 1)$.
6. The slope changes to zero at $t = 2$. Therefore, a ramp with slope +1 must be introduced at this point to eliminate the slope. The required sixth term is thus $r(t - 2)$.
7. Finally, a step of -0.5 unit is encountered at $t = 3$ Thus, the seventh term is $-0.5u(t - 3)$.

   The function is therefore represented by the equation

   $$x(t) = u(t + 2) + u(t + 1) - 2r(t + 1) + r(t - 0.5)$$
   $$+ 3u(t - 1) + r(t - 2) - 0.5u(t - 3)$$

In the next example, the signal is nonzero prior to the leftmost boundary between piecewise-defined segments. In this case, the representation contains a term that is neither a step function nor a ramp function, but merely the signal expression that exists prior to the first boundary.

Example 3.7

Find a sum of step and ramp functions which represents the signal

$$x(t) = \begin{cases} -0.5t & t < -2 \\ 2 & -2 < t < 1 \\ 1 & 1 < t < 2 \\ 3 - t & 2 < t < 3 \\ 0 & 3 < t \end{cases}$$

Solution

We first plot the signal, as shown in Figure 3.15.

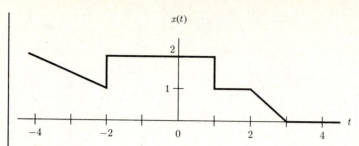

Figure 3.15 Plot of Signal for Example 3.7

1. Prior to the boundary furthest left between piecewise-defined segments (at $t = -2$), the function is $-0.5t$. Therefore, the first term is $-0.5t$.
2. A step of height +1 is encountered at $t = -2$. Therefore, the second term is $u(t + 2)$.
3. The slope also changes from -0.5 to 0 at $t = -2$. Therefore, a ramp with slope 0.5 must be introduced at this point to eliminate the slope. Thus, the third term is $0.5r(t + 2)$.
4. A step of -1 unit is encountered at $t = 1$. Therefore, the fourth term is $-u(t - 1)$.
5. The slope changes from 0 to -1 at $t = 2$. Therefore, a ramp with slope -1 must be introduced at this point to change the slope, giving a fifth term of $-r(t - 2)$.
6. The slope changes from -1 to 0 at $t = 3$. This slope change is introduced with the sixth term, $r(t - 3)$.

   There are no further changes in the signal; therefore, it is represented by

   $$x(t) = -0.5t + u(t + 2) + 0.5r(t + 2) - u(t - 1) - r(t - 2) + r(t - 3)$$

In Examples 3.6 and 3.7, the structural method used to generate the step and ramp function representations of signals yielded unique results. A different sum of step and ramp functions is generated if the signal is constructed by adding terms as needed from the right instead of from the left. For Example 3.7, this leads to

$$x(t) = r(3 - t) - r(2 - t) + u(1 - t) - u(-2 - t) + 0.5r(-2 - t) \quad (3.40)$$

It is also possible to represent signals with other combinations of step, ramp, and unit amplitude pulse functions if products are used as well as sums. For example, the signal of Example 3.7 can be written as

$$x(t) = u(2 - t)[2u(t + 2) - u(t - 1)] + 0.5r(-t)u(-2 - t) + r(3 - t)u(t - 2) \quad (3.41)$$

or as

$$x(t) = -\frac{t}{2}u(-2 - t) + 2\Pi\left(\frac{t - 0.5}{3}\right) + \Pi(t - 1.5) + (3 - t)\Pi(t - 2.5) \quad (3.42)$$

Equations (3.41) and (3.42) for the signal of Example 3.7 are less useful representations than the one determined in the example. We will see that this is particularly true when Laplace transforms of signals are considered in Chapter 7. Signals that extend to the left and cease at some time, and/or contain products, are more difficult to handle than signals that begin at some time, extend to the right, and contain sums but not products of functions.

We have already generated step and ramp functions with the first two integrals of the impulse function. We can generate additional singularity functions by further repeated integration of the impulse function, with the following general result:

$$\int_{-\infty}^{t} \int_{-\infty}^{\alpha_n} \cdots \int_{-\infty}^{\alpha_2} A\delta(\alpha_1) \ d\alpha_1 \ \ldots \ d\alpha_{n-1} d\alpha_n = \frac{At^{n-1}}{(n-1)!} u(t) \quad (3.43)$$

Derivatives of Impulse Functions

Derivatives of all orders of the impulse function are also singularity functions. The first derivative $\delta'(t)$ is referred to as a *doublet function*. When shifted by t_0 units from the origin, it is characterized by

$$A\delta'(t - t_0) = 0 \qquad t \neq t_0 \quad (3.44)$$

$$\int_{-\infty}^{\infty} A\delta'(t - t_0) \ dt = 0 \quad (3.45)$$

$$\int_{-\infty}^{\infty} x(t)A\delta'(t - t_0) \ dt = -Ax'(t_0) \quad (3.46)$$

where $x(t)$ is any continuous function having a continuous first derivative at $t = t_0$.

Higher order derivatives of the impulse function give

$$\int_{-\infty}^{\infty} x(t)\delta^{(k)}(t - t_0) \ dt = (-1)^k x^{(k)}(t_0) \quad (3.47)$$

where the superscript in parentheses indicates the order of the derivative. Doublets and higher order derivatives of the impulse function are not often used as signal models in system analysis.

3.4 The Ideal Sampling Signal

The continuous-time signal

$$\delta_s(t) = \sum_{i=-\infty}^{\infty} \delta(t - iT_s) \quad (3.48)$$

consists of a set of unit impulses that are equally spaced along the entire independent-variable axis. It is called the *ideal sampling signal* and is illustrated in Figure 3.16. Note that $\delta_s(t)$ is periodic with period T_s. This period is referred to as the *sampling interval* of the ideal sampling signal.

The ideal sampling signal is so called because of one of its properties; that is, if $\delta_s(t)$ multiplies the function $x(t)$, then it selects values (samples) of $x(t)$ at equally spaced intervals of t. These samples of $x(t)$ appear as the strengths of impulses located at the values of t where the sample values are extracted. The signal product $x(t)\delta_s(t)$ can be written as

$$x_s(t) = x(t)\delta_s(t) = x(t)\left[\sum_{i=-\infty}^{\infty} \delta(t - iT_s)\right]$$

$$= \sum_{i=-\infty}^{\infty} x(t)\delta(t - iT_s)$$

$$= \sum_{i=-\infty}^{\infty} x(iT_s)\delta(t - iT_s) \qquad (3.49)$$

where $x_s(t)$ is called the ideally sampled version of $x(t)$, $x(t)$ is continuous at all values of iT_s, and the last step is a result of Property V of the impulse function. An example of an ideally sampled version of a signal is shown in Figure 3.17.

The ideally sampled version of a signal is useful as a continuous-time model of a signal that consists of narrow pulses, the amplitudes of which are proportional to uniformly spaced samples of a continuous-time signal. The quantized version of a signal of this type appears at the output of an analog-to-digital (A/D) converter used to generate a digital signal from samples of an analog (continuous-time) signal. Since a signal consisting of narrow pulses with varying amplitudes can also be modeled as a discrete-time signal, ideal sampling provides a bridge between continuous-time and discrete-time representations of sampled signals; it is used in this context in Part III of this text.

We can also use the ideal sampling signal for other signal-modeling tasks. One of these tasks is the representation of periodic signals. This representation

Figure 3.16
The Ideal Sampling
Signal

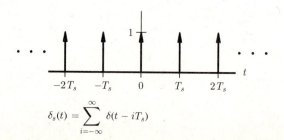

$$\delta_s(t) = \sum_{i=-\infty}^{\infty} \delta(t - iT_s)$$

Figure 3.17
Example of
Ideal-Signal
Sampling

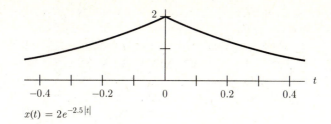

$$x(t) = 2e^{-2.5|t|}$$

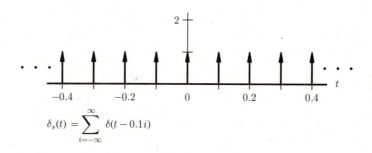

$$\delta_s(t) = \sum_{i=-\infty}^{\infty} \delta(t - 0.1i)$$

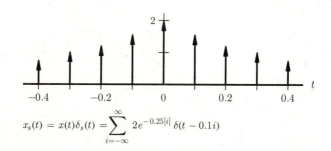

$$x_s(t) = x(t)\delta_s(t) = \sum_{i=-\infty}^{\infty} 2e^{-0.25|i|}\,\delta(t - 0.1i)$$

follows from Property VI of the impulse function, since, if $x(t)$ is a periodic signal with period T_0, $T_s = T_0$, and

$$x_p(t) = \begin{cases} x(t) & t_1 \leq t < t_1 + T_0 \\ 0 & \text{elsewhere} \end{cases} \tag{3.50}$$

then

$$x_p(t) * \delta_s(t) = x_p(t) * \left[\sum_{i=-\infty}^{\infty} \delta(t - iT_0) \right]$$

$$= \sum_{i=-\infty}^{\infty} x_p(t) * \delta(t - iT_0)$$

$$= \sum_{i=-\infty}^{\infty} x_p(t - iT_0)$$

$$= x(t) \tag{3.51}$$

Figure 3.18
Example of
Periodic Signal
Representation Using
the Ideal Sampling
Signal

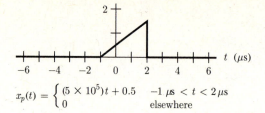

$$x_p(t) = \begin{cases} (5 \times 10^5)t + 0.5 & -1\,\mu\text{s} < t < 2\,\mu\text{s} \\ 0 & \text{elsewhere} \end{cases}$$

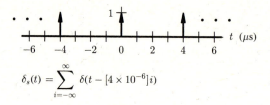

$$\delta_s(t) = \sum_{i=-\infty}^{\infty} \delta(t - [4 \times 10^{-6}]i)$$

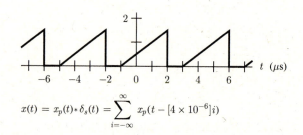

$$x(t) = x_p(t) * \delta_s(t) = \sum_{i=-\infty}^{\infty} x_p(t - [4 \times 10^{-6}]i)$$

where the last step follows from the representation of a periodic continuous-time signal given by eq. (1.6) in Chapter 1. That is, a periodic signal $x(t)$ with period T_0 can be represented by the convolution of an ideal sampling signal with period T_0 and a signal that is equal to $x(t)$ for one period and equal to zero everywhere else. This representation will prove useful when the Fourier transform of periodic signals is considered in the next chapter. It is illustrated in Figure 3.18 for the horizontal sweep signal $x(t)$ of a cathode-ray tube, which is modeled as a periodic signal with a period of $T_0 = 4\,\mu\text{s}$.

3.5 Signal Energy and Power

The terms *signal energy* and *signal power* describe signal characteristics. They are not actually measures of energy and power, since the energy or power absorbed or supplied by a system component is a function of the component, as well as of the signal that passes through it or that is measured across it. We provide motivation here for defining signal energy and signal power by considering the energy absorbed in a resistor in an electrical system. Assuming this resistor has resistance R, then the total energy absorbed in the resistor over all time is determined by

$$E_R = \lim_{T \to \infty} \int_{-T}^{T} i^2(t)R \, dt \tag{3.52}$$

or

$$E_R = \lim_{T \to \infty} \int_{-T}^{T} \frac{v^2(t)}{R} \, dt \tag{3.53}$$

where the energy, E_R, has units of joules (J) when the resistance has units of ohms (Ω), the current has units of amperes (A), and the voltage has units of volts (V). We define the signal energies associated with the signals $i(t)$ and $v(t)$ to be

$$E_i = \lim_{T \to \infty} \int_{-T}^{T} i^2(t) \, dt \tag{3.54}$$

and

$$E_v = \lim_{T \to \infty} \int_{-T}^{T} v^2(t) \, dt \tag{3.55}$$

That is, these signal energies are those that would be absorbed from the current signal $i(t)$ if it were passing through a 1-Ω resistor or from the voltage signal if it were measured across a 1-Ω resistor. The fact that signal energy is not an actual energy in the general case is shown in the following example.

Example 3.8

A 12-V pulse that is 4 s wide and centered at $t = 5$ s is applied across the terminals of an automotive seat-belt warning buzzer. We can model the buzzer as a pure resistance of 20 Ω. The resulting circuit and signal model for the buzzer and actuating signal are shown in Figure 3.19. Find the energy absorbed in the buzzer and the signal energies for the signals $v(t)$ and $i(t)$.

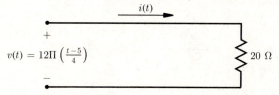

Figure 3.19 Electric Circuit and Signals for Example 3.8

Solution

$$i(t) = \frac{v(t)}{R} = 0.6\Pi\left(\frac{t-5}{4}\right)$$

The energy absorbed in the resistor is

$$E_R = \lim_{T \to \infty} \int_{-T}^{T} i^2(t)R\, dt$$

$$= \lim_{T \to \infty} \int_{-T}^{T} \left[0.36\Pi \frac{(t-5)}{4} \right] [20]\, dt$$

$$= \int_{3}^{7} 7.2\, dt = 28.8 \text{ J}$$

The signal energy for $v(t)$ is

$$E_v = \lim_{T \to \infty} \int_{-T}^{T} v^2(t)\, dt$$

$$= \lim_{T \to \infty} \int_{-T}^{T} \left[144\Pi \frac{(t-5)}{4} \right] dt$$

$$= \int_{3}^{7} 144\, dt = 576$$

The signal energy for $i(t)$ is

$$E_i = \lim_{T \to \infty} \int_{-T}^{T} i^2(t)\, dt$$

$$= \lim_{T \to \infty} \int_{-T}^{T} \left[0.36\Pi \frac{(t-5)}{4} \right] dt$$

$$= \int_{3}^{7} 0.36\, dt = 1.44$$

We now extend the preceding definition of signal energy to any signal $x(t)$, including signals that take on complex values.

Definition _____

The signal energy in the signal $x(t)$ is

$$E_x = \lim_{T \to \infty} \int_{-T}^{T} |x(t)|^2\, dt \qquad (3.56)$$

Since average power in a time interval is the energy in that interval divided by the length of the time interval, we arrive at the following definition.

Definition

The average signal power over all time in the signal $x(t)$ is

$$P_x = \lim_{T \to \infty} \frac{1}{2T} \int_{-T}^{T} |x(t)|^2 \, dt \qquad (3.57)$$

where P_x is called the *signal power* for short.

We often characterize signals as being *energy signals* or *power signals*.

Definition

An energy signal contains finite signal energy. That is, $x(t)$ is an energy signal if $0 < E_x < \infty$.

Note that the signal power for an energy signal is zero (that is, $P_x = 0$), since finite energy is divided by infinite time.

Definition

A power signal contains finite signal power. That is, if $0 < P_x < \infty$, then the signal is called a power signal.

Note that the signal energy for a power signal is always infinite ($E_x = \infty$), since finite average power is multiplied by infinite time.

Some signals are neither energy nor power signals. Examples are signals that contain infinite energy and zero power or infinite energy and infinite power. An example of a signal containing infinite energy and zero power is $x(t) = t^{-0.1}u(t - 1)$. An example of a signal containing infinite energy and infinite power is $x(t) = r(t)$. The proof of these assertions is left for the end-of-chapter problems. The impulse function also is neither an energy nor a power signal because it has infinite signal energy and power. This is demonstrated (see Problem 3.34) by considering the impulse function as the limit of the rectangular pulse $\frac{A}{\tau}\Pi(t/\tau)$ as $\tau \to 0$.

If a signal is periodic, then it cannot be an energy signal, since its signal energy is always either zero or infinite. We can compute its signal power without having to perform a limiting operation, as we will now demonstrate. The signal energy in one period of a periodic signal with period T_0 is

$$E_{x1} = \int_{t_1}^{t_1+T_0} |x(t)|^2 \, dt \qquad (3.58)$$

where t_1 is an arbitrary time. Since all periods are alike, the energy in n periods is

$$E_{xn} = n \int_{t_1}^{t_1+T_0} |x(t)|^2 \ dt \qquad (3.59)$$

The average signal power for all time (that is, over all periods) is the signal power

$$P_x = \lim_{n\to\infty} \frac{E_n}{nT_0} = \lim_{n\to\infty} \frac{1}{T_0} \int_{t_1}^{t_1+T_0} |x(t)|^2 \ dt$$

$$= \frac{1}{T_0} \int_{t_1}^{t_1+T_0} |x(t)|^2 \ dt \qquad (3.60)$$

We see that the signal power of a periodic signal is equal to the average signal power in any period of the signal. If the signal energy in one period of a periodic signal is finite, then the total energy of the periodic signal is infinite, because it contains an infinite number of periods, and the signal power of the periodic signal is finite. Therefore, the signal is a power signal. If the signal energy in one period is either zero or infinite, then the signal's total energy and power are zero or infinite, respectively, and the signal is neither an energy signal nor a power signal.

We illustrate computation of signal energy and signal power in the following two examples. We also indicate whether each signal is an energy signal or a power signal.

Example 3.9

A capacitor is discharged by connecting a resistor across its terminals at $t = 0$. The voltage measure across the terminals is

$$v(t) = e^{-3t} u(t)$$

Compute the signal energy and signal power for $v(t)$.

Solution

$$E_v = \lim_{T\to\infty} \int_{-T}^{T} \left| e^{-3t} u(t) \right|^2 \ dt$$

$$= \lim_{T\to\infty} \int_{-T}^{T} e^{-6t} u(t) \ dt$$

$$= \lim_{T\to\infty} \int_{0}^{T} e^{-6t} \ dt$$

$$= \lim_{T\to\infty} \left[-\frac{1}{6} e^{-6t} \right]_{0}^{T} = \frac{1}{6}$$

The signal power $P_v = 0$, since E_v is finite. The signal $v(t)$ is an energy signal, since E_v is finite.

The result of Example 3.9 is not surprising in that E_v is proportional to the energy supplied by the capacitor to the resistor; we know that this energy must be finite for a practical capacitor.

Example 3.10

Compute the signal energy and signal power for the following complex valued signal and indicate whether it is an energy signal or a power signal.

$$y(t) = Ae^{j2\pi\alpha t}$$

Solution

Since $y(t)$ is a periodic signal (see Section 3.1), it cannot be an energy signal. Therefore, we compute its signal power first. The signal period is $T_0 = 1/\alpha$.

$$P_y = \alpha \int_{t_1}^{t_1+(1/\alpha)} \left| Ae^{j2\pi\alpha t} \right|^2 \, dt$$

$$= \alpha \int_{t_1}^{t_1+(1/\alpha)} A^2 \, dt$$

$$= \alpha \left[A^2 t \right]_{t_1}^{t_1+(1/\alpha)} = A^2$$

Since the signal has finite power, it is a power signal and $E_y = \infty$.

3.6 Signal Representation by Generalized Fourier Series

For signal and system analysis, we sometimes find that it is convenient to approximate a signal, $x(t)$, over the time interval $t_1 < t < t_2$ by a linear combination (weighted sum) of a set of other signals, that is, by the series of weighted signals

$$\hat{x}(t) = \sum_{n=1}^{N} X_n \phi_n(t) \tag{3.61}$$

where $\hat{x}(t)$ is a continuous-time signal that approximates the signal $x(t)$ according to some approximation criterion over the time interval $t_1 < t < t_2$. Initially, we will consider only real functions of time for the signals $x(t)$ and $\phi_n(t)$ to make the results easy to follow. At the end of our discussion, we will summarize the extension to signals that are complex functions of time.

Examples of possible criteria of approximation are

$$\min\{\max|\hat{x}(t) - x(t)|\} \tag{3.62}$$

$$\min\left\{\int_{t_1}^{t_2}|\hat{x}(t) - x(t)|\ dt\right\} \tag{3.63}$$

and

$$\min\left\{\int_{t_1}^{t_2}[\hat{x}(t) - x(t)]^2\ dt\right\} \tag{3.64}$$

where eq. (3.62) produces the smallest possible maximum-error-signal magnitude, eq. (3.63) produces the smallest area under the magnitude of the error signal, and eq. (3.64) produces the smallest area under the square of the error signal. Since the last two approximation criteria shown consider errors at all points of time in the interval, they are more attractive than the first. The last approximation criterion is the most desirable, since it weights larger errors more heavily and is easier to use. Therefore, we will use eq. (3.64) as the criterion of approximation in the discussion that follows. Thus, we minimize the integral square error given by

$$\epsilon_N = \int_{t_1}^{t_2}[\hat{x}(t) - x(t)]^2\ dt = \int_{t_1}^{t_2}\left[\sum_{n=-\infty}^{N} X_n\phi_n(t) - x(t)\right]^2\ dt \tag{3.65}$$

for a sum of N-terms.

The signals $\phi_n(t)$ are called the *basis signals*. They are the set of continuous-time signals that we select to generate the weighted sum and are often chosen to highlight specific signal characteristics of interest. The X_n's are the weighting constants that we choose to make $\hat{x}(t)$ approximate $x(t)$ as closely as possible in the minimum-integral-square-error sense over the chosen interval, given the selected set of basis signals. For some choices of basis signals it is convenient to let n range from $-N$ to N rather than from 1 to N. It may be convenient to let N be infinity, so that the approximation will have an infinite number of terms. The choice of basis signals and corresponding constants should allow for a better approximation as more terms are added to the sum.

It is not possible for us to choose a set of values for the X_n's that minimizes the integral square error for every signal and set of basis signals combination. That is, a series may not exist that approximates a particular signal in the integral-square-error sense with a given set of basis signals. In the present discussion, we will not consider the derivation of conditions that must be satisfied for a series to exist and, for the remainder of this chapter, we will assume that the signals and basis signals satisfy any required conditions. In the following chapter, we will state conditions to be satisfied by signals when the set of basis signals is one that is particularly useful in signal analysis.

Note that $\hat{x}(t)$ need not approximate $x(t)$ in any sense outside the interval of interest, $t_1 < t < t_2$, since, at this point in the discussion, the range of time outside this interval is not of interest. Approximations that are valid over the entire time axis are of interest in many cases, and concepts that are useful in some of these cases will be discussed in the next chapter.

In Example 3.11, only two basis signals are used to illustrate the preceding concepts.

Example 3.11

The oscilloscope horizontal sweep signal $x(t)$ and the two basis signals $\phi_1(t)$ and $\phi_2(t)$ shown in Figure 3.20 are given. Find the constants X_n so that $\hat{x}(t) = \sum_{n=1}^{N} X_n \phi_n(t)$ approximates $x(t)$ over the interval $0 < t < 2$ with minimum integral square error in this interval. Do this for $N = 1$ and $N = 2$.

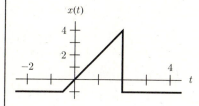

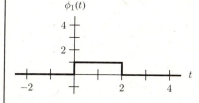

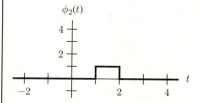

Figure 3.20 Signals for Example 3.11

Solution
For $N = 1$,

$$\hat{x}(t) = x_1 \phi_1(t)$$

and

$$\epsilon_N = \int_0^2 [X_1 - 2t]^2 \, dt = 2X_1^2 - 8X_1 + 32/3$$

To minimize ϵ_N, compute

$$\frac{d\epsilon_N}{dX_1} = 4X_1 - 8 = 0$$

which gives $X_1 = 2$ and $\epsilon_N = 8/3$. The signal and approximate signal are plotted in Figure 3.21.

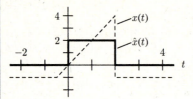

Figure 3.21 Signal Approximation for $N = 1$ in Example 3.11

For $N = 2$,

$$\hat{x}(t) = X_1 \phi_1(t) + X_2 \phi_2(t)$$

and

$$\epsilon_N = \int_0^1 [X_1 - 2t]^2 \, dt + \int_1^2 [X_1 + X_2 - 2t]^2 \, dt$$

$$= 2X_1^2 + X_2^2 + 2X_1 X_2 - 8X_1 - 6X_2 + 32/3$$

To minimize ϵ_N, compute

$$\frac{d\epsilon_N}{dX_1} = 4X_1 + 2X_2 - 8 = 0 \quad \text{and} \quad \frac{d\epsilon_N}{dX_2} = 2X_2 + 2X_1 - 6 = 0$$

and solve these two equations simultaneously to give $X_1 = 1$ and $X_2 = 2$. The resulting integral square error is $\epsilon_N = 2/3$; the signal and approximate signal are plotted in Figure 3.22.

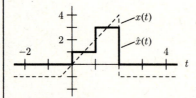

Figure 3.22 Signal Approximation for $N = 2$ in Example 3.11

Note that $\hat{x}(t)$ matches $x(t)$ more closely than it did when $N = 1$. This is also shown by the decrease of ϵ_N from 8/3 to 2/3.

We note in Example 3.11 that the coefficient of the first term changes when the second term is added to the approximation for the particular basis signals selected. It would be desirable that the basis signals be chosen so that each of the coefficients would not change as more terms were added to the approximation. This fixed nature of the coefficients is known as *finality of coefficients*. We illustrate the concept in Example 3.12 by using the signal of Example 3.11, but with a different set of basis signals.

Example 3.12

Repeat Example 3.11 using the two basis signals $\phi_1(t)$ and $\phi_2(t)$ shown in Figure 3.23, rather than those shown in Figure 3.20.

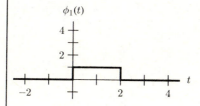

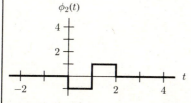

Figure 3.23 Basis Signals for Example 3.12

Solution
For $N = 1$,

$$\hat{x}(t) = X_1 \phi_1(t)$$

and

$$\epsilon_N = \int_0^2 [X_1 - 2t]^2 \, dt = 2X_1^2 - 8X_1 + 32/3$$

To minimize ϵ_N, compute

$$\frac{d\epsilon_N}{dX_1} = 4X_1 - 8 = 0$$

which gives $X_1 = 2$ and $\epsilon_N = 8/3$. The signal and approximate signal are plotted in Figure 3.24.

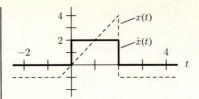

Figure 3.24 Signal Approximation for $N = 1$ in Example 3.12

For $N = 2$,
$$\hat{x}(t) = X_1\phi_1(t) + X_2\phi_2(t)$$
and
$$\epsilon_N = \int_0^1 [X_1 - X_2 - 2t]^2 \, dt + \int_1^2 [X_1 + X_2 - 2t]^2 \, dt$$
$$= 2X_2^2 + 2X_1^2 - 8X_1 - 4X_2 + 32/3$$

To minimize ϵ_N, compute
$$\frac{d\epsilon_N}{dX_1} = 4X_1 - 8 = 0, \text{ which yields } X_1 = 2$$
and
$$\frac{d\epsilon_N}{dX_2} = 4X_2 - 4 = 0, \text{ which yields } X_2 = 1$$

The resulting integral square error is $\epsilon_N = 2/3$; the signal and approximate signal are plotted in Figure 3.25.

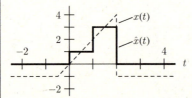

Figure 3.25 Signal Approximation for $N = 2$ in Example 3.12

Note that finality of coefficients holds for the pair of basis signals selected, since the value of X_1 is unchanged when the second term is added. The resulting approximate signals and integral square errors are the same as those of Example 3.11. This is attributable to the choice of basis functions. Do not interpret it as a general result!

Now, let us consider the integral of the product of the two basis signals over the interval of interest in the preceding two examples. For Example 3.11,
$$\int_0^2 \phi_1(t)\phi_2(t) \, dt = \int_1^2 (1)(1) \, dt = 1$$

For Example 3.12,

$$\int_0^2 \phi_1(t)\phi_2(t)\, dt = \int_0^1 (1)(-1)\, dt + \int_1^2 (1)(1)\, dt = 0$$

When the criterion of approximation is the minimum integral square error, and the integral of the product of two real basis signals is zero over the approximation interval, then we say that these two basis signals are *orthogonal* over the interval. The basis signals of Example 3.12 that produce finality of coefficients are orthogonal. In general, a set of basis signals that are mutually orthogonal over an interval produce the finality of coefficients property when a signal approximation is carried out over the interval.

The general definition of basis-function orthogonality follows. We then present the derivation of the expression required to compute the weighting coefficients X_n for the approximating series referred to as the *generalized Fourier series*.

Definition

The real basis signals, $\phi_i(t)$, are mutually orthogonal over the time interval $t_1 < t < t_2$ if and only if

$$\int_{t_1}^{t_2} \phi_n(t)\phi_m(t)\, dt = \begin{cases} \lambda_n & n = m \\ 0 & n \neq m \end{cases} \tag{3.66}$$

Definition

A generalized Fourier series is a series of weighted orthogonal basis signals that approximates a signal by minimizing the integral square error, ϵ_N, over the interval.

To find the coefficient values X_n required to minimize ϵ_N for a given set of basis signals, we first rewrite eq. (3.65) as

$$\epsilon_N = \int_{t_1}^{t_2} \left\{ x^2(t) - 2x(t)\left[\sum_{n=1}^N X_n\phi_n(t)\right] \right. $$
$$\left. + \left[\sum_{n=1}^N X_n\phi_n(t)\right]\left[\sum_{m=1}^N X_m\phi_m(t)\right] \right\} dt \tag{3.67}$$

Then, since the integral of a sum is the sum of the integrals,

$$\epsilon_N = \int_{t_1}^{t_2} x^2(t)\, dt - 2\sum_{n=1}^N X_n \int_{t_1}^{t_2} x(t)\phi_n(t)\, dt$$
$$+ \sum_{n=1}^N \sum_{m=1}^N X_n X_m \int_{t_1}^{t_2} \phi_n(t)\phi_m(t)\, dt \tag{3.68}$$

Since the integral in the last term is equal to λ_n when $n = m$ and is equal to zero when $n \neq m$, then

$$\epsilon_N = \int_{t_1}^{t_2} x^2(t)\, dt - 2 \sum_{n=1}^{N} X_n \int_{t_1}^{t_2} x(t)\phi_n(t)\, dt + \sum_{n=1}^{N} \lambda_n X_n^2 \qquad (3.69)$$

Completing the square gives

$$\epsilon_N = \int_{t_1}^{t_2} x^2(t)\, dt - \sum_{n=1}^{N} \lambda_n \left[\frac{1}{\lambda_n} \int_{t_1}^{t_2} x(t)\phi_n(t)\, dt \right]^2$$

$$+ \sum_{n=1}^{N} \lambda_n \left[\frac{1}{\lambda_n} \int_{t_1}^{t_2} x(t)\phi_n(t)\, dt - X_n \right]^2 \qquad (3.70)$$

Now, ϵ_N is positive, since it is the integral of a squared function. Also, the first two terms of eq. (3.70) do not depend on X_n and the last term is greater than or equal to zero. Therefore, ϵ_N is minimized if we choose X_n so that the last term is zero. That is, the value of X_n that minimizes ϵ_N is

$$X_n = \frac{1}{\lambda_n} \int_{t_1}^{t_2} x(t)\phi_n(t)\, dt \qquad (3.71)$$

where X_n must exist for all n of interest if the series is to exist. Since the computation of the value of X_n depends only on $x(t)$ and the basis signal $\phi_n(t)$, and not on any of the other basis signals, the finality of coefficients holds. The integral square error that remains when N terms are used to approximate $x(t)$ is

$$\epsilon_N = \int_{t_1}^{t_2} x^2(t)\, dt - \sum_{n=1}^{N} \lambda_n \left[\frac{1}{\lambda_n} \int_{t_1}^{t_2} x(t)\phi_n(t)\, dt \right]^2$$

$$= \int_{t_1}^{t_2} x^2(t)\, dt - \sum_{n=1}^{N} \lambda_n X_n^2 \qquad (3.72)$$

Example 3.13

In Example 3.12, a generalized Fourier series approximation to the signal $x(t)$ shown in Figure 3.20 was found using the basis functions defined in Figure 3.23. Recalculate the series coefficients, X_n, and integral square error using eqs. (3.71) and (3.72).

Solution

We first compute λ_1 and λ_2 as

$$\lambda_1 = \int_0^2 \phi_1^2(t)\, dt = \int_0^2 (1)\, dt = 2$$

and

$$\lambda_2 = \int_0^2 \phi_2^2(t) \, dt = \int_0^2 (1) \, dt = 2$$

Then we compute X_1 and X_2 by using eq. (3.71) to give

$$X_1 = \frac{1}{\lambda_1} \int_0^2 x(t)\phi_1(t) \, dt$$

$$= \frac{1}{2} \int_0^2 (2t)(1) \, dt = 2$$

and

$$X_2 = \frac{1}{\lambda_2} \int_0^2 x(t)\phi_2(t) \, dt$$

$$= \frac{1}{2} \int_0^1 (2t)(-1) \, dt + \frac{1}{2} \int_1^2 (2t)(1) \, dt = 1$$

Finally, we compute the integral square error as

$$\epsilon_N = \int_0^2 (2t)^2 \, dt - \sum_{n=1}^2 \lambda_n X_n^2$$

$$= \frac{32}{3} - (2)(2)^2 - (2)(1)^2 = \frac{2}{3}$$

If an infinite set of basis signals is defined, and $\epsilon_N \to 0$ as $N \to \infty$ for all signals for which a series exists, then we say that the basis-signal set is complete. In other words, with a complete set of basis signals, we can make the integral square error arbitrarily small by using more and more terms in the approximation. Note that an arbitrarily small integral square error does not imply that $\hat{x}(t)$ approaches $x(t)$ for all or any t as $N \to \infty$. Conditions under which $\hat{x}(t)$ approaches $x(t)$ for almost all values of t when a particular set of basis signals is used is considered in the next chapter.

In general, it is difficult to prove whether a set of basis functions is complete. Therefore, in this introductory treatment, such proofs are not considered. Completeness is simply stated for those sets of basis functions known to be complete.

For a complete set of basis functions,

$$\int_{t_1}^{t_2} x^2(t) \, dt - \sum_{n=1}^{\infty} \lambda_n X_n^2 = 0 \tag{3.73}$$

Therefore,

$$\int_{t_1}^{t_2} x^2(t) \, dt = \sum_{n=1}^{\infty} \lambda_n X_n^2 \tag{3.74}$$

The left side of eq. (3.74) is the energy contained in the signal in the time interval for which the approximation to the signal is generated. The energy contained in the nth term of the generalized Fourier series in the time interval $t_1 < t < t_2$ is

$$\int_{t_1}^{t_2} [X_n \phi_n(t)]^2 \, dt = X_n^2 \int_{t_1}^{t_2} \phi_n(t)\phi_n(t) \, dt = \lambda_n X_n^2 \qquad (3.75)$$

which means that the right side of eq. (3.74) is the sum of the energies contained by the terms of the generalized Fourier series in the approximation interval. Therefore, in the approximation interval, the energy contained in the signal is equal to the sum of energies contained in each of the individual components of the generalized Fourier series approximation. This result, which is defined by eq. (3.74), is called *Parseval's theorem* and is a consequence of the orthogonality of the basis signals.

In general, signals and basis signals may be complex valued. The following modifications to the real-valued generalized Fourier series of the preceding discussion are made to extend it to complex-valued signals and/or basis signals.

1. Basis signal mutual orthogonality definition:

$$\int_{t_1}^{t_2} \phi_n(t)\phi_m^*(t) \, dt = \begin{cases} \lambda_n & n = m \\ 0 & n \neq m \end{cases} \qquad (3.76)$$

2. Coefficient computation:

$$X_n = \frac{1}{\lambda_n} \int_{t_1}^{t_2} x(t)\phi_n^*(t) \, dt \qquad (3.77)$$

3. Integral square error:

$$\epsilon_N = \int_{t_1}^{t_2} |\hat{x}(t) - x(t)|^2 \, dt = \int_{t_1}^{t_2} |x(t)|^2 \, dt - \sum_{n=1}^{N} \lambda_n |X_n|^2 \qquad (3.78)$$

Equations (3.77) and (3.78) are easily derived by steps that parallel those of the preceding discussion. For a complete set of basis functions, Parseval's theorem is stated as

$$\int_{t_1}^{t_2} |x(t)|^2 \, dt = \sum_{n=1}^{\infty} \lambda_n |X_n|^2 \qquad (3.79)$$

Note that when the signal or the basis signals or both are complex valued, then the coefficients X_n are complex numbers, since they result from the integral of a complex-valued function.

We frequently refer to the generalized Fourier series $\hat{x}(t)$ that corresponds to the signal $x(t)$ over the interval $t_1 < t < t_2$ as the *generalized Fourier series expansion of the signal on this interval*. Likewise, we refer to the interval $t_1 < t < t_2$ as the *expansion interval*.

In the preceding discussion of the generalized Fourier series expansion of a signal over a time interval, we have shown the approximation rationale and general characteristics of the series. The only basis functions we will consider in this text are the complex exponentials and trigonometric functions. These basis functions can be used to determine some useful signal characteristics.

3.7 Summary

Analysis of signals and linear systems requires mathematical expressions that effectively model the signals when quantitative results are to be obtained. Useful signal expressions include the exponential and sinusoidal signals.

Sinusoidal signals can be used directly when analyzing power systems. They also provide background and insight into the important concepts of the frequency content of signals, bandwidth, and the frequency response of systems discussed in later chapters. These signals are defined by the three parameters of frequency, amplitude, and phase and can be represented as rotating phasors through the use of the complex-exponential signal. The phase of a sinusoid signal is directly related to the signal's time delay.

A frequently occurring group of signals in signal and system analysis are known as singularity functions. They are all based on a generalized function called the impulse function, that is defined by the sifting integral. An impulse-function input signal is very useful in generating a system response that provides fundamental information about the system characteristics, as will be discussed in Chapter 5. The impulse function can also be integrated to produce the step and ramp functions that are quite useful in modeling practical signals for use in system analysis.

The ideal sampling signal is a set of unit impulses that are equally spaced along the entire independent-variable axis. One use of the ideal sampling signal is in modeling the sampling of a continuous-time signal and in providing a bridge between continuous-time and discrete-time representations of sampled signals. We will consider this application in more detail in Part III of the text.

Although signal energy and signal power of a continuous-time signal are not measures of actual energy or power being transferred by the signal, they do provide useful information about a signal's characteristics. Signals with finite energy or finite power are of particular importance.

A signal can be represented approximately by a weighted sum (series) of basis signals. The selection of the basis signals is often made to highlight signal characteristics of interest. A minimum-integral-square-error approximation criterion is used to compute the basis signal weighting coefficients. Use of this approximation criterion permits the definition of orthogonal basis signals that lead to coefficients that do not change as more terms are added to the approximating series. The nonchanging coefficient property is referred to as finality of coefficients. When the sense of the approximation to a signal is that the integral square error is minimized and the basis functions are orthogonal, then the resulting weighted sum (series) is referred to as the generalized Fourier series representation of the signal.

Problems

3.1 Find the time delay of the following cosine functions with respect to the reference cosine function:

  a. $v(t) = 4\cos(20\pi t - 0.5)$

  b. $i(t) = 12\cos(365\pi t - 0.11)$

  c. $x(t) = \cos(0.1\pi t + 0.1)$

  d. $y(t) = 1.5\cos(2000\pi t + 1.6)$.

3.2 Find $z(t) = x(t) + y(t)$ when

  a. $x(t) = 3\cos(14\pi t - 0.2)$ and

  $y(t) = 2\cos(14\pi t + 0.1)$.

  b. $x(t) = 6\cos(10\pi t - 0.25)$ and

  $y(t) = 12\cos(10\pi t - 0.2)$.

  c. $x(t) = 5\cos(4\pi t)$ and

  $y(t) = 2\sin(4\pi t - 0.3)$.

  d. $x(t) = 2\cos(4\pi t + 0.4)$ and

  $y(t) = 3\cos(6\pi t + 0.6)$.

3.3 In Example 1.10 we indicated that the electric circuit shown in Figure 3.26 was a simple form of a filter circuit designed to reduce high-frequency interference that occurs along with a lower frequency message signal at the input. Find the output signal $y(t)$ when the input signal is $x(t) = x_m(t) + x_i(t)$, where $x_m(t) = 25\cos(t - 0.25)$ is the input message signal and $x_i = 2\cos(100t - 0.5)$ is the input interference signal. Does the filter circuit perform the function it was designed to perform?

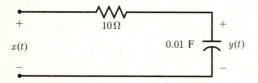

Figure 3.26

3.4 The circuit shown in Figure 3.27 models the output portion of a certain transistor amplifier. Find the current $i_1(t)$ through the load resistor and the current $i_2(t)$ through the capacitor.

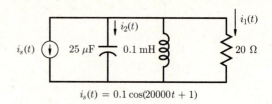

$i_s(t) = 0.1\cos(20000t + 1)$

Figure 3.27

3.5 For the circuit node shown in Figure 3.28, find $i_4(t)$ when $i_1(t) = 3\cos(40t - 0.1)$, $i_2(t) = 5\cos(40t + 0.2)$, and $i_3(t) = 3\cos(40t - 2)$.

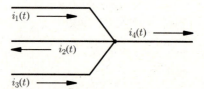

Figure 3.28

3.6 A sinusoidal signal measured in a television amplifier is a cosine that is delayed by 0.01 ms from the reference cosine signal. Find the phase of the signal if its frequency is (a) 1 KHz, (b) 5 KHz, (c) 10 KHz, (d) 40 KHz.

3.7 A sinusoidal signal that appears in the feedback loop of a control system is a cosine that is delayed by −50 ms from the reference cosine signal. Find the phase of the signal if its frequency is (a) 0.25 Hz, (b) 0.5 Hz, (c) 2 Hz, (d) 25 Hz.

3.8 A sinusoidal signal is a cosine with a phase angle of −1.4 rad. Find the time delay of this signal with respect to the reference cosine signal if its frequency is (a) 0.5 Hz, (b) 1 Hz, (c) 10 Hz.

3.9 Write the household voltage signal $v(t) = 115\sqrt{2}\cos[120\pi t + \pi/8]$ as a sum of counterrotating phasors and plot these phasors and their sum for (a) $t = 0$, (b) $t = 0.8$ ms, (c) $t = 5.0$ ms, (d) $t = 9.4$ ms.

3.10 The shock absorber on the front wheel of an automobile is worn out, causing the wheel to oscillate up and down as the automobile is driven. The vertical displacement $y(t)$ of the wheel with respect to its average vertical location is given by the signal $y(t) = 0.75 \cos[1.25\pi t - \pi/4]$. Write $y(t)$ as a sum of counterrotating phasors and plot these phasors and their sum for (a) $t = 0$, (b) $t = 1/10$ s, (c) $t = 1/5$ s, (d) $t = 2/5$ s.

3.11 Evaluate $\int_{-\infty}^{\infty} x(\tau)\delta(\tau - t_0)$ for

a. $x(\tau) = 3e^{-\tau}$ and $t_0 = 2$.

b. $x(\tau) = 1/(1 + \tau^2)$ and $t_0 = -0.5$.

c. $x(\tau) = \cos 2\pi\tau$ and $t_0 = 0.25$.

d. $x(\tau) = (\tau - 1)^3$ and $t_0 = 3$.

3.12 Evaluate $\int_{t_1}^{t_2} 2\delta(t - t_0) \, dt$ for

a. $t_1 = 0.25$, $t_2 = 2.5$, $t_0 = 1$.

b. $t_1 = -4$, $t_2 = 2$, $t_0 = -2$.

c. $t_1 = 0.5$, $t_2 = 1.5$, $t_0 = 2.5$.

d. $t_1 = 0$, $t_2 = \infty$, $t_0 = 100$.

3.13 Show that Property IV for impulse functions is valid.

3.14 Plot the following signals:

a. $y(t) = 3\delta(t + 1) - 4\delta(t - 0.5) + 2\delta(t - 2)$

b. $x(t) = 10\delta(t - 1) + 5\delta(t + 1) - 5\delta(t - 1)$

3.15 Write the expressions for and plot $y(t) = x(t)\delta(t - t_0)$ when

a. $x(t) = 2e^{-2|t|}$ and $t_0 = 0.25$.

b. $x(t) = 24/(2 + t^2)$ and $t_0 = 2$.

c. $x(t) = 3\cos(2\pi t - \pi/4)$ and $t_0 = 0.5$.

d. $x(t) = (t - 0.5)^2$ and $t_0 = 0.5$.

3.16 Write the expressions for and plot $z(t) = x(t) * \delta(t - t_0)$ for the four combinations of $x(t)$ and t_0 given in Problem 3.15.

3.17 Write the expressions for and plot $w(t) = y(t)[A\delta(t - t_0)]$ when

a. $y(t) = t^2$, $A = 2$, $t_0 = 2$.

b. $y(t) = 3 + \sin 10\pi t$, $A = 1.5$, $t_0 = -0.1$.

c. $y(t) = 1/(1 + t^2)$, $A = -5$, $t_0 = 2$.

d. $y(t) = |t - 1|$, $A = 3$, $t_0 = 0.75$.

3.18 Write the expressions for and plot $x(t) = y(t) * [A\delta(t - t_0)]$ for the four combinations of $y(t)$, A, and t_0 given in Problem 3.17.

3.19 Plot $z(t) = x(t) * y(t)$ for the following signals:

a. $x(t) = 3\delta(t - 2.5) - 2\delta(t + 2.5)$

$y(t) = 4\delta(t) + 2\delta(t - 0.5)$

b. $x(t) = \delta(t - 2) + 2\delta(t - 4)$

$y(t) = 3\delta(t + 0.5) + w(t)$

where $w(t) = \begin{cases} |t| & |t| < 1 \\ 0 & \text{elsewhere} \end{cases}$

c. $x(t) = -2\delta(t + 3) + 4\delta(t + 5)$

$y(t) = \delta(t - 3) + \delta(t - 4)$

3.20 Write single-equation expressions for the following signals using the unit step function and then plot the signals:

a. $x(t) = \begin{cases} 1/(1 + t^2) & t > 3 \\ 0 & t < 3 \end{cases}$

b. $x(t) = \begin{cases} e^{-|t|} & t < -2 \\ 0 & t > -2 \end{cases}$

c. $x(t) = \begin{cases} 1 - \cos 3t & 1 < t < 3 \\ 0 & \text{elsewhere} \end{cases}$

d. $x(t) = \begin{cases} e^{-(t-1)} & t > 1 \\ \cos 3(t - 1) & t < 1 \end{cases}$

e. $x(t) = \begin{cases} 2 & |t| < 1 \\ (3 - t) & 1 < t < 2 \\ 0 & \text{elsewhere} \end{cases}$

3.21 Plot the signals (a) $x(t) = r(t + 2)$, (b) $x(t) = 3r(t - 1.5)$, (c) $x(t) = 2.5r(0.5t + 1)$, (d) $x(t) = 4r(6 - t)$, (e) $x(t) = r(3 - 0.5t)$.

3.22 Plot the following signals:

a. $x(t) = 2u(t + 2) - u(t - 0.5) + r(t - 0.5)$
$- r(t - 2)$

b. $x(t) = 2r(t + 1) - 4r(t - 1) + u(t - 2)$
$+ 2r(t - 3) - u(t - 4)u(5 - t)$

c. $x(t) = 2u(t) - u(t - 1) - u(t + 1)$

d. $x(t) = r(1 - t)u(t + 1) - 2u(t + 1) + 2u(t)$
$+ 2r(t + 1)u(-t)$

Can you find a simpler expression for the signal of part (d)?

3.23 Plot the following signals:

a. $x(t) = 2r(t+2) - 2r(t+1) - 2r(t) + 2r(t-1)$

b. $x(t) = 2u(t-1.5) - r(t-1.5) + r(t-2.5)$
$\quad - u(t-4)$

c. $x(t) = r(t) \Pi[(t-3)/2] + 2u(2-t)$
$\quad + 4u(t-4)$

d. $x(t) = 2 + r(t-2) - r(t-4)$

e. $x(t) = 2 \Pi[(t-1)/3] + (2t+3) \Pi(t+1)$
$\quad + (7-2t) \Pi(t-3)$

3.24 Write equations for the test signals shown in Figure 3.29 in terms of a sum of step and/or ramp functions.

a.

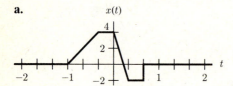
$x(t)$

b.

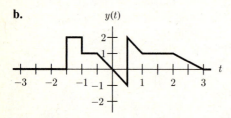
$y(t)$

Figure 3.29

3.25 Write equations for the test signals shown in Figure 3.30 in terms of a sum of step and/or ramp functions.

a.

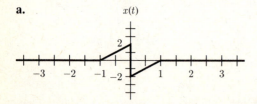
$x(t)$

Figure 3.30

b.

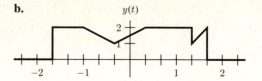
$y(t)$

c.

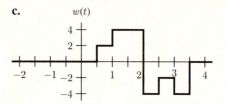
$w(t)$

Figure 3.30 *(continued)*

3.26 Plot $y(t) = x(t)\delta_s(t)$ for

a. $x(t) = (1-t)/(2+t)$, $T_s = 1$ s.

b. $x(t) = e^{-t/3}u(t+0.1)$, $T_s = 0.5$ s.

c. $x(t) = t^2$, $T_s = 0.1$ s.

d. $x(t) = 1 - |t - 0.5|$, $T_s = 0.25$ s.

3.27 Plot $x(t) = \cos(10\pi t) \, \delta_s(t)$ for

a. $T_s = 0.025$ s,

b. $T_s = 0.05$ s,

c. $T_s = 0.1$ s,

d. $T_s = 0.15$ s,

e. $T_s = 0.2$ s,

f. $T_s = 0.25$ s,

For what range of values of T_s does there appear to be enough samples of $\cos(10\pi t)$ to portray its characteristics? This concept will be pursued further in Chapter 11, in our discussion of the sampling theorem.

3.28 Repeat Problem 3.27 with $\cos(10\pi t)$ replaced by $\sin(10\pi t)$.

3.29 Plot $x(t)$, $\delta_s(t)$, and $y(t) = x(t) * \delta_s(t)$ for

a. $x(t) = \Pi(t/2)$, $T_s = 4$.

b. $x(t) = t \Pi[(t-1)/2]$, $T_s = 2$.

c. $x(t) = |t| \Pi(t/1.5)$, $T_s = 1.5$.

d. $x(t) = \cos(0.5\pi t) \Pi[(t-3)/4]$, $T_s = 4$.

e. $x(t) = 1/(1+t^3)$, $T_s = 4$.

3.30 Plot $x(t)$, $\delta_s(t - t_1)$, and $y(t) = x(t) * \delta_s(t - t_1)$ for

a. $x(t) = r(t) - 2r(t-1) + r(t-2)$, $T_s = 4$, $t_1 = 0$

b. $x(t) = r(t+1) - 2r(t) + r(t-1)$, $T_s = 4$, $t_1 = 1$

c. $x(t) = \sin(0.2\pi t) \Pi(t/10)$, $T_s = 10$, $t_1 = -2.5$

d. $x(t) = \Pi[(t-1)/2]$, $T_s = 1.5$, $t_1 = 1$

3.31 Determine the signal energy and signal power for each signal given and indicate whether it is an energy signal or a power signal.

a. $x(t) = e^{-2|t|}$

b. $x(t) = (3 - t)\Pi\left[(t - 2)/2\right]$

c. $x(t) = 4\cos(30t - 0.25)$

d. $x(t) = 4\cos(30t - 0.25)u(t)$

e. $x(t) = r(t)$

3.32 Determine the signal energy and signal power for each signal given and indicate whether it is an energy signal or power signal.

a. $x(t) = r(t) - r(t - 1) - u(t - 4)$

b. $x(t) = 0.5r(t + 1)u(t - 2)u(3 - t)$

c. $x(t) = 1/\sqrt{1 + t^2}$

d. $x(t) = t^{-0.1}u(t - 1)$

e. $x(t) = e^{-|t|}\cos 10\pi t$

3.33 The signal $y(t) = e^{-\alpha t}u(t)$ occurs in many systems. Determine the values of α, if any, that will produce (a) an energy signal, (b) a power signal, (c) a signal that is neither an energy signal nor a power signal.

3.34 Considering the signal

$$x(t) = \frac{A}{\tau}\Pi\left(\frac{t}{\tau}\right)$$

and taking the limit as τ approaches zero, show that the impulse function is neither an energy signal nor a power signal.

3.35 Determine which of the following pairs of signals are orthogonal over the time interval $1 < t < 3$:

a. $x(t) = t - 2$ and $y(t) = \Pi\left[(t - 3)/6\right]$

b. $x(t) = t$ and $y(t) = e^{-|t|}$

c. $x(t) = \cos 5\pi t$ and $y(t) = \sin 5\pi t$

d. $x(t) = \cos 2.5\pi t$ and $y(t) = \sin 2.5\pi t$

e. $x(t) = e^{j3\pi t}$ and $y(t) = e^{j2\pi t}$

f. $x(t) = e^{j2.5\pi t}$ and $y(t) = e^{-j2\pi t}$

g. $x(t) = e^{j5\pi t}$ and $y(t) = e^{-j5\pi t}$

3.36 Consider the signal $x(t)$ and the five basis signals $\phi_i(t)$, $1 \le n \le 5$, shown in Figure 3.31.

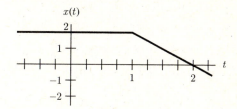

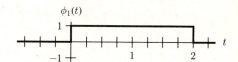

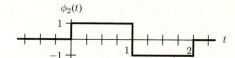

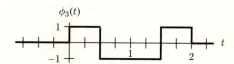

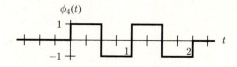

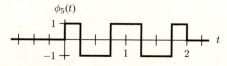

Figure 3.31

a. Show that the basis signals are mutually orthogonal over the interval $0 < t < 2$ and find the value of λ_n associated with each one.

b. Find and plot the generalized Fourier series approximation to $x(t)$ over the interval $0 < t < 2$ using the five basis signals.

c. Compute the integral square error over the interval of the approximation.

d. Compute the signal energy and the sum of the energies in each term of the generalized Fourier series approximation over the interval $0 < t < 2$. What is the ratio of the energy contained in the series approximation to the energy contained in the signal in the interval?

Note: The basis signals used in this problem are the first five of an infinite, complete set known as *Walsh functions*.

4 Spectra of Continuous-Time Signals

In the previous chapter we considered various time functions that are useful in modeling continuous-time signals. The time-function signal model portrays the characteristics of a continuous-time signal by showing its variation as a function of time. The model is called the *signal waveform* or the *time-domain representation* of the signal.

A cosine signal is completely characterized by the three quantities of frequency, amplitude, and phase, as defined in Section 3.1. These quantities are sometimes referred to as the *frequency-domain representation* of the cosine signal. In this chapter, we extend the concept of frequency-domain representation to a wide range of signals that contain energy at more than one frequency by considering the signal amplitude and the signal phase as functions of frequency. This representation, also referred to as the *frequency spectrum* or, simply as the *spectrum* of the signal, is useful in portraying the range of frequencies for which a signal has significant energy content, as well as the nature of the signal amplitude and phase in different portions of the frequency range. These characteristics are not easily visualized from time-domain representations for most signals. For example, the waveform of a voltage signal shown in Figure 4.1 is the sum of two sine functions. The frequencies, amplitudes, and phases of the two sine functions are 1 Hz, 2 V, $-\pi/4$ rad and 2 Hz, 0.75 V, $-\pi/3$ rad, respectively. We see that this information is not directly apparent from the time waveform shown in Figure 4.1.

4.1 Amplitude- and Phase-Spectra Concepts

For simplicity, let us consider the cosine signal

$$x(t) = A_x \cos(2\pi f_x t + \theta_x) \tag{4.1}$$

Figure 4.1
Example Voltage
Signal

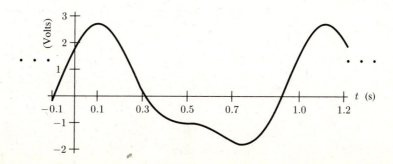

to introduce the concept of an amplitude spectrum and a phase spectrum. The cosine signal is characterized by the values of the three parameters: (1) frequency, f_x; (2) amplitude, A_x; and (3) phase, θ_x. Thus, we can plot the values of the amplitude and phase as a function of frequency to show the characteristics of the cosine signal, as illustrated in Figure 4.2. The amplitude as a function of frequency and the phase as a function of frequency are called, respectively, the *amplitude spectrum* and *phase spectrum* of the cosine signal. As mentioned previously, together they are the frequency-domain representation or the spectrum, of the cosine signal. Spectra of the form shown in Figure 4.2 are referred to as *line spectra*, since they are discrete-frequency functions plotted as lines. The units of the amplitude spectrum are the same as the units of the signal. For example, a voltage signal and its amplitude spectrum have units of volts. The units of the phase spectrum are usually radians. If degrees are used, it is specifically indicated.

Figure 4.2
Signal Waveform and
Spectrum for $x(t) =$
$3\cos\left[2\pi(10)t - \pi/3\right]$

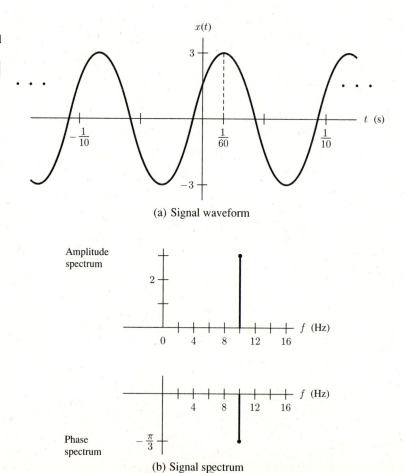

(a) Signal waveform

(b) Signal spectrum

We indicated in Chapter 3 that the cosine signal $x(t)$ can be represented by the sum of two counterrotating phasors with complex-conjugate magnitudes, the frequencies of rotation of which are f_x rotations per second. This representation is

$$x(t) = \left[\frac{A_x}{2} e^{j\theta_x} \right] e^{j2\pi f_x t} + \left[\frac{A_x}{2} e^{-j\theta_x} \right] e^{-j2\pi f_x t} \qquad (4.2)$$

We can write eq. (4.2) as

$$x(t) = \left[\frac{A_x}{2} e^{j\theta_x} \right] e^{j2\pi f_x t} + \left[\frac{A_x}{2} e^{-j\theta_x} \right] e^{j2\pi(-f_x)t} \qquad (4.3)$$

which shows that the cosine signal can also be represented as the sum of two rotating phasors with complex-conjugate magnitudes, the frequencies of rotation of which are f_x rotations per second and $-f_x$ rotations per second. The cosine signal representation eq. (4.3), and thus the cosine signal, is completely characterized by the amplitude and phase of the complex magnitudes as a function of the frequencies of the rotating phasors. These functions of frequency are referred to as the *double-sided amplitude and phase spectra* of the cosine signal. We illustrate them by example in Figure 4.3 for the same cosine signal shown in Figure 4.2.

The double-sided amplitude and phase spectra are even and odd functions of frequency, respectively. This fact is a consequence of the complex-conjugate nature of the complex magnitudes of the rotating phasors, which, in turn, is a consequence of the fact that $x(t)$ is real, so that imaginary parts of the terms summed must cancel. Note that the cosine function and its double-sided spectrum are easily related directly, since the amplitude of the cosine is twice the value of the double-sided amplitude spectrum at the positive frequency and the phase of the cosine is the value of the double-sided phase spectrum at the positive frequency.

We must emphasize that the double-sided spectrum does not imply that negative frequencies exist. The negative frequencies present are merely a mathematical abstraction resulting from the interpretation given in eq. (4.3) of the representation of a cosine by the sum of the rotating phasors. Such a mathematical abstraction is introduced because complex exponentials are easier to use than cosines, in many cases, and because the use of double-sided spectra makes analysis of systems, such as modulation systems, more convenient and easier to comprehend.

The amplitude and phase spectra shown in Figure 4.2 are referred to as *single-sided* amplitude and phase spectra to distinguish them from double-sided spectra. In most cases, we can refer to either the single-sided or double-sided spectra as simply the amplitude and phase spectra for simplicity. We will use the double-sided spectra in almost all cases in this text.

The amplitude and phase spectra of a single-cosine signal are not particularly interesting, since the information they provide is readily determined from the time waveform. However, we showed in Figure 4.1 that information about the frequencies that are present in a signal, as well as the amplitudes and

Figure 4.3
Signal Waveform and
Double-Sided
Spectrum for $x(t) =$
$\left(1.5e^{-j\pi/3}\right)e^{j2\pi(10)t} +$
$\left(1.5e^{j\pi/3}\right)e^{j2\pi(-10)t} =$
$3\cos\left[2\pi(10)t - \pi/3\right]$

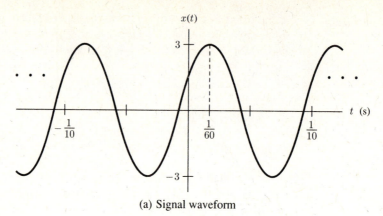

(a) Signal waveform

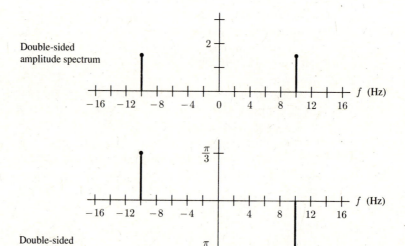

Double-sided
amplitude spectrum

Double-sided
phase spectrum

(b) Double-sided spectrum of signal

phases associated with these frequencies, is not readily apparent from the time
waveform if the signal is a sum of cosines of different frequencies. To indicate
this information, we can express the signal either as a sum of cosines with
different frequencies or as a sum of terms, such as those in eq. (4.3), with
different rotation frequencies. Thus, the single-sided or double-sided spectrum
of the signal is the sum of the single-sided or double-sided spectra for all cosine
terms with different frequencies. For example, we can express the signal

$$x(t) = \left(3/\sqrt{2}\right)\cos\left[2\pi(10)t + \pi/4\right] + \left(3/\sqrt{2}\right)\cos\left[2\pi(10)t - \pi/4\right]$$

$$+ \left(1/\sqrt{2}\right)\cos[2\pi(15)t] + \left(1/\sqrt{2}\right)\sin[2\pi(15)t]$$

$$+ 2\sin\left[2\pi(20)t + 5\pi/6\right] \tag{4.4}$$

shown in Figure 4.4 by the following sum of cosines of different frequencies:

$$x(t) = 3\cos[2\pi(10)t] + \cos[2\pi(15)t - \pi/4] + 2\cos[2\pi(20)t + \pi/3] \quad (4.5)$$

The double-sided spectrum of the signal expressed by eq. (4.4) is easily obtained from eq. (4.5); it is also illustrated in Figure 4.4. Note that the signal waveform is periodic with period of 0.2 s, and that one period is shown. The spectrum clearly indicates that the range of frequencies contained in the signal extends from 10 Hz to 20 Hz, and that the signal contains only three frequencies in this range. These facts are not readily apparent from the waveform plot of the signal.

Figure 4.4
Waveform and
Spectrum for the
Signal Expressed by
Eq. (4.4)

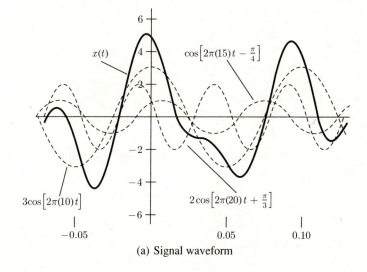

(a) Signal waveform

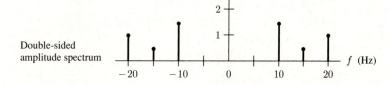

Double-sided
amplitude spectrum

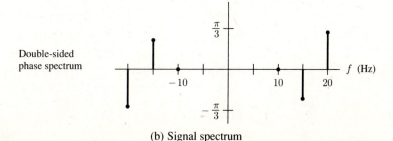

Double-sided
phase spectrum

(b) Signal spectrum

4.2 Fourier Series Representation of Signals

In the previous section we showed that the signal spectrum provides useful information about signal amplitude and phase characteristics as a function of frequency, and that the spectrum is easily obtained for a signal that is a sum of cosines, expressed either as a sum of cosines or as a sum of complex-conjugate pairs of complex exponentials. Thus, it is reasonable for us to approximate a signal with a generalized Fourier series that uses complex-exponential basis signals in order to easily obtain the spectrum of the signal. For real-valued signals, we will see that the resulting *complex-exponential Fourier series* (that is, the weighted sum of complex exponentials) can be written as a sum of cosines. This sum of the cosines is referred to as the *cosine Fourier series*. Either the complex exponential Fourier series or the cosine Fourier series is frequently called the *Fourier series* for simplicity.

We will first consider use of the complex-exponential Fourier series to represent a signal over some time interval. Then, we will turn to the special case where the signal is periodic and the interval is one period of the signal, since this leads to a representation of the signal over all time from which we can easily determine the signal spectrum.

Representation of Signals over an Interval by Fourier Series

Let us consider the approximation of the signal $x(t)$ over the interval $t_1 < t < t_1 + T_1$ by a generalized Fourier series constructed from an infinite set of complex-exponential basis signals. The basis signals chosen have harmonically related rotation frequencies and are orthogonal. They are

$$\phi_n(t) = e^{j2\pi(nf_1)t} \qquad n = 0, \pm 1, \pm 2, \ldots \qquad (4.6)$$

where $f_1 = 1/T_1$. The resulting complex-exponential Fourier series approximation of the signal $x(t)$ over the interval $t_1 < t < t_1 + T_1$ is thus

$$\hat{x}(t) = \sum_{n=-\infty}^{\infty} X_n e^{j2\pi n f_1 t} \qquad (4.7)$$

A truncated version of the series containing $2N + 1$ terms is

$$\hat{x}_N(t) = \sum_{n=-N}^{N} X_n e^{j2\pi n f_1 t} \qquad (4.8)$$

We see that $\phi_0(t) = 1$ and that $\phi_n(t)$ for $n > 0$ has a rotation frequency of $|n|f_1$ revolutions per second; thus, it is periodic with period $T_1/|n|$. The frequency f_1 produces one revolution or one cycle in the expansion interval, and is referred to as the *fundamental frequency*. The frequency $|n|f_1$, corresponding to the nth basis function, is called the nth *harmonic frequency* and produces n cycles in the expansion interval. Since $\phi_0(t)$ is constant, and the period $T_1/|n|$

of $\phi_n(t)$ divides into the period T_1 of $\phi_1(t)$ an integer number of times for all $|n| \geq 1$, then $\hat{x}(t)$ is periodic with period T_1. That is, $\hat{x}(t)$ repeats outside the expansion interval.

The proof that the harmonically related complex-exponential basis signals are mutually orthogonal over the expansion interval is left for the reader in the Problem section. The set of harmonically related complex-exponential basis signals can also be shown to be a complete set. Thus, as indicated in Section 3.6, the energy in the error signal $\hat{x}_N(t) - x(t)$ in the expansion interval approaches zero as $N \to \infty$ if the complex-exponential Fourier series exists (that is, if finite coefficients can be found for all terms).

The conditions a signal must satisfy over an interval in order for a Fourier series expansion for the signal to exist over the interval are rigorously considered in mathematics texts on Fourier analysis.[†] They are merely stated here to enable identification of the types of signals that can be expanded into a Fourier series.

A Fourier series approximation over the expansion interval exists for all signals having finite energy in the interval; that is, signals that satisfy

$$\int_{t_1}^{t_1+T_1} |x(t)|^2 \, dt < \infty \tag{4.9}$$

As we have already stated, $\hat{x}_N(t)$ approximates $x(t)$ in the expansion interval in the sense that the energy in the error signal $\hat{x}_N(t) - x(t)$ approaches zero as $N \to \infty$.

In addition, if the signal $x(t)$ satisfies what are known as the *Dirichlet conditions*, then, as $N \to \infty$, $\hat{x}_N(t)$ approaches $x(t)$ at every point in the expansion interval where $x(t)$ is continuous. At discontinuities, $\hat{x}_N(t)$ approaches the average of the limits of $x(t)$ taken from the right and from the left. Thus, if $x(t)$ satisfies the Dirichlet conditions and is defined to be equal to the average of the limits from the right and the left at any discontinuity points in the expansion interval, then

$$x(t) = \hat{x}(t) = \sum_{n=-\infty}^{\infty} X_n e^{j2\pi n f_1 t} \qquad t_1 < t < t_1 + T_1 \tag{4.10}$$

The Dirichlet conditions, developed by P. L. Dirichlet, are as follows:

1. $\displaystyle\int_{t_1}^{t_1+T_1} |x(t)| < \infty$
2. The signal $x(t)$ must have a finite number of maxima and minima in the expansion interval.
3. The signal $x(t)$ must have a finite number of finite discontinuities in the expansion interval.

[†] See, for example, R. V. Churchill, *Fourier Series and Boundary Value Problems* (New York: McGraw Hill, 1941).

Almost all signals of interest satisfy these conditions. Three examples of such signals and their corresponding Fourier series are shown in Figure 4.5. Note that $\hat{x}(t) = x(t)$ inside the expansion interval and repeats outside the expansion interval. Thus, the Fourier series converges to a periodic signal that we call the *periodic extension of the portion of the signal in the expansion interval.*

We compute the required coefficients for the complex-exponential Fourier series representation of $x(t)$ over the interval $t_1 < t < t_1 + T_1$ by using eq. (3.77). This gives

$$X_n = \frac{1}{\lambda_n} \int_{t_1}^{t_1+T_1} x(t) e^{-j2\pi n f_1 t}\, dt \tag{4.11}$$

where the parameter λ_n is computed with eq. (3.76) to be

$$\lambda_n = \int_{t_1}^{t_1+T_1} e^{j2\pi n f_1 t} e^{-j2\pi n f_1 t}\, dt = \int_{t_1}^{t_1+T_1} dt = T_1 \tag{4.12}$$

for all n.

Figure 4.5
Examples of Fourier Series Corresponding to Signals for Given Expansion Intervals

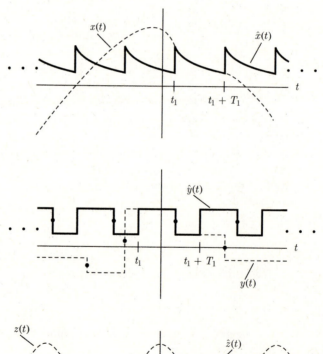

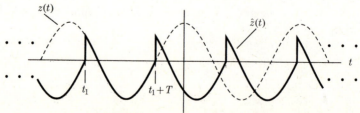

We now turn our attention to some characteristics of the coefficients of the complex-exponential Fourier series. To begin, let us consider the coefficient for the term produced when $n = 0$. This term is

$$X_0 = \frac{1}{T_1} \int_{t_1}^{t_1+T_1} x(t) e^{-j2\pi(0)f_1 t} \, dt$$

$$= \frac{1}{T_1} \int_{t_1}^{t_1+T_1} x(t) \, dt \tag{4.13}$$

Note that X_0 is the average value of $x(t)$ over the expansion interval. Next let us consider the coefficient for any other term in the series (that is, $n \neq 0$). Using Euler's theorem, we can write X_n as

$$X_n = \frac{1}{T_1} \int_{t_1}^{t_1+T_1} x(t) \left[\cos 2\pi n f_1 t - j \sin 2\pi n f_1 t \right] dt$$

$$= \frac{1}{T_1} \int_{t_1}^{t_1+T_1} x(t) \cos\left(2\pi n f_1 t\right) dt - j \frac{1}{T_1} \int_{t_1}^{t_1+T_1} x(t) \sin\left(2\pi n f_1 t\right) dt \tag{4.14}$$

Thus, the coefficients for $n \neq 0$ are complex, and we write them symbolically as

$$X_n = Re[X_n] + jIm[X_n] = |X_n| e^{j\underline{/X_n}} \tag{4.15}$$

In most cases, signals that are real functions of time are of the greatest interest to us. If $x(t)$ is real, then the integrands in eq. (4.14) are real; hence, the values of the integrals are real. Therefore, if $x(t)$ is real, then

$$Re\left[X_n\right] = \frac{1}{T_1} \int_{t_1}^{t_1+T_1} x(t) \cos\left(2\pi n f_1 t\right) dt \tag{4.16}$$

and

$$Im\left[X_n\right] = -\frac{1}{T_1} \int_{t_1}^{t_1+T_1} x(t) \sin\left(2\pi n f_1 t\right) dt \tag{4.17}$$

Note that when $x(t)$ is real, then

$$Re[X_{-n}] = Re[X_n] \tag{4.18}$$

and

$$Im[X_{-n}] = -Im[X_n] \tag{4.19}$$

since $\cos(-2\pi n f_1 t) = \cos(2\pi n f_1 t)$ and $\sin(-2\pi n f_1 t) = -\sin(2\pi n f_1 t)$. Equations (4.18) and (4.19) imply that

$$X_{-n} = X_n^* \tag{4.20}$$

which means that

$$|X_{-n}| = |X_n| \tag{4.21}$$

and

$$\underline{/\,X_{-n}} = -\underline{/\,X_n} \tag{4.22}$$

Therefore, the magnitude and angle of the complex-exponential Fourier series coefficients are, respectively, even and odd functions of n for signals that are real functions of time.

The complex-exponential Fourier series is easily converted into a cosine Fourier series when the signal that it represents over the expansion interval is a real function of time. To perform this conversion, we first rewrite eq. (4.7) as

$$\hat{x}(t) = X_0 + \sum_{n=1}^{\infty} \left[X_n e^{j2\pi n f_1 t} + X_{-n} e^{j2\pi(-n)f_1 t} \right] \tag{4.23}$$

Since

$$X_n = |X_n| e^{j\underline{/X_n}} \tag{4.24}$$

and

$$X_{-n} = |X_{-n}| e^{j\underline{/X_{-n}}} = |X_n| e^{-j\underline{/X_n}} \tag{4.25}$$

then

$$\hat{x}(t) = X_0 + \sum_{n=1}^{\infty} |X_n| \left[e^{j(2\pi n f_1 t + \underline{/X_n})} + e^{-j(2\pi n f_1 t + \underline{/X_n})} \right]$$

$$= X_0 + \sum_{n=1}^{\infty} 2|X_n| \cos\left(2\pi n f_1 t + \underline{/\,X_n}\right) \tag{4.26}$$

This is the desired cosine Fourier series. The complex-exponential Fourier series is usually easier to use than the series of cosine Fourier series in performing signal and system analysis. However, the cosine Fourier series is useful if we wish to indicate the terms to be added in constructing the Fourier series representation of a signal. It also helps us to visualize the meaning of the amplitude and phase spectra of a periodic signal.

We can further modify the cosine Fourier series by using the trigonometric identity for the cosine of a sum of angles to give

$$\hat{x}(t) = X_0 + \sum_{n=1}^{\infty} 2|X_n| [\cos(\underline{/\,X_n}) \cos(2\pi n f_1 t) - \sin(\underline{/\,X_n}) \sin(2\pi n f_1 t)]$$

$$= X_0 + \sum_{n=1}^{\infty} a_n \cos(2\pi n f_1 t) + \sum_{n=1}^{\infty} b_n \sin(2\pi n f_1 t) \tag{4.27}$$

where

$$a_n = 2|X_n| \cos(\underline{/\,X_n}) \quad \text{and} \quad b_n = -2|X_n| \sin(\underline{/\,X_n}) \tag{4.28}$$

It is not difficult to show that

$$a_n = \frac{2}{T_1} \int_{t_1}^{t_1+T_1} x(t) \cos(2\pi n f_1 t) \, dt \qquad (4.29)$$

and

$$b_n = \frac{2}{T_1} \int_{t_1}^{t_1+T_1} x(t) \sin(2\pi n f_1 t) \, dt \qquad (4.30)$$

Note that $a_0 = 2X_0$ and $b_0 = 0$. Therefore,

$$\hat{x}(t) = \frac{a_0}{2} + \sum_{n=1}^{\infty} a_n \cos(2\pi n f_1 t) + \sum_{n=1}^{\infty} b_n \sin(2\pi n f_1 t) \qquad (4.31)$$

is a generalized Fourier series, the basis signals of which are the harmonic sinusoids $\cos(2\pi n f_1 t)$ and $\sin(2\pi n f_1 t)$ for $n = 0, 1, 2, \dots$. This form of the Fourier series is called the trigonometric Fourier series. It is not easy use in signal and system analysis; morever, it does not provide ready interpretation of amplitude and phase characteristics of the components at different frequencies. It is included here only because it has historical significance in the development of Fourier series.

Example 4.1

Find the complex-exponential Fourier series expansion of $x(t) = e^{-0.1t}$ over the interval $-5 < t < 15$ s. Plot $x(t)$ and the truncated Fourier series for $N = 4$ — that is, $\hat{x}_4(t)$.

Solution

The complex-exponential Fourier series coefficients are

$$X_n = \frac{1}{T_1} \int_{t_1}^{t_1+T_1} x(t) e^{-j2\pi n f_1 t} \, dt$$

where

$$T_1 = 20 \quad \text{and} \quad f_1 = 1/20.$$

Therefore,

$$X_n = \frac{1}{20} \int_{-5}^{15} e^{-t/10} e^{-j\pi n t/10} \, dt = \frac{1}{20}\left[-\frac{10}{(1+j\pi n)} \right] \left[\exp\left\{ -\frac{t}{10}(1+j\pi n) \right\} \right]_{-5}^{15}$$

$$= \frac{0.5}{(1+j\pi n)} \left[1.649 e^{j\pi n/2} - 0.223 e^{-j3\pi n/2} \right]$$

and the complex-exponential Fourier series is

$$\hat{x}(t) = \sum_{n=-\infty}^{\infty} \frac{0.5}{(1+j\pi n)} \left[1.649 e^{j\pi n/2} - 0.223 e^{-j3\pi n/2} \right] e^{j\pi n t/10}$$

Evaluating the Fourier series coefficients for $0 \leq n \leq 4$ gives $X_0 = 0.713$, $X_1 = 0.216e^{j0.308}$, $X_2 = 0.112e^{j1.729}$, $X_3 = 0.075e^{-j3.036}$, and $X_4 = 0.057e^{-j1.491}$, where all angles are expressed in radians.

Since $x(t)$ is real, $X_{-n} = X_n^*$; therefore,

$$\hat{x}_4(t) = \left[0.057e^{j1.491}\right] e^{-j4\pi t/10} + \left[0.075e^{j3.036}\right] e^{-j3\pi t/10}$$

$$+ \left[0.112e^{-j1.729}\right] e^{-j2\pi t/10} + \left[0.216e^{-j0.308}\right] e^{-j\pi t/10}$$

$$+ 0.713 + \left[0.216e^{j0.308}\right] e^{j\pi t/10} + \left[0.112e^{j1.729}\right] e^{j2\pi t/10}$$

$$+ \left[0.075e^{-j3.036}\right] e^{j3\pi t/10} + \left[0.057e^{-j1.491}\right] e^{j4\pi t/10}$$

Furthermore, it is more convenient to write $\hat{x}_4(t)$ as the following cosine series in order to plot $\hat{x}_4(t)$:

$$\hat{x}_4(t) = X_0 + \sum_{n=1}^{4} 2|X_n| \cos(0.1\pi nt + \underline{/X_n})$$

$$= 0.713 + 0.432 \cos(0.1\pi t + 17.66°) + 0.224 \cos(0.2\pi t + 99.04°)$$

$$+ 0.150 \cos(0.3\pi t - 173.94°) + 0.114 \cos(0.4\pi t - 85.45°)$$

The plots of $x(t)$ and $\hat{x}_4(t)$ are shown in Figure 4.6.

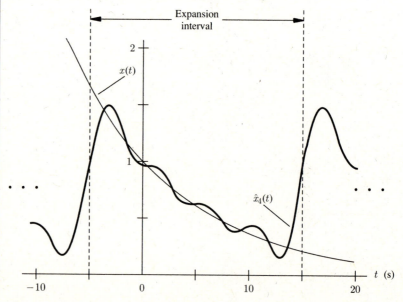

Figure 4.6 Signal $x(t)$ and Truncated Fourier Series Approximation for Example 4.1

We see in Figure 4.6 that the truncated Fourier series $x_4(t)$ provides a reasonable approximation to $x(t)$ in Example 4.1 over most of the expansion interval. The greatest difference is at the ends of the interval, since the Fourier series is converging to the periodic extension of the portion of $x(t)$ inside the expansion interval. This periodic extension has large discontinuities, or jumps, at the ends of the interval because the value of $x(t)$ is significantly different at the ends of the interval. These jumps represent rapid changes in the periodic extension of the signal, requiring rapidly changing (high-frequency) terms in the Fourier series.

If we use more terms in the truncated Fourier series (that is, increase N), then the number of ripples in $\hat{x}_N(t)$ in the expansion interval increases. Thus, the number of crossings of $\hat{x}_N(t)$ and $x(t)$ increases inside the expansion interval. The ripples decrease in size in the center. Those on the ends do not decrease in size, but just get narrower. This effect, called the *Gibbs phenomenon*, always occurs at discontinuities in the periodic extension of the signal being approximated by a truncated Fourier series. The effect is named after Josiah Gibbs, who first demonstrated that it would always occur. The Gibbs phenomenon does not signify that $\hat{x}_N(t)$ will not approach $x(t)$ at all values in the interval when N approaches infinity. It merely means that a larger N is required to decrease the error to some specified value as points closer to the discontinuity are considered.

Fourier Series Representation of Periodic Signals

In the preceding discussion, we indicated that the Fourier series corresponding to a signal that satisfies the Dirichlet conditions in the expansion interval converges to the periodic extension of the portion of the signal in the expansion interval. We illustrated this fact with the three examples in Figure 4.5. When we consider the periodic signal of the third example in Figure 4.5, we are led to conclude that the Fourier series obtained would converge to the signal for all time if we increased the expansion-interval length to equal the length of the period of the signal. This result holds for all periodic signals, since the periodic extension of one period of a periodic signal is the periodic signal itself. Therefore, given that the periodic signal $x(t)$ satisfies the Dirichlet conditions in all intervals of length equal to one period, then

$$x(t) = \hat{x}(t) = \sum_{n=-\infty}^{\infty} X_n e^{j2\pi n f_0 t} \qquad \text{for all } t \qquad (4.32)$$

where

$$X_n = \frac{1}{T_0} \int_{t_1}^{t_1+T_0} x(t) e^{-j2\pi n f_0 t} \, dt \qquad (4.33)$$

and $T_0 = 1/f_0$ is the period of $x(t)$.

Example 4.2

A waveform generator produces the square wave $x(t)$ shown in Figure 4.7. Find the complex-exponential Fourier series that converges to $x(t)$ for all time.

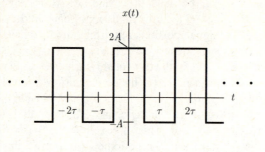

Figure 4.7 Square-Wave Signal for Example 4.2

Solution

The expansion-interval length must be one period so that $\hat{x}(t)$ will converge to $x(t)$ for all t. Since the period of $x(t)$ is $T_0 = 2\tau$, then

$$X_n = \frac{1}{2\tau} \int_{t_1}^{t_1+2\tau} x(t)e^{-j(\pi nt/\tau)} \, dt$$

We chose $t_1 = -\tau/2$, since this choice permits evaluation of the integral in two parts rather than three. This evaluation gives

$$X_n = \frac{1}{2\tau} \int_{-\tau/2}^{\tau/2} (2A)e^{-j(\pi nt/\tau)} \, dt + \frac{1}{2\tau} \int_{\tau/2}^{3\tau/2} (-A)e^{-j(\pi nt/\tau)} \, dt$$

$$= \frac{A}{\tau} \left[-\frac{\tau}{j\pi n} e^{-j(\pi nt/\tau)} \right]_{-\tau/2}^{\tau/2} - \frac{A}{2\tau} \left[-\frac{\tau}{j\pi n} e^{-j(\pi nt/\tau)} \right]_{\tau/2}^{3\tau/2}$$

$$= \frac{A}{j\pi n} \left[e^{j(\pi n/2)} - e^{-j(\pi n/2)} \right] - \frac{A}{j2\pi n} \left[e^{-j\pi n/2} - e^{-j(3\pi n/2)} \right]$$

$$= \frac{A}{j2\pi n} \left[2 - e^{-j\pi n} \right] \left[e^{-j\pi n/2} - e^{-j(\pi n/2)} \right]$$

$$= \frac{A}{\pi n} \left[2 - (-1)^n \right] \sin\left(\frac{\pi n}{2} \right)$$

Now, X_0 is of the form $0/0$. We could evaluate X_0 by using l'Hospital's rule. However, it is simpler to just reevaluate the integral for $n = 0$, which gives

$$X_0 = \frac{1}{2\tau} \int_{-\tau/2}^{\tau/2} (2A) \, dt + \frac{1}{2\tau} \int_{\tau/2}^{3\tau/2} (-A) \, dt$$

$$= \frac{1}{2\tau}(2A\tau) + \frac{1}{2\tau}(-A\tau) = \frac{A}{2}$$

This result is as expected because X_0 is the average or dc value of $x(t)$, which we can see is $A/2$ in Figure 4.7. If n is even, then $\sin(\pi n/2) = 0$, which means that $X_n = 0$. If $n = \ldots -11, -7, -3, 1, 5, 9, \ldots,$

$$X_n = \frac{A}{\pi n}[2 - (-1)](1) = \frac{3A}{\pi n}$$

If $n = \ldots -9, -5, -1, 3, 7, 11, \ldots,$

$$X_n = \frac{A}{\pi n}[2 - (-1)](-1) = -\frac{3A}{\pi n}$$

Therefore,

$$X_n = \begin{cases} A/2 & n = 0 \\ 3A/\pi|n| & n = \pm1, \pm5, \pm9, \ldots \\ -3A/\pi|n| & n = \pm3, \pm7, \pm11, \ldots \\ 0 & n = \pm2, \pm4, \pm6, \ldots \end{cases}$$

and the complex exponential Fourier series is

$$\hat{x}(t) = \sum_{n=-\infty}^{\infty} X_n e^{j(\pi n t/\tau)} = x(t) \qquad \text{for all } t$$

where the values of X_n are as indicated.

4.3 Amplitude and Phase Spectra of Periodic Signals

Let us now consider a periodic real signal, $x(t)$, that satisfies the Dirichlet conditions on any interval of length equal to one period. We define this signal at any discontinuity to have a value equal to the average of the limits of the signal from the right and left. Given these limitations, which are not very restrictive, the signal $x(t)$ is equal to the Fourier series, $\hat{x}(t)$, obtained by using an expansion interval equal to one period of the signal. Thus, the spectrum of $x(t)$ is the same as the spectrum of $\hat{x}(t)$. Actually, it is not important that the values defined at discontinuities be equal to the average of the limits from the right and left. If we choose different finite values, then $\hat{x}(t)$ differs from $x(t)$ only at the discontinuity points. This difference would not affect the value of the Fourier series coefficients, and thus would not affect the values of the signal spectrum components because we obtain the coefficients with an integral and $\hat{x}(t) - x(t)$ at one point in time has no area under it. Therefore, the spectrum of $\hat{x}(t)$ would still equal the spectrum of $x(t)$. Since $\hat{x}(t) = x(t)$ under the above conditions, then we can write the periodic signal $x(t)$ as

$$x(t) = X_0 + \sum_{n=1}^{\infty} 2|X_n| \cos(2\pi n f_0 t + \underline{/X_n}) \tag{4.34}$$

or

$$x(t) = \sum_{n=-\infty}^{\infty} X_n e^{j2\pi n f_0 t} \qquad (4.35)$$

where

$$X_n = \frac{1}{T_0} \int_{t_1}^{t_1+T_0} x(t) e^{-j2\pi n f_0 t} \, dt \qquad (4.36)$$

and $T_0 = 1/f_0$ is the period of the periodic signal.

Periodic-Signal Spectra

Since a periodic signal can be written as a sum of cosines of different frequencies (eq. [4.34]), or as a sum of complex-conjugate pairs of complex exponentials of different frequencies (eq. [4.35]), then we can obtain its spectrum easily from the parameters of the individual cosine terms or from the complex coefficients of the individual complex exponentials. Single-sided amplitude and phase spectra obtained from eq. (4.34) are illustrated in Figure 4.8 for a general periodic signal, $x(t)$. Double-sided amplitude and phase spectra obtained from eq. (4.35) are shown in Figure 4.9 for the same general periodic signal. Recall from Section 4.1 that either the single-sided or double-sided spectra can be used to represent the signal because the amplitude and phase of the cosine terms that make up the signal can be determined from either. Also recall that

Figure 4.8
Single-Sided
Spectrum of the
Periodic Signal $x(t)$

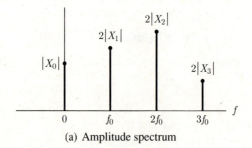

(a) Amplitude spectrum

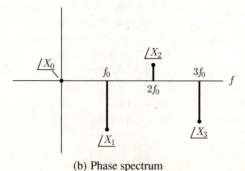

(b) Phase spectrum

Figure 4.9
Double-Sided
Spectrum of the
Periodic Signal $x(t)$

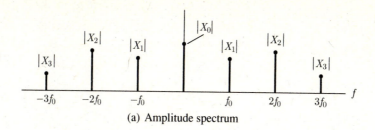

(a) Amplitude spectrum

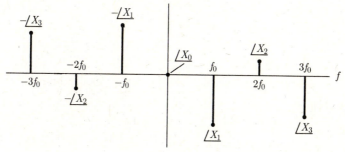

(b) Phase spectrum

the double-sided spectrum is typically used in signal and system analysis for mathematical simplicity, even though it contains the mathematical abstraction of negative frequencies.

Since the phase angle $\underline{/X_n} \pm 2\pi$ rad (or $\underline{/X_n} \pm 360°$, if angles are expressed in degrees) produces the same cosine as the phase angle $\underline{/X_n}$, then we can express all phase angles so that $-\pi \leq \underline{/X_n} \leq \pi$. However, signal characteristics are sometimes shown more clearly if we allow $|\underline{/X_n}|$ to be greater than π. Thus, we will follow this practice in many of the following examples. Note that the phase angle associated with the zero frequency or dc term must have a value that is a multiple of π. That is, $\underline{/X_0} = \pm m\pi$ rad, where m is even when X_0 is positive, and where m is odd when X_0 is negative. For example, $3 = 3e^{j0} = 3e^{j2\pi} = 3e^{-j2\pi}$ and $-3 = 3e^{j\pi} = 3e^{-j\pi} = 3e^{j3\pi}$.

Example 4.3

Plot the double-sided amplitude and phase spectra over the frequency range $|f| < 475$ Hz for the square-wave output signal from the waveform generator in Example 4.2 when the generator is adjusted so that $A = 10$ and $\tau = 0.01$ s.

Solution
The signal period is $T_0 = 2\tau = 0.02$ s. The fundamental frequency is $f_0 = 1/T_0 = 50$ Hz.

$$\frac{475}{50} = 9.5 \rightarrow \text{Plot to the ninth harmonic } (n = 9)$$

From Example 4.2,

$$X_2 = X_4 = X_6 = X_8 = 0$$

$$X_0 = A/2 = 5 = 5e^{j0}$$

$$X_1 = 3A/\pi = 9.549 = 9.549e^{j0}$$

$$X_3 = -A/\pi = -3.183 = 3.183e^{-j\pi}$$

$$X_5 = 3A/5\pi = 1.910 = 1.910e^{j0}$$

$$X_7 = -3A/7\pi = -1.364 = 1.364e^{-j\pi}$$

$$X_9 = 3A/9\pi = 1.061 = 1.061e^{j0}$$

We do not need to compute X_n for negative n, since $X_{-n} = X_n^*$. The amplitude and phase plots are shown in Figure 4.10.

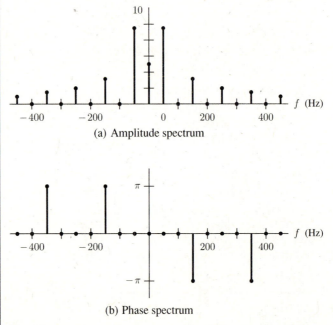

(a) Amplitude spectrum

(b) Phase spectrum

Figure 4.10 Amplitude and Phase Spectra for the Square-Wave Signal of Example 4.2

The phase-spectrum plot constructed in Example 4.3 includes phase-angle values of zero for those harmonic frequencies which have an amplitude of zero. We included these zero values for completeness. They are actually meaningless values, however, since the zero amplitude indicates that the harmonic frequencies do not exist in the signal. Note further that the phase angles associated with the harmonic frequencies of 150 Hz and 350 Hz could have been chosen to be π rad rather than $-\pi$ rad, since these two phase angles produce the same cosine function.

The function $\sin(\pi x)/\pi x$ occurs in the next example and sufficiently often in various analyses that we give it the following special notation:

$$\mathrm{sinc}(x) = \frac{\sin(\pi x)}{\pi x} \tag{4.37}$$

The area under this function is unity. The $\mathrm{sinc}(x)$ function and its amplitude and angle are plotted in Figure 4.11. Since the positive portions of $\mathrm{sinc}(x)$ have

Figure 4.11
The $\mathrm{sinc}(x)$ Function and Its Amplitude and Angle

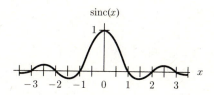

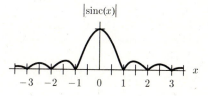

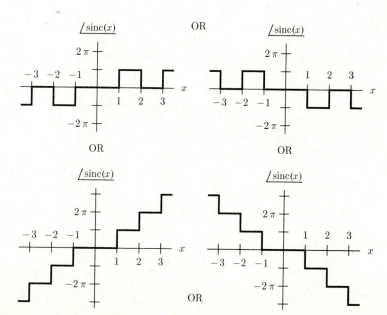

angles of $\pm n\pi$ where n is an even integer, and the negative portions of $\mathrm{sinc}(x)$ have angles of $\pm m\pi$ where m is odd, then there are a number of possible angle plots. Four of these plots are shown. In all cases the angles are plotted to be odd functions for ease of interpretation.

Example 4.4

The clock signal for a personal computer is the periodic rectangular-pulse voltage signal $x(t)$ shown in Figure 4.12.

a. Find the complex-exponential Fourier series representation for the clock signal.
b. Plot the amplitude and phase spectra for the clock signal when the computer has a clock rate of $f_c = 8$ MHz, the clock pulses have an amplitude of 4 V and a width of 0.05 μs, and $t_0 = 0$.
c. Repeat part (b) with $t_0 = 0$ replaced by $t_0 = 0.05$ μs.

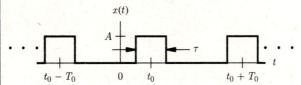

Figure 4.12 Periodic Rectangular-Pulse Voltage Signal

Solution

a. We require an expansion interval of length equal to one period of $x(t)$ to determine the values of the Fourier series coefficients. It is desirable to choose an expansion interval that contains only one pulse of the signal rather than two part pulses to reduce the integration labor required. A particularly convenient expansion interval, because of the symmetry involved, is the interval centered on the pulse closest to the time origin. Therefore, we choose $t_1 = t_0 - T_0/2$, which gives

$$X_n = \frac{1}{T_0} \int_{t_0 - T_0/2}^{t_0 + T_0/2} x(t) e^{-j2\pi n f_0 t}\, dt = \frac{1}{T_0} \int_{t_0 - \tau/2}^{t_0 + \tau/2} A e^{-j2\pi n f_0 t}\, dt$$

$$= \frac{A}{T_0} \left[\frac{e^{-j2\pi n f_0 t}}{-j2\pi n f_0} \right]_{t_0 - \tau/2}^{t_0 + \tau/2} = \frac{A}{\pi n f_0 T_0} \left[\left(e^{j\pi n f_0 \tau} - e^{-j\pi n f_0 \tau} \right)/2j \right] e^{-j2\pi n f_0 t_0}$$

$$= \frac{A\tau}{T_0} \left[\frac{\sin \pi n f_0 \tau}{\pi n f_0 \tau} \right] e^{-j2\pi n f_0 t_0} = \frac{A\tau}{T_0} \, \mathrm{sinc}(n f_0 \tau)\, e^{-j2\pi n f_0 t_0}$$

The complex-exponential Fourier series representation of the clock signal is therefore

$$x(t) = \sum_{n=-\infty}^{\infty} \left[\frac{A\tau}{T_0} \operatorname{sinc}(nf_0\tau) e^{-j2\pi n f_0 t_0} \right] e^{j2\pi n f_0 t}$$

b. $A = 4\text{V}$, $\tau = 0.05\ \mu\text{s}$, $T_0 = \dfrac{1}{f_c} = 0.125\ \mu\text{s}$, and $t_0 = 0$.

$$f_0 = 1/T_0 = f_c = 8\ \text{MHz}$$

$$X_n = 1.6\ \operatorname{sinc}\left(0.05 n f_0 \times 10^{-6}\right)$$

The nth term is at frequency $f = nf_0 = 8n$ MHz. The amplitude of the nth term is

$$|X_n| = \left| 1.6\ \operatorname{sinc}\left(0.05 n f_0 \times 10^{-6}\right) \right| = |1.6| \left| \operatorname{sinc}\left(0.05 n f_0 \times 10^{-6}\right) \right|$$

$$= 1.6 \left| \operatorname{sinc}\left(0.05 n f_0 \times 10^{-6}\right) \right| \qquad \text{(volts)}$$

The phase of nth term is

$$\underline{/X_n} = \underline{/1.6\ \operatorname{sinc}(0.05 n f_0 \times 10^{-6})}$$

$$= \underline{/1.6} + \underline{/\operatorname{sinc}(0.05 n f_0 \times 10^{-6})}$$

$$= 0 + \underline{/\operatorname{sinc}(0.05 n f_0 \times 10^{-6})} = \underline{/\operatorname{sinc}(0.05 n f_0 \times 10^{-6})} \qquad \text{(radians)}$$

The amplitude and phase spectra of the clock signal with the parameters of part (b) are plotted in Figure 4.13 for $|f| \le 48$ MHz.

c. $A = 4\text{V}$, $\tau = 0.05\ \mu\text{s}$, $T_0 = \dfrac{1}{f_c} = 0.125\ \mu\text{s}$, and $t_0 = 0.05\ \mu\text{s}$.

$$f_0 = 1/T_0 = f_c = 8\ \text{MHz}$$

$$X_n = 1.6\ \operatorname{sinc}\left(0.05 n f_0 \times 10^{-6}\right) e^{-j0.1\pi n f_0 \times 10^{-6}}$$

The nth term is at frequency $f = nf_0 = 8n$ MHz. The amplitude of the nth term is

$$|X_n| = \left| 1.6\ \operatorname{sinc}\left(0.05 n f_0 \times 10^{-6}\right) e^{-j0.1\pi n f_0 \times 10^{-6}} \right|$$

$$= |1.6| \left| \operatorname{sinc}\left(0.05 n f_0 \times 10^{-6}\right) \right| \left| e^{-j0.1\pi n f_0 \times 10^{-6}} \right|$$

$$= 1.6 \left| \operatorname{sinc}\left(0.05 n f_0 \times 10^{-6}\right) \right| \qquad \text{(volts)}$$

since $\left| e^{j0} \right| = 1$. The phase of the nth term is

$$\underline{/X_n} = \underline{/1.6\ \operatorname{sinc}(0.05 n f_0 \times 10^{-6}) e^{-j0.1\pi n f_0 \times 10^{-6}}}$$

$$= \underline{/1.6} + \underline{/\operatorname{sinc}(0.05 n f_0 \times 10^{-6})} + \underline{/e^{-j0.1\pi n f_0 \times 10^{-6}}}$$

$$= 0 + \underline{/\ \text{sinc}(0.05nf_0 \times 10^{-6})} + (-0.1\pi nf_0 \times 10^{-6})$$

$$= \underline{/\ \text{sinc}(0.05nf_0 \times 10^{-6})} - 0.1\pi nf_0 \times 10^{-6} \qquad \text{(radians)}$$

The amplitude and phase spectra of $x(t)$ with the parameters of part (c) are plotted in Figure 4.14 for $|f| \le 48$ MHz.

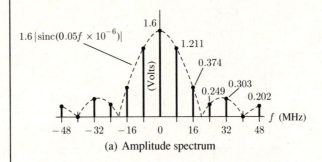

(a) Amplitude spectrum

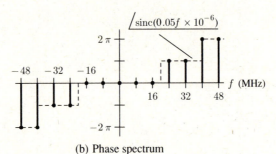

(b) Phase spectrum

Figure 4.13 Amplitude and Phase Spectra for the Clock Signal of Figure 4.12 with $A = 4$ V, $\tau = 0.05$ μs, $T_0 = 0.125$ μs, and $t_0 = 0$

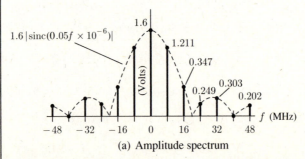

(a) Amplitude spectrum

Figure 4.14 Amplitude and Phase Spectra for the Clock Signal of Figure 4.12 with $A = 4$ V, $\tau = 0.05$ μs, $T_0 = 0.125$ μs, and $t_0 = 0.05$ μs

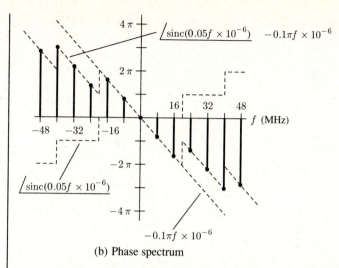

(b) Phase spectrum

Figure 4.14 Amplitude and Phase Spectra for the Clock Signal of Figure 4.12 with $A = 4$ V, $\tau = 0.05$ μs, $T_0 = 0.125$ μs, and $t_0 = 0.05$ μs *(continued)*

Continuous-frequency function envelopes (curves passing through all spectrum points) are shown on the plots of Figures 4.13 and 4.14 in Example 4.4. These envelopes help portray the variation of the amplitude and phase spectra with frequency. We can also use them in the approximate sketching of the spectra, if we are not using computer-generated sketches. In arriving at the phase-spectrum envelope for such sketching, it is convenient to add graphically the envelopes corresponding to each individual phase term. Therefore, they are also shown in Figure 4.14. We did not substitute the value of f_0 into the expressions for X_n in parts (b) and (c) of Example 4.4 in order to permit easy identification of the continuous-frequency envelopes. Obviously, the value of f_0 must be substituted to compute the specific values of amplitude and phase for different values of n.

Note that the phase-spectrum envelope in Figures 4.13 and 4.14 has a discontinuity at $f = 40$ MHz. The value plotted at this discontinuity was chosen so as to maximize its magnitude. Again, any finite value could have been chosen because the amplitude of the Fourier series component at $f = 40$ MHz is equal to zero; hence, it does not exist.

The amplitude spectra are the same in Figures 4.13 and 4.14 because the signals are the same, except for a time shift of 0.05 μs. This is always the case for a periodic signal, regardless of how the signal is located with respect to the time origin. The shift in time merely produces an additional phase term, the envelope of which is a linear function of frequency, as illustrated in Figure 4.14. We see that the additional linear envelope function is

$$\theta = -0.1\pi f \times 10^{-6} = -2\pi f t_0 \tag{4.38}$$

where t_0 is the signal shift to the right, or the time delay of the signal. Equation (4.38) matches the relation between cosine phase shift and time delay of a cosine shown by Eq. (3.4) in Chapter 3, as it must.

In addition to the clock signal, the pulse-train signal of Example 4.4 is an important signal for many applications. For example, a pulse train is used in a radar system, where the distance to a target is determined by the time interval between the transmission of a pulse and the reception of the reflection of the pulse from the target.

Periodic-Signal Power from Spectral Information

We showed in Chapter 3 (eq. [3.60]) that the average power in the periodic signal $x(t)$ is

$$P_x = \frac{1}{T_0} \int_{t_1}^{t_1+T_0} |x(t)|^2 \, dt \tag{4.39}$$

Since $\lambda_n = T_0$ for the complex-exponential Fourier series, we can use Parseval's theorem, as expressed by eq. (3.79), to rewrite the average power in a periodic signal as

$$P_x = \frac{1}{T_0} \left[\sum_{n=-\infty}^{\infty} T_0 |X_n|^2 \right] = \sum_{n=-\infty}^{\infty} |X_n|^2 = X_0^2 + \sum_{n=1}^{\infty} 2 |X_n|^2 \tag{4.40}$$

The last term of eq. (4.40) follows, since $|X_{-n}| = |X_n|$. We can see from the cosine form of the Fourier series in eq. (4.26) that the average power in a signal is the sum of the power in the constant, or dc, term and the average power in each cosine term.

The average power in the computer clock signal of parts (b) and (c) of Example 4.4 is $P_x = A^2 \tau / T_0 = 6.4$ (volts^2). Now let us assume that the clock signal is passed through a system (filter) that removes all components with frequencies greater than the fourth harmonic. The resulting output signal is $\hat{x}_4(t)$, and the power in this signal is

$$P_{x4} = |X_0|^2 + 2 \sum_{n=1}^{4} |X_n|^2 = |1.6|^2 + 2 \sum_{n=1}^{4} |1.6 \operatorname{sinc}(0.4n)|^2$$

$$= 1.6^2 + 2[(1.211)^2 + (0.374)^2 + (0.249)^2 + (0.303)^2]$$

$$= 6.080 \quad \text{(volts)}^2$$

Thus, the terms through the fourth harmonic contain 95% of the signal power. Plots of $\hat{x}_4(t)$ and $x(t)$ are shown in Figure 4.15. We can see that the output signal from the filter, $\hat{x}_4(t)$, is a distorted version of the rectangular-pulse-train input signal, $x(t)$, even though it contains 95% of the input-signal power. The Gibbs phenomenon is also readily apparent.

Figure 4.15
Signal of
Example 4.4 and
Fourier Series
Approximation
through the Fourth
Harmonic

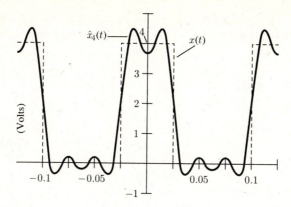

Bandwidth of Periodic Signals

An important property of a periodic signal is its *frequency bandwidth*. The bandwidth can be determined from either the single- or double-sided spectrum of the signal.

Definition _____

> The frequency bandwidth (called bandwidth) of a periodic signal is the width of the range of positive frequencies for which the signal has power.

Recall that the positive- and negative-frequency parts of a double-sided spectrum at a given frequency absolute value combine to produce the signal characteristics at that frequency. Thus, the negative-frequency part of a double-sided spectrum is not additional signal bandwidth.

We can see from the amplitude spectrum in Figures 4.13 and 4.14 that the bandwidth of the pulse-train signal of Example 4.4 is infinite. However, the signal power is concentrated at low frequencies because the amplitudes of the low-frequency components are much larger than those of the higher frequency components. We could define the significant bandwidth of the signal to encompass frequencies from zero up to the second zero value of the amplitude spectrum envelope, which, in effect, would include the signal components with the largest magnitudes, and thus most of the signal power. For the signal of Example 4.4, the bandwidth so defined would be $B_x = 2/\tau$. Other significant bandwidth definitions could include frequencies from zero up to the first zero value of the amplitude spectrum envelope ($B_x = 1/\tau$ for Example 4.4), or from zero up to the frequency at which the amplitude spectrum envelope is cX_0 where c is some fraction. In all of these definitions, the significant bandwidth of the pulse-train signal is inversely proportional to the width of the pulses. Therefore, if the pulses are made narrower, a wider bandwidth system is required to pass the same portion of the signal power.

4.4 The Fourier Transform and Spectra of Aperiodic Energy Signals

We have seen that the Fourier series is useful in the analysis of periodic signals. As long as the signal satisfies the Dirichlet conditions, the Fourier series equals the signal for all time. Furthermore, it readily provides information about the frequency content of the signal through the coefficients that specify the amplitude and phase spectra.

Because the Fourier series is a periodic function of time, it is not desirable to use it to represent aperiodic signals. The Fourier series is equal to the aperiodic signal only over the interval chosen for Fourier series expansion; outside this interval, it repeats. Therefore, the Fourier series is not the same signal as the aperiodic signal over all time; consequently, it cannot produce the true spectrum for the aperiodic signal. This leads to our discussion in this section on how to find the spectrum of an aperiodic signal. The method we will consider is called the *Fourier transform*.

Since the complex-exponential Fourier series provides an easy method for determining the spectrum of a periodic signal, it is reasonable for us to start with the complex-exponential Fourier series representation of an aperiodic signal over the interval $-T/2 < t < T/2$, and then let the interval increase in size until the entire time axis is encompassed. From eq. (4.9) we know that the Fourier series of $x(t)$ exists for any T if

$$\int_{-T/2}^{T/2} |x(t)|^2 \quad dt < \infty \tag{4.41}$$

In the limit, as $T \to \infty$ to cover the entire time axis,

$$\int_{-\infty}^{\infty} |x(t)|^2 \quad dt < \infty \tag{4.42}$$

which indicates that the spectrum given by the complex-exponential Fourier series coefficients in the limit exists if the aperiodic signal is an energy signal. We previously indicated that a signal could be recovered from its spectrum, as given by its Fourier series coefficients, as long as the signal satisfied the Dirichlet conditions. In that case, the Fourier series would equal the signal for all time. Similar conditions, called the *Dirichlet conditions for Fourier transforms* apply to aperiodic signals. They guarantee that the aperiodic signal can be recovered from the spectrum given by the complex-exponential Fourier series coefficients obtained in the limit as $T \to \infty$. The Dirichlet conditions for Fourier transforms are as follows:

1. $\int_{-\infty}^{\infty} |x(t)| \, dt < \infty$

2. The signal $x(t)$ must have a finite number of maxima and minima in any finite interval.

3. The signal $x(t)$ must have a finite number of finite discontinuities in any finite interval.

Most practical energy signals satisfy these conditions.

Development of the Fourier Transform

Let us consider the aperiodic function, $x(t)$, that satisfies the Dirichlet conditions for Fourier transforms. Thus, it satisfies the Dirichlet conditions for Fourier series for an expansion interval of any length. Then,

$$\hat{x}(t) = \sum_{n=-\infty}^{\infty} X_n e^{-j2\pi n f_0 t} \tag{4.43}$$

is equal to $x(t)$ over the interval $-T/2 < t < T/2$ and repeats outside this interval if $f_0 = 1/T$ and

$$X_n = \frac{1}{T} \int_{-T/2}^{T/2} x(t) e^{-j2\pi n f_0 t} \, dt \tag{4.44}$$

Therefore, $\hat{x}(t) = x_p(t)$ where $x_p(t)$ is the periodic extension of the portion of the aperiodic signal that is inside the expansion interval. We illustrate $x_p(t)$ in Figure 4.16 for an example aperiodic signal and two different values of T. The amplitude and phase spectra of $x_p(t)$ are obtained from the coefficients X_n; they are illustrated in Figure 4.17 for one value of T.

To obtain the amplitude and phase spectra for the original signal $x(t)$ from the amplitude and phase spectra for $x_p(t)$, we take the limit as $T \to \infty$, so that $x_p(t) = x(t)$ over the entire time axis. Since $f_0 = 1/T$, as T becomes larger, the lines in the spectra become closer together. The amplitude-spectrum lines also become smaller, since the finite-energy condition ensures that

$$\lim_{T \to \infty} \int_{-T/2}^{T/2} x(t) e^{-j2\pi n f_0 t} \, dt < \infty \tag{4.45}$$

which means that X_n, as given by eq. (4.44), approaches zero as $T \to \infty$. Therefore, as $T \to \infty$, an infinite number of infinitesimally small amplitude-spectrum lines occur in any finite region of the frequency axis.

To circumvent the problem of the spectral lines approaching zero amplitude and to illustrate more clearly the nature of the continuous frequency amplitude-density and phase spectra that result, we can redefine the amplitude and phase spectral plots for $x_p(t)$ as shown in Figure 4.18, where $\Delta f \equiv f_0$.

The redefined amplitude spectrum is the continuous-frequency function

$$|X_R(f)| = \sum_{n=-\infty}^{\infty} \frac{|X_n|}{\Delta f} \Pi[(f - n\Delta f)/\Delta f] \tag{4.46}$$

Figure 4.16
Example Signal and
Periodic Extensions
of Portions of the
Signal

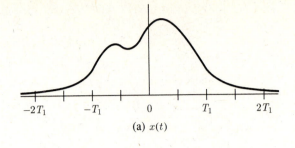

(a) $x(t)$

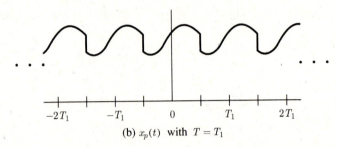

(b) $x_p(t)$ with $T = T_1$

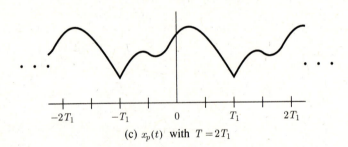

(c) $x_p(t)$ with $T = 2T_1$

Therefore, as illustrated in Figure 4.18a, the areas under each of the rectangles are equal to the amplitude of the spectral component that occurs within the width of the rectangle. Note that the height of a rectangle (that is, $|X_n|/\Delta f$) has units of amplitude per unit frequency. The redefined amplitude spectrum is thus an *amplitude-density spectrum* showing the signal amplitude contribution per unit frequency.

The redefined phase spectrum is the continuous-frequency function

$$\underline{/\,X_R(f)} = \sum_{n=-\infty}^{\infty} \underline{/\,X_n} \, \Pi[(f - n\Delta f)/\Delta f] \qquad (4.47)$$

We can see from Figure 4.18b that the height of each of the rectangles is equal to the phase of the spectral component that occurs within the width of the rectangle.

Figure 4.17
Amplitude and Phase
Spectra of $x_p(t)$ for
One Value of T.

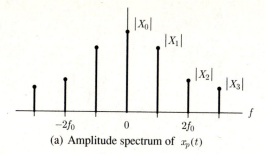

(a) Amplitude spectrum of $x_p(t)$

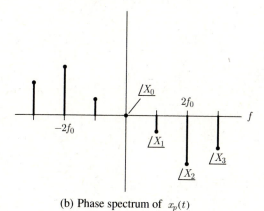

(b) Phase spectrum of $x_p(t)$

We can now proceed to perform the limiting operation $T \to \infty$ on the redefined continuous-frequency spectrum for $x_p(t)$ given by

$$X_R(f) = |X_R(f)| \exp[j \underline{/X_R(f)}] \tag{4.48}$$

When $T \to \infty$, then $x_p(t) \to x(t)$ for all t, so that the redefined amplitude and phase spectra of $x_p(t)$ are the spectra of $x(t)$. To compute these spectra for the arbitrary frequency f_a, we let $n \to \infty$ as $T \to \infty$ in such a manner that $n\Delta f = f_a$ for any T. We then define

$$X(f_a) \equiv \lim_{T \to \infty} X_R(f_a) = \lim_{\substack{T \to \infty \\ n\Delta f = f_a}} X_R(n\Delta f) \tag{4.49}$$

Note that

$$|X_R(n\Delta f)| = |X_n|/\Delta f = T|X_n| \tag{4.50}$$

and

$$\underline{/X_R(n\Delta f)} = \underline{/X_n} \tag{4.51}$$

Figure 4.18
Redefined Amplitude
and Phase Spectra of
$x_p(t)$

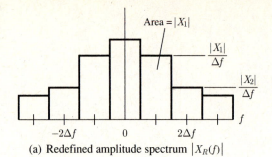

(a) Redefined amplitude spectrum $|X_R(f)|$

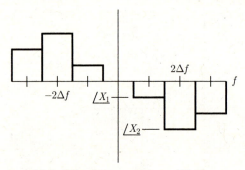

(b) Redefined phase spectrum $\underline{/X_R(f)}$

Substituting eqs. (4.48), (4.50), and (4.51) in eq. (4.49) gives

$$X(f_a) = \lim_{\substack{T \to \infty \\ n\Delta f = f_a}} T|X_n| \exp(j/X_n) = \lim_{\substack{T \to \infty \\ n\Delta f = f_a}} TX_n$$

$$= \lim_{\substack{T \to \infty \\ n\Delta f = f_a}} \int_{-T/2}^{T/2} x(t)e^{-j2\pi n\Delta f t}\, dt \qquad (4.52)$$

Since f_a is arbitrary, we can write eq. (4.52) in terms of the general frequency f as

$$X(f) = \int_{-\infty}^{\infty} x(t)e^{-j2\pi f t}\, dt \qquad (4.53)$$

The function $X(f)$, called the *Fourier transform* of $x(t)$, is referred to as the *spectrum* of $x(t)$. The amplitude of $X(f)$ is the double-sided amplitude-*density* spectrum (called *amplitude spectrum* for short) of $x(t)$, and the angle of $X(f)$ is the double-sided *phase spectrum of $x(t)$*.

Now, let us turn our attention to the limit of the Fourier series representation of $x_p(t)$. From eq. (4.48), (4.50), and (4.51),

$$\Delta f X_R(n\Delta f) = |X_n| \exp(j\underline{/X_n}) = X_n \qquad (4.54)$$

Therefore, using the equivalence $f_0 = \Delta f$,

$$x_p(t) = \sum_{n=-\infty}^{\infty} X_n e^{j2\pi n \Delta f t} = \sum_{n=-\infty}^{\infty} X_R(n\Delta f) e^{j2\pi n \Delta f t} \Delta f \qquad (4.55)$$

Taking the limit as $T \to \infty$, or equivalently as $\Delta f \to 0$,

$$x(t) = \lim_{T\to\infty} x_p(t) = \lim_{\Delta f\to 0} \sum_{n=-\infty}^{\infty} X_R(n\Delta f) e^{j2\pi n \Delta f t} \Delta f \qquad (4.56)$$

The values $X_R(n\Delta f) \exp(j2\pi n\Delta f t)$ are the values of the continuous-frequency function $X_R(f) \exp(j2\pi f t)$ at regularly spaced intervals of width Δf, and thus occur within adjacent intervals of Δf along the f axis. Therefore, the limit in eq. (4.56) is a *Riemann integral*, which we can write as

$$x(t) = \int_{-\infty}^{\infty} X(f) e^{j2\pi f t} \, df \qquad (4.57)$$

since

$$\lim_{T\to\infty} X_R(f) = X(f) \qquad (4.58)$$

We refer to eq. (4.57) as the *inverse Fourier transform*; it provides a method for determining an energy signal from its Fourier transform or spectrum. Note that the Fourier transform of any energy signal that satisfies the Dirichlet conditions is unique because the time function can be uniquely recovered from it. The two functions defined by the Fourier transform and inverse Fourier transform integrals,

$$X(f) = \int_{-\infty}^{\infty} x(t) e^{-j2\pi f t} \, dt \equiv \mathcal{F}[x(t)] \qquad (4.59)$$

and

$$x(t) = \int_{-\infty}^{\infty} X(f) e^{j2\pi f t} \, df \equiv \mathcal{F}^{-1}[X(f)] \qquad (4.60)$$

are called a *Fourier transform pair*. We frequently use a double-headed arrow to indicate Fourier transform pairs as

$$x(t) \leftrightarrow X(f) \qquad (4.61)$$

Sometimes the Fourier transform is defined as a function of $\omega = 2\pi f$, rather than of f; that is,

$$\int_{-\infty}^{\infty} x(t) e^{-j\omega t} \, dt = X\left(\frac{\omega}{2\pi}\right) \equiv X_\omega(\omega) \qquad (4.62)$$

Substituting $f = \omega/2\pi$ in eq. (4.60), we can recover the signal $x(t)$ from $X_\omega(\omega)$:

$$x(t) = \int_{-\infty}^{\infty} X\left(\frac{\omega}{2\pi}\right) e^{j\omega t} d\left(\frac{\omega}{2\pi}\right) = \frac{1}{2\pi} \int_{-\infty}^{\infty} X_\omega(\omega) e^{j\omega t} \, d\omega \qquad (4.63)$$

However, it is generally desirable to represent a spectrum as a function of frequency in hertz rather than in radians per second; consequently, $X(f)$ will be used in this text.

The Energy-Density Spectrum and Spectrum-Computation Examples

The concept of signal amplitude per unit range of frequency is somewhat nebulous. The signal energy per unit range of frequency has more intuitive appeal because the signal energy in a given frequency interval can be computed as the area under the signal energy per unit frequency curve in that interval. A plot of the signal energy per unit frequency curve is called the *energy-density spectrum*, or *energy spectral density* of the signal. We use eqs. (3.56), (4.59), and (4.60) to write the energy in the signal $x(t)$ as

$$
\begin{aligned}
E_x &= \int_{-\infty}^{\infty} |x(t)|^2 \, dt = \int_{-\infty}^{\infty} x^*(t)x(t) \, dt \\
&= \int_{-\infty}^{\infty} x^*(t) \left[\int_{-\infty}^{\infty} X(f)e^{j2\pi ft} \, df \right] dt \\
&= \int_{-\infty}^{\infty} X(f) \left[\int_{-\infty}^{\infty} x^*(t)e^{j2\pi ft} \, dt \right] df \\
&= \int_{-\infty}^{\infty} X(f) \left[\int_{-\infty}^{\infty} \left[x(t)e^{-j2\pi ft} \right]^* dt \right] df \\
&= \int_{-\infty}^{\infty} X(f) \left[\int_{-\infty}^{\infty} x(t)e^{-j2\pi ft} \, dt \right]^* df \\
&= \int_{-\infty}^{\infty} X(f)X^*(f) \, df = \int_{-\infty}^{\infty} |X(f)|^2 \, df
\end{aligned}
\tag{4.64}
$$

Thus, the signal energy is

$$
E_x = \int_{-\infty}^{\infty} |x(t)|^2 \, dt = \int_{-\infty}^{\infty} |X(f)|^2 \, df
\tag{4.65}
$$

This is called *Rayleigh's energy theorem* or *Parseval's theorem for Fourier transforms*. The function

$$
G_x(f) = |X(f)|^2
\tag{4.66}
$$

gives the total energy in the signal when integrated over all frequencies; therefore, $G_x(f)$ is defined as the energy-density spectrum of $x(t)$. We can show that

$$
\int_{-f_2}^{-f_1} G_x(f) \, df + \int_{f_1}^{f_2} G_x(f) \, df
$$

is the energy contained by the signal in the frequency range $f_1 < f < f_2$; hence, $G_x(f)$ is an appropriate definition for the energy-density spectrum. Recall that the positive and negative portions of a double-sided spectrum combine to produce the signal characteristics in a given frequency range.

Example 4.5

In Chapter 3 we indicated that the exponential signal $y(t) = e^{-\alpha t}u(t)$, where $\alpha > 0$, often occurs in systems when energy transfers from energy-storage components to energy-using components. Find the amplitude, phase, and energy-density spectra for $y(t)$ if it is a current signal in an electrical system.

Solution
The signal $y(t)$ is shown in Figure 4.19.

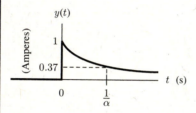

Figure 4.19 Exponential Current Signal for Example 4.5

$$Y(f) = \int_{-\infty}^{\infty} \left[e^{-\alpha t}u(t)\right] e^{-j2\pi ft}\, dt = \int_{0}^{\infty} e^{-(\alpha+j2\pi f)t}\, dt$$

$$= \frac{1}{\alpha + j2\pi f}\left[-e^{-(\alpha+j2\pi f)t}\right]_0^{\infty} = \frac{1}{\alpha + j2\pi f}$$

The amplitude spectrum is

$$|Y(f)| = \left|\frac{1}{\alpha + j2\pi f}\right| = \frac{1}{\sqrt{\alpha^2 + 4\pi^2 f^2}} \qquad \text{(amperes/hertz)}$$

The phase spectrum is

$$\angle Y(f) = \angle \frac{1}{\alpha+j2\pi f} = \angle 1 - \angle \alpha + j2\pi f$$

$$= 0 - \tan^{-1}(2\pi f/\alpha) = -\tan^{-1}(2\pi f/\alpha) \qquad \text{(radians)}$$

The energy-density spectrum is

$$G_y(f) = |Y(f)|^2 = \frac{1}{\alpha^2 + 4\pi^2 f^2} \qquad \text{([amperes]}^2\text{/hertz)}$$

The plots of these spectra are shown in Figure 4.20.

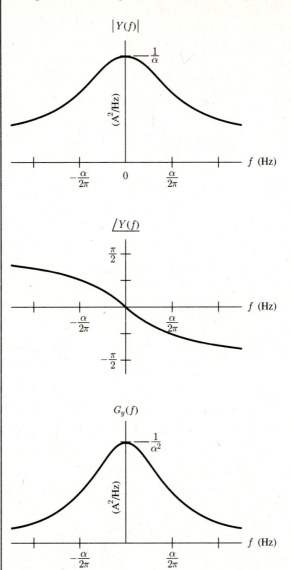

Figure 4.20 Spectra of the Exponential Current Signal in Example 4.5

We see in Figure 4.20 that the energy in an exponential signal is concentrated at lower frequencies. In fact, half of the signal energy occurs at frequencies lower than $\alpha/2\pi$ Hz, as is easily verified by integration of $G_Y(f)$.

Example 4.6

A rectangular pulse voltage signal can be used as a trigger pulse to initiate a single horizontal sweep across the screen of an oscilloscope. Find the amplitude, phase, and energy-density spectra for the rectangular-pulse voltage signal $x(t) = A\Pi(t/\tau)$ shown in Figure 4.21, where A may be positive or negative.

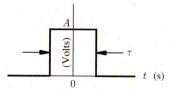

Figure 4.21 Rectangular-Pulse Voltage Signal for Example 4.6

Solution

$$X(f) = \int_{-\infty}^{\infty} A\Pi(t/\tau)e^{-j2\pi ft}\, dt$$

$$= \int_{-\tau/2}^{\tau/2} Ae^{-j2\pi ft}\, dt = \frac{A}{j2\pi f}\left[-e^{-j2\pi ft}\right]_{-\tau/2}^{\tau/2}$$

$$= \frac{A}{j2\pi f}[e^{j\pi f\tau} - e^{-j\pi f\tau}]$$

$$= A\tau\left[\frac{\sin(\pi f\tau)}{\pi f\tau}\right] = A\tau\ \text{sinc}(f\tau)$$

The amplitude spectrum is

$$|X(f)| = |A\tau\ \text{sinc}(f\tau)| = \tau|A|\,|\text{sinc}(f\tau)| \qquad \text{(volts/hertz)}$$

The phase spectrum is

$$\underline{/X(f)} = \underline{/A\tau\ \text{sinc}(f\tau)} = \underline{/A\tau} + \underline{/\text{sinc}(f\tau)}$$

$$= \begin{cases} 0 + \underline{/\text{sinc}(f\tau)} & A > 0 \\ \pm\pi + \underline{/\text{sinc}(f\tau)} & A < 0 \end{cases} \qquad \text{(radians)}$$

The energy-density spectrum is

$$G_x(f) = |A\tau\ \text{sinc}(f\tau)|^2 = A^2\tau^2\ \text{sinc}^2(f\tau) \qquad \text{([volts]}^2\text{/hertz)}$$

The plots of these spectra are shown in Figure 4.22. In plotting the phase for $A < 0$, we used the angle $-\pi$ for positive frequencies and the angle $+\pi$ for negative frequencies. The reason for this choice will be discussed later.

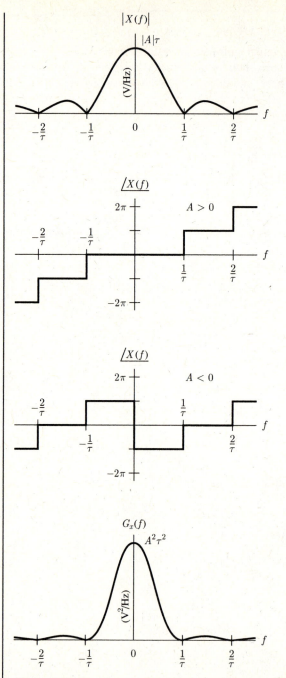

Figure 4.22 Spectra of the Rectangular Pulse in Example 4.6

We see from Figure 4.22 that the energy in a rectangular pulse is concentrated at the lower frequencies. Integration of $G_Y(f)$ shows that 90% of the signal energy is contained in frequencies below $1/\tau$.

Computation of the Fourier transform and plotting of the spectra for the very useful exponential and rectangular-pulse signals were shown in Examples 4.5 and 4.6. But for space limitations, we could continue such computation and spectrum plotting for many more useful signals. However, Table C.2 in Appendix C shows the Fourier transforms for a number of signals of interest. The reader is asked to verify some of these Fourier transform pairs in the Problem section of this chapter.

One of the signals in the table of Fourier transforms is the triangular pulse $\Lambda(t/\tau)$, defined by

$$\Lambda(t/\tau) \equiv \begin{cases} 1 - |t|/\tau & |t| < \tau \\ 0 & \text{elsewhere} \end{cases} \tag{4.67}$$

where τ is positive. This signal illustrated in Figure 4.23. Those signals in the table that are not energy signals are discussed in Section 4.5.

Bandwidth of Aperiodic Signals

As with periodic signals, frequency bandwidth is an important property of signals that are aperiodic.

Definition _____

The frequency bandwidth of an aperiodic signal is the width of the range of positive frequencies for which the signal has energy.

Again recall that the positive- and negative-frequency parts of a double-sided spectrum at a given frequency absolute value combine to produce the signal characteristics at that frequency; hence, the negative-frequency part of a double-sided spectrum is not additional signal bandwidth. The bandwidth of

Figure 4.23
The Triangular
Signal $\Lambda(t/\tau)$

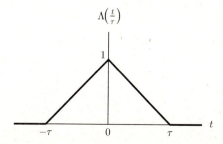

the pulse signal in Example 4.6 is infinite, as shown by the energy spectrum in Figure 4.22. However, the signal energy is concentrated at low frequencies. One definition of the significant bandwidth of the signal is $B_x = 1/\tau$, since most of the energy is contained in the large lobe of $G_x(f)$ centered at $f = 0$. Thus, the bandwidth of a pulse signal is inversely proportional to the width of the pulse. A wider bandwidth system is required to pass the pulse if it is made narrower, as we previously noted for a pulse-train signal at the end of Section 4.3.

Fourier Transform and Spectrum Properties

We will now present several properties associated with the Fourier transform, and thus with the spectrum of a signal. Before stating the first property, however, we must consider the definitions for a *time-limited signal* and a *bandlimited signal*.

Definition

A time-limited signal is one that is nonzero only for a finite-length time interval.

An example of a time-limited signal is the rectangular pulse of Example 4.6.

Definition

A bandlimited signal is one that has a nonzero spectrum only for a finite-length frequency interval.

The signal $x(t) = \text{sinc}(at)$ has the spectrum $X(f) = \frac{1}{a}\Pi(f/a)$ (see Table C.2). Since this spectrum is nonzero only for the frequency interval $|f| < a$, the signal is bandlimited.

Property I   A signal that is time limited cannot be bandlimited and a signal that is bandlimited cannot be time limited.

This property is a corollary of the *Paley Wiener theorem.*[†] The proof is beyond the scope of this text.

Property II   The inverse Fourier transform of the truncated spectrum of a signal exhibits the Gibbs phenomenon (see p. 129) at any signal discontinuities.

[†]A. Papoulis, *The Fourier Integral and its Applications* (New York: McGraw-Hill, 1962), p. 215.

The next three properties to be presented are Fourier transform- or spectrum-symmetry properties that occur for real-valued signals. To determine these properties, we first note that any real signal $x(t)$ can be written as the sum of a real and even signal $x_e(t)$ and a real and odd signal $x_0(t)$. That is,

$$x(t) = x_e(t) + x_0(t) \tag{4.68}$$

where

$$x_e(t) = [x(t) + x(-t)]/2 \tag{4.69}$$

and

$$x_0(t) = [x(t) - x(-t)]/2 \tag{4.70}$$

Example 4.7

Decompose the signal $x(t)$ shown in Figure 4.24 into its even and odd parts.

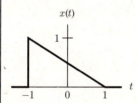

Figure 4.24 Signal for Example 4.7

Solution
The signal $x(t)$ is decomposed in Figure 4.25.

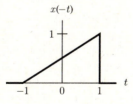

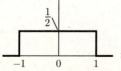

Figure 4.25 Construction of Even and Odd Parts of the Signal in Figure 4.24

$x_0(t) = [x(t) - x(-t)]/2$

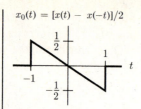

Figure 4.25 Construction of Even and Odd Parts of the Signal in Figure 4.24 *(continued)*

The Fourier transform of the real signal $x(t)$ is

$$X(f) = \int_{-\infty}^{\infty} x(t)e^{-j2\pi ft} \, dt$$

$$= \int_{-\infty}^{\infty} x_e(t)e^{-j2\pi ft} \, dt + \int_{-\infty}^{\infty} x_0(t)e^{-j2\pi ft} \, dt$$

$$= X_e(f) + X_o(f) \tag{4.71}$$

Now, Euler's theorem gives

$$X_e(f) = \int_{-\infty}^{\infty} x_e(t)\cos 2\pi ft \, dt - j\int_{-\infty}^{\infty} x_e(t)\sin 2\pi ft \, dt$$

$$= \int_{-\infty}^{\infty} x_e(t)\cos 2\pi ft \, dt \tag{4.72}$$

where the second step follows from the fact that $x_e(t)\sin 2\pi ft$ is an odd function. Likewise,

$$X_o(f) = \int_{-\infty}^{\infty} x_o(t)\cos 2\pi ft \, dt - j\int_{-\infty}^{\infty} x_o(t)\sin 2\pi ft \, dt$$

$$= j\left[-\int_{-\infty}^{\infty} x_o(t)\sin 2\pi ft \, dt\right] \tag{4.73}$$

where the second step follows from the fact that $x_o(t)\cos 2\pi ft$ is an odd function.

Since $x_e(t)$ and $x_o(t)$ are real, then the integrals in eqs. (4.72) and (4.73) are real. Therefore,

$$Re[X(f)] = \int_{-\infty}^{\infty} x_e(t)\cos 2\pi ft \, dt \tag{4.74}$$

and

$$Im[X(f)] = -\int_{-\infty}^{\infty} x_o(t)\sin 2\pi ft \, dt \tag{4.75}$$

Note that the real part of the transform results from the even part of the signal, and that the imaginary part of the transform results from the odd part of the signal. Also, $Re[X(f)]$ and $Im[X(f)]$ are, respectively, even and odd functions of f, for the reason that $\cos 2\pi ft$ and $\sin 2\pi ft$ are, respectively, even and odd functions of f. Therefore, we come to Properties III and IV.

> **Property III** Real and even signals produce real and even Fourier transforms.

> **Property IV** Real and odd signals produce imaginary and odd Fourier transforms.

Property III shows that we can plot the *total spectrum* $X(f)$ of a real and even signal $x(t)$, rather than the amplitude and phase spectra, because $X(f)$ is real. This could have been done in Example 4.6. Property IV shows that all phase angles corresponding to the spectrum of a real and odd signal are odd multiples of $\pi/2$. Note further that

$$|X(f)| = \sqrt{\{Re[X(f)]\}^2 + \{Im[X(f)]\}^2} \tag{4.76}$$

is even and that

$$\underline{/\,X(f)} = \tan^{-1}\{Im[X(f)]/Re[X(f)]\} \tag{4.77}$$

is odd, for the reason that $Re[X(f)]^2$ and $Im[X(f)]^2$ are even and that $Im[X(f)]/Re[X(f)]$ is odd. Therefore, we state Property V.

> **Property V** The amplitude and phase spectra of a real signal are even and odd functions respectively.

We arrived at Property V earlier for Fourier series, which is logical given that the Fourier transform derives from the Fourier series. In Example 4.6, there was a choice of $+\pi$ or $-\pi$ for a term in the equation for the phase spectrum. Since the signal in question was real, we chose $-\pi$ for positive frequencies and $+\pi$ for negative frequencies to preserve the odd-phase characteristic in the phase plot.

Fourier Transform Theorems

We will now state some Fourier transform characteristics in the form of theorems. These theorems are useful in computing Fourier transforms of more complicated signals. Proofs of some of the theorems are shown or indicated; proofs of others are left for the reader in the Problem section. For a summary of these theorems, see Table C.1 of Appendix C.

Theorem 1   Linearity

If

$$x(t) \leftrightarrow X(f) \quad \text{and} \quad y(t) \leftrightarrow X(f)$$

then

$$a(t) + by(t) \leftrightarrow aX(f) + bY(f) \tag{4.78}$$

Since an integral is a linear operation, this theorem is easily proved.

Example 4.8

Find the spectrum of the signal $x(t)$ shown in Figure 4.26.

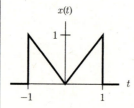

Figure 4.26 Signal for Example 4.8

Solution

$$x(t) = x_1(t) - x_2(t)$$

where $x_1(t)$ and $x_2(t)$ are shown in Figure 4.27.

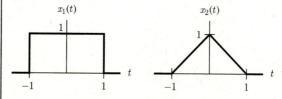

Figure 4.27 Decomposition of Signal for Example 4.8

From Table C.2 and the linearity theorem,

$$X(f) = X_1(f) - X_2(f) = 2 \text{ sinc } 2f - \text{sinc}^2 f$$

The spectra of $x(t)$ are shown in Figure 4.28. Note that the total spectrum can be plotted because $X(f)$ is real and even.

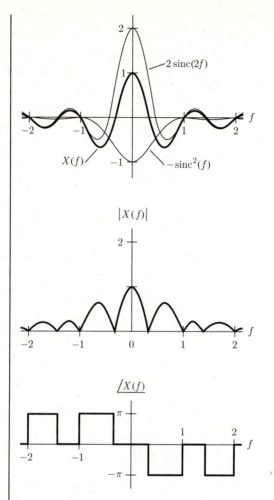

Figure 4.28 Spectra of the Signal in Example 4.8

Theorem 2   Scale Change
If

$$x(t) \leftrightarrow X(f)$$

then

$$x(at) \leftrightarrow \frac{1}{|a|} X(f/a) \qquad (4.79)$$

The proof of this theorem is left to the reader in the Problem Section.

Example 4.9

Use the scale-change theorem to find the spectrum of the rectangular pulse $y(t)$ from the spectrum of the rectangular pulse $x(t)$. The two signals are shown in Figure 4.29.

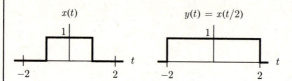

Figure 4.29 Signals for Example 4.9

Solution

From Table C.2, $X(f) = 2 \text{ sinc } 2f$; therefore, using the scale-change theorem,

$$Y(f) = \frac{1}{|1/2|}X(2f) = 4 \text{ sinc } 4f$$

This spectrum matches the spectrum shown for $y(t)$ in Table C.2. The spectra of $x(t)$ and $y(t)$ are plotted in Figure 4.30.

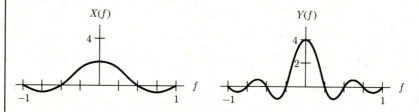

Figure 4.30 Signal Spectra for Example 4.9

Again we note that an increase in signal pulse width corresponds to a decrease in significant bandwidth.

Theorem 3   Time Reversal

If

$$x(t) \leftrightarrow X(f)$$

then

$$x(-t) \leftrightarrow X(-f) \tag{4.80}$$

This theorem is a special case of the scale-change theorem that occurs when $a = -1$. If the signal is real, then time reversal affects only the phase spectrum, the amplitude spectrum being an even function of frequency.

Theorem 4   Complex Conjugation
If
$$x(t) \leftrightarrow X(f)$$
then
$$x^*(t) \leftrightarrow X^*(-f) \tag{4.81}$$

The proof of this theorem is required in the Problem section.

Theorem 5   Duality
If
$$x(t) \leftrightarrow X(f) \quad \text{and} \quad y(t) = X(t)$$
then
$$Y(f) \leftrightarrow x(-f) \tag{4.82}$$

We see that the Fourier transform and inverse Fourier transform integrals are the same, except for a change of sign on the complex-exponential exponent; hence, the proof of this theorem is readily apparent. It is a very useful theorem, essentially doubling the size of a table of Fourier transforms.

Example 4.10

Use the duality theorem to find the spectrum of the complex signal $y(t) = 1/(3 + j2\pi t)$.

Solution
Define $X(f) = 1/(3 + j2\pi f)$. From Table C.2, $x(t) = e^{-3t}u(t)$; therefore, by the duality theorem, the spectrum of $y(t)$ is
$$Y(f) = x(-f) = e^{3f}u(-f)$$
The plot of $Y(f)$ is shown in Figure 4.31.

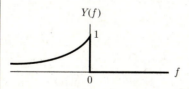

Figure 4.31 Spectrum of $y(t) = 1/(3 + j2\pi t)$

Theorem 6   Time Shift
If
$$x(t) \leftrightarrow X(f)$$
then
$$x(t - t_0) \leftrightarrow X(f)e^{-j2\pi f t_0} \tag{4.83}$$

Direct computation of the Fourier transform of $x(t - t_0)$ gives

$$\mathcal{F}[x(t - t_0)] = \int_{-\infty}^{\infty} x(t - t_0)e^{-j2\pi ft} \, dt \tag{4.84}$$

Letting $t - t_0 = \tau$ implies that $dt = d\tau$, $\tau \to \infty$ as $t \to \infty$, and $\tau \to -\infty$ as $t \to -\infty$. Therefore,

$$\mathcal{F}[x(t - t_0)] = \int_{-\infty}^{\infty} x(\tau) \, e^{-j2\pi f(\tau + t_0)} \, d\tau$$

$$= \left[\int_{-\infty}^{\infty} x(\tau) \, e^{-j2\pi f\tau} \, d\tau \right] e^{-j2\pi ft_0}$$

$$= X(f)e^{-j2\pi ft_0} \tag{4.85}$$

which completes the proof. Note that a similar theorem applies to Fourier series, as was illustrated in Example 4.4.

Example 4.11

The oscilloscope trigger signal of Example 4.6 is delayed by $\tau/2$ seconds as shown in Figure 4.32. Compute and plot its spectrum.

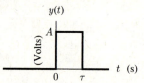

Figure 4.32 Rectangular-Pulse Voltage Signal for Example 4.11

Solution

$$y(t) = x(t - \tau/2) \quad \text{where} \quad x(t) = A\Pi(t/\tau)$$

From Example 4.6, $X(f) = A\tau \, \text{sinc}(f\tau)$; therefore, by the time-shift theorem,

$$Y(f) = X(f)e^{-j2\pi f(\tau/2)} = A\tau \, \text{sinc}(f\tau)e^{-j\pi f\tau}$$

Since $|e^{-j\pi ft}| = 1$ and $\underline{/e^{-j\pi f\tau}} = -\pi ft$, then

$$|Y(f)| = \tau|A| \, |\text{sinc}(f\tau)| \quad \text{(volts)}$$

and

$$\underline{/Y(f)} = \begin{cases} \underline{/\text{sinc}(f\tau)} + (-\pi f\tau) & A > 0 \\ \pm\pi + \underline{/\text{sinc}(f\tau)} + (-\pi f\tau) & A < 0 \end{cases} \quad \text{(radians)}$$

where the results of Example 4.6 have been used. The plots of the amplitude spectrum $|Y(f)|$ and the phase spectrum $\underline{/Y(f)}$ of $y(t)$ are shown in Figure 4.33.

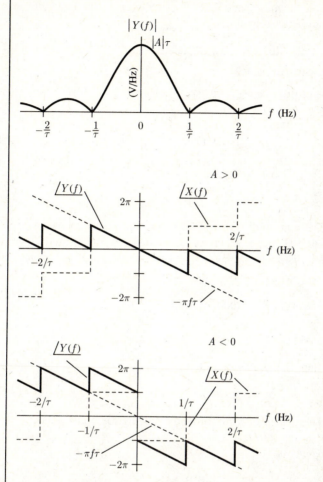

Figure 4.33 Spectra of the Delayed Trigger Signal in Example 4.11

The signal of Example 4.11 is of the same shape as the signal of Example 4.6, but the former is shifted to the right by $\tau/2$ units. Note that its amplitude spectrum is unchanged and that its phase spectrum has the additional linear term of $-2\pi f(\tau/2)$. This is the same result as indicated for Fourier series in the discussion following Example 4.4.

Theorem 7 Frequency Translation
If

$$x(t) \leftrightarrow X(f)$$

then

$$x(t)e^{j2\pi f_0 t} \leftrightarrow X(f - f_0) \qquad (4.86)$$

This theorem is a dual of the time-shift theorem. Therefore, the duality theorem can be used to prove it.

Example 4.12

Find the spectrum of the signal $w(t) = e^{-(\alpha - j2\pi f_0)t}u(t)$, where $\alpha > 0$.

Solution

$$w(t) = \left[e^{-\alpha t}u(t)\right]e^{j2\pi f_0 t} \equiv x(t)e^{j2\pi f_0 t}$$

From Example 4.5, $X(f) = 1/[\alpha + j2\pi f]$; therefore, by the frequency-shift theorem,

$$W(f) = 1/[\alpha + j2\pi(f - f_0)]$$

The amplitude spectrum $|W(f)|$ and phase spectrum $\underline{/W(f)}$ of $w(t)$ are the same as those shown in Figure 4.20, except shifted f_0 units to the right. Since $w(t)$ is complex, $|W(f)|$ is not even and $\underline{/W(f)}$ is not odd.

Theorem 8   Modulation
If

$$x(t) \leftrightarrow X(f)$$

then

$$x(t)\cos 2\pi f_0 t \leftrightarrow \frac{1}{2}X(f - f_0) + \frac{1}{2}X(f + f_0) \qquad (4.87)$$

Theorem 8 is easily proved by using the equivalence

$$\cos 2\pi f_0 t = \frac{1}{2}e^{j2\pi f_0 t} + \frac{1}{2}e^{-j2\pi f_0 t} \qquad (4.88)$$

and the frequency shift-theorem.

Example 4.13

Find the spectrum of the signal $x(t) = 2\,\text{sinc}(t/2)\cos 2\pi t$.

Solution
Define $z(t) = 2\,\text{sinc}(t/2)$ so that $x(t) = z(t)\cos 2\pi t$. From Table C.2,

$$Z(f) = 4\Pi(2f)$$

Therefore, using the modulation theorem,

$$X(f) = 2\Pi[2(f - 1)] + 2\Pi[2(f + 1)]$$

Plots of $z(t)$ and its spectrum $Z(f)$ and of $x(t)$ and its spectrum $X(f)$ are shown in Figure 4.34.

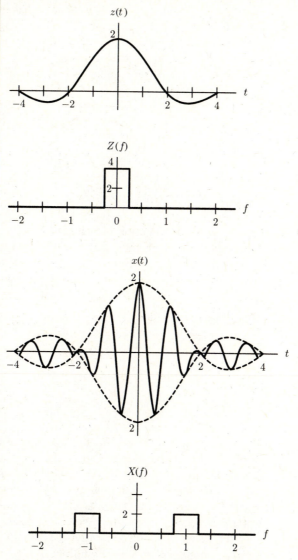

Figure 4.34 Signals and Their Spectra for Example 4.13

Note that the signal $x(t)$ of Example 4.13 is a cosine, the amplitude of which is $z(t)$; hence, the amplitude changes with time. A signal such as $x(t)$ is referred to as a double-sideband amplitude-modulated signal. It transmits the message signal $z(t)$ by using the sinusoidal carrier signal $\cos 2\pi t$. The double-sideband designation follows from the fact that the message information is contained in bands of frequencies both above and below the carrier frequency. This is shown in Figure 4.34 by the spectrum of the modulated signal $x(t)$. Note that the bandwidth of the modulated signal is double that of the message signal (0.5 Hz versus 0.25 Hz for the example). The carrier frequency used in the example is too low for a practical communications system, however. Carrier signals are typically in the megahertz range or larger to provide good signal-radiation efficiency with reasonable antenna lengths.

Theorem 9   Time Differentiation

If

$$x(t) \leftrightarrow X(f)$$

then

$$\frac{d^n x(t)}{dt^n} \leftrightarrow (j2\pi f)^n X(f) \qquad\qquad (4.89)$$

The proof of this theorem for $n = 1$ is considered in the Problem section.

Since the first Dirichlet condition indicates that $X(f)$ exists for all frequencies for a finite-energy signal, then $X(0)$ is finite for a finite-energy signal. Thus, the spectra of derivatives of finite-energy signals are zero at $f = 0$. Note that differentiation enhances the amplitude of high-frequency components in a signal because the spectrum of the original signal is multiplied by f.

Theorem 10   Time Integration

If

$$x(t) \leftrightarrow X(f)$$

then

$$\int_{-\infty}^{t} x(\lambda)\, d\lambda \leftrightarrow (j2\pi f)^{-1} X(f) + \frac{1}{2} X(0)\delta(f) \qquad\qquad (4.90)$$

The time-integration theorem shows that the spectrum of the integral of a signal contains an impulse function. To this point, we have not discussed spectra with impulse functions, but we will shortly. The impulse function arises because the integral of a finite-energy signal may produce a signal with infinite energy (for example, the integral of a rectangular pulse). We previously noted that the derivative of a finite-energy signal has a spectrum that is zero at $f = 0$. In this case, because the integral has finite energy, the differentiation and integration theorems produce reciprocal spectrum multipliers of $j2\pi f$ and $1/j2\pi f$.

Theorem 11   Convolution

If

$$x(t) \leftrightarrow X(f)$$

then

$$x(t) * y(t) \leftrightarrow X(f)Y(f) \qquad\qquad (4.91)$$

The proof of this important theorem follows. For notational simplicity, we define $x(t) * y(t) = z(t)$. Then,

$$z(t) = \int_{-\infty}^{\infty} x(\lambda)y(t - \lambda)\, d\lambda \tag{4.92}$$

and

$$Z(f) = \int_{-\infty}^{\infty} \left[\int_{-\infty}^{\infty} x(\lambda)y(t - \lambda)\, d\lambda \right] e^{-j2\pi ft}\, dt \tag{4.93}$$

Interchanging the order of integration gives

$$Z(f) = \int_{-\infty}^{\infty} x(\lambda) \left[\int_{-\infty}^{\infty} y(t - \lambda)e^{-j2\pi ft}\, dt \right] d\lambda \tag{4.94}$$

Letting $t - \lambda = \tau$ implies that $dt = d\tau$, $\tau \to \infty$ as $t \to \infty$, and $\tau \to -\infty$ as $t \to -\infty$. Therefore,

$$Z(f) = \int_{-\infty}^{\infty} x(\lambda) \left[\int_{-\infty}^{\infty} y(\tau)e^{-j2\pi f(\tau+\lambda)}\, d\tau \right] d\lambda$$

$$= \left[\int_{-\infty}^{\infty} x(\lambda)e^{-j2\pi f\lambda}\, d\lambda \right]\left[\int_{-\infty}^{\infty} x(\tau)e^{-j2\pi f\tau}\, d\tau \right]$$

$$= X(f)Y(f) = \mathcal{F}[x(t) * y(t)] \tag{4.95}$$

Note that the somewhat complicated operation of convolution of two signals corresponds to the simple operation of multiplication of their spectra.

Example 4.14

Find the signal $w(t) = x(t)*y(t)$ and its spectrum when $x(t) = y(t) = A\Pi(t/\tau)$.

Solution
From Example 4.6, $X(f) = Y(f) = A\tau \,\mathrm{sinc}(\tau f)$; therefore, using the convolution theorem,

$$W(f) = A^2\tau^2 \,\mathrm{sinc}^2(\tau f)$$

The spectra of $x(t)$, $y(t)$, and $w(t)$ are plotted in Figure 4.35 for $A\tau = 1$.

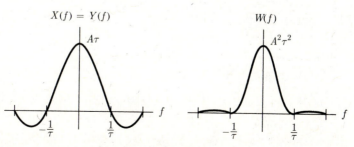

Figure 4.35 Spectra of Signals in Example 4.14

The signal $w(t)$ can be found by computing the inverse Fourier transform of its spectrum. This gives

$$w(t) = \mathcal{F}[W(f)] = A^2\tau\Lambda(t/\tau)$$

where Table C.2 was used to obtain the inverse transform. The plot of $w(t)$ is shown in Figure 4.36.

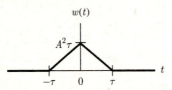

Figure 4.36 Signal $w(t)$ in Example 4.14

It is easily seen that the triangular signal $w(t)$ is the convolution of the rectangular pulse with itself.

Convolutions occur in the spectral analysis of systems; therefore, the convolution theorem is important. For example, we indicated in Chapter 1 that the output of a linear time-invariant system with zero initial conditions is the input signal convolved with a function determined by the system. This application of the convolution theorem will be discussed in Chapter 6.

Theorem 12   Multiplication
If

$$x(t) \leftrightarrow X(f) \quad \text{and} \quad y(t) \leftrightarrow Y(f)$$

then

$$x(t)y(t) \leftrightarrow X(f) * Y(f) \tag{4.96}$$

This theorem is a dual to the theorem for signal convolution; therefore, its proof is similar. It is required of the reader in the Problem section.

Example 4.15

Find the spectrum of $z(t) = x(t)y(t)$ when $x(t) = \operatorname{sinc}(t)$ and $y(t) = 2\operatorname{sinc}(2t)$.

Solution
The spectra for $x(t)$ and $y(t)$ are obtained from Table C.2 and are shown in Figure 4.37.

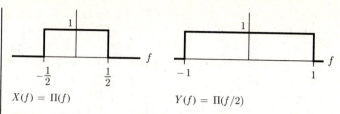

$$X(f) = \Pi(f) \qquad\qquad Y(f) = \Pi(f/2)$$

Figure 4.37 Spectra for Signals $x(t)$ and $y(t)$ in Example 4.15

Therefore, using the multiplication theorem,

$$Z(f) = X(f) * Y(f)$$

which is plotted in Figure 4.38. The plot is easily verified.

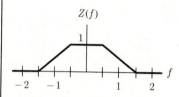

Figure 4.38 Spectrum for Signal $z(t)$ in Example 4.15

4.5 The Fourier Transform and Spectra of Nonenergy Signals

We have developed the Fourier transform and have discussed its properties and relation to the signal spectrum. Although the theory developed applies only to energy signals, generalized function theory can be used with Fourier transforms in the limit to show that the techniques developed are valid for nonenergy signals if impulses are permitted in the transforms. We will now illustrate this concept by means of an example.

The Fourier Transform in the Limit

First we compute the Fourier transform of the energy signal $x(t) = A\exp(-a^2t^2)$ to obtain its spectrum $X(f) = [\sqrt{\pi}\exp(-\pi^2 f^2/a^2)]/a$. Then, we obtain a Fourier transform in the limit by letting $a \to 0$. As $a \to 0$, $X(f)$ becomes concentrated at $f = 0$ with an amplitude approaching infinity and an area that remains constant with a value of A; therefore $X(f) \to A\delta(f)$. Likewise, $x(t) \to A$ for all t. Therefore, the Fourier transform in the limit has produced the important Fourier transform pair

$$A \leftrightarrow A\delta(f) \tag{4.97}$$

This Fourier transform pair shows that the spectrum of a constant, or dc, signal of amplitude A is an impulse located at $f = 0$ with area A. This result is reasonable, given that the area in a range of frequencies under the amplitude-density spectrum produced by the Fourier transform is the amplitude in the signal in that range of frequencies. Thus, the spectrum indicates that the amplitude of the signal is A and is concentrated at $f = 0$, or dc. We know this is true, since $x(t) = A$ is a dc signal with amplitude A.

The Fourier transform in the limit obeys all the properties of the Fourier transform. Therefore, we can illustrate the derivation of the spectra for other nonenergy signals using the properties and theorems already shown. The resulting Fourier transforms in the limit are referred to simply as *Fourier transforms* for simplicity.

Fourier Transforms and Spectra of Selected Signals

We can use the Fourier transform to find the spectra of several useful nonenergy signals. We will now consider four signals—impulse, signum function, unit step, and cosine.

Signal 1   Impulse

$$A\delta(t) \leftrightarrow A \tag{4.98}$$

To show the validity of the Fourier transform for signal 1, we compute the Fourier integral to give

$$\mathcal{F}[A\delta(t)] = \int_{-\infty}^{\infty} A\delta(t)e^{-j2\pi ft}\, dt = A \tag{4.99}$$

where the sifting integral accounts for the final result. Note an impulse signal contains equal amplitude density in all frequency ranges. Note, too, that since $G_x(f) = A^2$ for all f, the signal contains infinite energy.

Example 4.16

Find the spectrum of the time-shifted impulse signal $x(t) = 10\delta(t - 2)$.

Solution
Since $10\delta(t) \leftrightarrow 10$, use of the time-shift theorem gives

$$x(t) = 10e^{-j2\pi f(2)} = 10e^{-j4\pi f}$$

The amplitude and phase spectra for $x(t)$ are plotted in Figure 4.39.

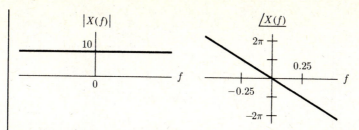

Figure 4.39 Amplitude and Phase Spectra for Time Shifted Impulse Signal

The next signal we will consider is modeled by the signum function, sgn(t), which is defined by

$$\text{sgn}(t) = \begin{cases} 1 & t > 0 \\ -1 & t < 0 \end{cases} \tag{4.100}$$

Signal 2 Signum Function

$$\text{sgn}(t) \leftrightarrow 1/j\pi f \tag{4.101}$$

The Fourier transform pair of eq. (4.101) is verified by recognizing that, for $a > 0$,

$$\text{sgn}(t) = \lim_{a \to 0} \left[e^{-at}u(t) - e^{-a(-t)}u(-t) \right] \tag{4.102}$$

and then by computing the Fourier transform of $e^{-at}u(t) - e^{-a(-t)}u(-t)$ in the limit as $a \to 0$.

Example 4.17

Find the spectrum of the signal $y(t) = 3\,\text{sgn}(t)$.

Solution
Using eq. (4.101), the spectrum of the signal is

$$Y(f) = 3/j\pi f = -j3/\pi f$$

The amplitude spectrum of the signal is

$$|Y(f)| = |-j3/\pi f| = 3/\pi|f|$$

The phase spectrum of the signal is

$$\angle Y(f) = \angle -j3/\pi f = \begin{cases} -\pi/2 & f > 0 \\ \pi/2 & f < 0 \end{cases}$$

The signal and its amplitude and phase spectra are plotted in Figure 4.40.

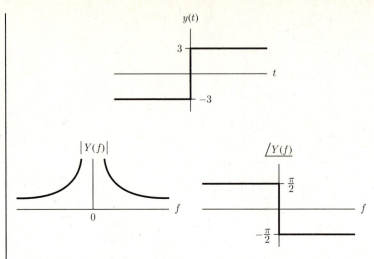

Figure 4.40 Signum Signal and Its Amplitude and Phase Spectra

Signal 3 Unit Step

$$u(t) \leftrightarrow \frac{1}{2}\delta(f) + 1/j2\pi f \tag{4.103}$$

The Fourier transform for a unit step signal is easily verified, recognizing that $u(t) = [1 + \text{sgn}\,(t)]/2$ and then using the linearity theorem.

Example 4.18

An automobile is driven off a 4-inch curb at $t = 1$ s. Thus the input signal to the automobile suspension system is $x(t) = -4u(t - 1)$. Find the spectrum of this signal.

Solution
Using eq. (4.103) and the time-shift theorem gives the spectrum of $x(t)$ as

$$X(f) = -[2\delta(f) - j2/\pi f]\,e^{-j2\pi f(1)} = -2\delta(f) + [j2/\pi f]e^{-j2\pi f}$$

The amplitude spectrum of $x(t)$ is

$$|X(f)| = 2\delta(f) + 2/\pi|f| \qquad \text{(inches/hertz)}$$

The phase spectrum of $x(t)$ is

$$\underline{/\,X(f)} = \begin{cases} -\pi/2 - 2\pi f & f < 0 \\ \pm\pi/2 & f = 0 \\ \pi/2 - 2\pi f & f > 0 \end{cases} \qquad \text{(radians)}$$

These spectra are plotted in Figure 4.41.

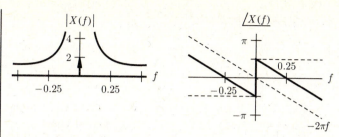

Figure 4.41 Amplitude and Phase Spectra for the Step Signal in Example 4.18

Note that the amplitude spectrum consists of two terms that cannot be added on the plot; thus, they are plotted separately. The vertical scale is amplitude density in inches/hertz for the continuous-frequency term $2/\pi|f|$ and area under amplitude density, or amplitude, in inches for the impulse term $2\delta(f)$.

Signal 5 Cosine

$$\cos\left(2\pi f_a t\right) \leftrightarrow \frac{1}{2}\delta(f - f_a) + \frac{1}{2}\delta(f + f_a) \tag{4.104}$$

Equation (4.104) can be verified by using eq. (4.97) and the modulation theorem.

Example 4.19

Wind currents cause a telephone wire to oscillate vertically. Midway between two supporting telephone poles, the vertical displacement of the wire from its at-rest position is $z(t) = 8\cos[2\pi(3)t + \pi/3]$ cm. Find the spectrum of the vertical-displacement signal.

Solution

$$z(t) = 8\cos\left[2\pi(3)(t + 1/18)\right]$$

Therefore, using eq. (4.104) and the time-shift theorem,

$$Z(f) = [4\delta(f - 3) + 4\delta(f + 3)]e^{-j2\pi f(1/18)}$$

$$= 4\delta(f - 3)e^{-j2\pi(3)(1/18)} + 4\delta(f + 3)e^{-j2\pi(-3)(1/18)}$$

$$= 4\delta(f - 3)e^{-j\pi/3} + 4\delta(f + 3)e^{j\pi/3}$$

The amplitude spectrum is

$$|Z(f)| = 4\delta(f - 3) + 4\delta(f + 3) \qquad \text{(centimeters/hertz)}$$

Since the frequency ranges for which the two terms are nonzero do not overlap, the amplitude of their sum equals the sum of their amplitudes.

The phase spectrum is

$$\underline{/\,Z(f)} = \begin{cases} -\pi/3 & f = -3 \\ \pi/3 & f = 3 \\ 0 & \text{elsewhere} \end{cases} \qquad \text{(radians)}$$

Actually, for $|f| \neq 3$ the phase is undefined and can be assigned any finite value because there is no signal amplitude at these frequencies. We define it to be zero for convenience. The amplitude and phase spectral plots are shown in Figure 4.42.

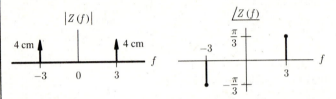

Figure 4.42 Amplitude and Phase Spectra for the Cosine Signal in Example 4.19

Note that the amplitude and phase spectra plots obtained for a cosine signal in Example 4.19 are like the double-sided spectra produced for a cosine signal with the complex-exponential Fourier series, except that in this case impulses occur in the amplitude spectrum. The impulses occur because the area under the amplitude-density spectrum produced by the Fourier transform in a given range of frequencies is equal to the amplitude in that range. For the cosine signal, all the amplitude is concentrated at the single frequency of the cosine. Thus, for the double-sided spectrum of the cosine of Example 4.19, the area at $f = \pm 3$ must equal 4 cm. This can be true only if impulses exist at these frequencies.

Fourier Transforms and Spectra of Periodic Signals

The Fourier transform of a single-cosine signal hints at what can be expected for the Fourier transform of any periodic signal that satisfies the Dirichlet conditions in each period, keeping in mind that such a signal can be written as a weighted sum of cosines. Letting $x(t)$ be a periodic signal with period equal to T_0, which satisfies the Dirichlet conditions in each period, then, the complex-exponential Fourier series representation of $x(t)$ is

$$x(t) = \sum_{n=-\infty}^{\infty} X_n e^{-j2\pi n f_0 t} \qquad (4.105)$$

where $f_0 = 1/T_0$ and X_n is the nth complex-exponential Fourier series coefficient defined by eq. (4.36). Using eq. (4.105) and the linearity theorem, we write the Fourier transform of $x(t)$ as

$$X(f) = \sum_{n=-\infty}^{\infty} \mathcal{F}\left[X_n e^{-j2\pi n f_0 t}\right] \qquad (4.106)$$

Equation (4.106) can be rewritten as

$$X(f) = \sum_{n=-\infty}^{\infty} X_n \delta(f - n f_0) \qquad (4.107)$$

using eq. (4.97) and the frequency-shift theorem. Equation (4.107) is the signal spectrum determined by computing the Fourier transform of the signal. Recall that the double-sided amplitude and phase spectra obtained from the Fourier series of a periodic signal are just the amplitude and angle of the Fourier series coefficients as a function of frequency. Therefore, the same amplitude and phase spectra for a periodic signal are obtained by both the Fourier series and the Fourier transform, except for the presence of impulses in the amplitude spectrum obtained by the latter. The impulses occur, rather than just values, for the reason described in reference to the cosine signal.

Example 4.20

Use the Fourier transform to find the spectrum of the periodic signal $x(t)$ shown in Figure 4.43.

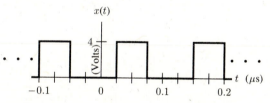

Figure 4.43 Periodic Signal for Example 4.20

Solution

Note that this is the same computer clock signal considered in part (c) of Example 4.4. It has a period of $T_0 = 0.125$ μs, and thus a fundamental frequency of $f_0 = 8$ Mhz. Using the results of Example 4.4, part (c),

$$X_n = 1.6 \operatorname{sinc}\left(0.05 n f_0 \times 10^{-6}\right) e^{-j0.1\pi n f_0 \times 10^{-6}}$$

Therefore, eq. (4.105) gives

$$X(f) = \sum_{n=-\infty}^{\infty} \left[1.6 \operatorname{sinc}\left(0.5 n f_0 \times 10^{-6}\right) e^{-j0.1\pi n f_0 \times 10^{-6}}\right] \delta(f - n f_0)$$

as the spectrum of $x(t)$.

The amplitude spectrum of $x(t)$ is

$$|X(f)| = \sum_{n=-\infty}^{\infty} 1.6 \left| \text{sinc}(0.5nf_0 \times 10^{-6}) \right| \delta(f - nf_0) \qquad \text{(volts/hertz)}$$

where, as for the cosine signal, the amplitude of the sum is the sum of the amplitudes of each term because the frequency ranges for which the terms are nonzero do not overlap. The value for f_0 is not substituted in the spectrum equation, so we can easily identify an envelope for the impulses to aid in plotting them. Recall that we did this when producing line spectra with Fourier series (see the discussion following Example 4.4).

The phase spectrum of $x(t)$ (in radians) is

$$
X(f) = \begin{cases} \underline{/X_n} & n = 0, \pm 1, \pm 2, \ldots \\ 0 & \text{elsewhere} \end{cases}
$$

$$
= \begin{cases} \underline{/ \text{sinc}(0.05nf_0 \times 10^{-6})} - 0.1\pi nf_0 \times 10^{-6} & n = 0, \pm 1, \pm 2 \\ 0 & \text{elsewhere} \end{cases}
$$

Usually, we plot the phase spectrum as the continuous-frequency envelope that passes through the specified values. Plotting the envelope is easier than plotting the individual lines and causes no problem in that the amplitude spectrum clearly indicates which frequencies are present. Thus, the phase spectrum plotted is

$$\underline{/X(f)} = \underline{/\text{sinc}(0.05f \times 10^{-6})} - 0.1\pi f \times 10^{-6}$$

The amplitude and phase spectra for $x(t)$ are plotted in Figure 4.44. Note that they do give the same information as the amplitude and phase plot obtained using the Fourier series that we showed in Figure 4.14.

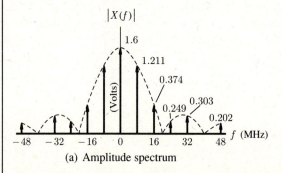

(a) Amplitude spectrum

Figure 4.44 Amplitude and Phase Spectra of the Periodic Signal in Example 4.20

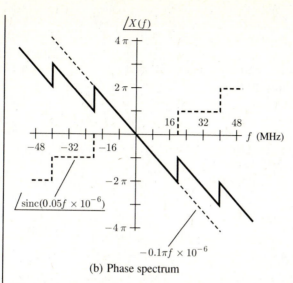

(b) Phase spectrum

Figure 4.44 Amplitude and Phase Spectra of the Periodic Signal in Example 4.20 (*continued*)

The ideal sampling signal defined by eq. (3.48) is a periodic signal with period T_s that does not satisfy the Dirichlet conditions (see Chapter 3, Figure 3.16). However, it does have a spectrum identical to the continuous-frequency function obtained by computing the Fourier series coefficients using eq. (4.36) and determining the spectrum with eq. (4.107). We will now perform these computations here and justify the result upon completion. The Fourier series coefficients for the ideal sampling signal

$$\delta_s(t) = \sum_{n=-\infty}^{\infty} \delta(t - nT_s) \tag{4.108}$$

are

$$\Delta_n = \frac{1}{T_s} \int_{t_1}^{t_1+T_s} \delta_s(t) e^{-j2\pi n f_s t} \, dt \tag{4.109}$$

where $f_s = 1/T_s$. Choosing $t_1 = -T_s/2$ gives an integration interval containing only the impulse located at the time origin. Therefore,

$$\Delta_n = \frac{1}{T_s} \int_{-T_s/2}^{T_s/2} \delta(t) e^{-j2\pi n f_s t} \, dt = \frac{1}{T_s} = f_s \tag{4.110}$$

where the last equality follows because the integral is a sifting integral. These Fourier series coefficients are substituted in eq. (4.107) to give the spectrum of the ideal sampling signal,

$$\Delta_s(f) = \sum_{n=-\infty}^{\infty} f_s\delta(f - nf_s) \tag{4.111}$$

The ideal sampling signal and its spectrum are illustrated in Figure 4.45.

The spectrum obtained for the ideal sampling signal shows that the signal contains a dc term of amplitude f_s and an infinite number of harmonically related cosines, each of which has an amplitude of $2f_s$. Figure 4.46, where the sum of the dc term and the first 10 cosines is shown, illustrates that this specrum is reasonable. Note that the most significant signal lobes produced are spaced by T_s with height $21f_s$ and that there are 9.5 small ripple cycles per period. When N cosines are added, the height of the most significant lobes are $(2N+1)$

Figure 4.45
Ideal Sampling
Signal and Its
Spectrum

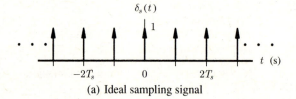

(a) Ideal sampling signal

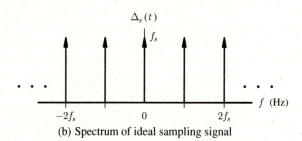

(b) Spectrum of ideal sampling signal

Figure 4.46
Fourier Series
Approximation to the
Ideal Sampling
Signal when $N = 10$

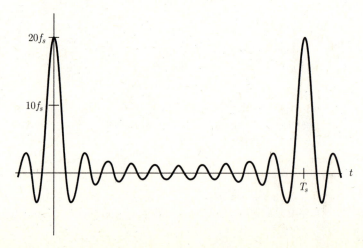

and there are $N - 0.5$ ripple cycles per period. As $N \to \infty$, the ripple cycles approach zero width; they have no area since their amplitudes approach constant values. The most significant lobes approach zero width, infinite amplitude, and unit area. Therefore, the cosine series converges to the ideal sampling signal.

Recall from eq. (3.51) that any periodic signal can be written as

$$x(t) = x_p(t) * \delta_s(t) \tag{4.112}$$

where $T_s = T_0$ and

$$x_p(t) = \begin{cases} x(t) & t_1 < t < t_1 + T_0 \\ 0 & \text{elsewhere} \end{cases} \tag{4.113}$$

Since, by the convolution theorem,

$$X(f) = X_p(f)\Delta_s(f) \tag{4.114}$$

the representation given by eq. (4.112) is useful in finding the spectrum of a perodic signal. Example 4.21 serves to illustrate this method.

Example 4.21

Find the spectrum of the periodic computer clock signal of Example 4.20 using eqs. (4.112) and (4.114).

Solution
First, we choose $t_1 = 0$ to give

$$x_p(t) = 4\Pi\left(\frac{t - 0.05 \times 10^{-6}}{0.05 \times 10^{-6}}\right)$$

We then use Table C.2 and the time-shift theorem to obtain

$$X_p(f) = \left(0.2 \times 10^{-6}\right)\text{sinc}\left(0.05 f \times 10^{-6}\right)e^{-j0.1\pi f \times 10^{-6}}$$

Since $T_s = T_0 = 0.125\ \mu s$, then $f_s = f_0 = 8$ MHz, and eq. (4.111) gives

$$\Delta_s(f) = \sum_{n=-\infty}^{\infty} 8 \times 10^6 \delta(f - nf_0)$$

Therefore, eq. (4.114) yields

$$X(f) = \left[0.2 \times 10^{-6}\ \text{sinc}\left(0.05 f \times 10^{-6}\right)e^{-j0.1\pi f \times 10^{-6}}\right]$$

$$\left[\sum_{n=-\infty}^{\infty} 8 \times 10^6 \delta(f - nf_0)\right]$$

$$= \sum_{n=-\infty}^{\infty} \left[1.6\ \text{sinc}\left(0.05 nf_0 \times 10^{-6}\right)e^{-j0.1\pi nf_0 \times 10^{-6}}\right]\delta(f - nf_0)$$

This spectrum is the same as that found for $x(t)$ in Example 4.20.

The method for computing the spectrum of periodic signals illustrated in Example 4.21 is convenient when we can easily find the transform for the function $x_p(t)$ from existing Fourier transform tables and theorems.

The Power-Density Spectrum

The distribution of signal power as a function of frequency in a nonenergy signal is of interest. It is called the *power density spectrum*, or *power spectral density*, and is developed in this section.

We defined the double-sided energy-density spectrum of the energy signal $x(t)$ as

$$G_x(f) = |X(f)|^2 \tag{4.115}$$

with eq. (4.66). We now define $\phi_x(\tau)$ by $\phi_x(\tau) \leftrightarrow G_x(f)$ where the time variable t is replaced by the time variable τ for convenience. Therefore,

$$\phi_x(\tau) = \mathcal{F}^{-1}[G_x(f)] = \mathcal{F}^{-1}\left[|X(f)|^2\right]$$
$$= \mathcal{F}^{-1}\left[X^*(f)X(f)\right] \tag{4.116}$$

Using the convolution, complex-conjugation, and time-reversal theorems, we find that

$$\phi_x(\tau) = \mathcal{F}^{-1}\left[X^*(f)\right] * \mathcal{F}^{-1}[X(f)]$$
$$= x^*(-\tau) * x(\tau)$$
$$= \int_{-\infty}^{\infty} x^*(-\alpha)x\left[(\tau - \alpha)\right]d\alpha$$
$$= \int_{-\infty}^{\infty} x^*(t)x(t + \tau)\,dt \tag{4.117}$$

where the last step follows by making the charge of variable $\alpha = -t$. Equation (4.117) can be written

$$\phi_x(\tau) = \lim_{T \to \infty} \int_{-T}^{T} x^*(t)x(t + \tau)\,dt \tag{4.118}$$

The function $\phi(\tau)$ is called the *time autocorrelation function* of the finite-energy signal $x(t)$. Note that

$$\phi_x(0) = \lim_{T \to \infty} \int_{-T}^{T} |x(t)|^2\,dt \tag{4.119}$$

which is the signal energy.

A similar time autocorrelation function, $R(\tau)$, can be defined for nonenergy signals by the equation

$$R_x(\tau) \equiv \lim_{T \to \infty} \frac{1}{2T} \int_{-T}^{T} x^*(t)x(t + \tau)\,d\tau \tag{4.120}$$

Note that $R_x(0)$ equals the signal power. Paralleling the preceding result for energy signals, we define

$$S_x(f) = \mathcal{F}[R_x(\tau)] \tag{4.121}$$

to be the double-sided power-density spectrum of the signal $x(t)$. Since

$$R_x(0) = \{\mathcal{F}^{-1}[S_x(f)]\}_{\tau=0}$$

$$= \int_{-\infty}^{\infty} S_x(f)e^{j2\pi f(0)} \, df$$

$$= \int_{-\infty}^{\infty} S_x(f) \, df \tag{4.122}$$

then the area under $S_x(f)$ is the power in the signal. It can be shown that

$$\int_{-f_2}^{-f_1} S_x(f) \, df + \int_{f_1}^{f_2} S_x(f) \, df$$

is the power contained in the signal in the range of frequencies $f_1 < f < f_2$, which further shows that $S_x(f)$ is an appropriate definition of the power-density spectrum.

Example 4.22

Find the power-density spectrum for the cosine signal of Example 4.19. Compute the power in the signal.

Solution

$$x(t) = 8 \cos[2\pi(3)t + \pi/3] \quad \text{(centimeters)}$$

$$R_x(\tau) = \lim_{T \to \infty} \frac{1}{2T} \int_{-T}^{T} 64 \cos[6\pi t + \pi/3]\cos[6\pi(t + \tau) + \pi/3] \, dt$$

Using the trigonometric identity,

$$\cos(x)\cos(y) = [\cos(x - y) + \cos(x + y)]/2$$

$$R_x(\tau) = \lim_{T \to \infty} \frac{16}{T} \int_{-T}^{T} [\cos(6\pi\tau) + \cos(12\pi t + 6\pi\tau + 2\pi/3)] \, dt$$

$$= \lim_{T \to \infty} \frac{16}{T} \cos[6\pi\tau] \int_{-T}^{T} dt + \lim_{T \to \infty} \frac{16}{T} \int_{-T}^{T} \cos[12\pi t + 6\pi\tau + 2\pi/3] \, dt$$

$$= 32 \cos[2\pi(3)\tau] \quad \text{(centimeters}^2\text{/second)}$$

where the limit of the second term is zero, since the integral is finite. The power-density spectrum is

$$S_x(f) = \mathcal{F}\{32\cos[2\pi(3)\tau]\}$$

$$= 16\delta(f-3) + 16\delta(f+3) \qquad \text{([centimeters}^2\text{/second]/hertz)}$$

and is plotted in Figure 4.47.

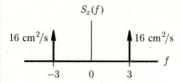

Figure 4.47 Power-Density Spectrum for the Signal in Example 4.22

The power-density spectrum for $x(t)$ shows that all power is concentrated at a frequency of 3 Hz, which we know to be true. The power in the signal is

$$P_x = \int_{-\infty}^{\infty} S_x(f)\,df = 32 \qquad \text{(centimeters}^2\text{/second)}$$

This is clearly correct for the cosine signal: its power is its amplitude squared divided by 2, which is $(8)^2/2 = 32$.

We see from the amplitude spectrum and power-density spectrum for the cosine signal shown in Examples 4.19 and 4.22 that both spectra contain impulses at the same locations, and that the strengths of the impulses in the power-density spectrum are the square of the strengths of the impulses in the amplitude spectrum.

Example 4.23

Find the power-density spectrum of the constant, or DC, signal $x(t) = A$. Compute the power in the signal.

Solution

$$R_x(\tau) = \lim_{T\to\infty} \frac{1}{2T} \int_{-T}^{T} (A)(A)\,dt = A^2$$

$$S_x(f) = \mathcal{F}[A^2] = A^2\delta(f)$$

The power-density spectrum for $x(t)$ shows that all power is concentrated at dc, which is correct. The power in the signal is

$$P_x = \int_{-\infty}^{\infty} S_x(f)\,df = A^2$$

which is clearly correct.

The power-density spectrum and amplitude spectrum for the constant signal again contain impulses at the same location, with the strength of the impulses in the power-density spectrum equal to the square of the strengths of the impulses in the amplitude spectrum. These results indicate that the power-density spectrum for any periodic signal can be easily found by squaring the strengths of the impulses in the amplitude spectrum.

4.6 Summary

The frequency spectrum of a signal is useful in portraying the range of frequencies for which the signal has significant energy content and the nature of the signal amplitude and phase in different portions of the frequency range. The spectrum consists of two parts: the amplitude spectrum, which displays amplitude characteristics of the signal, and the phase spectrum, which displays phase characteristics of the signal. Either single-sided or double-sided spectrum representations can be used. They arise from cosine signals and the complex-exponential representation of cosine signals, respectively.

The complex-exponential Fourier series and cosine Fourier series provide representations of a signal over the expansion interval $t_1 < t < t_1 + T_1$ in terms of the sum of complex exponentials or cosines. A complex-exponential or cosine Fourier series represents a periodic signal for all time if the expansion interval length is chosen to be equal to the period length of the signal. Thus, the terms in the cosine or complex-exponential Fourier series representation of a periodic signal permit easy determination of the single-sided or double-sided frequency spectrum of the signal.

If we construct the periodic extension of the portion of an aperiodic signal in the interval $|t| < T/2$ and let $T \to \infty$, then the Fourier series coefficients for the periodic extension produce the frequency spectrum for an aperiodic signal. The resulting spectrum-producing integral is called the Fourier integral, or Fourier transform, and produces an amplitude-*density* spectrum and a phase spectrum. The inverse Fourier transform is used to compute an aperiodic signal from its frequency spectrum. The energy-density spectrum of an aperiodic signal, which can be obtained from the amplitude spectrum, shows the density of the signal energy as a function of frequency.

There are several properties and a number of theorems for Fourier transforms that relate signal properties to spectrum properties. We can also use the theorems and Fourier transforms of simple signals to simplify the task of computing Fourier transforms of more complicated signals.

The Fourier transform method for computing the frequency spectrum of aperiodic energy signals extends to nonenergy signals through the concept of the Fourier transform in the limit. The amplitude spectrum of a nonenergy signal contains impulses. When the nonenergy signal is periodic, its amplitude spectrum contains only impulses. The power-density spectrum of nonenergy signals shows the density of the signal power as a function of frequency.

Use of the Fourier transform in the limit and the Fourier transform permits evaluation of the frequency spectrum for both aperiodic and periodic signals by Fourier transform techniques. Therefore, the Fourier transform is useful for determining the frequency spectrum of a broad class of signals.

Problems

In the problems that follow, all times are in seconds unless otherwise noted.

4.1 Plot the single-sided and double-sided amplitude and phase spectra for the following signals.

a. $x(t) = 6\cos(1000\pi t - 0.5)$

b. $y(t) = 0.5\cos(200\pi t + 0.3) + \cos(400\pi t)$

c. $v(t) = 10\sin(2\pi t) - 3\cos(6\pi t + 0.2)$

d. $i(t) = 4\cos(30\pi t)$

$\qquad + 6\cos(20\pi t + 0.5)\cos(30\pi t - 0.5)$

4.2 Plot the single-sided and double-sided amplitude and phase spectra for the following signals:

a. $z(t) = 12\cos(10000\pi t + \pi/4)$

b. $v(t) = 14\cos(15\pi t - 0.8) + 18\cos(20\pi t + 2.5)$

c. $i(t) = 3\sin(100\pi t - 1.25) - 4\sin(150\pi t + 0.75)$

d. $q(t) = 6\sin(3.5\pi t - 0.5)\cos(2\pi t + 1.5)$

$\qquad + 2\cos(7\pi t)$

4.3 Show that the complex-exponential basis signals in the set defined by eq. (4.6) are mutually orthogonal over the interval $t_1 < t < t_1 + T_1$.

4.4 Find the complex-exponential Fourier series representations for the signal $x(t)$ over the interval $0 \le t \le 2$ when

a. $x(t) = \begin{cases} 1 & t < 1 \\ t & t \ge 1 \end{cases}$

b. $x(t) = 2t^2$ for all t

c. $x(t) = \begin{cases} 1 + 2t/3 & -1.5 < t \le 1.5 \\ 5 - 2t & 1.5 < t \le 2.5 \\ 0 & \text{elsewhere} \end{cases}$

4.5 Plot the truncated Fourier series approximation, $\hat{x}_N(t)$, to the signals of Problem 4.4 over the time interval $-2 \le t \le 4$ when $N = 10$. Also plot the signals and indicate the expansion interval on

the same set of axes as each approximation (refer to Figure 4.6). Due to the large number of computations required to compute enough points to adequately portray the Fourier series approximation curves, we recommend using a computer for the computations in this problem.

4.6 Find the complex-exponential Fourier series representation for the signal $x(t) = t$ over the interval $0 \le t \le 1$.

4.7 Find the complex-exponential Fourier series representation for the signal

$$y(t) = \begin{cases} t & t \ge 0 \\ t + 1 & t < 0 \end{cases}$$

over the interval $|t| < 1$.

4.8 Compare the Fourier series obtained in Problems 4.6 and 4.7. Indicate reasons for similarities and/or differences.

4.9 As an illustration of the Gibbs phenomenon, find the complex-exponential Fourier series representation for the signal $x(t) = 4\Pi(t/4)$ over the interval $|t| < 4$. Plot $\hat{x}_5(t)$, $\hat{x}_{10}(t)$, and $\hat{x}_{20}(t)$ over this same interval. Use a computer to generate the data for the plots to obtain a sufficient number of points.

4.10 Given the periodic signal

$$x(t) = \sum_{i=-\infty}^{\infty} x_p(t - 20i)$$

where

$$x_p(t) = e^{-0.1t}\Pi\left[(t - 5)/20\right]$$

find the single-sided amplitude and phase spectra and the double-sided amplitude and phase spectra of $x(t)$. Plot the spectra for $f \le 0.3$ Hz.

4.11 The signal model, $x(t)$, for the output of the horizontal sweep oscillator in a monochrome television set is shown in Figure 4.48. The sweep-return time is very small; hence, it has been modeled as negligible. The sweep time is $t_s = 1/15.75$ ms.

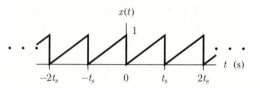

Figure 4.48

a. Plot the double-sided amplitude and phase spectra of the sweep signal up to a frequency of 65 kHz.

b. What fraction of the signal power is contained in the approximate signal corresponding to the portion of the spectrum plotted?

4.12 A full-wave rectifier converts a sinusoidal signal with frequency of 5 Hz into the signal $x(t)$ shown in Figure 4.49.

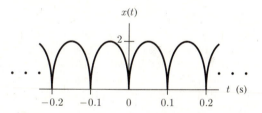

Figure 4.49

a. Plot the double-sided amplitude and phase spectra for the full-wave rectified sinusoidal signal. Plot for $|f| < 50$ Hz.

b. What fraction of the signal power is contained in the approximate signal corresponding to the portion of the spectrum plotted?

4.13 The periodic signal

$$y(t) = \sum_{i=-\infty}^{\infty} y_p(t - 0.1i)$$

where

$$y_p(t) = 3\Pi\left[(t + 0.04)/0.04\right]$$

is a train of pulses used in a multiplexer to periodically turn on a signal. Find and plot the double-

sided amplitude and phase spectra of $y(t)$ for $|f| \leq 75$ Hz.

4.14 Verify the Fourier transform pair

$$\Lambda(t/\tau) \leftrightarrow \tau \sin^2 \tau f$$

4.15 Verify the Fourier transform pair

$$e^{-a|t|} \leftrightarrow 2a \Big/ \left[a^2 + (2\pi f)^2\right]$$

where $a > 0$.

4.16 Verify the Fourier transform pair

$$e^{-a^2 t^2} \leftrightarrow \left[\sqrt{\pi} e^{-(\pi f/a)^2}\right]/a$$

where $a > 0$.

4.17 Verify the Fourier transform pair

$$\sin 2\pi f_0 t \leftrightarrow \frac{1}{2j}\left[\delta(f - f_0) - \delta(f + f_0)\right]$$

4.18 Plot the following signals and their even and odd parts:

a. $x(t) = e^{-|t-1|}$

b. $y(t) = 2\Pi\left[(t + 1)/4\right]$

c. $z(t) = 3\sin 6\pi t$

d. $w(t) = 4\cos(2\pi t - \pi/4)$

4.19 Find and plot the amplitude, phase, and energy spectra for the signal $y(t) = e^{3t}u(-t)$.

4.20 Find and plot the amplitude and phase spectra or, if possible, the total spectrum for the following time-limited signals:

a. $x(t) = e^{-0.1t}\Pi(t/8)$

b. $v(t) = e^{-0.1|t|}\Pi(t/8)$

c. $i(t) = 2t^2\Pi(t/2)$

d. $i(t) = 2t^2\Pi\left[(t - 1)/2\right]$

4.21 Prove the scale-change theorem (Theorem 2).

4.22 Prove the complex-conjugation theorem (Theorem 4).

4.23 Prove the time-differentiation theorem (Theorem 9), for $n = 1$.

4.24 Prove the multiplication theorem (Theorem 12).

4.25 Use the table of Fourier transforms (Table C.2) and the theorems to find the spectra for the signals shown in Figure 4.50. Plot the amplitude and phase spectra or, if possible, the total spectrum for each signal.

a.

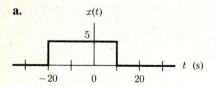

b.

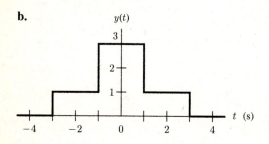

c.

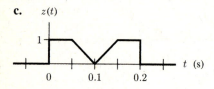

d. $v(t) = 2 + e^{-0.1t} u(t - 2)$

Figure 4.50

4.26 The signal $x(t)$ has a spectrum

$$X(f) = 3/(2 + j\pi f)$$

Use the theorems to find the spectra for the following signals:

a. $x_a(t) = 2\dfrac{dx(t)}{dt}$

b. $x_b(t) = x(-2 - t)$

c. $x_c(t) = x(-t/4)$

d. $x_d(t) = e^{-j3t} x(t - 5)$

e. $x_e(t) = 3x(t) \cos 6\pi t$

f. $x_f(t) = x(4t - 3)$

4.27 Use the table of Fourier transforms and the theorems to find the spectra for the signals shown in Figure 4.51. Plot the amplitude and phase spectra or, if possible, the total spectrum for each signal.

a.

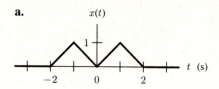

b.

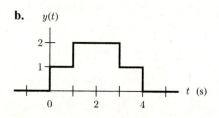

c. $v(t) = 1 - e^{-0.2|t+1|}$

Figure 4.51

4.28 A cosine carrier signal with frequency of 1000 Hz is used with double-sideband amplitude modulation to transmit the single-pulse message signal $m(t) = \Pi(t/0.01)$. Sketch the transmitted signal and plot its amplitude and phase spectra or, if possible, its total spectrum.

4.29 Use the Fourier transform to find the amplitude and phase spectra for the following signals:

a. $x(t) = 3 + \cos(1000\pi t + \pi/3)$

b. $y(t) = 4\cos(30\pi t - \pi/6) + 10\sin(50\pi t + \pi/4)$

c. $z(t) = 2\cos(4\pi t) + 4\sin(4\pi t)$
$$+ 6\cos(8\pi t - \pi/3)$$

4.30 An infrared sensor mounted on the belly of an aircraft sweeps linearly back and forth over the angle range from vertical to 1 radian off of vertical to produce an infrared image of the ground on one side of the aircraft flight path. The angular sweep signal $\theta(t)$ used to drive the sensor is shown in Figure 4.52. Use the Fourier transform table and theorems to find the spectrum of the sweep signal. Plot the amplitude and phase spectra or, if possi-

Figure 4.52

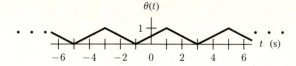

ble, the total spectrum for frequencies lower than or equal to 1 Hz.

4.31 A video camera used to monitor a scene is mounted on the corner of a building and scans by continual rotation. Since the camera points at the building walls for 1/4 of its scan, we want to blank its video signal for this portion of time. Figure 4.53 shows the signal that turns the video on and off. Use the Fourier transform table and theorems to find the spectrum of x(t). Plot the amplitude and phase spectra or, if possible, the total spectrum for frequencies lower than or equal to 0.2 Hz.

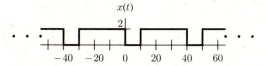

Figure 4.53

4.32 Use the Fourier transform table and theorems to find the spectrum of the periodic signal shown in Figure 4.54. Plot the amplitude and phase spectra or, if possible, the total spectrum for frequencies lower than or equal to 1 Hz.

4.33 Use the Fourier transform table and theorems to find the spectrum of the signal

$$x(t) = A\Pi\left(f_0 t / N\right) \cos 2\pi f_0 t$$

for (a) $N = 1$, (b) $N = 2$, and (c) $N = 4$. Plot the three signals and their spectra for frequencies lower than $3f_0$.

4.34 A pulse radar transmits a cosine carrier signal of frequency 4 MHz modulated by the pulse-train signal

$$x(t) = \sum_{n=-\infty}^{\infty} \Pi\left(\frac{t - nT}{\tau}\right)$$

where $t = 1$ μs and $T = 20$ μs. Use the Fourier transform table and theorems to find the spectrum of the transmitted radar signal. Plot the transmitted signal and its amplitude and phase spectra or, if possible, its total spectrum. (Note: These radar parameters are not practical; however, the result does show the nature of a pulsed radar signal and its spectrum).

4.35 Find the power-density spectrum of the infrared sensor-scanning signal of Problem 4.30.

4.36 Find the power-density spectrum for the periodic signal of Problem 4.32. What bandwidth of frequencies contains at least 95% of the signal power?

4.37 Find the power-density-spectrum for the transmitted radar signal of Problem 4.34. What band of frequencies contains at least 90% of the signal power?

Figure 4.54

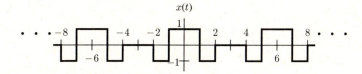

5 Time-Domain Analysis of Continuous-Time Systems

A continuous-time system can be represented by a block diagram indicating signal flow in the system and operations performed on signals by the system. An alternate representation for an electric circuit system—one that is usually more convenient—is a circuit diagram, in which component symbols, such as resistors, capacitors, and inductors, are shown with interconnecting conductors.

We can use the operations and interconnections specified by a block diagram or apply Kirchoff's current and voltage laws to a circuit diagram to produce a set of integro-differential equations that relate the signals in the system. Since in this text we emphasize linear time-invariant systems, the equations produced are linear integro-differential equations with constant coefficients. We can combine the equations in the set to eliminate internal signals, thus producing a single linear integro-differential equation with constant coefficients that relates the output signal (system response) to the input signal for the single-input, single-output type of system considered here.

We indicated in the discussion following Example 1.3 that we can differentiate the integro-differential equation relating the output signal to the input signal to produce a linear differential equation with constant coefficients that relates the output signal to the input signal. This differential equation is referred to as the *system differential equation*. We can use the system differential equation, the initial conditions specified by the energy stored in the system at the initial time of interest, t_0, and classical solution techniques for differential equations to determine the system response to the stored energy and a specified input signal for $t \geq t_0$. This approach was illustrated in Example 1.10 for an RC filter circuit.

In this chapter, we will consider an alternate mathematical model that relates the system response to the input signal when the continuous-time system is linear, time-invariant, and has zero initial conditions (that is, zero initial stored energy). This mathematical model, the *superposition integral*, is often more convenient to use than the differential equation. The solution of the superposition integral requires that the impulse response of the system be known.

5.1 System Impulse Response

Definition

The impulse response of a linear time-invariant system is the system's response to a unit impulse input signal located at $t = 0$ when the system initial conditions are zero.

We use the notation $h(t)$ for the impulse response. Thus, if $x(t)$ is the system input and $y(t)$ is the system response, then

$$x(t) = \delta(t) \quad \text{yields} \quad y(t) = h(t) \tag{5.1}$$

Direct determination of the impulse response for a system characterized by a system differential equation requires solution of the differential equation with a unit impulse input signal. As a simple example, consider the RC filter circuit analyzed in Example 1.10.

Example 5.1

Find the impulse response for the simple RC filter considered in Example 1.10 and shown in Figure 5.1.

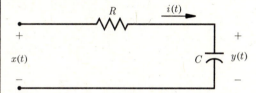

Figure 5.1 RC Filter Circuit for Example 5.1

Solution

From Example 1.10, the differential equation that relates the system response $y(t)$, and input signal $x(t)$ is

$$RC\frac{dy(t)}{dt} + y(t) = x(t)$$

This differential equation was solved in Example 1.10 to find the system response to a general input signal for $t \geq t_0$. The result is eq. (1.40), as follows.

$$y(t) = \int_{t_0}^{t} x(\tau)\left[\frac{1}{RC}\,e^{-(t-\tau)/RC}\right]d\tau + y(t_0)e^{-(t-t_0)/RC} \qquad t \geq t_0$$

because $Y_0 \equiv y(t_0)$ in Example 1.10. To find the impulse response for all time, we set $y(t_0) = 0$ so that initial conditions are zero, $t_0 = -\infty$, and $x(t) = \delta(t)$. This gives

$$h(t) = \int_{-\infty}^{t} \delta(\tau)\left[\frac{1}{RC}e^{-(t-\tau)/RC}\right]d\tau$$

Since the integral is a sifting integral,

$$h(t) = \begin{cases} 0 & t < 0 \\ \frac{1}{RC}e^{-t/RC} & t > 0 \end{cases}$$

$$= \frac{1}{RC}e^{-t/RC}u(t)$$

We can also determine the impulse response of a system by finding the output of the system to the rectangular pulse $x_\epsilon(t) = (1/\epsilon)\Pi(t/\epsilon)$ and then taking the limit of the resulting system response, $y_\epsilon(t)$ as $\epsilon \to 0$. This solution is outlined in the following example.

Example 5.2

Find the impulse response for the RC filter circuit of Example 5.1 using the input signal $x_\epsilon(t) = \frac{1}{\epsilon}\Pi(t/\epsilon)$ and taking the limit as $\epsilon \to 0$.

Solution

From Example 5.1, the system response at time t is

$$y_\epsilon(t) = \int_{t_0}^t \frac{1}{\epsilon}\Pi(\tau/\epsilon)\left[\frac{1}{RC}e^{-(t-\tau)/RC}\right]d\tau + y(t_0)e^{-(t-t_0)/RC} \qquad t \geq t_0$$

For $t \leq -\epsilon/2$,

$$y_\epsilon(t) = 0$$

since $t_0 = -\infty$, $y(t_0) = 0$, and $\Pi(t/\epsilon) = 0$.
For $-\epsilon/2 \leq t \leq \epsilon/2$,

$$y_\epsilon(t) = \int_{-\epsilon/2}^t \left[\frac{1}{\epsilon}\right]\left[\frac{1}{RC}e^{-(t-\tau)/RC}\right]d\tau + y_\epsilon(-\epsilon/2)e^{-(t+\epsilon/2)/RC}$$

$$= \frac{1}{\epsilon}\left[1 - e^{-(t+\epsilon/2)/RC}\right]$$

since $y_\epsilon(-\epsilon/2) = 0$.
For $t \geq \epsilon/2$,

$$y_\epsilon(t) = \int_{\epsilon/2}^t [0]\left[\frac{1}{RC}e^{-(t-\tau)/RC}\right]d\tau + y_\epsilon(\epsilon/2)e^{-(t-\epsilon/2)/RC}$$

$$= \frac{1}{\epsilon}\left[e^{\epsilon/2RC} - e^{-\epsilon/2RC}\right]e^{-t/RC}$$

since $y_\epsilon(\epsilon/2) = \frac{1}{\epsilon}[1 - e^{-\epsilon/RC}]$.
Computing the limit as $\epsilon \to 0$ produces

$$\delta(t) = \lim_{\epsilon \to 0} x_\epsilon(t)$$

Therefore, the impulse response is

$$h(t) = \lim_{\epsilon \to 0} y_\epsilon(t) = \begin{cases} 0 & t < 0 \\ \frac{1}{RC}e^{-t/RC} & t > 0 \end{cases}$$

$$= \frac{1}{RC}e^{-t/RC}u(t)$$

where we used l'Hospital's rule in evaluating the limit for $t > 0$. At $t = 0$, $h(t)$ has a step discontinuity and its value is undefined. The impulse response obtained is the same as that found in Example 5.1, as it must be.

Because the unit impulse signal can be interpreted as the derivative of the unit step signal, it follows that we can also find the impulse response of a system by computing the derivative of the system's step response.

The determination of the impulse response for the simple first-order RC filter circuit was not difficult. It is much more complicated to solve the differential equation for the impulse response of higher order systems. Fortunately, transform techniques can be used to simplify finding the system impulse response. We will consider these techniques in the next two chapters.

5.2 Response of Linear Time-Invariant Systems

In considering the response of a linear time-invariant system to a general input signal, our emphasis will be on those systems having zero initial conditions. However, we will also briefly outline a method for determining the system response when initial conditions are nonzero.

Systems with Zero Initial Conditions

When we compare the result of Example 5.1 with eq. (1.42) in Chapter 1, we note that the impulse response of the RC filter circuit is exactly the function that can be convolved with any input signal to produce the corresponding system response when the system initial condition is zero (that is, when there is zero initial energy stored in the capacitor). This result is not confined to just the RC filter circuit. We will now show that the system response (output signal) of *any* linear, time-invariant system with zero initial conditions is the input signal convolved with the system impulse response.

To begin, we define $x(t)$ as the input signal and $y(t)$ as the system response of a system with an impulse response of $h(t)$. We then consider the rectangular-pulse input signal

$$x(t) = x_{\Delta\tau}(t) = \left[\Pi(t/\Delta\tau) \right] \big/ \Delta\tau \tag{5.2}$$

The system response to this input signal is

$$y(t) = y_{\Delta\tau}(t) \tag{5.3}$$

where

$$\lim_{\Delta\tau \to 0} y_{\Delta\tau}(t) = h(t) \tag{5.4}$$

since

$$\lim_{\Delta\tau \to 0} x_{\Delta\tau}(t) = \delta(t) \tag{5.5}$$

The system is time-invariant; hence,

$$x(t) = x_{\Delta\tau}(t - \alpha) \quad \text{yields} \quad y(t) = y_{\Delta\tau}(t - \alpha) \tag{5.6}$$

We now consider a general input signal, $x(t)$, and express it as

$$x(t) = \lim_{\Delta\tau \to 0} \tilde{x}(t, \Delta\tau) \tag{5.7}$$

where $\tilde{x}(t, \Delta\tau)$ is the stairstep approximation to $x(t)$

$$\tilde{x}(t, \Delta\tau) = \sum_{n=-\infty}^{\infty} x(n\Delta\tau)\Pi\left[(t - n\Delta\tau)/\Delta\tau\right] \tag{5.8}$$

shown in Figure 5.2. We can write eq. (5.8) as

$$\tilde{x}(t, \Delta\tau) = \sum_{n=-\infty}^{\infty} [x(n\Delta\tau)\Delta\tau]x_{\Delta\tau}(t - n\Delta\tau) \tag{5.9}$$

by using eq. (5.2). Equation (5.6) shows that the system response to $x_{\Delta\tau}(t - n\Delta\tau)$ is $y_{\Delta\tau}(t - n\Delta\tau)$. Therefore, the system response to $\tilde{x}_{\Delta\tau}(t, \Delta\tau)$ is

$$\tilde{y}(t, \Delta\tau) = \sum_{n=-\infty}^{\infty} [x(n\Delta\tau)\Delta\tau]y_{\Delta\tau}(t - n\Delta\tau) \tag{5.10}$$

because the system is linear and superposition holds. Now the system response, $y(t)$, to the general input signal $x(t)$ is

$$y(t) = \lim_{\Delta\tau \to 0} \tilde{y}(t, \Delta\tau) = \lim_{\Delta\tau \to 0} \sum_{n=-\infty}^{\infty} x(n\Delta\tau)y_{\Delta\tau}(t - n\Delta\tau)\Delta\tau \tag{5.11}$$

since $\tilde{y}(t, \Delta\tau)$ is the system response to the input signal $\tilde{x}(t, \Delta\tau)$ and

$$x(t) = \lim_{\Delta\tau \to 0} \tilde{x}(t, \Delta\tau)$$

When we perform the limit in eq. (5.11) the integral

$$y(t) = \int_{-\infty}^{\infty} x(\tau)h(t - \tau)\, d\tau = x(t) * h(t) \tag{5.12}$$

Figure 5.2
Stairstep
Approximation to
Signal $x(t)$

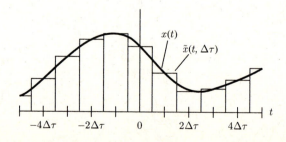

is produced because, by eq. (5.4), $y_{\Delta\tau}(t) \to h(t)$ as $\Delta\tau \to 0$. We refer to the integral in eq. (5.12) as the *superposition integral*, because it constructs the system response at time t by adding the system responses at time t resulting from input-signal values at all times τ (that is, for $-\infty \le \tau \le \infty$). Thus, it is apparent that the output response of a linear time-invariant system with zero initial conditions to a general input signal is, indeed, the convolution of the input signal and the impulse response, as was indicated at the beginning of this section. Since convolution is commutative, $y(t)$ can also be written as

$$y(t) = h(t) * x(t) = \int_{-\infty}^{\infty} h(\tau)x(t - \tau)\,d\tau \qquad (5.13)$$

The system differential equation completely specifies a system's response characteristics when the system initial conditions are zero because it can be used to find the system response to any input signal. The same can be said for the system impulse response when the system has zero initial conditions and is linear and time-invariant. Thus, the system impulse response is an alternate characterization for a linear time-invariant system with zero initial conditions.

If a specific input signal and the system differential equation is given for a linear time-invariant system with zero initial conditions, then we can find the system response to the signal by using this signal and solving the system differential equation. Alternatively, we can solve the system differential equation with the unit impulse input signal to find the system impulse response, and then compute the convolution of this impulse response and the given input signal to find the system response to that signal. One might wonder why the impulse-response characterization and superposition integral are of interest, if the differential equation must be solved anyway as part of the solution. One reason for our interest in this system-response solution technique is that the system differential equation needs to be solved only once if the system responses to a number of different input signals are desired. A second reason is that we can often find the impulse response from block diagrams or circuit diagrams through the use of transforms, without solving the system differential equation. Only algebra and an inverse transform are required. (As mentioned previously, we will consider transform methods in the following two chapters.) A third reason for our interest in the impulse response and superposition integral is that it may be possible to measure experimentally an approximate impulse response for a system and then use it in the superposition integral to find the approximate system response to various input signals.

Example 5.3

The circuit shown in Figure 5.3 is a filter that reduces high-frequency interference. It can be shown that this circuit has the impulse response $h(t) = 6[e^{-2t} - e^{-3t}]u(t)$ when $x(t)$ and $y(t)$ are the input and output, respectively. Find the system response to the input signal $x(t) = \Pi(t - 0.5)$ when no energy is initially stored in the system. Plot $x(t)$, $h(t)$ and $y(t)$.

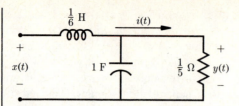

Figure 5.3 Filter Circuit for Example 5.3

Solution

$$y(t) = \int_{-\infty}^{\infty} h(\tau)x(t - \tau)d\tau$$

Step 1   Write equations and draw sketches for $h(t)$ and $x(t)$, as shown in Figure 5.4.

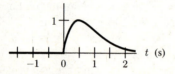

$$h(t) = \begin{cases} 6\left[e^{-2t} - e^{-3t}\right] & t > 0 \\ 0 & t < 0 \end{cases}$$

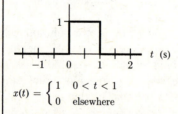

$$x(t) = \begin{cases} 1 & 0 < t < 1 \\ 0 & \text{elsewhere} \end{cases}$$

Figure 5.4 Step 1 for Example 5.3

Step 2   Write equations and draw sketches for $h(\tau)$ and $x(t - \tau)$, as shown in Figure 5.5.

$$h(\tau) = \begin{cases} 6\left[e^{-2t} - e^{-3\tau}\right] & \tau > 0 \\ 0 & \tau < 0 \end{cases}$$

Figure 5.5 Step 2 for Example 5.3

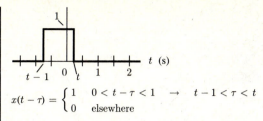

$$x(t - \tau) = \begin{cases} 1 & 0 < t - \tau < 1 \\ 0 & \text{elsewhere} \end{cases} \quad \rightarrow \quad t - 1 < \tau < t$$

Figure 5.5 Step 2 for Example 5.3 *(continued)*

Step 3   Perform integrations to find $y(t)$ for all values of t.

Case 1   For $t \leq 0$,

$$y(t) = \int_{-\infty}^{\infty} h(\tau)x(t - \tau)\,d\tau = 0$$

Case 2   For $0 < t \leq 1$,

$$y(t) = \int_{0}^{t} 6\big[e^{-2\tau} - e^{-3\tau}\big][1]\,d\tau = 2e^{-3t} - 3e^{-2t} + 1$$

Case 3   For $1 < t$,

$$y(t) = \int_{t-1}^{t} 6\big[e^{-2\tau} - e^{-3\tau}\big][1]\,d\tau = 3e^{-2t}\left(e^2 - 1\right) - 2e^{-3t}\left(e^3 - 1\right)$$

$$= 19.167e^{-2t} - 38.171e^{-3t}$$

Summarizing,

$$y(t) = \begin{cases} 0 & t \leq 0 \\ 2e^{-3t} - 3e^{-2t} + 1 & 0 < t \leq 1 \\ 19.167e^{-2t} - 38.171e^{-3t} & 1 < t \end{cases}$$

The input, $x(t)$, impulse response, $h(t)$; and system response, $y(t)$, are plotted in Figure 5.6.

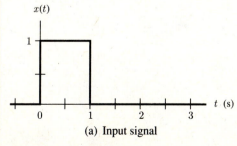

(a) Input signal

Figure 5.6 Signals for Example 5.3

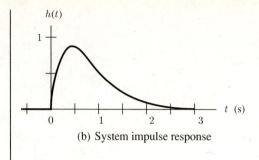

(b) System impulse response

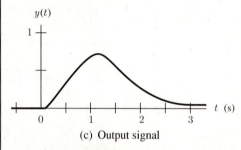

(c) Output signal

Figure 5.6 Signals for Example 5.3 *(continued)*

Example 5.4

A feedback control system is used to control a room's temperature about a nominal value. A simple model for this system is represented by the block diagram shown in Figure 5.7, where $x(t)$ is a signal representing the commanded temperature change from nominal, $y(t)$ is a signal representing the produced temperature change from nominal, and t is the time in minutes. Find (a) the differential equation relating $x(t)$ and $y(t)$, (b) the impulse response for the system, and (c) the temperature change from nominal produced by the system when the gain K is 0.5 and a step change of 2° is commanded at $t = 4$ min. Plot the temperature change produced.

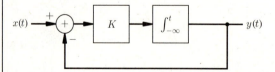

Figure 5.7 Temperature-Control Loop for Example 5.4

Solution

a. From the block diagram,

$$y(t) = \int_{-\infty}^{t} K\left[x(\tau) - y(\tau)\right] d\tau + y(-\infty)$$

Therefore,

$$\frac{dy(t)}{dt} + Ky(t) = Kx(t)$$

b. The differential equation can be rewritten as

$$\frac{1}{K}\frac{dy(t)}{dt} + y(t) = x(t)$$

This is the same differential equation we encountered in Example 5.1, but now $1/K$ has replaced RC. Therefore, from Example 5.1, the system impulse response is

$$h(t) = Ke^{-Kt}u(t)$$

c. The temperature change produced is

$$y(t) = \int_{-\infty}^{\infty} h(\tau)x(t - \tau) d\tau$$

Step 1   To solve this integral, first write equations and draw sketches for $x(t)$ and $h(t)$, as shown in Figure 5.8.

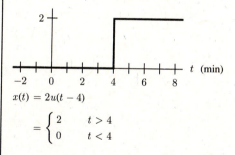

$$x(t) = 2u(t - 4)$$

$$= \begin{cases} 2 & t > 4 \\ 0 & t < 4 \end{cases}$$

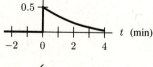

$$h(t) = \begin{cases} 0.5e^{-0.5t} & t > 0 \\ 0 & t < 0 \end{cases}$$

Figure 5.8 Step 1 for Example 5.4

Step 2   Next, write equations and draw sketches for $x(\tau)$ and $h(t - \tau)$, as shown in Figure 5.9.

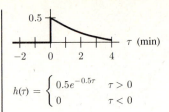

$$h(\tau) = \begin{cases} 0.5e^{-0.5\tau} & \tau > 0 \\ 0 & \tau < 0 \end{cases}$$

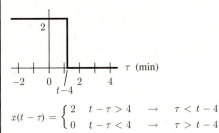

$$x(t - \tau) = \begin{cases} 2 & t - \tau > 4 & \rightarrow & \tau < t - 4 \\ 0 & t - \tau < 4 & \rightarrow & \tau > t - 4 \end{cases}$$

Figure 5.9 Step 2 for Example 5.4

Step 3 Finally, perform integrations to find the temperature change produced for all values of t.

Case 1 For $t - 4 \le 0 \quad \rightarrow \quad t \le 4$,

$$y(t) = \int_{-\infty}^{\infty} h(\tau)x(t - \tau)\, d\tau = 0$$

Case 2 For $t - 4 > 0 \rightarrow t > 4$,

$$y(t) = \int_{0}^{t-4} \left[0.5e^{-0.5\tau}\right][2]\, d\tau = 2\left[1 - e^{-0.5(t-4)}\right]$$

Therefore, the temperature change produced is

$$y(t) = 2\left[1 - e^{-0.5(t-4)}\right]u(t - 4)$$

and is plotted in Figure 5.10.

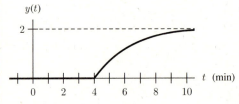

Figure 5.10 Temperature Change Produced in Example 5.4

Note that it takes approximately 6 min before the temperature change produced is close to the commanded $2°$ change.

Causal Systems

We recall from Chapter 1 that a causal system is one for which the value of the output at a specified time depends on values of the input signal only at times preceding or equal to the specified time. The portion of a linear system's output signal that depends on the input signal is the system response with zero initial conditions. For a linear time-invariant system, the zero-initial-condition response at time t is

$$y(t) = \int_{-\infty}^{\infty} h(\tau)x(t - \tau)\,d\tau \qquad (5.14)$$

In this integral, input-signal values that occur after time t are those for which $\tau < 0$. The system response does not depend on these input-signal values if $h(\tau)$ is zero for $\tau < 0$. Therefore, a linear time-invariant system is causal if its impulse response is zero for $t < 0$. This condition for system causality is actually both a necessary and sufficient condition. In other words, a linear time-invariant system is causal if and only if $h(t) = 0$ for $t < 0$. For a causal system, eq. (5.14) is equivalent to

$$y(t) = \int_{0}^{\infty} h(\tau)x(t - \tau)\,d\tau \qquad (5.15)$$

Also, since convolution is commutative, then

$$y(t) = \int_{-\infty}^{\infty} x(\tau)h(t - \tau)\,d\tau = \int_{-\infty}^{t} x(\tau)h(t - \tau)\,d\tau \qquad (5.16)$$

for a causal system because $h(t - \tau) = 0$ for $\tau > t$. Note that the systems of Examples 5.1, 5.2, 5.3, and 5.4 are causal; in these systems $h(t) = 0$ for $t < 0$.

Systems with Nonzero Initial Conditions

We have limited the preceding discussion to linear time-invariant systems with zero initial conditions. However, it is not conceptually difficult to find the additional response that is due to nonzero initial conditions. Since the system is linear, the zero-input response that results from initial energy stored in each energy-storage element can be determined separately and added to the system response from the input signal to yield the total output signal. These zero-input responses are equivalent to system responses to impulse input signals that are directly introduced into the energy-storage elements. We will not present a detailed discussion on this equivalency or its use, for the reason that systems with zero initial conditions are emphasized throughout this text. Moreover, systems with nonzero initial conditions are more easily analyzed using the transform techniques discussed in Chapter 7. At this point, however, we will illustrate the concept by finding the initial-condition response for the simple RC circuit previously considered.

For the RC filter circuit of Example 5.1, the initial energy stored in the capacitor at $t = 0^+$ is specified by the voltage across the capacitor $y(0^+) =$

$Y_0 \neq 0$. We can produce this voltage with the current input $i_i(t) = CY_0\delta(t)$ to the capacitor because

$$y\left(0^+\right) = \frac{1}{C} \int_{0^-}^{0^+} CY_0\delta(t)\, dt + y\left(0^-\right) = Y_0 \qquad (5.17)$$

when the capacitor is uncharged at $t = 0^-$; hence, $y(0^-) = 0$. Note that 0^- and 0^+ indicate approaches to zero from the left and right, respectively. With the input signal $x(t)$ set to zero, the initial-condition response (that is, the zero-input response) is

$$y_i(t) = \frac{1}{C} \int_{0^-}^{t} \left[i_i(t) - y_i(t)/R\right]\, dt + y\left(0^-\right)$$

$$= \frac{1}{C} \int_{0^-}^{t} \left[CY_0\delta(t) - y_i(t)/R\right]\, dt \qquad t \geq 0 \qquad (5.18)$$

Differentiating this equation gives

$$RC\frac{dy_i(t)}{dt} + y_i(t) = RCY_0\delta(t) \qquad (5.19)$$

This differential equation is like the one in Example 5.1; the initial condition is zero since $y_i(0^-) = 0$. Therefore, from Example 5.1,

$$y_i(t) = RCY_0 \left[\frac{1}{RC}e^{-t/RC}u(t)\right] = Y_0 e^{-t/RC}u(t) \qquad (5.20)$$

which is the response of the RC filter circuit to the initial condition corresponding to energy stored in the capacitor. This result matches that obtained in Example 1.10, being equal to the second term in eq. (1.40).

5.3 Summary

In Chapter 1, we indicated that a single-input, single-output linear time-invariant system can be modeled by a linear differential equation with constant coefficients. This equation can be solved, subject to initial conditions, to determine the system response (output signal) to a specified input signal and set of initial conditions. An alternate method for determining the system response to a specified input signal for a linear time-invariant system with zero initial conditions involves the convolution of the specified input with the system impulse response. The resulting convolution integral, referred to as the superposition integral, constructs the system response at time t by adding the system responses at time t that are due to input-signal values at all times τ (that is, $-\infty \leq \tau \leq \infty$).

The system impulse response, $h(t)$, is the system response to a unit impulse input signal applied at $t = 0$ when system initial conditions are all zero. The impulse response of a system characterizes the system, except for initial energy storage. A system is causal if and only if $h(t) = 0$ for $t < 0$.

Impulse-response methods can be extended to the analysis of systems with nonzero initial conditions. We considered only one example to illustrate this concept for the reason that systems with zero initial conditions are emphasized throughout the text. Also, the transform method of Chapter 7 provides an easier method for performing such analysis.

Problems

5.1 To reduce high-frequency interference, the simple RC filter of Example 5.1 is used in transferring a voltage signal that is proportional to pressure in a tank to a voltmeter used to indicate the pressure. The voltmeter can be modeled as a resistance of 2 $k\Omega$. The complete system representation is shown in Figure 5.11. Find and plot the impulse response for the pressure-measurement system, where $x(t)$ is the voltage signal proportional to pressure and $y(t)$ is the voltage indicated by the voltmeter.

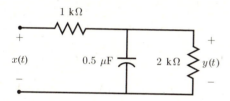

Figure 5.11

5.2 Find and plot the impulse response for the resistive circuit shown in Figure 5.12. The input signal is $v(t)$ and the output signal is $i(t)$.

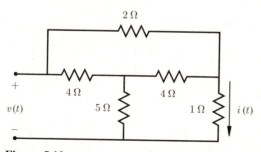

Figure 5.12

5.3 Figure 5.13 shows another filter circuit that reduces high-frequency noise. In this case, the input signal is a current and the output signal is a volt-

age. Find and plot the impulse response for the filter circuit.

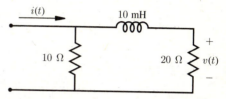

Figure 5.13

5.4 The flow rate $y(t)$, in a pipeline is proportional to a valve-opening area. The rate of change of this area is proportional to the signal $e(t)$ that actuates the valve. Figure 5.14 shows a feedback control system for controlling the flow rate to match a specified flow rate, $x(t)$. The system uses control signals proportional to the flow rate and the rate of change of flow rate. Find and plot the impulse response of the system.

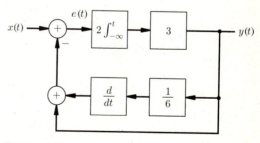

Figure 5.14

5.5 Show that the impulse response given for the circuit in Example 5.3 is correct.

5.6 A system has the impulse response $h(t) = 2e^{-3t}u(t)$. Find the system response to the input signal $x(t) = 3\cos(4\pi ft)u(t)$.

5.7 A system has the impulse response $h(t) = e^{-2t}u(t)$. Find and plot the system response when the input signal is $x(t) = -\Pi(t/2)$.

5.8 Rework part (c) for the temperature control system of Example 5.4 with different values of the gain K. Comment on the effect of the gain value.

5.9 A system has the impulse response $h(t) = e^{-|t|}$. Find and plot the system response to the input signal $x(t) = 3\Pi[(t+2)/2]$.

5.10 A system has the impulse response $h(t) = 2e^{-2t}u(t)$. What function of time does the output signal of the system approach as t becomes large when the input signal is (a) $x(t) = u(t)$, (b) $x(t) = \cos(10\pi t)u(t)$.

5.11 A system has the impulse response $h(t) = e^{-t}\cos(10\pi t)u(t)$. What function of time does the output signal approach as t becomes large when the input signal is (a) $x(t) = u(t)$, (b) $x(t) = \cos(10\pi t)u(t)$.

5.12 The impulse responses shown in Figure 5.15 characterize certain systems. Which ones char-

acterize causal systems? Give reasons for your answers.

5.13 A system has the impulse response
$$h(t) = 3e^{-0.5t}u(t) - 2e^{-t}u(t) + e^{-2(t-2)}u(t-2)$$
The input signal is $x(t) = 2\Pi[t-1)/0.5]$. Find the output signal. Plot the input signal, impulse response, and output signal.

5.14 A system has the impulse response
$$h(t) = \begin{cases} 0 & t < 0 \\ t & 0 \le t < 1 \\ 1 & 1 \le t \end{cases}$$
Find and plot the system response to the following input signals:

a. $x(t) = u(t)$

b. $x(t) = u(t+3)$

c. $x(t) = \cos(4\pi t)u(t)$

d. $x(t) = \Pi[(t-0.5)]$

e. $x(t) = \Pi[(t+1)/0.5]$

Figure 5.15

a.

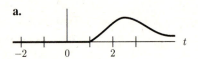

b.

c.

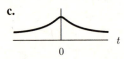

d.

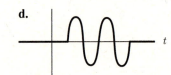

6 Spectral Analysis of Continuous-Time Systems

In Chapter 5, we considered time-domain analysis of single-input–single-output, linear, time-invariant, continuous-time systems. Time-domain analysis produces the time waveform of the system output signal (system response) from the time waveform of the input signal, the system characteristics and the initial energy stored in the system. We emphasized systems with zero initial stored energy (zero initial conditions).

When we want to know the frequency range over which the output signal has significant energy content and the distribution of the output-signal energy in this frequency range, the spectrum of the output signal, rather than its time waveform, is of interest to us. For example, the frequency ranges encompassed by the output signals of two neighboring radio stations must not overlap, so that each can be received without interference from the other.

The spectrum of the zero-initial-energy response of a system is of the greatest interest to us. The measure of its difference from the spectrum of the input signal indicates the effect of passing the signal through the system. Thus, we will consider only systems with zero initial conditions in developing spectral-analysis techniques for single-input–single-output, linear, time-invariant, continuous-time systems. These techniques are useful in determining the spectrum of the output signal from the system characteristics and the spectrum of the input signal. They permit us to see the effect of a system on such signal characteristics as bandwidth and to produce a system characterization as a function of signal frequency.

6.1 System Frequency Response

A fundamental concept in the spectral analysis of systems is the concept of system frequency response.

Definition

The system frequency response $H(f)$ of a continuous-time system is the function of frequency that produces the spectrum $Y(f)$ of the output signal $y(t)$ when multiplied by the spectrum $X(f)$ of the input signal $x(t)$ if system initial conditions are zero.

Not all systems possess a frequency response. A system has a frequency response if $Y(f)$ exists when $X(f)$ exists; for example, if a finite-energy input signal produces a finite-energy output signal.

In equation form, the frequency-response definition is

$$Y(f) = H(f)X(f) \tag{6.1}$$

If no impulses exist in $X(f)$ or $Y(f)$ (that is, if the input and output signals are not power signals), and if $X(f)$ is nonzero for all f, then we can compute the frequency response with

$$H(f) = Y(f)/X(f) \tag{6.2}$$

Equation (6.2) shows that the frequency response of a system is the ratio of the spectrum of the output signal to the spectrum of the input signal in this case. For power signals, eq. (6.1) shows that the strength of an impulse located at $f = f_1$ in the spectrum of the input signal, multiplied by the frequency response at $f = f_1$, produces the strength of an impulse located at $f = f_1$ in the spectrum of the output signal. However, just knowing the strengths of impulses in the spectra of the input and output power signals does not define the frequency response for all frequencies.

Frequency Response Characteristics

If the system frequency response is known, the system zero-initial-energy response to any input signal can be found. Therefore, the system frequency response completely characterizes a linear time-invariant system, except for initial conditions. We take the following steps to produce the system response:

1. Compute the Fourier transform of the input signal to find the spectrum of the input signal.
2. Multiply the spectrum of the input signal by the system frequency response to find the spectrum of the system response.
3. Compute the inverse Fourier transform of the spectrum of the system response to find the system response.

Applying the Fourier transform convolution theorem (Theorem 11 in Chapter 4) to eq. (6.1) yields

$$y(t) = \int_{-\infty}^{\infty} \left\{ \mathcal{F}^{-1} \left[H(f) \right]_{t=\tau} \right\} x(t - \tau)\, d\tau \tag{6.3}$$

When we compare this equation with the superposition integral (eq. [5.13] in Chapter 5), we note that

$$h(t) = \mathcal{F}^{-1} \left[H(f) \right] \tag{6.4}$$

or equivalently,

$$H(f) = \mathcal{F} \left[h(t) \right] \tag{6.5}$$

Therefore, the system frequency response is the Fourier transform of the system impulse response.

We showed in Chapter 4 that the Fourier transform of a time function is in general complex. Thus, the system frequency response is also complex, since it is the Fourier transform of the time function that is the system impulse response. Therefore, we can rewrite eq. (6.1) as

$$Y(f) = |Y(f)|e^{j\underline{/Y(f)}} = \left[|H(f)|e^{j\underline{/H(f)}}\right]\left[|X(f)|e^{j\underline{/X(f)}}\right]$$

$$= |H(f)||X(f)|e^{j[\underline{/H(f)}+\underline{/X(f)}]} \tag{6.6}$$

Equation (6.6) shows that

$$|Y(f)| = |H(f)||X(f)| \tag{6.7}$$

and

$$\underline{/Y(f)} = \underline{/H(f)} + \underline{/X(f)} \tag{6.8}$$

Because $|H(f)|$ relates the amplitude spectrum of the output signal to the amplitude spectrum of the input signal, we call it the *system amplitude response*. Likewise, we call $\underline{/H(f)}$ the *system phase response* because it relates the phase spectrum of the output signal to the phase spectrum of the input signal.

Example 6.1

A system has the frequency response

$$H(f) = \Lambda(f/5)$$

Find (a) the spectrum $Y(f)$ of the output signal $y(t)$ and (b) the output signal when the input signal is

$$x(t) = 3\cos(4\pi t) + 4\cos(6\pi t)$$

Solution

a. From Table C.2, the spectrum of the input signal is

$$X(f) = 1.5\delta(f - 2) + 1.5\delta(f + 2) + 2\delta(f - 3) + 2\delta(f + 3)$$

The system frequency response is

$$H(f) = \begin{cases} 1 - |f|/5 & |f| < 5 \\ 0 & \text{elsewhere} \end{cases}$$

The spectrum of the output signal is

$$Y(f) = H(f)X(f)$$

$$= 1.5H(2)\delta(f - 2) + 1.5H(-2)\delta(f + 2)$$

$$+ 2H(3)\delta(f - 3) + 2H(-3)\delta(f + 3)$$

$$= 0.9\delta(f - 2) + 0.9\delta(f + 2) + 0.8\delta(f - 3) + 0.8\delta(f + 3)$$

The plots of $X(f)$, $H(f)$, and $Y(f)$ are shown in Figure 6.1.

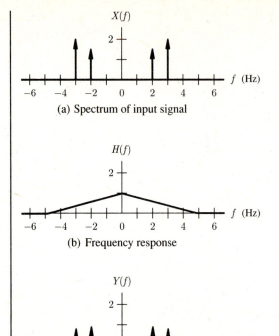

Figure 6.1 Signal Spectra and Frequency Response for Example 6.1

b. From Table C.2, the output signal is

$$y(t) = 1.8 \cos(4\pi t) + 1.6 \cos(6\pi t)$$

The impulse response for the system of Example 6.1 is

$$h(t) = \mathcal{F}^{-1}[H(f)] = 5 \operatorname{sinc}^2(5t) \tag{6.9}$$

To determine the spectrum of the output signal, it is much easier to use spectral-analysis techniques rather than to compute the convolution of $5 \operatorname{sinc}^2(5t)$ and $3 \cos(4\pi t) + 4 \cos(6\pi t)$ and then compute the Fourier transform of the result.

Example 6.2

The simple RC filter circuit of Example 1.10 is constructed with a $0.5 - \Omega$ resistor and $(2\pi)^{-1} - F$ capacitor, as shown in Figure 6.2. Find (a) the system frequency response, amplitude response, and phase response; (b) the spectrum $Y(f)$ of the output signal $y(t)$; and (c) the output signal when the input signal is $x(t) = 4\cos(4\pi t) + 4\cos(12\pi t)$, where the first term is the message signal and the second term is an additive interference signal.

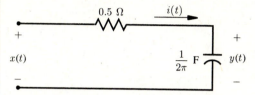

Figure 6.2 RC Filter Circuit for Example 6.2

Solution

a. From Example 5.1, the simple RC filter in Figure 6.2 has the system impulse response

$$h(t) = 4\pi e^{-4\pi t} u(t)$$

Therefore, referring to Table C.2, the system frequency response is

$$H(f) = \mathcal{F}[h(t)] = \frac{4\pi}{4\pi + j2\pi f} = \frac{1}{1 + j(f/2)}$$

The system amplitude response is

$$|H(f)| = 1\Big/\sqrt{1 + (f/2)^2}$$

and the system phase response is

$$\underline{/H(f)} = -\tan^{-1}(f/2)$$

The amplitude and phase responses are illustrated in Figure 6.3.

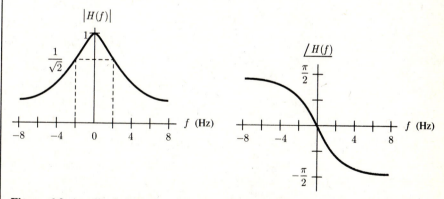

Figure 6.3 Amplitude Response and Phase Response for the Simple RC Filter

b. From Table C.2, the spectrum of the input signal is

$$X(f) = 2\delta(f - 2) + 2\delta(f + 2) + 2\delta(f - 6) + 2\delta(f + 6)$$

which is illustrated in Figure 6.4. The spectrum of the output signal is

$$Y(f) = H(f)X(f)$$

$$= 2H(2)\delta(f-2) + 2H(-2)\delta(f+2)$$

$$+ 2H(6)\delta(f-6) + 2H(-6)\delta(f+6)$$

$$= \left(2\big/\sqrt{2}\right)e^{-j0.25\pi}\delta(f-2) + \left(2\big/\sqrt{2}\right)e^{j0.25\pi}\delta(f+2)$$

$$+ \left(2\big/\sqrt{10}\right)e^{-j0.398\pi}\delta(f-6) + \left(2\big/\sqrt{10}\right)e^{j0.398\pi}\delta(f+6)$$

This spectrum is shown in Figure 6.4.

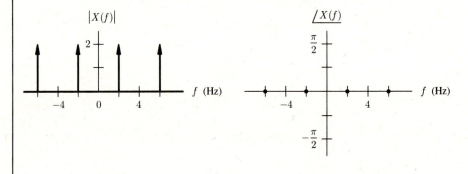

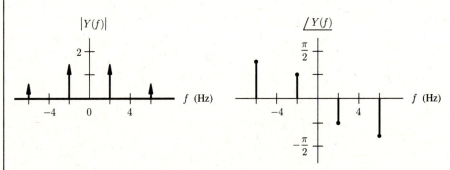

Figure 6.4 Spectra of Input and Output Signals for Example 6.2, part (b)

c. Using Table C.2, the output signal is found to be

$$y(t) = \left(4\big/\sqrt{2}\right)\cos(4\pi t - 0.25\pi) + \left(4\big/\sqrt{10}\right)\cos(12\pi t - 0.398\pi)$$

We see from the amplitude response in Figure 6.3 and the input- and output-signal amplitude spectra in Figure 6.4 that the simple RC filter does indeed reduce the amplitude of higher frequency interference signals more than that of lower frequency message signals, as asserted in Example 1.10.

The systems considered in Examples 6.1 and 6.2 have even amplitude responses and odd phase responses. Here, "system amplitude response" is especially descriptive. As we see from the examples, the value of the system amplitude response at the frequency of an input-signal sinusoidal component is the value by which the amplitude of that sinusoidal component is multiplied to produce the amplitude of an output-signal sinusoidal component at the same frequency. Similarly, "system phase response" is especially descriptive because the value of the system phase response at the frequency of an input-signal sinusoidal component is the value by which the phase of that signal component is shifted to produce the phase of an output-signal sinusoidal component at the same frequency. That is,

$$y(t) = |X||H(f_1)| \cos\left(2\pi f_1 t + \theta + \underline{/\,H(f_1)}\right) \tag{6.10}$$

when

$$x(t) = |X| \cos(2\pi f_1 t + \theta) \tag{6.11}$$

and the system frequency response is

$$H(f) = |H(f)| e^{j\underline{/H(f)}} \tag{6.12}$$

We frequently refer to the values of the amplitude response and phase response at a particular frequency as the *system gain* and *system phase shift* at that frequency. Note that the gain may be less than unity. The system in Example 6.1 does not produce any phase shift of the input-signal components, the phase response being zero at all frequencies.

Practical systems produce output signals that are real functions of time when input signals are real functions of time. These systems are the only type considered in this text. The system impulse response for such systems is a real function of time when the system is linear and time-invariant because the output signal is the input signal convolved with the impulse response. The corresponding system amplitude response is an even function of frequency and the system phase response is an odd function of frequency because they are the amplitude and angle associated with the Fourier transform of a real function of time. That is,

$$|H(-f)| = |H(f)| \tag{6.13}$$

and

$$\underline{/\,H(-f)} = -\underline{/\,H(f)} \tag{6.14}$$

as in the two examples.

If a system is also causal, then the phase shift produced by the system cannot be zero for all frequencies, except for the trivial case in which the output signal is merely a constant times the input signal. Causality implies that $h(t) = 0$ for $t < 0$, which implies that $h(t)$ is neither an even nor an odd function of time. This, in turn, implies that $H(f)$ is complex; hence, the phase characteristic just mentioned. Also, the nonzero portion of $|H(f)|$ cannot extend for only a finite range of frequencies for a causal system. This would require $h(t)$ to extend over the entire time axis, as indicated in Chapter 4 by Property I for Fourier transforms. Note that $|H(f)|$ for the system considered in Example 6.1 is nonzero only over the frequency range $|f| < 5$ and that the system is noncausal, having an impulse response of $h(t) = 5 \, \text{sinc}^2(5t)$.

In Examples 6.1 and 6.2, we found that it was easier to find the output signal by using spectral-analysis techniques than by using convolution directly. In many cases, however, this is not so, owing to the difficulty of the computation of the inverse Fourier transform.

Example 6.3

Find the spectrum of the output signal from the RC filter of Example 6.2 when the input signal is $x(t) = 4\Pi(t - 1)$.

Solution
From Table C.2, the spectrum of the input signal is

$$X(f) = 4 \, \text{sinc}(f)e^{-j2\pi f}$$

We found in Example 6.2 that the system frequency response is

$$H(f) = 1/\left[1 + j(f/2)\right]$$

Therefore, the spectrum of the output signal is

$$Y(f) = H(f)X(f) = \left[4 \, \text{sinc}(f)e^{-j2\pi f}\right]/\left[1 + j(f/2)\right]$$

We made no attempt to determine the output-signal waveform $y(t)$ in Example 6.3 because the inverse Fourier transform computation is difficult. It is easier to compute the convolution of the input signal and impulse response to produce the output signal. The input signal and impulse response are simple time functions, the convolution of which is easily computed. A transform technique that is easier to use in finding $y(t)$ for the system and input signal in Example 6.3 is the Laplace transform, to be discussed in the following chapter.

System Bandwidth

It is useful to define both total *system bandwidth* and *significant system bandwidth*, the latter being of more practical value.

Definition

The total bandwidth of a system is the width of the range of positive frequencies for which the amplitude response is nonzero.

Unfortunately, many systems (including all causal systems) have an infinite total bandwidth. Consequently, we use the following additional definition.

Definition _____

The significant system bandwidth is the width of the range of frequencies over which the amplitude response is greater than some specified fraction of its maximum value.

Usually, the significant system bandwidth is called simply the *system bandwidth*. The frequency at which the specified fractional amplitude response occurs is called the *cutoff frequency*.

A typical specified fraction is the value $1/\sqrt{2} = 0.707$, which we illustrated on the plot of the amplitude response shown in Figure 6.3. The resulting system bandwidth and cutoff frequency are referred to as the *half-power bandwidth* and *cutoff frequency of the system*, respectively. The designation half-power is a consequence of the fact that a sinusoidal signal amplitude ratio of $1/\sqrt{2}$ corresponds to a sinusoidal signal power ratio of 1/2.

We refer to the system considered in Example 6.1 as a *low-pass system* for the reason that its bandwidth encompasses the range of frequencies from $f = 0$ to the cutoff frequency. The system in Example 6.2 is also a low-pass system. We can see that the half-power cutoff frequencies and bandwidths for the systems in Examples 6.1 and 6.2 are 1.464 Hz and 2 Hz, respectively.

Figure 6.5 illustrates an amplitude response for a bandpass system. In this case, there are cutoff frequencies at each end of the range of frequencies encompassed by the system bandwidth. For the example shown, the half-power cutoff frequencies are 24 Hz and 40 Hz and the half-power system bandwidth is $40 - 24 = 16$ Hz.

Figure 6.5
Example Amplitude
Response for a
Bandpass System

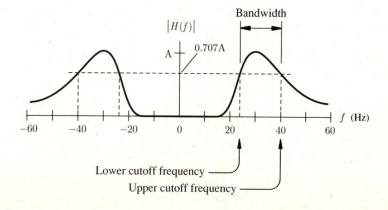

6.2 Frequency-Response Determination

The system frequency response is very useful in specifying system characteristics and in evaluating output-signal spectra; hence, we need methods for computing it. We could first solve the system differential equation for the system impulse response and then compute the Fourier transform of the impulse response. However, there are techniques with which we can compute the system frequency response without solving the system differential equation. We will now examine two of them.

Frequency-Response Determination from the System Differential Equation

We can determine the system frequency response directly from the system differential equation when the initial conditions are zero. This is accomplished by using the differentiation and linearity theorems (Theorems 9 and 1 in Chapter 4) to compute the Fourier transforms of the signals represented by each side of the system differential equation. Since these signals are equal and the Fourier transform is unique, then the Fourier transforms of these signals are equal. In future discussions, we will refer to this Fourier transform computation as the *Fourier transform of the differential equation* for simplicity. After we compute the Fourier transform of the differential equation, we then have an algebraic equation in terms of the spectra of the input signal and corresponding output signal. We solve this algebraic equation for the ratio of the spectrum of the output signal to the spectrum of the input signal, which is the frequency response. The following examples serve to illustrate this technique.

Example 6.4

Consider once again the simple RC filter circuit of Example 1.10, shown here in Figure 6.6. Find the frequency response of this circuit.

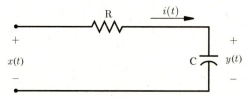

Figure 6.6 RC Filter Circuit for Example 6.4

Solution
Using circuit theory, we find that

$$Ri(t) + y(t) = x(t) \quad \text{and} \quad i(t) = C\frac{dy(t)}{dt}$$

Thus, the system differential equation is

$$RC\frac{dy(t)}{dt} + y(t) = x(t)$$

Using the time differentiation and linearity theorems, we compute the Fourier transform of the differential equation to be

$$(j2\pi fRC)Y(f) + Y(f) = X(f)$$

Therefore, the frequency response for the RC filter circuit is

$$H(f) = \frac{Y(f)}{X(f)} = \frac{1}{1 + j2\pi fRC}$$

We showed in Example 5.1 that the impulse response for the RC circuit of Example 6.4 is

$$h(t) = \frac{1}{RC}e^{-t/RC}u(t) \tag{6.15}$$

Thus, using Table C.2, we compute the system frequency response to be

$$H(f) = \mathcal{F}[h(t)] = \frac{1/RC}{1/RC + j2\pi f} = \frac{1}{1 + j2\pi fRC} \tag{6.16}$$

which verifies the result of Example 6.4.

Example 6.5

A linear, time-invariant, continuous-time control system is characterized by the system differential equation

$$\frac{d^3y(t)}{dt^3} + A\frac{d^2y(t)}{dt^2} + B\frac{dy(t)}{dt} + Cy(t) = D\frac{dx(t)}{dt} + Ex(t)$$

Find the system frequency response.

Solution
Using the time-differentiation theorem,

$$(j2\pi f)^3Y(f) + A(j2\pi f)^2Y(f) + B(j2\pi f)Y(f) + CY(f)$$
$$= D(j2\pi f)X(f) + EX(f)$$

Therefore,

$$H(f) = \frac{Y(f)}{X(f)} = \frac{E + j2\pi Df}{(C - 4\pi^2Af^2) + j2\pi f(B - 4\pi^2f^2)}$$

Frequency-Response Determination Using Phasor Concepts

We saw in Section 6.1 that the system frequency response $H(f)$ indicates the gain and phase shift supplied by a linear system with real impulse response to a sinusoidal input signal of frequency f. Therefore, we can determine a system's frequency response by computing the response of the system to a cosine signal, the frequency of which is f, where f is left as a variable, and thus encompasses all frequencies. Problems involving linear systems and sinusoidal signals are easily solved by using the phasor concept; therefore, as we will now illustrate, phasor concepts can be used to find the frequency response.

The output signal of a linear time-invariant system, the differential equation of which has real coefficients or, equivalently, the impulse response of which is real, is of the form

$$y(t) = |Y| \cos(2\pi f_1 t + \underline{/Y}) \tag{6.17}$$

when the input signal is of the form

$$x(t) = |X| \cos(2\pi f_1 t + \underline{/X}) \tag{6.18}$$

as we know from previous circuit-analysis studies. The phasor representations of these two signals are

$$\mathbf{X}_{f1} = |X| e^{j\underline{/X}} \tag{6.19}$$

and

$$\mathbf{Y}_{f1} = |Y| e^{j\underline{/Y}} \tag{6.20}$$

where the subscript $f1$ is used to indicate that the phasors in eqs. (6.19) and (6.20) represent sinusoidal signals with frequency $f = f_1$. From Table C.2, the spectrum of the input signal is

$$X(f) = \frac{|X|}{2} [e^{j\underline{/X}} \delta(f - f_1) + e^{-j\underline{/X}} \delta(f + f_1)] \tag{6.21}$$

The system frequency response is

$$H(f) = |H(f)| e^{j\underline{/H(f)}} \tag{6.22}$$

Therefore, the spectrum of the output signal is

$$
\begin{aligned}
Y(f) &= X(f)H(f) \\
&= \frac{|X||H(f_1)|}{2} e^{[j\underline{/X} + \underline{/H(f_1)}]} \delta(f - f_1) \\
&\quad + \frac{|X||H(-f_1)|}{2} e^{j[-\underline{/X} + \underline{/H(-f_1)}]} \delta(f + f_1)
\end{aligned} \tag{6.23}
$$

Since $h(t)$ is real, then eqs. (6.13) and (6.14) are true, and the spectrum of the output signal can be rewritten as

$$Y(f) = \frac{|X||H(f_1)|}{2}$$
$$\times \left\{ e^{[j\angle X + \angle H(f_1)]}\delta(f - f_1) + e^{-j[\angle X + \angle H(f_1)]}\delta(f + f_1) \right\} \qquad (6.24)$$

From Table C.2, we find that the output signal is

$$y(t) = |X||H(f_1)| \cos[2\pi f_1 t + \angle X + \angle H(f_1)] \qquad (6.25)$$

which means that the phasor representing the output signal is

$$\mathbf{Y}_{f1} = |X||H(f_1)|e^{j[\angle X + \angle H(f_1)]} \qquad (6.26)$$

Substituting eq. (6.19) in (6.26) and dividing both sides by $\mathbf{X}_{f1}$ yields

$$\mathbf{Y}_{f1}/\mathbf{X}_{f1} = |H(f_1)|e^{j\angle H(f_1)} = H(f_1) \qquad (6.27)$$

Since eq. (6.27) holds for any frequency, f, then

$$H(f) = \mathbf{Y}_f / \mathbf{X}_f \qquad (6.28)$$

and we can determine the system frequency response by computing the ratio of the phasor representing the output sinusoidal signal to the phasor representing the input sinusoidal signal for sinusoidal signals with the general frequency f.

Using the phasor concept for determining the frequency response eliminates the need for finding the system differential equation. However, we do need a technique for finding the ratio of the phasors representing the input and output signals.

If the system is represented by a block diagram, then the output–input phasor ratio is found by first transforming the block diagram. We replace all signals with their phasor representations and all differentiations and integrations with the multiplicative coefficients $j2\pi f$ and $1/j2\pi f$, respectively. We then solve the resulting algebraic equation for $\mathbf{Y}_f/\mathbf{X}_f$. The validity of this solution method is apparent. If $\mathbf{W}_f$ and $\mathbf{Z}_f$ are phasors that represent the cosine signals $w(t) = A\cos(2\pi ft + \theta)$ and $z(t) = B\cos(2\pi ft + \phi)$, respectively, then $\mathbf{W}_f + \mathbf{Z}_f$ represents $w(t) + z(t)$, $b\mathbf{W}_f$ represents $bw(t)$, $j2\pi f\mathbf{W}_f$ represents $dw(t)/dt = d[A\cos(2\pi ft + \theta)]/dt = -2\pi fA\sin(2\pi ft + \theta) = 2\pi fA\cos(2\pi ft + \theta + \pi/2)$, and $\mathbf{W}_f/j2\pi f$ represents $\int w(t)\,dt = \int A\cos[2\pi ft + \theta]\,dt = A\sin[2\pi ft + \theta]/2\pi f = A\cos[2\pi ft + \theta - \pi/2]/2\pi f$.

Example 6.6

Find the frequency response of the feedback temperature control system of Example 5.4 by using the phasor-analysis technique. Plot the amplitude and phase response for $K = 0.5$ and $K = 1.5$.

Solution

The system block diagram obtained from Example 5.4 and the transformed system block diagram are shown in Figure 6.7.[†]

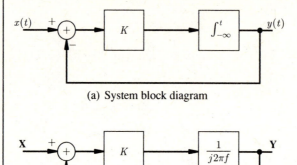

(a) System block diagram

(b) Transformed system block diagram

Figure 6.7 System Block Diagram and Transformed System Block Diagram for Example 6.6

Using the transformed block diagram to write the equation for the phasor representing the output signal yields $\mathbf{Y} = K(\mathbf{X} - \mathbf{Y})/j2\pi f$. Therefore, $\mathbf{Y}(K + 2j\pi f) = K\mathbf{X}$ and the system frequency response is

$$H(f) = \mathbf{Y}/\mathbf{X} = K/(K + 2j\pi f)$$

The amplitude response $|H(f)|$ and phase response $\underline{/H(f)}$ of the temperature control system are plotted in Figure 6.8.

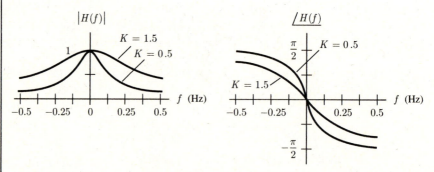

Figure 6.8 Amplitude and Phase Response for the Temperature Control System of Example 6.6

[†]The subscript f on the phasors representing the signals has been omitted for simplicity.

Solution

The transformed circuit diagram for the amplifier circuit is shown in Figure 6.12.

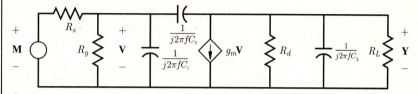

Figure 6.12 Transformed Circuit for Example 6.8

The two node equations for the transformed circuit are

$$(V - M)/R_s + (V - Y)j2\pi f C_2 + V\left[(1/R_g) + j2\pi f C_1\right] = 0$$

and

$$(Y - V)j2\pi f C_2 + g_m V + Y[(1/R_d) + (1/R_L) + j2\pi f C_3] = 0$$

For simplicity, we define $C_a = C_1 + C_2$, $C_b = C_2 + C_3$, $R_i = R_s R_g/(R_s + R_g)$, and $R_0 = R_d R_L/(R_d + R_L)$. The node equations then simplify to

$$[(1/R_i) + j2\pi f C_a]V - j2\pi f C_2 Y = M/R_s$$

and

$$[g_m - j2\pi f C_2]V = -[(1/R_0) + j2\pi f C_b]Y$$

Solving these two equations simultaneously for **Y/M** gives the amplifier frequency response

$$H(f) = Y/M = N(f)/D(f)$$

where

$$N(f) = -R_0 R_i [g_m - j2\pi f C_2]$$

and

$$D(f) = \left[R_s - 4\pi^2 f^2 R_s R_i R_0 \left(C_2^2 - C_a C_b\right)\right] + j2\pi f R_s [R_i C_a + R_0 C_b + g_m R_i R_0 C_2]$$

In the circuit model shown for the transistor amplifier in Example 6.8, G, D, and S are the locations of the transistor's gate, drain, and source. The circuit elements represent the following circuit properties: R_s is the resistance of the message source, R_g and R_d are the gate-to-source and drain-to-source circuit resistances, R_L is the load resistance, C_1 is the gate-to-source capacitance in the transistor and the input circuit-wiring capacitance, C_2 is the gate-to-drain capacitance in the transistor, and C_3 is the drain-to-source capacitance in the transistor and the output circuit-wiring capacitance.

6.4 Phase Delay and Group Delay

The phase response of a system supplies very useful information about the time delay of signals that are passed through the system. The reason is that the time delay of a sinusoidal signal corresponds to a phase shift of the signal—a correspondence we showed with eq. (3.4) in Chapter 3.

Phase Delay

We will first define the *system phase delay*, then show how it relates to the system phase response.

Definition ——————————————————————————————————

The time delay experienced by a single-frequency signal (that is, a sinusoidal signal) when the signal passes through a system is referred to as the system phase delay.

Now, let us assume that the system input signal is the single-frequency signal

$$x(t) = A\cos(2\pi f t + \theta) \tag{6.29}$$

and that the system frequency response is

$$H(f) = |H(f)|e^{j\underline{/H(f)}} \tag{6.30}$$

Then, from eqs. (6.10) and (6.11), the system output signal is

$$y(t) = A|H(f)|\cos(2\pi f t + \theta + \underline{/\,H(f)} \tag{6.31}$$

This signal can be written as

$$y(t) = A|H(f)|\cos\left\{2\pi f\left[t - t_p(f)\right] + \theta\right\} \tag{6.32}$$

where

$$t_p(f) = -\underline{/\,H(f)}/2\pi f \tag{6.33}$$

is the time delay experienced by the single-frequency signal with frequency f when it passes through the system. Thus, $t_p(f)$ is the system phase delay for a signal with frequency f. Its determination from the system phase response is illustrated in Figure 6.13. Note that a negative system phase response at positive frequencies indicates that a signal is delayed in time when it passes through the system, whereas a positive system phase response at positive frequencies indicates that a signal is advanced in time when it passes through the system.

If the system phase response for a range of frequencies lies on a straight line through the origin, then all signal components with frequencies in this range

Figure 6.13
Determination of
Phase Delay from
Phase Response

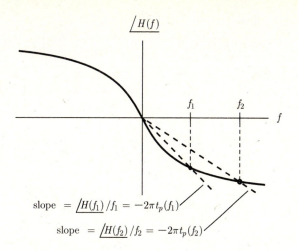

$$\text{slope} \;=\; \underline{/H(f_1)}\,/f_1 \;=\; -2\pi t_p(f_1)$$

$$\text{slope} \;=\; \underline{/H(f_2)}\,/f_2 \;=\; -2\pi t_p(f_2)$$

experience the same phase delay when passed through the system. That is, they are delayed by the same amount of time. We illustrate a phase response of this type in Figure 6.14. If an input signal contains only components with frequencies in the range lying on the straight line through the origin, then each signal component is delayed by the same amount of time in passing through the system. If, in addition, the amplitude response is constant over the range of frequencies, then the output signal is the equivalent of the input signal, only changed in amplitude and delayed in time.

Actually, the phase response illustrated in Figure 6.14 is not achievable in an actual system. However, the phase response illustrated in Figure 6.13 is achievable and produces an approximately constant phase delay at low frequencies, being nearly linear through the origin at low frequencies.

Figure 6.14
Example Phase
Response for System
with Constant Phase
Delay for a Range of
Frequencies

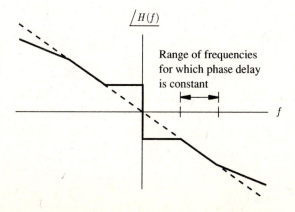

Range of frequencies
for which phase delay
is constant

Group Delay

When a bandlimited signal, $m(t)$, that contains only components with frequencies lower than f_1 is used to amplitude-modulate a cosine carrier signal, $x_c(t)$, with frequency f_c, then the modulated signal $x_m(t)$ contains the group of frequencies in the range $f_c - f_1$ to $f_c + f_1$, as illustrated in Figure 4.34. The time delay experienced by the modulating signal $m(t)$ when it passes through a system with the phase response $\angle H(f)$ is

$$t_g(f_c) = -\frac{1}{2\pi}\frac{d\angle H(f)}{df} \tag{6.34}$$

if the system amplitude response and the slope of the system phase response are constant over the range of frequencies encompassed by the modulated signal. Since the modulating signal is contained in a group of frequencies surrounding the carrier frequency, we call $t_g(f)$ the *system group delay*.

We now derive the equation that defines group delay (eq. [6.34]) by using single-frequency modulating and carrier signals with zero phase for simplicity. The assumed input signal is

$$x_m(t) = 2A\cos(2\pi\Delta f t)\cos(2\pi f t) \tag{6.35}$$

where f is the carrier frequency and Δf is the frequency of the single-frequency modulating signal. This input signal is illustrated in Figure 6.15. We assume that the system amplitude response and phase response are $|H(f)| = 1$ and $\angle H(f) = \phi(f)$, respectively. Unity system gain is assumed so that we can see the effect of phase response alone. The system phase response is shown in Figure 6.16.

We use the cosine-product trigonometric identity (see Appendix B, Table B.1) to rewrite the input signal as

$$x_m(t) = A\cos\left[2\pi(f - \Delta f)t\right] + A\cos\left[2\pi(f + \Delta f)t\right] \tag{6.36}$$

Figure 6.15
Input Signal for
Group-Delay
Derivation

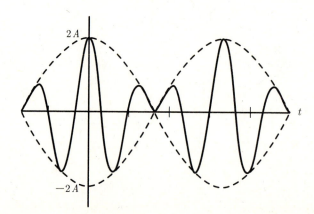

Figure 6.16
System Phase
Response for
Group-Delay
Derivation

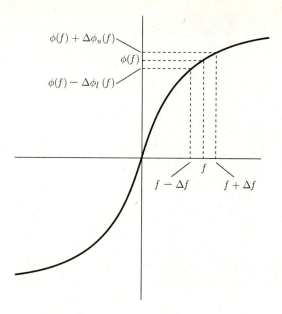

Equations (6.10) and (6.11) and the system phase response shown in Figure 6.16 are used to compute the system output signal

$$y(t) = A\cos\left[2\pi(f - \Delta f)t + \phi(f) - \Delta\phi_\ell(f)\right]$$

$$+ A\cos\left[2\pi(f + \Delta f)t + \phi(f) + \Delta\phi_u(f)\right] \qquad (6.37)$$

Regrouping terms in eq. (6.37) yields

$$y(t) = A\cos\left\{[2\pi ft + \phi(f)] - [2\pi\Delta ft + \Delta\phi_\ell(f)]\right\}$$

$$+ A\cos\left\{[2\pi ft + \phi(f)] + [2\pi\Delta ft + \Delta\phi_u(f)]\right\} \qquad (6.38)$$

For small Δf, the phase response is approximately linear from $f - \Delta f$ to $f + \Delta f$, and thus $\Delta\phi_u(f) \cong \Delta\phi\ell(f) \equiv \Delta\phi(f)$. We use the trigonometric identity again to write the approximate system output signal

$$y(t) \cong 2A\cos\left[2\pi\Delta ft + \Delta\phi(f)\right]\cos\left[2\pi ft + \phi(f)\right]$$

$$\cong 2A\cos\left\{2\pi\Delta f\left[t - t_m(f)\right]\right\}\cos\left\{2\pi f\left[t - t_c(f)\right]\right\} \qquad (6.39)$$

where

$$t_c(f) = -\phi(f)/2\pi f \qquad (6.40)$$

is the time delay of the carrier signal and

$$t_m(f) = -\Delta\phi(f)/2\pi\Delta f \qquad (6.41)$$

is the time delay of the modulating signal. These time delays are illustrated by the system output signal shown in Figure 6.17.

Since $\phi(f) = \underline{/\,H(f)}$, then

$$t_c(f) = -\underline{/\,H(f)}/2\pi f = t_p(f) \qquad (6.42)$$

Figure 6.17
Output Signal for
Group-Delay
Derivation

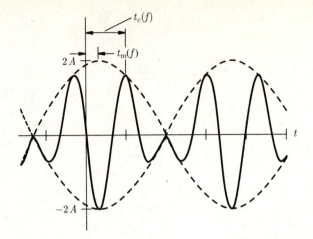

Therefore, since the carrier signal is a single-frequency signal, the carrier-signal time delay is the system phase delay at the carrier frequency. When $\Delta f \to 0$, then

$$t_m(f) \to -\frac{1}{2\pi}\frac{d\phi(f)}{df} = -\frac{1}{2\pi}\frac{d\,\underline{/H(f)}}{df} = t_g(f) \qquad (6.43)$$

where $t_g(f)$ is the group delay defined earlier. If the slope of the system phase response is constant over the frequency range $f - \Delta f$ to $f + \Delta f$, then the group delay is the time delay of the modulating signal, as asserted earlier. In most actual systems, the phase response is curved so that $t_m(f)$ is approximately equal to $t_g(f)$ only for values of Δf that are sufficiently small to allow for an approximately constant slope of the phase response over the frequency range $f - \Delta f$ to $f + \Delta f$. If the modulating signal contains more than one frequency, then the slope of the system phase response must be approximately constant over the frequency range encompassed by the resulting modulated signal (that is, the frequency range $f_c - f_1$ to $f_c + f_1$ for the modulated signal illustrated in Figure 4.34).

Example 6.9

Find the phase delay and group delay for the simple RC filter of Example 6.4 for the frequencies $f = 1/2\pi RC$ and $f = 1/\pi RC$.

Solution
From Example 6.4, the frequency response of the RC circuit is

$$H(f) = 1/(1 + j2\pi fRC)$$

Thus, the phase response is

$$\underline{/H(f)} = -\tan^{-1} 2\pi fRC$$

The phase delay is

$$t_p(f) = -\underline{/\,H(f)}\,/2\pi f = [\tan^{-1}(2\pi f RC)]/2\pi f$$

The group delay is

$$t_g(f) = -\frac{1}{2\pi}\frac{d\underline{/\,H(f)}}{df} = \frac{1}{2\pi}\frac{d[\tan^{-1}(2\pi f RC)]}{df}$$

$$= RC/\sqrt{1 + (2\pi f RC)^2}$$

At $f = 1/2\pi RC$,

$$t_p(1/2\pi RC) = (\pi/4)/(1/RC) = 0.785RC$$

and

$$t_g(1/2\pi RC) = RC/\sqrt{2} = 0.707RC$$

At $f = 1/\pi RC$

$$t_p(1/\pi RC) = (1.107)/(2/RC) = 0.554RC$$

and

$$t_g(1/\pi RC) = RC/\sqrt{5} = 0.447RC$$

6.5 Bode Plots of Amplitude and Phase Response

After we determine the frequency-response expression for a system, we can program it on a computer for computation and plotting of the system amplitude and phase responses. Particularly convenient forms of these plots are called *Bode plots* after H.W. Bode who first used them in system analysis.[†] They are plotted only for positive frequencies, because doing so produces all the frequency-response information for the systems with real impulse responses considered here. For simplicity, we choose the frequency variable $\omega = 2\pi f$ in radians/second, rather than f in hertz. By following this normally used convention, we avoid a number of 2π factors, at the same time maintaining consistency with most of the literature students are likely to encounter. Thus, we use the notation

$$H_\omega(\omega) = H(f)\big|_{f=\omega/2\pi} \tag{6.44}$$

The Bode plots for amplitude and phase responses are both plotted as a function of $\log(\omega) \equiv \log_{10}(\omega)$ which facilitates plotting and provides a compact representation of the frequency response over a wide frequency range. For

[†]H. W. Bode, *Network Analysis and Feedback Amplifier Design* (Princeton, N.J.: D. Van Nostrand, 1945).

ease in reading, the scale shown is ω rather than log (ω), meaning that it is logarithmic. We show the log (ω) and ω frequency scales for an example amplitude-response Bode plot in Figure 6.18.

The amplitude-response Bode plot (or gain Bode plot) is plotted as $|H_\omega(\omega)|_{dB}$ where

$$|H_\omega(\omega)|_{\mathrm{dB}} = 20\log|H_\omega(\omega)| = 20\log\left|H\left(\frac{\omega}{2\pi}\right)\right| \qquad (6.45)$$

is the system amplitude response expressed in decibels (dB). We illustrate this in Figure 6.18. The use of a decibel amplitude scale permits compact representation over a wide range of amplitude-response values. It also permits easier plotting because the decibel amplitude of multiplicative terms in the frequency response can be graphically added rather than multiplied to generate the complete amplitude response. Furthermore, the Bode plot of phase response is plotted directly because it does not vary over a large range of values and also because the phases of multiplicative terms in the frequency response add. It is usually plotted in degrees.

The frequency response of a lumped-parameter, linear, time-invariant system as a function of ω is

$$H_\omega(\omega) = \frac{a_\alpha(j\omega)^\alpha + a_{\alpha-1}(j\omega)^{\alpha-1} + \cdots + a_\gamma(j\omega)^\gamma}{b_\beta(j\omega)^\beta + b_{\beta-1}(j\omega)^{\beta-1} + \cdots + b_\epsilon(j\omega)^\epsilon} \qquad (6.46)$$

where $\alpha > \gamma$ and $\beta > \epsilon$. This can be readily shown by computing the Fourier transform of the system differential equation and solving for $H_\omega(\omega)$. Systems with real impulse responses are the only type with which we are concerned; therefore, the coefficients a_α through a_γ and b_β through b_ϵ are all real. We can factor $a_\gamma(j\omega)^\gamma$ and $b_\epsilon(j\omega)^\epsilon$ from the numerator and denominator, respectively. The remaining numerator and denominator polynomials in $j\omega$ can then be fac-

Figure 6.18
Example
Amplitude-Response
Bode Plot Illustrating
log(ω) and ω Scales

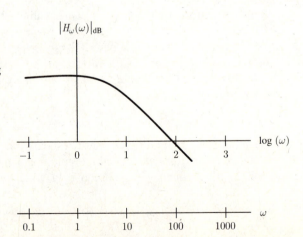

tored. It is possible that some of the factors may correspond to complex roots of the polynomials, and thus contain a complex coefficient. We can avoid complex coefficients because they always occur as complex-conjugate pairs in the set of factors, which results from the original polynomial's coefficients being real. Therefore, the factors containing the complex-conjugate coefficients produce a quadratic factor with real coefficients when multiplied together, which allows us to write the frequency response as

$$H_\omega(\omega) = [C_1](j\omega)^{\pm N} \prod_{i=1}^{I} L_i(\omega) \prod_{k=1}^{K} Q_k(\omega) \tag{6.47}$$

where C_1 is constant, $N = |\gamma - \epsilon|$, the linear factors are

$$L_i(\omega) = [(j\omega/\omega_{\ell i}) + 1]^{\pm M_i} \tag{6.48}$$

and the quadratic factors are

$$Q_k(\omega) = [(j\omega/\omega_{qk})^2 + 2\zeta_k(j\omega/\omega_{qk}) + 1]^{\pm P_k} \tag{6.49}$$

Positive exponents correspond to numerator factors and negative exponents correspond to denominator factors. Also $\zeta_k > 0$ if the frequency response exists (that is, if the system is stable) and $\zeta_k < 1$ because otherwise the quadratic factor could be factored into two linear factors with real coefficients. Note that repeated factors occur when M_i or P_k are greater than 1.

Example 6.10

Write the following frequency response in factored form.

$$H_\omega(\omega) = \frac{10(j\omega)^3 + 40(j\omega)^2 + 40(j\omega)}{(j\omega)^5 + 11(j\omega)^4 + 46(j\omega)^3 + 36(j\omega)^2}$$

Solution

Factoring,

$$H_\omega(\omega) = \frac{40(j\omega)[(j\omega/2)^2 + (j\omega) + 1]}{36(j\omega)^2[(j\omega) + 1][(j\omega/6)^2 + (10/36)(j\omega) + 1]}$$

$$= \frac{40[(j\omega/2) + 1]^2}{36(j\omega)[(j\omega) + 1][(j\omega/6)^2 + (10/6)(j\omega/6) + 1]}$$

Therefore,

$$H_\omega(\omega) = \left(\frac{10}{9}\right)(j\omega)^{-1}[(j\omega/2) + 1]^{+2}$$

$$\times [(j\omega + 1)]^{-1}[(j\omega/6)^2 + 2(5/6)(j\omega/6) + 1]^{-1}$$

From eq. (6.47) the amplitude response in decibels is

$$|H_\omega(\omega)|_{dB} = 20 \log |H_\omega(\omega)|$$

$$= 20 \log |C_1| + 20 \log |(j\omega)^{\pm N}|$$

$$+ \sum_{i=1}^{I} 20 \log |L_i(\omega)| + \sum_{k=1}^{K} 20 \log |Q_k(\omega)| \qquad (6.50)$$

and the phase response is

$$\underline{/\,H_\omega(\omega)} = \underline{/\,C_1} + \underline{/\,(j\omega)^{\pm N}} + \sum_{i=1}^{I} \underline{/\,L_i(\omega)} + \sum_{k=1}^{K} \underline{/\,Q_k(\omega)} \qquad (6.51)$$

To produce the Bode plots of amplitude and phase response, we compute and plot eqs. (6.50) and (6.51) as a function of $\log(\omega)$, considering each of the four types of factors separately. Then we obtain the total by graphical addition. In fact, we can obtain straight-line approximations to the plots for each factor. These straight-line approximate plots are easily added to produce the straight-line approximations $|H_\omega(\omega)|_s$ and $\underline{/\,H_\omega(\omega)}_s$ to the amplitude- and phase-response Bode plots, respectively. We can then produce smooth and approximate plots for the Bode plots by identifying key points of deviation from the straight lines and plotting a smooth curve through them. This procedure is illustrated in the following discussion and example. It provides a simple method for quickly sketching approximate Bode plots of the system amplitude and phase responses without point-by-point plotting and permits easy determination of which frequency-response factors significantly affect the frequency response at each frequency.

Constant Factor

The first factor we will consider is the *constant factor*

$$C(\omega) = C_1 \qquad (6.52)$$

For this factor,

$$|C(\omega)|_{dB} = 20 \log |C(\omega)| = 20 \log |C_1| \qquad (6.53)$$

$$\underline{/\,C(\omega)} = \begin{cases} 0 & C_1 > 0 \\ \pm 180^\circ & C_1 < 0 \end{cases} \qquad (6.54)$$

The amplitude- and phase-response Bode plots for this factor are shown in Figure 6.19.

Power-of-$j\omega$ Factor

The second type of frequency-response factor we will consider is the *power-of-$j\omega$ factor*

$$\Omega(\omega) = (j\omega)^{\pm N} \qquad (6.55)$$

Figure 6.19
Amplitude- and
Phase-Response
Bode Plots for a
Constant Factor

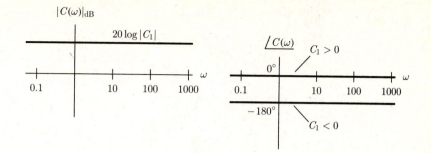

where a positive exponent corresponds to a numerator factor and a negative exponent corresponds to a denominator factor. For this factor,

$$|\Omega(\omega)|_{dB} = 20 \log |(j\omega)^{\pm N}| = \pm 20 N \log \omega$$

$$\underline{/\Omega(\omega)} = \underline{/(j\omega)^{\pm N}} = \pm 90 N°$$

(6.56)

The amplitude-response Bode plot is a straight line passing through 0_{dB} when $\omega = 1$ with a slope of $\pm 20N$ dB per decade (factor-of-10) change in frequency. The phase response is constant for all frequencies. The amplitude- and phase-response Bode plots for this factor are shown in Figure 6.20.

Linear Factors

The number of different *linear factors* is I, as indicated by eqs. (6.47) and (6.50). These factors are of the form

$$L_i(\omega) = [(j\omega/\omega_{\ell i}) + 1]^{\pm M_i}$$

(6.57)

as shown in eq. (6.48).

Figure 6.20
Amplitude- and
Phase-Response
Bode Plots for a
Power-of-$j\omega$ Factor

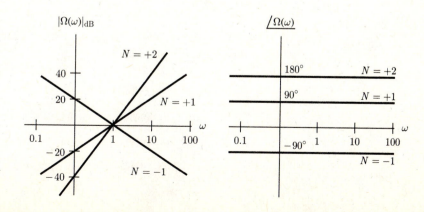

The amplitude response for a linear factor is given by

$$|L_i(\omega)|_{dB} = 20\log\left|[(j\omega/\omega_{\ell i}) + 1]^{\pm M_i}\right|$$
$$= \pm 20\,M_i\log\left|(j\omega/\omega_{\ell i}) + 1\right| \tag{6.58}$$

where the plus and minus signs correspond to numerator and denominator factors, respectively. For $\omega \ll \omega_{\ell i}$, $|L_i(\omega)|_{dB}$ asymptotically approaches the straight line

$$\pm 20\,M_i\log|1| = 0 \tag{6.59}$$

For $\omega \gg \omega_{\ell i}$, $|L_i(\omega)|_{dB}$ asymptotically approaches the straight line

$$\pm 20\,M_i\log|j\omega/\omega_{\ell i}| = \pm 20\,M_i[\log\omega - \log\omega_{\ell i}] \tag{6.60}$$

which has a slope of $\pm 20\,M_i$ dB per decade of frequency change. It passes through 0 dB when $\omega = \omega_{\ell i}$. The two straight-line asymptotes to the amplitude-response Bode plot for a linear factor intersect at $\omega = \omega_{\ell i}$ because their values are both zero at this frequency. The asymptotes form the straight-line approximation $|L_1\omega|_s$ to the amplitude-response Bode plot. The slope of $|L_1\omega|_s$ changes at the frequency $\omega = \omega_{\ell i}$. We refer to this frequency as the *break* (or *corner*) *frequency* for the linear factor.

The actual amplitude-response Bode plot for a linear factor differs most from the asymptotic straight-line approximation to it at the break frequency. The actual amplitude-response value at the break frequency is

$$\pm 20\,M_i\log|j1 + 1| = \pm 20\,M_i\log\left(\sqrt{2}\right) \cong \pm 3M_i \text{ dB} \tag{6.61}$$

We use the actual response value at the break frequency and the two straight-line asymptotes to easily sketch the amplitude-response Bode plot for the linear factor. The sketched amplitude response $|L_i\omega|_{dB}$ and straight-line approximation to it $|L_i\omega|_s$ are illustrated in Figure 6.21 for both a numerator factor and a denominator factor.

The phase response for a linear factor is given by

$$\underline{/L_i(\omega)} = \underline{/[(j\omega/\omega_{\ell i}) + 1]^{\pm M_i}}$$
$$= \pm M_i\underline{/(j\omega/\omega_{\ell i}) + 1} \tag{6.62}$$

where the plus and minus signs correspond to numerator and denominator terms, respectively. For $\omega \ll \omega_{\ell i}$, $\underline{/L_i(\omega)}$ asymptotically approaches the straight line

$$\pm M_i\underline{/j0 + 1} = 0° \tag{6.63}$$

For $\omega \gg \omega_{\ell i}$, $\underline{/L_i(\omega)}$ asymptotically approaches the straight line

$$\pm M_i\underline{/j\omega/\omega}_i = \pm 90 M_i° \tag{6.64}$$

At the break frequency, $\omega_{\ell i}$, the phase-response value for the factor is

$$\pm M_i\underline{/j1 + 1} = \pm 45 M_i° \tag{6.65}$$

Figure 6.21
Amplitude- and
Phase-Response
Bode Plots and
Straight-Line
Approximate Plots
for a Linear Factor

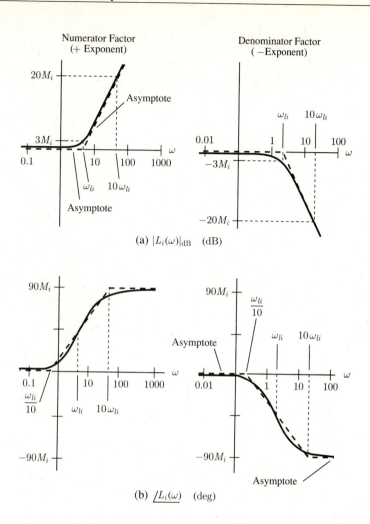

(a) $|L_i(\omega)|_{dB}$ (dB)

(b) $\underline{/L_i(\omega)}$ (deg)

We connect the two phase-response asymptotes by a straight line from $0°$ at $\omega = \omega_{\ell i}/10$ to $\pm 90 M_i^{\circ}$ at $\omega = 10\omega_{\ell i}$ to produce a complete straight-line approximation $\underline{/L_i(\omega)}_s$ to the phase-response Bode plot $\underline{/L_i(\omega)}$ for a linear factor. The connecting line passes through the proper value at $\omega = \omega_{\ell i}$ and approximately minimizes the maximum phase-difference magnitude between the actual phase-response plot and its straight-line approximation. The straight-line approximation produced gives a phase difference magnitude with respect to the phase response of less than $6M_i^{\circ}$. The maximum phase difference occurs at $\omega = \omega_{\ell i}/10$, $\omega = 10\omega_{\ell i}$, and at two points between these frequency values. This maximum phase difference is apparent in Figure 6.21, where the phase-response Bode plot $\underline{/L_i(\omega)}$ and straight-line approximation $\underline{/L_i(\omega)}_s$ for both numerator and denominator linear factors are shown.

Quadratic Factors

The number of different *quadratic factors* is K, as indicated by eqs. (6.47) and (6.50). Equation (6.49) shows that these factors are of the form

$$Q_k(\omega) = [(j\omega/\omega_{qk})^2 + 2\zeta_k(j\omega/\omega_{qk}) + 1]^{\pm P_k} \qquad (6.66)$$

The parameter ζ_k is called the *damping ratio* of the quadratic factor. The parameter ω_{qk} is the break frequency and is called the *undamped natural frequency* of the quadratic factor.

We will consider only single-degree factors ($P_k = 1$) here. If multiple-degree factors occur, then the results from single-degree factors are added.

The amplitude-response Bode plot for a quadratic factor is given by

$$|Q_k(\omega)|_{dB} = 20 \log |[(j\omega/\omega_{qk})^2 + 2\zeta_k(j\omega/\omega_{qk}) + 1]^{\pm 1}|$$

$$= \pm 20 \, \log |(j\omega/\omega_{qk})^2 + 2\zeta_k(j\omega/\omega_{qk}) + 1| \qquad (6.67)$$

where the plus and minus signs correspond to numerator and denominator factors, respectively. For $\omega \ll \omega_{qk}$, $|Q_k(\omega)|_{dB}$ asymptotically approaches the straight line

$$\pm 20 \log |1| = 0 \qquad (6.68)$$

For $\omega \gg \omega_q k$, $|Q_k(\omega)|_{dB}$ asymptotically approaches the straight line

$$\pm 20 \, \log |(j\omega/\omega_{qk})^2| = \pm 40[\log \omega - \log \omega_{qk}] \qquad (6.69)$$

which has a slope of ± 40 dB per decade of frequency change. It passes through 0 dB when $\omega = \omega_{qk}$. The two straight-line asymptotes to the amplitude-response Bode plot for a quadratic factor intersect at the break frequency $\omega = \omega_{qk}$, their values both being zero at this frequency. They form the straight-line approximation $|Q_k(\omega)|_s$ to the amplitude-response Bode plot $|Q_k(\omega)|_{dB}$ for the quadratic factor.

The actual amplitude-response Bode plot for a quadratic factor can differ significantly from the straight-line approximate plot in the vicinity of the break frequency. How much it differs depends on the damping ratio ζ_k (recall that $0 < \zeta_k < 1$). In Figure 6.22, we show amplitude-response Bode plots $|Q_k(\omega)|_{dB}$ and the straight-line asymptotic approximation to them $|Q_k(\omega)|_s$ for denominator factors with several values of ζ_k as a function of frequency normalized by ω_{qk}. The use of normalized frequency allows us to use these curves for denominator quadratic factors with any break frequency. If these curves are inverted, they then represent a numerator factor. If we differentiate eq. (6.67) and set it equal to zero, we find that the amplitude-response peak shown in Figure 6.22 occurs at

$$\omega_{pk} = \omega_{qk}\sqrt{1 - 2\zeta_k^2} \qquad (6.70)$$

where ω_{pk} is called the *maximum-response resonance frequency* or *peak frequency* of the quadratic factor. No peak occurs for $\zeta_k > 1/\sqrt{2}$, the result of

Figure 6.22
Amplitude-Response
Bode Plot and
Straight-Line
Approximate Plot for
Denominator
Quadratic Factors

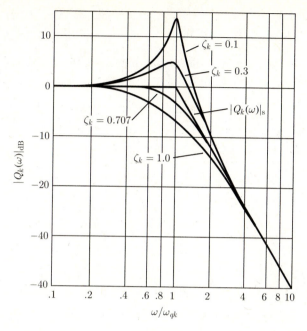

eq. (6.70) being the square root of a negative number. Solving for the amplitude-response Bode plot at $\omega = \omega_{pk}$ shows that the height of the peak with respect to 0 dB is

$$M_{pk} = \pm 20 \log \left[2\zeta_k \sqrt{1 - \zeta_k^2} \right] \qquad \text{(decibels)} \qquad (6.71)$$

where $0 \leq \zeta_k \leq 1/\sqrt{2}$ and the plus and minus signs correspond to numerator and denominator factors, respectively. The argument of the logarithm in eq. (6.71) is less than unity for $0 < \zeta_k < 1/\sqrt{2}$; therefore, the logarithm is negative. This means that a numerator factor has a negative peak and a denominator factor has a positive peak, as shown in Figure 6.22. Knowing the peak point and the asymptotes is usually sufficient to define an approximate sketch of the amplitude-response Bode plot for a quadratic factor. If no peak occurs (that is, if $\zeta_k > 1/\sqrt{2}$), then we can use the value of the amplitude response at $\omega = \omega_{qk}$ for sketching purposes. This value is

$$|Q_k(\omega_{qk})|_{dB} = \pm 20 \log |j2\zeta_k| = \pm 20 \log(2\zeta_k) \qquad (6.72)$$

The phase response for a quadratic factor is given by

$$\angle Q_k(\omega) = \angle [(j\omega/\omega_{qk})^2 + 2\zeta_k(j\omega/\omega_{qk}) + 1]^{\pm 1}$$
$$= \pm \angle (j\omega/\omega_{qk})^2 + 2\zeta_k(j\omega/\omega_{qk}) + 1 \qquad (6.73)$$

where the plus and minus signs correspond to numerator and denominator terms, respectively. We show the Bode plot of this phase response as a function of frequency normalized by ω_{qk} in Figure 6.23 for denominator factors with several different values of ζ_k. Once again, inversion of the curves produces the phase-response curves for a numerator factor.

We find the straight-line approximation to the phase-response Bode plot for a quadratic factor as follows. For $\omega \ll \omega_{qk}$, $\underline{/\,Q_k(\omega)}$ asymptotically approaches the straight line

$$\pm \underline{/\,j0 + 1} = 0° \tag{6.74}$$

For $\omega \gg \omega_{qk}$, $\underline{/\,Q_k(\omega)}$ asymptotically approaches the straight line

$$\pm \underline{/\,(j\omega/\omega_{qk})^2} = \pm \underline{/\,-\omega^2/\omega_{qk}^2} = \pm 180° \tag{6.75}$$

These two asymptotes form two segments of the straight-line approximation. We connect them with a center straight-line segment that passes through $0°$ at $\omega = \omega_L \equiv (10)^{-a}\omega_{qk}$ and $\pm 180°$ at $\omega = \omega_U \equiv (10)^a\omega_{qk}$, where

$$a = \begin{cases} 1.410\zeta_k - 0.150\zeta_k^2 & \zeta_k < 0.2 \\ 1.475\zeta_k - 0.475\zeta_k^2 & \zeta_k > 0.2 \end{cases} \tag{6.76}$$

This straight line passes through $\pm 90°$ at $\omega = \omega_{qk}$, which is the correct value of $\underline{/\,Q_k(\omega_{qk})}$. Thus, we need only to compute either ω_L or ω_U to define the

Figure 6.23
Phase-Response
Bode Plots for
Denominator
Quadratic Factor

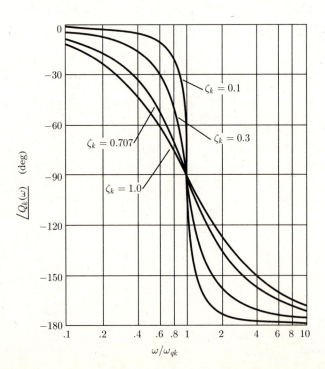

straight line required to complete the straight-line approximation because only one intersection point and the point at $\pm 90°$ and $\omega = \omega_{qk}$ specify the straight line. We selected the expressions in eq. (6.76) for parameter a, so that the defined straight-line approximation to the phase-response Bode plot would approximately minimize the maximum phase-difference magnitude between the actual phase-response curve and the straight-line approximation to it. Examples of two phase-response Bode plots $\underline{/\,Q_k(\omega)}$ for quadratic factors and their straight-line approximations $\underline{/\,Q_k(\omega)}_s$ are shown in Figure 6.24.

Bode Plots from Straight-Line Approximations

We will now present an example in which Bode plots of amplitude and phase responses and their straight-line approximations are found and plotted. In the discussion following the example, we will explain how the straight-line approximations can be used to deduce information about the amplitude and phase responses and to generate quick sketches of the amplitude- and phase-response Bode plots.

Figure 6.24
Phase-Response
Bode Plots and
Straight-Line
Approximate Plots
for Two Denominator
Quadratic Factors

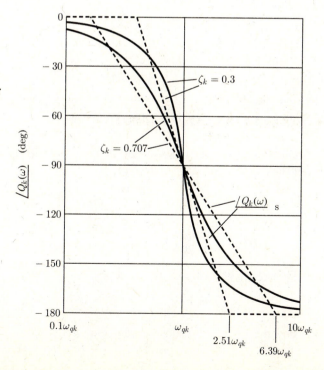

Example 6.11

Find the amplitude- and phase-response Bode plots for an amplifier with the following frequency response:

$$H_\omega(\omega) = \frac{45000[(j\omega)^2 + 18(j\omega) + 900]}{[j\omega + 90]^2[j\omega + 1000]}$$

Solution

$$H_\omega(\omega) = \frac{(45000)(900)[(j\omega/30)^2 + (18/30)(j\omega/30) + 1]}{(90)^2(1000)[(j\omega/90) + 1]^2[(j\omega/1000) + 1]}$$

$$= 5[(j\omega/30)^2 + 2(0.3)(j\omega/30) + 1][(j\omega/90) + 1]^{-2}[(j\omega/1000) + 1]^{-1}$$

For the constant factor, $C(\omega) = 5$,

$$|C(\omega)|_{dB} = 20\log|5| = 13.979 \text{ dB} \qquad \underline{/C(\omega)} = \underline{/5} = 0°$$

For the first linear factor, $L_1(\omega) = [(j\omega/90) + 1]^{-2}$,

- Break frequency: $\omega_{\ell1} = 90$ rad/s
- Amplitude-asymptote slope for $\omega > \omega_{\ell1}$: $-20(2) = -40$ dB/decade
- Amplitude response at $\omega = \omega_{\ell1}$: $-3(2) = -6$ dB
- Phase asymptote for $\omega > 10\omega_{\ell1}$: $-90(2) = -180°$

For the second linear factor, $L_2(\omega) = [(j\omega/1000) + 1]^{-1}$,

- Break frequency: $\omega_{\ell2} = 1000$ rad/s
- Amplitude-response slope for $\omega > \omega_{\ell2}$: $-20(1) = -20$ dB/decade
- Amplitude response at $\omega = \omega_{\ell2}$: $-3(1) = -3$ dB
- Phase asymptote for $\omega > 10\omega_{\ell2}$: $-90(1) = -90°$

For the quadratic factor, $Q_1(\omega) = [(j\omega/30)^2 + 2(0.3)(j\omega/30) + 1]$,

- Break frequency: $\omega_{q1} = 30$ rad/s
- $\zeta_1 = 0.3$
- Amplitude-asymptote slope for $\omega > 10\omega_{q1}$: $+40$ dB/decade
- Location of amplitude-response peak:

$$\omega_{p1} = 30\sqrt{1 - 2(0.3)^2} = 27.166 \text{ rad/s}$$

- Height of amplitude-response peak:

$$M_{p1} = 20\log\left[2(0.3)\sqrt{1 - (0.3)^2}\right] = -4.847 \text{ dB}$$

- Phase asymptote for $\omega > 10^a\omega_{q1}$: $180°$
- Exponent a: $1.475(0.3) - 0.475(0.3)^2 = 0.4$
- Straight-line phase approximation corner: $\omega_L = 10^{(-0.4)}(30) = 11.94$ rad/s

Using the preceding data, we plot the straight-line approximations to the amplitude- and phase-response Bode plots for each frequency-response factor.

We then add these straight-line approximations to produce the straight-line approximations $|H_\omega(\omega)|_s$ and $\underline{/\,H_\omega(\omega)}_s$ to the Bode plots. Straight lines are being added, simplifying the addition. We begin from the left and add or subtract slope, as indicated by the additional straight lines encountered. This is similar to the construction of piecewise-defined signals from step and ramp signals that we discussed in Chapter 3, Section 3.3. Straight-line approximations for the individual factors and $|H_\omega(\omega)|_s$ and $\underline{/\,H_\omega(\omega)}_s$ are shown in Figure 6.25. We also show the actual plots of amplitude response and phase response, which we compute directly from the frequency response, eq. (6.46). The amplifier gain in the frequency range from approximately $\omega_{\ell 1}$ to approximately $\omega_{\ell 2}$ is referred to as the *amplifier midband gain*.

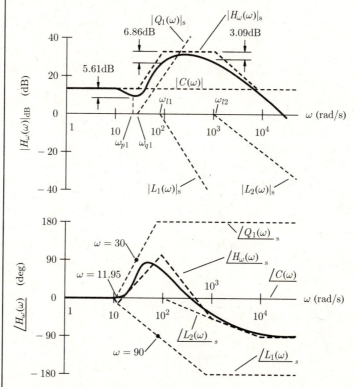

Figure 6.25 Amplitude- and Phase-Response Bode Plots and Straight-Line Approximate Plots for Example 6.11

Example 6.11 shows us that the straight-line approximate plots give a reasonable indication of the amplitude- and phase-response characteristics. From these plots, we also see that the break frequencies indicate the approximate frequencies at which changes occur in the slope of the amplitude-response Bode plot. In addition, the frequencies corresponding to the ends of the center straight-line segment of the straight-line approximations for the individual factors indicate the approximate frequencies at which changes occur in the slope of the phase-response Bode plot.

At the break frequency $\omega_{\ell2} = 1000$ rad/s in Figure 6.25, we observe that $|H_\omega(\omega)|_{dB} - |H_\omega(\omega)|_s = -3.09$ dB. This difference is the sum of $|L_2(\omega)|_{dB} - |L_2(\omega)|_s = -3$ dB, $|L_1(\omega)|_{dB} - |L_1(\omega)|_s$, and $Q_1(\omega)|_{dB} - |Q_1(\omega)|_s$ at $\omega = \omega_{\ell2}$. Similar sums apply at $\omega = \omega_{\ell1}$ and $\omega = \omega_{p1}$.

In Example 6.11, the value of $|H_\omega(\omega)|_{dB} - |H_\omega(\omega)|_s$ at each break or peak frequency is nearly the same as the value calculated for the individual factor corresponding to the break or peak frequency because the frequencies are far enough apart. Therefore, we can make a reasonably accurate, quick sketch of the amplitude-response Bode plot for this example by first plotting points at the break and peak frequencies that differ from the straight-line approximation by the calculated individual-factor amplitude differences and then using these points and the straight-line asymptotes to guide the sketch.

If the break and peak frequencies are fairly close together, a reasonably accurate, quick sketch of the amplitude response can be made using the straight-line approximation together with actual amplitude-response values computed at the break and peak frequencies. Thus, only a few computed values are required rather than values for the entire curve.

In many cases, however, quick sketching of the phase-response Bode plot from its straight-line approximation is not so easy to perform with any degree of accuracy. The problem is that the phase deviation from the straight-line approximation remains larger for a wider frequency range and is related to the straight-line approximation in a more complicated fashion. But, as mentioned previously, we can obtain a fairly good indication of the shape of the phase-response curve from the straight-line approximation. If we look at it carefully, we can usually note a few frequencies at which the actual phase response can be computed for use with the straight-line approximation to provide a reasonably accurate sketch.

Example 6.12

Find the straight-line approximations and approximate Bode plots of the amplitude and phase response for a system with the following frequency response:

$$H_\omega(\omega) = \frac{20(j\omega)^2 + 2000j\omega}{(j\omega)^2 + 210j\omega + 2000}$$

Solution

$$H_\omega(\omega) = \frac{20(j\omega)(j\omega + 100)}{(j\omega + 10)(j\omega + 200)}$$

$$= (j\omega)[(j\omega/10) + 1]^{-1}[(j\omega/100) + 1][(j\omega/200) + 1]^{-1}$$

For the $j\omega$ factor,

$$|\Omega(\omega)|_{dB} = 20\log\omega \qquad \underline{/\,\Omega(\omega)} = 90°$$

For the first linear factor, $L_1(\omega) = [(j\omega/10) + 1]^{-1}$,

- Break frequency: $\omega_{\ell 1} = 10$ rad/s
- Amplitude-asymptote slope for $\omega > \omega_{\ell 1}$: $-20(1) = -20$ dB/decade
- Amplitude response at $\omega = \omega_{\ell 1}$: $-3(1) = -3$ dB
- Phase asymptote for $\omega > 10\omega_{\ell 1}$: $-90(1) = -90°$

For the second linear factor, $L_2(\omega) = [(j\omega/100) + 1]$,

- Break frequency: $\omega_{\ell 2} = 100$ rad/s
- Amplitude-asymptote slope for $\omega > \omega_{\ell 2}$: $20(1) = 20$ dB/decade
- Amplitude response at $\omega = \omega_{\ell 2}$: $3(1) = 3$ dB
- Phase asymptote for $\omega > 10\omega_{\ell 2}$: $90(1) = 90°$

For the third linear factor, $L_3(\omega) = [(j\omega/200) + 1]^{-1}$,

- Break frequency: $\omega_{\ell 3} = 200$ rad/s
- Amplitude-asymptote slope for $\omega > \omega_{\ell 2}$: $-20(1) = -20$ dB/decade
- Amplitude response at $\omega = \omega_{\ell 3}$: $-3(1) = -3$ dB
- Phase asymptote for $\omega > 10\omega_{\ell 3}$: $-90(1) = -90°$

We first form the straight-line approximation sketches to the Bode plots by adding those due to each factor. The factor contributions and sum are shown in Figure 6.26. It is clear that the largest two break frequencies are too close together to permit us to use the computed amplitude-response values (shown by x's) for individual factors at their break frequencies as aids in sketching the amplitude-response Bode plot. Thus, we compute the actual amplitude response at these frequencies. The actual phase response is computed at frequencies where the straight-line approximation to the phase response changes slope to aid in sketching the phase-response Bode plot. The computed values are

$$|H_\omega(100)|_{dB} = 22.0 \text{ dB} \qquad |H_\omega(200)|_{dB} = 24.0 \text{ dB}$$

and $\underline{/H_\omega(1)} = 84.6°$, $\underline{/H_\omega(10)} = 47.8°$, $\underline{/H\omega(20)} = 32.2°$, $\underline{/H_\omega(100)} = 24.1°$, $\underline{/H_\omega(1000)} = 6.2°$.

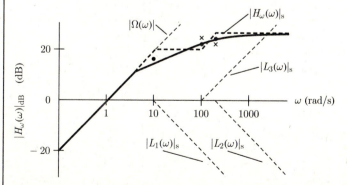

Figure 6.26 Approximate Amplitude- and Phase-Response Bode Plots and Straight-Line Approximate Plots for Example 6.11

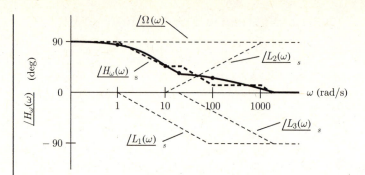

Figure 6.26 Approximate Amplitude- and Phase-Response Bode Plots and Straight-Line Approximate Plots for Example 6.11 *(continued)*

The approximate Bode plots obtained with these values and the asymptotes are shown in Figure 6.26.

Actually, for Example 6.12, the amplitude- and phase-response Bode plots are smooth enough for us to sketch them with reasonable accuracy directly from the straight-line approximations without the aid of additional specific points.

6.6 Summary

We often wish to determine the spectrum of the output signal from a system for which the spectrum of the input signal is known. Spectral-analysis methods allow us to find the output-signal spectrum without determining the signal time waveforms and system impulse response.

Important to spectral analysis is the concept of system frequency response; that is, the function of frequency that produces the output-signal spectrum when multiplied by the input-signal spectrum. The amplitude and phase of the system frequency response are referred to as the amplitude response and phase response of the system. For a sinusoidal signal and a system with real impulse response, the values of the amplitude and phase response at the frequency of the sinusoidal signal give the amplitude change (gain) and the phase shift experienced by the signal in passing through the system. The system frequency response is the Fourier transform of the system impulse response.

The significant bandwidth of a system is defined to be the width of the range of frequencies over which the system's amplitude response is greater than a specified fraction of its maximum value. It is usually called simply the system bandwidth. A typical specified fraction is $1/\sqrt{2}$, which produces the half-power bandwidth.

The system frequency response can be found without first finding the system impulse response (1) by computing the Fourier transform of the system differential equation and solving for the ratio of the output-signal spectrum to the input-signal spectrum or (2) by using phasor concepts to solve for the ratio between the phasors representing output and input sinusoidal signals of frequency f. Phasor concepts are particularly convenient for electric circuits for which a wealth of steady-state, sinusoidal-signal circuit-analysis tools are available.

System phase delay gives us information concerning the time delay experienced by single-frequency sinusoids in passing through a system. System group delay gives us information concerning the time delay of the amplitude modulation on a sinusoidal carrier in passing through a system. We can determine phase delay and group delay from the system phase response.

Amplitude- and phase-response Bode plots use a logarithmic frequency scale. Also, the amplitude-response Bode plot is plotted in terms of gain in decibels. These two choices of scale permit us to easily draw straight-line approximations to the Bode plots by summing straight-line approximations to the Bode plots for individual multiplicative factors in the frequency response. The straight-line approximations provide quick determination of basic frequency-response characteristics. Approximate amplitude- and phase-response Bode plots are easily sketched from the straight-line approximations and a few calculated points.

Problems

6.1 A system has the frequency response

$$H(f) = 500 \Big/ \Big[\Big(500 - f^2\Big) + j45f \Big]$$

and an input signal

$$x(t) = \cos(10\pi t) + \cos(20\pi t + \pi/3)$$
$$+ \cos(80\pi t - \pi/4)$$

a. Plot the system amplitude and phase responses.
b. Find and plot the amplitude and phase spectra for the input and output signals.

6.2 A system has the frequency response

$$H(f) = 10000 \Big/ \Big[\Big(10000 - f^2\Big) + j20f \Big]$$

and an input signal

$$x(t) = 2 + 2\cos(160\pi t + \pi/2)$$
$$+ 2\cos(400\pi t - \pi/4)$$

a. Plot the system amplitude and phase responses.
b. Find and plot the amplitude and phase spectra for the input and output signals.

6.3 A system has the frequency response

$$H(f) = 900 f^2 \Big/ \Big[\Big(-f^4 + 2500 f^2 - 64{,}000\Big) + j\Big(60 f^3 - 48{,}000 f\Big) \Big]$$

and an input signal

$$x(t) = 4\cos(20\pi t) + 4\cos(50\pi t - \pi/3)$$
$$+ 4\cos(70\pi t) + 4\cos(100\pi t + \pi/2)$$

a. Plot the system amplitude and phase responses.
b. Find and plot the amplitude and phase spectra for the input and output signals.

6.4 The system of Problem 6.3 has the following input signal:

$$x(t) = 40 \, \text{sinc}(20t) \cos(60\pi t)$$

a. Plot the system amplitude and phase responses.
b. Find and plot the amplitude and phase spectra for the input and output signals.

6.5 Plot the amplitude and phase responses and approximate the system half-power bandwidth and half-power cutoff frequency or frequencies for the systems with the frequency responses listed. Indicate whether the systems are low-pass or bandpass systems.

a. $H(f) = 100/(100 + jf)$

b. $H(f) = 200/[(10 + jf)(15 + jf)]$

c. $H(f) = 200f/[200f + j(f^2 - 240{,}000)]$

d. The frequency response of Problem 6.1

e. The frequency response of Problem 6.3

6.6 Find the approximate half-power bandwidth for a low-pass electric filter with the following frequency response:

$$H(f) = 1500\Big/\Big[\Big(3000 - 45f^2\Big) + jf\Big(350 - f^2\Big)\Big]$$

6.7 The input signal to the system of Problem 6.2 is a rectangular pulse centered at $t = 0$ with a width of 8 ms.

a. Plot the system amplitude and phase responses.

b. Find and plot the amplitude and phase spectra for the input and output signals.

6.8 A signal-repeater system in a satellite provides signal filtering and delays the signal before retransmission. The repeater system has the frequency response

$$H(f) = \exp(-j0.1\pi f)/[1 + j(f/2)]$$

a. Plot the system amplitude and phase responses.

b. Find and sketch the system impulse response.

6.9 An electric filter system has the frequency response

$$H(f) = j0.02f/(1 + j0.02f)$$

a. Plot the system amplitude and phase responses.

b. Find and sketch the system impulse response. (Hint: use the time-differentiation theorem.)

c. Approximate the system half-power cutoff frequency or frequencies.

d. From its amplitude response, what type of filter is this?

6.10 A system has the frequency response

$$H(f) = [1/(1 + j0.5f)] + [j0.05f/(1 + j0.05f)]$$

a. Plot the system amplitude and phase responses.

b. Find and sketch the system impulse response. (Hint: Use the linearity and time-differentiation theorems.)

c. Approximate the system half-power cutoff frequency or frequencies.

d. From its amplitude response, what type of system is this?

6.11 A control system is characterized by the following system differential equation:

$$\frac{d^3y(t)}{dt^3} + 0.5\frac{d^2y(t)}{dt^2} + 0.75\frac{dy(t)}{dt} + 2y(t) = x(t)$$

Find the frequency response for this system.

6.12 A system is characterized by the system differential equation

$$\frac{dy(t)}{dt} + 2y(t) = \frac{dx(t)}{dt} + x(t)$$

a. Find the system frequency response.

b. Plot the system amplitude and phase responses.

c. Find the system impulse response.

6.13 Consider the pipeline-flow-rate control system of Problem 5.4. It is represented by the block diagram in Figure 6.27.

a. Find the system differential equation.

b. Find the system frequency response.

Figure 6.27

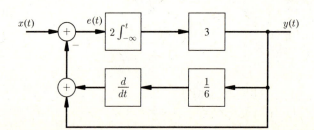

c. Plot the system amplitude and phase responses.
d. Find the system impulse response.

6.14 Work Problem 6.13 using the phasor concept instead of finding and using the system differential equation.

6.15 Consider the electric signal transmission circuit shown in Figure 6.28.

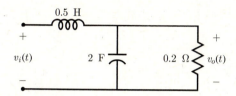

Figure 6.28

where $v_i(t)$ and $v_o(t)$ are the input and output signals, respectively.

a. Find the system differential equation and use it to find the system frequency response.
b. Find the system frequency response using phasor techniques.

6.16 A simple model of an operational amplifier used as a filter circuit is shown in Figure 6.29.

a. Find the frequency response of the circuit.
b. Plot the amplitude and phase responses, when (a) $R_1 = 1$ MΩ, (b) $R_2 = 2$ MΩ, (c) $C = 0.1$ μF, and (d) $A = 1000$.

6.17 Consider the electric circuit shown in Figure 6.30, where the input signal is $v(t)$ and the output signal is $i(t)$.

a. Find the frequency response of this circuit.
b. Plot the amplitude and phase response.

6.18 Find the phase delay and group delay for the electric signal transmission circuit of Problem 6.15 at the following frequencies: (a) 0.25 Hz, (b) 1 Hz, (c) 2 Hz, and (d) 5 Hz.

6.19 Find the phase and group delay for the system of Problem 6.2 at the following frequencies: (a) 5 Hz, (b) 80 Hz, (c) 125 Hz, and (d) 200 Hz.

6.20 A system has the frequency response

$$H(f) = j0.01f/(1 + j0.01f)$$

Find the phase delay and group delay for this system at the following frequencies: (a) 10 Hz, (b) 50 Hz, (c) 100 Hz, (d) 200 Hz, and (e) 1000 Hz.

Figure 6.29

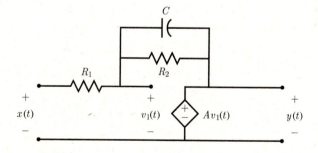

Figure 6.30

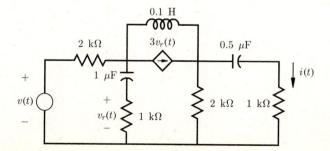

6.21 Given the system frequency response

$$H_\omega(\omega) = \frac{20(j\omega + 25)^2}{j\omega(j\omega + 500)}$$

a. Find and plot the straight-line approximations to the amplitude- and phase-response Bode plots for the system.

b. Sketch the approximate amplitude-response Bode plot for the system using the straight-line approximation and approximate values at break frequencies and/or peak frequencies.

c. Sketch the approximate phase-response Bode plot for the system using the straight-line approximation and a few calculated points.

6.22 Work Problem 6.21 with the frequency response

$$H_\omega(\omega) = \frac{500(j\omega + 15)}{j\omega(j\omega + 350)}$$

6.23 Work Problem 6.21 with the frequency response

$$H_\omega(\omega) = \frac{600j\omega + 600}{0.25(j\omega)^2 + 51(j\omega) + 200}$$

6.24 Work Problem 6.21 with the frequency response

$$H_\omega(\omega) = \frac{180(j\omega) + 360}{0.006(j\omega)^2 + 3.06(j\omega) + 30}$$

6.25 Work Problem 6.21 with the frequency response

$$H_\omega(\omega) = \frac{(j\omega)[(2 \times 10^6)(j\omega) + 10^7]}{(j\omega + 35)^2[(j\omega)^2 + 200(j\omega) + 10^6]}$$

7 Analysis of Continuous-Time Systems Using the Laplace Transform

Once again, we consider single-input–single-output, linear, time-invariant, continuous-time systems. We indicated in Chapter 1 that the system response (output signal) of such a system can be determined for $t \geq t_o$ from the input signal and the initial energy stored (initial conditions) by solving the system differential equation. Since the system is linear and time-invariant, the system differential equation is linear and has constant coefficients.

In Chapter 5, we saw that the system response when the initial conditions are zero (that is, the zero initial energy response) can also be found by computing the convolution of the input and system impulse response. We noted in Chapter 6 that the system impulse response required in the convolution computation is the inverse Fourier transform of the system frequency response and that it can be determined from the system representation without solving the system differential equation. However, the inverse Fourier transform is often difficult to evaluate, even for simple impulse responses of practical systems.

When the system initial conditions are zero, then we can obtain the output signal without convolution by computing the inverse Fourier transform of the spectrum of the output signal. We showed in Chapter 6 that the output signal spectrum is readily found from the input spectrum-signal using only algebraic operations. But once again it is often difficult to evaluate the inverse Fourier transform required, even for some relatively simple output signals (see Example 6.3).

In this chapter, we will consider the system response of a causal system to an input that is zero for $t < 0$ when initial conditions are either zero or nonzero. The tool we will use to determine the system response is the *single-sided Laplace transform*. This tool is particularly attractive for continuous-time system analysis for the following reasons:

1. It replaces the differential equation with an algebraic equation.
2. It finds the total solution (homogeneous plus particular) directly, meaning that initial-condition responses are automatically included.
3. It can be used with signals that do not possess a spectrum (that is, signals that do not have a Fourier transform), such as the ramp signal.

In addition, the single-sided Laplace transform of the impulse response of a causal, linear, time-invariant system produces the *system transfer function*—a function that is useful in characterizing the system.

7.1 The Laplace Transform

In this section, we will first define the *double-sided*, or *bilateral, Laplace transform* of a signal and the corresponding inverse Laplace transform. We will then consider the *single-sided Laplace transform*, which is a special, very useful, case of the double-sided Laplace transform. Conditions for the existence of the single-sided Laplace transform and some of its general properties are included in the discussion. Because the single-sided Laplace transform is so useful in continuous-time system analysis, we will consider it further in subsequent sections, where we refer to it as simply the Laplace transform.

The Double-Sided Laplace Transform

The *double-sided Laplace transform* is defined by an integral that we obtain from the Fourier transform integral by including the convergence factor $e^{-\sigma t}$. For the signal $x(t)$, the Fourier transform of $e^{-\sigma t}x(t)$ is, from eq. (4.62),

$$\mathcal{F}_\omega\left[e^{-\sigma t}x(t)\right] = \int_{-\infty}^{\infty} e^{-\sigma t}x(t)e^{-j\omega t}\, dt = \int_{-\infty}^{\infty} x(t)e^{-(\sigma+j\omega)t}\, dt \qquad (7.1)$$

where we chose σ to make $e^{-\sigma t}x(t)$ absolutely integrable; thus, the Fourier transform exists if $x(t)$ has a finite number of maxima and minima and a finite number of finite discontinuities in any finite interval (see Section 4.4). There are signals for which it is not possible to choose σ so that $e^{-\sigma t}x(t)$ is absolutely integrable. However, it is possible to do so for a wide variety of useful signals, including some for which $x(t)$ is not absolutely integrable; for example, the ramp signal $r(t)$.

The variable $\sigma + j\omega$ is a complex variable that we define as

$$s \equiv \sigma + j\omega \qquad (7.2)$$

for simplicity.

Definition

The double-sided Laplace transform $\mathcal{L}_D[x(t)]$ of the signal $x(t)$ is

$$\mathcal{L}_D[x(t)] = \mathcal{F}_\omega\left[e^{-\sigma t}x(t)\right]\Big|_{\sigma+j\omega=s} = \int_{-\infty}^{\infty} x(t)e^{-st}\, dt \equiv X_D(s) \qquad (7.3)$$

That is, the double-sided Laplace transform of $x(t)$ is the Fourier transform of $e^{-\sigma t}x(t)$ written as a function of the complex variable s.

To develop the inverse double-sided Laplace transform, we use the inverse Fourier transform given by eq. (4.63). Thus, we write

$$e^{-\sigma t}x(t) = \frac{1}{2\pi} \int_{-\infty}^{\infty} \mathcal{F}_\omega\left[e^{-\sigma t}x(t)\right]e^{j\omega t}\, d\omega \qquad (7.4)$$

and, therefore,

$$x(t) = \frac{1}{2\pi} \int_{-\infty}^{\infty} \mathcal{F}_\omega\left[e^{-\sigma t}x(t)\right]e^{(\sigma+j\omega)t}\, d\omega \qquad (7.5)$$

If we let $s = \sigma + j\omega$, then $d\omega = ds/j$, $s \to \sigma \pm j\infty$ as $\omega \to \pm\infty$, and

$$x(t) = \frac{1}{2\pi j} \int_{\sigma - j\infty}^{\sigma + j\infty} X_D(s)e^{st} \, ds \equiv \mathcal{L}^{-1}[X_D(s)] \qquad (7.6)$$

which is the inverse double-sided Laplace transform of $X_D(s)$.

Note that the evaluation of the inverse double-sided Laplace transform requires the integration of a function of the complex variable s along a line specified by $s = \sigma + j\omega$ in a plane (called the s-plane) on which the values of s are plotted. The s-plane and the integration line are illustrated in Figure 7.1. We must choose a value of σ that causes the line to be in the region of the s-plane where eq. (7.3) exists and converges to $X_D(s)$. We can find such a region if $X_D(s)$ exists for $x(t)$. It is referred to as the *region of convergence of the double-sided Laplace transform of the signal $x(t)$*. In many cases, more than one signal can produce the same double-sided Laplace transform function. The difference is the region in the s-plane where the double-sided Laplace transform converges. For example, $e^{-at}u(t)$ and $-e^{-at}u(-t)$ both produce the double-sided Laplace transform $1/(s+a)$, but the region of convergence is to the right of the line $a + j\omega$ for $e^{-at}u(t)$ and to the left of the same line for $-e^{-at}u(-t)$. Thus, the double-sided Laplace transform that corresponds to a particular signal consists of the transform function and its region of convergence.

The generality of the double-sided Laplace transform and the insight provided by its tie with the Fourier transform serves well in introducing the single-sided Laplace transform, to which we will now turn our attention. As mentioned previously, the single-sided Laplace transform is very useful in continuous-time system analysis. It is also simpler than the double-sided Laplace transform.

Figure 7.1
Line of Integration
for Inverse Laplace
Transform

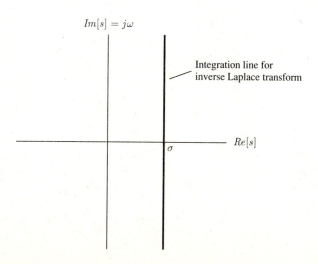

$Im[s] = j\omega$

Integration line for
inverse Laplace transform

$Re[s]$

σ

The Single-Sided Laplace Transform

The Laplace transform is very useful for evaluating the response of causal, linear, time-invariant systems when the time reference is chosen so that the input signal begins at $t \geq 0$. In this case, time functions of interest (that is, signals and impulse responses) are all zero for $t < 0$ and the double-sided Laplace transform integral is equivalent to

$$X_S(s) = \int_{0^-}^{\infty} x(t)e^{-st}dt \qquad (7.7)$$

This integral is referred to as the *single-sided Laplace transform of the signal* $x(t)$.

We set the lower limit of the integral in eq. (7.7) to 0^- to include the time origin in the integration in order to include discontinuities and impulses at the time origin. In most of the discussion to follow, the lower limit is designated 0 for simplicity; however, it should be kept in mind that 0^- is what we mean. Incidentally, mathematicians and some engineers use 0^+ as the lower limit. Since we are concerned with impulse functions, it is more convenient to use 0^-, so that the single-sided Laplace transform of an impulse at the origin can be computed and used.

Note that the single-sided Laplace transform of $x_a(t)$ equals the single-sided Laplace transform of $x_a(t)u(t)$. This is because the difference in $x_a(t)$ and $x_a(t)u(t)$ in the range of the transform integral is only at the single point $t = 0^-$, and thus does not produce a change in the value of the integral.

The region of convergence for a single-sided Laplace transform, if it exists, is always the region in the s-plane to the right of the vertical straight line given by $s = c + j\omega$, as illustrated by the shaded regions in Figure 7.2. The convergence region is more properly defined in terms of the existence theorem for the single-sided Laplace transform of a time function. The theorem is stated here without proof; its proof is beyond the scope of this text.[†]

> **Existence Theorem for the Single-Sided Laplace Transform:**
> If $x(t)$ is absolutely integrable (the integral of $|x(t)|$ is finite) over the interval $0^- \leq t < T$ for any $T > 0$, and if the real parameters c and K can be chosen so that $|e^{-ct}x(t)| \leq K$ for $t \geq T$, then the single-sided Laplace transform integral converges absolutely and uniformly for all s for which $Re[s] > c$.

The smallest possible value of c that we can choose is referred to as the *abscissa of absolute convergence*. This value of c defines the vertical line that is the left-hand boundary of the region of convergence in the s-plane, as illustrated in Figure 7.2. It restricts the value that we can use for the parameter σ in the convergence factor $e^{-\sigma t}$ to obtain a valid (convergent) single-sided Laplace

[†]For detailed discussion of the single-sided Laplace transform existence theorem, see W. M. Brown, *Analysis of Linear Time Invariant Systems* (New York: McGraw-Hill, 1963) and D. V. Widder, *The Laplace Transform* (Princeton, N.J.: Princeton University Press, 1941).

Figure 7.2

Example Regions of Convergence for Single-Sided Laplace Transforms

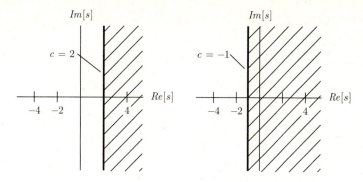

transform. The parameter σ must be larger than c in order that $e^{-\sigma t}x(t)$ approach zero rapidly enough at large values of t for

$$\int_{0-}^{\infty} \left| e^{-st}x(t) \right| \, dt = \int_{0-}^{\infty} \left| e^{-\sigma t} \right| \left| e^{-j\omega t} \right| |x(t)| \, dt = \int_{0-}^{\infty} \left| e^{-\sigma t}x(t) \right| \, dt \quad (7.8)$$

to exist. The existence theorem indicates that the Laplace transform of an impulse function exists because, in this case, the existence conditions are satisfied for any T greater than zero and any value of c.

The inverse double-sided Laplace transform and the inverse single-sided Laplace transform are given by the same integral for the reason that the double-sided Laplace transform of a signal $x(t)$ that is equal to zero for $t < 0$ is the same as the single-sided Laplace transform of that signal. That is, the inverse single-sided Laplace transform is

$$x(t) = \frac{1}{2\pi j} \int_{\sigma-j\infty}^{\sigma+j\infty} X_s(s)e^{st} \, ds \equiv \mathcal{L}_s^{-1}[X_s(s)] \quad (7.9)$$

where the integration line lies in the convergence region for $x(t)$. Since the region of convergence for every single-sided Laplace transform that exists lies to the right of a vertical line in the s-plane, each single-sided Laplace transform corresponds to a single convergence region. Therefore, all single-sided Laplace transforms uniquely correspond to one and only one signal that is zero for $t < 0$, and the region of convergence does not need to be stated to completely characterize the single-sided Laplace transform. The uniqueness of single-sided Laplace transforms makes them a particularly convenient tool.

Tables of time functions and corresponding single-sided Laplace transforms (that is, single-sided Laplace transform pairs) are readily available. Recall that a single-sided Laplace transform of the signal $x(t) = x_a(t)$ equals the single-sided transform of the signal $x(t) = x_a(t)u(t)$, where $x_a(t)$ may be defined for all t. Thus many tables, including those in this text, are expressed in terms of signals of the form $x(t) = x_a(t)u(t)$ to call attention to the fact that the single-sided Laplace transform is computed only over positive t. Recall when using these tables that $x_a(t)u(t)$ may differ from $x_a(t)$ at $t = 0^-$, as we indicated earlier. Some tables of single-sided Laplace transform pairs indicate the time function as $x(t) = x_a(t)$. In this case, one must be aware that $x(t)$ is actually assumed to be zero for $t < 0^-$, even though the signal expression in the table does not so indicate.

Since we are concerned with only the single-sided Laplace transform, in subsequent sections we will refer to it as simply the Laplace transform, and we will drop the subscript S on $X_S(s)$. This notation is consistent with most literature and Laplace transform tables.

The use of $X(s)$ for the Laplace transform of $x(t)$ is a similar notation to the one used for the Fourier transform; however, for the most part there should be no confusion if it is recalled that the notation indicates a Fourier transform when the argument is f and a Laplace transform when the argument is s. In cases where confusion may be possible, subscripts of f and l are used to indicate the Fourier and Laplace transforms, respectively.

We will use the same shorthand notation for Laplace transforms as we used for Fourier transforms; that is, $x(t) \leftrightarrow X(s)$ indicates a Laplace transform pair.

7.2 Laplace Transform Evaluations and Theorems

We will now compute the Laplace transforms of some useful simple signals to illustrate the direct computation of the Laplace transform of a signal. We will also present several Laplace transform theorems and illustrate them with examples. One use of these theorems is in the computation of the Laplace transform of more complicated signals.

Laplace Transforms of Simple Signals

In the following examples, we consider several useful simple signals to illustrate the direct computation of the Laplace transform.

Example 7.1

Find the Laplace transform and abscissa of convergence for the signal $x(t) = e^{-\alpha t}u(t)$.

Solution

$$X(s) = \int_0^\infty \left[e^{-\alpha t}u(t) \right] e^{-st}\, dt = \int_0^\infty e^{-(s+\alpha)t}\, dt$$

$$= \left[-\frac{e^{-(s+\alpha)t}}{s+\alpha} \right]_0^\infty = \left[-\frac{e^{-(\sigma+\alpha)t}e^{-j\omega t}}{s+\alpha} \right]_0^\infty$$

$$= \lim_{t \to \infty} \left[-\frac{e^{-(\sigma+\alpha)t}e^{-j\omega t}}{s+\alpha} \right] + \left[\frac{1}{s+\alpha} \right]$$

As $t \to \infty$, $e^{-j\omega t} = \cos \omega t - j \sin \omega t$ is bounded and

$$e^{-(\sigma+\alpha)t} \quad \text{approaches} = \begin{cases} \infty \text{ when } \sigma < -\alpha \\ 0 \text{ when } \sigma > -\alpha \end{cases}$$

Therefore, the Laplace transform integral converges if $\sigma > -\alpha$. The abscissa of convergence is thus $c = -\alpha$ and the Laplace transform is

$$X(s) = \frac{1}{s+\alpha}$$

Note that a region of convergence exists even if α has a finite negative value. Consequently, the Laplace transform exists for an increasing exponential signal that starts at $t = 0$. This signal does not have a spectrum because its Fourier transform does not exist.

Example 7.2

Find the Laplace transform and abscissa of convergence for the unit step signal.

Solution
The unit step signal is the special case of the signal in Example 7.1 that occurs when $\alpha = 0$. Therefore, if $x(t) = u(t)$, then

$$X(s) = \frac{1}{s} \quad \text{and} \quad c = 0$$

Example 7.3

Find the Laplace transform and region of convergence for the unit impulse signal.

Solution

$$x(t) = \delta(t)$$

$$X(s) = \int_0^\infty \delta(t)e^{-st} \, dt = 1$$

by the sifting integral because the lower limit of the integral is actually 0^-; thus, the integral includes the impulse function. The region of convergence is the entire s plane because the sifting integral selects only the value of e^{-st} at $t = 0$, which is 1 for all s.

Example 7.4

Find the Laplace transform of the unit ramp signal.

Solution

$$x(t) = r(t) = tu(t)$$

$$X(s) = \int_0^\infty tu(t)e^{-st}\,dt = \int_0^\infty te^{-st}\,dt$$

$$= \left[\frac{e^{-st}}{s^2}(-st - 1)\right]_0^\infty = \left[0 - \left(-\frac{1}{s^2}\right)\right] = \frac{1}{s^2}$$

Although we did not determine the abscissa of convergence in Example 7.4, it can be easily shown to be $c = 0$.

Table C.4 in Appendix C lists the Laplace transforms for a number of additional useful signals. Some of these Laplace transforms can be obtained from those already computed and a set of Laplace transform theorems. Let us now consider these theorems.

Laplace Transform Theorems

The theorems stated here are illustrated with examples. The proofs are similar to those for the corresponding Fourier transform theorems. Only a few proofs are shown, but the others can be constructed easily. The reader is asked to construct some of them in the end-of-chapter problems. The theorems are summarized in Table C.3 in Appendix C.

Theorem 1   Linearity
If

$$x(t) \leftrightarrow X(s) \quad \text{and} \quad y(t) \leftrightarrow Y(s)$$

then

$$ax(t) + by(t) \leftrightarrow aX(s) + bY(s) \tag{7.10}$$

This theorem is easily proved, the integral being a linear operation.

Example 7.5

Find the Laplace transform of the signal $x(t) = \sin(\omega_o t)\,u(t)$.

Solution

$$x(t) = \sin(\omega_o t)u(t) = \frac{1}{2j}e^{j\omega_o t}u(t) - \frac{1}{2j}e^{-j\omega_o t}u(t)$$

Example 7.1 showed that

$$e^{-\alpha t}u(t) \leftrightarrow \frac{1}{s+\alpha}$$

Therefore, by the linearity theorem,

$$X(s) = \frac{1}{2j}\left[\frac{1}{s-j\omega_o}\right] - \frac{1}{2j}\left[\frac{1}{s+j\omega_o}\right] = \frac{\omega_o}{s^2+\omega_o^2}$$

Theorem 2 Scale Change

If

$$x(t) \leftrightarrow X(s)$$

then

$$x(mt) \leftrightarrow \frac{1}{m}X\left(\frac{s}{m}\right) \qquad \text{for } m > 0 \qquad\qquad (7.11)$$

The parameter m cannot be less than or equal to zero for single-sided Laplace transforms. A negative m would result in a reversal of $x(t)$, meaning that $x(mt)$ would be nonzero only for $t < 0$, since $x(t) = 0$ for $t < 0$. Thus, the Laplace transform of $x(mt)$ is $X(s) = 0$ when m is negative, and the scaling theorem is not valid.

Example 7.6

Find the Laplace transform of the signal $y(t) = r(3t)$.

Solution

Letting $z(t) = r(t)$,

$$y(t) = z(3t)$$

From Table C.4,

$$Z(s) = 1/s^2$$

Thus, the scale-change theorem gives

$$X(s) = \frac{1}{3}Z\left(\frac{s}{3}\right) = \frac{1}{3}\left[1/(s/3)^2\right] = 3/s^2$$

The Laplace transform could also be computed by noting that

$$y(t) = r(3t) = 3tu(3t) = 3tu(t) = 3r(t) \leftrightarrow 3/s^2$$

However, our intent here was to illustrate the scale-change theorem.

Theorem 3 Time Delay

If

$$x(t) \leftrightarrow X(s) \quad \text{and} \quad x(t) = 0 \qquad \text{for } t < 0$$

then

$$x(t - t_0) \leftrightarrow X(s)e^{-st_o} \qquad \text{for} \quad t_o > 0 \qquad (7.12)$$

This theorem is similar to the time-shift theorem for Fourier transforms. However, t_0 must be greater than zero because a shift to the left causes a portion of the signal $x(t)$ to be located at $t < 0$. This portion is cut off when the single-sided Laplace transform is computed, so that more than time shifting occurs. Therefore, for the single-sided Laplace transform, Theorem 3 is a time-delay theorem rather than a time-shift theorem.

Example 7.7

A zero-order hold system has the impulse response shown in Figure 7.3. Find the Laplace transform of this impulse response.

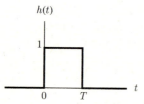

Figure 7.3 Impulse Response for Zero-Order Hold System

Solution

$$h(t) = u(t) - u(t - T)$$

We use the time-delay and linearity theorems and the transform $u(t) \leftrightarrow 1/s$ found in Example 7.2 to give

$$H(s) = \frac{1}{s} - \frac{1}{s}e^{-sT} = \left[1 - e^{-sT}\right]\Big/s$$

The system in this example is referred to as a *hold system*; it holds the value of the strength of an impulse for the time interval T.

Theorem 4 s-Shift

If

$$x(t) \leftrightarrow X(s)$$

then

$$e^{-at}x(t) \leftrightarrow X(s + a) \qquad (7.13)$$

This theorem corresponds to the frequency-translation theorem for Fourier transforms.

Example 7.8

Find the Laplace transform of the damped sinusoid signal $z(t) = e^{-at} \sin \omega_o t$.

Solution
From Table C.4,

$$\mathcal{L}[\sin \omega_o t] = \frac{\omega_o}{s^2 + \omega_o^2}$$

Therefore, the s-shift theorem gives

$$Z(s) = \frac{\omega_o}{(s+a)^2 + \omega_o^2}$$

Damped sinusoids frequently occur as signals in systems with energy-absorbing components; consequently, Example 7.8 has derived an important Laplace transform pair.

Theorem 5: t Multiplication
If

$$x(t) \leftrightarrow X(s)$$

then

$$tx(t) \leftrightarrow -\frac{dX(s)}{ds} \qquad (7.14)$$

The validity of this theorem is easily shown by differentiating the integral that defines $X(s)$ with respect to s.

Example 7.9

Find the Laplace transform of the signal $x(t) = te^{-at}u(t)$.

Solution
Letting $y(t) = e^{-at}u(t)$, then, from Table C.4,

$$Y(s) = \frac{1}{s+a}$$

Therefore,

$$X(s) = -\frac{d}{ds}\left[\frac{1}{s+a}\right] = \frac{1}{(s+a)^2}$$

Extending Theorem 5 by induction gives

$$t^n x(t) \leftrightarrow (-1)^n \frac{d^n X(s)}{ds^n} \qquad (7.15)$$

Theorem 6   Time Differentiation

If

$$x(t) \leftrightarrow X(s)$$

then

$$\frac{d^n x(t)}{dt^n} \leftrightarrow s^n X(s) - \sum_{i=0}^{n-1} s^{n-1-i} x^{(i)}\left(0^-\right) \tag{7.16}$$

where

$$x^{(i)}\left(0^-\right) = d^i x(t)/dt^i\big|_{t=0^-} \tag{7.17}$$

The proof of this theorem begins with the first derivative, the Laplace transform of which is

$$\mathcal{L}\left[x^{(1)}(t)\right] = \int_{0^-}^{\infty} x^{(1)}(t) e^{-st}\, dt \tag{7.18}$$

Using integration by parts,

$$\mathcal{L}\left[x^{(1)}(t)\right] = x(t) e^{-st}\big|_{0^-}^{\infty} - \int_{0^-}^{\infty} x(t) \left[-s e^{-st}\right] dt \tag{7.19}$$

Now, $x(t)e^{-st} \to 0$ as $t \to \infty$ for any s in the region of convergence of $X(s)$. Therefore,

$$\mathcal{L}\left[x^{(1)}(t)\right] = -x(0^-) + s \int_{0^-}^{\infty} x(t) e^{-st}\, dt = sX(s) - x(0^-) \tag{7.20}$$

Next, we consider the second derivative $x^{(2)}(t) = y^{(1)}(t)$, where $y(t) = x^{(1)}(t)$. Therefore,

$$\mathcal{L}\left[x^{(2)}(t)\right] = sY(s) - y(0^-) = s\left[sX(s) - x(0^-)\right] - x^{(1)}(0^-)$$

$$= s^2 X(s) - sx(0^-) - x^{(1)}(0^-) \tag{7.21}$$

We continue this procedure for higher order derivatives to inductively prove Theorem 6.

Example 7.10

Find the Laplace transform of the derivative of the signal $x(t) = e^{-t}u(t)$. Invert the transform to find the signal derivative.

Solution

By the time-differentiation theorem,

$$\mathcal{L}\left[x^{(1)}(t)\right] = sX(s) - x(0^-)$$

But $X(s) = \frac{1}{s+1}$ from Example 7.1, and $x(0^-) = 0$ because $u(t) = 0$ for $t < 0$. Therefore,

$$\mathcal{L}\big[x^{(1)}(t)\big] = \frac{s}{s+1}$$

Dividing s by $s+1$ yields

$$\mathcal{L}\big[x^{(1)}(t)\big] = 1 - \frac{1}{s+1}$$

By using transform pairs found in Table C.4, we find that the derivative of the signal is

$$x^{(1)}(t) = \delta(t) - e^{-t}u(t)$$

Now, let us compute the signal derivative found in Example 7.10 directly to check the result. The signal is shown in Figure 7.4 to help visualize the source of the two parts of the derivative.

$$x^{(1)}(t) = \frac{d}{dt}\big[e^{-t}u(t)\big] = e^{-t}\frac{du(t)}{dt} + u(t)\frac{d\big[e^{-t}\big]}{dt}$$

$$= e^{-t}\delta(t) + u(t)\big[-e^{-t}\big] = \delta(t) - e^{-t}u(t) \qquad (7.22)$$

Equation (7.22) checks the result found in Example 7.10. Note that the negative exponential term is the slope of the signal for $t > 0$. The impulse results from the step discontinuity of height 1 at $t = 0$.

Theorem 7   Time Integration

If

$$x(t) \leftrightarrow X(s)$$

then

$$y(t) = \int_{0-}^{t} x(\lambda)\, d\lambda + y(0^-) \leftrightarrow \frac{X(s)}{s} + \frac{y(0^-)}{s} \qquad (7.23)$$

Figure 7.4
Exponential Signal
for Example 7.9

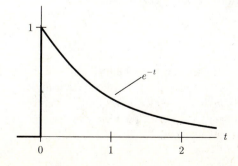

Example 7.11

Find the Laplace transform of $y(t) = \int_{0_-}^{t} x(\lambda) \, d\lambda + y(0^-)$ when $x(t) = r(t)$ and $y(0^-) = 2$.

Solution

From Table C.4,

$$X(s) = 1/s^2$$

Thus, the time-integration theorem gives

$$Y(s) = \frac{(1/s^2)}{s} + \frac{2}{s} = \frac{1}{s^3} + \frac{2}{s} = \frac{2s^2 + 1}{s^3}$$

Theorem 8   Convolution

If

$$x(t) \leftrightarrow X(s), \ y(t) \leftrightarrow Y(s), \quad \text{and} \quad x(t) = y(t) = 0 \text{ for } t < 0$$

then

$$x(t) * y(t) \leftrightarrow X(s)Y(s) \tag{7.24}$$

The proof of the convolution theorem is as follows:

$$\mathcal{L}[x(t) * y(t)] = \int_{0}^{\infty} \left[\int_{-\infty}^{\infty} x(\lambda) y(t - \lambda) \, d\lambda \right] e^{-st} \, dt$$

$$= \int_{0}^{\infty} \left[\int_{0}^{\infty} x(\lambda) y(t - \lambda) \, d\lambda \right] e^{-st} \, dt \tag{7.25}$$

where the second step follows because $x(\lambda) = 0$ for $\lambda < 0$. Interchanging the order of integration gives

$$\mathcal{L}[x(t) * y(t)] = \int_{0}^{\infty} x(\lambda) \left[\int_{0}^{\infty} y(t - \lambda) e^{-st} \, dt \right] d\lambda \tag{7.26}$$

Since $\lambda > 0$, the time-delay theorem can be used to give

$$\mathcal{L}[x(t) * y(t)] = \int_{0}^{\infty} x(\lambda) Y(s) e^{-s\lambda} \, d\lambda = \left[\int_{0}^{\infty} x(\lambda) e^{-s\lambda} \, d\lambda \right] Y(s) = X(s)Y(s)$$
$$\tag{7.27}$$

The output of a linear time-invariant system with zero initial conditions is the convolution of the input signal and the system impulse response; hence, the convolution theorem is especially important. When the system is causal and the input signal is zero for $t < 0$, then we can use the convolution theorem to solve for the system response.

Example 7.12

Consider the simple RC circuit shown in Figure 7.5 and the input signal $x(t) = 4\Pi(t-1)$. Find the output signal $y(t)$.

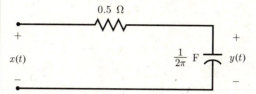

Figure 7.5 Simple RC Filter Circuit for Example 7.12

Solution

The circuit and input signal in this example are the same circuit and input signal considered in Example 6.3. We also considered the circuit in Example 6.2, where we found its impulse response to be

$$h(t) = 4\pi e^{-4\pi t} u(t)$$

The Laplace transform of this impulse response is

$$H(s) = \frac{4\pi}{s + 4\pi}$$

We express the input signal as

$$x(t) = 4u(t - 0.5) - 4u(t - 1.5)$$

the Laplace transform of which is found by the time-shift and linearity theorems to be

$$X(s) = \frac{4}{s}e^{-0.5s} - \frac{4}{s}e^{-1.5s}$$

Since $y(t) = x(t) * h(t)$ and $x(t) = h(t) = 0$ for $t < 0$, then we can use the convolution theorem to give

$$Y(s) = X(s)H(s) = 4\left[\frac{1}{s}\right]\left[\frac{4\pi}{s + 4\pi}\right]e^{-0.5s} - 4\left[\frac{1}{s}\right]\left[\frac{4\pi}{s + 4\pi}\right]e^{-1.5s}$$

Now $y(t) = x(t) * h(t) = 0$ for $t < 0$, since $x(t) = h(t) = 0$ for $t < 0$. Therefore, the integration, time-shift, and linearity theorems and the Laplace transform pair

$$\frac{4\pi}{s + 4\pi} \leftrightarrow 4\pi e^{-4\pi t} u(t)$$

yield

$$y(t) = \left\{ 4\left[\int_0^\tau 4\pi e^{-4\pi\lambda} u(\lambda)\, d\lambda \right] u(\tau) \right\}_{\tau = t - 0.5}$$

$$- \left\{ 4\left[\int_0^\tau 4\pi e^{-4\pi\lambda} u(\lambda)\, d\lambda \right] u(\tau) \right\}_{\tau = t - 1.5}$$

$$= \left\{ 4\left[-e^{-4\pi\lambda} \right]_0^\tau u(\tau) \right\}_{\tau = t - 0.5} - \left\{ 4\left[-e^{-4\pi\lambda} \right]_0^\tau u(\tau) \right\}_{\tau = t - 1.5}$$

$$= 4[1 - e^{-4\pi(t - 0.5)}]u(t - 0.5) - 4[1 - e^{-4\pi(t - 1.5)}]u(t - 1.5)$$

where $u(\tau)$ is included since the value of the integrals are only valid for $\tau > 0$. The output signal $y(t)$ is shown in Figure 7.6. Note that this is the output signal for which the spectrum was computed in Example 6.3. We did not attempt there to compute the inverse Fourier transform of the spectrum to find $y(t)$ because such a computation is difficult. The output $y(t)$ can actually be determined from $Y(s)$ without performing integration. This concept will be generalized later, when inverse Laplace transform evaluations are considered.

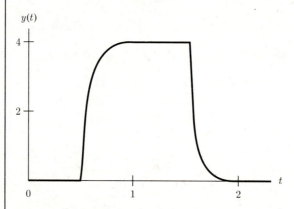

Figure 7.6 Output Signal for Example 7.12

Recall from Example 6.3 that the frequency response for the simple RC filter indicates that amplitudes of high-frequency components in the input signal are reduced more than amplitudes of low-frequency components in the input signal when the signal passes through the filter. This amplitude reduction of higher frequencies produces a decreased rate of change between signal levels and the rounding of pulse corners for the rectangular-pulse signal of Example 7.12. This effect is to be expected, because higher frequencies correspond to more rapid rates of change.

In Example 7.12, we computed the output signal by finding the inverse of the Laplace transform of the output signal. At times, however, we are interested only in the final value, $y(\infty)$, or initial value, $y(0^+)$ of the output signal. In many cases, we can find these values directly from the Laplace transform by using the final-value and initial-value theorems, avoiding the inverse transformation procedure. These theorems are also useful in checking the output-signal transform obtained for possible errors before proceeding with the transform inversion. This check is possible if the initial and final values of the output signal can be found easily by considering the system characteristics.

Theorem 9   Final Value

If (1) the Laplace transforms of $x(t)$ and $dx(t)/dt$ exist, and (2) $sX(s)$ has no singularities (points where $sX(s)$ approaches ∞) for $\sigma \geq 0$, where $s = \sigma + j\omega$, then

$$\lim_{s \to 0} sX(s) = \lim_{t \to \infty} x(t) \equiv x(\infty) \tag{7.28}$$

The proof of this theorem is left as an exercise for the reader.

Example 7.13

The final value of $w(t) = [2 + e^{-3t}]u(t)$ is obviously $w(\infty) = 2$. Show that this final value can be found with the final-value theorem.

Solution
From Table C.4 and the linearity theorem,

$$W(s) = \frac{2}{s} + \frac{1}{s+3} = \frac{3s+6}{s(s+3)}$$

Thus, $sW(s) = (3s + 6)/(s + 3)$ and it has a single singularity at $s = -3 + j0$. We can easily show that the Laplace transforms of $w(t)$ and $dw(t)/dt$ exist. Conditions for the final-value theorem are therefore satisfied and

$$w(\infty) = \lim_{s \to 0} \left(\frac{3s+6}{s+3} \right) = 2$$

Example 7.14

The signal $x(t) = \sin(2t)$ oscillates between $+1$ and -1 as $t \to \infty$; hence, it does not have a final value. Show that the application of the final-value theorem gives an incorrect result for this signal.

Solution
From Table C.4,

$$X(s) = \frac{2}{s^2 + 4}$$

Therefore,

$$
\lim_{s \to 0} sX(s) = \lim_{s \to 0} \frac{2s}{s^2 + 4} = 0
$$

This is not the value approached by $x(t)$ as $t \to \infty$. The problem is that $sX(s)$ has singularities located at $s = 0 \pm j2$. Therefore, $\sigma = 0$ for the singularities and $\lim_{s \to 0} sX(s)$ cannot be expected to give a correct result.

Theorem 10   Initial Value

If (1) the Laplace transforms of $x(t)$ and $dx(t)/dt$ both exist, and
(2) $\lim_{s \to \infty} sX(s)$ exists, then

$$
\lim_{s \to \infty} sX(s) = \lim_{t \to 0^+} x(t) \tag{7.29}
$$

We illustrate this theorem with the following example.

Example 7.15

Use the initial-value theorem to find the initial value of the signal corresponding to the Laplace transform

$$
Y(s) = \frac{s + 1}{s(s + 2)}
$$

Verify that the answer obtained is correct.

Solution

$$
\lim_{s \to \infty} sY(s) = \lim_{s \to \infty} \left(\frac{s + 1}{s + 2} \right) = 1 = y(0^+)
$$

The Laplace transform can be written as

$$
Y(s) = \frac{1}{2s} + \frac{1}{2}\left[\frac{1}{s + 2} \right]
$$

We use transform pairs from Table C.4 to obtain the corresponding signal

$$
y(t) = \frac{1}{2}u(t) + \frac{1}{2}e^{-2t}u(t)
$$

Therefore,

$$
y(0^+) = \frac{1}{2}u(0^+) + \frac{1}{2}e^{-2(0^+)}u(0^+) = \frac{1}{2} + \frac{1}{2} = 1
$$

which verifies the result obtained with the initial-value theorem.

7.3 Evaluation of Inverse Laplace Transforms

We showed in Section 7.1 that computation of the signal $x(t)$ from its single-sided Laplace transform $X(s)$ involves the evaluation of the complex integral

$$x(t) = \frac{1}{2\pi j} \int_{\sigma - j\infty}^{\sigma + j\infty} X(s)e^{st}\, ds \tag{7.30}$$

along the vertical line $s = \sigma + j\omega$ in the s-plane. Evaluation of this integral involves knowledge of complex-variable theory and is beyond the scope of this text.[†]

In the examples considered, we evaluated several inverse Laplace transforms, which were obtained using known Laplace transforms of simple signals and the Laplace transform theorems. This procedure can be extended to additional Laplace transforms obtained in the analysis of many systems and signals. Such extension is possible because linear, time-invariant, lumped-parameter systems and many useful signals produce Laplace transforms that are rational functions of s. That is, they produce Laplace transforms of the form

$$X(s) = \frac{b_m s^m + b_{m-1} s^{m-1} + \cdots + b_1 s + b_o}{a_n s^n + a_{n-1} s^{n-1} + \cdots + a_1 s + a_o} \equiv \frac{P(s)}{Q(s)} \tag{7.31}$$

Since we are considering only real-valued signals and systems that produce real-valued signals in response to real-valued signals, then all the coefficients in eq. (7.31) are real numbers. This is apparent from the single-sided Laplace transform integral (eq. [7.7]), where the only complex quantity is the parameter s, meaning that the integral produces a real function of the parameter s when $x(t)$ is a real-valued signal.

We can factor the numerator and denominator polynomials of eq. (7.31) into m and n linear factors, respectively, to give

$$X(s) = \frac{b_m(s - \beta_1)(s - \beta_2)\cdots(s - \beta_m)}{a_n(s - \alpha_1)(s - \alpha_2)\cdots(s - \alpha_n)} \tag{7.32}$$

Some of the factors may be repeated. That is, several of the β_i's may be equal and/or several of the α_i's may be equal. Also, if $b_o = 0$, then one β_i is zero; if $b_o = b_1 = 0$, then two β_i's are zero; and so on. The same is true for the a_i's and α_i's.

The numerator of $X(s)$ is zero when $s = \beta_i$ for any value of i between 1 and m. These values of s are called the *zeros of* $X(s)$, since they produce $X(s) = 0$. The denominator of $X(s)$ is zero when $s = \alpha_i$ for any value of i between 1 and n. These values of s, which produce $X(s) = \infty$, are called the *poles of* $X(s)$. The number of poles that exist at one location is referred to as the *order of the pole* at that location. The *order of a zero* is defined in the same way.

[†]R. V. Churchill and J. W. Brown, *Complex Variables and Applications*, 5th ed., (New York: John Wiley, 1990), Ch. 4.

A specific zero, say β_n, may be complex; that is, $\beta_n = c + jd$. In this case, another zero, say β_m, must equal $\beta_n^* = c - jd$, so that the product of the two factors is

$$(s - \beta_n)(s - \beta_m) = (s - c - jd)(s - c + jd) = s^2 + 2cs + (c^2 + d^2) \quad (7.33)$$

and contains only real coefficients. The product in eq. (7.33) must contain all real coefficients in order that the numerator polynomial of $X(s)$ contain only real coefficients. The same reasoning applies to the poles of $X(s)$. Thus, any complex poles or zeros of the Laplace transform of a real signal must occur in complex-conjugate pairs.

Inverse Transforms of Rational Functions of s

Now let us consider the evaluation of the inverse Laplace transform of a rational function of s. We can perform the inverse transformation by using partial-fraction expansion and tables of simple transforms if the rational function is a proper rational function; that is, if the degree of the numerator of the rational function is less than the degree of the denominator. If the degree of the numerator is greater than or equal to the degree of the denominator, then we must divide the numerator by the denominator to put the rational function in the form of a sum of various powers of s and a rational function with a numerator of lower degree than its denominator. We will illustrate this concept shortly, by means of an example.

A proper rational function can be expanded into a sum of rational functions of lower degree by the previously mentioned *partial-fraction-expansion* technique. We assume that the reader has studied partial-fraction expansion in previous mathematics courses; therefore, the procedures are only briefly stated here and then illustrated by example. To simplify the general discussion, the rational function

$$X(s) = \frac{P(s)}{(s - \alpha_1)^{n1}(s - \alpha_2)^{n2}} \quad (7.34)$$

is used for illustration. Note that eq. (7.34) has a denominator of degree $n = n1 + n2$ and has poles in two locations. The partial-fraction expansion of $X(s)$ is

$$X(s) = \frac{P(s)}{(s - \alpha_1)^{n1}(s - \alpha_2)^{n2}}$$

$$= \sum_{i=1}^{n1} A_{1,i}/(s - \alpha_1)^i + \sum_{i=1}^{n2} A_{2,i}/(s - \alpha_2)^i \quad (7.35)$$

We can find the $n1 + n2 = n$ desired coefficients $A_{1,i}$ and $A_{2,i}$ by any of three methods.

Method 1 Produce n different equations in the n desired coefficients by substituting n values, other than poles, for s. Solve the resulting linear equations simultaneously for the n desired coefficients.

For the partial-fraction expansion example of eq. (7.35), the equation produced when we set $s = s_a$ is

$$\frac{P(s_a)}{(s_a - \alpha_1)^{n1}(s_a - \alpha_2)^{n2}} = \sum_{i=1}^{n1} A_{1,i} / (s_a - \alpha_1)^i + \sum_{i=1}^{n2} A_{2,i} / (s_a - \alpha_2)^i \quad (7.36)$$

After numerical evaluation, eq. (7.36) is the linear algebraic equation

$$c_o = \sum_{i=1}^{n1} c_{1,i} A_{1,i} + \sum_{i=1}^{n2} c_{2,i} A_{2,i} \quad (7.37)$$

where the c's are constants. We produce $n-1$ additional equations like eq. (7.37) by substituting $n - 1$ other values for s. We then solve the n equations simultaneously for the values of the $A_{1,i}$'s and $A_{2,i}$'s.

Method 2 Multiply both sides of the partial-fraction-expansion equation (e.g. [7.35]) by all denominator factors to clear fractions. Simplify and equate the parameters that multiply each power of s on the two sides of the equation to produce n linear equations for the n desired coefficients. Solve the equations simultaneously for the n coefficients.

The example partial-fraction expansion of eq. (7.35) produces the following equation after we multiply each term by the denominator factors:

$$P(s) = \sum_{i=1}^{n1} A_{1,i}(s - \alpha_1)^{n1-i}(s-\alpha_2)^{n2} + \sum_{i=1}^{n2} A_{2,i}(s - \alpha_2)^{n2-i}(s-\alpha_1)^{n1} \quad (7.38)$$

We simplify eq. (7.38) to produce polynomials in s on both sides of the equation. The parameters that multiply the powers in s on the right side of the simplified eq. (7.38) are linear combinations of the n coefficients. We then equate coefficients of the powers of s on both sides of the simplified equation to produce n equations in the n coefficients. We solve these n equations for the n coefficients.

Method 3 Compute the $n\ell$ coefficients for the partial-fraction-expansion terms corresponding to the pole at location $s = \alpha_\ell$ by

$$A_{\ell,(n\ell-k)} = \frac{1}{k!} \frac{d^k}{ds^k} [X_\ell(s)] \Big|_{s=\alpha_\ell} \quad (7.39)$$

where $0 \le k \le n\ell - 1$ and $X_\ell(s) = (s - \alpha_\ell)^{n\ell} X(s)$. Repeat for each pole location.

Method 3 is the most elegant. It produces the coefficients directly without simultaneous-equation solution. However, computation of the repeated derivatives is usually messy. We frequently use Method 3 to determine as many of the coefficients as possible without differentiation (for example, $A_{1,n1}$ and $A_{2,n2}$ for the example given by eq. [7.35]) and then use one of the other two methods to find the remaining coefficients. This procedure is illustrated in the examples that follow.

Nonrepeated Factors

Method 3 is the most convenient when there are no repeated denominator factors (that is, when there are only single-order poles), no differentiation being required ($n\ell = 1$ for all ℓ). Moreover, the coefficients are produced directly by

$$A_{\ell,1} = (s - \alpha_\ell)X(s)\big|_{s=\alpha_\ell} \qquad (7.40)$$

This expression for the coefficients of nonrepeated factors is referred to as *Heaviside's expansion theorem*.

Example 7.16

Find the signal $y(t)$, the Laplace transform of which is

$$Y(s) = \frac{s^3 + 7s^2 + 18s + 20}{s^2 + 5s + 6}$$

Solution
Since the numerator is of higher degree than the denominator, we must first divide the numerator by the denominator until the remainder is one degree less than the denominator. This division appears as

$$
\begin{array}{r}
s + 2 \\
s^2 + 5s + 6 \overline{)\, s^3 + 7s^2 + 18s + 20} \\
\underline{s^3 + 5s^2 + 6s } \\
2s^2 + 12s + 20 \\
\underline{2s^2 + 10s + 12} \\
2s + 8
\end{array}
$$

and yields

$$Y(s) = s + 2 + \frac{2s + 8}{s^2 + 5s + 6}$$

The last term in this equation is a proper rational function. Performing partial fraction expansion of it gives

$$\frac{2s + 8}{s^2 + 5s + 6} = \frac{2s + 8}{(s+2)(s+3)} = \frac{A_1}{(s+2)} + \frac{A_2}{(s+3)}$$

Since the second subscript on the coefficients is always 1 for nonrepeated factors, we have dropped it for simplicity. We then obtain

$$A_1 = (s+2)\frac{2s+8}{(s+2)(s+3)}\bigg|_{s=-2} = \frac{2(-2)+8}{(-2)+3} = 4$$

$$A_2 = (s+3)\frac{2s+8}{(s+2)(s+3)}\bigg|_{s=-3} = \frac{2(-3)+8}{(-3)+2} = -2$$

by using Heaviside's expansion theorem. Therefore,

$$Y(s) = s+2+4\left(\frac{1}{s+2}\right) - 2\left(\frac{1}{s+3}\right)$$

Finally, we use Table C.4 and the differentiation and linearity theorems to produce

$$y(t) = \delta'(t) + 2\delta(t) + 4e^{-2t}u(t) - 2e^{-3t}u(t)$$

Repeated Factors

In general, it is not convenient to use Method 3 to find all the partial-fraction-expansion coefficients when repeated factors (multiple-order poles) exist because differentiation is required. As we indicated earlier, Method 3 is frequently used to determine as many coefficients as possible without differentiation, and then the remaining coefficients are found by either Method 1 or Method 2.

Example 7.17

Find the inverse Laplace transform of

$$X(s) = \frac{2s^3 + 8s^2 + 11s + 3}{(s+2)(s+1)^3}$$

Solution

$$X(s) = \frac{2s^3 + 8s^2 + 11s + 3}{(s+2)(s+1)^3}$$

$$= \frac{A_1}{s+2} + \frac{A_{2,1}}{s+1} + \frac{A_{2,2}}{(s+1)^2} + \frac{A_{2,3}}{(s+1)^3}$$

We find coefficients A_1 and $A_{2,3}$ first by using Heaviside's expansion theorem to give

$$A_1 = (s+2)X(s)\big|_{s=-2} = \frac{2(-2)^3 + 8(-2)^2 + 11(-2) + 3}{[(-2)+1]^3} = 3$$

$$A_{2,3} = (s+1)^3 X(s)\big|_{s=-1} = \frac{2(-1)^3 + 8(-1)^2 + 11(-1) + 3}{(-1)+2} = -2$$

Thus,

$$X(s) = \frac{3}{s+2} + \frac{A_{2,1}}{s+1} + \frac{A_{2,2}}{(s+1)^2} - \frac{2}{(s+1)^3}$$

We now use Method 1 with the substitutions $s = 0$ and $s = 1$ to compute $A_{2,1}$ and $A_{2,2}$. Note that neither 0 nor 1 is a pole of $X(s)$.

For $s = 0$,

$$\frac{2(0)^3 + 8(0)^2 + 11(0) + 3}{(0+2)(0+1)^3} = \frac{3}{0+2} + \frac{A_{2,1}}{0+1} + \frac{A_{2,2}}{(0+1)^2} - \frac{2}{(0+1)^3}$$

Therefore,

$$\frac{3}{2} = \frac{3}{2} + A_{2,1} + A_{2,2} - 2$$

which gives

$$A_{2,1} + A_{2,2} = 2$$

For $s = 1$,

$$\frac{2(1)^3 + 8(1)^2 + 11(1) + 3}{(1+2)(1+1)^3} = \frac{3}{1+2} + \frac{A_{2,1}}{1+1} + \frac{A_{2,2}}{(1+1)^2} - \frac{2}{(1+1)^3}$$

Therefore,

$$\frac{24}{24} = \frac{3}{3} + \frac{A_{2,1}}{2} + \frac{A_{2,2}}{4} - \frac{2}{8}$$

which gives

$$2A_{2,1} + A_{2,2} = 1$$

We then simultaneously solve the two equations in variables $A_{2,1}$ and $A_{2,2}$ to obtain

$$A_{2,1} = -1 \quad \text{and} \quad A_{2,2} = 3$$

Therefore,

$$X(s) = 3\left[\frac{1}{s+2}\right] - \left[\frac{1}{s+1}\right] + 3\left[\frac{1}{(s+1)^2}\right] - 2\left[\frac{1}{(s+1)^3}\right]$$

From Table C.4 and the s-shift and linearity theorems,

$$x(t) = 3e^{-2t}u(t) - e^{-t}u(t) + 3te^{-t}u(t) - t^2 e^{-t}u(t)$$

which completes the solution. However, we will now also illustrate the use of Method 2 to compute the coefficients $A_{2,1}$ and $A_{2,2}$.

Using Method 2, we start with

$$X(s) = \frac{2s^3 + 8s^2 + 11s + 3}{(s+2)(s+1)^3} = \frac{3}{s+2} + \frac{A_{2,1}}{s+1} + \frac{A_{2,2}}{(s+1)^2} - \frac{2}{(s+1)^3}$$

and multiply both sides of the equation by the denominator factors to clear fractions. The result is

$$2s^3 + 8s^2 + 11s + 3 = 3(s+1)^3 + A_{2,1}(s+2)(s+1)^2$$
$$+ A_{2,2}(s+2)(s+1) - 2(s+2)$$

We then perform the factor multiplications and collect terms to obtain

$$2s^3 + 8s^2 + 11s + 3 = (3 + A_{2,1})s^3 + (9 + 4A_{2,1} + A_{2,2})s^2$$
$$+ (7 + 5A_{2,1} + 3A_{2,2})s + (-1 + 2A_{2,1} + 2A_{2,2})$$

We equate the coefficients of the powers of s on both sides of the equation to produce the four equations

$$2 = 3 + A_{2,1}$$
$$8 = 9 + 4A_{2,1} + A_{2,2}$$
$$11 = 7 + 5A_{2,1} + 3A_{2,2}$$
$$3 = -1 + 2A_{2,1} + 2A_{2,2}$$

Solving the first two equations simultaneously, we find that

$$A_{2,1} = 2 - 3 = -1$$
$$A_{2,2} = -1 - 4A_{2,1} = 3$$

These are the same results obtained using Method 1. The values of $A_{2,1}$ and $A_{2,2}$ found by this method also satisfy the other two equations in $A_{2,1}$ and $A_{2,2}$. Any pair of the four equations can be solved simultaneously for $A_{2,1}$ and $A_{2,2}$. Thus, the four equations in the two unknowns are not independent. This nonindependence occurs because we solved for two of the unknown coefficients previously using Heaviside's expansion theorem. Had we not solved for A_1 and $A_{2,3}$ earlier, then the four equations obtained with Method 2 would have been independent equations in four unknowns, which could have been solved for the four unknowns.

Complex Poles

Thus far, in all our examples of partial-fraction-expansion evaluation we considered transforms for which roots of the denominator (poles) were real values. The poles can also be complex, in which case they occur in complex-conjugate pairs, as we indicated earlier. Since the previously defined techniques have not been specifically developed for real poles, they also can be used for complex poles. The only difference is that complex numbers occur in the coefficient solutions; thus, complex algebra must be used.

Example 7.18

Find the signal $w(t)$ that corresponds to the Laplace transform.

$$W(s) = \frac{3s^2 + 22s + 27}{s^4 + 5s^3 + 13s^2 + 19s + 10}$$

Solution

$$W(s) = \frac{3s^2 + 22s + 27}{s^4 + 5s^3 + 13s^2 + 19s + 10} = \frac{3s^2 + 22s + 27}{(s+1)(s+2)(s^2 + 2s + 5)}$$

$$= \frac{3s^2 + 22s + 27}{(s+1)(s+2)(s+1+j2)(s+1-j2)}$$

$$= \frac{A_1}{s+1} + \frac{A_2}{s+2} + \frac{A_3}{s+1+j2} + \frac{A_4}{s+1-j2}$$

$$A_1 = (s+1)W(s)|_{s=-1} = \frac{3(-1)^2 + 22(-1) + 27}{[(-1)+2][(-1)^2 + 2(-1) + 5]} = 2$$

$$A_2 = (s+2)W(s)|_{s=-2} = \frac{3(-2)^2 + 22(-2) + 27}{[(-2)+1][(-2)^2 + 2(-2) + 5]} = 1$$

$$A_3 = (s+1+j2)W(s)|_{s=-1-j2}$$

$$= \frac{3(-1-j2)^2 + 22(-1-j2) + 27}{[(-1-j2)+1][(-1-j2)+2][(-1-j2)+1-j2]} = -1.5 + j1$$

We could also compute A_4 in a similar manner; however, it is easy to show that $A_4 = A_3^*$ because $\alpha_4 = \alpha_3^*$ and all coefficients of the numerator and denominator polynomials of the transform are real. Therefore,

$$A_4 = -1.5 - j1$$

and

$$W(s) = \frac{2}{s+1} + \frac{1}{s+2} - \frac{1.5 - j1}{s+1+j2} - \frac{1.5 + j1}{s+1-j2}$$

Using Table C.4 and the linearity theorem, we obtain

$$w(t) = 2e^{-t}u(t) + e^{-2t}u(t) - (1.5 - j1)e^{-(1+j2)t}u(t) - (1.5 + j1)e^{-(1-j2)t}u(t)$$

$$w(t) = 2e^{-t}u(t) + e^{-2t}u(t)$$

$$- 3e^{-t}[(e^{j2t} + e^{-j2t})/2]u(t)$$

$$+ 2e^{-t}[(e^{j2t} + e^{-j2t})/2j]u(t)$$

$$w(t) = [2e^{-t} + e^{-2t} - 3e^{-t}\cos(2t) + 2e^{-t}\sin(2t)]u(t)$$

As we indicated in Example 7.18, the two partial-fraction-expansion terms corresponding to a pair of complex-conjugate poles $s = c + jd$ and $s = c - jd$ are of the form

$$X_c(s) = \frac{C + jD}{[s - (c + jd)]} + \frac{C - jD}{[s - (c - jd)]} \tag{7.41}$$

Combining these two terms into a single term gives

$$X_c(s) = \frac{(C + jD)(s - c + jd) + (C - jD)(s - c - jd)}{s^2 - 2cs + (c^2 + d^2)}$$

$$= \frac{2Cs - 2(Cc + Dd)}{s^2 - 2cs + (c^2 + d^2)} \equiv \frac{Es + F}{s^2 + es + f} \tag{7.42}$$

where $E = 2C$, $F = -(Cc + Dd)$, $e = 2c$, and $f = -2(c^2 + d^2)$. Thus, we can alternately perform partial-fraction expansion by including terms of the form of eq. (7.42) for quadratic denominator factors that cannot be factored without producing complex quantities. By so doing, we avoid complex numbers and complex arithmetic.

A term of the form of eq. (7.42) does not appear in the Laplace transform table in Appendix C. What is to be gained, then, by obtaining it in a partial-fraction expansion? The answer is that we can easily separate this term into two terms that are in the table. This separation is achieved by first completing the square in the denominator of eq. (7.42) to give

$$X_c(s) = \frac{Es + F}{(s + e/2)^2 + (f - e^2/4)} \tag{7.43}$$

We then define $\alpha = e/2$ and $\omega_0^2 = f - e^2/4$ and rewrite the numerator of eq. (7.43) so that s appears only in the form $s + \alpha$. The result is

$$X_c(s) = \frac{E(s + \alpha) + (F - E\alpha)}{(s + \alpha)^2 + \omega_o^2}$$

$$= E\left[\frac{s + \alpha}{(s + \alpha)^2 + \omega_o^2}\right] + \frac{F - E\alpha}{\omega_o}\left[\frac{\omega_o}{(s + \alpha)^2 + \omega_o^2}\right] \tag{7.44}$$

We note from Table C.4, that eq. (7.44) is the Laplace transform of the signal

$$x_c(t) = Ee^{-\alpha t}\cos(\omega_o t)u(t) + \left(\frac{F - E\alpha}{\omega_o}\right)e^{-\alpha t}\sin(\omega_o t)u(t) \tag{7.45}$$

The coefficients E and F in eq. (7.45) are computed in conjunction with the computation of coefficients of other terms not found with Heaviside's expansion theorem by using either Method 1 or Method 2. However, if all other poles are of order one (that is, nonrepeated factors) and we use Method 1, then it is more convenient to find E by computing the limit

$$\lim_{s \to \infty} sX(s) = \lim_{s \to \infty} \{s[\text{expansion of } X(s)]\} \tag{7.46}$$

This computation produces the value of E directly. We then find the coefficient F by substituting a convenient value of s, which is not equal to a pole, in the equation $X(s) =$ expansion of $X(s)$. This procedure is illustrated in the following example.

Example 7.19

Repeat Example 7.18 using partial-fraction expansion with quadratic factors.

Solution

$$X(s) = \frac{3s^2 + 22s + 27}{(s+1)(s+2)(s^2+2s+5)} = \frac{A_1}{s+1} + \frac{A_2}{s+2} + \frac{Es+F}{s^2+2s+5}$$

From Example 7.18, $A_1 = 2$ and $A_2 = 1$. Therefore,

$$X(s) = \frac{2}{s+1} + \frac{1}{s+2} + \frac{Es+F}{s^2+2s+5}$$

We find the coefficient E by computing

$$\lim_{s\to\infty}[sX(s)] = \lim_{s\to\infty}\left[\frac{3s^3 + 22s^2 + 27s}{s^4 + 5s^3 + 13s^2 + 19s + 10}\right]$$

$$= \lim_{s\to\infty}\left[\frac{2s}{s+1} + \frac{s}{s+2} + \frac{Es^2 + Fs}{s^2+2s+5}\right]$$

to give $0 = 2 + 1 + E$ which, in turn, yields $E = -3$.

To find F, we set $s = 0$, which is not a pole of $X(s)$, in the partial function expansion equation to give

$$\frac{3(0)^2 + 22(0) + 27}{(0)^4 + 5(0)^3 + 13(0)^2 + 19(0) + 10} = \frac{2}{0+1} + \frac{1}{0+2} + \frac{(-3)(0)+F}{(0)^2 + 2(0) + 5}$$

$$\frac{27}{10} = 2 + \frac{1}{2} + \frac{F}{5}$$

which yields $F = 1$. Therefore,

$$X(s) = \frac{2}{s+1} + \frac{1}{s+2} + \frac{-3s+1}{s^2+2s+5}$$

We then complete the square in the denominator of the quadratic term and rewrite its numerator so that s appears as $s + 1$. The result is

$$X(s) = \frac{2}{s+1} + \frac{1}{s+2} + \frac{-3(s+1)+4}{(s+1)^2 + 4}$$

By separating the quadratic term into two terms, we obtain

$$X(s) = 2\left[\frac{1}{s+1}\right] + \left[\frac{1}{s+2}\right] - 3\left[\frac{s+1}{(s+1)^2+4}\right] + 2\left[\frac{2}{(s+1)^2+4}\right]$$

From Table C.4 and the linearity theorem,

$$w(t) = [2e^{-t} + e^{-2t} - 3e^{-t}\cos(2t) + 2e^{-t}\sin(2t)]u(t)$$

which is the same result as that obtained in Example 7.18.

We can handle repeated quadratic factors corresponding to repeated complex poles in the same manner as repeated linear factors were handled before. Since a number of coefficients must be found by either Method 1 or Method 2, the algebra is tedious; thus, we will not include an example. Also, the table of transforms used to invert the partial-fraction terms must be sufficiently complete to include quadratic denominators raised to the highest power that appears.

Exponential in the Numerator

When we analyze systems, some of the input signals may start later than $t = 0$. As a result, the Laplace transform of these signals and output signals resulting from them may contain exponentials in s in the numerator. (Recall that $x(t - t_o)u(t - t_o) \leftrightarrow X(s)e^{-st_o}$. The inverse Laplace transform of such transforms is obtained by using the partial-fraction-expansion techniques on portions of the transform and then using the time-shift and linearity theorems to complete the inverse transformation.

Example 7.20

Find $x(t)$ that corresponds to

$$X(s) = \frac{s^2 e^{-2s} + e^{-3s}}{s(s^2 + 3s + 2)}$$

Solution

First, we separate the transform into a sum of rational functions multiplied by exponentials in s:

$$X(s) = \left[\frac{s}{s^2 + 3s + 2}\right]e^{-2s} + \left[\frac{1}{s^3 + 3s^2 + 2s}\right]e^{-3s}$$

Next, we construct the partial-fraction expansion of each rational function:

$$X_1(s) = \frac{s}{s^2 + 3s + 2} = \frac{s}{(s+2)(s+1)} = \frac{A_1}{s+2} + \frac{A_2}{s+1}$$

$$A_1 = (s+2)X_1(s)\big|_{s=-2} = \frac{(-2)}{(-2)+1} = 2$$

$$A_2 = (s+1)X_1(s)\big|_{s=-1} = \frac{(-1)}{(-1)+2} = -1$$

$$X_2(s) = \frac{1}{s^3 + 3s^2 + 2s} = \frac{1}{s(s+2)(s+1)} = \frac{A_1}{s} + \frac{A_2}{s+2} + \frac{A_3}{s+1}$$

$$A_1 = sX_2(s)\big|_{s=0} = \frac{1}{[(0)+2][(0)+1]} = 0.5$$

$$A_2 = (s+2)X_2(s)\big|_{s=-2} = \frac{1}{[-2][(-2)+1]} = 0.5$$

$$A_3 = (s+1)X_2(s)\big|_{s=-1} = \frac{1}{[-1][(-1)+2]} = -1$$

Substituting these results in the equation for $X(s)$ gives

$$X(s) = 2\left[\frac{1}{s+2}\right]e^{-2s} - \left[\frac{1}{s+1}\right]e^{-2s} + 0.5\left[\frac{1}{s}\right]e^{-3s}$$

$$+ 0.5\left[\frac{1}{s+2}\right]e^{-3s} - \left[\frac{1}{s+1}\right]e^{-3s}$$

We then use Table C.4 and the time-shift and linearity theorems to find the signal

$$x(t) = 2e^{-2(t-2)}u(t-2) - e^{-(t-2)}u(t-2)$$

$$+ 0.5u(t-3) + 0.5e^{-2(t-3)}u(t-3) - e^{-(t-3)}u(t-3)$$

7.4 Laplace Transform Solutions of Linear Integro-Differential Equations

An important application of Laplace transforms is in solving a set of constant-coefficient linear, integro-differential equations that characterize a system for $t \geq 0$. We perform this solution by first using the differentiation and linearity theorems to compute the Laplace transform of the signals represented by both sides of each integro-differential equation. Since these signals are equal and the Laplace transform is unique, then the Laplace transforms of the two signals are equal. In future discussions, we will refer to this Laplace transform computation as the *Laplace transform of the integro-differential equation* for simplicity. After we compute the Laplace transform of the set of integro-differential equations, we are left with a set of algebraic equations in terms of the Laplace transforms of system signals. We solve these algebraic equations simultaneously for the Laplace transform of the desired signal and invert this transform to find the signal.

Example 7.21

A second-order control system is characterized by the differential equation

$$y^{(2)}(t) + 5y^{(1)}(t) + 6y(t) = x(t)$$

Solve for $y(t)$ for $t \geq 0$ when $x(t) = u(t)$ and the initial conditions are $y(0^-) = 2$ and $y^{(1)}(0^-) = -12$.

Solution

Using the time-differentiation and linearity theorems, we find that the Laplace transform of the differential equation is

$$s^2 Y(s) - sy(0^-) - y^{(1)}(0^-) + 5[sY(s) - y(0^-)] + 6Y(s) = 1/s$$

since $u(t) \leftrightarrow 1/s$. Substituting the initial conditions and rearranging yields

$$(s^2 + 5s + 6)Y(s) = (2s^2 - 2s + 1)/s$$

Solving this algebraic equation for $Y(s)$ and constructing the partial-fraction expansion gives

$$Y(s) = \frac{2s^2 - 2s + 1}{s(s^2 + 5s + 6)} = \frac{2s^2 - 2s + 1}{s(s+2)(s+3)} = \frac{A_1}{s} + \frac{A_2}{s+2} + \frac{A_3}{s+3}$$

where

$$A_1 = (s)Y(s)|_{s=0} = \frac{2(0)^2 - 2(0) + 1}{[(0+2)(0+3)]} = \frac{1}{6}$$

$$A_2 = (s+2)Y(s)|_{s=-2} = \frac{2(-2)^2 - 2(-2) + 1}{(-2)[(-2)+3]} = -\frac{13}{2}$$

and

$$A_3 = (s+3)Y(s)|_{s=-3} = \frac{2(-3)^2 - 2(-3) + 1}{(-3)[(-3)+2]} = \frac{25}{3}$$

Therefore,

$$Y(s) = \frac{1}{6}\left[\frac{1}{s}\right] - \frac{13}{2}\left[\frac{1}{s+2}\right] + \frac{25}{3}\left[\frac{1}{s+3}\right]$$

From Table C.4 and the linearity theorem, the system response is

$$y(t) = \frac{1}{6}u(t) - \frac{13}{2}e^{-2t}u(t) + \frac{25}{3}e^{-3t}u(t)$$

Note that the response to initial conditions is automatically included in the system-response solution of Example 7.21; hence, the total system response for $t \geq 0$ is obtained directly.

Example 7.22

Consider the electric circuit shown in Figure 7.7. The switch is closed for all $t < 0$, so that the circuit is in equilibrium at $t = 0^-$. Find $v_o(t)$ for $t \geq 0$.

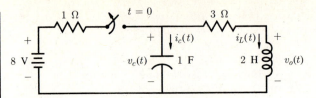

Figure 7.7 Electric Circuit for Example 7.22

Solution

We first obtain the three integro-differential equations that characterize the circuit for $t \geq 0$ (with the switch open) by using Kirchoff's laws and the equations characterizing the circuit elements. The resulting equations are

$$v_c(t) = \int_0^t i_c(t)dt + v_c(0^-) = 3i_L(t) + 2\frac{di_L(t)}{dt}$$

$$i_L(t) = -i_c(t)$$

and

$$v_o(t) = 2\frac{di_L(t)}{dt}$$

We then compute the Laplace transforms of these equations using the time-differentiation, time-integration, and linearity theorems. The results are

$$\left[I_c(s) + v_c(0^-)\right] / s = 3I_L(s) + 2\left[sI_L(s) - i_L(0^-)\right]$$

$$I_c(s) = -I_L(s)$$

and

$$V_o(s) = 2\left[sI_L(s) - i_L\left(0^-\right)\right]$$

where $v_c(0^-)$ and $i_L(0^-)$ are initial conditions that specify the initial energy stored in the capacitor and inductor, respectively. Solving these three algebraic equations simultaneously for $V_o(s)$, we obtain

$$V_o(s) = \frac{2sv_c\left(0^-\right) - 6si_L\left(0^-\right) - 2i_L\left(0^-\right)}{2s^2 + 3s + 1}$$

To find the initial conditions, note that the capacitor acts as an open circuit and the indicator acts as a short circuit at $t = 0^-$ because all voltages and currents are constant. Thus, the circuit representation at $t = 0^-$ is as shown in Figure 7.8.

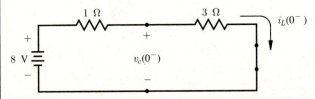

Figure 7.8 Electric Circuit for Example 7.22 at $t = 0^-$

Using Kirchoff's current law and Ohm's law, we find that the initial conditions are

$$i_L\left(0^-\right) = \frac{8}{1+3} = 2\text{A}$$

$$v_c\left(0^-\right) = 3i_L\left(0^-\right) = 6\text{V}$$

Substituting these initial conditions in the expression for $V_o(s)$ and constructing the partial-fraction expansion gives

$$V_o(s) = \frac{-4}{2s^2 + 3s + 1} = \frac{-4}{(2s+1)(s+1)} = \frac{-2}{(s+0.5)(s+1)}$$

$$= \frac{A_1}{s+0.5} + \frac{A_2}{s+1}$$

where we divided the numerator and denominator by 2 so that the partial-fraction terms would correspond to tabulated transforms. Using Heaviside's expansion theorem to calculate the partial-fraction-expansion coefficients gives

$$A_1 = (s+0.5)V_o(s)|_{s=-0.5} = \frac{-2}{(-0.5)+1} = -4$$

and

$$A_2 = (s+1)V_o(s)|_{s=-1} = \frac{-2}{(-1)+0.5} = 4$$

Therefore,

$$V_o(s) = -4\left[\frac{1}{s+0.5}\right] + 4\left[\frac{1}{s+1}\right]$$

Using Table C.4 and the linearity theorem to find $v_o(t)$ for $t \geq 0$, we obtain

$$v_o(t) = -4e^{-0.5t}u(t) + 4e^{-t}u(t)$$

which is plotted in Figure 7.9.

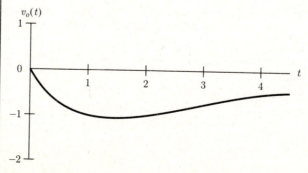

Figure 7.9 Output Signal for the System of Example 7.22

In Example 7.22 we first computed the Laplace transforms of the three integro-differential equations and then solved the resulting algebraic equations simultaneously for $V_o(s)$. For this example, this procedure was more convenient than combining the integro-differential equations to produce an integro-differential equation for $v_o(t)$ and then computing the Laplace transform of the resulting equation. Also, the most convenient initial conditions in Example 7.22 were the capacitor voltage and inductor current. They define the energy stored in the two energy storage elements in the circuit. Since $v_c(t)$ and $i_L(t)$ cannot change instantaneously when the switch is opened, $v_c(0^+) = v_c(0^-)$ and $i_L(0^+) = i_L(0^-)$. Thus, by applying Kirchoff's voltage law to the circuit at $t = 0^+$ we see that $v_o(0^+) = 0$, as indicated in the solution. Also, $v_o(t)$ must approach zero as $t \to \infty$ because all energy stored in the capacitor and inductor leaks off and is dissipated in the resistor. This result also checks the value of the solution as $t \to \infty$.

7.5 Laplace Transform Solutions for Electric Circuits

In Example 7.22, we obtained integro-differential equations for the electric circuit and solved for a circuit voltage signal by using Laplace transform techniques. The solution included the portion of the signal due to initial conditions. We will now see that the solutions for signals in an electric circuit can be found without writing integro-differential equations if the circuit operations and signals are represented with their Laplace transform equivalents. We refer to a circuit diagram produced from these equivalents as a *transformed circuit diagram*. We use this transformed circuit diagram, transforms of input signals, initial conditions, and ordinary techniques of circuit analysis to obtain transforms of signals of interest. The procedure is similar to the one used in Section 6.3 to determine the frequency response of an electric circuit. In that case, a transformed representation of the circuit corresponding to Fourier transformed operations was used.

In order to use the technique we have just described, we require the Laplace transform models for individual circuit elements. The circuit-element representations for these models are shown in Figure 7.10. Those corresponding to the original circuit are referred to as the *t-domain representations* and those corresponding to the transformed circuit are referred to as the *s-domain representations*.

The Laplace transform models of signal sources are obvious, in that all signals are represented by their Laplace transforms. The representations of these models are shown in Figure 7.10 as the first two circuit elements.

Resistance in a circuit is modeled by the equation

$$v(t) = Ri(t) \tag{7.47}$$

when voltage and current directions are defined as in Figure 7.10 and when R is a constant representing the resistance in ohms. Computing the Laplace

Figure 7.10
Representation of
Laplace Transform
Circuit-Element
Models

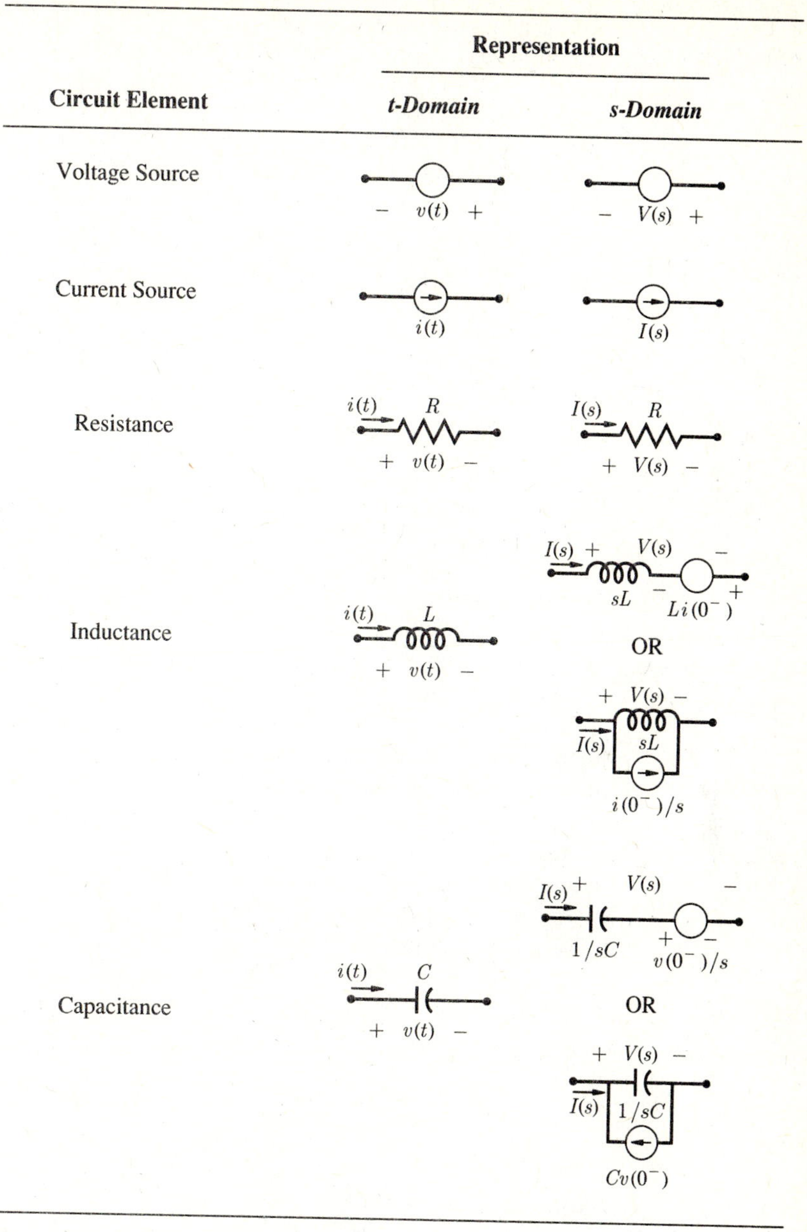

Circuit Element	Representation	
	t-Domain	*s-Domain*
Voltage Source	$-\quad v(t)\quad +$	$-\quad V(s)\quad +$
Current Source	$i(t)$	$I(s)$
Resistance	$i(t)\quad R$ $+\quad v(t)\quad -$	$I(s)\quad R$ $+\quad V(s)\quad -$
Inductance	$i(t)\quad L$ $+\quad v(t)\quad -$	$I(s)\ +\quad V(s)\quad -$ $sL\qquad Li(0^-)$ OR $+\quad V(s)\ -$ $I(s)\quad sL$ $i(0^-)/s$
Capacitance	$i(t)\quad C$ $+\quad v(t)\quad -$	$I(s)\ +\quad V(s)\quad -$ $1/sC\qquad v(0^-)/s$ OR $+\quad V(s)\ -$ $I(s)\quad 1/sC$ $Cv(0^-)$

transform of this equation yields eq. (7.48), which leads to the transformed circuit-element representation shown in Figure 7.10.

$$V(s) = RI(s) \tag{7.48}$$

Since the quantity R produces the transform of the voltage across the resistance when multiplied by the transform of the current through the resistance, we refer to it as the *complex impedance associated with the resistance symbol.*

Inductance in a circuit is modeled by the equation

$$v(t) = L\frac{di(t)}{dt} \tag{7.49}$$

when voltage and current directions are defined as shown in Figure 7.10 and L represents the inductance in henrys. Computing the Laplace transform of this equation gives

$$V(s) = sLI(s) - Li(0^-) \tag{7.50}$$

Direct implementation of eq. (7.50) leads us to the first transformed circuit representation for inductance shown in Figure 7.10, where sL is the complex impedance associated with the inductance symbol. The second representation shown in Figure 7.10 can be easily obtained from the first by source transformation.

Note that the first representation for inductance contains an ideal voltage source, the voltage of which is proportional to the initial current through the inductor. Thus, the initial condition corresponding to this energy-storage element is automatically included in its transformed representation. Actually, since the transformed initial-condition voltage source is constant, it corresponds to an impulse-function source that instantaneously establishes the initial current in the inductor at $t = 0$. Note that this voltage source is equivalent to the additional initial-condition source introduced when solution of systems with nonzero initial conditions was briefly discussed in Chapter 5.

The equation that models the capacitance circuit element is

$$i(t) = C\frac{dv(t)}{dt} \tag{7.51}$$

when voltage and current directions are defined as shown in Figure 7.10 and when C represents the capacitance in farads. Computing the Laplace transform of eq. (7.51) gives

$$I(s) = sCV(s) - Cv(0^-) \tag{7.52}$$

which leads us to the second transformed circuit representation shown for capacitance in Figure 7.10. In this representation, the capacitance symbol is labeled with the complex impedance $1/sC$, corresponding to the complex admittance sC present in eq. (7.52). The first representation for capacitance shown in Figure 7.10 is again easily obtained by performing a source transformation. Note from the second transformed capacitance representation that the initial-capacitance voltage is included by means of a source current that sends an impulse of current through the capacitance at $t = 0$ to charge it to its initial voltage at $t = 0$. Thus, the initial energy-storage condition is once again included in the transformed representation.

Laplace transform models and s-domain representations can be developed for other circuit elements, such as ideal transformers and mutual inductors. The procedures we have outlined apply to these cases as well.

Example 7.23

Consider the simple RC filter circuit and input signal shown in Figure 7.11. Find the output voltage $v_2(t)$ for $t \geq 0$.

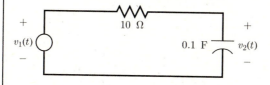

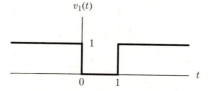

Figure 7.11 Simple RC Filter Circuit and Input Signal for Example 7.23

Solution

For $t < 0$, no current flows because $v_1(t)$ is constant. Therefore, $v_2(0^-) = v_1(0^-) = 1$. For $t \geq 0$,

$$v_1(t) = u(t-1) \leftrightarrow V_1(s) = \frac{e^{-s}}{s}$$

The transformed circuit is shown in Figure 7.12.

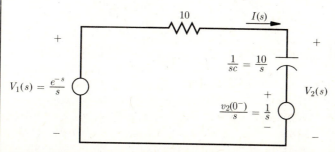

Figure 7.12 Transformed Circuit for Example 7.23

Using Kirchoff's voltage law,

$$\frac{e^{-s}}{s} - 10I(s) - \frac{10}{s}I(s) - \frac{1}{s} = 0$$

Solving this equation for $I(s)$ yields

$$I(s) = \frac{e^{-s} - 1}{10(s + 1)}$$

The transform of the output voltage is

$$V_2(s) = \frac{10}{s}I(s) + \frac{1}{s} = \left[\frac{1}{s(s + 1)}\right]\left[e^{-s} - 1\right] + \frac{1}{s}$$

Letting $Y(s) = \dfrac{1}{s(s + 1)} = \dfrac{A_1}{s} + \dfrac{A_2}{s + 1}$,

$$A_1 = sY(s)|_{s=0} = \frac{1}{0 + 1} = 1$$

$$A_2 = (s + 1)Y(s)|_{s=-1} = \frac{1}{(-1)} = -1$$

and

$$V_2(s) = \left[\frac{1}{s}\right]e^{-s} - \left[\frac{1}{s + 1}\right]e^{-s} + \frac{1}{s + 1}$$

We then use Table C.4 and the time-delay and linearity theorems to obtain the output voltage

$$v_2(t) = u(t - 1) - e^{-(t-1)}u(t - 1) + e^{-t}u(t)$$

The voltage $v_2(t)$ is plotted for all t in Figure 7.13.

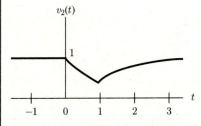

Figure 7.13 Circuit Output Voltage for Example 7.23

If we solve the differential equation directly for $v_2(t)$ in Example 7.23, then two solutions are necessary. These solutions are for $0 \leq t < 1$ and for $t \geq 1$. Also, we must determine a second initial condition at $t = 1^-$ for the second part of the solution. The Laplace transform technique automatically solves for $v_2(t)$ for all $t \geq 0$ in one step.

Example 7.24

Repeat Example 7.22 using the transformed-circuit-and-signals technique.

Solution

At $t = 0^-$, the capacitor acts as an open circuit and the inductor acts as a short circuit because all voltages and currents are constant. Thus, the circuit representation at $t = 0^-$ is as shown in Figure 7.14.

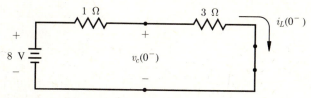

Figure 7.14 Electric Circuit at $t = 0^-$ for Example 7.24

As we see from the figure, the initial conditions required for the transformed capacitance and inductance models are

$$i_L\left(0^-\right) = \frac{8}{1+3} = 2\text{ A} \quad \text{and} \quad v_c\left(0^-\right) = 3i_L\left(0^-\right) = 6\text{V}$$

respectively. The transformed circuit for $t \geq 0$ is shown in Figure 7.15.

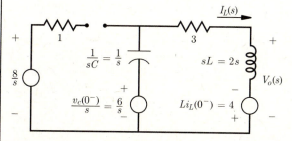

Figure 7.15 Transformed Electric Circuit for $t \geq 0$ for Example 7.24

Using Kirchhoff's voltage law around the circuit loop, we obtain the loop equation

$$\frac{6}{s} - \frac{1}{s}I_L(s) - 3I_L(s) - 2sI_L(s) + 4 = 0$$

The solution of this equation for $I_L(s)$ is

$$I_L(s) = \frac{4s + 6}{2s^2 + 3s + 1}$$

From the transformed circuit in Figure 7.15, we see that the transform of the output voltage is

$$V_o(s) = 2sI_L(s) - 4 = \frac{-4}{2s^2 + 3s + 1}$$

This transform checks the transform of $V_o(s)$ found in Example 7.22. As shown in that example, inversion of this transform gives the output signal

$$v_o(t) = -4e^{-0.5t}u(t) + 4e^{-t}u(t)$$

Comparing Examples 7.24 and 7.22 we see that by using a transformed circuit and signals we can avoid writing integro-differential equations in the solution for $v_o(t)$. Another nice feature of using the transformed circuit model is that all the tools of circuit analysis previously learned, such as Kirchhoff's laws, loop analysis, node analysis, voltage division, current division, and Thevenin's theorem, can be applied. The only difference here is that signals and impedances are a function of s. The following more complicated problem provides an additional example of the use of the transformed-circuit-and-signals technique.

Example 7.25

Consider the circuit and input signal shown in Figure 7.16. Find $v_2(t)$ for $t \geq 0$.

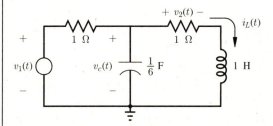

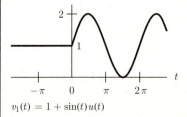

$v_1(t) = 1 + \sin(t)u(t)$

Figure 7.16 Electric Circuit and Input Signal for Example 7.25

Solution

At $t = 0^-$, the capacitance acts like an open circuit and the inductance acts like a short circuit; therefore, $i_L(0^-) = 0.5$ and $v_c(0^-) = 0.5$. For $t \geq 0$, the input signal is equivalent to $v_1(t) = u(t) + \sin(t)u(t)$; therefore,

$$V_1(s) = \frac{1}{s} + \frac{1}{s^2 + 1} = \frac{s^2 + s + 1}{s(s^2 + 1)}$$

The transformed circuit for $t \geq 0$ is shown in Figure 7.17.

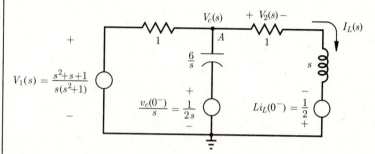

Figure 7.17 Transformed Electric Circuit for $t \geq 0$ for Example 7.25

The nodal equation written at node A is

$$\left[V_c(s) - \frac{s^2 + s + 1}{s(s^2 + 1)} \right] + \left[V_c(s) - \frac{1}{2s} \right]\frac{s}{6} + \left[V_c(s) - \left(-\frac{1}{2}\right) \right]\left[\frac{1}{s+1}\right] = 0$$

Solving the nodal equation for $V_c(s)$, we obtain

$$V_c(s) = \frac{s^4 + 7s^3 + 25s^2 + 19s + 12}{2s(s + 3)(s + 4)(s^2 + 1)}$$

We use Kirchoff's voltage law around the right-hand circuit loop and the impedance relation for the resistor in this loop to obtain the equations

$$V_c(s) - V_2(s) - sI_L(s) + 0.5 = 0 \quad \text{and} \quad I_L(s) = (1)V_2(s)$$

These two equations combine to give the equation

$$V_2(s) = \frac{V_c(s) + 0.5}{s + 1}$$

We now substitute the expression found earlier for $V_c(s)$ in this equation and perform partial-fraction expansion, the result of which is

$$V_2(s) = \frac{s^5 + 8s^4 + 20s^3 + 32s^2 + 31s + 12}{2s(s + 1)(s + 3)(s + 4)(s^2 + 1)}$$

$$= \frac{A_1}{s} + \frac{A_2}{s + 1} + \frac{A_3}{s + 3} + \frac{A_4}{s + 4} + \frac{Es + F}{s^2 + 1}$$

Then, from Heaviside's expansion, we obtain the coefficient values

$$A_1 = 0.5 \qquad A_2 = 0 \qquad A_3 = 0.6 \qquad A_4 = -0.353$$

Substitution of the computed A_i values in the partial fraction expression for $V_2(s)$, multiplication of both sides by s, and evaluation of the limit as $s \to \infty$ gives

$$E = -0.247$$

We then substitute the values of E and the A_i's in the partial-fraction expression for $V_2(s)$, set $s = 1$ (1 is not a pole) and solve the resulting equation for the parameter F to give $F = 0.388$. Therefore,

$$V_2(s) = \frac{0.5}{s} + \frac{0.6}{s+3} - \frac{0.353}{s+4} - \frac{0.247s}{s^2+1} + \frac{0.388}{s^2+1}$$

Finally, we can use Table C.4 and the linearity theorem to obtain

$$v_2(t) = [0.5 + 0.6e^{-3t} - 0.353e^{-4t} - 0.247\cos(t) + 0.388\sin(t)]u(t)$$

In Example 7.25, we found the partial-fraction-expansion coefficient A_2 to be zero. This means that the factor associated with this coefficient (that is, $s + 1$) is a factor of the numerator also. For the circuit and signal of Example 7.25, we note that when $t \to \infty$, $v_2(t)$ approaches

$$v_2(t) = 0.5 - 0.247\cos(t) + 0.388\sin(t) = 0.5 + 0.460\sin(t - 32.48°)$$

This signal is the sum of a constant term and a sinusoidal term. We refer to these two signal terms as the *steady-state response* of the circuit. They result after the constant plus a sinusoid input signal has been applied to the circuit for a long period of time. The steady-state response terms can be found with dc and ac circuit analysis, in which it is assumed that signals have been applied for long periods of time preceding the analysis time and, therefore, that components due to signal switching have died out. The signal components that die out are referred to as the *transient response* of the circuit.

7.6 The System Transfer Function

The use of the *system transfer function* to represent physical systems is an outgrowth of the use of the Laplace transform. The system transfer function is one of the principal tools in determining output signals from input signals and in determining system properties. In this section, we will define the transfer function and discuss some of its characteristics.

Transfer-Function Definition and Computation

In Chapter 5, we demonstrated that

$$y(t) = \int_{-\infty}^{\infty} x(\lambda)h(t - \lambda)\,d\lambda \qquad (7.53)$$

for linear time-invariant systems with zero initial conditions, where $x(t)$, $h(t)$, and $y(t)$ are the input signal, impulse response, and output signal, respectively. In addition, if the system is causal (that is, if $h(t) = 0$ for $t < 0$) and if $x(t) = 0$ for $t < 0$, then eq. (7.53) reduces to

$$y(t) = \int_0^t x(\lambda)h(t - \lambda)\, d\lambda \tag{7.54}$$

since $x(\lambda) = 0$ for $\lambda < 0$ and $h(t - \lambda) = 0$ for $\lambda > t$. Applying the convolution theorem to eq. (7.54), we obtain

$$Y(s) = X(s)H(s) \tag{7.55}$$

as the expression for the Laplace transform of the system output. Equation (7.55) can be expressed as

$$\mathcal{L}[h(t)] = H(s) = Y(s)/X(s) \tag{7.56}$$

to show that the Laplace transform of the impulse response of a causal system is the ratio of the Laplace transform of the output signal to the Laplace transform of the input signal when initial conditions are zero.

Definition ──

The system transfer function is

$$H(s) = Y(s)/X(s) \tag{7.57}$$

where $Y(s)$ is the Laplace transform of the output signal that corresponds to zero initial conditions when the Laplace transform of the input signal is $X(s)$.

Except for initial conditions, the system transfer function completely characterizes the system because the impulse response completely characterizes the system, except for initial conditions, and because the single-sided Laplace transform is unique.

When the initial conditions are zero, we can use the system transfer function to find the output of a causal system to any input that is zero for $t < 0$. This procedure consists of the following steps:

1. Compute the Laplace transform of the input.
2. Multiply this transform by the transfer function.
3. Compute the inverse transform of the result.

We indicated in Chapter 1 that the differential equation for an nth order, lumped-parameter, linear, time-invariant system is of the form

$$\sum_{i=0}^{n} a_i \frac{d^i y(t)}{dt^i} = \sum_{k=0}^{m} b_k \frac{d^k x(t)}{dt^k} \tag{7.58}$$

The coefficients of eq. (7.58) are all real and constant for systems for which the output is real when the input is real. With zero initial conditions, the Laplace transform of eq. (7.58) is

$$\sum_{i=0}^{n} a_i s^i Y(s) = \sum_{k=0}^{m} b_k s^k X(s) \tag{7.59}$$

which gives the transfer function

$$H(s) = Y(s)/X(s) = \left[\sum_{k=0}^{m} b_k s^k\right] \bigg/ \left[\sum_{i=0}^{n} a_i s^i\right] \tag{7.60}$$

As we indicated earlier, the transfer function is a rational function of s.

The two examples that follow illustrate the determination and use of transfer functions. For electric circuits, transfer functions are easily found using transformed circuits and the usual circuit-analysis techniques.

Example 7.26

Find the transfer function for the bridged-T filter supplying the 2-Ω resistor load shown in Figure 7.18.

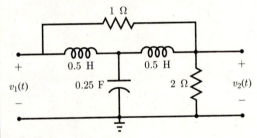

Figure 7.18 Bridged-T Filter Circuit for Example 7.26

Solution
The transformed circuit corresponding to the circuit of Figure 7.18 is shown in Figure 7.19.

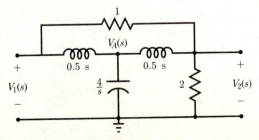

Figure 7.19 Transformed Circuit for Example 7.26

Note that because the transfer function is defined for a system with zero initial conditions, no initial-condition sources are shown.

For the node equations at the two independent nodes, we write

$$[V_A(s) - V_1(s)][2/s] + [V_A(s) - 0][s/4] + [V_A(s) - V_2(s)][2/s] = 0$$

and $\quad [V_2(s) - V_1(s)][1] + [V_2(s) - V_A(s)][2/s] + [V_2(s) - 0][1/2] = 0$

We then collect terms, combine the equations to eliminate $V_A(s)$, and solve the resulting equation for $V_2(s)/V_1(s)$. From these steps follows

$$H(s) = \frac{V_2(s)}{V_1(s)} = \frac{2s^3 + 32s + 32}{3s^3 + 4s^2 + 48s + 32}$$

The next example illustrates transfer-function determination and use of the transfer function for a system that is not an electric circuit.

Example 7.27

The rotor system of a television antenna is shown in Figure 7.20. Neglecting the inductance and resistance of the motor and the friction and inertia of the motor and antenna, find (a) the transfer function that relates the transform of the angular position of the antenna to the transform of the angular position of the adjusting knob and (b) the angular position of the antenna for $t > 0$ when a step input of C degrees is applied to the adjusting knob at $t = 0$ and $\theta_i(0^-) = \theta_o(0^-) = 0$.

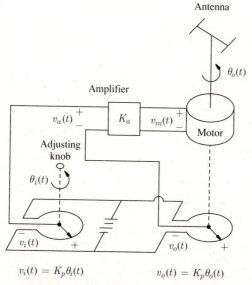

$$v_i(t) = K_p\theta_i(t) \qquad\qquad v_o(t) = K_p\theta_o(t)$$

Figure 7.20 Television-Antenna Rotor System

Solution

a. Since the resistance and inductance of the motor and the friction and inertia of the motor and antenna are all assumed to be zero, the shaft velocity of the motor is proportional to the motor input voltage. That is, $K_m v_m(t) = d\theta_o(t)/dt$, where K_m is the motor constant in degrees/second per volt. Therefore,

$$\theta_0(t) = K_m \int_0^t v_m(\lambda)\, d\lambda + \theta_0(0^-) = K_m \int_0^t v_m(\lambda)\, d\lambda$$

The system is represented by the system block diagram shown in Figure 7.21, where K_p and K_a are the potentiometer factor and amplifier gain, respectively.

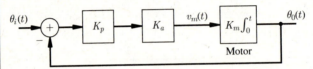

Figure 7.21 Block-Diagram Representation of Antenna Rotor System

We transform the system block diagram by transforming the signals and the operations corresponding to each block. The result is the transformed system block diagram shown in Figure 7.22.

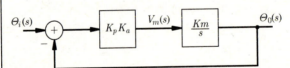

Figure 7.22 Transformed Block Diagram for Antenna Rotor System

From the transformed block diagram, we obtain

$$\Theta_0(s) = K_p K_a K_m [\Theta_i(s) - \Theta_0(s)]/s$$

We solve this equation to find the system transfer function. The result is

$$H(s) = \Theta_0(s)/\Theta_i(s) = \frac{K_p K_a K_m}{s + K_p K_a K_m} \equiv \frac{K}{s + K}$$

where $K = K_p K_a K_m$.

b. For a step input signal of amplitude C (that is, a sudden turn of adjusting knob by C degrees to a new position),

$$\theta_i(t) = C u(t)$$

Therefore,

$$\theta_i(s) = C/s$$

and

$$\theta_0(s) = \theta_i(s)H(s) = \frac{CK}{s(s+K)} = \frac{A_1}{s} + \frac{A_2}{s+k}$$

where

$$A_1 = \frac{CK}{(0)+K} = C \quad \text{and} \quad A_2 = \frac{CK}{(-K)} = -C$$

The partial-fraction expansion of the output-signal transform is thus

$$\theta_0(s) = \frac{C}{s} - C\left(\frac{1}{s+K}\right)$$

We then apply transforms found in Table C.4 and the linearity theorem to the transformed output signal to obtain the output signal

$$\theta_0(t) = C\left[1 - e^{-Kt}\right]u(t)$$

which is the television antenna angular position in degrees. This angular position is illustrated in Figure 7.23a.

Figure 7.23
Step Response for
Antenna Rotor
System

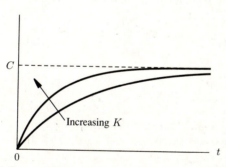

(a) Step response for system model of Example 7.26

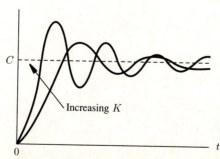

(b) Step response for a more complete system model

We see from Example 7.27 that the angular location of the antenna approaches the commanded angular location if we wait long enough, and that increasing the rotor-system gain, $K = K_p K_a K_m$, increases the speed with which the system responds. In a practical antenna rotor system, however, the system gain cannot be increased indefinitely to improve performance. The neglected terms produce a higher degree denominator with complex poles, so that the output to a step input is of the form

$$\theta_0(t) = C\big[1 - e^{-\alpha t}\cos\omega_0 t - \beta e^{-\alpha t}\sin\omega_0 t\big]u(t) \qquad (7.61)$$

An output of this form is as shown in Figure 7.23b. Very large overshoot of and oscillation about the commanded antenna position can occur if the system gain is made too large, producing excessive strain on the antenna.

Transfer-Function Characteristics

We showed with eq. (7.60) that the transfer function of an nth-order, lumped-parameter, linear, time-invariant system is the rational function

$$H(s) = \frac{b_o + b_1 s + \cdots + b_m s^m}{a_o + a_1 s + \cdots + a_n s^n} = \frac{N(s)}{D(s)} \qquad (7.62)$$

When we factor the transfer-function numerator and denominator polynomials, we obtain

$$H(s) = \left(\frac{b_m}{a_n}\right)\frac{(s - z_1)(s - z_2)\cdots(s - z_m)}{(s - p_1)(s - p_2)\cdots(s - p_n)} = \frac{N(s)}{D(s)} \qquad (7.63)$$

where the z_i's and p_i's are the zeros and poles, respectively, of the transfer function. Note that the transfer function is characterized, except for a gain factor, by its zeros and poles. Thus, a system is characterized, except for initial conditions and a gain factor, by the location of the zeros and poles corresponding to its transfer function. We frequently plot these zeros and poles on the s plane to indicate this characterization, using O's for zeros and X's for poles. We refer to the plot as the *pole-zero plot* or, alternatively, the *pole-zero diagram* or the *pole-zero constellation*.

Example 7.28

Find the pole-zero plots for the transfer functions

$$H_1(s) = \frac{2s^2 + 6s - 8}{s^3 + 4s^2 + 9s + 10}$$

and

$$H_2(s) = \frac{s^2 + 10s + 34}{s^4 + 6s^3 + 38s^2 + 62s + 29}$$

Solution

We factor the transfer functions specified to give

$$H_1(s) = \frac{2(s-1)(s+4)}{(s+2)(s+1+j2)(s+1-j2)}$$

and

$$H_2(s) = \frac{(s+3)(s+5+j3)(s+5-j3)}{(s+1)^2(s+2+j5)(s+2-j5)}$$

The corresponding pole-zero plots are shown in Figure 7.24, where the number ② beside the pole at $s = -1$ for $H_2(s)$ indicates a second-order pole.

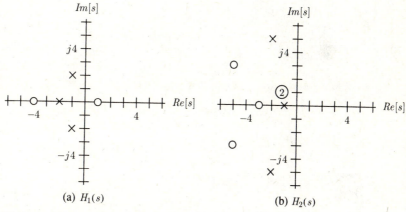

(a) $H_1(s)$ (b) $H_2(s)$

Figure 7.24 Pole-Zero Plots for Example 7.28

The Laplace transform of the output of a system when its initial conditions are zero (that is, the zero-initial-energy response of the system) is

$$Y(s) = H(s)X(s) = \frac{N(s)}{D(s)}X(s) \tag{7.64}$$

Therefore,

$$Y(s) = \frac{G_y(s-z_1)\cdots(s-z_m)(s-\beta_1)\cdots(s-\beta_k)}{(s-p_1)\cdots(s-p_n)(s-\alpha_1)\cdots(s-\alpha_\ell)} \tag{7.65}$$

where the β_i's and α_i's are the zeros and poles, respectively, of the input signal and G_y is a real constant.

If $Y(s)$ contains only single-order poles, then its partial-fraction expansion contains components of the form

$$Y_c(s) = \frac{A}{s-\alpha} \tag{7.66}$$

$$Y_c(s) = \frac{A}{(s - \alpha + j\omega_0)(s - \alpha - j\omega_0)} = \frac{A}{(s - \alpha)^2 + \omega_0^2} \qquad (7.67)$$

or

$$Y_c(s) = \frac{A(s - \alpha)}{(s - \alpha + j\omega_0)(s - \alpha - j\omega_0)} = \frac{A(s - \alpha)}{(s - \alpha)^2 + \omega_0^2} \qquad (7.68)$$

The corresponding output-signal components are

$$y_c(t) = Ae^{\alpha t}u(t) \qquad (7.69)$$

$$y_c(t) = Ae^{\alpha t}\sin(\omega_0 t)u(t) \qquad (7.70)$$

or

$$y_c(t) = Ae^{\alpha t}\cos(\omega_0 t)u(t) \qquad (7.71)$$

where the value of A depends on G_y and the pole and zero locations and α and ω_0 are specified by the pole locations.

We can use the same reasoning for repeated poles. In this case, the output-signal components that result are of the form

$$y_c(t) = f_1(t)e^{\alpha t}u(t) \qquad (7.72)$$

$$y_c(t) = f_1(t)e^{\alpha t}\sin(\omega_o t)u(t) \qquad (7.73)$$

or

$$y_c(t) = f_1(t)e^{\alpha t}\cos(\omega_o t)u(t) \qquad (7.74)$$

where $f_1(t)$ is a polynomial in t, the degree of which is one less than the order of the pole and the coefficient values of which depend on G_y and the pole and zero locations. Note that eqs. (7.69) through (7.71) are special cases of eqs. (7.72) through (7.74) where the polynomial $f_1(t)$ is a constant.

We note from eq. (7.65) that the output signal contains some components, the forms of which are determined by the poles of the transfer function. Other output-signal components have forms determined by the poles of the transform of the input signal.

We can see from eqs. (7.72) through (7.74) that the value of the real part, α, of a pole determines how quickly the signal component corresponding to the pole grows in time if $\alpha > 0$ or decays in time if $\alpha < 0$. Poles close to the imaginary axis produce slowly growing or decaying components. Furthermore, eqs. (7.72) through (7.74) show that the value of the imaginary parts of complex-conjugate pairs of poles of a signal transform determines the frequency of oscillation of oscillatory components in the signal.

If we return to eq. (7.58) and compute the Laplace transform when system initial conditions are not zero, then

$$\sum_{i=0}^{n} a_i s^i Y(s) + I(s) = \sum_{k=0}^{m} b_k s^k X(s) \qquad (7.75)$$

where $I(s)$ is a polynomial in s resulting from the initial conditions. Substituting eq. (7.62) in eq. (7.75) and rearranging, we obtain

$$Y(s) = \frac{N(s)}{D(s)}X(s) + \frac{I(s)}{D(s)} \tag{7.76}$$

as the expression for the Laplace transform of the output signal. The first component in eq. (7.76) is the Laplace transform of the zero-initial-energy response previously discussed. The second component in eq. (7.76) is the Laplace transform of the system output when the initial conditions are nonzero and the input signal is zero (that is, the zero-input response). Note that the forms of the components of the system zero-input response depend on the location of the poles corresponding to the transfer-function denominator polynomial $D(s)$. The numerator of the system transfer function has no effect on the zero-input response; hence, the zeros of the system transfer function have no effect on the zero-input response.

7.7 Use of the Transfer Function for Determining System Stability and Frequency Response

With the system transfer function, we can perform extensive system analysis and design directly, without having to apply specific signals and invert the transform of the system response to determine the output signal. Nor do we have to perform a separate spectral analysis. For example, control-system design and compensation is frequently performed by finding the pole and zero locations of the system transfer function and adding compensating system components to modify these locations to produce desired changes in system performance characteristics. Another example is in the determination of system stability without actually computing the output signal. Still another example is in the determination of the system frequency response without performing a separate spectral analysis.

System Stability

In most cases, we want a system to produce a bounded output to any bounded input. That is, the system must have bounded-input bounded-output (BIBO) stability, as defined in Section 1.4. We can determine whether a system is BIBO stable directly from the system transfer function. If the system transfer function is a rational function, it must satisfy two conditions. The first condition is that the degree of the numerator must be no larger than the degree of the denominator. If the numerator degree is k larger then the denominator degree, then we can divide to produce

$$H(s) = \frac{N(s)}{D(s)} = C_k s^k + \cdots + C_1 s + C_0 + \frac{N_1(s)}{D(s)} \tag{7.77}$$

where the degree of $N_1(s)$ is less than the degree of $D(s)$ in the remainder term. If we apply a unit step input signal (which is bounded) to the system, then the transform of the system response is

$$Y(s) = \frac{1}{s}H(s) = C_k s^{k-1} + \cdots + C_1 + \frac{C_0}{s} + \frac{N_1(s)}{sD(s)} \qquad (7.78)$$

Assuming that $N_1(s)/sD(s)$ corresponds to a bounded signal, then

$$y(t) = C_k \delta^{(k-1)}(t) + \cdots + C_1 \delta(t) + C_0 u(t) + y_1(t) \qquad (7.79)$$

where $C_0 u(t)$ and $y_1(t)$ are bounded, but the rest of the components are not. Since we found a bounded input that produces an unbounded output when $k > 0$, then the system does not possess BIBO stability when $k > 0$.

The second condition the system transfer function must satisfy for the system to possess BIBO stability is that its poles must lie in the left half of the s-plane (that is, at $s = \alpha \pm j\omega_0$ where $\alpha < 0$). To show that this condition is required, we recall that the system response contains components of the form shown in eqs. (7.72) through (7.74), corresponding to poles of the transform of the input signal and poles of the transfer function. Note that these components are bounded if $\alpha \leq 0$ for single-order poles and if $\alpha < 0$ for repeated poles.

Let us first consider the case where none of the poles of the input-signal transform are equal to a transfer-function pole. In this case, the output-signal components corresponding to the poles of the input-signal transform have the same form as the input-signal components; thus, they are bounded if the input signal is bounded. Note that the transform of a bounded input signal may have repeated poles for which $\alpha < 0$, single-order poles for which $\alpha = 0$, and no poles for which $\alpha > 0$. The output-signal components corresponding to the transfer-function poles are bounded if $\alpha \leq 0$ for single-order poles and if $\alpha < 0$ for repeated poles.

In the case where the input-signal transform and transfer-function poles satisfy the preceding conditions, but an input-signal pole is equal to a single-order transfer-function pole, then a repeated pole is generated in the output-signal transform at this pole location. In this case, a bounded output is produced only if the pole does not lie on the imaginary axis (that is, if $\alpha < 0$).

We see from the two cases described that a system produces a bounded output to any bounded input if the transfer-function poles lie in the left half of the s-plane (that is, $\alpha < 0$). In this case, the system is BIBO stable. We also see that a system produces a bounded output to most bounded inputs if the transfer-function poles lie in the left half of the plane or are single-order on the imaginary axis. We say, in this case, that the system is marginally BIBO stable. In all other cases, the system is unstable. Therefore, we have shown that a system possesses BIBO stability if its transfer function contains poles only in the left half of the s-plane.

If the system impulse response is absolutely integrable, then c in the existence theorem for single-sided Laplace transforms can be chosen to be 0^-. This implies that the transfer function converges for $Re[s] > 0^-$ and consequently that the system poles are in the left half of the s-plane and that the system possesses BIBO stability. Therefore, an equivalent condition for a system to possess BIBO stability is that its impulse response be absolutely integrable.

A technique such as the *Routh-Hurwitz test* can be used to determine whether a transfer function has poles that are not in the left half of the s-plane without actually factoring the denominator of the transfer function. Although this technique is beyond the scope of this text, it is discussed in elementary control-system texts.[†]

It is possible for a system to produce a bounded output signal for any bounded input signal but not produce a bounded output to initial conditions. This result occurs if $N(s)$ contains a zero in the right half of the s-plane and $D(s)$ contains a pole at this same location in addition to one or more poles in the left half of the s-plane. We can see from eq. (7.76) that pole-zero cancellation occurs in the transfer function in this case, so that the transfer function does not contain the pole in the right half of the s-plane. To produce a bounded output signal for any bounded input signal and any set of initial conditions, the poles corresponding to $D(s)$ must lie in the left half of the s-plane. This being the case, we say that the system is asymptotically stable; both the zero-initial-energy response to bounded input signals and the zero-input response are bounded. In fact, the zero-input response approaches zero as t approaches infinity. Note that asymptotic stability implies BIBO stability, but that the converse is not true. System stability is treated in detail in advanced texts on linear systems.[††] Problem 7.35 at the end of the chapter considers a system that is BIBO stable but not asymptotically stable.

System Frequency Response

We showed in Chapter 6 that the frequency response of a linear time-invariant system with zero initial conditions shows how the signal spectrum changes when the signal passes through the system. We also showed that the frequency response can be obtained as the Fourier transform of the impulse response.

The system transfer function of a causal system is the unique single-sided Laplace transform of the system impulse response. It is computed without resorting to transforms in the limit. Thus, we should be able to obtain the system frequency response from the system transfer function for a causal system, if the frequency response exists in a nonlimiting sense; that is, if the frequency response can be computed as the Fourier transform of the system impulse response without resorting to Fourier transforms in the limit. The frequency response is

$$H(f) = \mathcal{F}[h(t)] = \int_{-\infty}^{\infty} h(t)e^{-j2\pi ft} \, dt \qquad (7.80)$$

[†]See for example R. C. Dorf, *Modern Control Systems*, 5th ed. (Reading, Mass.: Addison-Wesley, 1989).

[††]See for example T. Kailath, *Linear Systems* (Englewood-Cliffs, N.J.: Prentice-Hall, 1980).

Since the system is causal, $h(t) = 0$ for $t < 0$ and

$$H(f) = \int_0^\infty h(t)e^{-j2\pi ft}\, dt = \mathcal{L}[h(t)]\Big|_{s=j2\pi f} \tag{7.81}$$

Thus, if the frequency response exists in the nonlimiting sense, then it is the transfer function with s replaced by $j2\pi f$. We see from eq. (7.81) that the frequency response is the transfer function evaluated along the imaginary axis in the s plane (that is, along the line $s = 0 + j2\pi f = 0 + j\omega$). This is illustrated in Figure 7.25a for a system with a single pole on the real axis and in Figure 7.25b for a system with a complex-conjugate pair of poles. We used a system with zero phase shift in these illustrations for simplicity. The system frequency response does not exist in a nonlimiting sense if the system has poles on the imaginary axis or in the right half of the s-plane. In such

Figure 7.25
Illustration of
Transfer
Function–Frequency
Response
Relationship

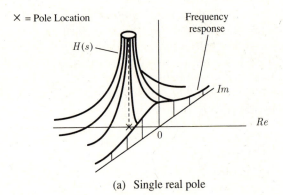

(a) Single real pole

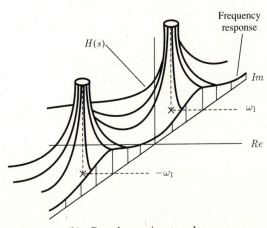

(b) Complex conjugate poles

cases, the transfer function does not converge on the imaginary axis; hence, the Fourier transform of the impulse response does not exist.

Figure 7.25 shows that the system pole location affects the system frequency response. Note from Figure 7.25a that a single pole on the negative real axis corresponds to a low-pass system because the corresponding frequency response shows that the gain applied to low-frequency signal components is greater than the gain applied to high-frequency signal components. More rapid gain decrease with increasing frequency (that is narrower system bandwidth) is obtained when the pole is moved closer to the imaginary axis. The complex pair of poles give a frequency response with a band-pass characteristic, with the peak response corresponding to the frequency ω_1 when $s = \alpha \pm j\omega_1$ are the pole locations. Again, the system bandwidth becomes narrower as the poles approach the imaginary axis.

7.8 Summary

The single-sided Laplace transform is an effective tool for finding the response of a causal system to an input signal that is zero for $t < 0$ because (1) it replaces integro-differential equations with algebraic equations, (2) it automatically includes the effects of nonzero initial conditions, and (3) it can be used with many signals that do not have a spectrum, such as the ramp signal. The single-sided Laplace transform is a special case of the double-sided Laplace transform, which is obtained from the Fourier transform through the use of a convergence factor. The term single-sided Laplace transform is shortened to Laplace transform for simplicity of reference. It produces unique transform pairs. A set of Laplace transform theorems proves useful in the evaluation of Laplace transforms of signals. These theorems and a number of Laplace transform pairs are summarized in Tables C.3 and C.4 of Appendix C.

Many signals of interest to us, including the impulse response of a causal, lumped-parameter, linear, time-invariant system, produce Laplace transforms which are rational functions of s. Partial-fraction expansion is a useful technique for performing the inverse transform of rational functions of s.

Laplace transforms can be used in the solution of sets of integro-differential equations. This transform solution requires (1) Laplace transformation of the equations and independent time functions, (2) simultaneous algebraic solution of the resulting transformed equations for the transform of a signal of interest, and (3) inverse transformation to determine the time function of the signal of interest. Initial conditions are automatically included when the equations are transformed. For electric circuits, it is easier to transform the circuit representation and use circuit-analysis techniques. This eliminates the necessity of finding the integro-differential equations.

The transfer function of a system is the ratio of the Laplace transform of the output signal to the Laplace transform of the input signal when initial conditions are zero. It is also the Laplace transform of the system impulse response.

The system transfer function completely characterizes a system, except for the system initial conditions. The poles and zeros of the system transfer function provide characterization of the system, except for gain and initial conditions. They indicate the form of some components of the system response and the characteristics of the system frequency response. The frequency response of a system and the stability of the system can be determined directly from the system transfer function without performing a separate spectral analysis or considering a specific input signal and inverting the transform of the system response to find the system response.

Problems

7.1 Use the Laplace transform integral to find the Laplace transforms of the following signals:

a. $x(t) = 3u(t) - 3u(t - 3)$

b. $y(t) = r(t) - r(t - 1) - u(t - 2)$

c. $w(t) = 10u(t) - 10e^{-t}u(t)$

d. $z(t) = 3\cos(4t)u(t)$

7.2 Use the Laplace transform integral to find the Laplace transforms of the following signals:

a. $x(t) = (t - 1)u(t)$

b. $y(t) = e^{-(t+4)}u(t)$

c. $w(t) = e^{(t-3)}u(t - 4)$

d. $z(t) = 10\sin(2t)u(t)$

7.3 Use tables and theorems to find the Laplace transforms of the signals in Problem 7.1.

7.4 Use tables and theorems to find the Laplace transforms of the signals in Problems 7.2.

7.5 Use tables and theorems to find the Laplace transforms of the following signals:

a. $x(t) = e^{-5t}[u(t) - u(t - 5)]$

b. $y(t) = [e^{-3t} - e^{-10t} + 2\cos(3t)]u(t)$

c. $w(t) = [2\cos(4t) + 3\sin(4t) - e^{-2t}]u(t)$

d. $z(t) = [4e^{-2t}\cos(5t) - 3e^{-2t}\sin(5t)]u(t)$

Simplify each transform found so that it is a ratio of functions of s.

7.6 Prove the linearity theorem (Theorem 1).

7.7 Prove the time-delay theorem (Theorem 3).

7.8 Prove the time-integration theorem (Theorem 7).

7.9 Prove the final-value theorem (Theorem 9).

7.10 Find the Laplace transform of the signal $x(t) = \Lambda(t - 1)$ by using the signal $y(t) = \Pi(t - 0.5)$ and the convolution theorem.

7.11 Use the initial- and final-value theorems to find the initial and final values for the time functions corresponding to the following Laplace transforms:

a. $X(s) = \dfrac{s}{(s + 1)(s + 4)}$

b. $Y(s) = \dfrac{2s + 1}{s(s + 2)}$

c. $Z(s) = \dfrac{e^{-2s}}{(s + 1)(s + 2)^2}$

7.12 Find the initial and final values for the time functions corresponding to the following Laplace transforms:

a. $X(s) = \dfrac{1}{s + 3}$

b. $Y(s) = \dfrac{5s - 12}{s^2 + 4s + 13}$

c. $W(s) = \dfrac{2 - e^{-4s}}{3s^2 + 2s}$

7.13 Find the inverse Laplace transforms of

a. $X(s) = \dfrac{s}{(s + 1)(s + 4)}$

b. $Y(s) = \dfrac{s^3 + 6s^2 + 6s}{s^2 + 6s + 8}$

c. $W(s) = \dfrac{s+1}{s^2+2s}$

7.14 Find the inverse Laplace transforms of

a. $X(s) = \dfrac{19s^2 + 25s + 16}{(s+3)^3(s+1)}$

b. $Y(s) = \dfrac{6s^2 - 2s + 2}{(s+1)(s^2+4s+13)}$

7.15 Find the inverse Laplace transforms of

a. $X(s)$

$$= \dfrac{-5s^5 + 20s^4 + 64s^3 + 103s^2 + 85s + 34}{(s+2)^2(s+6)(s+1)^3}$$

b. $Y(s) = \dfrac{-s^2 + 7s + 10}{s^3 + 2s^2 + 5s}$

7.16 Find the inverse Laplace transforms of

a. $X(s) = \dfrac{1 + e^{-2s}}{3s^2 + 2s}$

b. $Y(s) = \dfrac{se^{-s} + 2s^2 + 9}{s(s^2 + 9)}$

7.17 Use Laplace transforms to solve the following differential equation for $x(t)$ for $t \geq 0$, subject to the given initial conditions:

$$x^{(2)}(t) + 2x^{(1)}(t) + x(t) = 3\delta(t)$$

$$x(0^-) = 1 \qquad x^{(1)}(0^-) = -2$$

7.18 Use Laplace transforms to solve the following differential equation for $y(t)$ for $t \geq 0$, subject to the given initial conditions:

$$y^{(2)}(t) + 6y^{(1)}(t) + 9y(t) = 9u(t)$$

$$y(0^-) = -1 \qquad y^{(1)}(0^-) = 9$$

7.19 Consider the electric circuit shown in Figure 7.26 where the switch is closed at $t = 0$.

a. Write a set of integro-differential equations that include the current $i(t)$ and characterize the circuit for $t \geq 0$.

b. Use Laplace transforms to solve for $i(t)$ for $t \geq 0$.

7.20 Repeat Problem 7.19 for the circuit shown in Figure 7.27.

7.21 The electric circuit shown in Figure 7.28 supplies the 4-Ω resistive load from one of two dc sources. The switch is instantaneously thrown from point A to point B at $t = 0$.

a. Write a set of integro-differential equations that include the voltage $v_o(t)$ and characterize the circuit for $t \geq 0$.

b. Use Laplace transforms to solve for $v_o(t)$ for $t \geq 0$.

7.22 The pipeline flow-rate control system of Problem 5.4 is represented by the block diagram shown in Figure 7.29, where time is expressed in minutes.

Figure 7.26

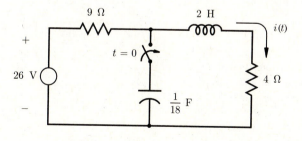

Figure 7.27

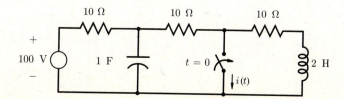

Figure 7.28

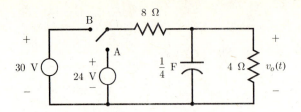

Figure 7.29

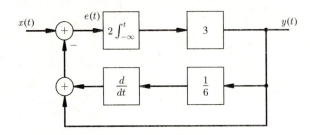

a. Write a set of integro-differential equations that includes the pipeline flow rate and specified flow rate ($y(t)$ and $x(t)$, respectively) and that characterizes the system for $t \geq 0$.

b. The specified flow rate has been 200 gal/min for a long time preceding $t = 0$, so that the pipeline flow is 200 gal/min at $t = 0^-$. Use Laplace transforms to solve for the pipeline flow rate for $t \geq 0$ when the specified flow rate is instantly changed to 300 gal/min at $t = 2$ min. Plot the flow rate for $0 \leq t \leq 4$ min.

c. Repeat parts (a) and (b) when the derivative feedback is removed. Comment on the effect of the derivative feedback.

7.23 Suppose you are traveling 40 mph in your car. At $t = 0$, you have traveled 10 mi and you depress the accelerator a sufficient amount to increase your commanded speed to 60 mph. The acceleration you experience is proportional to the difference in commanded speed and the car's speed. The block-diagram representation of the system that relates the commanded speed $v_c(t)$, in miles per hour, to

the distance traveled $y(t)$, in miles, is shown in Figure 7.30, where time is expressed in hours.

a. Write a set of integro-differential equations that includes the commanded speed and traveled distance ($v_c(t)$ and $y(t)$, respectively) for $t \geq 0$.

b. Use the Laplace transform to find the total distance traveled and the car's speed at $t = 3$ s, 10 s, 30 s, 1 min, and 10 min.

7.24 Transform the circuit of Problem 7.19 and solve for $i(t)$ for $t \geq 0$.

7.25 Transform the circuit of Problem 7.20 and solve for $i(t)$ for $t \geq 0$.

7.26 Transform the circuit of Problem 7.21 and solve for $v_o(t)$ for $t \geq 0$.

7.27 The electric circuit shown in Figure 7.31 reduces ripple in the current from a power supply to the 2-Ω resistor load. Find the transfer function for the electric circuit.

7.28 Repeat Problem 7.27 for the circuit shown in Figure 7.32.

Figure 7.30

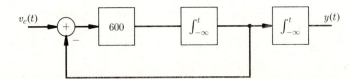

Figure 7.31

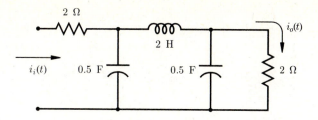

Figure 7.32

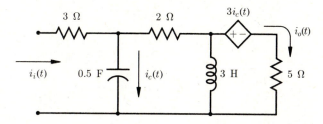

7.29 For the electric circuit shown in Figure 7.33, find each of the following transfer functions:

a. $H_a(s) = V_2(s)/V_i(s)$

b. $H_b(s) = I_2(s)/I_i(s)$

c. $H_c(s) = I_2(s)/V_i(s)$

d. $H_d(s) = V_2(s)/I_i(s)$

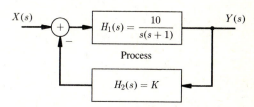

Figure 7.34

7.31 Draw the transformed block diagram and find the transfer function for the pipeline flow-rate control system of Problem 7.22.

7.30 A feedback control system for a manufacturing process is represented by the transformed block diagram shown in Figure 7.34, where $X(s)$ and $Y(s)$ are the transforms of the input and output signals, respectively.

a. Find the transfer function for the system.

b. Find the system response $y(t)$ for $K = 0$ and for $K = 0.125$ when the input signal $x(t)$ is a unit step function.

7.32 Draw the transformed block diagram and find the transfer function for the system relating commanded car speed and distance traveled in Problem 7.23.

7.33 A system has the transformed block diagram shown in Figure 7.35, where $X(s)$ and $Y(s)$ are the transforms of the input and output signals, respectively. Find the system transfer function. Draw the system pole-zero plot.

Figure 7.33

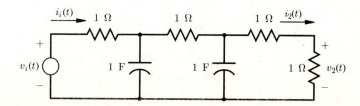

Figure 7.35

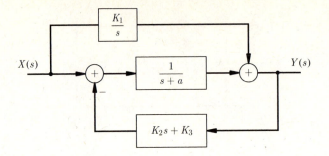

7.34 Determine whether the systems defined by the following transfer functions are stable, marginally stable, or unstable:

a. $H(s) = \dfrac{1}{s^2 + s - 2}$

b. $H(s) = \dfrac{s - 3}{(s + 1)(s + 4)}$

c. $H(s) = \dfrac{s}{s^2 - 2s + 5}$

d. $H(s) = \dfrac{s + 2}{s^2 + s + 3}$

7.35 A control system is represented by the block diagram shown in Figure 7.36.

a. Show that the system possess BIBO stability, but not asymptotic stability.
b. Find the response of the system to a unit step input for $t \geq 0$ when initial conditions are zero.
c. Find the response of the system to a unit step input for $t \geq 0$ when the initial conditions are $y(0^-) = 0$ and $y^{(1)}(0^-) = 0.1$.

7.36 Find and plot the unit step responses of the systems defined by the following transfer functions:

a. $H(s) = \dfrac{1}{s - 2}$

b. $H(s) = \dfrac{s - 1}{(s^2 + 3s + 2)}$

c. $H(s) = \dfrac{1}{s^2 + 4}$

d. $H(s) = \dfrac{1}{s^2 - s + 1.25}$

7.37 For the systems defined by the following transfer functions, find and plot the responses to the input signal $x(t) = \cos(2t)u(t)$.

a. $H(s) = \dfrac{1}{s^2 + 2s + 5}$

b. $H(s) = \dfrac{1}{s^2 + 4}$

c. $H(s) = \dfrac{1}{s^2 - 2s + 5}$

Comment on the results.

7.38 Find and plot the amplitude response and phase response for the systems defined by the following transfer functions:

Figure 7.36

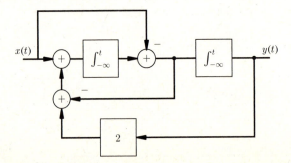

a. $H(s) = \dfrac{2}{s+2}$

b. $H(s) = \dfrac{10}{s+10}$

c. $H(s) = \dfrac{10}{s-10}$

7.39 Repeat Problem 7.38 for the systems defined by the following transfer functions:

a. $H(s) = \dfrac{s}{s^2+9}$

b. $H(s) = \dfrac{s-5}{s^2+3s+2}$

c. $H(s) = \dfrac{1}{s^2+2s+5}$

Also plot the pole-zero plots for these systems.

8 Continuous-Time Filters

The analysis techniques for continuous-time signals and systems developed in the preceding chapters can be applied to a wide range of systems, including control systems and communications systems. We illustrated some of these applications in the example problems.

One of the important systems used in several examples, but without formal definition, is known as a *filter*. It is apparent from the examples that, in general, a filter lets input-signal components with some frequencies pass with little change, while essentially eliminating input-signal components with other frequencies. Filters for continuous-time signals are frequently referred to as *analog filters*.

An example of an electric-signal filter is the station-selection filter in a radio receiver. The effect of this filter is illustrated in Figure 8.1 for three ampli-

Figure 8.1
Illustration of Signal-Selection Filtering in an AM Radio Receiver

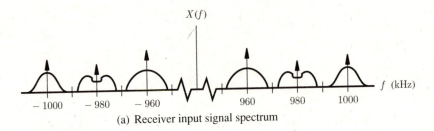

(a) Receiver input signal spectrum

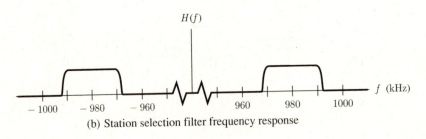

(b) Station selection filter frequency response

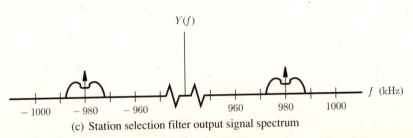

(c) Station selection filter output signal spectrum

tude modulation (AM) radio stations—1, 2, and 3—transmitting simultaneously at assigned carrier frequencies of 960, 980, and 1000 kHz, respectively. The station-selection filter in the receiver is centered at 980 kHz, has constant gain over the bandwidth of a station, and gain that goes to zero outside this bandwidth, so that station 2 is received and stations 1 and 3 are rejected.

In this chapter, we will briefly discuss filtering concepts and filters for electric signals. The techniques of spectral analysis and Laplace transform analysis developed in the preceding chapters are relevant to this discussion. Our intention here is not to provide an in-depth treatment of filter characteristics or design. For further reference, the reader can consult a wide variety of excellent texts devoted to this subject.[†] Various companies have even developed computer software packages to perform the computations required to produce filters with desired characteristics.[††]

8.1 Distortionless Transmission

As background to the discussion of filters, let us consider a system that transmits (passes) any signal with no changes to the signal, except for amplification and time delay; that is, the signal shape remains unchanged. Such a system is referred to as a *distortionless transmission system*.

Definition

> A distortionless transmission system is one that passes any signal with no change, except possibly for amplification and time delay.

Note that the amplification factor may be less than one. In this case, we frequently say that the system provides *signal attenuation*. A distortionless transmission system preserves all information in the signal, except for its overall amplitude and location in time. As long as the output signal possesses an amplitude that is not too small and a time delay that is not excessive (as is the case with commercial radio signals or public address systems, for example), this amplitude and time location are often not important.

The output signal from the defined distortionless transmission system is

$$y(t) = Kx(t - \tau) \tag{8.1}$$

[†]Among these texts are L. Weinberg, *Network Analysis and Synthesis* (New York: McGraw-Hill, 1962); M. E. Van Valkenberg, *Analog Filter Design* (New York: Holt, Rinehart and Winston, 1982); L. P. Huelsman and P. E. Allen, *Introduction to the Theory and Design of Active Filters* (New York: McGraw-Hill, 1980); E. A. Guilleman, *Synthesis of Passive Networks* (New York: Wiley, 1957); and IIT Staff, *Reference Data for Radio Engineers*, 6th ed. (Indianapolis, Ind.: Howard W. Sams, 1975).

[††]Two such companies are DGS Associates, Inc., 1353 Sarita Way, Santa Clara, CA 95051 and RLM Research, P.O. Box 3630, Boulder, CO 80307.

when the input signal is $x(t)$. Computing the Fourier transform of eq. (8.1), we obtain the expression for the output-signal spectrum

$$Y(f) = KX(f)e^{-j2\pi f\tau} \tag{8.2}$$

We then compute the frequency response of the distortionless transmission system as

$$H(f) = Y(f)/X(f) = Ke^{-j2\pi f\tau} \tag{8.3}$$

The amplitude and phase response corresponding to this frequency response are illustrated in Figure 8.2. Note that a distortionless transmission system has constant gain at all frequencies and a phase shift given by a straight line through the origin. This phase-shift characteristic is as expected, because it produces constant phase delay and constant group delay (see Section 6.4).

A practical transmission system may have amplitude and phase responses such as those shown in Figure 8.3. Referring to this figure, we see that approximately distortionless transmission occurs if the input signal contains only components with frequencies lower than f_1 Hz. Over this range of frequencies, system gain is approximately constant and the phase response is approximately a straight line through the origin. A signal containing components with a wider range of frequencies has phase distortion because the phase is no longer ap-

Figure 8.2
Frequency Response
for a Distortionless
Transmission System

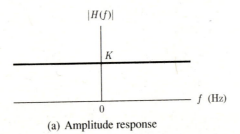

(a) Amplitude response

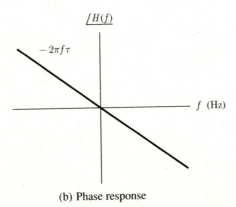

(b) Phase response

Figure 8.3
Example Frequency
Response for a
Practical
Transmission System

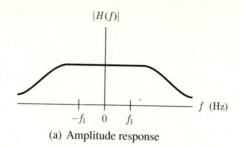

(a) Amplitude response

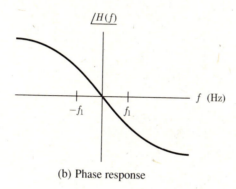

(b) Phase response

proximately a straight line through the origin. It also has amplitude distortion because the gain is not constant for all signal frequencies.

8.2 Ideal Filters

We would like to be able to construct a system (filter) that would pass the portion of the input signal corresponding to some spectral components without distortion and completely attenuate the remainder of the spectral components. The reason is that the desired signal may occupy only a certain spectral width; thus, anything in the input signal outside this frequency range is not desired signal, but unwanted interference or noise.

The frequency ranges for which the input-signal spectral components are passed and not passed by the filter are called the *passband* and *stopband* of the filter, respectively. Filter types of interest include the *low-pass filter* (LPF), *high-pass filter* (HPF), *band-pass filter* (BPF), and *band-rejection* (or *notch*) *filter* (BRF). The names are descriptive of the filter nature.

Definition

An ideal low-pass filter passes all signal components with frequencies lower than B Hz with no distortion and completely attenuates signal components with frequencies greater than B Hz.

From the definitions of the ideal low-pass filter and distortionless transmission, we see that the frequency response for an ideal low-pass filter is

$$H_L(f) = K\Pi\left(\frac{f}{2B}\right)e^{-j2\pi f\tau} \tag{8.4}$$

The corresponding amplitude and phase response are shown in Figure 8.4. Also shown in Figure 8.4 is the impulse response of the ideal low-pass filter. We derive the impulse response by computing the inverse Fourier transform of $H_L(f)$; that is,

$$h_L(t) = \mathcal{F}^{-1}[H_L(f)] = \mathcal{F}^{-1}\left[K\Pi\left(\frac{f}{2B}\right)e^{-j2\pi f\tau}\right] \tag{8.5}$$

Figure 8.4
Frequency Response and Impulse Response for an Ideal Low-Pass Filter

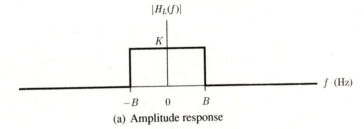

(a) Amplitude response

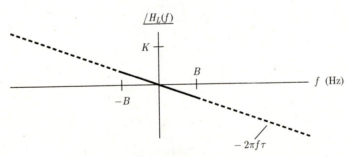

(b) Phase response

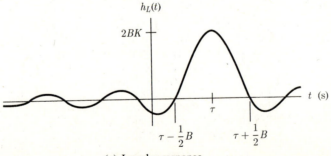

(c) Impulse response

From Table C.2 in Appendix C, we find that

$$\mathcal{F}^{-1}[K\Pi\left(\frac{f}{2B}\right)] = 2BK \ \mathrm{sinc}(2Bt) \tag{8.6}$$

We then use the time-shift theorem to obtain the impulse-response expression

$$h_L(t) = 2BK \ \mathrm{sinc}[2B(t - \tau)] \tag{8.7}$$

Note that the ideal low-pass filter is noncausal because $h_L(t) \neq 0$ for $t < 0$; thus, we must know the input signal at all future times to produce a filtered output signal at any time instant. Therefore, in practice we cannot implement the ideal low-pass filter. At best, we can tolerate only a finite time delay; thus, only a finite time portion of the signal that occurs at times later than the time of interest can be known.

Definition

> An ideal band-pass filter passes all signal components with frequencies in a band of width B Hz centered at the frequency f_o Hz with no distortion and completely attenuates signal components with frequencies outside this band.

The preceding definition indicates that the frequency response of an ideal band-pass filter is

$$H_B(f) = K\left[\Pi\left(\frac{f - f_o}{B}\right) + \Pi\left(\frac{f + f_o}{B}\right)\right]e^{-j2\pi f\tau} \tag{8.8}$$

The corresponding amplitude response and phase response for the ideal band-pass filter are shown in Figure 8.5. The impulse response is

$$h_B(t) = 2BK \ \mathrm{sinc}\left[B(t - \tau)\right]\cos\left[2\pi f_o(t - \tau)\right] \tag{8.9}$$

and is computed as the inverse Fourier transform of eq. (8.8). This inverse transform can be computed easily using Table C.2, the modulation theorem, and the time-shift theorem. The ideal band-pass filter is noncausal; thus it cannot be implemented as a practical filter.

Similar definitions can be made for ideal high-pass and ideal band-rejection filters and their corresponding impulse responses determined. This is requested of the reader in the end-of-chapter problems.

8.3 Approximation of Ideal Filters

Several approximations to ideal filters have been devised over the years. Some attempt to match the amplitude response as closely as possible without considering the phase response (for example, the Butterworth, Chebyshev, and

Figure 8.5
Frequency Response
and Impulse
Response for an Ideal
Band-Pass Filter

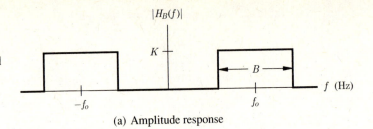

(a) Amplitude response

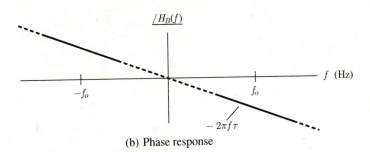

(b) Phase response

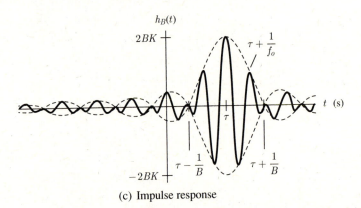

(c) Impulse response

elliptic filters). This is appropriate for such systems as audio filters because the human ear is not sensitive to signal-component phase shift. Other approximations attempt to match the phase response as closely as possible without considering the amplitude response (for example, the Bessel filter). It is not possible, however, to optimize both the amplitude-response approximation and the phase-response approximation at the same time. This is because the phase response of a stable causal filter with a given amplitude response cannot be chosen arbitrarily, and vice versa.[†]

[†]M. E. Van Valkenberg, *Introduction to Modern Network Synthesis* (New York: Wiley, 1960), Ch. 8 and D. F. Tuttle, *Network Synthesis* (New York: Wiley, 1958), Ch. 8.

In this section, we will briefly consider only two approximation techniques. In both cases, an approximation to the amplitude response is optimized in some sense.

For notational simplicity, the frequency response of filters will be expressed as a function of $j\omega$ in the remainder of this chapter. That is, the frequency response of a filter appears as

$$H(j\omega) = H_\omega(\omega) = \int_{-\infty}^{\infty} h(t)e^{-j\omega t}\, dt \qquad (8.10)$$

where $h(t)$ is the filter impulse response. This notation makes it easy to express the relation between the frequency response and the transfer function because $H(j\omega) = H(s)|_{s=j\omega}$.

If we want to construct a filter with a given amplitude response, then we must find a transfer function that corresponds to the amplitude response to identify a physical-component realization of the filter. More than one transfer function that produces the given amplitude response may exist. However, each different transfer function produces a different phase response.

We write the given amplitude response squared as

$$|H(j\omega)|^2 = H(j\omega)H^*(j\omega) = H(j\omega)H(-j\omega) \qquad (8.11)$$

The last step in eq. (8.11) follows because we want a system that produces real output-signal responses to real input signals. Therefore, $h(t)$ must be real, which implies that the coefficients of $H(j\omega)$ are real, so that the only way j enters $H(j\omega)$ is through $j\omega$. Continuing, we write eq. (8.11) as

$$|H(j\omega)|^2 = H(s)H(-s)\big|_{s=j\omega} \qquad (8.12)$$

where $H(s)$ is the filter transfer function. The poles and zeros of $H(-s)$ exist at values of s that are the negatives of those for $H(s)$. Thus, the poles and zeros of $H(s)H(-s)$ occur in pairs that are symmetric with respect to the origin. This symmetry is illustrated in Figure 8.6 for two examples. Due to the pole and zero symmetry,

$$H(s)H(-s) = \frac{C \prod_{i=1}^{m} (z_i + s)(z_i - s)}{\prod_{k=1}^{n} (p_k + s)(p_k - s)} \qquad (8.13)$$

where z_i and p_k are the zero and pole locations, respectively, for which $Re[z_i] \geq 0$ and $Re[p_k] \geq 0$. Multiplying factors gives

$$H(s)H(-s) = \frac{C \prod_{i=1}^{m} (z_i^2 - s^2)}{\prod_{k=1}^{n} (p_k^2 - s^2)} \qquad (8.14)$$

Figure 8.6
Pole-Zero Plot for
Two Examples of
$H(s)H(-s)$

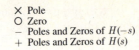

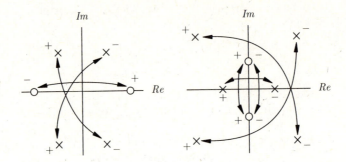

which shows that s appears only as even powers (that is, as s^2, s^4, s^6, etc.) in $H(s)H(-s)$. When we replace s with $j\omega$ in eq. (8.12), $s^2 \rightarrow -\omega^2$, $s^4 \rightarrow (-\omega^2)^2$, $s^6 \rightarrow (-\omega^2)^3$, etc. Therefore, we can rewrite eq. (8.12) as

$$|H(j\omega)|^2 = H(s)H(-s)\big|_{s^2=-\omega^2} \tag{8.15}$$

which yields

$$H(s)H(-s) = |H(j\omega)|^2\big|_{\omega^2=-s^2} = \frac{C\displaystyle\prod_{i=1}^{m}(z_i+s)(z_i-s)}{\displaystyle\prod_{k=1}^{n}(p_k+s)(p_k-s)} \tag{8.16}$$

when combined with eq. (8.13).

We must now split $H(s)H(-s)$ to identify $H(s)$. The split is not unique; all that is necessary is to assign one of each of the negatively paired poles and zeros to $H(s)$ and to give $H(s)$ a multiplicative constant of $\sqrt{C}$ so that the product of $H(s)$ and $H(-s)$ is the right side of eq. (8.16). Normally, $H(s)H(-s)$ does not contain poles on the imaginary axis, and we usually want a stable filter. Therefore, we choose the poles in the left half of the s plane for $H(s)$. The remaining poles are the negatives of those selected for $H(s)$, and thus are appropriate for $H(-s)$.

There is no stability constraint on the selection of zeros for $H(s)$. However, we quite often select zeros for $H(s)$ that are not in the right half of the s plane to produce a system that provides minimum phase shift (that is, minimum time delay) to the input signal. Therefore, we split $H(s)H(-s)$ in the following manner:

$$H(s)H(-s) = \left[\frac{\sqrt{C} \prod\limits_{i=1}^{m} (z_i + s)}{\prod\limits_{k=1}^{n} (p_k + s)} \right] \left[\frac{\sqrt{C} \prod\limits_{i=1}^{m} (z_i - s)}{\prod\limits_{k=1}^{n} (p_k - s)} \right] \tag{8.17}$$

Since $Re[z_i] \geq 0$ and $Re[p_k] > 0$, then the stable minimum-phase transfer function that produces the desired amplitude response is

$$H(s) = \left[\frac{\sqrt{C} \prod\limits_{i=1}^{m} (z_i + s)}{\prod\limits_{k=1}^{n} (p_i + s)} \right] \tag{8.18}$$

Example 8.1

The feedback signal in an antenna-pointing control system contains components with frequencies in the range $0 \leq \omega \leq 8.5$ rad/s. System performance is improved if feedback signal components with frequencies near 2.75 rad/s are increased in amplitude by a factor of approximately 1.5 with respect to components with frequencies near the ends of the signal frequency range. To produce this effect, the feedback signal is to be passed through a filter, the amplitude response of which is

$$H(j\omega) = \sqrt{\frac{7.5 + 7.5\omega^2}{25 + 7.25\omega^2 + 0.25\omega^4}}$$

Find the transfer function for a stable minimum-phase filter that corresponds to the desired amplitude response. Plot the filter amplitude response and pole-zero plot.

Solution

$$|H(j\omega)|^2 = \frac{7.5 + 7.5\omega^2}{25 + 7.25\omega^2 + 0.25\omega^4} = \frac{30(1 + \omega^2)}{100 + 29\omega^2 + \omega^4}$$

$$H(s)H(-s) = |H(j\omega)|^2 \Big|_{\omega^2 = -s^2} = \frac{30(1 - s^2)}{100 - 29s^2 + s^4}$$

$$= \frac{30(1 - s^2)}{(25 - s^2)(4 - s^2)} = \frac{30(1 + s)(1 - s)}{(5 + s)(5 - s)(2 + s)(2 - s)}$$

Therefore,

$$H(s) = \frac{\sqrt{30}\,(1+s)}{(5+s)(2+s)} = \frac{\sqrt{30}\,s + \sqrt{30}}{s^2 + 7s + 10}$$

The amplitude response and pole-zero plot for the filter are shown in Figure 8.7.

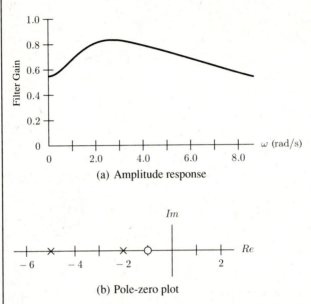

(a) Amplitude response

(b) Pole-zero plot

Figure 8.7 Amplitude Response and Pole-Zero Plot for the Filter in Example 8.1

8.4 Ideal Amplitude-Response Approximations

We are now ready to discuss two methods for approximating the amplitude response of an ideal filter by amplitude responses that can be produced by practical filters. The discussion is limited to low-pass filters for which the boundary between the passband and stopband of the filter is defined as the cutoff frequency ω_c and the maximum amplitude response (that is, the maximum gain) of the filter is defined as G_m. At the end of this section, we show frequency transformations that are used to generate band-pass, high-pass, and band-rejection amplitude-response approximations from the low-pass filter amplitude-response approximations.

The Low-Pass Butterworth Approximation

The first approximation we will consider for the amplitude response of an ideal low-pass filter is the *Butterworth approximation*. The resulting filters are referred to as *Butterworth filters*.

Definition

A low-pass Butterworth filter is one that has the amplitude response

$$|H_{bn}(j\omega)| = G_m\Big/\sqrt{1 + (\omega/\omega_c)^{2n}} \tag{8.19}$$

where $n \geq 1$ is the filter order and the subscript b denotes a Butterworth filter.

Two criteria are used in generating the Butterworth approximation, eq. (8.19), to the ideal low-pass filter amplitude response. The first criterion is that the gain of the filter at the cutoff frequency ω_c is $G_m/\sqrt{2}$, where G_m is the maximum gain of the filter. Equivalently, the power gain at $\omega = \omega_c$ is $G_m^2/2$, which is one-half the maximum power gain of the filter, or 3.01 dB less than the maximum power gain of the filter. Thus, the filter cutoff frequency is the half-power cutoff frequency, also called the 3-dB cutoff frequency. Note that the maximum filter gain is at $\omega = 0$ (that is, at dc).

The second criterion used in generating the Butterworth approximation is that the amplitude response be maximally flat at $\omega = 0$. That is, $|H(j\omega)|$ is to have as many derivatives as possible equal to zero at $\omega = 0$. The Butterworth amplitude response has $2n - 1$ derivatives equal to zero at $\omega = 0$.

We obtain the transfer function for the nth-order Butterworth low-pass filter from

$$H_{bn}(s)H_{bn}(-s) = |H_{bn}(j\omega)|^2\big|_{\omega^2 = -s^2}$$

$$= \frac{G_m^2}{1 + (\omega^2/\omega_c^2)^n}\bigg|_{\omega^2 = -s^2} = \frac{G_m^2\omega_c^{2n}}{\omega_c^{2n} + (-s^2)^n} \tag{8.20}$$

For first- and second-order filters ($n = 1, 2$),

$$H_{b1}(s)H_{b1}(-s) = \frac{G_m^2\omega_c^2}{\omega_c^2 - s^2} = \frac{G_m^2\omega_c^2}{(\omega_c + s)(\omega_c - s)} \tag{8.21}$$

and

$$H_{b2}(s)H_{b2}(-s) = \frac{G_m^2\omega_c^4}{\omega_c^4 + s^4} = \frac{G_m^2\omega_c^4}{\displaystyle\prod_{k=1}^{4}(p_k - s)} \tag{8.22}$$

where

$$p_k = \omega_c e^{j(0.25\pi + 0.5\pi k)} \tag{8.23}$$

It can be easily verified that the denominator factors shown are correct by performing the polynomial factorization. The pole locations for $H_{b1}(s)H_{b1}(-s)$ and $H_{b2}(s)H_{b2}(-s)$ are shown in Figure 8.8a. Note that the pole locations are equally spaced around a circle of radius ω_c and symmetric with respect to the imaginary axis. This equal spacing is a general result for $H_{bn}(s)H_{bn}(-s)$

Figure 8.8
$H_{bn}(s)H_{bn}(-s)$ Pole
Locations for
Low-Pass
Butterworth Filters

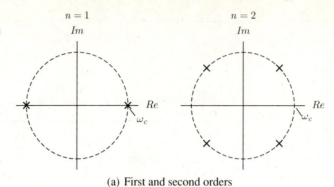

(a) First and second orders

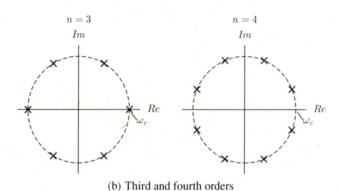

(b) Third and fourth orders

and is illustrated further in Figure 8.8b for third- and fourth-order Butterworth filters. Note that $H_{bn}(s)H_{bn}(-s)$ contains two real poles for odd-order filters and no real poles for even-order filters. We select the poles in the left half of the s-plane to be the poles of the Butterworth filter transfer function. From eqs. (8.21) and (8.22), this gives

$$H_{b1}(s) = \frac{G_m\omega_c}{\omega_c + s} \tag{8.24}$$

and

$$H_{b2}(s) = \frac{G_m\omega_c^2}{\omega_c^2 + \sqrt{2}\omega_c s + s^2} \tag{8.25}$$

as the transfer functions for the first- and second-order Butterworth low-pass filters. Circuits that realize these two orders of Butterworth filters with unity maximum gain (that is, $G_m = 1$) for voltage signals are shown in Figure 8.9, where $v_i(t)$ is the input voltage signal and $v_o(t)$ is the output voltage signal.

We show the amplitude response and phase response for the first three orders of Butterworth filters in Figure 8.10. It is clear that the ideal amplitude response is approached as the order, n, becomes large. However, a large number of filter

Figure 8.9
Circuit Realizations
of Low-Pass
Butterworth Filters

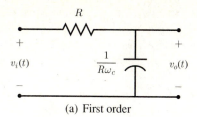

(a) First order

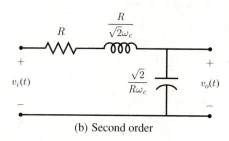

(b) Second order

Figure 8.10
Amplitude and Phase
Response for
Low-Pass
Butterworth Filters

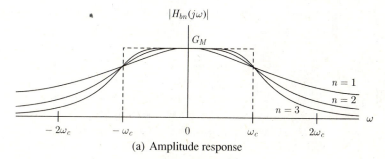

(a) Amplitude response

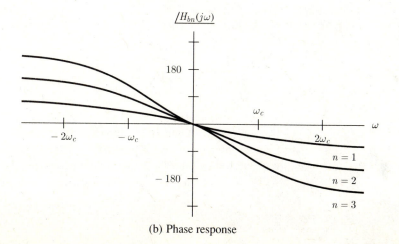

(b) Phase response

components is required, the size of the components becomes large because poles approach the imaginary axis, and the time delay of signals passed by the filter increases as n increases. The increased time delay is expected because the impulse response of an ideal low-pass filter can be approximated closely for all time only if the portion of it that occurs for $t < 0$ is small, this portion being set to zero by any causal approximation.

The Low-Pass Chebyshev Approximation

We now come to the second approximation for the amplitude response of an ideal low-pass filter—the *Chebyshev approximation*. The resulting filters are referred to as *Chebyshev filters*.

Definition _____

A low-pass Chebyshev filter is one that has the amplitude response

$$|H_{cn}(j\omega)| = G_m \Big/ \sqrt{1 + \epsilon^2 C_n^2(\omega/\omega_c)} \qquad (8.26)$$

where $C_n(x)$ is the nth-order Chebyshev polynomial, ϵ is a constant, $n \geq 1$ is the order of the filter, and the subscript c denotes a Chebyshev filter.

Except for the first-order filter, the Chebyshev filter gives a sharper cutoff and less gain in the stopband than an equal-order Butterworth filter, at the expense of variation (amplitude-response ripples) in the passband. The amount of variation is determined by the constant ϵ.

The Chebyshev polynomials can be obtained with a recursion relation.[†] The result for the Chebyshev polynomials of orders one through four are $C_1(\omega/\omega_c) = \omega/\omega_c$, $C_2(\omega/\omega_c) = 2(\omega/\omega_c)^2 - 1$, $C_3(\omega/\omega_c) = 4(\omega/\omega_c)^3 - 3(\omega/\omega_c)$, and $C_4(\omega/\omega_c) = 8(\omega/\omega_c)^4 - 8(\omega/\omega_c)^2 + 1$. These are illustrated in Figure 8.11. In compact form, the Chebyshev polynomial of the nth order is

$$C_n(\omega/\omega_c) = \begin{cases} \cos[n\cos^{-1}(\omega/\omega_c)] & (\omega/\omega_c) \leq 1 \\ \cosh[n\cosh^{-1}(\omega/\omega_c)] & (\omega/\omega_c) > 1 \end{cases} \qquad (8.27)$$

The value of a Chebyshev polynomial of any order varies from -1 to $+1$ when $w < w_c$, is equal to $+1$ at $w = w_c$, and increases monotonically for $\omega > \omega_c$. The rapidity with which a Chebyshev polynomial increases for $\omega > \omega_c$ increases as the order increases. Thus, from eq. (8.26), the amplitude response or gain for a Chebyshev filter varies between G_m and $G_m/\sqrt{1 + \epsilon^2}$ for $\omega < \omega_c$, is equal to $G_m/\sqrt{1 + \epsilon^2}$ for $\omega = \omega_c$, and decreases monotonically for $\omega > \omega_c$.

[†]National Bureau of Standards, *Tables of Chebyshev Polynomials* (Washington, D.C.: U.S. Government Printing Office, 1952).

Figure 8.11
Chebyshev
Polynomials

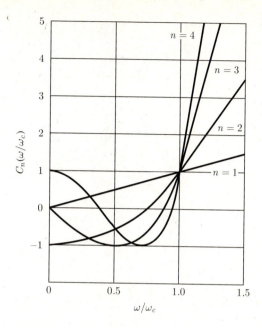

Note that the ratio of the gain of the filter at the cutoff frequency to the maximum gain depends on the value that we chose for the constant ϵ.

From Figure 8.11, we see that the filter gain at $\omega = 0$ (that is, at dc) is the maximum gain for odd-order Chebyshev filters. Even-order Chebyshev filters have a gain at $\omega = 0$ that is equal to the gain at $\omega = \omega_c$. Again, we see this from Figure 8.11.

There are n maxima and minima in the Chebyshev filter amplitude response when $\omega < \omega_c$. This, too, is apparent from Figure 8.11, because $|H_{cn}(j/\omega)|$ has maximum and minimum values when $C_n(\omega/\omega_c) = 0$ and $C_n(\omega/\omega_c) = \pm 1$, respectively. The amount of variation between these maximum and minimum values (that is, amplitude-response ripple in the passband) is determined by the value of the parameter ϵ because the ripple, in decibels, is

$$r = 10\log[|H_{cn}(j\omega)|^2_{\max}/|H_{cn}(j\omega)|^2_{\min}] \qquad (8.28)$$
$$= 10\log(1 + \epsilon^2)$$

We can solve eq. (8.28) to obtain the value of ϵ required to produce a given passband ripple. This solution is

$$\epsilon = \sqrt{10^{r/10} - 1} \qquad (8.29)$$

Figure 8.12 shows the effect of changing ϵ when the filter order is fixed and the effect of changing the filter order when ϵ is fixed. For a fixed-order filter, the gain in the stopband decreases with increasing ϵ, but the gain ripple in the passband increases. For a fixed ϵ, the gain ripple in the passband does not depend on filter order. However, with increasing order, more gain maxima

Figure 8.12
Chebyshev Filter
Amplitude-Response
Dependence on
Filter Order and
Parameter ϵ

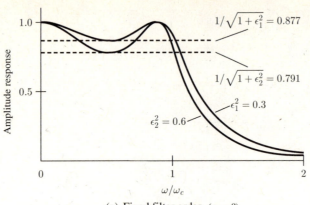

(a) Fixed filter order $(n = 3)$

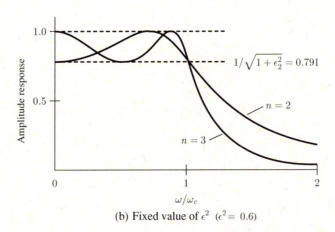

(b) Fixed value of ϵ^2 $(\epsilon^2 = 0.6)$

and minima occur in the passband, and the stopband gain decreases. Thus, the sharpness of amplitude-response cutoff is improved for a Chebyshev filter by allowing more gain variation in the passband or by increasing the complexity (order) of the filter. We emphasize that the definition of the location of the cutoff frequency for a Chebyshev filter is not the same as that for a Butterworth filter, except when ϵ is chosen to give 3.01 dB of gain ripple in the passband.

Applying the technique discussed in Section 8.3 to a Chebyshev filter amplitude response produces the transfer function for the filter. The poles of the transfer function lie on an ellipse.[†] The transfer function for the first two orders of low-pass Chebyshev filters with unity dc gain and a passband ripple of 2 dB ($\epsilon^2 = 0.5849$) are

$$H_{c1}(s) = \frac{1.3076\omega_c}{1.3076\omega_c + s} \tag{8.30}$$

[†]Van Valkenburg, *Introduction to Modern Network Synthesis*, pp. 379–384.

and

$$H_{c2}(s) = \frac{0.6368\omega_c^2}{0.6368\omega_c^2 + 0.8038\omega_c s + s^2} \tag{8.31}$$

These two filters are realized by the circuits shown in Figure 8.9, with the capacitance of the capacitor replaced by $1/1.3076R\omega_c$ for the first-order filter and with the capacitance of the capacitor and inductance of the inductor replaced by $0.8038/0.6368R\omega_c$ and $R/0.8038\omega_c$, respectively, for the second-order filter. It can be easily shown that the Chebyshev filters are produced by the circuit transfer functions using the specified component values.

Other Low-Pass Approximations

Other approximations to the ideal low-pass filter amplitude response are also used. In general, they improve the approximation in some sense by increasing the filter complexity. Examples are the *inverse Chebyshev filter*, also called the *Type II Chebyshev filter*, and the *elliptic filter*, both of which include zeros on the imaginary axis. The Type II Chebyshev filter obtains sharper frequency cutoff than the Butterworth filter by permitting gain ripple in the stopband rather than in the passband. The elliptic filter achieves even sharper frequency cutoff by permitting ripple in both the passband and stopband. At high frequencies, it has greater gain (less attenuation) than filters with no finite zeros.

Band-Pass, High-Pass, and Band Reject Approximations

Approximations of ideal band-pass, high-pass, and band-rejection filters of the same type discussed for the low-pass filter are obtained from low-pass filters by nonlinear frequency transformations. We make the transformations by substituting for the frequency variable ω_L in the frequency response $H_L(j\omega_L)$ of the low-pass filter. Note that we use the frequency variable ω_L for the low-pass filter to avoid confusion. The transformation equations are

$$H_B\left(j\omega\right) = \left. H_L\left(j\omega_L\right)\right|_{\omega_L = \omega_c\left(\omega^2 - \omega_u\omega_\ell\right)/\omega(\omega_u - \omega_\ell)} \tag{8.32}$$

for a band-pass filter,

$$H_H\left(j\omega\right) = \left. H_L\left(j\omega_L\right)\right|_{\omega_L = \omega_c\omega_{ch}/\omega} \tag{8.33}$$

for a high-pass filter, and

$$H_R\left(j\omega\right) = \left. H_L\left(j\omega_L\right)\right|_{\omega_L = \omega_c\omega(\omega_u - \omega_\ell)/\left(\omega^2 - \omega_u\omega_\ell\right)} \tag{8.34}$$

for a band-rejection filter, where ω_c is the cutoff frequency (passband/stopband boundary) for the low-pass filter, ω_{ch} is the cutoff frequency (passband/stopband boundary) for the high-pass filter, and ω_u and ω_ℓ are the upper and lower cut-off frequencies (passband/stopband boundaries) for the band-pass and band-rejection filters. The sense of the definition of the passband/stopband boundaries is the same as that of the low-pass filter from which the transformation is made (that is, the frequency at which gain $= G_m/\sqrt{2}$ for a Butterworth filter and gain $= G_m/\sqrt{1+\epsilon^2}$ for a Chebyshev filter). Figure 8.13 shows the frequency transformations specified in eqs. (8.32), (8.33), and (8.34).

Figure 8.13
Frequency
Transformations
Used to Convert a
Low-Pass Filter to
a Band-Pass,
High-Pass, or
Band-Rejection Filter

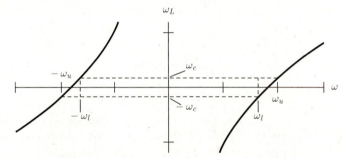

(a) Frequency transformation for band-pass filter

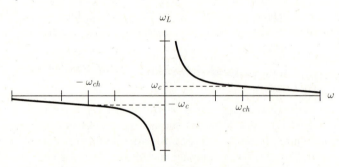

(b) Frequency transformation for high-pass filter

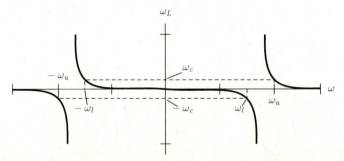

(c) Frequency transformation for band-rejection filter

8.5 Design of Low-Pass Filters

In this section we will address the application of Butterworth and Chebyshev approximations of an ideal filter in the design of a low-pass filter to satisfy specified requirements. The general form of the transfer function for an nth-order Butterworth or Chebyshev low-pass filter is

$$H_L(s) = \frac{A_o}{B_o + B_1 s + \cdots + B_{n-1} s^{n-1} + s^n} \tag{8.35}$$

Note that the nth-order low-pass filter has n poles.

Coefficients have been tabulated for Butterworth and Chebyshev low-pass filters with a cutoff frequency of 1 rad/s. We refer to filters of this type as *normalized low-pass filters*, the transfer functions of which can be expressed as

$$H_{LN}(s_{LN}) = \frac{a_o}{b_o + b_1 s_{LN} + \cdots + b_{n-1} s_{LN}^{n-1} + s_{LN}^n} \tag{8.36}$$

where the complex variable is written as s_{LN} to avoid confusion in later equations relating the transfer functions for the normalized and nonnormalized low-pass filters. Table 8.1 consists of an abbreviated listing of denominator polynomial coefficients for normalized Butterworth and Chebyshev filters of order up to $n = 4$.

Transfer functions for low-pass filters with other cutoff frequencies are obtained from eq. (8.36) by frequency scaling. Thus, we eliminate the need to regenerate denominator coefficients each time a Butterworth or Chebyshev low-pass filter is designed with a different cutoff frequency. We will discuss this frequency-scaling technique shortly.

If factors of the denominator polynomial are left as quadratic factors when complex-conjugate poles exist, then there is one linear factor, $\ell_o + s$, and $(n-1)/2$ quadratic factors, $q_{oi} + q_{1i} s + s^2$, $1 \le i \le (n-1)/2$, if n is odd. If n is even, then the denominator polynomial contains $n/2$ quadratic factors. The coefficients for these factors are shown in Table 8.2 and are used in Chapter 15 when discrete time filters are considered.

Sometimes we choose the numerator coefficient, a_o, of eq. (8.36) to give a specified filter gain at $\omega = 0$ (that is, dc gain). Since

$$H_{LN}(j\omega_{LN}) = H_{LN}(s_{LN})\big|_{s_{LN}=j\omega_{LN}}$$

$$= \frac{a_o}{b_o + b_1(j\omega_{LN}) + \cdots + (j\omega_{LN})^n} \tag{8.37}$$

where we define ω_{LN} as the frequency variable for the normalized low-pass filter, then the dc gain, G_{DC}, is

$$G_{DC} = |H_{LN}(j0)| = \left|\frac{a_o}{b_o}\right| \tag{8.38}$$

Table 8.1 Denominator Polynomial Coefficients for
Normalized Low-Pass Filters

n	b_0	b_1	b_2	b_3
		Butterworth		
1	1.0000	—	—	—
2	1.0000	1.4142	—	—
3	1.0000	2.0000	2.0000	—
4	1.0000	2.6131	3.4142	2.6131
	0.5-dB Passband Ripple Chebyshev $\left(\epsilon^2 = 0.1220\right)$			
1	2.8628	—	—	—
2	1.5162	1.4256	—	—
3	0.7157	1.5349	1.2529	—
4	0.3791	1.0255	1.7169	1.1974
	1.0-dB Passband Ripple Chebyshev $\left(\epsilon^2 = 0.2589\right)$			
1	1.9652	—	—	—
2	1.1025	1.0977	—	—
3	0.4913	1.2384	0.9883	—
4	0.2756	0.7426	1.4539	0.9528
	3.0-dB Passband Ripple Chebyshev $\left(\epsilon^2 = 0.9953\right)$			
1	1.0024	—	—	—
2	0.7079	0.6449	—	—
3	0.2506	0.9283	0.5972	—
4	0.1770	0.4048	1.1691	0.5816

Since b_o is positive, then a specified dc gain can be obtained by choosing

$$|a_o| = b_o G_{DC} \tag{8.39}$$

We choose the sign of a_o to be positive unless signal inversion is desired, in which case we choose it to be negative.

More often, we choose the numerator coefficient to give a specified maximum gain, G_M. This is the dc gain for Butterworth and odd-order Chebyshev filters. Thus, for these filters, $|a_o|$ can be calculated by eq. (8.39) with G_{DC} replaced by G_M. For even-order Chebyshev filters,

$$|a_o/b_o| = G_{DC} = G_M/\sqrt{1 + \epsilon^2} \tag{8.40}$$

Therefore, we obtain the specified maximum gain by choosing

Table 8.2 Factored-Denominator Polynomial Coefficients for Normalized Low-Pass Filters

n	ℓ_0	q_{01}	q_{11}	q_{02}	q_{12}
Butterworth					
1	1.0	—	—	—	—
2	—	1.0	1.4142	—	—
3	1.0	1.0	1.0	—	—
4	—	1.0	0.7654	1.0	1.8478
0.5-dB Passband Ripple Chebyshev $\left(\epsilon^2 = 0.1220\right)$					
1	2.8628	—	—	—	—
2	—	1.5162	1.4256	—	—
3	0.6265	1.1424	0.6265	—	—
4	—	1.0635	0.3507	0.3564	0.8467
1.0-dB Passband Ripple Chebyshev $\left(\epsilon^2 = 0.2589\right)$					
1	1.9652	—	—	—	—
2	—	1.1025	1.0977	—	—
3	0.4942	0.9942	0.4942	—	—
4	—	0.9865	0.2791	0.2794	0.6737
3.0-dB Passband Ripple Chebyshev $\left(\epsilon^2 = 0.9953\right)$					
1	1.0024	—	—	—	—
2	—	0.7079	0.6449	—	—
3	0.2986	0.8392	0.2986	—	—
4	—	0.9031	0.1703	0.1960	0.4112

$$|a_o| = b_o G_M \left/ \sqrt{1 + \epsilon^2} \right.$$

(8.41)

The sign of a_o is chosen as before.

We usually determine the order of a low-pass Butterworth or Chebyshev filter required for a particular application from the stopband gain characteristics required. Thus, curves of the gain (amplitude response) in the stopband of the normalized low-pass filters are useful in determining the required filter order. Such curves are shown in Figures 8.14 through 8.17 for the normalized Butterworth low-pass filter and the normalized Chebyshev low-pass filters with passband ripples of 0.5 dB, 1.0 dB, and 3.0 dB. The gain is shown in decibels below the maximum gain. The use of these curves in the design of low-pass filters is illustrated in the examples at the end of this section.

Figure 8.14
Stopband Gain
for Normalized
Low-Pass
Butterworth Filter

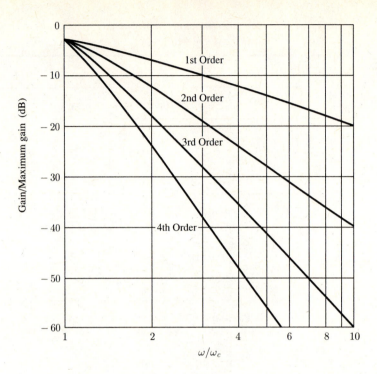

Figure 8.15
Stopband Gain
for Normalized
Low-Pass Chebyshev
Filter with 0.5-dB
Passband Ripple

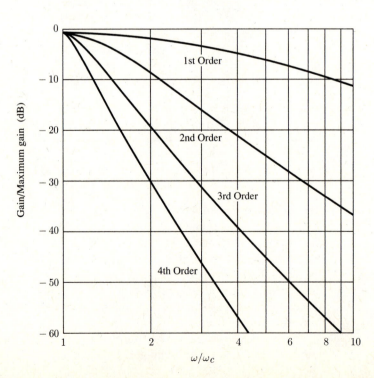

Figure 8.16
Stopband Gain
for Normalized
Low-Pass Chebyshev
Filter with 1.0-dB
Passband Ripple

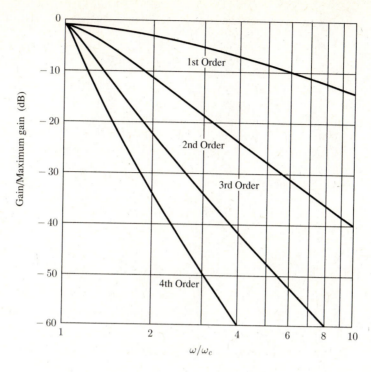

Figure 8.17
Stopband Gain
for Normalized
Low-Pass Chebyshev
Filter with 3.0-dB
Passband Ripple

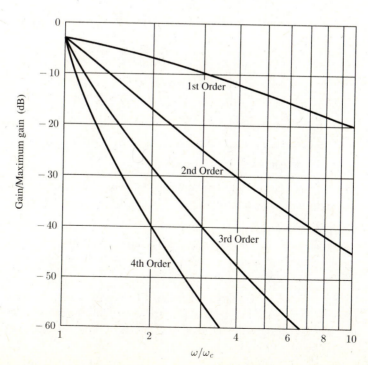

A frequency-scaled version of the frequency response of the normalized low-pass filter is produced by the frequency transformation

$$H_L(j\omega) = \left. H_{LN}(j\omega_{LN}) \right|_{\omega_{LN}=\omega_2\omega/\omega_1}$$

$$= \frac{a_o}{b_o + b_1(j\omega_2\omega/\omega_1) + \cdots + (j\omega_2\omega/\omega_1)^n} \qquad (8.42)$$

This transformation compresses or expands the frequency response along the frequency axis, so that the value of $H_L(j\omega_1)$ is the same as the value of $H_{LN}(j\omega_2)$. Multiplying both sides of the frequency transformation by j (that is, $j\omega_{LN} = j\omega_2\omega/\omega_1$) and replacing $j\omega_{LN}$ by s_{LN} and $j\omega$ by s gives the following expression for the transfer function of the frequency-scaled filter in terms of the transfer function of the normalized filter:

$$H_L(s) = \left. H_{LN}(s_{LN}) \right|_{s_{LN}=s\omega_2/\omega_1}$$

$$= \frac{a_o}{b_o + b_1(s\omega_2/\omega_1) + \cdots + (s\omega_2/\omega_1)^n} \qquad (8.43)$$

In the design of low-pass filters, we want to scale the cutoff frequency of the normalized low-pass filter (1 rad/s) to the desired cutoff frequency ω_c. Therefore, $\omega_2 = 1$ and $\omega_1 = \omega_c$ and the transfer function for a low-pass filter with cutoff frequency ω_c rad/s is

$$H_L(s) = \left. H_{LN}(s_{LN}) \right|_{s_{LN}=s/\omega_c}$$

$$= \frac{a_o}{b_o + b_1(s/\omega_c) + \cdots + (s/\omega_c)^n}$$

$$= \frac{a_o\omega_c^n}{b_o\omega_c^n + b_1\omega_c^{n-1}s + \cdots + s^n} \qquad (8.44)$$

We now illustrate the use of the preceding normalized low-pass filter data and frequency scaling for designing Butterworth or Chebyshev low-pass filters to satisfy design specifications.

Example 8.2

A low-pass filter is to be designed to reduce higher frequency noise on a signal from a temperature sensor. The filter is to satisfy the following specifications:

1. maximum gain of 2;
2. gain variation less than or equal to 3 dB from dc to 50 Hz; and
3. gain less than or equal to -50 dB with respect to maximum gain for $f \geq$ 250 Hz.

For both a Butterworth and a Chebyshev filter, find the minimum order that can be used and the transfer function. Plot the amplitude responses for the two filter designs for $0 \leq f \leq 250$ Hz. Use a dB amplitude scale and a linear frequency scale.

Solution

From specification 2, the required filter cutoff frequency is

$$\omega_c = 2\pi(50) = 100\pi \text{ rad/s}$$

From specification 3, the stopband frequency of interest (the frequency above which the gain is less than -50 dB with respect to the maximum gain) is

$$\omega_1 = 2\pi(250) = 500\pi \text{ rad/s}$$

The scaled stopband frequency of interest for the corresponding normalized low-pass filter is

$$\omega_{LN1} = \frac{\omega_1}{\omega_c} = \frac{500\pi}{100\pi} = 5$$

From Figures 8.14 and 8.17, a fourth-order Butterworth filter or a third-order Chebyshev filter is required to satisfy specification 3.

For the Butterworth filter, from Table 8.1, the transfer function for the fourth-order normalized low-pass filter is

$$H_{LN}(s_{LN}) = \frac{a_o}{1 + 2.6131 s_{LN} + 3.4142 s_{LN^2} + 2.6131 s_{LN^3} + s_{LN^4}}$$

To satisfy specification 1,

$$a_o = b_o G_M = (1)(2) = 2$$

Scaling the transfer function to obtain the desired cutoff frequency gives the Butterworth filter transfer function

$$H_L(s) = \frac{2}{1 + 2.6131\left(\frac{s}{100\pi}\right) + 3.4142\left(\frac{s}{100\pi}\right)^2 + 2.6131\left(\frac{s}{100\pi}\right)^3 + \left(\frac{s}{100\pi}\right)^4}$$

$$= \frac{1.948 \times 10^{10}}{9.741 \times 10^9 + (8.102 \times 10^7)s + (3.370 \times 10^5)s^2 + 506.8s^3 + s^4}$$

The amplitude response for the Butterworth filter is

$$|H_L(j2\pi f)|$$

$$= \frac{2}{\sqrt{\left\{1 - 3.4142\left(f/50\right)^2 + \left(f/50\right)^4\right\}^2 + \left\{2.6131\left(f/50\right)\left[1 - \left(f/50\right)^2\right]\right\}^2}}$$

For the Chebyshev filter, from Table 8.1, the transfer function for the third-order normalized low-pass filter with a 3-dB passband ripple is

$$H_{LN}(s_{LN}) = \frac{a_o}{0.2506 + 0.9283 s_{LN} + 0.5972 s_{LN^2} + s_{LN^3}}$$

Since the filter is of odd order,

$$a_o = b_o G_M = (0.2506)(2) = 0.5012$$

will satisfy specification 1. Scaling the transfer function to obtain the desired cutoff frequency gives the Chebyshev filter transfer function

$$H_L(s) = \frac{0.5012}{0.2506 + 0.9283\left(\frac{s}{100\pi}\right) + 0.5972\left(\frac{s}{100\pi}\right)^2 + \left(\frac{s}{100\pi}\right)^3}$$

$$= \frac{1.554 \times 10^7}{7.770 \times 10^6 + (9.162 \times 10^4)s + 187.6s^2 + s^3}$$

The amplitude response for the Chebyshev filter is

$$|H_L(j2\pi f)|$$

$$= \frac{0.5012}{\sqrt{\left\{0.2506 - 0.5972\left(f/50\right)^2\right\}^2 + \left\{0.9283\left(f/50\right) - \left(f/50\right)^3\right\}^2}}$$

The amplitude responses for the two filter designs are shown in Figure 8.18. Note that the filter designs do indeed meet the desired specifications.

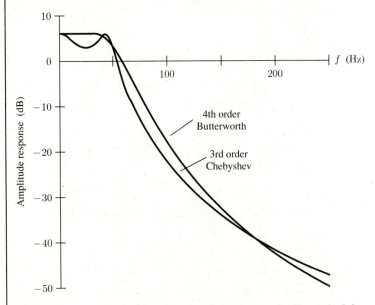

Figure 8.18 Low-Pass Filter Amplitude Responses for Example 8.2

8.6 Design of Band-Pass Filters

We indicated in Section 8.4 that a low-pass filter can be converted to a band-pass filter with similar characteristics by using the frequency transformation

$$\omega_L = \omega_c(\omega^2 - \omega_u\omega_\ell)/\omega(\omega_u - \omega_\ell) \qquad (8.45)$$

in the expression for the frequency response of the low-pass filter. Recall that ω_L is the frequency variable for the low-pass filter. The frequencies ω_u and ω_ℓ are the upper and lower cutoff frequencies (passband/stopband boundaries) of the resulting band-pass filter when ω_c is the cutoff frequency of the low-pass filter from which the transformation is made. The bandwidth of the band-pass filter is

$$\omega_b = \omega_u - \omega_\ell \qquad (8.46)$$

and its geometric-center frequency is

$$\omega_g = \sqrt{\omega_u\omega_\ell} \qquad (8.47)$$

Therefore,

$$H_B(j\omega) = H_{LN}(j\omega_{LN})\big|_{\omega_{LN}=(\omega^2-\omega_g^2)/\omega\omega_b} \qquad (8.48)$$

expresses the frequency response of the band-pass filter in terms of the frequency response of the normalized low-pass filter, because $\omega_c = 1$ for the normalized low-pass filter and ω_{LN} is the frequency variable for the normalized low-pass filter. Note that the frequency response of the band-pass filter at its geometric-center frequency, $\omega = \omega_g$, equals the dc response of the normalized low-pass filter (that is, the response at $\omega_{LN} = 0$).

The transformation shown in eq. (8.48) is nonlinear. Therefore, the resulting frequency response of the band-pass filter is not symmetric about its center frequency, which is defined as

$$\begin{aligned}
\omega_o &= (\omega_u + \omega_\ell)/2 \\
&= \omega_\ell + \omega_b/2 \\
&= \omega_u - \omega_b/2
\end{aligned} \qquad (8.49)$$

This lack of symmetry is illustrated in Figure 8.19. The lack of symmetry becomes less apparent when ω_g approaches ω_o as ω_o becomes much larger than ω_b. This is often the case for practical filters. Note that gains at frequencies below the geometric-center frequency of the band-pass filter correspond to gains at negative frequencies in the double-sided frequency response of the low-pass filter.

We multiply both sides of the transformation in eq. (8.48) by j and perform some algebra to give

$$\begin{aligned}
j\omega_{LN} &= j(\omega^2 - \omega_g^2)/\omega\omega_b \\
&= -[-(j\omega)^2 - \omega_g^2]/j\omega\omega_b \\
&= [(j\omega)^2 + \omega_g^2]/j\omega\omega_b
\end{aligned} \qquad (8.50)$$

Then, we use eq. (8.50) in eq. (8.48), replacing $j\omega$ with s and $j\omega_{LN}$ with s_{LN} to give the following expression for the transfer function of the band-pass filter

Figure 8.19
Amplitude Responses
of Example
Low-Pass Filter and
Corresponding
Band-Pass Filter
Obtained by
Frequency
Transformation

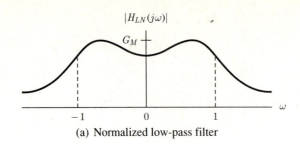

(a) Normalized low-pass filter

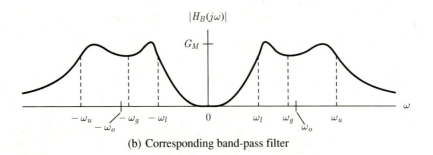

(b) Corresponding band-pass filter

in terms of the transfer function of the normalized low-pass filter:

$$H_B(s) = H_{LN}(s_{LN})\big|_{s_{LN}=(s^2+\omega_g^2)s\omega_b} \tag{8.51}$$

Since s_{LN} is replaced by a term that contains s^2, then the transfer function of the band-pass filter has twice as many poles as the normalized low-pass filter from which it is generated. That is, the band-pass filter has $2n$ poles, where n is the order of the normalized low-pass filter.

The procedure for designing a Butterworth or Chebyshev band-pass filter is similar to the one for designing the low-pass filter, except that a different frequency transformation is used. Also, we must determine the filter gain for various filter orders at frequencies of interest (that is, frequencies for which gain specifications are given) in the stopbands both above and below the pass-band to determine the required filter order. For these gain determinations, the frequencies of interest in the stopband below the filter passband transform into negative frequencies for the corresponding normalized low-pass filter. Since the double-sided amplitude response has even symmetry, in this case we can use the curves of Figures 8.14 through 8.17 with the frequency scale replaced by its negative.

Example 8.3

We need a filter that will pass the portion of the audio frequency band near 10 kHz. The filter specifications are as follows:

1. Chebyshev filter with 1 dB of ripple in the passband;
2. maximum gain: unity;
3. center frequency: 10 kHz;
4. bandwidth: 1 kHz; and
5. stopband gain: less than or equal to -10 dB for $f \geq 13$ kHz and $f \leq 7$ kHz.

Find the filter with the fewest number of poles that will satisfy the specifications. Determine the filter transfer function and plot its amplitude response for $0 \leq f \leq 15$ kHz using linear amplitude and frequency scales.

Solution

From specifications 3 and 4, the upper and lower cutoff frequencies and the geometric-center frequency are

$$\omega_u = \omega_o + \omega_b/2 = 2\pi(10) + \pi(1) = 21\pi \text{ krad/s}$$

$$\omega_\ell = \omega_o - \omega_b/2 = 2\pi(10) - \pi(1) = 19\pi \text{ krad/s}$$

$$\omega_g = \sqrt{\omega_u \omega_\ell} = \sqrt{21\pi(19)\pi} = 19.975\pi \text{ krad/s}$$

From specification 5 and the frequency transformation in eq. (8.48), the stop-band frequencies of interest for the corresponding normalized low-pass filter are

$$\omega_{1N} = \frac{[2\pi(7)]^2 - (19.975\pi)^2}{[2\pi(7)](2\pi)} = -7.250$$

$$\omega_{2N} = \frac{[2\pi(13)]^2 - (19.975\pi)^2}{[2\pi(13)](2\pi)} = 5.327$$

We see from Figure 8.16 that the normalized low-pass Chebyshev filter of the first order with 1 dB of ripple in the passband has a gain of less than -10 dB with respect to the maximum gain at $\omega_{LN} = \omega/\omega_c = -7.250$, but not at $\omega_{LN} = \omega/\omega_c = 5.327$. A second-order filter has a gain of less than -10 dB with respect to the maximum gain at both normalized frequencies. Thus, the minimum filter order required is second order, which implies that the normalized low-pass filter has two poles. This means that the minimum number of poles required for the band-pass filter is four. The resulting filter has a gain with respect to the maximum gain of less than approximately -29 dB for $f \geq 13$ kHz and $f \leq 7$ kHz. Actually, the -10-dB specification given does not produce a very selective filter. However, this specification reduced the required filter order and thereby kept the example sample.

From Table 8.1, we find that the second-order normalized low-pass Chebyshev filter with 1 dB of ripple in the passband has the transfer function

$$H_{LN}(s_{LN}) = \frac{a_o}{1.1025 + 1.0977 s_{LN} + s_{LN^2}}$$

To satisfy specification 2,

$$a_o = b_o G_M / \sqrt{1 + \epsilon^2}$$

$$= (1.1025)(1) / \sqrt{1 + (0.2589)} = 0.9826$$

To convert $H_{LN}(s_{LN})$ to the transfer function of the desired band-pass filter, we replace s_{LN} by

$$s_{LN} = (s^2 + \omega_g^2)/\omega_b s = [s^2 + (19975\pi)^2]/2000\pi s$$

The resulting transfer function of the band-pass filter is

$$H_B(s)$$

$$= \frac{3.8791 \times 10^7 s^2}{(1.5508 \times 10^{19}) + (2.7160 \times 10^{13})s + (7.9195 \times 10^9)s^2 + 6897.1 s^3 + s^4}$$

The amplitude response of the band-pass filter is given by $|H_B(j2\pi f)|$ and is plotted in Figure 8.20.

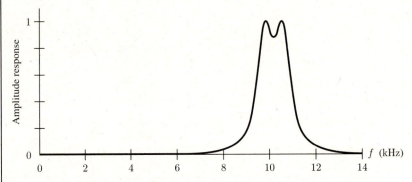

Figure 8.20 Amplitude Response of the Band-Pass Filter for Example 8.3

8.7 Summary

One of the important continuous-time systems is the filter. In general, a filter lets input-signal components with some frequencies pass with little change, while essentially eliminating input-signal components with other frequencies. We use the analysis tools developed in the preceding chapters to analyze and design filters.

Distortionless transmission is defined as transmission that changes the input signal only with respect to amplification or attenuation and/or time delay. Ideal low-pass, band-pass, high-pass, and band-rejection filters provide distortionless transmission for signals with components having certain ranges of frequencies, at the same time completely eliminating signals with components having other ranges of frequencies.

To construct filters that approximate the amplitude response or phase response of the ideal filter, we determine the filter transfer function that produces an approximation of the desired amplitude or phase response. In this chapter, we considered only those filters constructed to approximate the desired amplitude response.

Two types of amplitude-response approximations are the Butterworth and Chebyshev filter approximations. These approximations are defined for low-pass filters. The Butterworth approximation gives a maximally flat amplitude response near dc. The Chebyshev approximation provides sharper amplitude-response cutoff by permitting amplitude-response ripple in the passband. The band-pass, high-pass, and band-rejection forms of Butterworth and Chebyshev filters are found by frequency transformations.

We define a normalized low-pass filter as a low-pass filter with a cutoff frequency of 1 rad/s. It is convenient to use a normalized low-pass filter as an integral part of the design procedure when designing Butterworth and Chebyshev filters. The design can then be achieved using a few tables of parameters for the normalized filter together with frequency scaling to obtain a desired cutoff frequency or frequency transformation with which to convert the filter to a band-pass, high-pass, or band-rejection filter.

We provided a brief introduction to filtering concepts and filter design. Techniques used in filter design depend on the technology available to implement the filters and on the response characteristics desired. For detailed discussion of various aspects of filter design and design techniques many excellent resources can be consulted. Computer software is also available to generate filter designs to satisfy given specifications.

Problems

8.1 Define the ideal high-pass filter in the manner in which the ideal low-pass and band-pass filters were defined in Section 8.2. Sketch the amplitude and phase responses.

8.2 Write the mathematical expression for the frequency response of the ideal high-pass filter defined in Problem 8.1. Find its impulse response.

8.3 Define the ideal band-rejection filter in the manner in which the ideal low-pass and band-pass filters were defined. Sketch the amplitude and phase responses.

8.4 Write the mathematical expression for the frequency response of the ideal band-rejection filter defined in Problem 8.3. Find its impulse response.

8.5 Find the transfer function for the minimum-phase

stable filter with the amplitude response

$$|H(j\omega)| = \frac{2(16 + \omega^2)}{\sqrt{6 + 5\omega^2 + \omega^4}}$$

Plot its poles and zeros on the s-plane.

8.6 Find the transfer function for the minimum-phase stable filter with the amplitude response

$$|H(j\omega)| = \sqrt{\frac{4 + 68\omega^2 + 64\omega^4}{4 + 17\omega^2 + 4\omega^4}}$$

Plot its poles and zeros on the s-plane.

Note: In Problems 8.7–8.22, computer solutions should be used to generate data for the requested plots to provide sufficient plot detail.

8.7 A filter with the amplitude response

$$|H(j\omega)| = \sqrt{\frac{1 + \omega^2}{1 - \omega^2 + \omega^4}}$$

is needed to compensate for some undesired characteristics of a signal sensor.

a. Find the transfer function for the minimum-phase stable filter having this amplitude response.

b. Plot the amplitude and phase response for the filter of part (a) and for the stable filter obtained by choosing the zero locations to be the negative of those selected in part (a). Comment on the effect of the zero locations.

8.8 Repeat Problem 8.7 for the amplitude response

$$|H(j\omega)| = \sqrt{\frac{4 + \omega^4}{169 + 159\omega^2 - 9\omega^4 + \omega^6}}$$

8.9 For $n = 1$, 2, and 3, find and plot the amplitude and phase response for the Chebyshev low-pass filter 3 dB ripple in the passband. Compare the results with those shown for the Butterworth low-pass filter in Figure 8.10.

8.10 Find the transfer function for a second-order normalized Chebyshev low-pass filter with unity gain and 2 dB of ripple in the passband. Express your answer in the form of eq. (8.36).

8.11 A low-pass filter for reducing noise on a satellite-position measurement signal is to be of third order, have a gain variation less than or equal to 3 dB in the frequency range from 0 Hz to 10 Hz, and have a maximum gain of 2. Consider a Butterworth and a Chebyshev filter and find the filter transfer functions. Also plot the amplitude responses and phase responses for the filters from 0 Hz to 20 Hz using linear scales.

8.12 Find the transfer function for a fourth-order Chebyshev low-pass filter having a maximum gain of unity and a gain variation of no more than 0.5 dB in the frequency range from 0 to 500 Hz.

8.13 A low-pass Chebyshev filter for reducing image-processing noise is to have three poles, unity maximum gain, and no more than 1 dB of gain variation in the frequency range from 0 to 200 Hz. Find the filter transfer function and the filter gain at a frequency of 500 Hz.

8.14 Specifications for a low-pass filter to restrict the bandwidth for a voice-transmission system are (1) maximum gain variation of 3 dB from 0 to 5 kHz and (2) gain less than or equal to -30 dB with respect to the maximum gain for $f \geq 15$ kHz. Determine the minimum order required for Butterworth and Chebyshev filters to satisfy the specifications.

8.15 A Chebyshev filter is to have a gain that varies no more than 1 dB from 0 to 10 kHz and gain less than or equal to -40 dB with respect to the maximum gain for $f \geq 25$ kHz. What order of filter is required?

8.16 The gain of a low-pass filter is to vary no more than 3 dB from 0 to 16 KHz and be less than or equal to one-tenth of the maximum gain for $f \geq 40$ kHz. Determine the minimum order of Butterworth and Chebyshev filters that will meet specifications.

8.17 A low-pass filter is to satisfy the following specifications: (1) Chebyshev type, (2) maximum gain of 4, (3) gain variation of no more than 0.5 dB from 0 to 100 Hz, and (4) gain less than or equal to -30 dB with respect to the maximum gain for $f \geq 800$ Hz. Find the minimum-order filter that will meet specifications and the transfer function for the filter.

8.18 Find the transfer function for a Butterworth band-pass filter that has two poles, a center frequency of 500 Hz, a bandwidth of 100 Hz, and a maximum gain of unity. Plot the amplitude and phase responses for the filter from 0 to 2000 Hz using linear scales.

8.19 A band-pass filter of the Chebyshev type is to be constructed. It is to have four poles, a center frequency of 100 Hz, a bandwidth of 80 Hz, unity maximum gain, and 3 dB of passband ripple. Find the required transfer function and plot the amplitude and phase responses from 0 to 250 KHz using linear scales.

8.20 A band-pass filter is to be constructed with unity maximum gain, a passband from 100 Hz to 200 Hz with ripple of 0.5 dB, and gain less than or equal to 0.03 for $f \leq 50$ Hz and $f \geq 300$ Hz. What is the minimum number of poles the filter must contain?

8.21 A high-pass filter is needed to reduce low-frequency interference, including power-source hum, in a transmitted voice signal. It is constructed by transforming a normalized Chebyshev low-pass filter with 1 dB of passband ripple. The resulting high-pass filter has a maximum gain of unity, a cutoff frequency of 200 Hz, and a gain less than or equal to 0.01 for $f \leq 50$ Hz. Find the transfer function for the filter and plot the amplitude response for 0 Hz $\leq f \leq$ 1000 Hz using linear scales.

8.22 A band-rejection filter is needed to reduce power-source hum (60-Hz hum) in an amplifier. It is constructed by transforming a normalized Butterworth low-pass filter. The resulting filter is to have unity maximum gain and a rejection band with lower and upper cutoff frequencies of 50 Hz and 70 Hz, respectively. The gain between 57 Hz and 61 Hz is to be less than or equal to 0.04. Find the filter transfer function and plot the amplitude response for 0 Hz $\leq f \leq$ 100 Hz using linear scales.

9 State-Variable Concepts for Continuous-Time Linear Systems

In the preceding chapters, we considered a number of techniques for analyzing continuous-time signals and systems. In particular, we defined Fourier transform and Laplace transform techniques. These techniques are useful in the analysis of signal spectra and, for linear time-invariant systems, in the analysis of system response signals and their spectra. To this point, our discussion has been limited to single-input, single-output systems.

A multiple-input, multiple-output system can often be modeled as an interconnection of a group of single-input, single-output systems, as shown by the general block diagram in Figure 9.1. Thus, the techniques previously discussed apply. However, because the additional system complexity makes analysis more difficult, it is desirable to have an organized method for writing and solving system equations. An organized method is also desirable for very complicated single-input, single-output systems that produce large-order differential equations. One method for writing and solving such equations is called the *state variable*, or *state space, method*. In this method, we define internal signals called *state variables*, system-specified equations called *state-variable equations* or *state equations*, and *output equations*. We use the equations to solve for the state variables and the output signals. The state variables provide in-

Figure 9.1
Multiple-Input, Multiple-Output System Modeled as an Interconnection of Single-Input, Single-Output Systems

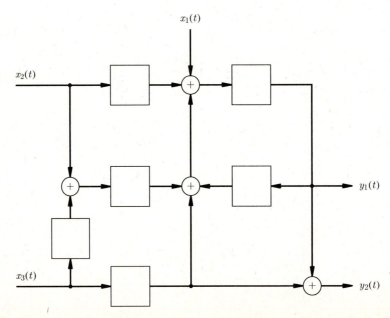

formation about the internal characteristics of the system—another motivation for using them in system analysis.

In this chapter we will briefly discuss the basic concepts associated with state variables and state-variable analysis of linear systems. This discussion will serve to introduce the basic concepts of analysis with state variables, enabling the reader to see how concepts such as impulse response and transfer function can be extended to multiple-input, multiple-output systems. Space does not permit a treatment of state-variable concepts in greater depth or breadth; indeed, entire texts are devoted to such treatments.[†]

9.1 State Variables and Equations

In Chapter 1, we indicated that if a system is an nth-order lumped-parameter, linear, continuous-time system, then an internal signal or an output signal is related to an input signal through an nth-order ordinary linear differential equation. For simplicity in the discussion here, we also assume that the differential equation contains no derivatives of the input signals of order higher than n. The system that corresponds to a differential equation of this form is referred to as a *proper system*. Most systems of interest are proper systems.

We can also express the relation between an internal signal or an output signal and an input signal in terms of n first-order ordinary differential equations in n unknown variables. To illustrate, consider a second-order, lumped-parameter, linear, time-varying, single-input–single-output, continuous-time system. Note that the system is permitted to be time-varying here for generality.

An internal signal $w(t)$ in the defined system can be found by solving the differential equation

$$a(t)\frac{d^2w(t)}{dt^2} + b(t)\frac{dw(t)}{dt} + c(t)w(t) = d(t)x(t) \tag{9.1}$$

for $w(t)$, where $x(t)$ is the input signal and $a(t)$, $b(t)$, $c(t)$, and $d(t)$ are time-varying system parameters. Let us assume that the output signal of the system is related to the input signal $x(t)$ and the internal signal $w(t)$ by the equation

$$y(t) = e(t)\frac{dw(t)}{dt} + f(t)w(t) + g(t)x(t) \tag{9.2}$$

Now we define the two signals

$$q_1(t) = \frac{dw(t)}{dt} \tag{9.3}$$

and

$$q_2(t) = w(t) \tag{9.4}$$

[†] See for example P. M. DeRusso, R. J. Roy, and C. M. Close, *State Variables for Engineers* (New York: Wiley, 1965) and L. K. Timothy and B. E. Bona, *State Space Analysis: An Introduction* (New York: McGraw-Hill, 1968).

Substituting eqs. (9.3) and (9.4) in eq. (9.1), and eq. (9.4) in eq. (9.3), we obtain the two first-order differential equations

$$\frac{dq_1(t)}{dt} = -\frac{b(t)}{a(t)}q_1(t) - \frac{c(t)}{a(t)}q_2(t) + \frac{d(t)}{a(t)}x(t) \tag{9.5}$$

and

$$\frac{dq_2(t)}{dt} = q_1(t) \tag{9.6}$$

We can simultaneously solve eqs. (9.5) and (9.6) for the two signals $q_1(t)$ and $q_2(t)$. The signals $q_1(t)$ and $q_2(t)$ are the state variables for the system and eqs. (9.5) and (9.6) are the state equations for the system. Note that the state equations express the derivatives of the state variables as weighted sums of the state variables and the input signal, where the weighting coefficients $-b(t)/a(t)$, $-c(t)/a(t)$, and $d(t)/a(t)$ depend on the system parameters. Also, the number of state variables is equal to the order of the system because we assumed the system to be proper.

We can express the output signal as a weighted sum of the state variables and the input signal by substituting eqs. (9.3) and (9.4) in eq. (9.2) to give

$$y(t) = e(t)q_1(t) + f(t)q_2(t) + g(t)x(t) \tag{9.7}$$

This is referred to as the *output equation* for the system.

The preceding example provides background for the definitions that follow.

Definition

The state equations for an nth-order, lumped-parameter, linear, continuous-time system are n first-order, linear, ordinary differential equations that express the derivatives of the n state variables of the system as weighted sums of the state variables and the input signals of the system.

Definition

The output equations for an nth-order, lumped-parameter, linear, continuous-time system are linear algebraic equations that express the output signals of the system as weighted sums of the n state variables and the input signals of the system.

The choice of state variables is not unique. The choice that we made in the preceding example was an internal signal and its derivative. In general, the most convenient state variables to use are those that characterize energy stored in the energy-storage components of the system. Solution of the state equations then provides information about the energy stored in the system. For electric circuits, it is convenient to use the voltages across capacitors and the currents through inductors. There are n of these for an nth-order electric circuit. They represent the state of the energy stored in the circuit because the energy stored in a capacitor is a function of the voltage across the capacitor, and the energy stored in an inductor is a function of the current through the inductor.

Example 9.1

Consider the electric circuit shown in Figure 9.2, where $x(t)$ and $y(t)$ are the input and output signals, respectively. Identify a set of system state variables and find the corresponding state equations and output equation for the system.

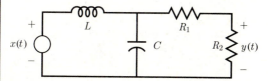

Figure 9.2 Electric Circuit for Example 9.1

Solution

The system is of the second order because it has two energy-storage elements. We define the required two state variables as the current $q_1(t)$ through the inductor and the voltage $q_2(t)$ across the capacitor. These state-variable definitions and other voltages and currents of interest are shown in Figure 9.3.

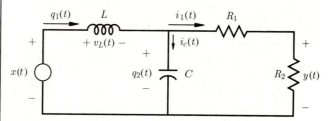

Figure 9.3 Definition of State Variables and Other Voltages and Currents for Example 9.1

The equations for the derivatives of the two state variables are

$$L\frac{dq_1(t)}{dt} = v_L(t) = x(t) - q_2(t)$$

and

$$C\frac{dq_2(t)}{dt} = i_C(t) = q_1(t) - i_1(t)$$

Also

$$i_1(t) = q_2(t)/(R_1 + R_2)$$

Combining and simplifying these three equations, we obtain the two state equations for the system

$$\frac{dq_1(t)}{dt} = \left[-\frac{1}{L}\right] q_2(t) + \left[\frac{1}{L}\right] x(t)$$

and

$$\frac{dq_2(t)}{dt} = \left[\frac{1}{C}\right] q_1(t) + \left[-\frac{1}{C(R_1 + R_2)}\right] q_2(t)$$

These state equations express the derivatives of the state variables as weighted sums of the state variables and input signals. Now, since

$$y(t) = i_1(t)R_1$$

then the output equation for the system is

$$y(t) = \left[\frac{R_1}{R_1 + R_2}\right] q_2(t)$$

With the particular choice of state variables as previously defined and as illustrated in the example, solution of the state equations and output equations provides not only the system output signal, but also information about the energy stored in the electric circuit as a function of time. Thus, as we stated earlier, state variables provide information about the internal characteristics of the system.

We now use the preceding definitions to write the state equations and output equations for an nth-order, lumped-parameter, linear, continuous-time system with m inputs and p outputs. These equations are

$$\frac{dq_i(t)}{dt} = \sum_{j=1}^{n} a_{ij}(t)q_j(t) + \sum_{k=1}^{m} b_{ik}(t)x_k(t) \qquad 1 \leq i \leq n \qquad (9.8)$$

and

$$y_\ell(t) = \sum_{j=1}^{n} c_{\ell j}(t)q_j(t) + \sum_{k=1}^{m} d_{\ell k}(t)x_k(t) \qquad 1 \leq \ell \leq p \qquad (9.9)$$

We show the block-diagram representation corresponding to eqs. (9.8) and (9.9) for a single state variable and single output signal in Figure 9.4. This block diagram is a portion of the complete system block diagram. The complete system block diagram contains n integrators, the inputs of which are weighted sums of the n state variables and the m input signals and the outputs of which are the state variables. The block diagram also contains p output summations of n weighted state variables and m weighted input signals. We see that the system block diagram becomes very complicated when the system contains very many states, inputs, and outputs.

Figure 9.4
Portion of Block
Diagram for a
Multiple-Input,
Multiple-Output
System

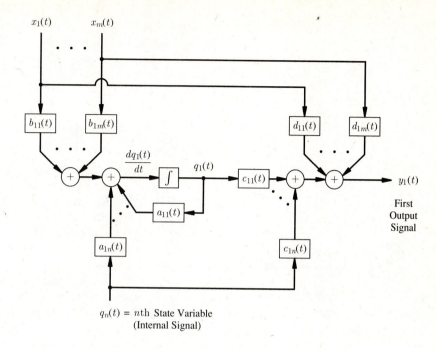

$q_n(t) = n$th State Variable
(Internal Signal)

For analysis purposes, it is convenient to express the state equations and output equations in matrix notation. (See Appendix A for a summary of matrix properties and operations.) The resulting matrix state equation and matrix output equation are

$$\frac{d\mathbf{q}(t)}{dt} = \mathbf{A}(t)\mathbf{q}(t) + \mathbf{B}(t)\mathbf{x}(t) \tag{9.10}$$

and

$$\mathbf{y}(t) = \mathbf{C}(t)\mathbf{q}(t) + \mathbf{D}(t)\mathbf{x}(t) \tag{9.11}$$

In eqs. (9.10) and (9.11),

$$\mathbf{x}(t) = \begin{bmatrix} x_1(t) \\ \vdots \\ x_m(t) \end{bmatrix} \qquad \mathbf{q}(t) = \begin{bmatrix} q_1(t) \\ \vdots \\ q_n(t) \end{bmatrix} \qquad \mathbf{y}(t) = \begin{bmatrix} y_1(t) \\ \vdots \\ y_p(t) \end{bmatrix} \tag{9.12}$$

are the $m \times 1$ input-signal matrix, the $n \times 1$ state-variable matrix, and the $p \times 1$ output-signal matrix, respectively. Since these matrices only contain one column, they are also referred to as *vectors*. The $n \times 1$ vector $d\mathbf{q}(t)/dt$ is defined

$$\frac{d\mathbf{q}(t)}{dt} = \begin{bmatrix} dq_1(t)/dt \\ \vdots \\ dq_n(t)/dt \end{bmatrix} \tag{9.13}$$

and is the state-variable derivative vector. The matrices

$$\mathbf{A}(t) = \begin{bmatrix} a_{11}(t) & \dots & a_{1n}(t) \\ \vdots & & \vdots \\ a_{n1}(t) & \dots & a_{nn}(t) \end{bmatrix} \qquad \mathbf{B}(t) = \begin{bmatrix} b_{11}(t) & \dots & b_{1m}(t) \\ \vdots & & \vdots \\ b_{n1}(t) & \dots & b_{nm}(t) \end{bmatrix}$$

$$\mathbf{C}(t) = \begin{bmatrix} c_{11}(t) & \dots & c_{1n}(t) \\ \vdots & & \vdots \\ c_{p1}(t) & \dots & c_{pn}(t) \end{bmatrix} \qquad \mathbf{D}(t) = \begin{bmatrix} d_{11}(t) & \dots & d_{1m}(t) \\ \vdots & & \vdots \\ d_{p1}(t) & \dots & d_{pm}(t) \end{bmatrix} \qquad (9.14)$$

are the parameter matrices of the system. Note that the sizes of the $\mathbf{A}(t)$, $\mathbf{B}(t)$, $\mathbf{C}(t)$, and $\mathbf{D}(t)$ matrices are $n \times n$, $n \times m$, $p \times n$, and $p \times m$, respectively. The block-diagram representation that corresponds to the matrix-equation representation of the system is frequently drawn as shown in Figure 9.5, where the double-line paths indicate vector-signal paths and the integral component corresponds to n integrations. We see that the use of matrix notation permits us to draw a simple system block diagram.

We allowed the system to be time-varying in the preceding development of state equations and output equations. We can perform sequential digital-computer approximate solutions of these equations to obtain successive samples of the state variables and output signals for $t \geq t_0$ if (1) the time functions characterizing the system parameters and the input signals are known for $t \geq t_0$, (2) the value of the state variables at $t = t_0$ are known, and (3) the time interval, Δt, between samples is sufficiently small to permit us to assume that derivatives of the state variables are constant over one time interval. Since this problem pertains to a discrete-time signal and system, we will postpone further discussion until Chapter 16.

In the remainder of this chapter, we will limit the linear systems considered to time-invariant systems. This limitation is in keeping with our practice in the preceding chapters and permits analytic solutions to be developed simply. For linear time-invariant systems, the matrix state equation and matrix output equation are

$$\frac{d\mathbf{q}(t)}{dt} = \mathbf{A}\mathbf{q}(t) + \mathbf{B}\mathbf{x}(t) \qquad (9.15)$$

Figure 9.5
Block-Diagram
System
Representation
Corresponding to
Matrix State and
Output Equations

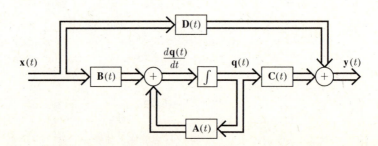

and

$$\mathbf{y}(t) = \mathbf{Cq}(t) + \mathbf{Dx}(t) \qquad (9.16)$$

where $\mathbf{A}$, $\mathbf{B}$, $\mathbf{C}$, and $\mathbf{D}$ are *constant matrices* (that is, matrices with elements that are constants).

The notation used in many texts that consider state-variable analysis techniques is $\mathbf{y}(t)$ for the output-signal vector, $\mathbf{x}(t)$ for the state vector, and $\mathbf{u}(t)$ for the input-signal vector. In this text, $\mathbf{y}(t)$ is used for the output-signal vector and $\mathbf{x}(t)$ for the input-signal vector to be consistent with the output- and input-signal notation used throughout and to avoid confusion with the unit step signal. Since $\mathbf{x}(t)$ is used for the input-signal vector, another notation is required for the state vector. We have chosen the notation $\mathbf{q}(t)$.

9.2 Concepts and Definitions

Before proceeding with a discussion of the solution of state and output equations, it is useful to consider some basic concepts and definitions concerning state variables and systems modeled by state and output equations.

Definition _____

The system state at time t is the value of the state vector at time t; that is, $\mathbf{q}(t)$.

_____

We can find the system output vector $\mathbf{y}(t)$, for $t \geq t_0$ from the state equations and output equations if the initial system state $\mathbf{q}(t_0)$ and the input signal vector $\mathbf{x}(t)$ for $t \geq t_0$ are known.

Definition _____

A zero state of a system, $\mathbf{q}_0(t_1)$, is a system state that produces the output-signal vector $\mathbf{y}(t) = \mathbf{0}$ for $t \geq t_1$ if the input-signal vector is $\mathbf{x}(t) = \mathbf{0}$ for $t \geq t_1$, where $\mathbf{0}$ is the zero matrix (vector, in this case) that has all components equal to zero.

_____

The zero vector, $\mathbf{0}$, is always a zero state of a system if the system is linear and is the only zero state for most linear systems of interest; however, some linear systems may also have other zero states. For some nonlinear systems, the zero vector may not be a zero state.

We are interested in two types of system response: the *zero-input response* and the *zero-state response*. These are defined in terms of the output-signal vector.

Definition _____

The zero-input response of a system is the output-signal vector $\mathbf{y}(t) = \mathbf{y}_{zi}(t)$ for $t \geq t_0$ that results from the initial state $\mathbf{q}(t_0)$ when the input-signal vector is $\mathbf{x}(t) = \mathbf{0}$ for $t \geq t_0$.

Definition _____

The zero-state response of a system is the output-signal vector $\mathbf{y}(t) = \mathbf{y}_{zs}(t)$ for $t \geq t_0$ that results from the input-signal vector $\mathbf{x}(t)$ when the system initial state at $t = t_0$ is a zero state; that is, when $\mathbf{q}(t_0) = \mathbf{q}_0(t_0)$.

The term *linear system* is used in the preceding discussions of a system modeled with state variables and equations. Now, let us define a linear system in terms of the state-variable model.

Definition _____

A system is linear if and only if:

1. its system response is decomposable into the sum of its zero-input response and its zero-state response;
2. superposition holds for its zero-state response; and
3. superposition holds for its zero-input response.

This definition of system linearity is a more general statement of the definition presented and illustrated in Chapter 1. Whereas the definition in Chapter 1 applied only to single-input, single-output systems, this definition applies to the general multiple-input, multiple-output system.

9.3 Time-Domain Analysis of Systems

We are now prepared to consider the solution of the state and output equations to determine the state vector and the output-signal vector, as functions of time, that correspond to a given initial state and a given input-signal vector. As mentioned previously, we are concerned with only linear and time-invariant systems; hence, the state and output equations are those given by eqs. (9.15) and (9.16). In addition, only the zero state $\mathbf{q}_0(t) = \mathbf{0}$ is considered for simplicity. Recall that $\mathbf{0}$ is always a zero state for a linear system and is the only zero state for most linear systems of interest.

Solutions for State and Output Signals

Let us first consider the solution of the state equations for $t \geq t_0$ when $\mathbf{x}(t) = \mathbf{0}$ for $t \geq t_0$. That is, the solution of

$$\frac{d\mathbf{q}_{zi}(t)}{dt} = \mathbf{A}\mathbf{q}_{zi}(t) \qquad (9.17)$$

where $\mathbf{q}_{zi}(t)$ is the zero-input state vector for $t \geq t_0$. The matrix differential equation of eq. (9.17) can be solved to yield a solution of the form

$$\mathbf{q}_{zi}(t) = e^{(t-t_0)\mathbf{A}} \mathbf{a} \qquad t \geq t_0 \qquad (9.18)$$

where $\mathbf{a}$ is an arbitrary $n \times 1$ constant matrix (that is, constant vector) and $e^{(t-t_0)\mathbf{A}}$ is an $n \times n$ matrix defined for $t \geq t_0$. The matrix $e^{(t-t_0)\mathbf{A}}$ is called the *matrix exponential*. The proof that eq. (9.18) is the solution of eq. (9.17) is beyond the scope of this text; however, note the similarity to the solution of a first-order scalar differential equation.

The $n \times n$ matrix exponential is defined in terms of its infinite series expansion as

$$e^{(t-t_0)\mathbf{A}} = \mathbf{I} + (t-t_0)\mathbf{A} + (t-t_0)^2 \mathbf{A}^2/2! + \ldots \qquad t \geq t_0 \qquad (9.19)$$

where $\mathbf{I}$ is the $n \times n$ identity matrix. Note the parallel with the infinite-series representation of the scalar exponential $e^{(t-t_0)a}$. From eq. (9.19), we see that

$$e^{(0)\mathbf{A}} = \mathbf{I} \qquad (9.20)$$

and, from eq. (A.17) in Appendix A,

$$\frac{de^{(t-t_0)\mathbf{A}}}{dt} = \mathbf{0} + \mathbf{A} + 2(t-t_0)\mathbf{A}^2/2! + 3(t-t_0)^2\mathbf{A}^3/3! + \cdots$$

$$= \mathbf{A}[\mathbf{I} + (t-t_0)\mathbf{A} + (t-t_0)^2\mathbf{A}^2/2! + \cdots]$$

$$= \mathbf{A}e^{(t-t_0)\mathbf{A}} \qquad t \geq t_0 \qquad (9.21)$$

The vector $\mathbf{q}_{zi}(t_0)$ is the initial state vector when the input signal is zero for $t \geq t_0$. However, the initial state vector $\mathbf{q}(t_0)$ for the system does not depend on the input signal that is present at $t \geq t_0$. Therefore,

$$\mathbf{q}(t_0) = \mathbf{q}_{zi}(t_0) = \mathbf{I}\mathbf{a} = \mathbf{a} \qquad (9.22)$$

which shows that the arbitrary constant vector $\mathbf{a}$ is equal to the initial state vector for the system. The zero-input state vector is thus

$$\mathbf{q}_{zi}(t) = e^{(t-t_0)\mathbf{A}}\mathbf{q}(t_0) \qquad t \geq t_0 \qquad (9.23)$$

The matrix exponential is called the *state transition matrix* because it relates the zero-input state vector at time t to the zero-input state vector at time t_0. It is a function of $t - t_0$, for which we use the notation

$$\mathbf{r}(t - t_0) = e^{(t-t_0)\mathbf{A}} \qquad t \geq t_0 \qquad (9.24)$$

Using this notation in eq. (9.23), we find that the zero-input state vector is

$$\mathbf{q}_{zi}(t) = \mathbf{r}(t - t_0)\mathbf{q}(t_0) \qquad t \geq t_0 \qquad (9.25)$$

The zero-input response of the system is found from the matrix output equation to be

$$\mathbf{y}_{zi}(t) = \mathbf{C}\mathbf{q}_{zi}(t) = \mathbf{C}\mathbf{r}(t - t_0)\mathbf{q}(t_0) \qquad t \geq t_0 \qquad (9.26)$$

The total state vector for $t \geq t_0$ is the zero-input state vector for $t \geq t_0$ plus the state vector obtained for $t \geq t_0$ from $x(t)$ when $\mathbf{q}(t_0) = \mathbf{0}$. We find it from the zero-input state vector by the method of variation of parameters, as we did for the single-input, single-output system considered in Example 1.10. In using variation of parameters, we replace the arbitrary constant vector $\mathbf{a}$ in the zero-input state vector given by eq. (9.18) with the time-varying vector $\mathbf{a}(t)$, and then solve for $\mathbf{a}(t)$. The details follow.

From eq. (9.18), the assumed solution for the state vector is

$$\mathbf{q}(t) = e^{(t-t_0)\mathbf{A}}\mathbf{a}(t) \qquad t \geq t_0 \qquad (9.27)$$

We then compute the derivative of eq. (9.27) and substitute in eq. (9.15), the matrix state equation, to obtain

$$\mathbf{A}e^{(t-t_0)\mathbf{A}}\mathbf{a}(t) + e^{(t-t_0)\mathbf{A}}\left[\frac{d\mathbf{a}(t)}{dt}\right] = \mathbf{A}e^{(t-t_0)\mathbf{A}}\mathbf{a}(t) + \mathbf{B}x(t) \qquad (9.28)$$

Simplifying eq. (9.28) gives

$$e^{(t-t_0)\mathbf{A}}\left[\frac{d\mathbf{a}(t)}{dt}\right] = \mathbf{B}x(t) \qquad t \geq t_0 \qquad (9.29)$$

The matrix exponential has the inverse

$$\left[e^{(t-t_0)\mathbf{A}}\right]^{-1} = e^{-(t-t_0)\mathbf{A}} \qquad t \geq t_0 \qquad (9.30)$$

The proof of eq. (9.30) is not shown; it is beyond the scope of this text. Premultiplying eq. (9.29) by the inverse of the matrix exponential, we obtain

$$\mathbf{I}\left[\frac{d\mathbf{a}(t)}{dt}\right] = \frac{d\mathbf{a}(t)}{dt} = e^{-(t-t_0)\mathbf{A}}\mathbf{B}x(t) \qquad t \geq t_0 \qquad (9.31)$$

Integration of this matrix equation produces

$$\mathbf{a}(t) = \int_{t_0}^{t} e^{-(\tau-t_0)\mathbf{A}}\mathbf{B}x(\tau)\,d\tau + \mathbf{a}(t_0) \qquad t \geq t_0 \qquad (9.32)$$

where

$$\mathbf{a}(t_0) = \mathbf{I}\mathbf{a}(t_0) = e^{(0)\mathbf{A}}\mathbf{a}(t_0) = \mathbf{q}(t_0) \qquad (9.33)$$

with the last step of (9.33) resulting from the application of eq. (9.27). Note that the integral indicated in eq. (9.32) is really n integrals, one for each component of the vector $\mathbf{a}(t)$. Substituting eqs. (9.32) and (9.33) in eq. (9.27), we obtain the state vector

$$\mathbf{q}(t) = e^{(t-t_0)\mathbf{A}} \int_{t_0}^{t} e^{-(\tau - t_0)\mathbf{A}} \mathbf{Bx}(\tau) \, d\tau + e^{(t-t_0)\mathbf{A}} \mathbf{q}(t_0)$$

$$= \int_{t_0}^{t} e^{(t-\tau)\mathbf{A}} \mathbf{Bx}(\tau) \, d\tau + e^{(t-t_0)\mathbf{A}} \mathbf{q}(t_0)$$

$$= \mathbf{r}(t - t_0)\mathbf{q}(t_0) + \int_{t_0}^{t} \mathbf{r}(t - \tau)\mathbf{Bx}(\tau) \, d\tau \qquad t \geq t_0 \qquad (9.34)$$

We then obtain the system response by substituting eq. (9.34) in eq. (9.16), the matrix output equation, to produce

$$\mathbf{y}(t) = \mathbf{Cr}(t - t_0)\mathbf{q}(t_0) + \int_{t_0}^{t} \mathbf{Cr}(t - \tau)\mathbf{Bx}(\tau) \, d\tau + \mathbf{Dx}(t) \qquad t \geq t_0 \quad (9.35)$$

By comparison with eq. (9.26), we see that the first term of eq. (9.35) is the zero-input response of the system. Therefore, the sum of the second and third terms of eq. (9.35) is the zero-state response because the system is linear, and thus its response is decomposable.

When we were considering single-input–single-output, linear, time-invariant systems, the importance of the impulse response became apparent. We saw that the system response to any input signal could be found by computing the convolution of the impulse response and the input signal, if initial conditions were zero. For a multiple-input, multiple-output system, the initial state of which is a zero state, the result is similar, as we will now demonstrate.

With $\mathbf{q}(t_0) = \mathbf{q}_0(t_0) = \mathbf{0}$, the system response is

$$\mathbf{y}(t) = \int_{t_0}^{t} \mathbf{Cr}(t - \tau)\mathbf{Bx}(\tau) \, d\tau + \mathbf{Dx}(t) \qquad t \geq t_0 \qquad (9.36)$$

Let us assume that the integration interval includes both limits; then, we can use the sifting property of the impulse function to write

$$\mathbf{y}(t) = \int_{t_0}^{t} \mathbf{Cr}(t - \tau)\mathbf{Bx}(\tau) \, d\tau + \int_{t_0}^{t} \delta(\tau - t)\mathbf{Dx}(\tau) \, d\tau \qquad t \geq t_0 \quad (9.37)$$

By Property I of the impulse function (given in Chapter 3), $\delta(\tau - t) = \delta(t - \tau)$. Therefore,

$$\mathbf{y}(t) = \int_{t_0}^{t} [\mathbf{Cr}(t - \tau)\mathbf{B} + \delta(t - \tau)\mathbf{D}]\mathbf{x}(\tau) \, d\tau \qquad t \geq t_0 \qquad (9.38)$$

which we can write as

$$\mathbf{y}(t) = \int_{t_0}^{t} \mathbf{h}(t - \tau)\mathbf{x}(\tau) \, d\tau \qquad t \geq t_0 \qquad (9.39)$$

if

$$\mathbf{Cr}(t)\mathbf{B} + \delta(t)\mathbf{D} \equiv \mathbf{h}(t) = \begin{bmatrix} h_{11}(t) & \cdots & h_{1m}(t) \\ \vdots & & \vdots \\ h_{p1}(t) & \cdots & h_{pm}(t) \end{bmatrix} \qquad t \geq 0 \qquad (9.40)$$

The $p \times m$ matrix $\mathbf{h}(t)$ is called the *impulse-response matrix* of the system because its components are the responses at an individual-output terminal for $t \geq 0$ to a unit-impulse input applied to a single-input terminal at $t = 0$. We illustrate this statement by considering the ith output-signal component given by eq. (9.39) when $t_0 = 0$. When all components of the input-signal vector $\mathbf{x}(t)$ are equal to zero, except for $x_j(t)$, which is the unit impulse $\delta(t)$, then

$$y_i(t) = \int_{t_0}^{t} h_{ij}(t - \tau)\delta(\tau)\, d\tau = h_{ij}(t) \qquad t \geq 0 \qquad (9.41)$$

Note that all system parameter matrices enter into the computation of the impulse-response matrix. The matrices $\mathbf{B}$, $\mathbf{C}$, and $\mathbf{D}$ enter directly and the matrix $\mathbf{A}$ enters through the state transition matrix $\mathbf{r}(t)$.

Computation of the State Transition Matrix

The state transition matrix must be computed before we can compute the solutions for the state and output vectors for $t \geq t_0$. This computation is nontrivial. The definition of the state transition matrix is the infinite series

$$\mathbf{r}(t - t_0) = \mathbf{I} + (t - t_0)\mathbf{A} + \frac{(t - t_0)^2}{2!}\mathbf{A}^2 + \cdots$$

$$\mathbf{r}(t - t_0) = \mathbf{I} + \sum_{i=1}^{\infty} \frac{(t - t_0)^i}{i!}\mathbf{A}^i \qquad t \geq t_0 \qquad (9.42)$$

and its closed-form sum must be found in order for it to be useful in the solution of eqs. (9.34) and (9.35).

In some cases, we can find a general form for the powers of $\mathbf{A}$. We can then sum the resulting series to obtain a closed-form expression for it. The following example illustrates a case of this type.

Example 9.2

The system parameter matrix

$$\mathbf{A} = \begin{bmatrix} -3 & 0 \\ 0 & -2 \end{bmatrix}$$

corresponds to a satellite-attitude control system. Find the state transition matrix for the system.

Solution

$$\mathbf{A}^i = \begin{bmatrix} -3 & 0 \\ 0 & -2 \end{bmatrix}^i = \begin{bmatrix} (-3)^i & 0 \\ 0 & (-2)^i \end{bmatrix}$$

Therefore,

$$\mathbf{r}(t - t_0) = \begin{bmatrix} 1 & 0 \\ 0 & 1 \end{bmatrix} + \sum_{i=1}^{\infty} \frac{(t - t_0)^i}{i!} \begin{bmatrix} (-3)^i & 0 \\ 0 & (-2)^i \end{bmatrix}$$

$$= \begin{bmatrix} \displaystyle\sum_{i=0}^{\infty} \frac{(-3)^i (t - t_0)^i}{i!} & 0 \\ 0 & \displaystyle\sum_{i=0}^{\infty} \frac{(-2)^i (t - t_0)^i}{i!} \end{bmatrix}$$

$$= \begin{bmatrix} e^{-3(t-t_0)} & 0 \\ 0 & e^{-2(t-t_0)} \end{bmatrix} \qquad t \geq t_0$$

In most cases, it is very difficult to identify the closed-form expression for the elements of the state transition matrix directly from the infinite sums. In these cases, it is easier to use an eigenfunction evaluation and the *Cayley-Hamilton theorem* associated with square matrices. We will now discuss this technique.

Let us first consider a square matrix $\mathbf{M}$ of size $n \times n$. The equation in the variable λ given by the determinant

$$|\mathbf{M} - \lambda\mathbf{I}| = m_0 + \left[\sum_{i=1}^{n-1} m_i \lambda^i\right] + \lambda^n = 0 \tag{9.43}$$

is called the *characteristic equation* of the matrix. The determinant is a polynomial of degree n with coefficients m_i. The n roots of this polynomial are called the *eigenvalues* of the matrix. According to the Cayley-Hamilton theorem, a square matrix satisfies its own characteristic equation when that equation is expressed as a matrix equation. That is,

$$m_0\mathbf{I} + \left[\sum_{i=1}^{n-1} m_i \mathbf{M}^i\right] + \mathbf{M}^n = \mathbf{0} \tag{9.44}$$

Now, let us consider a linear time-invariant system with n states. The system's parameter matrix $\mathbf{A}$ has dimension $n \times n$ and n eigenvalues $\lambda = \lambda_k$, where $1 \leq k \leq n$, that satisfy its characteristic equation

$$|\mathbf{A} - \lambda\mathbf{I}| = a_0 + \left[\sum_{i=1}^{n-1} a_i \lambda^i\right] + \lambda^n = 0 \tag{9.45}$$

Since $\mathbf{A}$ is square, it satisfies its own characteristic equation; therefore,

$$a_0\mathbf{I} + \left[\sum_{i=1}^{n-1} a_i \mathbf{A}^i\right] + \mathbf{A}^n = \mathbf{0} \tag{9.46}$$

Solving eq. (9.46) for $\mathbf{A}^n$ gives

$$\mathbf{A}^n = -a_0\mathbf{I} - \left[\sum_{i=1}^{n-1} a_i\mathbf{A}^i\right] \tag{9.47}$$

Now, multiplying eq. (9.47) by the matrix $\mathbf{A}$, we obtain

$$\mathbf{A}^{n+1} = -a_0\mathbf{A} - \left[\sum_{i=1}^{n-2} a_i\mathbf{A}^{i+1}\right] - a_{n-1}\mathbf{A}^n \tag{9.48}$$

Then, substituting eq. (9.47) in eq. (9.48), we obtain

$$\mathbf{A}^{n+1} = -a_0\mathbf{A} - \left[\sum_{i=1}^{n-2} a_i\mathbf{A}^{i+1}\right] + a_{n-1}a_0\mathbf{I} + \left[\sum_{i=1}^{n-1} a_{n-1}a_i\mathbf{A}^i\right] \tag{9.49}$$

Finally, combining terms of eq. (9.49) yields

$$\mathbf{A}^{n+1} = a_{n-1}a_0\mathbf{I} + \left[\sum_{i=1}^{n-1} (a_{n-1}a_i - a_{i-1})\mathbf{A}^i\right] \tag{9.50}$$

Note from eqs. (9.47) and (9.50) that the matrices $\mathbf{A}^n$ and $\mathbf{A}^{n+1}$ can both be expressed as the weighted sum of the n matrices $\mathbf{I}, \mathbf{A}, \ldots, \mathbf{A}^{n-1}$. We can repeat the steps given by eqs. (9.48) through (9.50) to produce similar expressions for higher powers of $\mathbf{A}$. Therefore, $\mathbf{A}^\ell$ can be expressed as a weighted sum of $\mathbf{I}$, $\mathbf{A}, \ldots, \mathbf{A}^{n-1}$ for all $\ell \geq n$. That is,

$$\mathbf{A}^\ell = w_{0\ell}\mathbf{I} + \sum_{i=1}^{n-1} w_{i\ell}\mathbf{A}^i = \sum_{i=0}^{n-1} w_{i\ell}\mathbf{A}^i \qquad \ell \geq n \tag{9.51}$$

where the last expression follows because $\mathbf{A}^0\mathbf{A}^k = \mathbf{A}^{(0+k)} = \mathbf{A}^k$, which implies that $\mathbf{A}^0 = \mathbf{I}$. The constants $w_{i\ell}$ are combinations of the characteristic-equation constants a_i, where $0 \leq i \leq n-1$. We substitute the equations of eq. (9.51) in the infinite-series representation of the state transition matrix $\mathbf{r}(t - t_0)$ (that is, eq. [9.42]) and collect like powers of $\mathbf{A}$ to obtain the finite-series representation of the state transition matrix

$$\mathbf{r}(t - t_0) = g_0(t - t_0)\mathbf{I} + \sum_{i=1}^{n-1} g_i(t - t_0)\mathbf{A}^i \qquad t \geq t_0 \tag{9.52}$$

where the functions $g_i(t - t_0)$ for $0 \leq i \leq n-1$ are n functions in the variable $t - t_0$. To complete the computation of the state transition matrix in closed form, we must evaluate the n functions $g_i(t - t_0)$, substitute them in eq. (9.52), and perform the necessary matrix multiplications and additions.

We will now illustrate a method for finding the n functions $g_i(t - t_0)$ when all the eigenvalues are distinct (that is, when they have different values). The method requires modifications if the eigenvalues are not distinct; however, we will not deal with that circumstance in this introductory treatment. To begin, we know from matrix theory that a matrix similarity transformation $\mathbf{PAP}^{-1}$ can be found that will transform the square matrix $\mathbf{A}$ into a diagonal matrix $\mathbf{A}_D$, with the eigenvalues on the diagonal.[†] That is, we can find the square matrix $\mathbf{P}$ so that

$$\mathbf{PAP}^{-1} = \mathbf{A}_D = \begin{bmatrix} \lambda_1 & 0 & \dots & 0 \\ 0 & \lambda_2 & & \vdots \\ \vdots & & \ddots & \vdots \\ 0 & \dots & \dots & \lambda_n \end{bmatrix} \qquad (9.53)$$

Since $\mathbf{P}^{-1}\mathbf{P} = \mathbf{I}$ (see Appendix A), then

$$\mathbf{PA}^2\mathbf{P}^{-1} = \mathbf{PAIAP}^{-1} = \mathbf{PAP}^{-1}\mathbf{PAP}^{-1} = \mathbf{A}_D^2$$

$$= \begin{bmatrix} \lambda_1^2 & 0 & \dots & 0 \\ 0 & \lambda_2^2 & & \vdots \\ \vdots & & \ddots & \vdots \\ 0 & \dots & \dots & \lambda_n^2 \end{bmatrix} \qquad (9.54)$$

This procedure can be repeated for higher powers of $\mathbf{A}$ to give

$$\mathbf{PA}^\ell\mathbf{P}^{-1} = \mathbf{A}_D^\ell = \begin{bmatrix} \lambda_1^\ell & 0 & \dots & 0 \\ 0 & \lambda_2^\ell & & \vdots \\ \vdots & & \ddots & \vdots \\ 0 & \dots & \dots & \lambda_n^\ell \end{bmatrix} \qquad (9.55)$$

Also, from eq. (9.42),

$$\mathbf{Pr}(t - t_0)\mathbf{P}^{-1} = \mathbf{P}\left[\mathbf{I} + \sum_{i=1}^{\infty} \frac{(t - t_0)^i}{i!}\mathbf{A}^i\right]\mathbf{P}^{-1}$$

$$= \mathbf{PIP}^{-1} + \sum_{i=1}^{\infty} \frac{(t - t_0)^i}{i!}\mathbf{PA}^i\mathbf{P}^{-1}$$

$$= I + \sum_{i=1}^{\infty} \frac{(t - t_0)^i}{i!}\mathbf{A}_D^i$$

$$= \begin{bmatrix} \beta_1 & 0 & \dots & 0 \\ 0 & \beta_2 & & 0 \\ \vdots & & \ddots & \vdots \\ 0 & 0 & \dots & \beta_n \end{bmatrix} \qquad t \geq t_0 \qquad (9.56)$$

[†]S. Perlis, *Theory of Matrices* (Reading, Mass.: Addison-Wesley, 1952), pp. 169–174.

where

$$\beta_k = 1 + \sum_{i=1}^{\infty} \frac{(\lambda_k)^i (t - t_0)^i}{i!} = e^{\lambda_k (t - t_0)} \qquad 1 \le k \le n \qquad (9.57)$$

Using eq. (9.52),

$$\mathbf{Pr}(t - t_0)\mathbf{P}^{-1} = \mathbf{P}[g_0(t - t_0)\mathbf{I} + \sum_{i=1}^{n-1} g_i(t - t_0)\mathbf{A}^i]\mathbf{P}^{-1}$$

$$= g_0(t - t_0)\mathbf{I} + \sum_{i=1}^{n-1} g_i(t - t_0)\mathbf{P}\mathbf{A}^i\mathbf{P}^{-1} \qquad t \ge t_0 \quad (9.58)$$

Substituting eqs. (9.55), (9.56), and (9.57) in eq. (9.58), we obtain the n equations

$$e^{\lambda_k (t - t_0)} = \sum_{i=0}^{n-1} \lambda_k^i g_i(t - t_0) \qquad t \ge t_0, \, 1 \le k \le n \qquad (9.59)$$

corresponding to the diagonal terms of the matrices in the matrix equation. All other terms in the matrices are zero. We then combine the n equations of eq. (9.59) to obtain expressions for the n functions $g_i(t - t_0)$ after first determining the eigenvalues of $\mathbf{A}$ by solving for the roots of the characteristic equation corresponding to $\mathbf{A}$.

Example System Analysis

We will now illustrate the techniques described for the time-domain analysis of systems. For convenience, our time reference is selected so that $t_0 = 0$. For the state variables, we select the outputs of the integrators in the system because the integrators are the system energy-storage components.

Example 9.3

In a chemical processing plant, the chemical flow rates $y_1(t)$, $y_2(t)$, and $y_3(t)$ in gallons per second in three pipes interconnecting intermediate storage tanks are controlled with two signals $x_1(t)$ and $x_2(t)$ derived from process characteristics. The system relating the flow rates to the control signals is the second-order system shown in Figure 9.6. The control signal $x_1(t)$ is a large-amplitude, extremely narrow pulse of unit area that can be modeled as a unit impulse. The control signal $x_2(t)$ is a unit step signal. Find the flow rates in the three pipes for $t \ge 0$ when the initial state values are (a) $q_1(0) = 1$, $q_2(0) = -2$, and (b) $q_1(0) = q_2(0) = 0$.

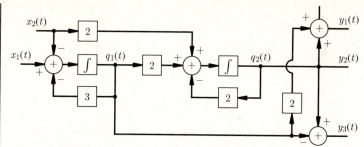

Figure 9.6 Flow-Rate Control System for Example 9.3

Solution
From the block diagram in Figure 9.6,

$$\frac{dq_1(t)}{dt} = (-3)q_1(t) + (0)q_2(t) + (1)x_1(t) + (-1)x_2(t)$$

$$\frac{dq_2(t)}{dt} = (2)q_1(t) + (-2)q_2(t) + (0)x_1(t) + (2)x_2(t)$$

$$y_1(t) = (2)q_1(t) + (1)q_2(t)$$

$$y_2(t) = (0)q_1(t) + (1)q_2(t)$$

$$y_3(t) = (-1)q_1(t) + (1)q_2(t)$$

Therefore,

$$\frac{d\mathbf{q}(t)}{dt} = \mathbf{A}\mathbf{q}(t) + \mathbf{B}\mathbf{x}(t)$$

$$\mathbf{y}(t) = \mathbf{C}\mathbf{q}(t) + \mathbf{D}\mathbf{x}(t)$$

where

$$\mathbf{A} = \begin{bmatrix} -3 & 0 \\ 2 & -2 \end{bmatrix} \qquad \mathbf{B} = \begin{bmatrix} 1 & -1 \\ 0 & 2 \end{bmatrix} \qquad \mathbf{C} = \begin{bmatrix} 2 & 1 \\ 0 & 1 \\ -1 & 1 \end{bmatrix} \qquad \mathbf{D} = \mathbf{0}$$

The characteristic equation for the system is

$$|\mathbf{A} - \lambda\mathbf{I}| = \begin{vmatrix} -3 - \lambda & 0 \\ 2 & -2 - \lambda \end{vmatrix} = (-3 - \lambda)(-2 - \lambda) = 0$$

The roots of the characteristic equation are the eigenvalues

$$\lambda_1 = -2 \quad \text{and} \quad \lambda_2 = -3$$

Since $n = 2$ and $t_0 = 0$,

$$\mathbf{r}(t) = g_0(t)\mathbf{I} + g_1(t)\mathbf{A} \qquad t \geq 0$$

The functions $g_0(t)$ and $g_1(t)$ are obtained from the two equations given by eq. (9.59). These equations are

$$e^{-2t} = g_0(t) - 2g_1(t) \qquad t \geq 0$$

and

$$e^{-3t} = g_0(t) - 3g_1(t) \qquad t \geq 0$$

Simultaneous solution of these equations for $g_0(t)$ and $g_1(t)$ gives

$$g_0(t) = 3e^{-2t} - 2e^{-3t} \qquad t \geq 0$$

and

$$g_1(t) = e^{-2t} - e^{-3t} \qquad t \geq 0$$

Substituting these equations in the equation for $\mathbf{r}(t)$, we obtain

$$\mathbf{r}(t) = (3e^{-2t} - 2e^{-3t})\begin{bmatrix} 1 & 0 \\ 0 & 1 \end{bmatrix} + (e^{-2t} - e^{-3t})\begin{bmatrix} -3 & 0 \\ 2 & -2 \end{bmatrix}$$

$$= \begin{bmatrix} e^{-3t} & 0 \\ 2e^{-2t} - 2e^{-3t} & e^{-2t} \end{bmatrix} \qquad t \geq 0$$

Then, from eq. (9.35),

$$\mathbf{y}(t) = \mathbf{Cr}(t)\mathbf{q}(0) + \int_0^t \mathbf{Cr}(t - \tau)\mathbf{Bx}(\tau)\,d\tau + \mathbf{Dx}(t) \qquad t \geq 0$$

Since

$$\mathbf{Cr}(t) = \begin{bmatrix} 2 & 1 \\ 0 & 1 \\ -1 & 1 \end{bmatrix} \begin{bmatrix} e^{-3t} & 0 \\ 2e^{-2t} - 2e^{-3t} & e^{-2t} \end{bmatrix}$$

$$= \begin{bmatrix} 2e^{-2t} & e^{-2t} \\ 2e^{-2t} - 2e^{-3t} & e^{-2t} \\ 2e^{-2t} - 3e^{-3t} & e^{-2t} \end{bmatrix} \qquad t \geq 0$$

and

$$\mathbf{Cr}(t)\mathbf{B} = \begin{bmatrix} 2e^{-2t} & e^{-2t} \\ 2e^{-2t} - 2e^{-3t} & e^{-2t} \\ 2e^{-2t} - 3e^{-3t} & e^{-2t} \end{bmatrix} \begin{bmatrix} 1 & -1 \\ 0 & 2 \end{bmatrix}$$

$$= \begin{bmatrix} 2e^{-2t} & 0 \\ 2e^{-2t} - 2e^{-3t} & 2e^{-3t} \\ 2e^{-2t} - 3e^{-3t} & 3e^{-3t} \end{bmatrix} \qquad t \geq 0$$

then, for part (a), for $t \geq 0$,

$$\mathbf{y}(t) = \begin{bmatrix} 2e^{-2t} & e^{-2t} \\ 2e^{-2t} - 2e^{-3t} & e^{-2t} \\ 2e^{-2t} - 3e^{-3t} & e^{-2t} \end{bmatrix} \begin{bmatrix} 1 \\ -2 \end{bmatrix}$$

$$+ \int_0^t \begin{bmatrix} 2e^{-2(t-\tau)} & 0 \\ 2e^{-2(t-\tau)} - 2e^{-3(t-\tau)} & 2e^{-3(t-\tau)} \\ 2e^{-2(t-\tau)} - 3e^{-3(t-\tau)} & 3e^{-3(t-\tau)} \end{bmatrix} \begin{bmatrix} \delta(\tau) \\ u(\tau) \end{bmatrix} d\tau$$

Therefore,

$$y_1(t) = 2e^{-2t} - 2e^{-2t} + \int_0^t 2e^{-2(t-\tau)}\delta(\tau)\, d\tau$$

$$= 2e^{-2t} \qquad t \geq 0$$

$$y_2(t) = 2e^{-2t} - 2e^{-3t} - 2e^{-2t} + \int_0^t 2e^{-2(t-\tau)}\delta(\tau)\, d\tau$$

$$- \int_0^t 2e^{-3(t-\tau)}\delta(\tau)d\tau + \int_0^t 2e^{-3(t-\tau)}u(\tau)\, d\tau$$

$$= -2e^{-3t} + 2e^{-2t} - 2e^{-3t} + \frac{2}{3}(1 - e^{-3t})$$

$$= \frac{2}{3} + 2e^{-2t} - \frac{14}{3}e^{-3t} \qquad t \geq 0$$

and

$$y_3(t) = 2e^{-2t} - 3e^{-3t} - 2e^{-2t} + \int_0^t 2e^{-2(t-\tau)}\delta(\tau)\, d\tau$$

$$- \int_0^t 3e^{-3(t-\tau)}\delta(\tau)\, d\tau + \int_0^t 3e^{-3(t-\tau)}u(\tau)\, d\tau$$

$$= -3e^{-3t} + 2e^{-2t} - 3e^{-3t} + (1 - e^{-3t})$$

$$= 1 + 2e^{-2t} - 7e^{-3t} \qquad t \geq 0$$

are the flow rates obtained with the initial state values of part (a).

To solve part (b), we will use the system impulse response $\mathbf{h}(t)$, for purposes of illustration. From eq. (9.40), we obtain

$$\mathbf{h}(t) = \mathbf{C}r(t)\mathbf{B} + \delta(t)\mathbf{D}$$

$$= \begin{bmatrix} 2e^{-2t} & 0 \\ 2e^{-2t} - 2e^{-3t} & 2e^{-3t} \\ 2e^{-2t} - 3e^{-3t} & 3e^{-3t} \end{bmatrix} \qquad t \geq 0$$

as the impulse-response matrix for the flow-rate control system. We then compute the output-signal (that is, flow-rate) vector by using eq. (9.39), which is

$$\mathbf{y}(t) = \int_0^t \mathbf{h}(t - \tau)\mathbf{x}(\tau)\, d\tau \qquad t \geq 0$$

The resulting flow rates are

$$y_1(t) = \int_0^t 2e^{-2(t-\tau)}\delta(\tau)\,d\tau = 2e^{-2t} \qquad t \geq 0$$

$$y_2(t) = \int_0^t 2e^{-2(t-\tau)}\delta(\tau)\,d\tau - \int_0^t 2e^{-3(t-\tau)}\delta(\tau)\,d\tau$$

$$+ \int_0^t 2e^{-3(t-\tau)}u(\tau)\,d\tau = \frac{2}{3} + 2e^{-2t} - \frac{8}{3}e^{-3t} \qquad t \geq 0$$

and

$$y_3(t) = \int_0^t 2e^{-2(t-\tau)}\delta(\tau)\,d\tau - \int_0^t 3e^{-3(t-\tau)}\delta(\tau)\,d\tau$$

$$+ \int_0^t 3e^{-3(t-\tau)}u(\tau)\,d\tau = 1 + 2e^{-2t} - 4e^{-3t} \qquad t \geq 0$$

The chemical flow rates corresponding to parts (a) and (b) of this example are plotted in Figure 9.7.

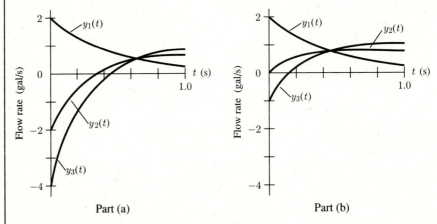

Part (a)   Part (b)

Figure 9.7 Flow Rates for Parts (a) and (b) of Example 9.3

9.4 System Analysis Using Laplace Transforms

If the time reference is chosen to be $t_0 = 0$, then we can assume that all signals are zero for $t < 0$ because the effects of signal values at $t < 0$ are contained in the state-variable values at $t = 0$. With this assumption, we can use the single-sided Laplace transform to find solutions for the state variables and the output signals for $t \geq 0$ because the system is linear and time-invariant. We will now describe the analysis techniques and illustrate them with an example.

To begin, recall that the matrix state and output equations are

$$\frac{d\mathbf{q}(t)}{dt} = \mathbf{A}\mathbf{q}(t) + \mathbf{B}\mathbf{x}(t) \tag{9.60}$$

and

$$\mathbf{y}(t) = \mathbf{C}\mathbf{q}(t) + \mathbf{D}\mathbf{x}(t) \tag{9.61}$$

Computing the Laplace transform of these equations gives

$$s\mathbf{Q}(s) - \mathbf{q}(0) = \mathbf{A}\mathbf{Q}(s) + \mathbf{B}\mathbf{X}(s) \tag{9.62}$$

and

$$\mathbf{Y}(s) = \mathbf{C}\mathbf{Q}(s) + \mathbf{D}\mathbf{X}(s) \tag{9.63}$$

where the Laplace transform of a matrix or vector is computed as the Laplace transform of each component (see eq. [A.19] in Appendix A). We can write eq. (9.62) as

$$s\mathbf{I}\mathbf{Q}(s) - \mathbf{q}(0) = \mathbf{A}\mathbf{Q}(s) + \mathbf{B}\mathbf{x}(s) \tag{9.64}$$

where $\mathbf{I}$ is an $n \times n$ identity matrix. Therefore,

$$(s\mathbf{I} - \mathbf{A})\mathbf{Q}(s) = \mathbf{q}(0) + \mathbf{B}\mathbf{X}(s) \tag{9.65}$$

The matrix $s\mathbf{I} - \mathbf{A}$ is an $n \times n$ matrix and has an inverse $(s\mathbf{I} - \mathbf{A})^{-1}$. Therefore,

$$\mathbf{Q}(s) = (s\mathbf{I} - \mathbf{A})^{-1}\mathbf{q}(0) + (s\mathbf{I} - \mathbf{A})^{-1}\mathbf{B}\mathbf{X}(s) \tag{9.66}$$

Substituting eq. (9.66) in eq. (9.63) yields

$$\mathbf{Y}(s) = \mathbf{C}(s\mathbf{I} - \mathbf{A})^{-1}\mathbf{q}(0) + \mathbf{C}(s\mathbf{I} - \mathbf{A})^{-1}\mathbf{B}\mathbf{X}(s) + \mathbf{D}\mathbf{X}(s) \tag{9.67}$$

as the Laplace transform of the output-signal vector.

Since $t_0 = 0$, eq. (9.34) gives the expression

$$\mathbf{q}(t) = \mathbf{r}(t)\mathbf{q}(0) + \int_0^t \mathbf{r}(t - \tau)\mathbf{B}\mathbf{x}(\tau)\,d\tau \qquad t \geq 0 \tag{9.68}$$

for the state-variable vector. We defined the state transition matrix $\mathbf{r}(t)$ only for $t \geq 0$; therefore, we can consider it to be zero for $t < 0$. Also $\mathbf{x}(t) = \mathbf{0}$ for $t < 0$ by prior assumption. Therefore, $\mathbf{r}(t - \tau) = \mathbf{0}$ for $\tau > t$ and $\mathbf{x}(\tau) = \mathbf{0}$ for $\tau < 0$, which means that we can write eq. (9.68) as

$$\mathbf{q}(t) = \mathbf{r}(t)\mathbf{q}(0) + \int_{-\infty}^{\infty} \mathbf{r}(t - \tau)\mathbf{B}\mathbf{x}(\tau)\,d\tau$$

$$= \mathbf{r}(t)\mathbf{q}(0) + \mathbf{r}(t) * [\mathbf{B}\mathbf{x}(t)] \qquad t \geq 0 \tag{9.69}$$

where the convolution of two matrices is performed in the manner of matrix multiplication, except that the individual-element multiplications are replaced with convolutions. This can be easily verified by performing the matrix multiplication and integration of the second term of eq. (9.69). Computation of the Laplace transform of eq. (9.69) yields

$$\mathbf{Q}(s) = \mathbf{R}(s)\mathbf{q}(0) + \mathbf{R}(s)\mathbf{B}\mathbf{X}(s) \tag{9.70}$$

as the Laplace transform of the state-variable vector. When we compare eq. (9.70) with eq. (9.66), we see that

$$\mathbf{R}(s) = (s\mathbf{I} - \mathbf{A})^{-1} \tag{9.71}$$

which means that the matrix $(s\mathbf{I} - \mathbf{A})^{-1}$ is the Laplace transform of the state transition matrix. We substitute eq. (9.71) in eq. (9.67) to obtain the expression

$$\mathbf{Y}(s) = \mathbf{CR}(s)\mathbf{q}(0) + \mathbf{CR}(s)\mathbf{BX}(s) + \mathbf{DX}(s) \tag{9.72}$$

for the Laplace transform of the output-signal vector.
  If the initial state is the zero state, $\mathbf{q}(0) = \mathbf{0}$, then

$$\mathbf{Y}(s) = [\mathbf{CR}(s)\mathbf{B} + \mathbf{D}]\mathbf{X}(s) \equiv \mathbf{H}(s)\mathbf{X}(s) \tag{9.73}$$

where we define

$$\mathbf{H}(s) = \mathbf{CR}(s)\mathbf{B} + \mathbf{D} \tag{9.74}$$

to be the transfer-function matrix of the system. Note that

$$\mathcal{L}^{-1}[\mathbf{H}(s)] = \mathbf{C}r(t)\mathbf{B} + \delta(t)\mathbf{D} = \mathbf{h}(t) \tag{9.75}$$

where $\mathbf{h}(t)$ is the impulse-response matrix for the system. Thus, the transfer-function matrix is the Laplace transform of the impulse-response matrix.
  Let us now reconsider the previous example, this time using the Laplace transform solution techniques.

Example 9.4

Use Laplace transforms to solve Example 9.3.

Solution
Recall from Example 9.3 that

$$\mathbf{A} = \begin{bmatrix} -3 & 0 \\ 2 & -2 \end{bmatrix} \qquad \mathbf{B} = \begin{bmatrix} 1 & -1 \\ 0 & 2 \end{bmatrix} \qquad \mathbf{C} = \begin{bmatrix} 2 & 1 \\ 0 & 1 \\ -1 & 1 \end{bmatrix} \qquad \mathbf{D} = \mathbf{0}$$

First, we solve for the Laplace transform of the state transition matrix

$$s\mathbf{I} - \mathbf{A} = \begin{bmatrix} s & 0 \\ 0 & s \end{bmatrix} - \begin{bmatrix} -3 & 0 \\ 2 & -2 \end{bmatrix} = \begin{bmatrix} s+3 & 0 \\ -2 & s+2 \end{bmatrix}$$

From Appendix A, the inverse of the 2×2 matrix

$$\mathbf{M} = \begin{bmatrix} m_{11} & m_{12} \\ m_{21} & m_{22} \end{bmatrix}$$

is

$$\mathbf{M}^{-1} = \frac{\begin{bmatrix} m_{22} & -m_{12} \\ -m_{21} & m_{11} \end{bmatrix}}{|\mathbf{M}|}$$

Therefore,

$$\mathbf{r}(s) = (s\mathbf{I} - \mathbf{A})^{-1} = \frac{\begin{bmatrix} s+2 & 0 \\ 2 & s+3 \end{bmatrix}}{(s+3)(s+2) - (-2)(0)}$$

$$= \begin{bmatrix} \dfrac{1}{s+3} & 0 \\ \dfrac{2}{(s+2)(s+3)} & \dfrac{1}{s+2} \end{bmatrix} = \begin{bmatrix} \dfrac{1}{s+3} & 0 \\ \dfrac{2}{s+2} - \dfrac{2}{s+3} & \dfrac{1}{s+2} \end{bmatrix}$$

From eq. (9.72),

$$\mathbf{Y}(s) = \mathbf{CR}(s)\mathbf{q}(0) + [\mathbf{CR}(s)\mathbf{B} + \mathbf{D}]\mathbf{X}(s)$$

$$= \mathbf{CR}(s)\mathbf{q}(0) + \mathbf{H}(s)\mathbf{X}(s)$$

where $\mathbf{X}(s) = \begin{bmatrix} 1 \\ 1/s \end{bmatrix}$ because $x_1(t) = \delta(t)$ and $x_2(t) = u(t)$. Now,

$$\mathbf{CR}(s) = \begin{bmatrix} 2 & 1 \\ 0 & 1 \\ -1 & 1 \end{bmatrix} \begin{bmatrix} \dfrac{1}{s+3} & 0 \\ \dfrac{2}{s+2} - \dfrac{2}{s+3} & \dfrac{1}{s+2} \end{bmatrix}$$

$$= \begin{bmatrix} \dfrac{2}{s+2} & \dfrac{1}{s+2} \\ \dfrac{2}{s+2} - \dfrac{2}{s+3} & \dfrac{1}{s+2} \\ \dfrac{2}{s+2} - \dfrac{3}{s+3} & \dfrac{1}{s+2} \end{bmatrix}$$

and

$$\mathbf{H}(s) = \mathbf{CR}(s)\mathbf{B}$$

$$= \begin{bmatrix} \dfrac{2}{s+2} & \dfrac{1}{s+2} \\ \dfrac{2}{s+2} - \dfrac{2}{s+3} & \dfrac{1}{s+2} \\ \dfrac{2}{s+2} - \dfrac{3}{s+3} & \dfrac{1}{s+2} \end{bmatrix} \begin{bmatrix} 1 & -1 \\ 0 & 2 \end{bmatrix}$$

$$= \begin{bmatrix} \dfrac{2}{s+2} & 0 \\ \dfrac{2}{s+2} - \dfrac{2}{s+3} & \dfrac{2}{s+3} \\ \dfrac{2}{s+2} - \dfrac{3}{s+3} & \dfrac{3}{s+3} \end{bmatrix}$$

Note that

$$\mathbf{h}(t) = \mathcal{L}^{-1}[\mathbf{H}(s)] = \begin{bmatrix} 2e^{-2t}u(t) & 0 \\ (2e^{-2t} - 2e^{-3t})u(t) & 2e^{-3t}u(t) \\ (2e^{-2t} - 3e^{-3t})u(t) & 3e^{-3t}u(t) \end{bmatrix}$$

which checks the result that we obtained in Example 9.3 for the impulse-response matrix.

For part (a),

$$\mathbf{Y}(s) = \begin{bmatrix} \dfrac{2}{s+2} & \dfrac{1}{s+2} \\ \dfrac{2}{s+2} - \dfrac{2}{s+3} & \dfrac{1}{s+2} \\ \dfrac{2}{s+2} - \dfrac{3}{s+3} & \dfrac{1}{s+2} \end{bmatrix} \begin{bmatrix} 1 \\ -2 \end{bmatrix}$$

$$+ \begin{bmatrix} \dfrac{2}{s+2} & 0 \\ \dfrac{2}{s+2} - \dfrac{2}{s+3} & \dfrac{2}{s+3} \\ \dfrac{2}{s+2} - \dfrac{3}{s+3} & \dfrac{3}{s+3} \end{bmatrix} \begin{bmatrix} 1 \\ 1/s \end{bmatrix}$$

$$= \begin{bmatrix} \dfrac{2}{s+2} \\ \dfrac{2}{s+2} - \dfrac{4}{s+3} + \dfrac{2}{s(s+3)} \\ \dfrac{2}{s+2} - \dfrac{6}{s+3} + \dfrac{3}{s(s+3)} \end{bmatrix} = \begin{bmatrix} \dfrac{2}{s+2} \\ \dfrac{2}{3}\left(\dfrac{1}{s}\right) + \dfrac{2}{s+2} - \dfrac{14}{3}\left(\dfrac{1}{s+3}\right) \\ \dfrac{1}{s} + \dfrac{2}{s+2} - \dfrac{7}{s+3} \end{bmatrix}$$

Since

$$Y_1(s) = \frac{2}{s+2}$$

$$Y_2(s) = \frac{2}{3}\left(\frac{1}{s}\right) + \frac{2}{s+2} - \frac{14}{3}\left(\frac{1}{s+3}\right)$$

and

$$Y_3(s) = \frac{1}{s} + \frac{2}{s+2} - \frac{7}{s+3}$$

then the flow rates are

$$y_1(t) = 2e^{-2t}u(t)$$

$$y_2(t) = \left(\frac{2}{3} + 2e^{-2t} - \frac{14}{3}e^{-3t}\right)u(t)$$

and

$$y_3(t) = \left(1 + 2e^{-2t} - 7e^{-3t}\right)u(t)$$

which checks those found in Example 9.3.

For Part (b),

$$\mathbf{Y}(s) = \mathbf{H}(s)\mathbf{X}(s) = \begin{bmatrix} \dfrac{2}{s+2} & 0 \\[2mm] \dfrac{2}{s+2} - \dfrac{2}{s+3} & \dfrac{2}{s+3} \\[2mm] \dfrac{2}{s+2} - \dfrac{3}{s+3} & \dfrac{3}{s+3} \end{bmatrix} \begin{bmatrix} 1 \\[1mm] 1/s \end{bmatrix}$$

$$= \begin{bmatrix} \dfrac{2}{s+2} \\[2mm] \dfrac{2}{s+2} - \dfrac{2}{s+3} + \dfrac{2}{s(s+3)} \\[2mm] \dfrac{2}{s+2} - \dfrac{3}{s+3} + \dfrac{3}{s(s+3)} \end{bmatrix} = \begin{bmatrix} \dfrac{2}{s+2} \\[2mm] = \dfrac{2}{3}\left(\dfrac{1}{s}\right) + \dfrac{2}{s+2} - \dfrac{8}{3}\left(\dfrac{1}{s+3}\right) \\[2mm] \dfrac{1}{s} + \dfrac{2}{s+2} - \dfrac{4}{s+3} \end{bmatrix}$$

Therefore, the flow rates are

$$y_1(t) = \mathcal{L}^{-1}[Y_1(s)] = 2e^{-2t}u(t)$$

$$y_2(t) = \mathcal{L}^{-1}[Y_2(s)] = \left(\frac{2}{3} + 2e^{-2t} - \frac{8}{3}e^{-3t}\right)u(t)$$

$$y_3(t) = \mathcal{L}^{-1}[Y_3(s)] = (1 + 2e^{-2t} - 4e^{-3t})u(t)$$

which, once again, checks the flow rates obtained in Example 9.3.

9.5 Summary

State variables of a system are internal signals, and possibly output signals. The number of state variables needed to produce solutions for the system signals is equal to the order, n, of the system. The derivative of each state variable can be expressed as a weighted sum of the state variables and input signals. The resulting expressions are called state equations. The output signals are computed with output equations that express the output signals as a weighted sum of the state variables and input signals.

Analysis techniques using state variables, state equations, and output equations provide an organized approach for finding output signals of multiple-input, multiple-output systems or of complicated single-input, single-output systems. They also provide information about internal system characteristics through the solutions found for the state variables.

The choice of state variables is not unique. In general, the most convenient set of state variables is the set that characterizes the energy stored in the n energy-storage elements in the system.

The state equations and output equations used in the state-variable method of system analysis are most easily expressed and solved in matrix format. We can use either time-domain or Laplace transform techniques to solve these matrix equations to find state variables and output signals for $t \geq t_0$ where t_0 is the initial time of interest. Use of these techniques produces the impulse-response matrix $\mathbf{h}(t)$, and the transfer-function matrix $\mathbf{H}(s)$ of the system. These two matrices are a Laplace transform pair; that is, $\mathbf{h}(t) \leftrightarrow \mathbf{H}(s)$.

We only briefly introduced the basic concepts associated with state variables and state-variable analysis of linear systems in this chapter. For detailed discussion of the full power of state-variable techniques for characterizing and analyzing systems, the readers is encouraged to consult texts devoted to state-variables.

Problems

9.1 Write a set of state equations corresponding to the differential equation

$$3\frac{d^3 y(t)}{dt^3} - 2\frac{dy^2(t)}{dt^2} - \frac{dy(t)}{dt} + 4y(t) = 2x(t)$$

Also express them as a matrix state equation.

9.2 Repeat Problem 9.1 for the differential equation

$$\frac{d^4 y(t)}{dt^4} - 2\frac{d^2 y(t)}{dt} - 5y(t) = -3x(t)$$

9.3 The temperatures required for two steps in the processing of steel are interdependent and depend on the initial thickness and temperature of the steel. The system that relates the processing-step tem-

peratures, $y_1(t)$ and $y_2(t)$, to the initial thickness and temperature, $x_1(t)$ and $x_2(t)$, is shown in Figure 9.8. Identify the system state variables and write the matrix state and output equations for the system.

9.4 The system shown in Figure 9.9 on page 367 is part of a flight-control system for aircraft. Identify system state variables and write the matrix state and output equations for the system.

9.5 Define state variables and find the corresponding state and output equations for the electric circuit shown in Figure 9.10 on page 367.

Figure 9.8

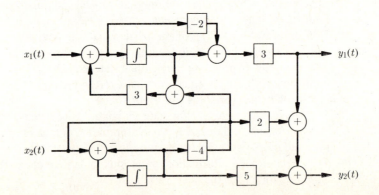

Figure 9.9

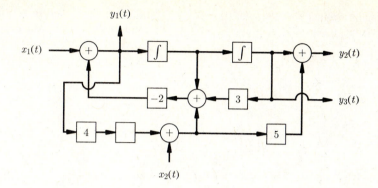

9.6 An engineer has designed the electric circuit shown in Figure 9.11 to supply power from the voltage source to the three resistive loads. Define state variables and find the corresponding state and output equations in matrix form for this electric circuit.

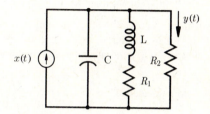

Figure 9.10

9.7 Define state variables and find the corresponding state and output equations in matrix form for the electric circuit shown in Figure 9.12.

9.8 Use the eigenvalue method to find the state transition matrix and impulse-response matrix for the system with the following parameter matrices:

$$\mathbf{A} = \begin{bmatrix} -4 & 1 \\ -2 & -1 \end{bmatrix} \quad \mathbf{B} = \begin{bmatrix} 1 & 2 & 0 \\ 0 & -1 & 1 \end{bmatrix}$$

$$\mathbf{C} = [-3 \quad 1] \quad \mathbf{D} = [-1 \quad 0 \quad 1]$$

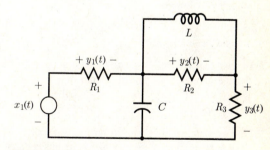

Figure 9.11

Figure 9.12

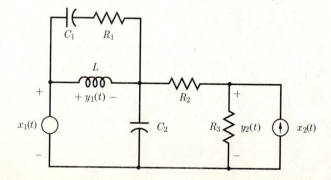

How many inputs, outputs, and states does this system have?

9.9 Repeat Problem 9.8 for the system with the following parameter matrices:

$$A = \begin{bmatrix} -3 & 2 \\ 1 & -4 \end{bmatrix} \qquad B = \begin{bmatrix} 2 \\ 1 \end{bmatrix}$$

$$C = \begin{bmatrix} 1 & 0 \\ 0 & 1 \\ 1 & 0 \end{bmatrix} \qquad D = \begin{bmatrix} 0 \\ 0 \\ 1 \end{bmatrix}$$

9.10 Repeat Problem 9.8 using Laplace transforms.

9.11 Repeat Problem 9.9 using Laplace transforms.

9.12 Use time-domain analysis to find the output signals for $t \geq 0$ for the system of Problem 9.8 if the input signals are all $u(t)$ and all initial state values equal 2.

9.13 Use time-domain analysis to find the output signals for $t \geq 0$ for the system of Problem 9.9 if the input signals are all impulse functions and the initial state values equal -1.

9.14 Repeat Problem 9.12 using Laplace transform analysis techniques.

9.15 Repeat Problem 9.13 using Laplace transform analysis techniques.

9.16 Use Laplace transforms to find the output signals for the temperature control system of Problem 9.3 if $x_1(t) = r(t)$, $x_2(t) = u(t)$ and all initial state values equal 0.

9.17 Two cameras are used in an industrial security system to monitor a scene. This scene has constant illumination, except for bright flashes from a beacon in one location, which can be modeled as impulses of light, and lights from turning vehicles in another location, which appear and then exponentially diminish in intensity as the vehicle turns. The cameras have an interconnected automatic-exposure control system. The response of the exposure control system to the rotating-beacon and turning-vehicle light sources is represented by the system block diagram shown in Figure 9.13. The signals $y_1(t)$ and $y_2(t)$ are proportional to the amount that the lens openings are closed in response to extra light, and the signals $x_1(t)$ and $x_2(t)$ are the light-source intensities of the beacon and turning vehicle, respectively. Use time-domain analysis to find $y_1(t)$ and $y_2(t)$ for $t \geq 0$ when $x_1(t) = 4\delta(t - 1)$, $x_2(t) = e^{-4t}u(t)$, and all state values at $t = 0$ equal 0.

9.18 Repeat Problem 9.17 using Laplace transform analysis techniques.

9.19 The circuit shown in Figure 9.14 supplies the $1/6$-Ω resistive load from one of two dc sources. A low-pass filter is included to reduce switching transients. The switch is at point A for all $t < 0$ and is instantaneously thrown from point A to point B at $t = 0$. Use state-variable techniques to find $y(t)$ for all $t \geq 0$.

9.20 Use state-variable techniques to find the output current $y(t)$ for $t \geq 0$ for the circuit considered in Problem 9.5 if $R_1 = 1\ \Omega$, $R_2 = 2\ \Omega$, $C = 0.125$ F, $L = 0.5$ H, $x(t) = 3u(t)$ amperes, 0.5 joule of energy is stored in the capacitor at $t = 0$, and 1.5 joules of energy are stored in the inductor at $t = 0$.

Figure 9.13

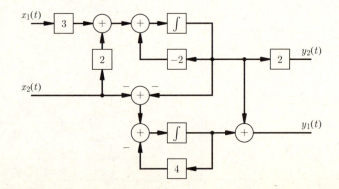

9.21 Use state-variable techniques to find the output signals for the portion of the flight-control system considered in Problem 9.4 when $x_1(t) = 2\delta(t)$, $x_2(t) = -u(t)$, and all state values equal -2 at $t = 0$.

Figure 9.14

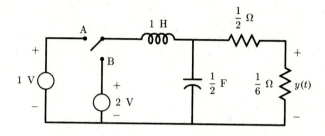

III

Discrete-Time Signals and Systems

10 Discrete-Time Signals

We indicated in Chapter 1 that discrete-time signals are defined only for discrete points in time (called sample times) and are specified by a sequence of values $x(n)$ (called samples) and the time separation T between the sample times (called sample spacing). Thus, we defined $x_T(n)$ and $[x(n), \quad T]$ as two notations to indicate discrete-time signals: The first represents a signal in general and the second represents it specifically by its particular sequence and sample spacing.

A discrete-time signal may occur directly from a computer output or it may be the result of the nature of the system considered. For example, in the repayment of a loan with monthly payments, certain events occur only at payment times; thus, the signal representing the payments is a discrete-time signal.

Alternatively, a discrete-time signal may consist of samples of a continuous-time signal generated from the continuous-time signal by analog-to-digital conversion (A/D conversion). An A/D converter samples the continuous-time signal at a sample spacing of T seconds and generates an output discrete-time sequence of values equal to the samples of the continuous-time signal. A/D conversion is required if we want to process a continuous-time signal by a discrete-time or digital system (for example, as in digital audio processing by compact disc players).

Actually, we can think of any discrete-time signal as samples of a continuous-time signal that is constructed to pass through all the discrete-time-signal samples. This conceptualization of discrete-time signals permits easy extension of many characteristics of continuous-time signals to discrete-time signals. Note that the continuous-time signal that can be constructed is not unique because an infinite number of different continuous-time signals can be constructed through a set of discrete-times-signal samples.

We can specify the defining sequence $x(n)$ for a deterministic discrete-time signal by listing the values for the sequence in order, using the notation of eq. (1.3) in Chapter 1. However, if the signal duration is long, this is rather cumbersome. If possible, then, it is more convenient for us to specify $x(n)$ by either a single mathematical function of n for all n (simply-defined sequence) or a set of mathematical functions of n, with each function valid over a defined range of n (piecewise-defined sequence). In this chapter we will consider several simple mathematical functions that are useful individually and in combination for specifying sequences corresponding to discrete-time signals. Most of these sequences are samples of the continuous-time signals discussed in Chapter 3. Note that the mathematical functions we will define can be used to represent discrete-variable functions of frequency, distance, temperature, and so forth, as well as time.

Another consideration in this chapter is the representation of a discrete-time signal as an ideally sampled continuous-time signal. This representation is useful in extending the analysis techniques for continuous-time signals and systems of Part II to discrete-time signals and systems, at the same time enhancing understanding of the nature of the results obtained.

10.1 Sinusoidal and Complex-Exponential Signals

The discrete-time sinusoidal and complex-exponential signals consist of samples of the corresponding continuous-time signals presented in Section 3.1. The sample spacing is T seconds. The sinusoidal sequence corresponding to the discrete-time sinusoidal signal is thus

$$x(n) = A\cos(2\pi f_0 nT + \theta)$$

$$\equiv A\cos(2\pi r_0 n + \theta) \qquad \text{for all } n \qquad (10.1)$$

where f_0 is the frequency in hertz, A is the amplitude, and θ is the phase in radians. Either a sine or a cosine function can be used. Both will produce the same sequence if the angle θ is $\pi/2$ radians greater for the sine because $\cos(\alpha) = \sin(\alpha + \pi/2)$. The parameter r_0 is the frequency of the sinusoid normalized by the frequency $f_s = 1/T$ because $r_0 = f_0 T = f_0/f_s$. The frequency f_s is the frequency or rate of sample occurrence (that is, the sample frequency or sample rate) because it is the reciprocal of the sample spacing.

The sinusoidal sequence defined by eq. (10.1) consists of samples of a continuous-time sinusoidal function that is periodic with period $T_0 = 1/f_0$. However, the sinusoidal sequence is not necessarily periodic (that is, it does not necessarily consist of sample values that occur periodically). It is periodic with period equal to N samples if and only if the argument of the sinusoid changes by an integer multiple of 2π when the sample number changes from n to $n + N$; that is, if

$$2\pi r_0(n + N) - 2\pi r_0 n = 2\pi i \qquad (10.2)$$

which reduces to

$$i/N = r_0 = f_0 T = T/T_0 \qquad (10.3)$$

where i and N are integers. Therefore, to produce a periodic sinusoidal sequence, the normalized frequency r_0 must be rational, which implies that the ratio of the sample spacing T to the period T_0 of the continuous-time sinusoidal function sampled must be rational. When $r_0 = T/T_0$ is rational and expressed as a fraction with all common factors removed from the numerator and denominator, then the denominator of the fraction, N, is the period of the sinusoidal sequence, and the numerator of the fraction, i, is the number of cycles of the continuous-time sinusoid that are spanned by these N samples. The aperiodic and periodic discrete-time sinusoidal signals corresponding to aperiodic and periodic sinusoidal sequences are illustrated in Figure 10.1.

Figure 10.1
Aperiodic and
Periodic
Discrete-Time
Sinusoidal Signals

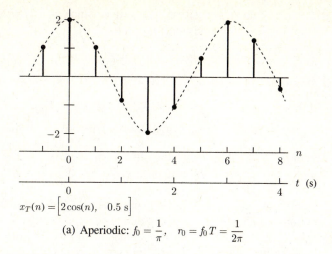

$$x_T(n) = \left[2\cos(n), \quad 0.5 \text{ s}\right]$$

(a) Aperiodic: $f_0 = \dfrac{1}{\pi}$, $r_0 = f_0 T = \dfrac{1}{2\pi}$

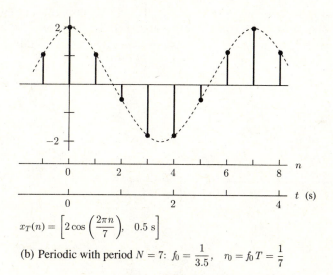

$$x_T(n) = \left[2\cos\left(\frac{2\pi n}{7}\right), \quad 0.5 \text{ s}\right]$$

(b) Periodic with period $N = 7$: $f_0 = \dfrac{1}{3.5}$, $r_0 = f_0 T = \dfrac{1}{7}$

Another property of a discrete-time sinusoidal signal is that it corresponds to samples of more than one continuous-time sinusoidal function, the functions differing only in frequency. To show this, let us consider the sinusoidal sequence

$$x_k(n) = A\cos(2\pi f_k n T + \theta)$$

$$\equiv A\cos(2\pi r_k n + \theta) \qquad \text{for all } n \qquad (10.4)$$

corresponding to the discrete-time sinusoidal signal $x_{kT}(n) = [x_k(n), T]$, where the frequency of the sampled continuous-time sinusoidal function is

$$f_k = f_0 + k/T \qquad (10.5)$$

and k is any integer, positive or negative. Since $r_k = f_k T$ and $r_0 = f_0 T$, then we can express the condition of eq. (10.5) as

$$r_k = r_0 + k \qquad (10.6)$$

Substituting eq. (10.5) in eq. (10.4) gives

$$x_k(n) = A\cos(2\pi f_0 nT + 2\pi kn + \theta)$$

$$= A\cos(2\pi f_0 nT + \theta) = x(n) \qquad \text{for all } n \qquad (10.7)$$

Equation (10.7) shows that the sinusoidal sequence $x_k(n)$ contains the same samples as the sinusoidal sequence given by eq. (10.1), regardless of the value of the integer k.

The ambiguity concerning the frequency of the continuous-time sinusoidal function sampled to produce a discrete-time sinusoidal signal is illustrated in Figure 10.2. This ambiguity is a fundamental property of signal sampling that will be discussed more extensively in the following chapter. We will show there that no ambiguity exists (that is, that unique discrete-time sinusoidal signals are obtained) if the continuous-time sinusoidal functions that are sampled are restricted to those with frequencies lower than $1/2T$. Equivalently, no ambiguity exists if we restrict the normalized frequency r_k to be less than 0.5.

Figure 10.2
Two Continuous-Time Sinusoids That Produce the Same Discrete-Time Sinusoidal Signal ($k = 0$ and $k = 1$)

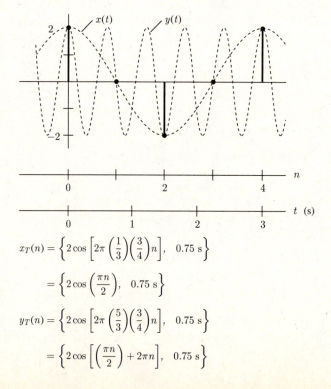

$$x_T(n) = \left\{ 2\cos\left[2\pi\left(\frac{1}{3}\right)\left(\frac{3}{4}\right)n\right], \quad 0.75\text{ s}\right\}$$

$$= \left\{ 2\cos\left(\frac{\pi n}{2}\right), \quad 0.75\text{ s}\right\}$$

$$y_T(n) = \left\{ 2\cos\left[2\pi\left(\frac{5}{3}\right)\left(\frac{3}{4}\right)n\right], \quad 0.75\text{ s}\right\}$$

$$= \left\{ 2\cos\left[\left(\frac{\pi n}{2}\right) + 2\pi n\right], \quad 0.75\text{ s}\right\}$$

Note that the value of f_k in eq. (10.4) does not uniquely define how rapidly the amplitudes of the discrete-time sinusoidal-signal samples vary, unless f_k is restricted to be less than $1/2T$. This is because the same set of samples, and thus the same rate of sample-amplitude variation, occurs for all values of f_k defined by eq. (10.5).

Using eqs. (10.4) and (10.6), we can write the discrete-time cosine sequence of eq. (10.1) as

$$x(n) = Re\left[Ae^{j(2\pi r_0 n + \theta)}\right] = Re\left[Ae^{j(2\pi r_k n + \theta)}\right] \qquad \text{for all } n \qquad (10.8)$$

since, by Euler's theorem, the complex-exponential sequence can be written

$$e^{j(2\pi r_0 n + \theta)} = \cos(2\pi r_0 n + \theta) + j\sin(2\pi r_0 n + \theta) \qquad (10.9)$$

Note that the complex-exponential sequence possesses the same periodicity characteristics and the same property of equality for the values of r_k given by eq. (10.6) as the sinusoidal sequence.

We can also write the cosine sequence of eq. (10.8) as

$$x(n) = \frac{A}{2}e^{j(2\pi r_k n + \theta)} + \frac{A}{2}e^{-j(2\pi r_k n + \theta)} \qquad \text{for all } n \qquad (10.10)$$

Thus, each sample value is the sum of two complex-conjugate vectors of equal length that rotate in opposite directions as n increases. This representation of a sinusoidal sequence corresponds to the counterrotating phasors used to represent a continuous-time sinusoidal signal in Section 3.1.

10.2 Exponential Signals

Discrete-time signals that correspond to exponential sequences occur in many discrete-time systems. For example, they represent growth or decay processes that occur in economic systems, population systems, or energy-storage-and-dissipation systems. The exponential sequence corresponding to a discrete-time exponential signal is

$$x(n) = \begin{cases} A(K)^n & n \geq n_1 \\ 0 & n < n_1 \end{cases} \qquad (10.11)$$

where K is a real number and the growth or decay process begins at sample index n_1 (that is, at signal time $n_1 T$). Growth or decay in the magnitude of the samples occurs when $|K| > 1$ or $|K| < 1$, respectively. The growth or decay is oscillatory if K is negative. Two exponential sequences are illustrated in Figure 10.3.

When K is positive, it can be expressed as $K = e^a$, where the parameter a is equal to $\ln(K)$. Thus, the exponential sequence can be written

$$x(n) = \begin{cases} Ae^{an} & n \geq n_1 \\ 0 & n < n_1 \end{cases} \qquad (10.12)$$

Figure 10.3
Examples of
Exponential
Sequences

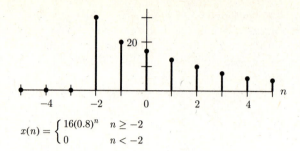

$$x(n) = \begin{cases} 16(0.8)^n & n \geq -2 \\ 0 & n < -2 \end{cases}$$

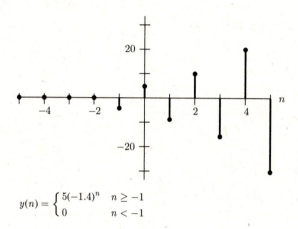

$$y(n) = \begin{cases} 5(-1.4)^n & n \geq -1 \\ 0 & n < -1 \end{cases}$$

when K is positive. This sequence consists of samples of the continuous-time exponential signal presented in Section 3.2 when $a = \alpha T$. Positive and negative exponents are used to permit modeling of both growing and decaying signal values.

We can obtain samples of exponentially growing or decaying sinusoidal functions by multiplying the discrete-time exponential sequence and the discrete-time sinusoidal sequence to give

$$x(n) = \begin{cases} Ae^{an} \cos(2\pi r_0 n + \theta) & n \geq n_1 \\ 0 & n < n_1 \end{cases} \tag{10.13}$$

An exponentially decaying (exponentially damped) sequence is illustrated in Figure 10.4.

10.3 Unit Step, Unit Ramp, and Unit Pulse Signals

Three other frequently occurring discrete-time signals are the unit step signal, unit ramp signal, and unit pulse signal. In this section, we will define and describe the sequences corresponding to these discrete-time signals.

Figure 10.4
Example of Damped
Sinusoidal Sequence

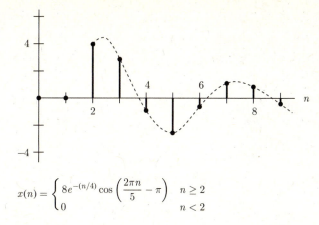

$$x(n) = \begin{cases} 8e^{-(n/4)} \cos\left(\dfrac{2\pi n}{5} - \pi\right) & n \geq 2 \\ 0 & n < 2 \end{cases}$$

The Unit Step and Unit Ramp Sequences

A special case of the exponential sequence of considerable interest is that in which $A = K = 1$ and $n_1 = 0$. This special case gives the *unit step sequence*

$$u(n) = \begin{cases} 1 & n \geq 0 \\ 0 & n < 0 \end{cases} \tag{10.14}$$

corresponding to the discrete-time unit step signal $u_T(n)$. Note that the unit step sequence consists of samples of the continuous-time unit step function defined in Section 3.3, except that $u(0)$ is given the value 1. To permit representation of step sequences that are not unity in value, and/or that are shifted in time by an integer number of sample spacings, and/or that are reversed in time, we write the general step sequence as

$$Au\,(cn + d) = \begin{cases} A & cn + d \geq 0 \\ 0 & cn + d < 0 \end{cases} \tag{10.15}$$

where A, c and d are constants, c is not equal to zero, and d/c is an integer. The general step sequence consists of samples of the continuous-time step function defined by eq. (3.36) when $c = aT, d = b$, and the continuous-time step function is defined to be unity at $t = -b/a$.

The unit step sequence is shown in Figure 10.5 along with two applications of the general step sequence to the representation of discrete-time signals that begin and/or cease at a finite time. The last discrete-time signal shown in Figure 10.5 consists of samples of a time-shifted version of the continuous-time unit rectangular pulse defined in Figure 3.11.

The *unit ramp sequence* is defined as

$$r(n) = nu(n) = \begin{cases} n & n \geq 0 \\ 0 & n < 0 \end{cases} \tag{10.16}$$

and corresponds to the discrete-time unit ramp signal $r_T(n)$. To permit representation of sequences that do not have unity slope as a function of the sample

Figure 10.5
Unit Step Sequence
and Applications
to Signal
Representation

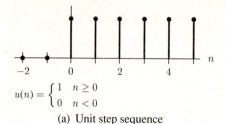

$$u(n) = \begin{cases} 1 & n \geq 0 \\ 0 & n < 0 \end{cases}$$

(a) Unit step sequence

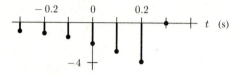

$$x_T(n) = \left[-2e^{0.35n} u(-n+2), \quad 0.1 \text{ s} \right]$$

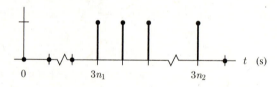

$$y_T(n) = [u(n+n_1) - u(n-n_2-1), \quad 3 \text{ s}]$$

(b) Applications to signal representation

number n, and/or are shifted in time by an integer number of sample spacings, and/or are reversed in time, we write the general ramp sequence as

$$Ar(cn + d) = A(cn + d)u(cn + d) = \begin{cases} A(cn + d) & cn + d \geq 0 \\ 0 & cn + d < 0 \end{cases} \quad (10.17)$$

where A, c, and d are constants, c is not equal to zero, and d/c is an integer. Note that the ramp sequence consists of samples of the continuous-time ramp function (see eq. [3.39]) when $c = aT$ and $d = b$. The unit ramp sequence is shown in Figure 10.6 along with two examples of the general ramp sequence.

Any piecewise-defined sequence that consists of values lying along straight-line segments can be represented as a sum of step and ramp sequences. This representation permits a compact notation for the sequence. The representation and compact notation are similar to the representation and compact notation described in Chapter 3 for continuous-time signals consisting of straight-line segments. We will see in Chapter 14 that the compact notation makes it easy for us to determine the z-transform of such a signal when it begins after $n = 0$. Using the techniques described in Chapter 3, one can easily verify that the sequence sketch and sequence expression shown in Figure 10.7 correspond.

Figure 10.6
Unit Ramp Sequence
and General Ramp
Sequences

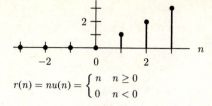

$$r(n) = nu(n) = \begin{cases} n & n \geq 0 \\ 0 & n < 0 \end{cases}$$

(a) Unit ramp sequence

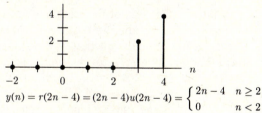

$$x(n) = 2r(-n+1) = 2(-n+1)u(-n+1) = \begin{cases} -2n+2 & n \leq 1 \\ 0 & n > 1 \end{cases}$$

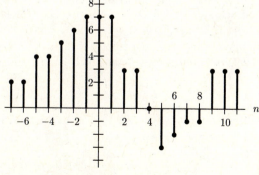

$$y(n) = r(2n-4) = (2n-4)u(2n-4) = \begin{cases} 2n-4 & n \geq 2 \\ 0 & n < 2 \end{cases}$$

(b) General ramp sequences

Figure 10.7
Sequence Definition
Using Step and
Ramp Sequences

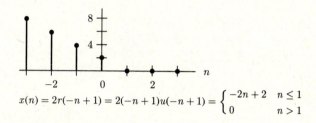

$$x(n) = 2 + 2u(n+5) + r(n+4) - r(n+1)$$
$$- 4u(n-2) - 3r(n-3) + 4r(n-5)$$
$$- r(n-7) + 4u(n-9)$$

The Unit Pulse Sequence

The *unit pulse sequence* is defined as

$$\delta(n) = \begin{cases} 1 & n = 0 \\ 0 & n \neq 0 \end{cases} \tag{10.18}$$

and corresponds to the discrete-time unit pulse signal $\delta_T(n)$. The unit pulse sequence is sometimes referred to as the *unit sample sequence*. It is illustrated in Figure 10.8.

The unit pulse sequence and its corresponding unit pulse signal are very important. Their roles in discrete-time system analysis are similar to the role of the impulse function in continuous-time system analysis. For example, the convolution of the sequence corresponding to the unit pulse response of a system with the sequence corresponding to the discrete-time input signal produces the sequence corresponding to the discrete-time output signal for a linear, time-invariant, discrete-time system with zero initial conditions. Also, the discrete-time Fourier transform of a discrete-time system's unit pulse response is the system's frequency response. These relationships will be derived in subsequent chapters.

As already mentioned, the roles of the unit pulse sequence and the discrete-time unit pulse signal corresponding to it in discrete-time system analysis are similar to the role of the impulse function in continuous-time system analysis; hence, we use the notation $\delta(n)$ for the unit pulse sequence. It should be noted, however, that the unit pulse sequence does not consist of samples of the continuous-time impulse function because the value of the impulse function is undefined at $t = 0$. Instead, the unit pulse sequence consists of samples of a signal, the magnitude of which is unity at $t = 0$ and zero at all other sample times (that is, at $t = nT$ for $n \neq 0$). This difference is important when constructing discrete-time systems that approximate the characteristics of continuous-time systems by attempting to match the impulse response. In these cases, it is convenient for us to interpret the unit pulse signal $\delta_T(n)$ as consisting of samples of the continuous-time unit rectangular pulse $\Pi(t/T)$. Then, $(1/T)\delta_T(n)$ is the sampled version of a continuous-time signal that approaches an impulse function as T approaches zero. This interpretation is shown in Figure 10.9. No confusion should result from the use of the notation $\delta(t)$

Figure 10.8
The Unit Pulse
Sequence

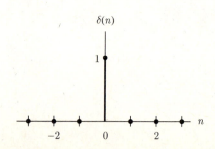

Figure 10.9
Unit Pulse Signal
$\delta_T(n)$ as a Sampled
Version of the Unit
Rectangular Pulse

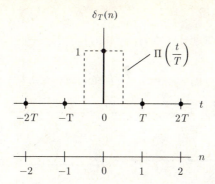

for the unit impulse function and $\delta(n)$ for the unit pulse sequence, because the form of the argument (that is, continuous variable t or sample number n) clearly indicates which is being considered.

We can represent a single-sample value that occurs at sample number $n = n_1$ by the single-pulse sequence (or single-sample sequence)

$$A\delta(n - n_1) = \begin{cases} A & n = n_1 \\ 0 & n \neq n_1 \end{cases} \tag{10.19}$$

Single-pulse sequences can be added so as to represent sequences of samples. We illustrate this addition with the following example.

Example 10.1

Write expressions for the sequences shown in Figure 10.10 by using sums of single-pulse sequences.

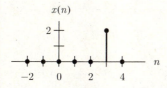

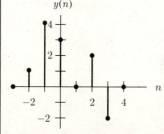

Figure 10.10 Sequences for Example 10.1

Solution

$$x(n) = 2\delta(n - 3)$$

$$y(n) = \delta(n + 2) + 4\delta(n + 1) + 3\delta(n) + 2\delta(n - 2) - 2\delta(n - 3)$$

Example 10.2

Cash flow between two large banks occurs once each day. For a ten-day period, with $t = 0$ at day one, the cash flow in millions of dollars is represented by the discrete-time signal

$$x_T(n) = \big[\{1.6, 2.3, -3.7, 0.5, -1.2, 5.2,$$
$$1.5, -2.3, -4.3, 1.0\}, \quad 1 \text{ day}\big]$$

Write the expression for the sequence $x(n)$ corresponding to the cash-flow signal $x_T(n)$

Solution
Since $x(n) = \{1.6, 2.3, -3.7, 0.5, -1.2, 5.2, 1.5, -2.3, -4.3, 1.0\}$, then, by inspection

$$x(n) = 1.6\delta(n) + 2.3\delta(n - 1) - 3.7\delta(n - 2) + 0.5\delta(n - 3)$$
$$- 1.2\delta(n - 4) + 5.2\delta(n - 5) + 1.5\delta(n - 6) - 2.3\delta(n - 7)$$
$$- 4.3\delta(n - 8) + 1.0\delta(n - 9)$$

The convolution of a single-pulse sequence with any other sequence is the original sequence multiplied by the value of the single pulse and shifted by an amount equal to the shift of the single pulse from the origin. This property parallels a similar property for the continuous-time impulse function and is easily shown by

$$A\delta(n - n_1) * x(n) = \sum_{m=-\infty}^{\infty} [A\delta(m - n_1)]x(n - m)$$

$$= A \sum_{m=-\infty}^{\infty} \delta(m - n_1)x(n - m)$$

$$= Ax(n - n_1) \tag{10.20}$$

Example 10.3

The two sequences

$$x(n) = 2\delta(n+1) - \delta(n-4)$$

$$y(n) = \delta(n+2) + 3\delta(n+1) + 2\delta(n)$$

are given. Plot $x(n)$, $y(n)$, and $z(n) = x(n) * y(n)$.

Solution

$$z(n) = [2\delta(n+1) - \delta(n-4)] * [\delta(n+2) + 3\delta(n+1) + 2\delta(n)]$$

$$= [2\delta(n+3) + 6\delta(n+2) + 4\delta(n+1)]$$

$$+ [-\delta(n-2) - 3\delta(n-3) - 2\delta(n-4)]$$

$$= 2\delta(n+3) + 6\delta(n+2) + 4\delta(n+1) - \delta(n-2)$$

$$- 3\delta(n-3) - 2\delta(n-4)$$

Plots of $x(n)$, $y(n)$, and $z(n)$ are shown in Figure 10.11.

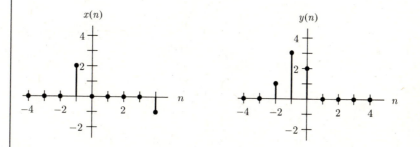

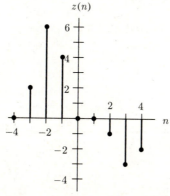

Figure 10.11 Plots of Sequences for Example 10.3

10.4 Signal Energy and Power

The definitions for *signal energy* and *signal power* in a discrete-time signal parallel the definitions of eqs. (3.56) and (3.57) in Section 3.5 for signal energy and signal power in a continuous-time signal.

Definition —————————————————————————————————————

The signal energy in the signal $x_T(n)$ is

$$E_{dx} = \lim_{N \to \infty} \left[T \sum_{n=-N}^{N} |x_T(n)|^2 \right] \qquad (10.21)$$

Definition —————————————————————————————————————

The signal power in the signal $x_T(n)$ is

$$P_{dx} = \lim_{N \to \infty} \left[\left(\frac{1}{2N+1} \right) \sum_{n=-N}^{N} |x_T(n)|^2 \right] \qquad (10.22)$$

For signal energy, we show the parallelism by using a discrete-time signal $x_T(n)$ that consists of samples $x(nT)$ of the continuous-time signal $x(t)$ and the definition of an integral to give

$$\lim_{T \to 0} \sum_{n=-\infty}^{\infty} |x_T(n)|^2 \, T = \lim_{T \to 0} \sum_{n=-\infty}^{\infty} |x(nT)|^2 T = \int_{-\infty}^{\infty} |x(t)|^2 \, dt \qquad (10.23)$$

As we indicated in Section 3.5, signal energy and power can be used to indicate characteristics of a discrete-time signal, but they are not actually measures of energy or power absorbed in or supplied by a system component when the signal passes through the component or is measured across it. A discrete-time energy signal is defined as one for which $0 < E_{dx} < \infty$, and a discrete-time power signal is defined as one for which $0 < P_{dx} < \infty$. Again, these definitions parallel similar definitions for continuous-time signals. It is possible for a discrete-time signal to be neither an energy signal nor a power signal; for example, $(nT)^{-0.1} u_T(n-1)$ and $r_T(n)$, which are sampled versions of the nonenergy and nonpower signals indicated in Section 3.5.

Example 10.4

Compute the signal energy and signal power for the following signals and indicate whether they are energy or power signals:

$$x_T(n) = \left[(-0.5)^n u(n), \qquad 0.01 \text{ s} \right]$$

$$y_T(n) = \left[2e^{3jn} u(n), \qquad 0.2 \text{ s} \right]$$

Solution

$$E_{dx} = \lim_{N \to \infty} T \sum_{n=-N}^{N} |x_T(n)|^2 = 0.01 \sum_{n=0}^{\infty} |(-0.5)^n|^2$$

$$= 0.01 \sum_{n=0}^{\infty} (-0.5)^{2n} = 0.01 \sum_{n=0}^{\infty} (0.25)^n$$

$$= \frac{0.01}{1 - 0.25} = \frac{0.04}{3}$$

where the last step follows from the sum of a geometric series.

Since E_{dx} is finite, $P_{dx} = 0$. Since $0 < E_{dx} < \infty$, the discrete-time signal $x_T(n)$ is an energy signal.

$$P_{dy} = \lim_{N \to \infty} \left(\frac{1}{2N+1} \right) \sum_{n=-N}^{N} |y_T(n)|^2$$

$$= \lim_{N \to \infty} \left(\frac{1}{2N+1} \right) \sum_{n=0}^{N} |2e^{j3n}|^2$$

$$= \lim_{N \to \infty} \left(\frac{1}{2N+1} \right) \sum_{n=0}^{N} 2^2$$

$$= \lim_{N \to \infty} \frac{4(N+1)}{2N+1} = 2$$

Since P_{dy} is finite, $E_{dy} = \infty$. Since $0 < P_{dy} < \infty$, the discrete-time signal $y_T(n)$ is a power signal.

10.5 Representation of Discrete-Time Signals as Ideally Sampled Continuous-Time Signals

As indicated previously, we can think of any discrete-time signal as samples of a continuous-time signal constructed so that it passes through all the values of the sequence corresponding to the discrete-time signal. Thus, a discrete-time signal $x_T(n)$ can be represented as a continuous-time signal by writing it as an ideally sampled version $x_s(t)$ of the continuous-time signal $x(t)$ constructed through the values specified by $x_T(n)$. This ideally sampled version of the constructed continuous-time signal is defined in Section 3.4 as

$$x_s(t) = \sum_{n=-\infty}^{\infty} x(nT)\delta(t - nT) \tag{10.24}$$

where $x(nT)$ is the value of $x(t)$ at $t = nT$ and the sample spacing is T seconds. Since $x(t)$ passes through the discrete-time-signal values $x_T(n)$ at

Figure 10.12
Representation of
Discrete-Time
Signals as Ideally
Sampled
Continuous-Time
Signals

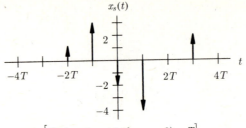

$$x_T(n) = \left[(\{1,3,-2,-4,0,2\}n_0 = -2), \quad T\right]$$
$$x_s(t) = \delta(t+2T) + 3\delta(t+T) - 2\delta(t)$$
$$\qquad - 4\delta(t-T) + 2\delta(t-3T)$$

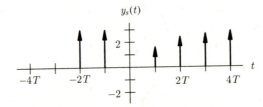

$$y_T(n) = \left[3u(n+2) - 3(0.5)^n u(n), \quad T\right]$$
$$y_s(t) = \sum_{n=-2}^{\infty} 3\delta(t-nT) - \sum_{n=0}^{\infty} 3(0.5)^n \delta(t-nT)$$

$t = nT$, then $x(nT) = x_T(n)$, and the strengths of the impulses in the ideally
sampled signal representation of a discrete-time signal are the values of the
sequence corresponding to the discrete-time signal. Thus, the representation of
a discrete-time signal as an ideally sampled, continuous-time signal is

$$x_s(t) = \sum_{n=-\infty}^{\infty} x_T(n)\delta(t-nT) \qquad (10.25)$$

Examples of discrete-time signals and their representations as ideally sampled,
continuous-time signals are shown in Figure 10.12. Note that we define $u(t)$ to
be 1 at $t = 0$ rather than leaving it undefined, as we did before. We indicated in
Chapter 3 that this is an acceptable definition. We will use it in the remainder
of the text because samples at $t = 0$ are of interest.

10.6 Summary

Discrete-time signals may be generated directly or as samples of a continuous-
time signal. Actually, any discrete-time signal can be thought of as samples
of a continuous-time signal. This conceptualization permits easy extension of
many characteristics of continuous-time signals to discrete-time signals.

Useful mathematical functions of n that model values for discrete-time-signal sequences (that is, samples) include sinusoidal and complex-exponential sequences, exponential sequences, step and ramp sequences, and the unit pulse sequence. The unit pulse sequence and its corresponding unit pulse signal are very important, playing roles in discrete-time system analysis similar to the role of the unit impulse signal in continuous-time system analysis. Despite the similar roles of the discrete-time unit pulse signal and the continuous-time unit impulse signal, the unit pulse signal does not consist of samples of the unit impulse signal.

The definitions of energy, E_{dx}, and power, P_{dx}, for a discrete-time signal parallel the definitions for continuous-time signals. Discrete-time energy and power signals are those for which $0 < E_{dx} < \infty$ and $0 < P_{dx} < \infty$, respectively.

The representation of a discrete-time signal as a continuous-time signal by using an ideally sampled signal with impulse strengths equal to the discrete-time-signal sequence is frequently useful. This representation is made possible by the fact that the sequence values for any discrete-time signal can be considered as samples of a continuous-time signal. The use of this representation permits us to use the concepts and techniques discussed in Part II to develop similar concepts and techniques for discrete-time signals and systems. By doing so, we enhance our understanding of discrete-time signal and system concepts.

Problems

10.1 Plot the following sinusoidal sequences:

a. $x(n) = 3\cos\left[(\pi n/4) - \pi/4\right]$

b. $y(n) = 3\sin(4\pi n/3)$

c. $z(n) = 3\sin\left[(\pi n/\sqrt{2}) + 5\pi/3\right]$

d. $w(n) = 3\cos\left[3n/\pi + 15\pi/4\right]$

Which are periodic sequences? Are any of the sequences the same?

10.2 Plot the following sinusoidal sequences:

a. $x(n) = 2\sin\left[(4n/\pi) - \pi\right]$

b. $y(n) = 2\cos\left[(3\pi n/8) - \pi/2\right]$

c. $z(n) = 2\sin\left[(3\pi n/8) - \pi/2\right]$

d. $w(n) = 2\sin\left[19\pi n/8\right]$

e. $v(n) = 2\cos\left[(5\pi n/8) + 3\pi/8\right]$

Which are periodic sequences? Are any of the sequences the same?

10.3 Without plotting the following sinusoidal sequences, determine which are periodic and state which produce the same sequence. Find the period of any sequences that are periodic.

a. $v(n) = 4\cos\left[(\pi n/3) + 3\pi/8\right]$

b. $w(n) = 4\sin\left[(3n/\pi) - 3\pi/8\right]$

c. $x(n) = 4\cos\left[(9n/4) + \pi/3\right]$

d. $y(n) = 4\sin\left[(13\pi n/3) - 9\pi/8\right]$

e. $z(n) = 4\sin\left[(8\pi n/3) + 7\pi/8\right]$

10.4 Without plotting the following sinusoidal sequences, determine which are periodic and state which produce the same sequence. Find the period of any sequences that are periodic.

a. $v(n) = 3\cos\left[(3n/2) - \pi/5\right]$

b. $w(n) = 3\cos\left[(17\pi n/7) - 13\pi/2\right]$

c. $x(n) = 3\cos\left[5n/2\right]$

d. $y(n) = 3\cos\left[(3\pi n/7) + 3\pi/2\right]$

e. $z(n) = 3\cos\left[(4\pi n/7) - \pi/2\right]$

10.5 Find four frequencies corresponding to continuous-time sinusoidal signals, the samples of which produce the following discrete-time signals:

a. $x_T(n) = [1.5\cos(0.2\pi n + 0.3),\quad 0.01\text{ s}]$

b. $y_T(n) = [2.6\sin(3.1\pi n - 0.1),\quad 0.01\text{ s}]$

10.6 Repeat Problem 10.5, but now assume the discrete-time signals produced are

a. $x_T(n) = [4\cos(1.1n - 0.35),\quad 0.2\text{ s}]$

b. $y_T(n) = [3\sin(13.3n + 0.15),\quad 0.2\text{ s}]$

10.7 Plot the two complex-conjugate vectors that can be added to produce the sequence values for $x(n) = 4\cos\left[(\pi n/3) + \pi/4\right]$ when $n = 0, 1, 2,$ and 3.

10.8 Work Problem 10.7 when $x(n) = 4\sin[(\pi n/4) - 3\pi/4]$.

10.9 Plot the following sequences for $|n| \le 4$:

a. $x(n) = 3(-0.2)^n u(n)$

b. $y(n) = 3(-0.2)^n u(-n + 2)$

c. $v(n) = 2e^{-0.2n} u(n - 1)$

d. $w(n) = 1.5e^{0.4n} u(n + 2)$

e. $z(n) = 6e^{(n/3)}\cos\left(\frac{2\pi n}{3}\right) u(n + 3)$

10.10 Plot the following discrete-time signals for $|n| \le 10$:

a. $x_T(n) = \left[-10(0.85)^n u(n - 4),\quad 2\text{ ms}\right]$

b. $y_T(n) = \left[20(-1.1)^n u(-n - 2),\quad 0.1\text{ s}\right]$

c. $z_T(n) = \left[e^{0.1n} u(n - 1),\quad 10\text{ s}\right]$

d. $w_T(n) = \{3e^{-0.2n}\sin\left[(9n/5) - \pi/3\right] u(n),$
    $0.5\text{ s}\}$

10.11 Plot the following sequences for $|n| \le 5$:

a. $u(3n - 6)$

b. $3u(n + 2)$

c. $4u(-n - 1)$

d. $2u(-5n + 15)$

10.12 Plot the following discrete-time signals for $|n| \le 5$:

a. $x_T(n) = [3r(n),\quad 1.6\text{ s}]$

b. $y_T(n) = [r(n - 2),\quad 0.35\text{ s}]$

c. $z_T(n) = [2r(-n + 1),\quad 3\text{ ms}]$

d. $w_T(n) = \{3r(7n - 14),\quad 5\text{ }\mu\text{s}\}$

10.13 Plot the following sequences for $|n| \le 10$.

a. $x(n) = -n + r(n + 5) - 5u(n + 1)$
    $+ 3r(n - 3) - 3r(n - 8)$

b. $y(n) = 4u(n + 7) - 2u(n + 2) - 2r(n + 2)$
    $+ r(n - 2) + r(n - 5) + 3u(n - 8)$

c. $z(n) = -2r(n + 1) + 4r(n - 3) - 2r(n - 7)$

10.14 Plot the following discrete-time signals for $|n| \le 10$:

a. $x_T(n) = [-2 + 4u(n + 6) + r(n + 6)$
    $- r(n) - 6u(n - 4) - 2u(n - 7),$
                                    $2\text{ s}]$

b. $y_T(n) = [3u(n + 4) + 3u(n + 1)$
    $+ 3u(n - 2) - 3r(n - 5) + 3r(n - 8),$
                                    $0.1\text{ s}]$

c. $z_T(n) = [n - r(n - 1) + u(n - 2)$
    $+ 2r(n - 5) - 4r(n - 7) + 2r(n - 9),$
                                    $15\text{ s}]$

10.15 The sequences shown in Figure 10.13 correspond to digitized signals that command a robot's position in x and y coordinates. The sequence $x(n)$ is equal to 1 for all $n \le -7$ and -2 for all $n \ge 7$. Write expressions for the sequences in terms of sums of step and ramp sequences.

10.16 The sequences shown in Figure 10.14 correspond to short digitized voice commands. Write expressions for the sequences using sums of step and ramp sequences.

10.17 Plot the signal

a. $w_T(n) = [\delta(n + 2)x(n),\quad 0.1\text{ s}]$

b. $z_T(n) = [\delta(n + 2) * y(n),\quad 2.5\text{ s}]$

c. $x_T(n) = [2\delta(n - 3)y(n),\quad 4\text{ s}]$

Figure 10.13

a.

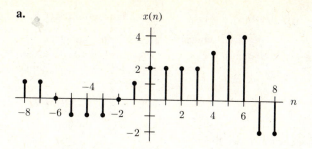

b.

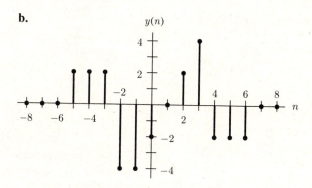

Figure 10.14

a.

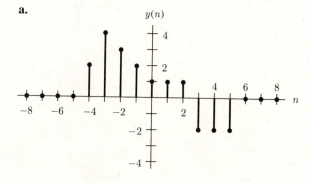

b.

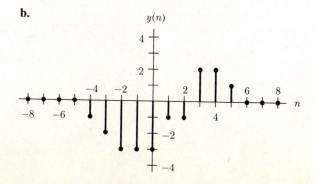

d. $a_T(n) = [2\delta(n-3) * x(n), \quad 0.01 \text{ s}]$

where $x(n)$ and $y(n)$ are the digitized voice-command sequences shown in Problem 10.16.

10.18 Plot the sequences

a. $w(n) = \delta(n-1)y(n)$

b. $z(n) = \delta(n-1) * y(n)$

where $y(n)$ is the second sequence shown in Problem 10.15.

10.19 Determine the signal energy and signal power for each of the given signals and indicate whether it is an energy signal or a power signal.

a. $x_T(n) = [2u(n+4) - 2r(n+1) + 3r(n-1)$
$\qquad - r(n-3), \quad 4 \text{ s}]$

b. $y_T(n) = [3(-0.2)^n u(n-3), \quad 2 \text{ ms}]$

c. $z_T(n) = [4(1.1)^n u(n+1), \quad 0.02 \text{ s}]$

d. $v_T(n) = [2 - u(n+2) - u(n-10), \quad 5 \text{ s}]$

e. $w_T(n) = \left[6\sin\left[(3\pi n/8) - \pi/3\right], \quad 1.5 \text{ s}\right]$

10.20 Determine the signal energy and signal power for each of the given signals and indicate whether it is an energy signal or a power signal.

a. Sequence $x(n)$ of Problem 10.15 with $T = 0.4$ s.

b. Sequence $y(n)$ of Problem 10.15 with $T = 25$ s.

c. $z_T(n) = [e^{-0.4|n|}, \quad 2.5 \text{ ms}]$.

d. $v_T(n) = \left[3\cos(5\pi n/2), \quad 0.05 \text{ s}\right]$.

10.21 Write the expression for the representations of the signals of Problem 10.19 as ideally sampled continuous-time signals. Plot them.

10.22 Write the expressions for the representations of the signals of Problem 10.20 as ideally sampled continuous-time signals. Plot them.

10.23 Write the sequences corresponding to the discrete-time signals represented by the following ideally sampled continuous-time signals and plot them:

a. $x_s(t) = -3\delta(t+2T) - 2\delta(t) + 2\delta(t-T)$
$\qquad + 2\delta(t-2T) + 2\delta(t-3T)$

b. $y_s(t) = \displaystyle\sum_{n=-4}^{\infty} (0.5)^n \delta(t-nT)$

c. $z_s(t) = \displaystyle\sum_{n=-\infty}^{n=3} (0.5)^{-n} \delta(t-nT)$

d. $w_s(t) = \displaystyle\sum_{n=-\infty}^{\infty} 2\delta(t-nT) + \sum_{n=-3}^{0} (n+3)$

$\qquad \delta(t-nT) + \displaystyle\sum_{n=1}^{3} (3-n)\delta(t-nT)$

10.24 Write the sequences corresponding to the discrete-time signals represented by the following ideally sampled continuous-time signals and plot them.

a. $x_s(t) = \displaystyle\sum_{n=-2}^{2} (n+2)\delta(t-nT)$

$\qquad + \displaystyle\sum_{n=3}^{8} 4\delta(t-nT)$

b. $y_s(t) = 4\delta(t-2T) + 3\delta(t-3T)$
$\qquad - 3\delta(t-5T) - 4\delta(t-6T)$

c. $z_s(t) = \displaystyle\sum_{n=2}^{\infty} e^{-0.4n}\delta(t-nT)$

d. $w_s(t) = \displaystyle\sum_{n=-3}^{6} (1.2)^n \delta(t-nT)$

11 Spectra of Discrete-Time Signals

Recall from the previous chapter that a discrete-time signal $x_T(n)$ is specified by a sequence of values $x(n)$, called samples, that occur at successive sample times and by the constant time T between sequence values, called the sample spacing. The sequence and sample spacing show how a discrete-time signal varies as a function of time in what we refer to as the time-domain representation of the discrete-time signal.

We can also use a frequency-domain representation to characterize a discrete-time signal, just as we did with continuous-time signals. The frequency-domain representation is the signal's spectrum, which consists of amplitude and phase spectra of the signal. It is useful for portraying the range of frequencies for which a signal has significant energy content and for describing the nature of the amplitude and phase of signal components at all frequencies. In this chapter, we will define the spectrum of a discrete-time signal. Our definition builds upon the spectrum concepts defined and considered in Chapter 4; thus, much of this chapter's discussion parallels that of Chapter 4.

The definition of the spectrum of a discrete-time signal provides us with an efficient method for discussing a property associated with discrete-time signals that are generated by sampling continuous-time signals. This property is formally stated by the sampling theorem and indicates the condition under which the resulting samples uniquely represent the continuous-time signal. The property is very important if discrete-time systems are used to process sampled continuous-time signals.

We found in Chapter 4 that the Fourier transform produces the spectrum of a continuous-time energy signal. Therefore, we will define a similar transform, called the *discrete-time Fourier transform*, to compute the spectrum of a discrete-time energy signal. We will also define the *inverse discrete-time Fourier transform* by which we determine a discrete-time energy signal from its spectrum. Properties and theorems associated with the discrete-time Fourier transform will be discussed to indicate spectrum properties.

We extend the discrete-time Fourier transform techniques to the computation of the spectrum of a discrete-time power signal by permitting impulse functions in the spectrum and by using Fourier transforms in the limit.

We indicated in Chapter 1 that the sequence $x(n)$ can represent different discrete-time signals that differ only by a scaling factor in time if the value of T is not specified. Thus, we can think of a sequence as a time-scale-normalized version of the corresponding discrete-time signal, where the time-scale normalization factor is the sample spacing T. We will consider a frequency-scale-normalized spectrum corresponding to the time-scale-normalized signal (that is, to the sequence). This frequency-scale-normalized spectrum portrays the spectrum characteristics for all discrete-time signals with the same sequence.

We obtain the actual spectrum for a discrete-time signal from the frequency-scale-normalized spectrum by performing a frequency scaling.

11.1 Spectra and Bandwidth of Discrete-Time Signals

We showed in Chapter 4 that the spectrum of a continuous-time signal portrays the signal's characteristics as a function of frequency, and that the signal's bandwidth indicates the range of frequencies that are present with significant amplitude in the signal. We will now apply these concepts to discrete-time signals.

Signal Spectrum

Since we can represent any discrete-time signal as an ideally sampled continuous-time signal, then we can find the spectrum of the discrete-time signal by computing the Fourier transform of its representation as an ideally sampled continuous-time signal. This representation for the discrete-time signal $x_T(n)$ is

$$x_s(t) = \sum_{n=-\infty}^{\infty} x_T(n)\delta(t - nT) = \sum_{n=-\infty}^{\infty} x(nT)\delta(t - nT)$$

$$= \sum_{n=-\infty}^{\infty} x(t)\delta(t - nT) = x(t)\left[\sum_{n=-\infty}^{\infty} \delta(t - nT)\right]$$

$$= x(t)\delta_s(t) \tag{11.1}$$

where $x(t)$ is a continuous-time signal constructed through the values of $x_T(n)$ and $\delta_s(t)$ is the ideal-sampling signal defined in Chapter 3. The second equivalence in eq. (11.1) follows from eqs. (10.24) and (10.25), the third equivalence follows from Property V of the impulse function (see Chapter 3), and the fourth equivalence results from the fact that $x(t)$ is not a function of n. We compute the Fourier transform of eq. (11.1) to obtain

$$X_d(f) \equiv X_s(f) = X(f) * \Delta_s(f) = X(f) * \sum_{m=-\infty}^{\infty} f_s\delta(f - mf_s)$$

$$= \sum_{m=-\infty}^{\infty} f_sX(f) * \delta(f - mf_s) = \sum_{m=-\infty}^{\infty} f_sX(f - mf_s) \tag{11.2}$$

as the spectrum $X_d(f)$ of the discrete-time signal $x_T(n)$, where the fourth equivalence occurs because the convolution operation is distributive and the fifth equivalence holds by Property VI of the impulse function. The parameter $f_s = 1/T$ is the frequency of sample occurrence in samples per second or, equivalently, hertz. We refer to it as the *sample frequency or sample rate*.

We noted in Chapter 10 that $x(t)$ is not unique for the given discrete-time signal $x_T(n)$. Therefore, $X(f)$ is not unique. However, $X_s(f)$ is unique because the Fourier transform of the ideally sampled continuous-time signal is unique. Thus, the sum in eq. (11.2) is the same regardless of which continuous-time signal that we construct through the values of $x_T(n)$.

Note that the spectrum of a discrete-time signal is the sample frequency, f_s, times the sum of the spectrum, $X(f)$, of the continuous-time signal, $x(t)$, and versions of this spectrum, $X(f - mf_s)$ for $m \neq 0$, that are shifted by frequencies that are multiples of the sample frequency. The equation

$$X_d(f - f_s) = \sum_{m=-\infty}^{\infty} f_s X(f - f_s - mf_s)$$

$$= \sum_{m=-\infty}^{\infty} f_s X[f - (m+1)f_s] = X_d(f) \qquad (11.3)$$

shows us that the spectrum $X_d(f)$ of a discrete-time signal is periodic with period f_s. Therefore, we need only to compute and plot $X_d(f)$ for $|f| \leq f_s/2$, since it repeats outside this range of frequencies. Throughout most of this chapter, we will plot several periods of $X_d(f)$ to highlight its periodic nature.

The spectrum $X_d(f)$ of a discrete-time signal is a continuous-frequency function, the value of which is complex for some or all values of f because $X(f)$ has this characteristic (see Chapter 4). Thus, $|X_d(f)|$ is the amplitude spectrum (actually, the amplitude-density spectrum, as indicated in Chapter 4) and $\angle X_d(f)$ is the phase spectrum of the discrete-time signal $x_T(n)$. Since the amplitude and phase spectra of a discrete-time signal are obtained in the same manner as the amplitude and phase spectra of a continuous-time signal, then they portray the nature of the discrete-time signal's amplitude and phase at frequencies over the entire frequency range.

Example 11.1

Suppose you hear the sound of a gunshot echoing between two walls. The echoes are separated by a 0.1-s interval. They exponentially decay in amplitude because only $e^{-0.5}$ (60.7%) of the amplitude is reflected at each subsequent reflection. The amplitude of the echoes relative to the first echo at $t = 0$ is given by the discrete-time signal

$$x_T(n) = \left[e^{-0.5n} u(n), \quad 0.1 \text{ s} \right]$$

Find the spectrum of the echo signal. Plot the amplitude and phase spectra of the signal.

Solution
Since $T = 0.1$, the discrete-time signal can be written as

$$x_T(n) = \left[e^{-5nT} u(n), \quad 0.1 \text{ s} \right]$$

The signal $x_T(n)$ is plotted in Figure 11.1.

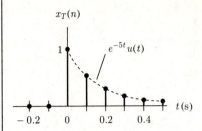

Figure 11.1 Echo Signal for Example 11.1

The representation of this discrete-time signal as an ideally sampled continuous-time signal is

$$x_s(t) = \sum_{n=-\infty}^{\infty} x_T(n)\delta(t - nT)$$

$$= \sum_{n=0}^{\infty} e^{-0.5n}\delta(t - 0.1n)$$

$$= e^{-5t} \sum_{n=0}^{\infty} \delta(t - 0.1n)$$

The plot of $x_s(t)$ is shown in Figure 11.2.

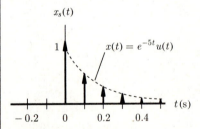

Figure 11.2 Ideally Sampled Continuous-Time Representation of Echo Signal in Example 11.1

From Table C.2 in Appendix C,

$$x(t) \leftrightarrow X(f) = \frac{1}{5 + j2\pi f}$$

The sample frequency is $f_s = 1/T = 10$ Hz; therefore, from eq. (11.2), the spectrum of the discrete-time signal is

$$X_d(f) = \sum_{m=-\infty}^{\infty} \frac{10}{5 + j2\pi(f - 10m)}$$

The amplitude and phase spectra are

$$|X_d(f)| = \left| \sum_{m=-\infty}^{\infty} \frac{10}{5 + j2\pi(f - 10m)} \right|$$

and

$$\underline{/X_d(f)} = \underline{/\sum_{m=-\infty}^{\infty} \frac{10}{5 + j2\pi(f - 10m)}}$$

respectively and are plotted in Figure 11.3. We see that the amplitude and phase spectra are periodic with period 10 Hz $= f_s = 1/T$, as expected. We will explain the dashed lines and associated numbers in Figure 11.3 shortly.

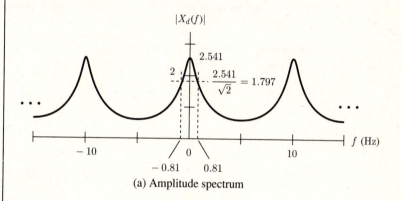

(a) Amplitude spectrum

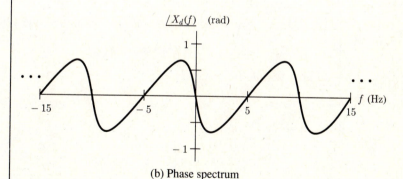

(b) Phase spectrum

Figure 11.3 Amplitude and Phase Spectra of the Discrete-Time Echo Signal of Example 11.1

The periodicity of the signal spectrum in Example 11.1 is a consequence of the regular 0.1-s spacing of the echoes. Note that, if the echo amplitudes decay more slowly, then the discrete-time signal is $x_T(n) = [e^{-\alpha n}u(n), \; 0.1 \text{ s}]$, where α is less than 0.5. In this case, the larger amplitude portions of the amplitude spectrum that are concentrated at the frequencies $\pm m f_s$ are narrower and higher, which means that the lower frequency components in the signal are more dominant. This dominance indicates that the amplitude of the signal-sequence values (that is, the echo amplitudes) varies more slowly.

In order to plot the amplitude and phase spectra for Example 11.1, we must find the amplitude and angle associated with an infinite sum of complex functions. We can find a reasonably accurate approximation to the infinite sum by summing a small number of its terms if $|X(f)|$ rapidly approaches zero and if $\underline{/\,X(f)}$ rapidly approaches a constant as f becomes larger.

Note that the spectrum of a discrete-time signal is similar to that of a continuous-time signal in that both are continuous-frequency functions. But because the spectrum of a discrete-time signal is a periodic function of frequency, it differs from the spectrum of a continuous-time signal.

Example 11.2

Find the spectrum of the discrete-time signal

$$x_T(n) = \begin{bmatrix} 3\cos(0.6\pi n - \pi/3), & 0.2\ \mathrm{s} \end{bmatrix}$$

Plot the amplitude and phase spectra of the signal.

Solution

Since $T = 0.2$, the discrete-time signal given can be written as

$$x_T(n) = \begin{bmatrix} 3\cos\left[2\pi(1.5)nT - \pi/3\right], & 0.2\ \mathrm{s} \end{bmatrix}$$

Thus, $x_T(n)$ consists of samples of the continuous-time signal

$$x(t) = 3\cos\left[(2\pi(1.5)t - \pi/3)\right]$$

and the sample frequency is

$$f_s = 1/T = 5\ \mathrm{Hz}$$

From Table C.2 in Appendix C,

$$x(t) \leftrightarrow X(f) = \left[\frac{3}{2}e^{-j\pi/3}\right]\delta(f - 1.5) + \left[\frac{3}{2}e^{j\pi/3}\right]\delta(f + 1.5)$$

and, from eq. (11.2), the spectrum of $x_T(n)$ is

$$X_d(f)$$

$$= \sum_{m=-\infty}^{\infty} \left[\left(7.5e^{-j\pi/3}\right)\delta(f - 1.5 - 5m) + \left(7.5e^{j\pi/3}\right)\delta(f + 1.5 - 5m)\right]$$

The amplitude and phase spectra corresponding to $x_T(n)$ are obtained from $X_d(f)$ and are plotted in Figure 11.4.

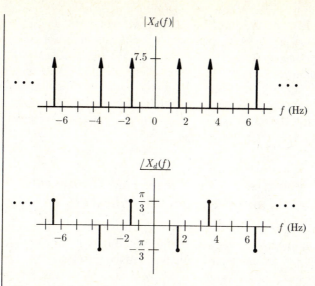

Figure 11.4 Amplitude and Phase Spectra for the Signal in Example 11.2

In Example 11.2, the contributions at $f = \pm 1.5$ Hz in the amplitude and phase spectra of $x_T(n)$ show the frequency content of the continuous-time cosine signal that is sampled. The remaining components in the spectrum of $x_T(n)$ are repeated versions of these contributions centered at multiples of the sampling frequency $f_s = 5$ Hz.

Signal Bandwidth

If we apply the bandwidth definition for continuous-time signals to discrete-time signals, we find that the bandwidth of all discrete-time signals is infinite. This is because their spectra are periodic, thus possessing significant amplitudes at frequencies near all multiples of the sample frequency f_s. The signal bandwidth should indicate the greatest time rate of amplitude variation in the signal samples. We showed in Section 10.1 of the previous chapter that the rates of variation of samples corresponding to sinusoidal-signal components of a discrete-time signal are uniquely specified for frequencies in the unambiguous frequency interval; that is, for frequencies lower than $1/2T = f_s/2$. Consequently, we define the bandwidth of a discrete-time signal as follows:

Definition _____

The bandwidth of a discrete-time signal is the range of positive frequencies within the frequency range $0 \leq f \leq f_s/2$ for which the amplitude-spectrum value is greater than or equal to α times its maximum value, where α is a selected constant.

Perhaps the most common choice of a value for α value is $1/\sqrt{2}$. This value produces the half-power bandwidth, as we indicated when we discussed continuous-time signals. For the discrete-time signal of Example 11.2, this definition of α produces a signal bandwidth of 0 Hz because the signal consists of samples of a single-frequency sinusoidal continuous-time signal. For the discrete-time signal of Example 11.1, the signal bandwidth is 0.81 Hz if we select the value of the constant α to be $1/\sqrt{2}$. This bandwidth is illustrated in Figure 11.3 by the dashed lines.

11.2 Spectral Analysis of Sampled Continuous-Time Signals and the Sampling Theorem

A frequent objective in signal processing and/or transmission is to process continuous-time signals with a digital signal processor or to transmit continuous-time signals as digital signals to reduce the effect of transmission noise. Examples are the processing of sensor data in an aircraft's flight-control system, the transmission of data from a satellite, and the storage and subsequent reading of audio signals using compact-disc technology. To perform these tasks, we must sample the continuous-time signal at discrete points in time to form a discrete-time signal that represents the continuous-time signal. This operation is performed in an A/D converter. We will now analyze the characteristics of the discrete-time representation of a continuous-time signal using spectral analysis made possible by the spectrum definition for discrete-time signals of the previous section.

If we represent a continuous-time signal $x(t)$ by samples $x(nT)$ taken at a time spacing of T seconds, then it is desirable that the samples represent $x(t)$ as well as possible. It is apparent that we will obtain more signal data if we take the samples closer together (that is, with T smaller). However, if $x(t)$ is bandlimited, then its amplitude-variation rate is less than a rate proportional to its bandwidth. Therefore, there is redundant information in adjacent amplitude values. For bandlimited signals, then, no additional information is obtained if samples are taken closer together than a maximum sample spacing determined by the signal's bandwidth. In other words, the signal is completely specified by samples that are at least as close together as the maximum sample spacing determined by the signal's bandwidth. We express this fact for equally spaced samples in the form of a theorem in which we use the sample rate $f_s = 1/T$ rather than the sample spacing T.

> **Uniform-Sampling Theorem** If $x(t)$ is bandlimited with no components at frequencies greater than f_h Hz, then it is completely specified by samples taken at the uniform rate $f_s > 2f_h$ Hz.

The minimum sampling rate, or minimum sampling frequency, $f_s = 2f_h$, for complete specification of the continuous-time-signal, is referred to as the

Nyquist rate, or *Nyquist frequency*. We can also express the sampling theorem in terms of maximum sample spacing rather than a minimum sampling rate because $T = 1/f_s < 1/2f_h$; hence, the maximum sample spacing for complete specification of the continuous-time-signal is $T = 1/2f_h$. Sampling a signal at a rate that is less than or greater than the Nyquist rate is referred to as *undersampling* or *oversampling*, respectively.

To illustrate the sampling theorem, let us now consider the spectrum of the discrete-time signal $x_T(n)$, the sequence values of which are the samples of the continuous-time signal $x(t)$ and the sample frequency of which is the continuous-time-signal sampling frequency. This discrete-time signal is

$$x_s(t) = \sum_{n=-\infty}^{\infty} x_T(n)\delta(t - nT) = \sum_{n=-\infty}^{\infty} x(nT)\delta(t - nT) \qquad (11.4)$$

when represented as an ideally sampled continuous-time-signal. Its spectrum, as given by eq. (11.2), is

$$X_d(f) = \sum_{m=-\infty}^{\infty} f_s X(f - mf_s) \qquad (11.5)$$

where $X(f)$ is the spectrum of $x(t)$.

For discussion purposes, we will assume that $x(t)$ is bandlimited and has the spectrum shown in Figure 11.5.

The corresponding spectrum of the discrete-time signal $x_T(n)$ formed from the samples of $x(t)$ is shown in Figure 11.6a for the case in which $f_s > 2f_h$. Figure 11.6b illustrates the case in which $f_s < 2f_h$. It is apparent from Figure 11.6 that the sampling theorem is valid. We can see that if an ideally sampled continuous-time signal $x_s(t)$ is constructed using the samples of $x(t)$ taken at a rate $f_s > 2f_h$, then $x(t)$ can be reconstructed from $x_s(t)$ with a reconstruction filter that is an ideal low-pass filter with gain equal to $1/f_s$ and bandwidth greater than f_h but less than $f_s - f_h$. On the other hand, if $f_s < 2f_h$, then no such reconstruction is possible because portions of $X(f - f_s)$ and $X(f + f_s)$ overlap $X(f)$, and thus add to $X(f)$ in producing $X_d(f)$. We refer to this overlap as *aliasing* because the components of $X(f - f_s)$ and $X(f + f_s)$ that correspond to signal frequencies greater than $f_s/2$ overlap $X(f)$ and appear as if they were signal components with frequencies lower

Figure 11.5
Spectrum of
Example Bandlimited
Signal $x(t)$

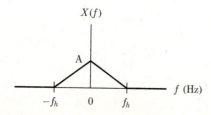

Figure 11.6
Spectrum of
Discrete-Time Signal
$x_T(n)$ Consisting of
Samples $x(t)$

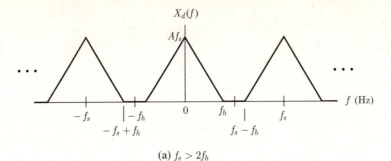

(a) $f_s > 2f_h$

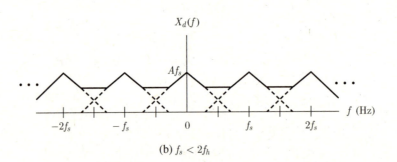

(b) $f_s < 2f_h$

than $f_s/2$.[†] An example of this effect is in movies, where the spoked wheels of fast-moving wagons appear to rotate more slowly than is consistent with the wagon speed, or even sometimes backward. The sampling operation in this case results from the fact that a movie is made up of a sequence of still frames (usually 18 to 24 frames per second).

Note that unique representation of a bandlimited continuous-time signal by samples occurs if $f_h < f_s/2 = 1/2T$; that is, if the frequencies of the signal's components are restricted to be lower than $1/2T$. These are the unambiguous frequencies defined in Chapter 10, and it is now clear why they are unambiguous. They are the set of continuous-time-signal frequencies that produce unique discrete-time-signal samples; thus, they uniquely define how rapidly the discrete-time-signal samples vary.

Example 11.3

The acoustic pulse received by a sonar receiver on a submarine can be represented with the signal

$$x(t) = 30 \ \text{sinc}^2(100t)$$

We want to process this signal with a discrete-time system. Find the minimum

---

[†]The sampling rate cannot equal the Nyquist rate if the signal contains a sinusoidal component with the frequency f_h because overlap of spectrum impulses occurs, which produces aliasing. This is why f_s is not permitted to equal $2f_h$ in the sampling theorem.

sampling rate that we must use in the A/D converter, the spectrum of the discrete-time signal formed from the samples, and the characteristics of the filter required to reconstruct $x(t)$ from the signal samples, if impulses are used to represent the signal samples.

Solution

From Table C.2 of Appendix C and the duality theorem for Fourier transforms,

$$X(f) = 0.3\Lambda(f/100)$$

Therefore $f_h = 100$ Hz and the minimum sampling rate is $f_s = 2f_h = 200$ Hz. The spectrum of the discrete-time signal formed from the samples is

$$X_d(f) = \sum_{m=-\infty}^{\infty} f_s X (f - mf_s)$$

$$= \sum_{m=-\infty}^{\infty} 60\Lambda\left(\frac{f - 200m}{100}\right)$$

and is plotted in Figure 11.7.

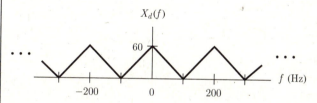

Figure 11.7 Spectrum of the Ideally Sampled Signal in Example 11.3

The reconstruction filter required is an ideal low-pass filter with gain of $1/f_s = 0.005$ and bandwidth of $f_h = 100$ Hz.

Impulse signals, truly bandlimited continuous-time signals, and ideal low-pass filters do not exist in practical continuous-time systems. Therefore, the reconstruction of a continuous-time signal from the discrete-time signal made up of its samples is affected by practical system considerations. Consider first the effect that results from not being able to use impulse signals. Assume that the signal samples are represented by rectangular pulses of height equal to the sequence values of the discrete-time signal (that is, sample values) and width equal to τ seconds. A signal of this type is generated by turning on the continuous-time signal $x(t)$ for a very short time each sampling period and then stretching the resulting signal pulse widths by using a zero-order hold system (see Example 7.7). The resulting signal is

$$x_{ps}(t) = \sum_{n=-\infty}^{\infty} x(nT)\Pi\left(\frac{t - \tau/2 - nT}{\tau}\right)$$

$$= \sum_{n=-\infty}^{\infty} \Pi\left(\frac{t-\tau/2}{\tau}\right) * x(nT)\delta(t-nT)$$

$$= \Pi\left(\frac{t-\tau/2}{\tau}\right) * \sum_{n=-\infty}^{\infty} x(nT)\delta(t-nT)$$

$$= \Pi\left(\frac{t-\tau/2}{\tau}\right) * x_s(t) \qquad (11.6)$$

and is illustrated in Figure 11.8a. The spectrum of this signal is

$$X_{ps}(f) = \mathcal{F}\left[\Pi\left(\frac{t-\tau/2}{\tau}\right)\right]\mathcal{F}[x_s(t)]$$

$$= \tau \ \mathrm{sinc}(\tau f)X_d(f)e^{-j2\pi f(\tau/2)} \qquad (11.7)$$

The amplitude spectrum of $x_{ps}(t)$ is illustrated in Figure 11.8c when the spectrum $X(f)$ of the continuous-time signal sampled is assumed to be as shown in Figure 11.8b. We have assumed a rectangular form for $X(f)$ for illustration

Figure 11.8
Signal $x_{ps}(t)$ and
Amplitude Spectrum
for Rectangular Pulse
Representation of
Signal Samples

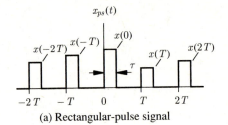

(a) Rectangular-pulse signal

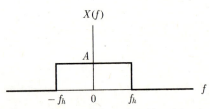

(b) Assumed spectrum of a continuous-time signal that is sampled

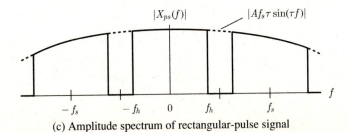

(c) Amplitude spectrum of rectangular-pulse signal

purposes due to its simplicity, even though it is not the actual spectrum of the continuous-time signal corresponding to the samples $x(nT)$ defined in Figure 11.8a. Except for a time delay of $\tau/2$ seconds, due to the $\exp[-j2\pi f(\tau/2)]$ term in $X_{ps}(f)$, the continuous-time signal $x(t)$ can be reconstructed from $x_{ps}(t)$ with a low-pass filter, the frequency response of which is

$$H(f) = \Pi\left(\frac{f}{2B}\right)\bigg/ f_s\tau \ \text{sinc}(\tau f) \tag{11.8}$$

where the filter bandwidth B is greater than f_h but less than $f_s - f_h$. A filter of this type is called an *ideal equalizer filter* because it equalizes the overall system gain for all frequencies in the filter passband and cuts off sharply. We do not need an equalizer filter if $\tau \ll T$ (that is, when the pulses are narrow with respect to pulse spacing). In that case, $\tau f \le \tau f_h < \tau f_s/2 = \tau/2T \ll 1$ over the low-pass-filter bandwidth, which implies that $\text{sinc}(\tau f) \cong 1$ over the filter bandwidth.

A second effect to be considered for practical systems is that which results from the fact that all practical signals are time-limited (that is, they have not existed since $t = -\infty$ and will not continue to exist until $t = \infty$). Therefore, practical signals cannot be truly bandlimited, as we indicated in the discussion of spectrum properties in Section 4.4. However, practical signals have amplitude spectra that approach zero at higher frequencies; otherwise, they would contain infinite energy (refer to the discussion of energy-density spectra in Section 4.4). Therefore, aliasing can be made negligible by choosing a large enough sampling rate, as illustrated in Figure 11.9. In the case illustrated, the small-amplitude ("negligible") higher frequency components of the continuous-time signal $x(t)$ being sampled and reconstructed are eliminated by the ideal low-pass reconstruction filter, and a small ("negligible") amount of aliasing is produced at all frequencies in the filter passband. Since the higher frequency components of $x(t)$ are lost anyway in reconstruction, we frequently reduce their amplitude with a low-pass filter before sampling $x(t)$. In this way, aliasing is further reduced.

Even if the signal $x(t)$ is bandlimited, we cannot reconstruct it exactly from samples taken at a rate $f_s > 2f_h$ because an ideal low-pass filter or ideal equalizer filter (that is, a sharp cutoff filter) cannot be constructed (refer to the

Figure 11.9

Reconstruction of Nonbandlimited Signal from Its Samples Using an Ideal Low-Pass Filter

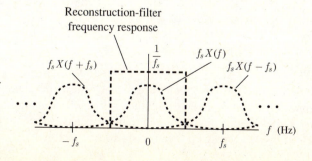

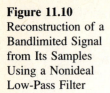

Figure 11.10
Reconstruction of a
Bandlimited Signal
from Its Samples
Using a Nonideal
Low-Pass Filter

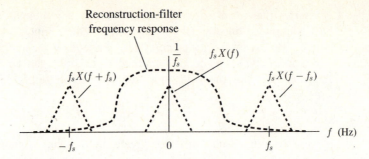

practical-filter discussion in Chapter 8). A causal approximation to an ideal
LPF has an amplitude response (gain) that approaches zero at higher frequencies, but never actually reaches zero (the time-limited impulse response cannot
produce a band-limited frequency response). Therefore, its output to a sampled
signal input will always contain contributions from the sampled signal spectrum components centered at multiples of the sampling frequency f_s. This is
illustrated in Figure 11.10. Note that we cannot use a sampling rate near the
minimum possible rate (that is, the Nyquist rate) if the continuous-time signal
is to be reconstructed from its samples with a high degree of accuracy. This is
because sufficient frequency space must exist between significant parts of the
$X(f)$ spectral component and the $X(f+f_s)$ and $X(f-f_s)$ spectral components
of the discrete-time-signal spectrum $X_d(f)$ to allow the practical low-pass filter to change from nearly constant passband gain over the significant part of
$X(f)$ to nearly zero stopband gain over the significant parts of $X(f-f_s)$ and
$X(f+f_s)$.

11.3 The Discrete-Time Fourier Transform and Spectra of Energy Signals

In computing the spectrum of a discrete-time signal in the previous section,
we used the representation of a discrete-time signal as an ideally sampled
continuous-time signal, continuous-time Fourier transforms, and the multiplication theorem for Fourier transforms. Note that both Fourier transforms and
Fourier transforms in the limit are admitted. The result is an infinite sum of
frequency-shifted continuous-frequency functions. An alternative method for
computing this spectrum for energy signals is the *discrete-time Fourier transform* (DTFT).

The Discrete-Time Fourier Transform

We derive the discrete-time Fourier transform by computing the Fourier transform of the ideally sampled continuous-time representation $x_s(t)$ of the discrete-
time signal $x_T(n)$ directly, rather than by using the multiplication theorem.
Performing the direct Fourier transform of $x_s(t)$ gives

$$X_d(f) = \int_{-\infty}^{\infty} x_s(t)e^{-j2\pi ft}\, dt$$

$$= \int_{-\infty}^{\infty} \left[\sum_{n=-\infty}^{\infty} x_T(n)\delta(t - nT) \right] e^{-j2\pi ft}\, dt$$

$$= \sum_{n=-\infty}^{\infty} x_T(n) \int_{-\infty}^{\infty} e^{-j2\pi ft}\delta(t - nT)\, dt$$

$$= \sum_{n=-\infty}^{\infty} x_T(n)e^{-j2\pi nfT} \tag{11.9}$$

where the last step follows by the sifting integral. Equation (11.9) defines the discrete-time Fourier transform of the discrete-time signal $x_T(n)$.

The discrete-time Fourier transform and thus the spectrum of the discrete-time signal exist if the summation in eq. (11.9) converges. The summation converges in the mean-square-error sense if the discrete-time signal $x_T(n)$ corresponds to an energy signal. That is, if

$$E_{dx} = T \sum_{n=-\infty}^{\infty} |x_T(n)|^2 < \infty \tag{11.10}$$

This is similar to the existence condition stated for Fourier transforms (eq. [4.42] in Section 4.4).

If, in addition, the signal is absolutely summable; that is, if

$$\sum_{n=-\infty}^{\infty} |x_T(n)| < \infty \tag{11.11}$$

then eq. (11.9) converges absolutely and uniformly to a continuous function of frequency that is the discrete-time Fourier transform and spectrum of the corresponding discrete-time signal. Eq. (11.11) is similar to the first Dirichlet condition for continuous-time signals (see Section 4.4). The second Dirichlet condition is always satisfied for discrete-time signals because the sample rate is finite, and thus there is a finite number of samples in any finite interval. The third Dirichlet condition is not applicable to discrete-time signals because discontinuities do not exist in discrete-time signals. Note that an absolutely summable discrete-time signal is also a finite-energy signal. Although the reverse is not always true, most finite-energy discrete-time signals of interest are also absolutely summable.

Example 11.4

The discrete-time echo signal of Example 11.1 defined by

$$x_T(n) = \left[e^{-0.5n}u(n), \quad 0.1 \text{ s} \right]$$

is absolutely summable because eq. (11.11) is the sum of a convergent geometric series for this signal. Use the discrete-time Fourier transform to find the amplitude and phase spectra for the echo signal.

Solution
Since $T = 0.1$, the discrete-time signal can be written as

$$x_T(n) = \left[e^{-5nT}u(n), \quad 0.1 \text{ s}\right]$$

and is plotted in Figure 11.11.

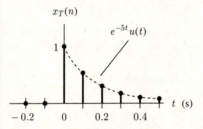

Figure 11.11 Echo Signal for Example 11.4

Using the discrete-time Fourier transform, we compute the spectrum of $x_T(n)$ and obtain

$$X_d(f) = \sum_{n=0}^{\infty} x_T(n)e^{-j2\pi nfT} = \sum_{n=0}^{\infty} e^{-(0.5+j0.2\pi f)n}$$

$$= \frac{1}{1 - \exp(-0.5 - j0.2\pi f)}$$

$$= \frac{1}{[1 - 0.607\cos(0.2\pi f)] + j0.607\sin(0.2\pi f)}$$

where the second step follows because the sum is the sum of a geometric series and because $\left|e^{-(0.5+j0.2\pi f)}\right| < 1$. Thus, the amplitude and phase spectra of $x_T(n)$ are

$$|X_d(f)| = 1 \Big/ \left\{[1 - 0.607\cos(0.2\pi f)]^2 + [0.607\sin(0.2\pi f)]^2\right\}^{0.5}$$

and

$$\underline{/X_d(f)} = -\tan^{-1}\{0.607\sin(0.2\pi f)/[1 - 0.607\cos(0.2\pi f)]\}$$

respectively. The amplitude and phase spectra produce the same plots as those shown in Figure 11.3, as they must.

Note that it is easier for us to plot the amplitude and phase spectra from the expressions found in Example 11.4 than from the expression found in Example 11.1. In Example 11.4, the infinite sum is not required. This easier plotting occurs when the convergent infinite sum of the discrete-time Fourier transform can be expressed in closed form. If the sum cannot be expressed in closed form, then an infinite sum is required. In some of these cases, if an $X(f)$ can be found easily, it may be convenient to use the expression found by the technique of Section 11.1 to obtain the data for spectral plots. This is particularly true if $X(f) = 0$ for $|f| \geq f_s/2$, in which case the infinite sum of eq. (11.2) consists of nonoverlapping terms.

Example 11.5

The discrete-time signal defined by

$$y_T(n) = \left[\left\{ \begin{matrix} 1 & |n| \leq 2 \\ 0 & |n| > 2 \end{matrix} \right\}, \ 0.2 \text{ s} \right]$$

consists of samples of the continuous-time rectangular pulse, the amplitude of which is 1 and the width of which is 1 s. Find the amplitude and phase spectra for this signal.

Solution
The discrete-time signal is plotted in Figure 11.12.

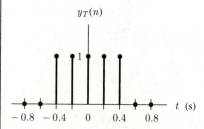

Figure 11.12 Discrete-Time Signal for Example 11.5

Using the discrete-time Fourier transform to compute the spectrum of $y_T(n)$, we obtain

$$Y_d(f) = \sum_{n=-\infty}^{\infty} y_T(n)e^{-j2\pi nfT}$$

$$= \sum_{n=-2}^{2} e^{-j0.4\pi nf}$$

$$= \sum_{n=-2}^{0} \left[e^{-j0.4\pi nf} \right] + \sum_{n=0}^{2} \left[e^{-j0.4\pi nf} \right] - 1$$

$$= \sum_{n=0}^{2} \left[e^{j0.4\pi nf} \right] + \sum_{n=0}^{2} \left[e^{-j0.4\pi nf} \right] - 1$$

$$= \frac{1 - e^{j1.2\pi nf}}{1 - e^{j0.4\pi nf}} + \frac{1 - e^{-j1.2\pi nf}}{1 - e^{-j0.4\pi nf}} - 1$$

We multiply the numerator and denominator of the first term by $-e^{-j0.2\pi f}$ and the numerator and denominator of the second term by $e^{j0.2\pi f}$ to obtain

$$Y_d(f) = \frac{-e^{-j0.2\pi nf} + e^{j\pi f}}{-e^{-j0.2\pi nf} + e^{j0.2\pi f}} + \frac{e^{j0.2\pi f} - e^{-j\pi f}}{e^{j0.2\pi f} - e^{-j0.2\pi f}} - 1$$

and simplify the result to give

$$Y_d(f) = \frac{e^{j\pi f} - e^{-j\pi f}}{e^{j0.2\pi f} - e^{-j0.2\pi f}} = \frac{\sin(\pi f)}{\sin(0.2\pi f)}$$

Therefore, the amplitude and phase spectra of $x_T(n)$ are

$$|Y_d(f)| = |\sin(\pi f) / \sin(0.2\pi f)|$$

and

$$\angle Y_d(f) = \angle \sin(\pi f) / \sin(0.2)\pi f)$$

The amplitude and phase spectra are plotted in Figure 11.13. If we select $\alpha = 1/\sqrt{2}$, then the bandwidth of $x_T(n)$ is 0.451 Hz. This signal bandwidth is illustrated in Figure 11.13.

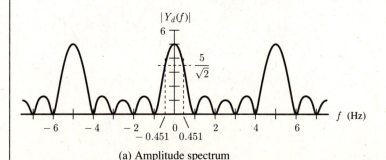

(a) Amplitude spectrum

Figure 11.13 Amplitude and Phase Spectra of the Discrete-Time Signal in Example 11.5

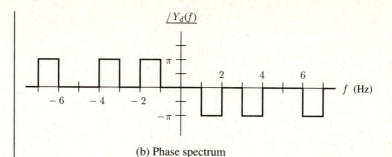

(b) Phase spectrum

Figure 11.13 Amplitude and Phase Spectra of the Discrete-Time Signal in Example 11.5 *(continued)*

Example 11.6

Find the amplitude and phase spectra for the single-pulse signal $x_T(n) = [A\delta(n), \quad T_1]$.

Solution
The single-pulse signal is plotted in Figure 11.14.

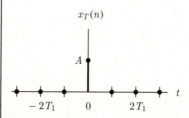

Figure 11.14 Single-Pulse Signal

The spectrum of the single-pulse signal is

$$X_d(f) = \sum_{n=-\infty}^{\infty} x_T(n)e^{-j2\pi nfT_1}$$

$$= \sum_{n=-\infty}^{\infty} A\delta(n)e^{-j2\pi nfT_1}$$

$$= A$$

which gives the amplitude and phase spectra

$$|X_d(f)| = A$$

and

$$\underline{/\,X_d(f)} = \underline{/\,A} = \left\{ \begin{array}{ll} 0 & A > 0 \\ \pm\pi & A < 0 \end{array} \right\}$$

In the previous three examples, we computed the discrete-time Fourier transform and the amplitude and phase spectra for three very useful signals. The discrete-time Fourier transforms for a number of discrete-time signals are listed in Table C.6 of Appendix C. Those signals in the table that are not energy signals will be discussed in a subsequent section.

If the discrete-time Fourier transform (spectrum) of a discrete-time signal exists, then we can find the discrete-time signal from its spectrum by using an inverse discrete-time Fourier transform, just as we found a continuous-time signal from its spectrum with the inverse Fourier transform. We will now show how the inverse discrete-time Fourier transform is derived.

Since the spectrum $X_d(f)$ of a discrete-time signal is periodic in f with period equal to f_s, then we can express it by a complex-exponential Fourier Series in the variable f. Therefore,

$$X_d(f) = \sum_{n=-\infty}^{\infty} a_n e^{-j2\pi n(1/f_s)f}$$

$$= \sum_{n=-\infty}^{\infty} a_n e^{-j2\pi n f T} \tag{11.12}$$

where

$$a_n = \frac{1}{f_s} \int_{-f_s/2}^{f_s/2} X_d(f) e^{j2\pi n f T}\, df \tag{11.13}$$

Since the complex-exponential Fourier series of a periodic function is unique, then comparison of eqs. (11.12) and (11.9) shows us that the discrete-time signal $x_T(n)$ corresponding to the spectrum (that is, the discrete-time Fourier transform) $X_d(f)$ is $x_T(n) = a_n$. Therefore, from eq. (11.13), the inverse discrete-time Fourier transform is

$$x_T(n) = \frac{1}{f_s} \int_{-f_s/2}^{f_s/2} X_d(f) e^{j2\pi n f T}\, df \tag{11.14}$$

Equations (11.9) and (11.14) are the discrete-time Fourier transform and the inverse discrete-time Fourier transform, respectively, of the discrete-time signal $x_T(n)$. These equations are referred to as the *discrete-time Fourier transform pair*. Thus, the discrete-time Fourier transform pair is

$$X_d(f) = \sum_{n=-\infty}^{\infty} x_T(n) e^{-j2\pi n f T} \equiv \mathcal{F}_d[x_T(n)] \tag{11.15}$$

and

$$x_T(n) = \frac{1}{f_s} \int_{-f_s/2}^{f_s/2} X_d(f) e^{j2\pi n fT} \, \equiv \mathcal{F}_d^{-1}\left[X_d(f)\right] \qquad (11.16)$$

We frequently use

$$x_T(n) \leftrightarrow X_d(f) \qquad (11.17)$$

to indicate a discrete-time Fourier transform pair. This notation parallels the form used for other transform pairs.

The Energy-Density Spectrum

To define the energy-density spectrum for a discrete-time energy signal, we express the energy in the signal as

$$E_{dx} = T \sum_{n=-\infty}^{\infty} |x_T(n)|^2 = T \sum_{n=-\infty}^{\infty} x_T^*(n) x_T(n)$$

$$= T \sum_{n=-\infty}^{\infty} x_T^*(n) \left[\frac{1}{f_s} \int_{-f_s/2}^{f_s/2} X_d(f) e^{j2\pi n fT} \, df \right]$$

$$= T^2 \int_{-f_s/2}^{f_s/2} X_d(f) \left[\sum_{n=-\infty}^{\infty} x_T^*(n) e^{j2\pi n fT} \right] df$$

$$= T^2 \int_{-f_s/2}^{f_s/2} X_d(f) \left[\sum_{n=-\infty}^{\infty} x_T(n) e^{-j2\pi n fT} \right]^* df$$

$$= T^2 \int_{-f_s/2}^{f_s/2} X_d(f) X_d^*(f) \, df = T^2 \int_{-f_s/2}^{f_s/2} |X_d(f)|^2 \, df \qquad (11.18)$$

Thus,

$$E_{dx} = T \sum_{n=-\infty}^{\infty} |x_T(n)|^2 = \int_{-f_s/2}^{f_s/2} T^2 |X_d(f)|^2 \, df \qquad (11.19)$$

Definition

The energy-density spectrum of the discrete-time energy signal $x_T(n)$ is

$$G_{dx}(f) = T^2 |X_d(f)|^2 \qquad (11.20)$$

We call $G_{dx}(f)$ the energy-density spectrum of the discrete-time signal because it gives the total energy in the signal when it is integrated over the unambiguous frequencies in the discrete-time signal; that is, over the interval $|f| \leq f_s/2 = 1/2T$. Since it can be shown that

$$\int_{-f_2}^{-f_1} G_{dx}(f) \, df + \int_{f_1}^{f_2} G_{dx}(f) \, df \tag{11.21}$$

is the energy contained in the signal in the range of unambiguous frequencies $f_1 < f < f_2$, where $f_2 \leq f_s/2$, this definition for the energy-density spectrum is appropriate. (Recall that the positive and negative portions of a double-sided spectrum go together to produce signal characteristics at a given frequency.) Note that eq. (11.19) is a form of Parseval's theorem (see eq. [3.79] in Chapter 3).

Example 11.7

Find and plot the energy-density spectra for the discrete-time echo signal of Example 11.1 and the discrete-time rectangular-pulse signal of Example 11.5. Note that the energy-density spectra exist because both signals are energy signals.

Solution

The spectrum for the echo signal of Example 11.1 was computed in closed form in Example 11.4. The signal spectrum is

$$X_d(f) = \frac{1}{1 - \exp(-0.5 - j0.2\pi f)}$$

Thus, the energy-density spectrum for the echo signal of Example 11.1 is

$$G_{dx}(f) = T^2 |X_d(f)|^2$$
$$= 0.01 / \left\{ [1 - 0.607 \cos(0.2\pi f)]^2 + [0.607 \sin(0.2\pi f)]^2 \right\}$$

This energy-density spectrum is plotted in Figure 11.15a for the range of frequencies $|f| \leq f_s/2 = 1/2T = 5$ Hz.

For the signal of Example 11.5, $T = 0.2$ s and the signal spectrum is

$$Y_d(f) = \frac{\sin(\pi f)}{\sin(0.2\pi f)}$$

Therefore, the energy-density spectrum is

$$G_{dy}(f) = T^2 |Y_d(f)|^2$$
$$= (0.2)^2 \left| \frac{\sin(\pi f)}{\sin(0.2\pi f)} \right|^2$$
$$= \frac{0.04 \sin^2(\pi f)}{\sin^2(0.2\pi f)}$$

This energy-density spectrum is plotted in Figure 11.15b for the range of frequencies $|f| \leq f_s/2 = 1/2T = 2.5$ Hz.

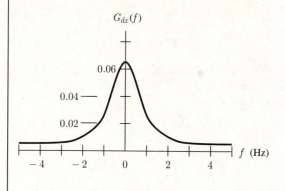

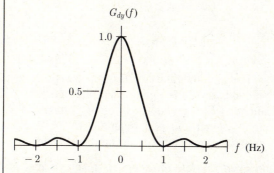

Figure 11.15 Energy-Density Spectra of the Discrete-Time Energy Signals in Examples 11.1 and 11.5

Discrete-Time Fourier Transforms and Spectrum Properties

The discrete-time Fourier transform possesses many properties and theorems that parallel or are similar to those for Fourier transforms. We will discuss these properties and theorems here as we did for the Fourier transform in Chapter 4, except that in this case we will omit the details and illustrations by example for properties and theorems that exactly parallel those for the Fourier transform.

Discrete-Time Fourier Transform Symmetry

Any real discrete-time signal can be written as the sum of a real and even signal and a real and odd signal. That is,

$$x_T(n) = x_{eT}(n) + x_{oT}(n) \tag{11.22}$$

where

$$x_{eT}(n) = [x_T(n) + x_T(-n)]/2 \qquad (11.23)$$

and

$$x_{oT}(n) = [x_T(n) - x_T(-n)]/2 \qquad (11.24)$$

Therefore, we can use a derivation that parallels the derivation in Chapter 4 to show that the discrete-time Fourier transform of any real discrete-time signal can be written as

$$X_d(f) = Re[X_d(f)] + jIm[X_d(f)] \qquad (11.25)$$

where

$$Re[X_d(f)] = \sum_{n=-\infty}^{\infty} x_{eT}(n)\cos(2\pi nfT) \qquad (11.26)$$

and

$$Im[X_d(f)] = -\sum_{n=-\infty}^{\infty} x_{oT}(n)\sin(2\pi nfT) \qquad (11.27)$$

Note that the real and imaginary parts of the transform result, respectively, from the even and odd parts of the signal. Also, since $\cos(2\pi nfT)$ is an even function of f, and $\sin(2\pi nfT)$ is an odd function of f, then $Re[X_d(f)]$ and $Im[X_d(f)]$ are, respectively, even and odd functions of f. Therefore, we can state the following properties associated with the discrete-time Fourier transform:

1. Real and even discrete-time signals result in real and even discrete-time Fourier transforms.
2. Real and odd discrete-time signals result in imaginary and odd discrete-time Fourier transforms.

In the first case, we can plot the total spectrum $X_d(f)$, rather than the amplitude and phase spectra, because $X_d(f)$ is real. A total spectrum plot for the signal of Example 11.5 is shown in Figure 11.16.

Figure 11.16
Total Spectrum of the Discrete-Time Signal in Example 11.5

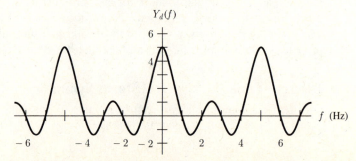

Since $Re[X_d(f)]$ and $Im[X_d(f)]$ are even and odd functions of f when $x_T(n)$ is real, then $|X_d(f)|$ and $\underline{/\,X_d(f)}$ are, respectively, even and odd functions of f when $x_T(n)$ is real. This symmetry parallels the symmetry of the Fourier transform of a real continuous-time signal.

Discrete-Time Fourier Transform Theorems

We now come to the theorems concerning discrete-time Fourier transforms. When the theorems are similar to those for Fourier transforms, but do not exactly parallel them, we will note the differences and, in some cases, illustrate the theorem with an example. These theorems are summarized in Table C.5 of Appendix C.

Theorem 1   Linearity
If

$$x_T(n) \leftrightarrow X_d(f) \quad \text{and} \quad y_T(n) \leftrightarrow Y_d(f)$$

then

$$ax_T(n) + by_T(n) \leftrightarrow aX_d(f) + bY_d(f) \tag{11.28}$$

The linearity theorem is an exact parallel of that for the Fourier transform; its proof is left for the reader in the Problem section of this chapter.

A change in time scale for a discrete-time signal is obtained by changing the sample spacing. Therefore, a time-scale change is indicated by changing $x_T(n)$ to $x_{T/a}(n)$, where a is a real constant.

Theorem 2   Scale Change
If

$$x_T(n) \leftrightarrow X_d(f)$$

then

$$x_{T/a}(n) \leftrightarrow X_d\left(f/a\right) \tag{11.29}$$

The proof of this theorem is given by

$$\mathcal{F}_d\big[x_{T/a}(n)\big] = \sum_{n=-\infty}^{\infty} x_{T/a}(n)e^{-j2\pi nf(T/a)}$$

$$= \sum_{n=-\infty}^{\infty} x_T(n)e^{-j2\pi n(f/a)T}$$

$$= X_d\left(f/a\right) \tag{11.30}$$

where the first step follows because the two signals have the same samples. This theorem is similar to the scale-change theorem for Fourier transforms, differing only by the absence of the factor $1/|a|$ in the transform of the time-scaled signal. The absence of $1/|a|$ is a consequence of two factors: (1) The sample spacing is scaled by the same factor as the discrete-time signal; thus, the samples are the same; and (2) the amplitude of the spectrum of a discrete-time signal is proportional to the inverse of the sample spacing. We will now illustrate these reasons for the absence of $1/|a|$ by an alternate proof of the scale-change theorem.

We can consider the discrete-time signal $x_T(n)$ to be samples of the continuous-time signal $x(t)$ taken at the spacing T. Thus, the discrete-time signal $x_{T/a}(n)$ consists of samples of the continuous-time signal $x(at)$ taken at the spacing T/a because the sequence values of $x_{T/a}(n)$ are the same as those for $x_T(n)$. The sample frequency for $x_T(n)$ is $f_s = 1/T$ and the sample frequency for $x_{T/a}(n)$ is $f_{sa} = |a/T|$, where the absolute value is necessary since a negative sign for the scaling factor a merely inverts the continuous-time signal $x(at)$ on the time axis and does not affect the sample frequency. Thus, by using eq. (11.2), we can write the discrete-time Fourier transform (or spectrum) of $x_{T/a}(n)$ as

$$\mathcal{F}_d[x_{T/a}(n)] = \sum_{m=-\infty}^{\infty} f_{sa}\left\{\mathcal{F}[x(at)]\right\}_{f_1=f-mf_{sa}}$$

$$= \sum_{m=-\infty}^{\infty} \left|\frac{a}{T}\right|\left\{\mathcal{F}[x(at)]\right\}_{f_1=f-m|a/T|} \qquad (11.31)$$

where f_1 is the frequency variable used here for $\mathcal{F}[x(at)]$. From Theorem 2 for Fourier transforms,

$$\mathcal{F}[x(at)] = \frac{1}{|a|}X(f_1/a) \qquad (11.32)$$

Therefore,

$$\mathcal{F}_d[x_{T/a}(n)] = \sum_{m=-\infty}^{\infty} \frac{1}{|T|}X[(f - m\,|a/T|)\,/a]$$

$$= \sum_{m=-\infty}^{\infty} \frac{1}{|T|}X\{[f - (ma/T)]\,/a\}$$

$$= \sum_{m=-\infty}^{\infty} f_s X[(f/a) - mf_s]$$

$$= \{\mathcal{F}_d[x_T(n)]\}_{f_2=f/a} = X_d(f/a) \qquad (11.33)$$

where T is positive and f_2 is the frequency variable used here for $\mathcal{F}_d[x_T(n)]$. The first step is valid for all $a \neq 0$ (that is, positive or negative), the order used in performing the summation being unimportant; the last step follows from eq. (11.2).

Example 11.8

Assume that the walls producing the gunshot echo in Example 11.1 are 1.25 times as far apart as they were in that example. Now, the echo signal is

$$z_T(n) = \left[e^{-0.5n}u(n), \quad 0.125 \text{ s}\right]$$

the echo amplitudes being the same, but in this case separated by 0.125 s rather than by 0.1 s. Find the spectrum of the changed echo signal.

Solution

We can write the discrete-time echo signal, defined $z_T(n)$, as

$$z_T(n) = \left[e^{-4nT}u(n), \quad 0.125 \text{ s}\right]$$

This signal is plotted in Figure 11.17.

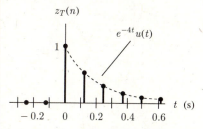

Figure 11.17 Echo Signal for Example 11.8

The discrete-time signal defined in Example 11.1 is

$$x_{T_1}(n) = \left[e^{-0.5n}u(n), \quad 0.1 \text{ s}\right] = [x(n), \quad 0.1 \text{ s}]$$

Therefore, we can write $z_T(n)$ as

$$z_T(n) = [x(n), \quad 0.125 \text{ s}]$$

or, equivalently, as

$$z_T(n) = \left[x(n), \quad \frac{0.1}{0.8} \text{ s}\right] = x_{T_1/a}(n)$$

where $T_1 = 0.1$ s and $a = 0.8$. Therefore, we use the scaling theorem with $a = 0.8$ and the spectrum $X_d(f)$ that we found for $x_T(n)$ in Example 11.4 to obtain

$$Z_d(f) = X_d(f/0.8)$$

$$= 1\big/\left[1 - \exp(-0.5 - j0.25\pi f)\right]$$

as the expression for the spectrum of $z_T(n)$.

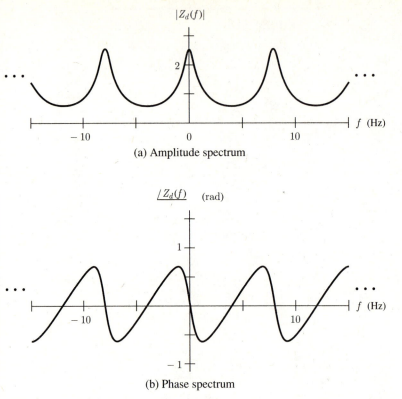

(a) Amplitude spectrum

(b) Phase spectrum

Figure 11.18 Amplitude and Phase Spectra of the Discrete-Time Echo Signal in Example 11.8

The corresponding amplitude and phase spectra for $z_T(n)$ are plotted in Figure 11.18. Note that they are much like the spectra in Figure 11.3, except these are compressed along the frequency axis by the factor 0.8. The discrete-time signal in this example is much like the discrete-time signal of Example 11.1, except that this signal is expanded along the time axis by the factor 1.25; hence, the compression along the frequency axis in this case.

Theorem 3   Time Reversal

If
$$X_T(n) \leftrightarrow X_d(f)$$

then
$$x_{-T}(n) \leftrightarrow X_d(-f) \tag{11.34}$$

This is a special case of the scale change that occurs when $a = -1$. Since the subscript $-T$ indicates that the sample spacing is unchanged but the signal is reversed in time, we can alternatively write this theorem in simpler form, as follows:

If

$$x_T(n) \leftrightarrow X_d(f)$$

then

$$x_T(-n) \leftrightarrow X_d(-f) \tag{11.35}$$

Theorem 4 Complex Conjugation
If

$$x_T(n) \leftrightarrow X_d(f)$$

then

$$x_T^*(n) \leftrightarrow X_d^*(-f) \tag{11.36}$$

This theorem parallels a Fourier transform theorem exactly; hence we leave its proof for the reader in the Problem section.

Following the complex-conjugation theorem for the Fourier transform in Chapter 4 is the duality theorem. No such theorem exists for the discrete-time Fourier transform, since the signal is discrete-time while the transform is continuous-frequency.

Theorem 5 Time Shift
If

$$x_T(n) \leftrightarrow X_d(f)$$

then

$$x_T(n - n_1) \leftrightarrow X_d(f)e^{-j2\pi n_1 fT} \tag{11.37}$$

Since the proof of this theorem parallels that for the similar theorem for Fourier transforms, it is left for the reader in the end-of-chapter problems. Note that theorem 5 differs from the like-named theorem for Fourier transforms in that it considers only time shifts that are integer multiples of the sample spacing.

Theorem 6 Frequency Translation
If

$$x_T(n) \leftrightarrow X_d(f)$$

then

$$e^{j2\pi n f_0 T} x_T(n) \leftrightarrow X_d(f - f_0) \tag{11.38}$$

This theorem is the exact parallel of the like-named Fourier transform theorem; its proof is required in the problems.

Theorem 7 Modulation
If

$$x_T(n) \leftrightarrow X_d(f)$$

then

$$x_T(n) \cos[2\pi n f_0 T] \leftrightarrow \tfrac{1}{2} X_d(f - f_0) \tfrac{1}{2} X_d(f + f_0) \qquad (11.39)$$

Again, this theorem is an exact parallel of a Fourier transform theorem and is proved in a similar fashion.

Time-differentiation is not a discrete-time-system operation. Therefore, a time-differentiation theorem does not exist for the discrete-time Fourier transform. However, since

$$\frac{dx(t)}{dt}\bigg|_{t=nT} = \lim_{T \to 0} \frac{x[nT] - x[(n - 1)T]}{T} \qquad (11.40)$$

then the *differencing operation* $x_T(n) - x_T(n - 1)$ for discrete-time signals plays a role similar to that of the differentiation operation for continuous-time signals. Therefore, a time-differencing theorem is of interest.

Theorem 8   Time Differencing

If

$$x_T(n) \leftrightarrow X_d(f)$$

then

$$x_T(n) - x_T(n - 1) \leftrightarrow \left(1 - e^{-j2\pi fT}\right) X_d(f) \qquad (11.41)$$

The proof of this theorem is given by

$$\mathcal{F}_d[x_T(n) - x_T(n - 1)] = \sum_{n=-\infty}^{\infty} [x_T(n) - x_T(n - 1)] e^{-j2\pi n fT}$$

$$= \sum_{n=-\infty}^{\infty} x_T(n) e^{-j2\pi n fT} - \sum_{n=-\infty}^{\infty} x_T(n - 1) e^{-j2\pi n fT} \qquad (11.42)$$

Using the linearity and time-shift theorems, we obtain

$$\mathcal{F}_d[x_T(n) - x_T(n - 1)] = \left(1 - e^{-j2\pi fT}\right) X_d(f) \qquad (11.43)$$

This step completes the proof. Since $X_d(f)$ exists for all f when the signal has finite energy, then $X_d(0)$ exists (that is, is finite). Thus, the spectrum of the signal corresponding to the sequence $[x_T(n) - x_T(n - 1)]$ is zero at $f = 0$ when $x_T(n)$ is an energy signal.

Example 11.9

Find the spectrum of the signal

$$y_T(n) = \{[x(n) - x(n - 1)], \quad 2 \text{ s}\}$$

when $x(n) = \delta(n) + \delta(n - 1) + \delta(n - 2)$.

Solution

$$X_d(f) = \sum_{n=-\infty}^{\infty} x_T(n)e^{-j2\pi nfT}$$

$$= 1 + e^{-j4\pi f} + e^{-j8\pi f}$$

$$= \left(1 - e^{-j12\pi f}\right) / \left(1 - e^{-j4\pi f}\right)$$

since

$$\sum_{n=0}^{N} K^n = \left(1 - K^{N+1}\right)/(1 - K)$$

Using the time-differencing theorem, we obtain

$$Y_d(f) = \left(1 - e^{-j4\pi f}\right)X_d(f) = 1 - e^{-j12\pi f}$$

As a check, note that

$$y_T(n) = \mathcal{F}_d^{-1}[Y_d(f)] = \delta_T(n) - \delta_T(n - 3)$$

and

$$x_T(n) - x_T(n - 1) = [\delta_T(n) + \delta_T(n - 1) + \delta_T(n - 2)]$$

$$- [\delta_T(n - 1) + \delta_T(n - 2) + \delta_T(n - 3)]$$

$$= \delta_T(n) - \delta_T(n - 3)$$

Sample summation plays a role for discrete-time signals similar to that of integration for continuous-time signals. Therefore, the time-integration theorem for Fourier transforms is replaced by the summation theorem for discrete-time Fourier transforms.

Theorem 9 Summation
If

$$x_T(n) \leftrightarrow X_d(f)$$

then

$$\sum_{i=-\infty}^{n} x_T(i) \leftrightarrow \frac{X_d(f)}{(1 - e^{-j2\pi fT})} + \frac{f_s X_d(0)}{2} \sum_{m=-\infty}^{\infty} \delta(f - mf_s) \qquad (11.44)$$

The impulse functions in the spectrum occur because the summation of a finite-energy signal may produce a signal with infinite energy.

We indicated previously that the spectrum of $y_T(n) = x_T(n) - x_T(n-1)$ is zero for $f = 0$ if $x_T(n)$ is an energy signal. Therefore, we have the following corollary to the summation theorem:

If

$$y_T(n) = x_T(n) - x_T(n-1) \quad \text{and} \quad \sum_{n=-\infty}^{\infty} |x_T(n)|^2 < \infty$$

then

$$\mathcal{F}_d\left[\sum_{i=-\infty}^{n} y_T(i)\right] = Y_d(f)/\left(1 - e^{-j2\pi fT}\right) \tag{11.45}$$

We see that if $y_T(n) = x_T(n) - x_T(n-1)$, then

$$\sum_{i=-\infty}^{n} y_T(i) = \sum_{i=-\infty}^{n} [x_T(i) - x_T(i-1)]$$

$$= \sum_{i=-\infty}^{n} x_T(i) - \sum_{i=-\infty}^{n} x_T(i-1)$$

$$= \sum_{i=-\infty}^{n} x_T(i) - \sum_{i=-\infty}^{n-1} x_T(i)$$

$$= x_T(n) \tag{11.46}$$

Therefore, the time-differencing and summation operations are inverse discrete-time operations.

Theorem 10   Convolution
If

$$x_T(n) \leftrightarrow X_d(f) \quad \text{and} \quad y_T(n) \leftrightarrow Y_d(f)$$

then

$$z_T(n) = x_T(n) * y_T(n) \leftrightarrow X_d(f)Y_d(f) \tag{11.47}$$

This important theorem exactly parallels a theorem for Fourier transforms. It can be proved in the manner of the Fourier transform theorem proof, the integrals being replaced by summations.

Example 11.10

A stepping motor adjusts the angular position of an airflow control vane. The discrete-time signal representing the angular location of the air vane in degrees is

$$x_T(n) = \left[\left\{\begin{matrix} 5(1 - |n/5|) & |n| \leq 5 \\ 0 & |n| > 5 \end{matrix}\right\}, \ 0.2 \ \text{s}\right]$$

Find the spectrum for $x_T(n)$ and plot the amplitude and phase spectra or, if possible, the total spectrum of $x_T(n)$.

Solution

We show the plot of the angular position signal $x_T(n)$ in Figure 11.19.

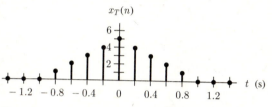

Figure 11.19 Air-Vane Position Signal for Example 11.10

Note that, if

$$y_T(n) = \left[\left\{ \begin{matrix} 1 & |n| \le 2 \\ 0 & |n| > 2 \end{matrix} \right\}, \quad 0.2 \text{ s} \right]$$

then

$$x_T(n) = y_T(n) * y_T(n) \quad \text{and} \quad X_d(f) = Y_d^2(f)$$

Since $y_T(n)$ is the signal of Example 11.5, then from Example 11.5

$$Y_d(f) = \frac{\sin(\pi f)}{\sin(0.2\pi f)} \quad \text{and} \quad X_d(f) = \frac{\sin^2(\pi f)}{\sin^2(0.2\pi f)}$$

The signal $x_T(n)$ is real and even; therefore, $X_d(f)$ is real and we can plot the total spectrum of $x_T(n)$. This plot is shown in Figure 11.20. If the definition of a significant amplitude-spectrum value is taken to be one that is greater than $1/\sqrt{2}$ times the maximum spectrum value, then $\alpha = 1/\sqrt{2}$ and the bandwidth of $x_T(n)$ is 0.325 Hz. Note that the bandwidth of $x_T(n)$ is narrower than the bandwidth of $y_T(n)$ (see Example 11.5) because $x_T(n)$ changes less rapidly than $y_T(n)$.

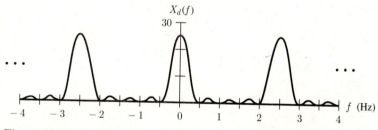

Figure 11.20 Total Spectrum of the Air-Vane Position Signal in Example 11.10

Theorem 11   Multiplication

If

$$x_T(n) \leftrightarrow X_d(f) \quad \text{and} \quad y_T(n) \leftrightarrow Y_d(f)$$

then

$$x_T(n)y_T(n) \leftrightarrow \frac{1}{f_s} \int_{-f_s/2}^{f_s/2} X_d(\alpha) Y_d(f - \alpha) \, d\alpha \qquad (11.48)$$

To prove the multiplication theorem, we compute the discrete-time Fourier transform of the product of signals. This computation gives

$$\mathcal{F}[x_T(n)y_T(n)] = \sum_{n=-\infty}^{\infty} x_T(n)y_T(n)e^{-j2\pi nfT}$$

$$= \sum_{n=-\infty}^{\infty} y_T(n) \left[\frac{1}{f_s} \int_{-f_s/2}^{f_s/2} X_d(\alpha)e^{j2\pi n\alpha T} \, d\alpha \right] e^{-j2\pi nfT}$$

$$= \frac{1}{f_s} \int_{-f_s/2}^{f_s/2} X_d(\alpha) \left[\sum_{n=-\infty}^{\infty} y_T(n)e^{-j2\pi n(f-\alpha)T} \right] d\alpha$$

$$= \frac{1}{f_s} \int_{-f_s/2}^{f_s/2} X_d(\alpha) Y_d(f - \alpha) \, d\alpha \qquad (11.49)$$

This proves the theorem. Note that the spectrum of $x_T(n)y_T(n)$ is produced by an operation similar to continuous-time convolution, except that in this case the integral is only over one period of the periodic-spectrum product.

11.4 The Discrete-Time Fourier Transform in the Limit and Spectra of Power Signals

We found in Chapter 4 that we could not use the Fourier transform to find the spectrum of continuous-time power signals because power signals do not have finite energy. This is also true for the discrete-time Fourier transform and discrete-time power signals. However, proceeding as we did for continuous-time power signals in Chapter 4, we can find the spectrum of a discrete-time power signal if impulse functions are permitted in the continuous-frequency spectrum and a discrete-time Fourier transform in the limit is used. The concept of a discrete-time Fourier transform in the limit is illustrated by the following example.

Example 11.11

Find the spectrum of a sequence of constant values using the discrete-time Fourier transform in the limit.

Solution

The discrete-time signal that is a sequence of constant values is

$$x_T(n) = [A, \quad T]$$

and is plotted in Figure 11.21.

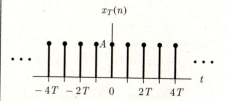

$x_T(n)$

$-4T \quad -2T \quad 0 \quad 2T \quad 4T$ t

Figure 11.21 Constant-Valued Discrete-Time Signal

To find the discrete-time Fourier transform in the limit of $x_T(n)$, we represent $x_T(n)$ by

$$x_T(n) = \lim_{a \to 0} y_T(n)$$

where $y_T(n) = [Ae^{-a|n|}, \quad T]$ and $a > 0$. Then, we compute the discrete-time Fourier transform in the limit of $x_T(n)$ with

$$X_d(f) = \lim_{a \to 0} Y_d(f)$$

Since

$$Y_d(f) = \sum_{n=-\infty}^{\infty} Ae^{-a|n|}e^{-j2\pi nfT}$$

$$= \sum_{n=-\infty}^{-1} Ae^{an}e^{-j2\pi nfT} + \sum_{n=0}^{\infty} Ae^{-an}e^{-j2\pi nfT}$$

$$= \sum_{n=1}^{\infty} Ae^{-an}e^{j2\pi nfT} + \sum_{n=0}^{\infty} Ae^{-an}e^{-j2\pi nfT}$$

$$= A + \sum_{n=1}^{\infty} 2Ae^{-an} \cos 2\pi nfT$$

then

$$X_d(f) = \lim_{a \to 0} Y_d(f) = A + \sum_{n=1}^{\infty} 2A \cos 2\pi nfT \qquad (11.50)$$

We can write eq. (11.50) as

$$X_d(f) = Af_s \left[T + \sum_{n=1}^{\infty} 2T \cos 2\pi nfT \right] \qquad (11.51)$$

since $f_s = 1/T$. If we replace the variable t by the variable f and the constant T_s by the constant $1/T$ in eq. (4.108) and in the discussion following eq. (4.108), then we see that the term in brackets in eq. (11.51) can be replaced by

$$\sum_{m=-\infty}^{\infty} \delta\left(f - \frac{m}{T}\right) = \sum_{m=-\infty}^{\infty} \delta(f - mf_s)$$

as the number of cosines summed approaches infinity. Therefore, the spectrum of a discrete-time signal that is a sequence of constant values is

$$X_d(f) = Af_s \sum_{m=-\infty}^{\infty} \delta(f - mf_s)$$

This spectrum is plotted in Figure 11.22.

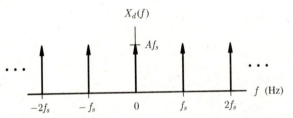

Figure 11.22 Spectrum of Constant-Valued Discrete-Time Signal

We can also find the spectrum for a discrete-time signal that is a sequence of constant values by using the signal's representation as an ideally sampled continuous-time signal

$$x_s(t) = \sum_{n=-\infty}^{\infty} x_T(n)\delta(t - nT) = A\delta_s(t) \qquad (11.52)$$

We find the spectrum of $x_s(t)$ to be

$$X_d(f) = \mathcal{F}[x_s(t)] = Af_s \sum_{n=-\infty}^{\infty} \delta(f - nf_s) \qquad (11.53)$$

from the Fourier transform for $\delta_s(t)$ in Table C.2 of Appendix C. This is the same spectrum we found by the discrete-time Fourier transform in the limit, as it must be.

Since the discrete-time Fourier transform in the limit obeys all the properties of the discrete-time Fourier transform, the spectra for other discrete-time power signals are derived from the properties and theorems already shown, even though we know that they were derived for energy signals. The discrete-time

Fourier transforms in the limit that are obtained are referred to as discrete-time Fourier transforms for simplicity.

Discrete-Time Fourier Transforms and Spectra for Basic Power Signals

We will now consider the discrete-time Fourier transforms (spectra) for two basic power signals—the unit step and the cosine.

Signal 1   Unit Step

$$u_T(n) \leftrightarrow \frac{1}{1 - e^{-j2\pi fT}} + \frac{f_s}{2} \sum_{m=-\infty}^{\infty} \delta(f - mf_s) \qquad (11.54)$$

The discrete-time Fourier transform pair given by eq. (11.54) can be easily verified by using the summation theorem, keeping in mind that $u_T(n) = \sum_{n=0}^{\infty} \delta_T(n)$ and $\delta_T(n) \leftrightarrow 1$.

Example 11.12

Find the spectrum of the time-shifted step signal

$$y_T(n) = 3u_T(n - 2)$$

Solution

We use eq. (11.54) and the time-shift theorem to find

$$Y_d(f) = 3\left[\frac{1}{1 - e^{-j2\pi fT}} + \frac{f_s}{2} \sum_{m=-\infty}^{\infty} \delta(f - mf_s) \right] e^{-j2\pi(2)fT}$$

$$= \frac{3e^{-j4\pi fT}}{1 - e^{-j2\pi fT}} + \frac{f_s}{2} \sum_{m=-\infty}^{\infty} e^{-j4\pi fT} \delta(f - mf_s)$$

$$= \frac{3e^{-j4\pi fT}}{1 - e^{-j2\pi fT}} + \frac{f_s}{2} \sum_{m=-\infty}^{\infty} e^{-j4\pi m} \delta(f - mf_s)$$

$$= \frac{3e^{-j4\pi fT}}{1 - e^{-j2\pi fT}} + \sum_{m=-\infty}^{\infty} \frac{f_s}{2} \delta(f - mf_s)$$

as the expression for the spectrum of $y_T(n)$. The amplitude and phase spectra of $y_T(n)$ are plotted in Figure 11.23. Note that these consist of two separate superimposed plots in each case. One of these is for the first term of $Y_d(f)$ and the other is for the impulses of the second term. The phase angle associated with the impulses is zero because $f_s/2$ is real and positive.

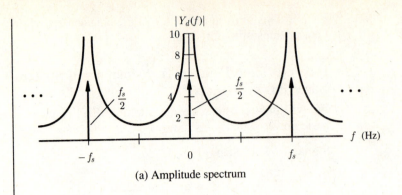

(a) Amplitude spectrum

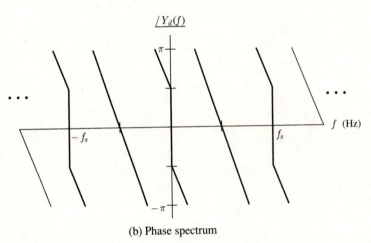

(b) Phase spectrum

Figure 11.23 Amplitude and Phase Spectra of the Discrete-Time Signal in Example 11.12

Signal 2   Cosine

$$\cos(2\pi n f_a T + \theta) \leftrightarrow \sum_{m=-\infty}^{\infty} \frac{f_s}{2}\left[e^{j\theta}\delta(f - f_a - mf_s)\right.$$

$$\left. + e^{-j\theta}\delta(f + f_a - mf_s)\right] \qquad (11.55)$$

We verify this discrete-time Fourier transform pair by using the equivalence

$$\cos(2\pi n f_a T + \theta) = \frac{1}{2}e^{j\theta}e^{2\pi n f_a T} + \frac{1}{2}e^{-j\theta}e^{-j2\pi n f_a T}$$

the discrete-time Fourier transform pair

$$\frac{1}{2}e^{\pm j\theta} \leftrightarrow \sum_{m=-\infty}^{\infty} \frac{f_s e^{\pm j\theta}}{2}\delta(f - mf_s)$$

from Example 11.11, and the frequency-translation theorem.

Example 11.13

Find the spectrum of the discrete-time cosine signal

$$x_T(n) = [3\cos(0.6\pi n - \pi/3), \quad 0.2 \text{ s}]$$

Note that this is the signal of Example 11.2.

Solution

Since $T = 0.2$ s, then we can write the discrete-time signal as

$$x_T(n) = \{3\cos[2\pi(1.5)nT - \pi/3], \quad 0.2 \text{ s}\}$$

Therefore, eq. (11.55) gives the spectrum of $x_T(n)$ as

$$X_d(f)$$

$$= \sum_{m=-\infty}^{\infty} \left[\left(7.5e^{-j\pi/3}\right)\delta(f - 1.5 - 5m) + \left(7.5e^{j\pi/3}\right)\delta(f + 1.5 - 5m)\right]$$

since $f_s = 1/T = 5$ Hz.

As expected, Example 11.13 gives the same spectrum for $x_T(n)$ as was obtained in Example 11.2 where we computed the Fourier transform of the signal as represented by an ideally sampled continuous-time signal. The amplitude and phase spectra of $x_T(n)$ are plotted in Example 11.2.

Discrete-Time Fourier Transforms and Spectra of Periodic Discrete-Time Signals

We can use the ideally sampled continuous-time representation to discuss the characteristics of the discrete-time Fourier transform of a periodic discrete-time signal. To begin, let us consider the discrete-time signal $x_T(n)$, the ideally sampled continuous-time representation of which is

$$x_s(t) = \sum_{n=-\infty}^{\infty} x_T(n)\delta(t - nT) \tag{11.56}$$

where $x_T(n)$ are samples taken with a spacing of T seconds from the periodic continuous-time signal $x(t)$. The period of $x(t)$ is T_0. Thus, the spectrum of the discrete-time signal $x_T(n)$ is

$$X_d(f) = \sum_{m=-\infty}^{\infty} f_s X(f - mf_s) \tag{11.57}$$

where $f_s = 1/T$ and $X(f) \leftrightarrow x(t)$. Since $x(t)$ is periodic with period T_0, then we can represent it with the Fourier series

$$x(t) = \sum_{k=-\infty}^{\infty} X_k e^{j2\pi k f_0 t} \tag{11.58}$$

where $f_0 = 1/T_0$ is the fundamental frequency. Therefore, from eq. (4.107), the spectrum of $x(t)$ is

$$X(f) = \sum_{k=-\infty}^{\infty} X_k \delta(f - kf_0) \qquad (11.59)$$

where impulses occur in $X(f)$ because a periodic signal is a power signal. We substitute eq. (11.59) in eq. (11.57) to obtain the spectrum

$$X_d(f) = \sum_{m=-\infty}^{\infty} \sum_{k=-\infty}^{\infty} f_s X_k \delta(f - kf_0 - mf_s) \qquad (11.60)$$

of the discrete-time signal $x_T(n)$.

The spectrum $X(f)$ of a typical real and even periodic signal $x(t)$ and some of the impulses in the spectrum $X_d(f)$ of the corresponding discrete-time signal $x_T(n)$ are shown in Figure 11.24, where the horizontal arrows indicate that the impulses continue indefinitely. We can see from this illustration that impulses will occur in $X_d(f)$ at all frequencies if $X(f)$ has impulses spaced by f_0 that span the entire frequency axis and if f_0/f_s is irrational. These impulses will have varying strengths. If f_0/f_s is rational, then, from eq. (10.3)

$$f_0/f_s = T/T_0 = i/N \qquad (11.61)$$

where i and N are integers. In this case, $x_T(n)$ is periodic, as we showed in Section 10.1. If i and N are in reduced form (that is, if all common factors

Figure 11.24
Spectra of a Periodic
Continuous-Time
Signal and the
Discrete-Time Signal
Formed from Its
Samples

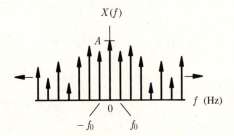

(a) Spectrum of the continuous-time signal

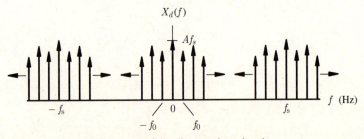

(b) Spectrum of the discrete-time signal

are removed), then the period of $x_T(n)$ is equal to N samples. Substituting eq. (11.61) in eq. (11.60), we obtain

$$X_d(f) = \sum_{m=-\infty}^{\infty} \sum_{k=-\infty}^{\infty} f_s X_k \delta\left[f - \left(\frac{ik + mN}{N}\right) f_s\right] \qquad (11.62)$$

Since i and N are integers with no common factors, then ik and mN do not have a common factor for all m and k. Also, the integers m and k span all possible integers. Therefore, $ik + mN$ spans all possible integers, and the individual impulses in the spectrum $X_d(f)$ are spaced uniformly by f_s/N Hz. If $X_k \neq 0$ for all $k \leq k_{max}$, where $k_{max} \geq N/2$, then all impulses have nonzero strength. In Figure 11.25 we show the spectra for a continuous-time periodic signal $x(t)$ and the discrete-time signal $x_T(n)$ constructed from samples of $x(t)$ for the case in which $i = 3, N = 8$, and $k_{max} = 4 = N/2$. The discrete-time-signal spectrum shown is for a periodic discrete-time signal with a period of $N = 8$ samples because $f_0/f_s = i/N = 3/8$ is rational.

If the highest frequency contained in $x(t)$ is lower than $if_s/2$, or if $X_k = 0$ for some $k < N/2$, then some of the impulse at a spacing of f_s/N Hz will be missing.

Figure 11.25

Spectra of a Periodic Continuous-Time Signal and the Periodic Discrete-Time Signal Formed from Its Samples When f_0/f_s Is Rational

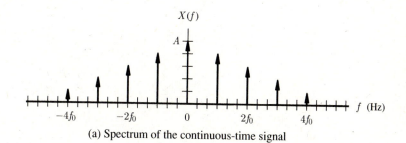

(a) Spectrum of the continuous-time signal

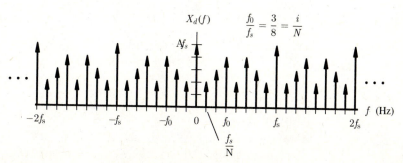

(b) Spectrum of the discrete-time signal

Computation of the spectrum of a periodic discrete-time signal is more easily accomplished with the discrete Fourier transform discussed in Chapter 17 than with eq. (11.62). The continuous-time signal $x(t)$ that passes through the discrete-time signal sample values does not need to be identified. The result is a line spectrum like those we obtained for continuous-time periodic signals with the Fourier series.

Power Density Spectrum

The power-density spectrum of a discrete-time power signal indicates the distribution of signal power as a function of frequency. Our definition for it is similar to the definition we used for the power-density spectrum for continuous-time signals in Chapter 4.

To begin the development of an expression for the power-density spectrum, we recall from eq. (11.20) that the energy-density spectrum for the discrete-time signal $x_T(n)$ is

$$G_{dx}(f) = T^2 \left| X_d(f) \right|^2 \tag{11.63}$$

We define $\phi_{xT}(k) \leftrightarrow G_{dx}(f)/T$, where we have replaced the sample index n with the sample index k for convenience. Therefore,

$$\phi_{xT}(k) = T\mathcal{F}_d^{-1}\left[X_d^*(f)X_d(f) \right] \tag{11.64}$$

We then use the convolution, complex-conjugation, and time-reversal theorems to obtain

$$\phi_{xT}(k) = T\mathcal{F}_d^{-1}\left[X_d^*(f) \right] * \mathcal{F}_d^{-1}\left[X_d(f) \right]$$

$$= Tx_T^*(-k) * x_T(k)$$

$$= T \sum_{\ell=-\infty}^{\infty} x_T^*(-\ell)\, x_T(k-\ell) \tag{11.65}$$

The variable change $\ell = -n$ gives

$$\phi_{xT}(k) = T \sum_{n=-\infty}^{\infty} x_T^*(n)x_T(n+k) \tag{11.66}$$

because the summation can be performed in either order. The function $\phi_{xT}(k)/T$ is called the *discrete-time autocorrelation of the energy signal* $x_T(n)$. Note that

$$\phi_{xT}(0) = \lim_{N\to\infty} T \sum_{n=-N}^{N} \left| x_T(n) \right|^2 = E_{dx} \tag{11.67}$$

where E_{dx} is the signal energy that we defined in Chapter 10.

A similar autocorrelation of $x_T(n)$ can be defined if $x_T(n)$ is a discrete-time power signal. This autocorrelation is

$$R_{xT}(k) = \lim_{N \to \infty} \frac{1}{2N+1} \sum_{n=-N}^{N} x_T^*(n) x_T(n+k) \tag{11.68}$$

Note that $R_{xT}(0)$ is equal to the signal power. Since $\phi_{xT}(0)$ gives the energy in an energy signal, then we can parallel the preceding result for energy signals to produce the definition of the power-density spectrum.

Definition ───

The power density spectrum $S_{dx}(f)$ of the power signal $x_T(n)$ is

$$S_{dx}(f) = \mathcal{F}_d[T R_{xT}(k)] \tag{11.69}$$

where $R_{xT}(k)$ is the autocorrelation of $x_T(n)$.

Note that

$$P_{dx} = R_{xT}(0) = \left\{ \mathcal{F}_d^{-1} \left[S_{dx}(f)/T \right] \right\}_{k=0}$$

$$= \frac{1}{f_s} \int_{-f_s/2}^{f_s/2} \frac{S_{dx}(f)}{T} e^{j2\pi(0)fT} \, df$$

$$= \int_{-f_s/2}^{f_s/2} S_{dx}(f) \, df \tag{11.70}$$

Therefore, it is reasonable to call $S_{dx}(f)$ the power density of the signal, because it represents the total signal power when integrated over the unambiguous frequencies in the discrete-time signal (that is, over $|f| \le f_s/2 = 1/2T$). Also, it can be shown that

$$\int_{-f_2}^{-f_1} S_{dx}(f) \, df + \int_{f_1}^{f_2} S_{dx}(f) \, df \tag{11.71}$$

is the signal power contained in the range of unambiguous frequencies $f_1 < f < f_2$, where $f_2 \le f_s/2$. This further verifies that $S_{dx}(f)$ is an appropriate definition for the power-density spectrum of the signal.

Example 11.14

Find the power-density spectrum of the discrete-time cosine signal

$$x_T(n) = [3 \cos(0.6\pi n - \pi/3), \quad 0.2 \text{ s}]$$

Also compute the power in the signal.

Solution

$$R_{xT}(k) = \lim_{N \to \infty} \frac{1}{2N+1} \sum_{n=-N}^{N} x_T^*(n) x_T(n+k)$$

$$= \lim_{N \to \infty} \frac{9}{2N+1} \sum_{n=-N}^{N} \cos\left[0.6\pi n - \pi/3\right] \cos\left[0.6\pi(n+k) - \pi/3\right]$$

We use the cosine-product trigonometric identity from Appendix B to obtain

$$R_{xT}(k) = \lim_{N \to \infty} \frac{4.5}{2N+1} \left[\sum_{n=-N}^{N} \left\{\cos(0.6\pi k) + \cos\left[0.6\pi(2n+k) - \frac{2\pi}{3}\right]\right\}\right]$$

$$= \lim_{N \to \infty} \frac{9N}{2N+1} \cos(0.6\pi k)$$

$$+ \lim_{N \to \infty} \frac{4.5}{2N+1} \sum_{n=-N}^{N} \cos\left[0.6\pi(2n+k) - \frac{2\pi}{3}\right]$$

$$= 4.5 \cos(0.6\pi k)$$

The limit of the second term is zero because the summation is always finite. Since $T = 0.2$ s and $f_s = 1/T = 5$ Hz, then

$$R_{xT}(k) = 4.5 \cos[2\pi(1.5)kT]$$

and

$$S_{dx}(f) = \mathcal{F}_d[T R_{xT}(k)]$$

$$= \sum_{m=-\infty}^{\infty} 2.25[\delta(f - 1.5 - 5m) + \delta(f + 1.5 - 5m)]$$

is the power-density spectrum of $x_T(n)$. The power in $x_T(n)$ is

$$P_{dx} = \int_{-f_s/2}^{f_s/2} S_{dx}(f)$$

$$= \int_{-2.5}^{2.5} \left\{\sum_{m=-\infty}^{\infty} 2.25[\delta(f - 1.5 - 5m) + \delta(f + 1.5 - 5m)]\right\} df$$

$$= \int_{-2.5}^{2.5} 2.25\delta(f - 1.5)\, df + \int_{-2.5}^{2.5} 2.25\delta(f + 1.5)\, df$$

$$= 2.25 + 2.25 = 4.5$$

The power in $x_T(n)$ can also be computed with

$$P_{dx} = R_{xT}(0) = 4.5 \cos[0.6\pi(0)] = 4.5$$

Figure 11.26
Amplitude Spectrum
and Power-Density
Spectrum of the
Discrete-Time
Cosine Signal in
Example 11.14

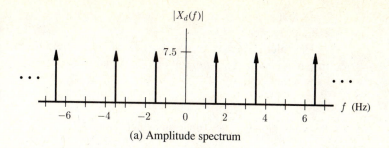

(a) Amplitude spectrum

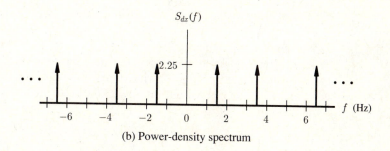

(b) Power-density spectrum

The cosine signal in Example 11.14 is the same signal that we considered Examples 11.13 and 11.2. The amplitude spectrum $|X_d(f)|$ of the signal is plotted in Figure 11.4. We repeat it in Figure 11.26 along with the power-density spectrum of the signal. When we compare the amplitude and power-density spectra of the cosine signal analyzed in Example 11.14, we see that they both consist of impulses located at the same set of frequencies. The strength of each impulse in the power-density spectrum is the square of T times the amplitude-spectrum impulse strength. This is true for all signals that are sampled versions of periodic signals.

11.5 Frequency-Scale-Normalized Spectra

The spectrum of the discrete-time signal $x_T(n)$ is

$$X_d(f) = \sum_{n=-\infty}^{\infty} x_T(n)e^{-j2\pi nfT} \qquad (11.72)$$

Since $T = 1/f_s$, we can define a frequency-scale-normalized version of the spectrum as

$$X_N(r) = \sum_{n=-\infty}^{\infty} x_T(n)e^{-j2\pi nr} = \sum_{n=-\infty}^{\infty} x(n)e^{-j2\pi nr} \qquad (11.73)$$

where $x(n)$ is the sequence corresponding to $x_T(n)$ and

$$r = fT = f/f_s \qquad (11.74)$$

Note that r is a normalized frequency, where the normalization factor is the sample frequency f_s.

The values of the sequence $x(n)$ for many discrete-time signals can be written in terms of just the sample index n. As an example, consider the discrete-time exponential signal

$$x_T(n) = \left[e^{-an}, \quad T \right] \tag{11.75}$$

the corresponding sequence of which is

$$x(n) = e^{-an} \tag{11.76}$$

This sequence depends only on the sample index n, not on the sample spacing T.

We can see from eq. (11.73) that the frequency-scale-normalized spectrum $X_N(r)$ corresponding to $x_T(n)$ depends only on the sequence values $x(n)$ and not on the sample spacing T when $x(n)$ does not depend on T. Thus, we can think of $X_N(r)$ as the discrete-time Fourier transform, or spectrum, of the sequence $x(n)$. That is

$$x(n) \leftrightarrow X_N(r) \tag{11.77}$$

Note that $x(n)$ is a time-scale-normalized version of the discrete-time signal $x_T(n)$, where the normalization factor is the sample spacing T.

In some cases, the values of the sequence $x(n)$ may depend on T; for example, the ramp signal $x_T(n) = [Tr(n), \quad T]$ or the cosine signal $x_T(n) = [A\cos(an + T), \quad T]$. In these cases, the spectrum of the sequence; that is, the frequency-scale-normalized spectrum $X_N(r)$ of $x_T(n)$, is a function of T.

Since the spectrum $X_d(f)$ is periodic in f with period f_s, and the frequency-scale-normalized spectrum $X_N(r)$ is obtained by normalizing the frequency by f_s, then $X_N(r)$ is periodic in r with period equal to one. If $X_N(r)$ is plotted, it is normally plotted only for $|r| \leq 1/2$. This is because $X_N(r)$ repeats outside this interval; thus, the interval contains all the spectrum information.

Example 11.15

Find the spectrum of the sequence

$$x(n) = \{1, \quad 2, \quad 1\} \qquad n_0 = -1$$

Plot the amplitude and phase spectra or, if possible, the total spectrum.

Solution

$$X_N(r) = \sum_{n=-\infty}^{\infty} x(n)e^{-j2\pi nr}$$

$$= (1)e^{j2\pi r} + (2)(1) + (1)e^{-j2\pi r}$$

$$= 2 + 2\cos(2\pi r)$$

The spectrum for the sequence $x(n)$ is shown in Figure 11.27.

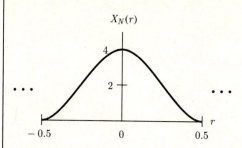

Figure 11.27 Spectrum of the Sequence $x(n)$ in Example 11.15

A sequence $x(n)$ that does not depend on T corresponds to many different discrete-time signals that differ only in their sample spacing or, equivalently, in their sample rate. Thus, we can obtain the spectra for these signals from the spectrum of $x(n)$ by appropriately scaling the frequency axis. This technique is illustrated in the following example.

Example 11.16

Find the spectra of the two signals shown in Figure 11.28. Plot the amplitude and phase spectra or, if possible, the total spectra of the signals.

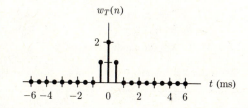

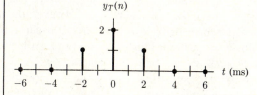

Figure 11.28 Signals for Example 11.16

Solution

$$w_T(n) = [x(n), \quad 0.5 \text{ ms}] \quad \text{and} \quad y_T(n) = [x(n), \quad 2 \text{ ms}]$$

where $x(n) = \{1, 2, 1\}$, $n_0 = -1$, and is the same sequence defined in Example 11.15. Therefore, using the result of Example 11.15, we obtain

$$W_d(f) = \left. X_N(r) \right|_{r=f(0.5 \times 10^{-3})} \qquad Y_d(f) = \left. X_N(r) \right|_{r=f(2 \times 10^{-3})}$$

and

$$= 2 + 2\cos(\pi f \times 10^{-3}) \qquad \qquad = 2 + 2\cos(4\pi f \times 10^{-3})$$

since T is 0.5 ms and 2 ms for $w_T(n)$ and $y_T(n)$, respectively. The plots of the signal spectra for $|f| \leq f_s/2$ are shown in Figure 11.29.

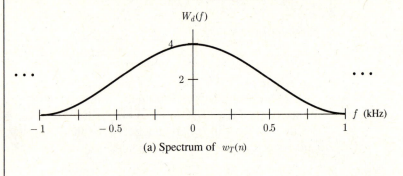

(a) Spectrum of $w_T(n)$

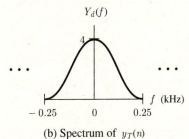

(b) Spectrum of $y_T(n)$

Figure 11.29 Spectra of the Signals $w_T(n)$ and $y_T(n)$ in Example 11.16

The spectra of $w_T(n)$ and $y_T(n)$ in Example 11.16 have the same shape because both signals have the same sequence. However, the bandwidth of $w_T(n)$ is four times wider than the bandwidth of $y_T(n)$ because the rate of change of sample amplitude is four times faster for $w_T(n)$ than it is for $y_T(n)$.

In the discussions of the following chapters, we will often use sequences and their associated spectra (that is, time-scale-normalized signals and their corresponding frequency-scale-normalized spectra). In this way, we avoid the necessity of explicitly indicating the sample spacing T. When specific sample-spacing values are of interest, then normalized results obtained are easily de-normalized by the appropriate time and frequency scaling, as we did in Example 11.16. Henceforth, we will use the term *normalized spectra* in reference to frequency-scale-normalized spectra for simplicity.

11.6 Summary

The Fourier transform of a continuous-time signal gives the spectrum of the signal. Therefore, the spectrum $X_d(f)$ of the discrete-time signal $x_T(n)$ can be defined as the Fourier transform of the ideally sampled continuous-time representation of $x_T(n)$. The spectrum is a continuous-frequency function that is periodic with period $f_s = 1/T$, where f_s is the frequency of sample occurrence (called sample frequency). The signal spectrum, which consists of its amplitude and phase spectra, portrays the range of frequencies for which the signal has significant energy content and the nature of the amplitude and phase of signal components at all frequencies. The bandwidth of the discrete-time signal is determined by the nature of its spectrum in the period covered by $|f| \le f_s/2$, this period being the interval of the unambiguous frequencies (defined in Section 10.1) that indicate the rate of time variation of sinusoidal-signal sequence values.

If the discrete-time signal is produced by taking samples of a bandlimited continuous-time signal, then the sampling rate must exceed a lower limit so that it will produce samples that uniquely characterize the bandlimited continuous-time signal. The limit is defined by the uniform sampling theorem to be $f_s > 2f_h$, where f_h is the highest frequency component of the continuous-time signal. However, practical constraints in generating and using the samples and in filtering the sampled signal to reconstruct the original signal cause the signal reconstruction to be imperfect.

Direct computation of the Fourier transform of the representation of the discrete-time signal as an ideally sampled continuous-time signal leads to the definition of the discrete-time Fourier transform. This transform computes the spectrum, $X_d(f)$ of a discrete-time energy signal $x_T(n)$ directly from the sample values and sample spacing of the signal. We use the inverse discrete-time Fourier transform to find a discrete-time signal from its spectrum.

The energy-density spectrum for the discrete-time signal $x_T(n)$ is $G_{dx}(f) = T^2|X_d(f)|^2$. It shows the energy distribution in the signal as a function of frequency.

Several properties and a number of theorems for the discrete-time Fourier transform are useful in relating signal and spectrum properties and in computing spectra for complicated signals. These properties and theorems are similar to or parallel exactly those for continuous-time signals, as presented in Chapter 4.

The use of the discrete-time Fourier transform to compute the spectrum of a discrete-time energy signal is extended to discrete-time power signals by permitting impulses in the spectrum and by using discrete-time Fourier transforms in the limit. This concept parallels a similar one for continuous-time signals.

The frequency-scale-normalized spectrum $X_N(r)$ of the discrete-time signal $x_T(n)$ is produced from the spectrum $X_d(f)$ by scaling the frequency by the sample frequency f_s. This normalized spectrum can be thought of as the spectrum of the sequence $x(n)$ corresponding to the discrete-time signal $x_T(n)$; it does not depend on the sample spacing T if $x(n)$ does not depend on T. If $x(n)$ does not depend on T, then we can determine the spectrum of any discrete-time signal with a sequence of $x(n)$ from $X_N(r)$ by appropriate scaling of the frequency axis.

Problems

In the following problems, all times are in seconds unless otherwise noted. To solve many of the problems, computer or programmable calculator computations should be used to generate sufficient data for the plots requested.

11.1 Use the technique described in Section 11.1 to find the spectrum for the following signals:

a. $x_T(n) = [6\cos(0.5\pi n), \quad 0.025]$

b. $w_T(n) = [6\cos(1.5\pi n), \quad 0.025]$

c. $z_T(n) = [2\sin(1.5n), \quad 4]$

d. $y_T(n) = [15\cos(0.8\pi n + \pi/6), \quad 0.025]$

11.2 Repeat Problem 11.1 for the signals

a. $x_T(n) = [10\cos(0.4\pi n), \quad 0.01]$

b. $y_T(n) = [5\cos(1.6\pi n), \quad 0.01]$

c. $w_T(n) = \left[3\cos\left(1.2\pi n - \pi/4\right)\right.$
$\left. + 2\cos\left(0.3\pi n + \pi/3\right), \quad 0.1\right]$

d. $z_T(n) = \left[4\cos\left(0.6\pi n - \pi/5\right)\right.$
$\left. + 2\cos\left(1.4\pi n + \pi/4\right), \quad 0.2\right]$

11.3 Use the technique of Section 11.1 to find the spectrum for the signals

a. $x_T(n) = [4\ \text{sinc}^2(0.48n), \quad 8\ \text{ms}]$

b. $y_T(n) = [4\ \text{sinc}^2(0.6n), \quad 10\ \text{ms}]$

c. $w_T(n) = [e^{-1.25n}u(n), \quad 0.025]$

d. $z_T(n) = \left[\left\{\begin{matrix} 2\left(1 - |n|/3\right) & |n| \le 3 \\ 0 & |n| > 3 \end{matrix}\right\}, \quad 0.1\right]$

11.4 Determine the bandwidth of the signals of Problem 11.2 if bandwidth is defined by $\alpha = 0.5$.

11.5 Repeat Problem 11.4 for the signals of Problem 11.3.

11.6 Assuming ideal systems, find the minimum sampling rate that can be used to obtain samples that completely specify the following signals:

a. $w(t) = \text{sinc}(400t)$

b. $x(t) = 10\cos(20\pi t) - 5\cos(100\pi t)$
$\qquad + 20\cos(400\pi t)$

c. $y(t) = 2\cos(20\pi t) + 4\sin(20\pi t - \pi/4)$
$\qquad + 5\cos(8\pi t)$

11.7 Repeat Problem 11.6 with the following signals:

a. $w(t) = 4\ \text{sinc}(10t)\text{sinc}(4t)$

b. $z(t) = 10\cos(50\pi t) - 20\cos(40\pi t)$

c. $y(t) = 15\cos\left(100\pi t - \pi/3\right)$
$\qquad + 20\sin\left(250\pi t - \pi/4\right)$

11.8 Plot the ideally sampled signal and its amplitude and phase spectra or, if possible, its total spectrum for signal (a) of Problem 11.6 for sampling rates of 300 Hz and 500 Hz. Are these sampling rates acceptable for an ideal system?

11.9 Plot the ideally sampled signal and its amplitude and phase spectra or, if possible, its total spectrum for signal (a) of Problem 11.7 for sampling rates of 10 Hz, 15 Hz, and 20 Hz. Are these sampling rates acceptable for an ideal system?

11.10 Plot the ideally sampled signal and its amplitude and phase spectra or, if possible, its total spectrum for signal (c) of Problem 11.6 for sampling rates of 15 Hz, 20 Hz, and 25 Hz. Are these sampling rates acceptable for an ideal system?

11.11 A continuous-time signal $x(t)$ has the spectrum shown in Figure 11.30.

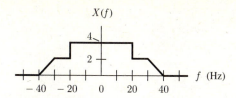

Figure 11.30

Perform ideal sampling on the signal and plot the spectrum of the resulting sampled signal over the frequency range $|f| \leq 120$ Hz for sampling rates of 60 Hz, 65 Hz, 75 Hz, and 85 Hz. Which of these sampling rates are acceptable if an ideal low-pass filter is available for signal reconstruction.

11.12 In signaling between two locations using pulses of light, a single pulse can be represented by the signal $y(t) = 8\pi e^{-4\pi|t|}$. The pulse is processed by a discrete-time system. What minimum sampling rate would you use in an ideal system so as to keep aliasing amplitude-spectrum values under 1% of the maximum amplitude-spectrum value?

11.13 Samples of the continuous-time signal $x(t)$ are transmitted over a digital communication link. Assume that $x(t)$ has the following spectrum:

$$X(f) = 10\Pi\left(\frac{f}{100}\right)$$

Ideal sampling is used on this signal and the ideal samples are transmitted; however, an ideal low-pass filter is not available for signal reconstruction from the samples. The low-pass filter that is available has the frequency response $H(f) = 4/\sqrt{1 + af^2}$. Choose the parameter a so as to keep the gain variation over the bandwidth of $x(t)$ to under 2%. Then choose a sampling rate so as to keep the filter gain for unwanted signal components to under 1% of the maximum filter gain for the desired signal components.

11.14 Verify the discrete-time Fourier transform pair

$$\left[a^n u(n), \quad T\right] \leftrightarrow 1/\left[1 - a\exp(-j2\pi fT)\right]$$

$$|a| < 1$$

11.15 Verify the discrete-time Fourier transform pair

$$\left[(n+1)a^n u(n), \quad T\right] \leftrightarrow 1/\left[1 - a\exp(-j2\pi fT)\right]^2$$

$$|a| < 1$$

11.16 Find and plot the amplitude and phase spectra or, if possible, the total spectrum for the following signals:

a. $x_T(n) = [\{3, -1, 0, 1, -2, 4\}, \quad 20$ ms]

b. $y_T(n) = [(\{-6, -4, 2, -2, 4, 8, 2\}n_0 = -2), \quad 0.1]$

c. $z_T(n) = \left[\left\{\begin{matrix} 3 - 0.5|n| & |n| \leq 4 \\ 0 & |n| > 4 \end{matrix}\right\}, \quad 5\right]$

Also, plot the signals.

11.17 Find and plot the amplitude and phase spectra or, if possible, the total spectrum for the signals shown in Figure 11.31.

11.18 Prove the linearity theorem (Theorem 1).

11.19 Prove the complex-conjugation theorem (Theorem 4).

11.20 Prove the time-shift theorem (Theorem 5).

11.21 Prove the frequency-translation theorem (Theorem 6).

11.22 Prove the convolution theorem (Theorem 10).

11.23 Use the table of discrete-time Fourier transforms (Table C.6) and theorems (Table C.5) to find the spectra of the system test signals shown in Figure 11.32. Plot the amplitude and phase spectra or, if possible, the total spectrum for each signal.

11.24 The signal $x_T(n) = [x(n), \quad 0.05]$ has the spectrum

$$X_d(f) = 1/\left[1 - 0.8\exp(-j\pi f/10)\right]$$

Use the theorems to find the spectra of the following signals:

a. $a_T(n) = \left[4\{x(n) - x(n-1)\}, \quad 0.05\right]$

b. $b_T(n) = x_T(-3 - n)$

c. $c_T(n) = x_{T/2}(n)$

d. $d_T(n) = e^{-j\pi n}x_T(n)$

e. $e_T(n) = 4x_T(n)\cos(0.5\pi n)$

f. $f_T(n) = x_{4T}(n - 1)$

g. $g_T(n) = \sum_{m=-\infty}^{n} x_T(m)$

Figure 11.31   **a.**

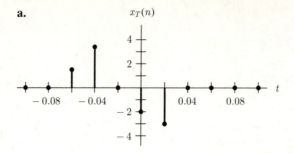

b.

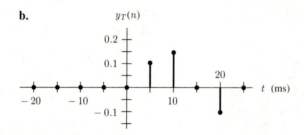

c.

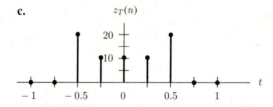

11.25 Use the table of discrete-time Fourier transforms (Table C.6) and the theorems (Table C.5) to find the spectra of the system test signals shown in Figure 11.33. Plot the amplitude and phase spectra or, if possible, the total spectrum for each signal. Also plot the signal $z_T(n)$.

11.26 Find the spectra of the power signals

a. $x_T(n) = \left[\left(1 - e^{-0.7n}\right)u(n), \quad 0.2\right]$

b. $y_T(n) = \left[\left\{\begin{matrix} 0 & -1 \le n \le 3 \\ 2 & \text{elsewhere} \end{matrix}\right\}, \quad 2\right]$

c. $z_T(n) = \left[\{u(n+2) + u(-n+2)\}, \quad 5 \text{ ms}\right]$

11.27 Use the discrete-time Fourier transform table and theorems to find the spectra of the periodic signals listed. Plot the amplitude and phase spectra or, if possible, the total spectrum for each signal over the frequency range $|f| \le 3f_s/2$.

a. $x_T(n) = \big[\{3 + 2\cos(\pi n/2) + \cos(\pi n)$
$$+ 2\cos(3\pi n/2)\}, \quad 10\big]$$

b. $y_T(n) = \big[\{2 + 3\cos(0.8\pi n) + 4\cos(1.6\pi n)$
$$+ 2\cos(2.4\pi n)\}, \quad 0.5\big]$$

c. $z_T(n) = \big[\{1 + 2\cos(\pi n/2)$
$$+ 3\cos(3\pi n/2)\}, \quad 0.01\big]$$

11.28 In a data-transmission system, the signal

$$x_T(n) = \Bigg[\Bigg\{ \sum_{i=-\infty}^{\infty} [u(n+4-27i)$$

$$- u(n-5-27i)]\Bigg\}, \quad 5 \text{ ms}\Bigg]$$

Figure 11.32   **a.**

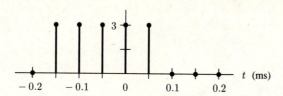

b.

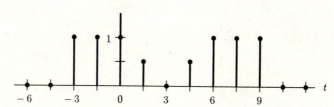

c.

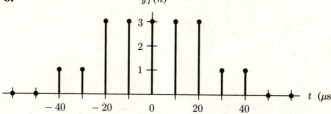

Figure 11.33   **a.**

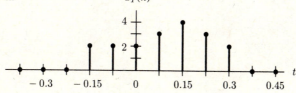

b.

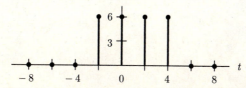

c. $z_T(n) = \left[\left\{e^{-0.4n}u(n) + e^{0.4n}u(-n) - (n)\right\}, \ 0.1\right]$

multiplies the incoming discrete-time signal to select portions of it that contain synchronization data.
(a) Plot $x_T(n)$ and (b) find the spectrum of $x_T(n)$
and plot the amplitude and phase spectra or, if possible, the total spectrum over the frequency range
$|f| \leq 3f_s/2$.

11.29 Consider the signal

$$y_T(n) = \left[x(n) \cos(\pi n/4), \quad 5 \text{ ms} \right]$$

where $x(n)$ is the sequence corresponding to the
signal $x_T(n)$ in Problem 11.28. (a) Plot $y_T(n)$ and
(b) find the spectrum of $y_T(n)$ and plot the amplitude and phase spectra or, if possible, the total spec-
trum over the range of frequencies $|f| \leq 3f_s/2$.

11.30 Find and plot the power-density spectra of the
signals in Problem 11.27.

11.31 Find and plot the power-density spectrum of
the signal in Problem 11.28.

11.32 Find the spectra of the sequences listed. Plot
the amplitude and phase spectra or, if possible, the
total spectrum of each sequence.

a. $x(n) = 0.5^n u(n)$

b. $u(n)$

c. $y(n) = u(n+3) - u(n-4)$

d. $\delta(n)$

e. $z(n) = 2 \cos(0.25\pi n - \pi/3)$

11.33 Find the spectra of the sequences listed. Plot
the amplitude and phase spectra or, if possible, the
total spectrum of each sequence.

a. $x(n) = 4\delta(n+2) + 2\delta(n+1) - \delta(n)$
$$+ \delta(n-1)$$

b. $y(n) = \{3, 2, -2, 2, 2\} \quad n_0 = 2$

c. $z(n) = \{-2, -2, 3, 4, 2\}$

d. $w(n) = \sum_{i=-\infty}^{\infty} [u(n+2-10i)$
$$- u(n-3-10i)]$$

11.34 The two signals $y_T(n) = [x(n), \quad 5 \text{ ms}]$ and
$z_T(n) = [x(n), \quad 0.02 \text{ ms}]$, where $x(n) = 0.2^n u(n)$,
are the output signals of two discrete-time process-
ing systems operating at different sample frequencies. Find the spectra of the two signals. Plot the
amplitude and phase spectra or, if possible, the total
spectrum of the signals.

12 Time-Domain Analysis of Discrete-Time Systems

In this chapter, we will consider time-domain analysis methods for linear, time-invariant, discrete-time systems. We will use these methods to determine time waveforms of the discrete-time-system output signal. Therefore, our discussion here parallels that of Chapter 5, where we considered continuous-time system time-domain analysis.

A single-input–single-output, linear, time-invariant, discrete-time system can be modeled by an ordinary linear difference equation with constant coefficients (see Section 1.3). As an example, consider a savings-plan system in which deposits are made by payroll deduction on the first day of each month. This savings plan also pays interest on the first day of each month on the amount in the plan during the previous month. Assume $y_T(n)$ to be the amount in the plan after the deposit at time nT, α to be the interest rate per month, and $x_T(n)$ to be the monthly deposit to the savings plan (note: $T =$ one month). We define the signals $x_T(n)$ and $y_T(n)$ to be the input and output signals of the savings-plan system. The difference equation that characterizes the savings-plan system is then

$$y_T(n) = y_T(n-1) + \alpha y_T(n-1) + x_T(n) \tag{12.1}$$

The general difference equation

$$y_T(n) + \sum_{i=1}^{k} a_i y_T(n-i) = \sum_{i=0}^{l} b_i x_T(n-i) \tag{12.2}$$

models a kth-order causal system with an output signal at sample number n (that is, sample time $t = nT$) that depends directly on the input signal at sample number n and on the previous l samples of the input signal. The output signal at sample number n also depends on the previous k output-signal samples. Through these samples, it depends on all previous input-signal samples. Note that the sample spacing of the signals is T; thus, the system must operate at a sample rate of $f_s = 1/T$ samples per second.

We can use the difference equation and initial-condition values of the system to determine the output-signal samples for $n \geq n_0$ (that is, for $t = nT \geq n_0 T$) corresponding to a specified input signal. The integer n_0 is the index number of the initial sample of interest. The initial-condition values are specified by the energy stored in the system components at $n = n_0$. For very simple difference equations, we may be able to find the solution for the output signal by using successive substitution, noting the emerging sequence pattern. This was the solution method used for eq. (1.21) in Chapter 1. In general, we can use classical solution methods for difference equations to solve the difference equation for the output signal.[†] However, as indicated in Chapter 1, these methods are not covered in this text.

The superposition sum that we will develop in this chapter is an alternate mathematical model that relates the output signal (system response) to the input signal for a discrete-time system when the system is linear and time-invariant and has zero initial conditions. We noted previously that this system response is called the zero-initial-energy response or, equivalently, the zero-initial condition-response. To use the superposition sum, we must know the system's unit pulse response.

We can analyze systems with nonzero initial conditions by treating the initial conditions as inputs, as we will illustrate by example. We will also discuss the recursive method of difference-equation solution. This method is useful in discrete-time-system simulation for specific input signals when general output-signal characteristics are not required.

In this and the following chapters, discrete-time signals and systems will be referred to as signals and systems, dropping the "discrete-time" adjective, unless it is needed for clarity. When block-diagram representations are shown, the sample delay operation defined by eq. (1.20) in Chapter 1 will be represented by a block with the letter D inside. When we need to represent a delay of m sample spacings, the notation D^m will appear in the block.

12.1 Unit Pulse Response

The *unit pulse response of a discrete-time system* is defined as follows:

Definition _____

The unit pulse response of a discrete-time system is the system's response to a unit pulse input signal located at $t = nT = 0$ [that is, to $\delta_T(n)$] when the system initial conditions are zero.

_____

We use the notation $h_T(n)$ for the unit pulse response; therefore,

$$x_T(n) = \delta_T(n) \quad \text{yields} \quad y_T(n) = h_T(n) \tag{12.3}$$

where $x_T(n)$ and $y_T(n)$ are the system input and output signals, respectively.

Direct determination of the unit pulse response of a system requires solution of the system difference equation for the output signal when the input signal is a unit pulse and initial conditions are zero. A simple illustration of direct determination of the unit pulse response is shown in the following example.

[†]See for example H. Levy and F. Lessman, *Finite Difference Equations* (New York: Macmillan, 1961).

Example 12.1

The savings-plan system we described at the beginning of the chapter is characterized by the difference equation

$$y_T(n) = (1 + \alpha)y_T(n - 1) + x_T(n)$$

where $x_T(n)$ is the deposit and $y_T(n)$ is the amount in the plan at $t = nT$. The parameter α is the monthly interest rate. Find the unit pulse response of the savings-plan system.

Solution

The system difference equation in this example is the difference equation (eq. [1.21]) in Section 1.3, with $a = 1 + \alpha$ and $b = 1$. We found the solution of eq. (1.21) for $n \geq n_0$ to be eq. (1.26). This solution, with $a = 1 + \alpha$ and $b = 1$, is

$$y_T(n) = (1+\alpha)^{(n-n_0+1)}y_T(n_0 - 1)+\sum_{i=n_0}^{n}(1 + \alpha)^{(n-i)}x_T(i) \qquad n \geq n_0 \quad (12.4)$$

where $x_T(n)$ is the input signal consisting of the monthly deposits. The value $y_T(n_0 - 1)$ is a constant value that is established by the system initial condition; that is, by the money in the savings plan one month prior to the initial time considered. To find the unit pulse response for all time, we set $y_T(n_0 - 1) = 0$, $n_0 = -\infty$, and $x_T(n) = \delta_T(n)$ where $y_T(n_0 - 1)$ is set equal to zero to give zero initial conditions. Having done this, $y_T(n)$ in eq. (12.4) is $h_T(n)$ and is given by

$$h_T(n) = \sum_{i=-\infty}^{n}(1 + \alpha)^{(n-i)}\delta_T(i) \qquad n \geq -\infty \quad (12.5)$$

Since the only nonzero term in the sum defined by eq. (12.5) is the term corresponding to $i = 0$, then

$$h_T(n) = \begin{cases} 0 & n < 0 \\ (1 + \alpha)^n & n \geq 0 \end{cases}$$

or, equivalently,

$$h_T(n) = (1 + \alpha)^n u_T(n) \quad (12.6)$$

is the unit pulse response of the savings-plan system.

An interesting property of money invested at compound interest can be determined from the unit pulse response of the savings-plan system. We begin by noting that the unit pulse response for the system gives the amount in the plan after n months when a single deposit of one unit is made at time $t = 0$. If we assume an interest rate of, say, 6% per year (that is, 0.5% per month), then

$\alpha = 0.005$, and a single deposit of one unit at $t = 0$ results in a savings-plan amount of $(1 + 0.005)^{139} = 2.0002$ units after 139 months (that is, 11 years, 7 months). Thus, a single-time deposit in the savings plan approximately doubles in size after 11 years, 7 months if the interest rate is 6% per year and the interest is compounded monthly. One can easily show that the time required for the single-time deposit to double reduces to 7 years if the interest rate is increased to 10%.

We found in Example 12.1 that it was not difficult to determine the unit pulse response for the system corresponding to the simple first-order difference equation. Solutions for the impulse responses of systems corresponding to higher-order difference equations require classical solution techniques for difference equations, although, as mentioned previously, these methods are not covered here. We can determine the unit pulse responses for these systems more easily using transform techniques. This method of determining the impulse response will be treated in the next two chapters, where we consider the use of transform techniques in the analysis of discrete-time systems.

12.2 General Response of Linear Time-Invariant Systems

In this section, we will consider the response of a linear time-invariant system to a general input signal. Those systems having zero initial conditions are of primary interest to us because very often we can choose the time origin so that initial conditions are zero. However, we do briefly outline one method for determining the system response for systems with nonzero initial conditions.

Systems with Zero Initial Conditions

Consider the system response for all values of n for the system of Example 12.1 when initial conditions are zero; that is, for the cases when $n_0 = -\infty$ and $y_T(n_0 - 1) = 0$. Using these assumptions, we obtain the system response to a general input signal $x_T(n)$ from eq. (12.4) The result is

$$y_T(n) = \sum_{i=-\infty}^{n} (1 + \alpha)^{(n-i)} x_T(i) \qquad (12.7)$$

Since $u_T(n - i) = 0$ for $i > n$, then we can write eq. (12.7) in the form

$$y_T(n) = \sum_{i=-\infty}^{\infty} (1 + \alpha)^{(n-i)} u_T(n - i) x_T(i) \qquad (12.8)$$

which we further simplify to

$$y_T(n) = \sum_{i=-\infty}^{\infty} x_T(i) h_T(n - i) = x_T(n) * h_T(n) \qquad (12.9)$$

by using eq. (12.6). Therefore, for this particular linear, time-invariant, first-order system, the system response is the input signal convolved with the unit pulse response when initial conditions are zero.

To show that the preceding relationship is true for any linear, time-invariant, discrete-time system with zero initial conditions, we consider the system with an input signal, output signal, and unit pulse response of $x_T(n)$, $y_T(n)$, and $h_T(n)$, respectively. Therefore,

$$x_T(n) = \delta_T(n) \quad \text{yields} \quad y_T(n) = h_T(n) \tag{12.10}$$

We can write the input signal as

$$x_T(n) = \sum_{m=-\infty}^{\infty} x_T(m)\delta_T(n-m) \tag{12.11}$$

where $x_T(m)$ is the signal value at the mth sample location. Since the system is time-invariant,

$$x_T(n) = \delta_T(n-m) \quad \text{yields} \quad y_T(n) = h_T(n-m) \tag{12.12}$$

This means that the output signal is

$$y_T(n) = \sum_{m=-\infty}^{\infty} x_T(m)h_T(n-m) = x_T(n) * h_T(n) \tag{12.13}$$

because the system is linear, and thus superposition holds for any weighted sum of input signals such as the unit pulses $x_T(m)\delta_T(n-m)$ in eq. (12.11). Equation (12.13) shows that the output signal is, indeed, the convolution of the input signal and the system unit pulse response when the system is linear and time-invariant with zero initial conditions. We see from eq. (12.13) that the output signal is a sum of time-shifted unit pulse responses that are individually scaled by the input-signal sample values. We refer to eq. (12.13) as the *superposition sum for a discrete-time system*. Since convolution is commutative, we can also write eq. (12.13) in the form

$$y_T(n) = h_T(n) * x_T(n) = \sum_{m=-\infty}^{\infty} h_T(m)x_T(n-m) \tag{12.14}$$

Note that the sample spacing must be the same for the input signal and the unit pulse response if we are to compute the system response by using the superposition sum. The sample spacing for the system response is also the same as that for the input signal and unit pulse response.

We indicated in Chapter 2 that we perform the convolution of two discrete-time signals by computing the convolution of their sequences, keeping in mind that the discrete-time signals corresponding to the sequences have a spacing of T between samples. Therefore,

$$y(n) = x(n) * h(n) = h(n) * x(n) \tag{12.15}$$

is the superposition sum expressing the relationship between the sequences corresponding to the input signal, unit pulse response, and output signal when system initial conditions are zero.

The difference equation that relates the system response to the input signal completely specifies the system-response characteristics when the initial conditions are zero, because we can use it to find the system response to any input signal. Likewise, the system unit pulse response completely characterizes the system-response characteristics when the system is linear and time-invariant with zero initial conditions for the same reason.

It appears at this point that we must solve the difference equation to find the unit pulse response of a system before we can find the zero-initial-condition response of the system to an input signal by convolving the input signal with the unit pulse response. However, we can often find the unit pulse response directly from the system representation without solving the difference equation. Only algebra and inverse transformation are required. This method of determining a unit pulse response will be developed in the following two chapters.

Example 12.2

A system has the unit pulse response

$$h_T(n) = \left[(1/3)^{n+1} u(n+1), \quad 0.1 \text{ s} \right]$$

Find the zero-initial-condition response of the system to a unit step input.

Solution
The sequences corresponding to the input signal and unit pulse response are

$$x(n) = u(n)$$

and

$$h(n) = (1/3)^{n+1} u(n+1)$$

The superposition sum gives the sequence corresponding to the output signal. It is

$$y(n) = x(n) * h(n) = h(n) * x(n)$$

$$= \sum_{i=-\infty}^{\infty} h(i)x(n-i)$$

Since $h(i) = 0$ for $i < -1$ and $x(n-i) = 0$ for $i > n$, then, from eq. (2.15) in Chapter 2,

$$y(n) = \begin{cases} 0 & n < -1 + 0 = -1 \\ \sum_{i=-1}^{n} h(i)x(n-i) & n \geq -1 + 0 = -1 \end{cases}$$

Computing $y(n)$ for $n \geq -1$ yields

$$y(n) = \sum_{i=-1}^{n} (1/3)^{i+1} = \sum_{k=0}^{n+1} (1/3)^k = \frac{1 - (1/3)^{n+2}}{1 - (1/3)}$$

$$= (3/2) - (3/2)(1/3)^{n+2} = (3/2) - (1/6)(1/3)^n$$

Therefore, the zero-initial-condition response for the system is

$$y_T(n) = \left\{ \left[(3/2) - (1/6)(1/3)^n \right] u(n+1), \quad 0.1 \text{ s} \right\}$$

This response is plotted in Figure 12.1.

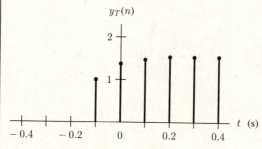

Figure 12.1 Zero-Initial-Condition Response for Example 12.2

Example 12.3

A process control signal in a chemical processing plant is smoothed by using a digital processor. That is, the control signal is converted to a discrete-time signal by an A/D converter, the discrete-time signal is smoothed with a discrete-time filter, and the smoothed discrete-time signal is converted to a continuous-time smoothed control signal by a D/A converter. The smoothing filter has the unit pulse response

$$h_T(n) = \left[(0.5^n - 0.25^n) u(n), \quad 0.25 \text{ s} \right]$$

Find the zero-initial-condition response of the discrete-time filter when the input signal samples are

$$x_T(n) = \left[\{1, 1, 1\}, \quad 0.25 \text{ s} \right]$$

Solution

The sequences corresponding to the input and unit pulse response are

$$x(n) = \{1, 1, 1\}$$

and

$$h(n) = (0.5^n - 0.25^n) u(n)$$

Using the superposition sum, we write the sequence corresponding to the filter response as

$$y(n) = x(n) * h(n)$$

$$= \sum_{i=-\infty}^{\infty} x(i)h(n-i)$$

Since $x(n) = 0$ for $n < 0$ and $h(n) = 0$ for $n < 1$, then, from eq. (2.15) in Chapter 2,

$$y(n) = \begin{cases} 0 & n < 0 + 1 = 1 \\ \sum_{i=0}^{n-1} x(i)h(n-i) & n \geq 0 + 1 = 1 \end{cases}$$

Computing $y(n)$ for $n = 1$ and $n = 2$ yields

$$y(1) = x(0)h(1) = (1)(0.5 - 0.25) = 0.25$$

and

$$y(2) = x(0)h(2) + x(1)h(1)$$

$$= (1)(0.5^2 - 0.25^2) + (1)(0.5 - 0.25)$$

$$= 0.4375$$

Since $x(n) = 0$ for $n > 2$, then

$$y(n) = \sum_{i=0}^{2} x(i)h(n-i)$$

for $n > 2$. Therefore

$$y(n) = (1)\left(0.5^n - 0.25^n\right) + (1)\left(0.5^{(n-1)} - 0.25^{(n-1)}\right)$$

$$+ (1)\left(0.5^{(n-2)} - 0.25^{(n-2)}\right)$$

$$= 0.5^n[1 + 2 + 4] - 0.25^n[1 + 4 + 16]$$

$$= 7(0.5)^n - 21(0.25)^n$$

In summary, the filter response is

$$y_T(n) = \left\{ \begin{bmatrix} 0 & n < 1 \\ 0.25 & n = 1 \\ 0.4375 & n = 2 \\ 7(0.5)^n - 21(0.25)^n & n > 2 \end{bmatrix}, \quad 0.25 \text{ s} \right\}$$

The input signal, unit pulse response, and system response of the discrete-time filter to the input signal are plotted in Figure 12.2. Note that the output sampled control signal is smoothed, not changing as rapidly as the input sampled control signal. It is also delayed in time.

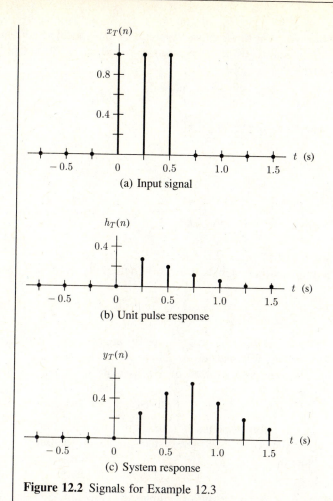

Figure 12.2 Signals for Example 12.3

In the preceding example, we processed a continuous-time signal with a discrete-time system by using a system consisting of a cascade of an A/D converter, the discrete-time system, and a D/A converter. The discrete-time unit pulse response was given.

In some cases, we may want to design the discrete-time system so that the output samples it produces from samples of a continuous-time signal are approximately equal to samples of the continuous-time output signal produced by a specified continuous-time system from the same continuous-time signal. In these cases, a possible choice for the unit pulse response contains sequence values that are T times the samples of the impulse response of the specified continuous-time system. To show this, we first consider the specified operation on the continuous-time signal

$$y(t) = h_c(t) * x(t) = \int_{-\infty}^{\infty} h_c(\tau)x(t - \tau)\,d\tau \qquad (12.16)$$

where $h_c(t)$ is the impulse response of the specified continuous-time system. From calculus, we use the limiting definition of the integral with $\Delta t = T$ to write

$$y(t) = \lim_{T \to 0} \sum_{m=-\infty}^{\infty} h_c(mT)x(t - mT)T \qquad (12.17)$$

Evaluating eq. (12.17) at $t = nT$ yields

$$y(nT) = \lim_{T \to 0} \sum_{m=-\infty}^{\infty} h_c(mT)x(nT - mT)T \qquad (12.18)$$

For small T,

$$y(nT) \doteq \sum_{m=-\infty}^{\infty} Th_c(mT)x(nT - mT) \qquad (12.19)$$

Since $y(nT)$, $h_c(mT)$, and $x(nT - mT)$ are sequences of samples of the continuous-time signals, we can write eq. (12.19) as

$$y_T(n) \doteq \sum_{m=-\infty}^{\infty} Th_{cT}(m)x_T(n - m) \qquad (12.20)$$

where $y_T(n), h_{cT}(m)$, and $x_T(n-m)$ are discrete-time signals. When we compare eq. (12.20) with eq. (12.14), we see that the unit pulse response of the discrete-time system used to approximate the specified continuous-time system is

$$h_T(n) = Th_{cT}(n) \qquad (12.21)$$

This unit pulse response is T times the discrete-time signal consisting of samples of the continuous-time system impulse response.

Note that

$$h_{cT}(n) = (1/T)\, h_T(n) \qquad (12.22)$$

Thus $h_{cT}(n)$, consisting of samples of the continuous-time system impulse response, is the response of the discrete-time system to the input signal $(1/T)\delta_T(n)$. The signal $(1/T)\delta_T(n)$ is the sampled version of a continuous-time signal that approaches an impulse function as T approaches zero (see Figure 10.9 and the accompanying discussion in Chapter 10).

The discrete-time system, the unit pulse response of which is $Th_{cT}(n)$ produces only approximations to the desired continuous-time output-signal samples from continuous-time input-signal samples because eq. (12.19) only approximates eq. (12.18). The approximation is better for some input signals than it is for others; it improves as T becomes smaller. We will consider it further when we discuss the design of discrete-time filters in Chapter 15.

Causal Systems

In Chapter 1, we defined a causal system as one for which the value of the output signal at a specified time depends only on values of the input signal at the specified time and/or at times preceding the specified time. For a linear time-invariant system, the portion of the output signal at sample number n produced by the input signal is the zero-initial-condition response of the system, which is given by the superposition sum as

$$y_T(n) = \sum_{i=-\infty}^{\infty} h_T(i) x_T(n-i) \qquad (12.23)$$

In eq. (12.23), the input-signal samples that follow sample number n are those for which $i < 0$. Therefore, a sufficient condition for a linear time-invariant, discrete-time system to be causal is that its unit pulse response be zero for $n < 0$. As we indicated for continuous-time systems, this condition is actually a necessary and sufficient condition. That is, a linear, time-invariant, discrete-time system is causal if and only if $h_T(n) = 0$ for $n < 0$. Since convolution is commutative, then

$$y_T(n) = \sum_{i=-\infty}^{\infty} x_T(i) h_T(n-i)$$

$$= \sum_{i=-\infty}^{n} x_T(i) h_T(n-i) \qquad (12.24)$$

for a causal system, where the second form follows from the fact that $h_T(n-i) = 0$ for $i > n$. Equation (12.24) clearly shows that the output signal at sample number n (that is, at $t = nT$) depends only on input-signal samples at sample number n and/or at preceding sample numbers (that is, at $t \le nT$). Note that the systems in Examples 12.1 and 12.3 are causal systems and that the system in Example 12.2 is noncausal.

Systems with Nonzero Initial Conditions

We can analyze discrete-time systems with nonzero initial conditions using a technique similar to the one we used for continuous-time systems with nonzero initial conditions. That is, we replace the initial conditions by single-pulse inputs directly at the delay (energy-storage) components and then analyze the resulting system with zero initial conditions. We will not discuss this procedure in detail here, for the reason that systems with nonzero initial conditions can generally be analyzed more easily using the transform techniques of Chapter 14. However, we include the following example to illustrate the concept and technique.

Example 12.4

A second-order system processes the difference between successive image-intensity samples $x_T(n)$ obtained from a satellite to produce a signal $y_T(n)$

that is used to analyze image-intensity variations. The system is characterized by the difference equation

$$y_T(n) + 0.3y_T(n-1) - 0.1y_T(n-2) = x_T(n) - x_T(n-1)$$

Find the system response to the general input signal $x_T(n)$ for $n \geq 0$ when the initial conditions are $y_T(-1) = Y_1$ and $y_T(-2) = Y_2$.

Solution
We can write the system difference equation as

$$y_T(n) = -0.3y_T(n-1) + 0.1y_T(n-2) + x_T(n) - x_T(n-1)$$

We can draw the system block diagram easily by first constructing $y_T(n)$ as the output of an adder and then constructing the proper delays, multiplies, and signal paths to provide the needed input signals to this adder. The resulting system block diagram is shown in Figure 12.3.

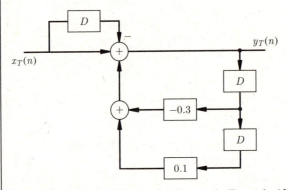

Figure 12.3 Block Diagram for System in Example 12.4

We then substitute the initial conditions $y_T(-1) = Y_1$ and $y_T(-2) = Y_2$ and the input-signal samples for $n = -1, 0,$ and 1 in the system difference equation to obtain the output-signal samples for $n = 0$ and $n = 1$. The samples obtained are

$$y_T(0) = -0.3y_T(-1) + 0.1y_T(-2) + x_T(0) - x_T(-1)$$
$$= -0.3Y_1 + 0.1Y_2 + x_T(0) - x_T(-1) \qquad (12.25)$$

and

$$y_T(1) = -0.3y_T(0) + 0.1y_T(-1) + x_T(1) - x_T(0)$$
$$= -0.3y_T(0) + 0.1Y_1 + x_T(1) - x_T(0) \qquad (12.26)$$

Next, we compute output-signal samples recursively for $n \geq 2$ by repeated use of the system difference equation; that is,

$$y_T(n) =$$

$$-0.3y_T(n-1) + 0.1y_T(n-2) + x_T(n) - x_T(n-1) \qquad n \geq 2 \qquad (12.27)$$

The same output-signal samples are obtained for the system with initial conditions changed to zero [that is, $y_T(-1) = y_T(-2) = 0$] if we write eqs. (12.25) through (12.27) as

$$y_T(0) = -0.3y_T(-1) + 0.1y_T(-2) - 0.3Y_1 + 0.1Y_2 + x_T(0) - x_T(-1)$$

$$y_T(1) = -0.3y_T(0) + 0.1y_T(-1) + 0.1Y_1 + x_T(1) - x_T(0)$$

$$y_T(n) = -0.3y_T(n-1) + 0.1y_T(n-2) + x_T(n) - x_T(n-1) \qquad n \geq 2$$

If we use unit pulses, then we can write these three equations in terms of the single equation

$$y_T(n) = -0.3y_T(n-1) + 0.1y_T(n-2) - 0.3Y_1\delta_T(n)$$

$$+ 0.1Y_2\delta_T(n) + 0.1Y_1\delta_T(n-1) + x_T(n) - x_T(n-1) \quad (12.28)$$

for $n \geq 0$, where

$$y_T(-1) = y_T(-2) = 0$$

Using this equation to redraw the system block diagram results in the equivalent system block diagram shown in Figure 12.4.

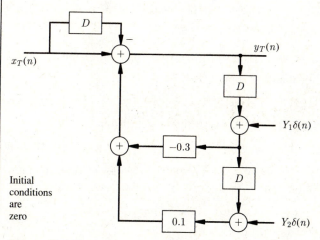

Figure 12.4 Equivalent System Block Diagram for Example 12.4

We see that the original initial conditions specifying energy stored in energy-storage components at $n = 0$ are replaced in the equivalent block diagram by inputs that establish this energy at the energy-storage-component outputs at $n = 0$.

We can write the difference equation (eq. [12.28]) in the form

$$y_T(n) + 0.3y_T(n-1) - 0.1y_T(n-2) = [-0.3Y_1 + 0.1Y_2]\,\delta_T(n)$$

$$+ 0.1Y_1\delta_T(n-1) + x_T(n) - x_T(n-1)$$

where

$$y_T(-1) = y_T(-2) = 0$$

Since the difference equation is linear, we can use superposition to solve it. For the first input-signal term,

$$y_{1T}(n) + 0.3y_{1T}(n-1) - 0.1y_{1T}(n-2) = [-0.3Y_1 + 0.1Y_2]\,\delta_T(n)$$

This difference equation has the solution

$$y_{1T}(n) = h_{bT}(n) * [-0.3Y_1 + 0.1Y_2]\,\delta_T(n)$$

$$= [-0.3Y_1 + 0.1Y_2]\,h_{bT}(n)$$

where $h_{bT}(n)$ is defined to be the unit pulse response of the system characterized by the difference equation

$$y_T(n) + 0.3y_T(n-1) - 0.1y_T(n-2) = x_T(n)$$

Note that the subscript b has been added to distinguish this unit pulse response from the system unit pulse response $h_T(n)$. The difference equation for the second input-signal term is

$$y_{2T}(n) + 0.3y_{2T}(n-1) - 0.1y_{2T}(n-2) = 0.1Y_1\delta_T(n-1)$$

and has the solution

$$y_{2T}(n) = h_{bT}(n) * 0.1Y_1\delta_T(n-1) = 0.1Y_1 h_{bT}(n-1)$$

For the third input-signal term,

$$y_{3T}(n) + 0.3y_{3T}(n-1) - 0.1y_{3T}(n-2) = x_T(n) - x_T(n-1)$$

The solution of this difference equation is

$$y_{3T}(n) = h_{bT}(n) * [x_T(n) - x_T(n-1)] = h_T(n) * x_T(n)$$

where $h_T(n)$ is the unit pulse response of the original system. Since superposition applies to the system, the total system response is

$$y_T(n) = y_{1T}(n) + y_{2T}(n) + y_{3T}(n)$$

$$= [-0.3Y_1 + 0.1Y_2]\,h_{bT}(n) + 0.1Y_1 h_{bT}(n-1)$$

$$+ x_T(n) * h_T(n) \tag{12.29}$$

The first two terms of the total system response given by eq. (12.29) for Example 12.4 are the system response to the initial conditions when the input signal is zero. This part of the response is the zero-input response. Note that the zero-input response for the second-order system in Example 12.4 consists of a weighted sum of $h_{bT}(n)$ and $h_{bT}(n-1)$, where $h_{bT}(n)$ is the unit pulse response for a second-order system corresponding to the system difference equation with all terms containing the input signal replaced by the single input-signal term $x_T(n)$. The weighting coefficients are, in turn, weighted sums of the initial-condition values $y_T(-1) = Y_1$ and $y_T(-2) = Y_2$. This characteristic of the zero-input response is true in general for a kth-order system. That is, for a kth-order system, the zero-input response is the weighted sum of $h_{bT}(n), \ldots, h_{bT}[n-(k-1)]$, where the weighting coefficients are weighted sums of the k initial-condition values $y_T(-1), \ldots, y_T(-k)$.

The last term of eq. (12.29) is the system response to the input signal when the initial conditions are zero. This part of the response is the zero-initial-condition response of the system.

12.3 Recursive Solution of Difference Equations

A useful analysis tool for a discrete-time system is a digital computer simulation of the system. The simulation is used to determine the system response to specific input signals and specific initial conditions. To implement the simulation, we successively introduce samples of the input signal and use these samples in the system difference equation to recursively compute successive samples of the output signal.

Equation (12.2) is the difference equation that relates the output signal to the input signal for a causal, linear, time-invariant, discrete-time system, when the system is of the kth order and the current output-signal sample depends directly on the current and previous l input-signal samples. It can be rearranged to yield

$$y_T(n) = \sum_{i=1}^{k} (-a_i)\, y_T(n-i) + \sum_{i=0}^{l} b_i x_T(n-i) \tag{12.30}$$

In performing the system simulation, we use eq. (12.30) to recursively compute the values of $y_T(n)$ for $n \geq n_0$ when the input-signal samples $x_T(n)$ and the k initial-condition values $y_T(n_0 - 1), \ldots, y_T(n_0 - k)$ are known. We obtain successive samples of the output signal rather than a general expression for the output signal. If the set of these samples is sufficiently large, it can be used to illustrate the response characteristics of the system.

Example 12.5

A system is represented by the block diagram shown in Figure 12.5. Find $y_T(n)$ for $0 \leq n \leq 3$ when the input signal is $x_T(n) = 2r_T(n)$ and the initial conditions are $y_T(-1) = -2$ and $y_T(-2) = 1$.

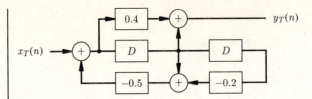

Figure 12.5 System Block Diagram for Example 12.5

Solution

To find the system difference equation, we first write equations for the signals at the outputs of all the adders in the system block diagram. To do so, we first define the the signal at the output of the leftmost adder as $w_T(n)$ and the signal at the output of the bottom adder as $z_T(n)$. From the block diagram, we write

$$z_T(n) = w_T(n-1) - 0.2w_T(n-2) \tag{12.31}$$

$$w_T(n) = -0.5z_T(n) + x_T(n) \tag{12.32}$$

and

$$y_T(n) = 0.4w_T(n) + w_T(n-1) \tag{12.33}$$

as the equations for the signals at each adder output. We substitute eq. (12.31) in eq. (12.32) and then substitute the result in eq. (12.33) to produce

$$
\begin{aligned}
y_T(n) &= 0.4\left\{-0.5\left[w_T(n-1) - 0.2w_T(n-2)\right] + x_T(n)\right\} \\
&\quad + \left\{-0.5\left[w_T(n-2) - 0.2w_T(n-3)\right] + x_T(n-1)\right\} \\
&= -0.5\left[0.4w_T(n-1) + w_T(n-2)\right] \\
&\quad + 0.1\left[0.4w_T(n-2) + w_T(n-3)\right] \\
&\quad + 0.4x_T(n) + x_T(n-1) \\
&= -0.5y_T(n-1) + 0.1y_T(n-2) + 0.4x_T(n) + x_T(n-1) \tag{12.34}
\end{aligned}
$$

as the system difference equation. The last step in producing eq. (12.34) uses the substitution $0.4w_T(n) + w_T(n-1) = y_T(n)$ obtained from eq. (12.33).

We recursively solve the system difference equation for the output signal samples to give

$$
\begin{aligned}
y_T(0) &= -0.5y_T(-1) + 0.1y_T(-2) + 0.4x_T(0) + x_T(-1) \\
&= -0.5(-2) + 0.1(1) + 0.4(0) + (0) \\
&= 1.1 \\
y_T(1) &= -0.5y_T(0) + 0.1y_T(-1) + 0.4x_T(1) + x_T(0) \\
&= -0.5(1.1) + 0.1(-2) + 0.4(2) + (0) \\
&= 0.05
\end{aligned}
$$

$$y_T(2) = -0.5y_T(1) + 0.1y_T(0) + 0.4x_T(2) + x_T(1)$$

$$= -0.5(0.05) + 0.1(1.1) + 0.4(4) + (2)$$

$$= 3.685$$

$$y_T(3) = -0.5y_T(2) + 0.1y_T(1) + 0.4x_T(3) + x_T(2)$$

$$= -0.5(3.685) + 0.1(0.05) + 0.4(6) + (4)$$

$$= 4.5625$$

Note that we can continue the recursive solution in Example 12.5 to produce additional output-signal samples. The result is shown in Figure 12.6 for $0 \leq n \leq 6$.

12.4 Summary

The determination of the time-waveform representation of the response of a discrete-time system to a specified input signal and specified initial conditions is performed using time-domain analysis of the system. In general, the determination of the system response requires the solution of the system difference equation with the specified input signal and initial conditions. An alternate method for determining the system response for a single-input–single-output, linear, time-invariant, discrete-time system with zero initial conditions involves using the superposition sum. The superposition sum is the discrete convolution of the input signal and the system unit pulse response. The system unit pulse response, $h_T(n)$, is the zero-initial-condition response of the system to a unit pulse input signal located at $t = 0$. It characterizes a single-input–single-output, linear, time-invariant, discrete-time system, except for initial conditions. A system is causal if and only if $h_T(n) = 0$ for $n < 0$ (that is, for $t < 0$).

We can extend the use of the superposition-sum method to systems with nonzero initial conditions. In the extension, we replace the initial conditions

Figure 12.6
Input and Output
Signals for Example
12.5

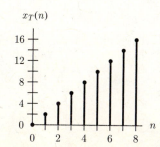

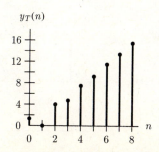

with single-pulse inputs. We then compute the output-signal components produced by the original input signal and single-pulse inputs by convolution of each input with the unit pulse response corresponding to the input location. The output-signal component produced by the original input signal is the zero-initial-condition response of the system. The sum of the output-signal components produced by single-pulse inputs is the zero-input response of the system. The sum of the zero-initial-condition response and the zero-input response is the response of the system to the original input signal and initial conditions.

A recursive solution of the system difference equation can be used to compute output-signal samples by successive substitution of input-signal samples, initial-condition values, and previously computed output-signal samples. We can simulate the operation of the discrete-time system by using a digital computer to perform the recursive solution of the system difference equation. This simulation produces the output-signal samples corresponding to specified input-signal samples and initial conditions rather than a general expression for the output signal.

Problems

12.1 A simple single-loop discrete-time feedback control system is shown in Figure 12.7. Find and plot the unit pulse response for the system.

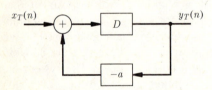

Figure 12.7

12.2 Find and plot the unit pulse response for the system shown in Figure 12.8. Show the block diagram of a much simpler equivalent system.

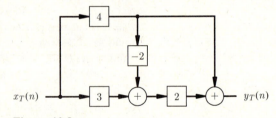

Figure 12.8

12.3 Find and plot the unit pulse response for the system shown in Figure 12.9.

12.4 Verify that the unit pulse response given for the system in Example 12.4 is

$$h_T(n) = \frac{1}{7}\left[15(-0.5)^n - 8(0.2)^n\right]u_T(n)$$

Figure 12.9

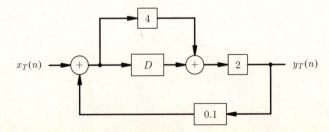

(Hint: Use solution substitution of the given solution in the difference equation.)

12.5 Verify that $h_{bT}(n)$ for Example 12.4 is

$$h_{bT}(n) = \frac{1}{7}\left[5(-0.5)^n + 2(0.2)^n\right]u_T(n)$$

(See the hint given for Problem 12.4.)

12.6 A discrete-time low-pass filter has the unit pulse response

$$h_T(n) = \left[0.85^n u(n), \quad 0.1 \text{ s}\right]$$

Find all output-signal sequence values for $|t| \leq 1.5$ s when the input signal is the cosine

$$x_T(n) = [4\cos 0.4\pi n, \quad 0.1 \text{ s}]$$

and initial conditions are zero. Repeat for a cosine with frequency twice as large. (Hint: Use a computer or programmable calculator to achieve sufficient precision.)

12.7 Electron impacts on a cathode-ray-tube surface cause it to glow. The unit pulse response of the glow intensity sampled at intervals of T seconds to an electron impact is $h_T(n) = (0.2)^n u_T(n)$. Find and plot the glow-intensity samples for the electron-impact signal $x_T(n) = u_T(n) - u_T(n - 5)$ when the glow intensity is initially zero.

12.8 A system has the unit pulse response $h_T(n) = [(0.4)^{-|n|}, \quad T]$. Find and plot the system zero-initial-condition response to the input signal

$$x_T(n) = u_T(n + 2) - u_T(n - 2)$$

12.9 A system has the impulse response $h_T(n) = (0.2)^{n+1}u_T(n)$. The input signal is $x_T(n) = 3u_T(n)$. What discrete-time function does the output signal approach for large time?

12.10 The unit pulse responses for a set of systems is shown in Figure 12.10. Which of the systems are causal systems?

a.

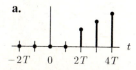

Figure 12.10

b.

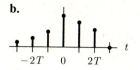

c.

d.

Figure 12.10 *(continued)*

12.11 A discrete-time filter has the unit pulse response

$$h_T(n) = 4(0.2)^n u_T(n) - 4(0.1)^n u_T(n)$$
$$+ (0.4)^n u_T(n - 2)$$

Find the zero-initial-condition response for the filter when the input signal is

$$x_T(n) = u_T(n + 1) - u_T(n - 1)$$

Plot the input signal, unit pulse response, and output signal.

12.12 It can be shown that $h_{bT}(n) = (0.8)^n u_T(n)$ and $h_T(n) = -\delta(n) + 2(0.8)^n u_T(n)$ for the system characterized by the difference equation

$$y_T(n) - 0.8y_T(n - 1) = x_T(n) + 0.8x_T(n - 1)$$

where $h_{bT}(n)$ is defined in Example 12.4. Use the technique of Example 12.4 to find the total response of the system when the input is $x_T(n) = \delta_T(n)$ and the initial condition is $y_T(-1) = -2$. Simplify the solution to the extent possible.

12.13 The third-order system with the difference equation

$$y_T(n) - 0.2y_T(n - 1) - 0.01y_T(n - 2)$$
$$+ 0.002y_T(n - 3)$$
$$= 5x_T(n) - 1.3x_T(n - 1)$$

yields

$$h_T(n) = \left[4(0.1)^n + 3(-0.1)^n - 2(0.2)^n\right]u_T(n)$$

and

$$h_{bT}(n) =$$

$$\left[-\frac{1}{2}(0.1)^n + \frac{1}{6}(-0.1)^n + \frac{4}{3}(0.2)^n\right]u_T(n)$$

where h_{bT} is defined in Example 12.4. Use the technique of Example 12.4 to find the total response of the system when the input signal is $x_T(n) = 3\delta_T(n)$ and the initial conditions are $y_T(-1) = -3$, $y_T(-2) = 0.5$, and $y_T(-3) = 2$. Simplify the solution to the extent possible.

12.14 Use a recursive solution to find the output-signal values at $0 \le t \le 6T$ for the system of Problem 12.12 if the input is changed to $x_T(n) = 2u_T(n)$.

12.15 Use a recursive solution to find the output-signal values at $0 \le t \le 6T$ for the system of Problem 12.13 if the input is changed to $x_T(n) = u_T(n) - u_T(n-5)$.

12.16 The distance of an iceberg from a ship is 2 km. Measurements of the distance to the iceberg are made with a radar on 10 successive pulses that are 10 ms apart. Noise in the measurement system causes the measurements to be in error. The sequence of 10 measurements is $x(n) = \{2.05, 2.10, 1.94, 2.03, 1.96, 1.91, 2.00, 1.99, 1.97, 2.01\}$ km. As they are received, these measurement values are smoothed with a smoothing filter characterized by the difference equation $y_T(n) = 0.9y_T(n-1) + 0.1535x_T(n)$. The output of the filter is read after the tenth measurement is received (that is, at $n = 9$). Use a recursive solution to find the filter output for $0 \le n \le 9$. The initial condition $y(-T)$ is zero. What is the reading for the smoothed measurement value?

13 Spectral Analysis of Discrete-Time Systems

The spectrum of the output signal of a discrete-time system is of interest to us because it indicates the signal's characteristics as a function of frequency. In particular, comparison of the spectrum of the portion of the output signal that is due to the input signal (that is, the zero-initial-condition response) with the spectrum of the input signal shows how components of the input signal at different frequencies are affected by the system when the signal passes through the system. Thus, in this chapter we will consider only systems with zero initial conditions as we develop spectral methods for analyzing single-input–single-output, linear, time-invariant, discrete-time systems.

The methods of analysis to be described are sometimes also referred to as *frequency-domain methods of analysis*. They parallel the spectral-analysis methods discussed in Chapter 6 for continuous-time systems. We can use them to characterize discrete-time systems and to analyze the effect of a discrete-time system on signal spectrum characteristics, just as we did for continuous-time systems in Chapter 6.

13.1 System Frequency Response

The concept of the discrete-time-system frequency response is fundamental to the frequency-domain method of analysis.

Definition _____

The frequency response $H_d(f)$ of a discrete-time system is the function of frequency that produces the spectrum $Y_d(f)$ of the output signal $y_T(n)$ when multiplied by the spectrum $X_d(f)$ of the input signal $x_T(n)$, if system initial conditions are zero.

Not all systems possess a frequency response. A system has a frequency response if $Y_d(f)$ exists when $X_d(f)$ exists; for example, if an absolutely summable input signal produces an absolutely summable output signal.

In equation form, the frequency-response definition is

$$Y_d(f) = H_d(f)X_d(f) \tag{13.1}$$

The frequency response $H_d(f)$ must be periodic in f with period equal to the input signal sample rate, $f_s = 1/T$, to produce an output signal with the same sample rate. That is, $H_d(f)$ must be periodic with period f_s when $X_d(f)$ is periodic with period f_s, so that $Y_d(f)$ will be periodic with period f_s. We illustrate this spectrum and frequency-response periodicity in Figure 13.1. The periodicity of $H_d(f)$ indicates that the inverse discrete-time Fourier transform of

Figure 13.1
Periodic Nature of
Discrete-Time-
System Frequency
Response

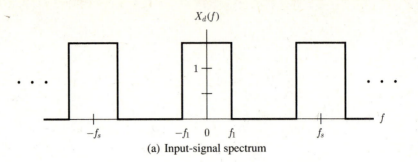

(a) Input-signal spectrum

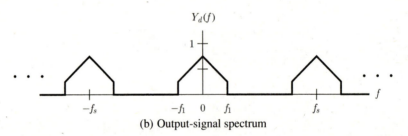

(b) Frequency response

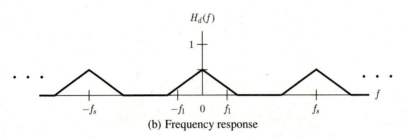

(b) Output-signal spectrum

$H_d(f)$ is a discrete-time signal with sample spacing $T = 1/f_s$. We usually plot the frequency response of a system only for the frequency interval $|f| \leq f_s/2$ (that is, the unambiguous frequency interval) for the reason that the frequency response repeats outside this interval. However, the frequency-response plots shown in Figure 13.1 and those of the first two examples in this chapter are plotted for a wider frequency interval to emphasize the periodic nature of the frequency response.

If no impulses exist in $X_d(f)$ or in $Y_d(f)$ (that is, if the input and output signals are not power signals), and if $X_d(f)$ is nonzero for all f, then we can compute the system frequency response with

$$H_d(f) = Y_d(f)/X_d(f) \tag{13.2}$$

Equation (13.2) shows that the frequency response of a system is the ratio of the spectrum of the output signal to the spectrum of the input signal in this case.

For power signals, eq. (13.1) shows that the strength of an impulse located at $f = f_1$ in the input-signal spectrum multiplied by the frequency response at

$f = f_1$ produces the strength of an impulse located at $f = f_1$ in the output-signal spectrum. However, just knowing the strengths of the impulses in the input and output signal spectra does not define the frequency response for all frequencies.

If we scale frequency values by the sample frequency $f_s = 1/T$, then we obtain the frequency-scale-normalized version, $H_N(r)$, of the frequency response. It is defined by

$$H_N(r) = Y_N(r)/X_N(r) \tag{13.3}$$

and is found from the frequency response by replacing fT by the normalized frequency r (see Section 11.5). Henceforth, we will refer to $H_N(r)$ as the *normalized frequency response* for simplicity. Following the usual convention, we will show plots of the normalized frequency response only for $|r| \leq 1/2$, just as we did for the scaled-frequency spectra in Chapter 11, because $H_N(r)$ repeats outside this interval.

Frequency-Response Characteristics

One characteristic of the frequency response of a discrete-time system we have already noted is that it is periodic in frequency with period equal to the sample frequency $f_s = 1/T$.

If we apply the discrete-time Fourier transform convolution theorem (Theorem 10 in Chapter 11) to eq. (13.1), we obtain

$$y_T(n) = \sum_{m=-\infty}^{\infty} \left\{ \mathcal{F}^{-1}[H_d(f)]\big|_{n=m} \right\} x_T(n-m) \tag{13.4}$$

Comparison of eq. (13.4) with the superposition sum given by eq. (12.14) in Chapter 12 shows that

$$h_T(n) = \mathcal{F}_d^{-1}[H_d(f)] \tag{13.5}$$

or, equivalently,

$$H_d(f) = \mathcal{F}_d[h_T(n)] \tag{13.6}$$

Therefore, we see that the system frequency response is the discrete-time Fourier transform of the system unit pulse response. Note that, except for initial conditions, the frequency response completely characterizes any discrete-time system because it is uniquely determined by the system unit pulse response which, in turn, characterizes a system except for initial conditions.

Example 13.1

A discrete-time data-filtering system operates with a sample spacing of $T = 0.1$ s (that is, a sample rate or frequency of $f_s = 10$ Hz). It has the frequency response

$$H_d(f) = \sum_{m=-\infty}^{\infty} \Lambda\left(\frac{f - 10m}{5}\right)$$

Find the output-signal spectrum, $Y_d(f)$, and the output signal, $y_T(n)$, when the input signal is

$$x_T(n) = \{[3\cos(0.4\pi n) + 4\cos(0.6\pi n)], \quad 0.1 \text{ s}\}$$

Note that the system frequency response does correspond to a system with sample rate of 10 Hz because it consists of individual portions summed together that have the same shape and are separated by 10 Hz.

Solution
Since $T = 0.1$ s, we can write the input signal in the form

$$x_T(n) = \{[3\cos(2\pi(2)nT) + 4\cos(2\pi(3)nT)], \quad 0.1 \text{ s}\}$$

From Table C.6, the input-signal spectrum is

$$X_d(f) = \sum_{m=-\infty}^{\infty} [15\delta(f - 2 - 10m) + 15\delta(f + 2 - 10m)$$

$$+ 20\delta(f - 3 - 10m) + 20\delta(f + 3 - 10m)]$$

The plots of the input-signal spectrum and system frequency response are shown in Figure 13.2.

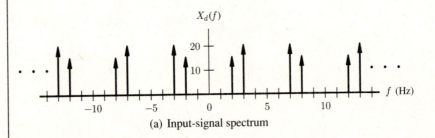

(a) Input-signal spectrum

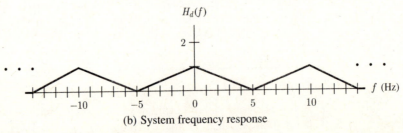

(b) System frequency response

Figure 13.2 Signal Spectra and System Frequency Response for Example 13.1

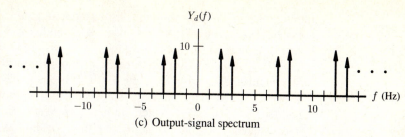

(c) Output-signal spectrum

Figure 13.2 Signal Spectra and System Frequency Response for Example 13.1 (*continued*)

The spectrum of the output signal is

$$Y_d(f) = H_d(f)X_d(f)$$

$$= \sum_{m=-\infty}^{\infty} [15H_d(2+10m)\delta(f-2-10m)$$

$$+ 15H_d(-2+10m)\delta(f+2-10m)$$

$$+ 20H_d(3+10m)\delta(f-3-10m)$$

$$+ 20H_d(-3+10m)\delta(f+3-10m)]$$

$$= \sum_{m=-\infty}^{\infty} [9\delta(f-2-10m)+9\delta(f+2-10m)$$

$$+ 8\delta(f-3-10m)+8\delta(f+3-10m)]$$

and is also plotted in Figure 13.2. From Table C.6 and the linearity theorem, we find that the output signal is

$$y_T(n) = \{[1.8\cos(2\pi(2)nT)+1.6\cos(2\pi(3)nT)], \quad 0.1 \text{ s}\}$$
$$= \{[1.8\cos(0.4\pi n)+1.6\cos(0.6\pi n)], \quad 0.1 \text{ s}\}$$

Since $h_T(n) = \mathcal{F}^{-1}[H_d(f)]$, then, from Table C.6, the unit pulse response for the system of Example 13.1 is

$$h_T(n) = 0.5 \sum_{n=-\infty}^{\infty} \text{sinc}^2(0.5n) \tag{13.7}$$

Comparing Example 13.1 with Example 6.1, we see that the input and output signals of Example 13.1 are sampled versions of the input and output signals of Example 6.1. We also see that the discrete-time-system frequency response in Example 13.1 is the sum of the continuous-time system frequency response in Example 6.1 and frequency-shifted versions of it. The unit pulse response of the discrete-time system of Example 13.1 is T times the sampled version of the impulse response of the continuous-time system of Example 6.1, as predicted by eq. (12.21) in Chapter 12. We can also see in another way why this relationship holds. Note that the frequency response of the system in Example 13.1, $H_d(f)$, is $T = 1/f_s$ times the discrete-time Fourier transform of the sampled version of the impulse response in Example 6.1. This is due to the fact that the f_s factor that occurs in the expression for the spectrum of a sampled continuous-time signal (see eq. [11.2] in Chapter 11) is not present in $H_d(f)$.

The frequency response of a discrete-time system has several properties in common with the frequency response of a continuous-time system. These properties are discussed in considerable detail in Chapter 6; here, we will treat them in brief.

The first properties are that $H_d(f)$ is in general complex and that $|H_d(f)|$ and $\underline{/H_d(f)}$ are the system amplitude response and phase response, respectively. Therefore,

$$|Y_d(f)| = |H_d(f)||X_d(f)| \tag{13.8}$$

and

$$\underline{/Y_d(f)} = \underline{/H_d(f)} + \underline{/X_d(f)} \tag{13.9}$$

Likewise, the frequency-scale-normalized amplitude and phase responses are $|H_N(r)|$ and $\underline{/H_N(r)}$, respectively. We will refer to them as the *normalized amplitude response* and *normalized phase response* for simplicity.

The amplitude response and phase response are, respectively, even and odd functions of frequency if the unit pulse response consists of real sequence values. That is,

$$|H_d(-f)| = |H_d(f)| \tag{13.10}$$

and

$$\underline{/H_d(-f)} = -\underline{/H_d(f)} \tag{13.11}$$

if $h_T(n)$ is real. We will consider only systems for which $h_T(n)$ is real because an output signal with real sequence values is desired if the input signal has real sequence values. If, in addition, a system is causal, then $\underline{/H_d(f)}$ cannot be equal to zero for all frequencies and $|H_d(f)|$ cannot be zero for any finite interval in the frequency interval $|f| \le f_s/2$. These properties are explained according to the reasoning presented in Chapter 6.

Example 13.2

A system operates with the sample spacing of $T = 0.1$ s. The block-diagram representation of this system is shown in Figure 13.3.

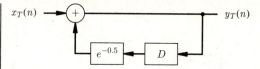

Figure 13.3 System Block Diagram for Example 13.2

Find the spectrum of the output signal and the output signal when the input signal is

$$x_T(n) = \{[3\cos(0.4\pi n) + 4\cos(0.6\pi n)], \quad 0.1 \text{ s}\}$$

Solution

We computed the spectrum of the input signal in Example 13.1. The corresponding amplitude and phase spectra of this signal are

$$|X_d(f)| = \sum_{n=-\infty}^{\infty} [15\delta(f - 2 - 10m) + 15\delta(f + 2 - 10m)$$

$$+ 20\delta(f - 3 - 10m) + 20\delta(f + 3 - 10m)]$$

and

$$\underline{/\,X_d(f)} = 0$$

These spectra are plotted in Figures 13.4a and 13.5a, respectively.

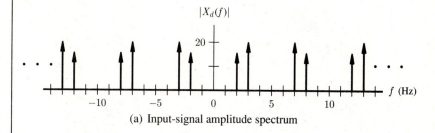

(a) Input-signal amplitude spectrum

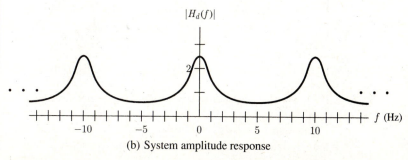

(b) System amplitude response

Figure 13.4 Amplitude Spectra and Amplitude Response for Example 13.2

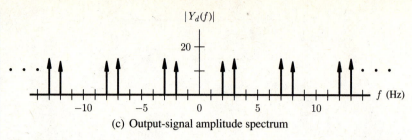

(c) Output-signal amplitude spectrum

Figure 13.4 Amplitude Spectra and Amplitude Response for Example 13.2 *(continued)*

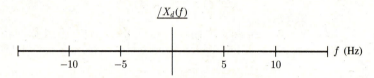

(a) Input-signal phase spectrum

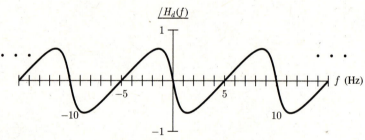

(b) System phase response

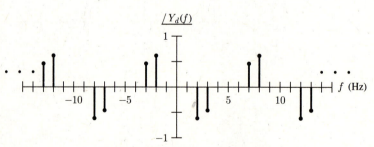

(c) Output-signal phase spectrum

Figure 13.5 Phase Spectra and Phase Response for Example 13.2

It can be easily shown from the system block diagram that the system difference equation is the first-order difference equation

$$y_T(n) = e^{-0.5}y_T(n-1) + x_T(n)$$

We found the system unit pulse response for a system described by a first-order system difference equation in Example 12.1. Using that result, we find that the unit pulse response for the system in this example is

$$h_T(n) = e^{-0.5n} u_T(n)$$

Computation of the discrete-time Fourier transform of $h_T(n)$ with $T = 0.1$ s produces the frequency response for the system. We computed the discrete-time Fourier transform of $y_T(n) = e^{-0.5n} u_T(n)$ in Example 11.4. Using those results, with $Y_d(f)$ replaced by $H_d(f)$, gives the system amplitude response

$$|H_d(f)| = 1/|1 + \exp(-0.5 - j0.2\pi f)|$$

$$= 1/\left\{ [1 - 0.607\cos(0.2\pi f)]^2 + [0.607\sin(0.2\pi f)]^2 \right\}^{0.5}$$

and system phase response

$$\underline{/\,H_d(f)} = \underline{/\,1/[1 + exp(-0.5 - j0.2\pi f)]}$$

$$= -\tan^{-1}\{0.607\sin(0.2\pi f)/[1 - 0.607\cos(0.2\pi f)]\}$$

The amplitude response and phase response are plotted in Figures 13.4b and 13.5b, respectively. Note that $|H_d(f)|$ is an even function of frequency and that $\underline{/\,H_d(f)}$ is an odd function of frequency, since $h_T(n)$ is real. Also note that $H_d(f)$ is periodic with period $f_s = 10$ Hz, as it must be for a system with a sample frequency of $f_s = 10$ Hz.

The spectrum of the output signal is

$$Y_d(f) = H_d(f)X_d(f)$$

$$= \sum_{m=-\infty}^{\infty} [15H_d(2 + 10m)\delta(f - 2 - 10m)$$

$$+ 15H_d(-2 + 10m)\delta(f + 2 - 10m)$$

$$+ 20H_d(3 + 10m)\delta(f - 3 - 10m)$$

$$+ 20H_d(-3 + 10m)\delta(f + 3 - 10m)]$$

From the equation for $H_d(f)$, we compute

$$H_d(2 + 10m) = H_d(2) = 1.004e^{-j0.617}$$

$$H_d(-2 + 10m) = H_d(-2) = 1.004e^{j0.617}$$

$$H_d(3 + 10m) = H_d(3) = 0.758e^{-j0.452}$$

$$H_d(-3 + 10m) = H_d(-3) = 0.758e^{j0.452}$$

Substituting the results in the equation for $Y_d(f)$ yields

$$Y_d(f) = \sum_{m=-\infty}^{\infty} [15(1.004)e^{-j0.617}\delta(f-2-10m)$$

$$+ 15(1.004)e^{j0.617}\delta(f+2-10m)$$

$$+ 20(0.758)e^{-j0.452}\delta(f-3-10m)$$

$$+ 20(0.758)e^{+j0.452}\delta(f+3-10m)]$$

Therefore, the amplitude spectrum of the output signal is

$$|Y_d(f)| = \sum_{m=-\infty}^{\infty} \{15.060[\delta(f-2-10m) + \delta(f+2-10m)]$$

$$+ 15.160[\delta(f-3-10m) + \delta(f+3-10m)]\}$$

and the phase spectrum of the output signal is

$$\underline{/Y_d(f)} = \begin{cases} -0.617 & f = 2+10m \\ 0.617 & f = -2+10m \\ -0.452 & f = 3+10m \\ 0.452 & f = -3+10m \end{cases} \quad \text{for all } m$$

The output-signal amplitude and phase spectra are plotted in Figures 13.4c and 13.5c, respectively.

From Table C.6 and the linearity theorem, the output signal is

$$y_T(n) = \{[3.012\cos(0.4\pi n - 0.617) + 3.032\cos(0.6\pi n - 0.452)], \quad 0.1 \text{ s}\}$$

This output signal can be expressed as

$$y_T(n)$$

$$= \{[3(1.004)\cos(0.4\pi n - 0.617) + 4(0.758)\cos(0.6\pi n - 0.452)], \quad 0.1 \text{ s}\}$$

Example 13.2 shows that if $h_T(n)$ is real, the amplitude- and phase-response values at the frequency $f = f_1$ are the amplitude multiplication (gain) and the phase shift supplied by the discrete-time system to a sampled sinusoidal input signal of frequency $f = f_1$ to produce a sampled sinusoidal output signal of frequency $f = f_1$. Thus, the amplitude and phase responses have the same characteristics as the amplitude and phase responses of a continuous-time system. Note once again that the gain may be less than unity. Typically, only the unambiguous frequencies (that is, $f < f_s/2$) are of interest because higher frequencies in a signal appear as one of these frequencies through aliasing.

We obtained the output signals in Examples 13.1 and 13.2 through spectral techniques by performing inverse discrete-time Fourier transforms. As for continuous-time systems, computation of this inverse transform is quite difficult

Figure 13.6
Example Amplitude
Response of a
Bandpass
Discrete-Time
System

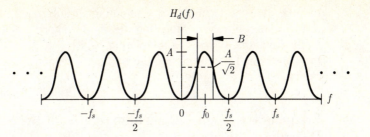

for many signals and systems of interest. Therefore, we normally use spectral-analysis techniques only to analyze spectral characteristics and the modification of these characteristics. In the next chapter, we will consider another transform technique that is more useful if output-signal time functions are of interest.

System Bandwidth

The definition for the bandwidth of a discrete-time system is of the same type as that used in Chapter 11 for a discrete-time signal. These bandwidth definitions must be compatible and, as mentioned previously, the system characteristics are really of interest only for unambiguous frequencies.

Definition _____

The bandwidth of a discrete-time system is that range of positive frequencies within the frequency range $0 \leq f \leq f_s/2$ for which the amplitude-response value is greater than α times its maximum value, where α is a selected constant.

We often select $1/\sqrt{2}$ as the value for α, just as we did for signal bandwidth in Chapter 11, to produce the half-power bandwidth. When we use this value for α, the bandwidths of the systems in Examples 13.1 and 13.2 are 1.464 Hz and 0.813 Hz, respectively. These systems are low-pass systems; thus, cutoff frequencies are the same as their bandwidths.

We illustrate the amplitude response of a bandpass discrete-time system in Figure 13.6. The bandpass system corresponding to this amplitude response has a bandwidth of B and a center frequency of f_0 when the selected value for α is $1/\sqrt{2}$.

13.2 Frequency-Response Determination

We can find the frequency response of a system by solving the system difference equation for the unit pulse response and then computing the discrete-time Fourier transform of the unit pulse response. However, the frequency response is more easily determined by computing the discrete-time Fourier transform of the system difference equation, if this difference equation is known, or directly from the system representation, if it is known. We will now discuss these methods for determining the frequency response and illustrate them with examples.

Frequency-Response Determination from the System Difference Equation

The system frequency response is the ratio of the output-signal spectrum to the input-signal spectrum of a system with zero initial conditions. Therefore, it is the ratio of the discrete-time Fourier transforms of the output and input signals when initial conditions are zero, which means that we can find it from the discrete-time Fourier transform of the system difference equation. To do so, we recall from eq. (12.2) in Chapter 12 that the general difference equation for a kth-order causal system is

$$y_T(n) + \sum_{i=1}^{k} a_i y_T(n-i) = \sum_{i=0}^{l} b_i x_T(n-i) \tag{13.12}$$

We use the linearity theorem and the time-shift theorem to compute the discrete-time Fourier transform of the signal represented by each side of the difference equation. These Fourier transforms are unique; thus, they are equal, the two sides of the equation being equal. Therefore,

$$Y_d(f) + \sum_{i=1}^{k} a_i e^{-j2\pi i fT} Y_d(f) = \sum_{i=0}^{l} b_i e^{-j2\pi i fT} X_d(f) \tag{13.13}$$

We will refer to eq. (13.13) as the *discrete-time Fourier transform of the difference equation* for simplicity. Equation (13.13) can be written

$$Y_d(f) \left[1 + \sum_{i=1}^{k} a_i e^{-j2\pi i fT} \right] = X_d(f) \sum_{i=0}^{l} b_i e^{-j2\pi i fT} \tag{13.14}$$

Since the system frequency response is $H_d(f) = Y_d(f)/X_d(f)$, then we can find it from eq. (13.14) by finding the ratio $Y_d(f)/X_d(f)$. That is,

$$H_d(f) = Y_d(f)/X_d(f)$$

$$= \left[\sum_{i=0}^{l} b_i e^{-j2\pi i fT} \right] \bigg/ \left[1 + \sum_{i=1}^{k} a_i e^{-j2\pi i fT} \right] \tag{13.15}$$

Equation (13.15) shows $H_d(f)$ as a ratio of nonfactored polynomials in $\exp(-j2\pi fT)$, the constant term in the denominator of which is equal to unity. Henceforth, we will refer to this form of $H_d(f)$ as the *standard form of the frequency response*. We can write it directly from the system difference equation because we can obtain the a_i and b_i coefficients for eq. (13.15) by inspection from the system difference equation written in the form of eq. (13.12). The normalized frequency response is obtained from eq. (13.15) by replacing fT by r to give

$$H_N(r) = \left[\sum_{i=0}^{l} b_i e^{-j2\pi ir}\right] \Bigg/ \left[1 + \sum_{i=1}^{k} a_i e^{-j2\pi ir}\right] \tag{13.16}$$

After we have determined the frequency response, we frequently wish to plot the corresponding amplitude and phase responses and/or compute their values at specific frequencies of interest to us. This is easily accomplished by

1. using Euler's theorem to write $e^{-j2\pi ifT}$ as $\cos(2\pi ifT) - j\sin(2\pi ifT)$;
2. collecting the real and imaginary terms in the numerator and denominator;
3. using rectangular-to-polar conversion of the numerator and denominator terms; and
4. computing the ratio of the numerator and denominator amplitudes and the difference of the numerator and denominator angles.

In equation form, the first three steps in the procedure are

$$H_d(f) = \frac{\displaystyle\sum_{i=0}^{l} b_i \left[\cos(2\pi ifT) - j\sin(2\pi ifT)\right]}{1 + \displaystyle\sum_{i=1}^{k} a_i \left[\cos(2\pi ifT) - j\sin(2\pi ifT)\right]}$$

$$= \frac{\left[\displaystyle\sum_{i=0}^{l} b_i \cos(2\pi ifT)\right] + j\left[\displaystyle\sum_{i=0}^{l} -b_i \sin(2\pi ifT)\right]}{\left[1 + \displaystyle\sum_{i=1}^{k} a_i \cos(2\pi ifT)\right] + j\left[\displaystyle\sum_{i=1}^{k} -a_i \sin(2\pi ifT)\right]}$$

$$= \frac{N_R(f) + jN_I(f)}{D_R(f) + jD_I(f)} \equiv \frac{|N(f)|e^{j\underline{/N(f)}}}{|D(f)|e^{j\underline{/D(f)}}} \tag{13.17}$$

and the fourth step in the procedure is

$$|H_d(f)| = |N(f)|/|D(f)| \tag{13.18}$$

and

$$\underline{/\,H_d(f)} = \underline{/\,N(f)} - \underline{/\,D(f)} \tag{13.19}$$

Equations (13.18) and (13.19) are, respectively, the amplitude response and phase response for the system. We can compute values of the scaled-frequency amplitude and phase responses in the same way by replacing fT with r.

Example 13.3

A satellite temperature measurement is sent to Earth each 5 s, as part of the data sent from the satellite. The temperature is known to vary only slowly. The data is noisy; therefore, it is processed with a smoothing filter characterized by the difference equation

$$y_T(n) = 1.3y_T(n-1) - 0.42y_T(n-2) + 3x_T(n) - x_T(n-1)$$

Find the frequency response of the data-smoothing filter. Plot the amplitude and phase responses for $|f| \le f_s/2$.

Solution
We compute the discrete Fourier transform of the difference equation to obtain

$$Y_d(f) = 1.3e^{-j2\pi fT}Y_d(f) - 0.42e^{-j4\pi fT}Y_d(f) + 3X_d(f) - e^{-j2\pi fT}X_d(f)$$

We then find the standard form of the frequency response for the smoothing filter from the transformed difference equation by inspection. The result is

$$H_d(f) = \frac{Y_d(f)}{X_d(f)} = \frac{3 - e^{-j2\pi fT}}{1 - 1.3e^{-j2\pi fT} + 0.42e^{-j4\pi fT}}$$

$$= \frac{3 - e^{-j10\pi f}}{1 - 1.3e^{-j10\pi f} + 0.42e^{-j20\pi f}}$$

To plot the smoothing-filter amplitude and phase responses for $|f| \le f_s/2 = 1/2T = 0.1$ Hz, we first write the frequency response as

$$H_d(f) = \frac{N_R(f) + jN_I(f)}{D_R(f) + jD_I(f)} = \frac{N(f)}{D(f)}$$

where

$$N_R(f) = 3 - \cos(10\pi f)$$

$$N_I(f) = \sin(10\pi f)$$

$$D_R(f) = 1 - 1.3\cos(10\pi f) + 0.42\cos(20\pi f)$$

and

$$D_I(f) = 1.3\sin(10\pi f) - 0.42\sin(20\pi f)$$

We then compute $|N(f)|$, $\underline{/N(f)}$, $|D(f)|$ and $\underline{/D(f)}$ for each value of f considered by computing $N_R(f)$, $N_I(f)$, $D_R(f)$ and $D_I(f)$ and using rectangular-to-polar conversion. This is followed by computing $|H_d(f)|$ and $\underline{/H_d(f)}$ using eqs. (13.18) and (13.19). As an illustration, for $f = 0.05$,

$$N_R(0.05) = 3 - \cos(0.5\pi) = 3$$

$$N_I(0.05) = \sin(0.5\pi) = 1$$

$$N(0.05) = 3 + j1 \rightarrow |N(0.05)| = 3.162 \qquad \underline{/N(0.05)} = 0.322 \text{ rad}$$

$$D_R(0.05) = 1 - 1.3\cos(0.5\pi) + 0.42\cos(\pi) = 0.58$$

$$D_I(0.05) = 1.3\sin(0.5\pi) - 0.42\sin(\pi) = 1.3$$

$$D(0.05) = 0.58 + j1.3 \rightarrow |D(0.05)| = 1.424 \qquad \underline{/D(0.05)} = 1.215 \text{ rad}$$

$$|H(0.05)| = |N(0.05)|/|D(0.05)| = 2.221$$

$$\underline{/H(0.05)} = \underline{/N(0.05)} - \underline{/D(0.05)} = -0.829$$

Actually, computations are best carried out by using a programmable calculator or a personal computer. The resulting amplitude and phase responses are plotted in Figure 13.7.

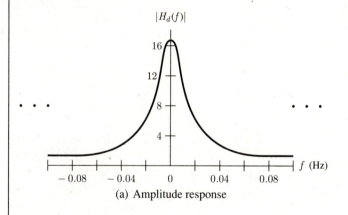

(a) Amplitude response

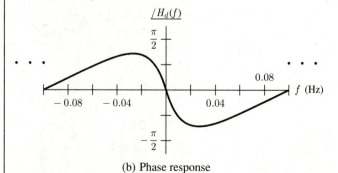

(b) Phase response

Figure 13.7 Amplitude and Phase Responses of the Data-Smoothing Filter in Example 13.3

The frequency response in Example 13.3 indicates that the data-smoothing filter is a low-pass filter that attenuates the higher frequency noise components. It can be easily determined that the filter bandwidth is 0.0091 Hz when the bandwidth definition used is the half-power bandwidth (that is, $\alpha = 1/\sqrt{2}$). The filter cutoff frequency corresponds to signal components that have a period of $1/0.0091 = 109.9$ s, or approximately 2 min. Thus, the smoothing filter is effective for temperature values that have no components with periods under 2 min.

Example 13.4

a. Find the normalized frequency response of the system represented by the block diagram shown in Figure 13.8. Also plot the normalized amplitude and phase responses of the system.

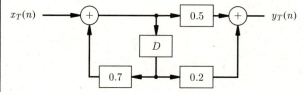

Figure 13.8 System Block Diagram for Example 13.4

b. Find the system output signal if the system input signal consists of samples of $2\cos[2\pi(10)t]$ taken with a sample spacing of 0.02 s.
c. Find the system output signal if the system input signal consists of samples of $2\cos[2\pi(25)t]$ taken with a sample spacing of 8 ms.

Solution

a. We find the system difference equation by writing the expression for the signal at the output of each adder and then combining these equations to eliminate internal signals from the resulting expression. Therefore, we first define the signal at the output of the left adder as $w_T(n)$. Then, the signals at the outputs of the two adders are

$$w_T(n) = x_T(n) + 0.7w_T(n-1)$$

and

$$y_T(n) = 0.5w_T(n) + 0.2w_T(n-1)$$

We substitute the first equation in the second equation to produce

$$y_T(n) = 0.5\left[x_T(n) + 0.7w_T(n-1)\right] + 0.2[x_T(n-1) + 0.7w_T(n-2)]$$

$$= 0.7\left[0.5w_T(n-1) + 0.2w_T(n-2)\right] + 0.5x_T(n) + 0.2x_T(n-1)$$

$$= 0.7y_T(n-1) + 0.5x_T(n) + 0.2x_T(n-1)$$

as the system difference equation. Computation of the discrete-time Fourier transform of the difference equation gives

$$Y_d(f) = 0.7e^{-j2\pi fT}Y_d(f) + 0.5X_d(f) + 0.2e^{-j2\pi fT}X_d(f)$$

We solve this equation for the system frequency response, the result being

$$H_d(f) = \frac{Y_d(f)}{X_d(f)} = \frac{0.5 + 0.2e^{-j2\pi fT}}{1 - 0.7e^{-j2\pi fT}}$$

Substitution of r for fT yields the normalized frequency response

$$H_N(r) = \frac{0.5 + 0.2e^{-j2\pi r}}{1 - 0.7e^{-j2\pi r}}$$

$$= \frac{[0.5 + 0.2\cos(2\pi r)] - j0.2\sin(2\pi r)}{[1 - 0.7\cos(2\pi r)] + j0.7\sin(2\pi r)}$$

The normalized amplitude and phase responses are

$$|H_N(r)| = \frac{|[0.5 + 0.2\cos(2\pi r)] - j0.2\sin(2\pi r)|}{|[1 - 0.7\cos(2\pi r)] + j0.7\sin(2\pi r)|}$$

and

$$\underline{/H_N(r)} = \underline{/[0.5 + 0.2\cos(2\pi r)] - j0.2\sin(2\pi r)}$$
$$- \underline{/[1 - 0.7\cos(2\pi r)] + j0.7\sin(2\pi r)}$$

They are plotted in Figure 13.9.

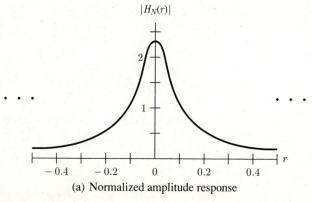

(a) Normalized amplitude response

Figure 13.9 Normalized Amplitude and Phase Responses of the System in Example 13.4

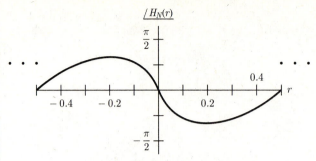

(b) Normalized phase response

Figure 13.9 Normalized Amplitude and Phase Responses of the System in Example 13.4 *(continued)*

b. The first input signal specified is

$$x_T(n) = \{2\cos[2\pi(10)nT], \quad 0.02 \text{ s}\}$$

The frequency of this signal is $f_1 = 10$ Hz and the sample frequency is $f_s = 1/T = 1/0.02 = 50$ Hz. The normalized frequency of the input signal is $r_1 = f_1/f_s = 10/50 = 0.2$. The normalized frequency is also apparent from

$$x_T(n) = \{2\cos[2\pi(0.2)n], \quad 0.02 \text{ s}\}$$

obtained by substituting the value for T in the preceding expression for the discrete-time input-signal sequence. The gain and phase shift supplied by the system to this input signal are $|H(0.2)|$ and $\underline{/\,H(0.2)}$, respectively. Therefore, the output signal is

$$y_T(n) = \{2|H_N(0.2)|\cos[2\pi(0.2)n + \underline{/\,H_N(0.2)}], \quad 0.02 \text{ s}\}$$

$$= \{1.154\cos[2\pi(0.2n - 0.328\pi)], \quad 0.02 \text{ s}\}$$

c. The second input signal specified is

$$x_T(n) = \{2\cos[2\pi(25)nT], \quad 8 \text{ ms}\}$$

$$= \{2\cos[2\pi(0.2)n], \quad 8 \text{ ms}\}$$

and once again the normalized frequency is 0.2. Therefore, the output signal is

$$y_T(n) = \{2|H_N(0.2)|\cos[2\pi(0.2)n + \underline{/\,H_N(0.2)}], \quad 8 \text{ ms}\}$$

$$= \{1.154\cos[2\pi(0.2n - 0.328\pi)], \quad 8 \text{ ms}\}$$

The sequence values for the input signals in parts (b) and (c) of Example 13.4 are exactly the same because both signals have the same normalized frequency. However, the samples of the first signal are 2.5 times as far apart as those of the second signal. The system difference equation does not depend on T, and thus the output-signal sequence values are the same for both output signals, but once again are 2.5 times as far apart for the first signal as for the second. The input and output signals are illustrated in Figure 13.10. For the second signal, the system has the same difference equation as for the first signal, but it is operating at a rate that is 2.5 times higher than that for the first signal. At this higher operating rate, the system bandwidth is 2.5 times wider; thus, the higher frequency second signal passes through the system with the same gain and phase shift as the first signal did with the lower operating rate. The wider bandwidth is illustrated in Figure 13.11, where the amplitude responses are shown for the system when operating rates correspond to the two sample rates of the signals.

Figure 13.10
Input and Output
Signals for
Example 13.4

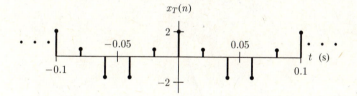

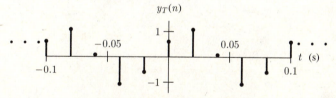

(a) Input and output signals for part b

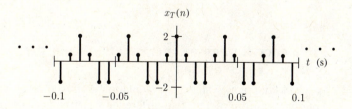

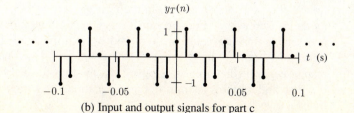

(b) Input and output signals for part c

Figure 13.11
Amplitude Response
of the System in
Example 13.4 at Two
Different Sample
Rates of Operation

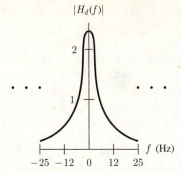

(b) Operating at a sample rate of 50 Hz
to correspond to first signal

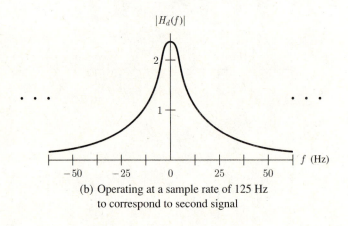

(b) Operating at a sample rate of 125 Hz
to correspond to second signal

Frequency-Response Determination from the System Representation

A block-diagram system representation corresponding to the discrete-time Fourier transform of a system difference equation is referred to as the system's *transformed block diagram*. It looks exactly like the block-diagram representation of the system, except that in the transformed version signals are replaced by their discrete-time Fourier transforms (spectra) and blocks corresponding to one sample-spacing delay are replaced by multiplication blocks with a multiplier of $e^{-j2\pi fT}$, since $x_T(n-1) \leftrightarrow e^{-j2\pi fT} X_d(f)$. The system frequency response can be determined directly from the transformed block diagram without first finding the system difference equation. This procedure is best shown by examples.

Example 13.5

Find the frequency response for the system represented by the block diagram shown in Figure 13.8 (that is, for the system of Example 13.4).

Solution
First we draw the transformed system block diagram, as shown in Figure 13.12.

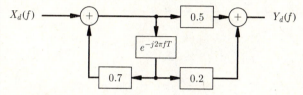

Figure 13.12 Transformed System Block Diagram for the System Shown in Figure 13.8

We then define the spectrum of the signal at the output of the first adder as $W_d(f)$ so that we can write the expressions for the spectra of the signals at the output of each adder. These expressions are

$$W_d(f) = X_d(f) + 0.7e^{-j2\pi fT}W_d(f)$$

and

$$Y_d(f) = 0.5W_d(f) + 0.2e^{-j2\pi fT}W_d(f)$$

Solving the first equation for $W_d(f)$, we obtain

$$W_d(f) = X_d(f)/\left(1 - 0.7e^{-j2\pi fT}\right)$$

Then we substitute this expression for $W_d(f)$ in the second equation and find the ratio $Y_d(f)/X_d(f)$ to determine the system frequency response

$$H_d(f) = \frac{Y_d(f)}{X_d(f)} = \frac{0.5 + 0.2e^{-j2\pi fT}}{1 - 0.7e^{-j2\pi fT}}$$

Since this system is the same as that in Example 13.4, then its frequency response is the same.

If we want to obtain the normalized frequency response, then the transformed block diagram will contain frequency-scale-normalized spectra for signals and $e^{-j2\pi r}$ multiplier blocks for T-second-delay blocks.

Example 13.6

Find the normalized frequency response for the system represented by the block diagram in Figure 13.13.

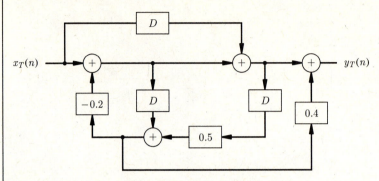

Figure 13.13 System Block Diagram for Example 13.6

Solution

We first draw the transformed system block diagram in terms of normalized frequency. It is shown in Figure 13.14.

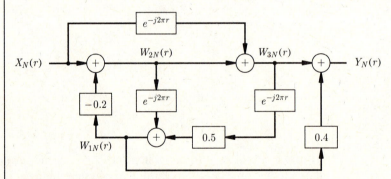

Figure 13.14 Transformed System Block Diagram for Example 13.6

The transformed equations for the system are given by the expressions obtained for the normalized spectra of the signals at the output of each adder. These equations are

$$W_{1N}(r) = e^{-j2\pi r}W_{2N}(r) + 0.5e^{-j2\pi r}W_{3N}(r) \tag{13.20}$$

$$W_{2N}(r) = -0.2W_{1N}(r) + X_N(r) \tag{13.21}$$

$$W_{3N}(r) = W_{2N}(r) + e^{-j2\pi r}X_N(r) \tag{13.22}$$

$$Y_N(r) = 0.4W_{1N}(r) + W_{3N}(r) \tag{13.23}$$

We solve these equations simultaneously to find the relationship between $Y_N(r)$ and $X_N(r)$. The solution is accomplished by first substituting eq. (13.21) in eq. (13.22) to give

$$W_{3N}(r) = -0.2W_{1N}(r) + \left(1 + e^{-j2\pi r}\right)X_N(r) \tag{13.24}$$

We then substitute eqs. (13.21) and (13.24) in eq. (13.20) to obtain

$$W_{1N}(r) = -0.2e^{-j2\pi r}W_{1N}(r) + e^{-j2\pi r}X_N(r)$$
$$-0.1e^{-j2\pi r}W_{1N}(r) + 0.5\left(e^{-j2\pi r} + e^{-j4\pi r}\right)X_N(r) \tag{13.25}$$

Collecting terms in eq. (13.25) yields

$$\left(1 + 0.3e^{-j2\pi r}\right)W_{1N}(r) = \left(1.5e^{-j2\pi r} + 0.5e^{-j4\pi r}\right)X_N(r) \tag{13.26}$$

Next, we solve eq. (13.26) for $W_{1N}(r)$. The result is

$$W_{1N}(r) = \left(1.5e^{-j2\pi r} + 0.5e^{-j4\pi r}\right)X_N(r)/\left(1 + 0.3e^{-j2\pi r}\right) \tag{13.27}$$

Substituting eq. (13.27) in eq. (13.24) gives

$$W_{3N}(r) = \left[\left(-0.3e^{-j2\pi r} - 0.1e^{-j4\pi r}\right)/\left(1 + 0.3e^{-j2\pi r}\right)\right.$$
$$\left. + \left(1 + e^{-j2\pi r}\right)\right]X_N(r)$$
$$= \left(1 + e^{-j2\pi r} + 0.2e^{-j4\pi r}\right)X_N(r)/\left(1 + 0.3e^{-j2\pi r}\right) \tag{13.28}$$

Equations (13.27) and (13.28) are now substituted in eq. (13.23) to find the normalized spectrum of the output signal in terms of the normalized spectrum of the input signal. The result is

$$Y_N(r) = \left[\left(0.6e^{-j2\pi r} + 0.2e^{-j4\pi r}\right)/\left(1 + 0.3e^{-j2\pi r}\right)\right.$$
$$\left. + \left(1 + e^{-j2\pi r} + 0.2e^{-j4\pi r}\right)/\left(1 + 0.3e^{-j2\pi r}\right)\right]X_N(r)$$
$$= \left(1 + 1.6e^{-j2\pi r} + 0.4e^{-j4\pi r}\right)X_N(r)/\left(1 + 0.3e^{-j2\pi r}\right) \tag{13.29}$$

Dividing both sides of eq. (13.29) by $X_N(r)$ gives the normalized frequency response

$$H_N(r) = \frac{Y_N(r)}{X_N(r)} = \frac{1 + 1.6e^{-j2\pi r} + 0.4e^{-j4\pi r}}{1 + 0.3e^{-j2\pi r}}$$

13.3 Summary

Spectral-analysis techniques for discrete-time systems are useful for determining the spectrum of the output signal from the spectrum of the input signal and in characterizing discrete-time systems. A fundamental concept that we use in the spectral analysis of systems is that of system frequency response. The system frequency response is defined to be the function of frequency that produces the output-signal spectrum when multiplied by the input-signal spectrum. For a discrete-time system, the system frequency response is a continuous-frequency function that is periodic with period equal to the sample frequency. The amplitude and angle of the frequency response as a function of frequency are referred to as the amplitude response and phase response, respectively. The frequency-scale-normalized frequency response is defined in the same way as the frequency-scale-normalized spectrum of a signal. That is, the frequency-normalization factor is the sample frequency. We refer to the frequency-scale-normalized frequency response as the normalized frequency response for simplicity.

When the system unit pulse response is a real sequence and the input signal consists of samples of a sinusoid, then the system amplitude and phase responses at the frequency of the sinusoidal signal give the amplitude change (gain) and phase shift experienced by the signal in passing through the system. This result shows that the system frequency response provides the gain and phase-shift characteristics of the system for input-signal components with any frequency. Also, the system frequency response is the discrete-time Fourier transform of the system unit pulse response. The discrete-time system bandwidth is determined by the nature of the system frequency response in the frequency period covered by $|f| \leq f_s/2$, just as the signal bandwidth is.

We can find the system frequency response without having to solve the system difference equation by using the discrete-time Fourier transform of the system difference equation or the discrete-time Fourier transform of the system representation. Theoretically, we could use the resulting frequency response to find the system unit pulse response by computing the inverse discrete-time Fourier transformation of the frequency response. This procedure would eliminate the need to solve the system difference equation directly. However, the inverse discrete-time Fourier transform often is not easy to compute; thus, transform techniques described in the next chapter are frequently used to find the system unit pulse response.

Problems

To solve many of the following problems, computer or programmable calculator solutions should be used to generate sufficient data for the plots requested.

13.1 A system is operating at a sample rate of $f_s = 50$ Hz and has the frequency response

$$H_d(f) = \sum_{m=-\infty}^{\infty} \Pi\left(\frac{f - 50m}{20}\right)$$

The system input signal is

$$x_T(n) = \{[4\cos(0.16\pi n) + 2\cos(0.32\pi n) + \cos(0.48\pi n)], \quad 0.02 \text{ s}\}$$

Find the amplitude and phase spectra of the input signal, the system amplitude and phase responses, and the amplitude and phase spectra of the output signal. Plot them for $|f| \leq 3f_s/2$.

13.2 Plot the input and output signals corresponding to Problem 13.1 for $|t| \leq 0.25$ s.

13.3 Find and plot the normalized amplitude and phase responses for the system in Problem 13.1.

13.4 Repeat Problem 13.1, but this time with the frequency response

$$H_d(f) = \sum_{m=-\infty}^{\infty}$$

$$\left[\Pi\left(\frac{f - 50m}{20}\right) + \Lambda\left(\frac{f - 50m}{10}\right)e^{-j0.16\pi(f-50m)}\right]$$

and the following input signal:

$$x_T(n) = \{[5\cos(0.16\pi n - 0.3)$$

$$+ 2\sin(0.32\pi n + 0.1)], \quad 0.02 \text{ s}\}$$

13.5 Find and plot the normalized amplitude and phase responses for the system in Problem 13.4.

13.6 Plot the input and output signals corresponding to Problem 13.4 for $|t| \le 0.25$ s.

13.7 Prove that a system's amplitude response and phase response are, respectively, even and odd functions of f when the system's unit pulse response is real.

13.8 A system used to transmit measurements of the liquid level in a tank produces an output signal with the spectrum

$$Y_d(f) = \sum_{m=-\infty}^{\infty} 6/[1 + j0.25(f - 40m)]$$

when the input measurement signal has the spectrum

$$X_d(f) = \sum_{m=-\infty}^{\infty} 6/[1 + j0.1(f - 40m)]$$

What sample spacing is being used? Plot the amplitude and phase spectra of the input and output signals and the system amplitude and phase responses for $|f| \le f_s/2$. The frequency response given by the amplitude and phase response is of interest because it can be used to estimate the rate of response and time delay of the measurement system. Hint: Use only those summation terms that contribute significantly over $|f| \le f_s/2$.

13.9 An image enhancement system is designed to emphasize the components of image-intensity samples that correspond to samples of sinusoids with frequencies near 25 Hz. The system produces an output signal with the spectrum

$$Y_d(f) = \sum_{m=-\infty}^{\infty} 1250 \Big/ \left\{ \left[625 - (f - 100m)^2\right] \right.$$

$$\left. + j12(f - 100m) \right\}$$

when the input image intensity sample signal has the spectrum

$$X_d(f) = \sum_{m=-\infty}^{\infty} 3/[1 + j0.04(f - 100m)]$$

What sample spacing is being used? Plot the amplitude and phase spectra of the input and output signals and the system amplitude and phase responses for $|f| \le f_s/2$. (The hint for Problem 13.8 also applies in this case.) Note that the amplitude response exhibits the desired characteristic.

13.10 A system produces an output signal with the spectrum

$$Y_d(f) = 2/\left[1 - \exp(-0.25 - j0.05\pi f)\right]$$

when its input signal has the spectrum

$$X_d(f) = 4/\left[1 - \exp(-0.5 - j0.05\pi f)\right]$$

What sample spacing is being used? Plot the amplitude and phase spectra of the input and output signals and the system amplitude and phase responses for $|f| \le f_s/2$.

13.11 Find and plot the normalized amplitude and phase responses for the system of Problem 13.9.

13.12 A system operating at a sample rate of $f_s = 100$ Hz has the frequency response

$$H_d(f) = \sum_{m=-\infty}^{\infty} \Pi\left(\frac{f - 100m}{50}\right)e^{-j0.2\pi(f-100m)}$$

and the input signal

$$x_T(n) = \left[4e^{-0.02n}u(n), \quad 0.01 \text{ s}\right]$$

Find the amplitude and phase spectra of the input signal, the system amplitude and phase responses, and the amplitude and phase spectra of the output signal. Plot them for $|f| \le 3f_s/2$.

13.13 A system has the frequency response

$$H_d(f) = 1/\left[1 - \exp(-0.2 - j0.002\pi f)\right]$$

The input signal to the system consists of samples of a cosine signal with amplitude of 10, frequency of 200 Hz, and phase shift of -0.1 rad. The signal samples are taken at a rate corresponding to the system sample rate. Find expressions for the input and output signals and plot them for $|t| \leq 0.005$ s.

13.14 A system has the frequency response

$$H_d(f) = \sum_{m=-\infty}^{\infty} 75 \Big/ \Big\{ \Big[25 - (f - 25m)^2 \Big]$$

$$+ j20(f - 25m) \Big\}$$

The input signal to the system consists of samples of a cosine signal with amplitude of 4, frequency of 6 Hz, and phase shift of 0.2 rad. The signal samples are taken at a rate corresponding to the system sample rate. Find expressions for the input and output signals and plot them for $|t| \leq 0.4$ s.

13.15 Determine the approximate bandwidth and cut-off frequency or frequencies for the system of Problem 13.4 if $\alpha = 1/\sqrt{2}$ is used to define the system bandwidth.

13.16 Repeat Problem 13.15 using the system of Problem 13.13.

13.17 Repeat Problem 13.15 using the system of Problem 13.14.

13.18 Find the frequency responses for the systems characterized by the following difference equations and sample spacings.

a. $y_T(n) + 0.2y_T(n-1) = x_T(n)$ $T = 0.001$ s

b. $y_T(n) - 0.3y_T(n-1)$
$$= 2x_T(n) - x_T(n-2) \quad T = 5 \text{ s}$$

c. $y_T(n) + 0.4y_T(n-1) - 0.25y_T(n-2)$
$$+ 0.01y_T(n-3)$$
$$= x_T(n) + 3x_T(n-2) - x_T(n-3)$$
$$T = 0.1 \text{ s}$$

13.19 Find the normalized frequency responses for the systems in Problem 13.18.

13.20 The system characterized by the difference equation

$$y_T(n) + 0.5y_T(n-1) = x_T(n) - x_T(n-1)$$

is used to process temperature measurements to analyze their variation. The measurement spacing (sample spacing) is $T = 0.2$ s. Find and plot the amplitude and phase responses for the system.

13.21 A system is characterized by the difference equation

$$y_T(n) - 1.5y_T(n-1) + y_T(n-2)$$
$$= x_T(n) + 0.5x_T(n-1)$$

and the sample frequency $f_s = 1000$ Hz.

a. Find and plot the system's amplitude and phase responses.

b. Find expressions for the discrete-time input and output signals when the input signal consists of samples of a cosine signal with amplitude = 4, frequency = 50 Hz, and phase = 0.15 rad.

13.22 The difference equation

$$y_T(n) + 1.2y_T(n-1) + 0.85y_T(n-2) = x_T(n)$$

characterizes a system that transmits discrete-time process-control signals having a sample spacing of $T = 0.1$ s.

a. Find and plot the system's amplitude and phase responses.

b. Find expressions for the discrete-time input and output signals when the input signal consists of samples of a sine signal with amplitude = 20, frequency = 2 Hz, and phase = -0.2 rad.

13.23 A system is characterized by the difference equation

$$y_T(n) - 0.25y_T(n-2) = x_T(n)$$

a. Find and plot the system's normalized amplitude and phase responses.

b. Given that the system input signal consists of samples of the continuous-time signal

$$x(t) = 5\cos(10\pi t - 0.25)$$

find the expression for the output discrete-time signal if the signal and system sample rate is $f_s = 15$ Hz.

c. Repeat part (b) with $f_s = 30$ Hz.

13.24 A system is characterized by the difference equation

$$y_T(n) - 1.55y_T(n-1) + 0.6y_T(n-2)$$

$$= 0.5\,[x_T(n) - x_T(n-1)]$$

a. Find and plot the system's normalized amplitude and phase responses.

b. Given that the system input signal consists of samples of the continuous-time signal

$$x(t) = 14\sin(250\pi t)$$

find the expression for the discrete-time output signal if the signal and system sample rate is $f_s = 1.5$ kHz.

c. Repeat part (b) with $f_s = 5$ kHz.

13.25 A system that provides smoothing for a spacecraft attitude-control signal is characterized by the block-diagram representation shown in Figure 13.15. The system sample rate is $f_s = 1$ kHz. Find and plot the system amplitude and phase responses.

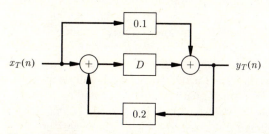

Figure 13.15

13.26 The block diagram shown in Figure 13.16 represents a discrete-time-signal transmission system. Find and plot the normalized amplitude and phase responses for the system. Also, write the expression for $y_T(n)$ if $x_T(n)$ consists of samples of the continuous-time signal

$$x(t) = 3\sin(500\pi t - 0.25)$$

and the signal and system sample rate is $f_s = 2$ kHz.

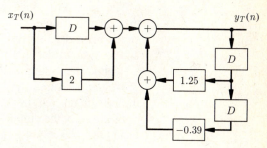

Figure 13.16

13.27 A system is represented by the block-diagram shown in Figure 13.17. The system input signal is

$$x_T(n) = \left[e^{-0.2n},\quad T\right]$$

Find and plot (a) the normalized amplitude and phase spectra of the input signal, (b) the normalized amplitude and phase responses of the system, and (c) the normalized amplitude and phase spectra of the output signal.

Figure 13.17

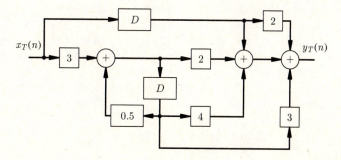

14 Analysis of Discrete-Time Systems Using the z-Transform

In Chapter 1, we indicated that a single-input–single-output, linear, time-invariant, discrete-time system can be characterized by an ordinary linear difference equation with constant coefficients. We can use the difference equation, a specified input signal, and initial conditions to solve for the corresponding output signal (system response).

The system response, when initial conditions are zero, can be found by using the superposition sum—a technique that relates the output signal to the input signal for a discrete-time system when the system is linear and time-invariant and has zero initial conditions. This technique, described in Chapter 12, requires us to find the system unit pulse response.

We can also find the system output signal from its spectrum by computing the inverse discrete-time Fourier transform of the spectrum. When initial conditions are zero, the output-signal spectrum is found by using the spectral-analysis techniques described in Chapter 13. These techniques require that we compute the spectrum of the input signal using the transform techniques described in Chapter 11 and determine the system frequency response. We can find the system frequency response directly from the system difference equation or system representation without solving the system difference equation. However, it may be difficult for us to evaluate the output signal through the inverse discrete-time Fourier transform of its spectrum, even for relatively simple output signals.

In this chapter, we will consider an alternate technique for finding the system response. This technique, called the *z-transform*, parallels the Laplace transform technique that we defined and discussed in Chapter 7 for continuous-time signals and systems. It is applicable to systems with either zero or nonzero initial conditions. After initial definition of the z-transform, we focus our attention on the single-sided z-transform, just as we did in discussing the Laplace transform in Chapter 7. The single-sided z-transform is adequate for analyzing causal systems having inputs that are zero for $t < 0$. These cases are of particular interest to us because we can usually choose the time reference so that input signals are zero for $t < 0$. Limiting our consideration to the single-sided z-transform produces unique transforms for each signal and simplifies transform-convergence considerations.

The single-sided z-transform of the system unit pulse response produces a function of z that we call the *system transfer function*. This transfer function parallels the transfer function defined for continuous-time systems in Chapter 7 and is useful in characterizing the system.

14.1 The z-Transform

In this section, we will first define the double-sided z-transform of a discrete-time signal and the corresponding inverse z-transform. We will then consider a special case of the double-sided z-transform, the single-sided z-transform, and state conditions for its existence and some of its general properties. We will also use the ideally sampled continuous-time-signal representation of a discrete-time signal in an alternate definition of the single-sided z-transform to call attention to parallels between the analysis of discrete-time and continuous-time signal and systems.

Because the single-sided z-transform is so useful in the analysis of discrete-time signals and systems, we will discuss it in considerable detail following this introductory section. In these discussions, the single-sided z-transform will be called simply the *z-transform*, except when its single-sided nature must be emphasized.

The Double-Sided z-Transform

To begin the development leading to the definition of the double-sided z-transform, consider the signal $r^{-n}x_T(n)$ where r^{-n} is a convergence factor with r chosen so that the signal is absolutely summable. Therefore, the discrete-time Fourier transform of $r^{-n}x_T(n)$ exists and is the convergent sum

$$\mathcal{F}_d\big[r^{-n}x_T(n)\big] = \sum_{n=-\infty}^{\infty} r^{-n}x_T(n)e^{-j2\pi nfT}$$

$$= \sum_{n=-\infty}^{\infty} x_T(n)\Big(re^{j2\pi fT}\Big)^{-n} \qquad (14.1)$$

We cannot choose r so that $r^{-n}x_T(n)$ will be absolutely summable for all signals $x_T(n)$. However, we can do so for a wide variety of useful signals, including some for which $x_T(n)$ is not absolutely summable; for example, the unit step signal $u_T(n)$.

The quantity

$$z \equiv re^{j2\pi fT} \qquad (14.2)$$

is a complex variable that varies with different values of r and f for any given sample spacing T. A plane surface with orthogonal coordinate axes along which the real and imaginary components of z are plotted is called the *z-plane*. The points in the z-plane correspond to values of z.

Definition ───

The double sided z-transform $\mathcal{Z}_D[x_T(n)]$ of the signal $x_T(n)$ is

$$\mathcal{Z}_D[x_T(n)] = \mathcal{F}_d\big[r^{-n}x_T(n)\big]\Big|_{re^{j2\pi fT}=z}$$

$$= \sum_{n=-\infty}^{\infty} x_T(n)z^{-n} \equiv X_D(z) \qquad (14.3)$$

That is, the double-sided z-transform of $x_T(n)$ is the discrete-time Fourier transform of $r^{-n}x_T(n)$ written as a function of z. Note that the z-transform is a function only of the signal sequence values and the complex variable z. Therefore, if the sequence values are not a function of the sample spacing T, then the z-transform does not depend on T. That is,

$$\mathcal{Z}_D\big[\{x(n),\quad 0.1\text{ s}\}\big] = \mathcal{Z}_D\big[\{x(n),\quad 5\text{ s}\}\big] \tag{14.4}$$

when the values of the sequence $x(n)$ are not a function of T.

The inverse double-sided z-transform is developed in the following manner. Using the inverse discrete-time Fourier transform given by eq. (11.14), we obtain

$$r^{-n}x_T(n) = T \int_{\frac{-1}{2T}}^{\frac{1}{2T}} \mathcal{F}_d\big[r^{-n}x_T(n)\big]e^{j2\pi nfT}\,df \tag{14.5}$$

since $f_s = 1/T$. Therefore,

$$x_T(n) = T \int_{\frac{-1}{2T}}^{\frac{1}{2T}} \mathcal{F}_d\big[r^{-n}x_T(n)\big]\left(re^{j2\pi fT}\right)^n\,df \tag{14.6}$$

For a given sample spacing T and a fixed value of r,

$$dz = d\Big[re^{j2\pi fT}\Big] = j2\pi Tre^{j2\pi fT}\,df = (j2\pi Tz)\,df \tag{14.7}$$

Therefore,

$$df = dz/j2\pi Tz \tag{14.8}$$

Also, the complex variable z traverses the circle of radius r in the z-plane from an angle of $2\pi(-\frac{1}{2T})T = -\pi$ to $2\pi(\frac{1}{2T})T = \pi$ as f varies from $-\frac{1}{2T}$ to $\frac{1}{2T}$. Therefore, making the change of variable $re^{j2\pi fT} = z$ in eq. (14.6) produces

$$x_T(n) = \frac{1}{2\pi j} \oint_r \Big\{ \mathcal{F}_d[r^{-n}x_T(n)]\big|_{re^{j2\pi fT}=z}\Big\} z^{n-1}\,dz$$

$$= \frac{1}{2\pi j} \oint_r X_D(z)z^{n-1}\,dz \equiv \mathcal{Z}_D^{-1}[X_D(z)] \tag{14.9}$$

where $\oint_r$ is an integral in the z-plane in the counterclockwise direction around a circle, the radius of which is r, centered at the origin. Since eq. (14.9) produces $x_T(n)$ from its double-sided z-transform $X_D(z)$, it is the inverse double-sided z-transform of $X_D(z)$, and is so indicated in eq. (14.9) by the appropriate notation.

The evaluation of the inverse double-sided z-transform requires the integration around the circle of radius r in the z-plane. This integration curve in the z-plane is illustrated in Figure 14.1. The radius of the circle must be chosen so that the circle will be within the region of the z-plane where eq. (14.3) exists and converges to $X_D(z)$. We can find this region of convergence if $X_D(z)$ exists for $x_T(n)$.

Figure 14.1
Circle of Integration
for Inverse
z-Transform

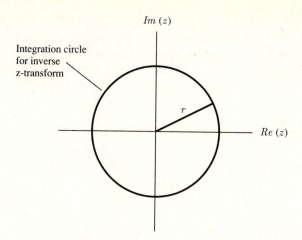

In many cases, more than one signal produces the same double-sided z-transform function.[†] The difference is the region in the z-plane where the double-sided z-transform converges. Thus, the double-sided z-transform that corresponds to a particular signal consists of both the transform function and its region of convergence. This result parallels the results we saw in Chapter 7 for the Laplace transform of continuous-time signals.

In developing the concept of the double-sided z-transform, we exploited its tie to the discrete-time Fourier transform. Now, because of its utility and simplicity, we will place our emphasis on the special case of the double-sided z-transform known as the single-sided z-transform.

The Single-Sided z-Transform

We use the z-transform most often in evaluating the responses of causal, linear, time-invariant systems for which the time reference can be chosen so that all signals are zero for $t < 0$. Thus, all time functions of interest to us (that is, signals and the unit pulse response) are zero for $t < 0$, and the double-sided z-transform of the signal $x_T(n)$ is equivalent to

$$X_S(z) = \sum_{n=0}^{\infty} x_T(n)z^{-n} = \mathcal{Z}_S[x_T(n)] \qquad (14.10)$$

Equation (14.10) is referred to as the *single-sided z-transform of the signal* $x_T(n)$.

The single-sided z-transform of a general signal $x_T(n)$ automatically sets $x_T(n) = 0$ for $n < 0$ because the summation begins at $n = 0$. Therefore, the single-sided z-transform of $x_T(n)$ equals the single-sided z-transform of $x_T(n)u_T(n)$.

[†]A. V. Oppenheim and R. W. Schafer, *Discrete-Time Signal Processing* (Englewood Cliffs, N.J.: Prentice-Hall, 1989), pp. 153–164.

If the region of convergence for a single-sided z-transform exists, then it is always the region in the z plane outside a circle of radius r_c, as illustrated with the shaded regions in the examples of Figure 14.2. The region of convergence is more properly defined in terms of the existence theorem for the single-sided z-transform. The proof of this existence theorem is beyond the scope of this book; it parallels the single-sided Laplace transform existence theorem stated in Chapter 7.

> **Single-Sided z-Transform Existence Theorem**   If $x_T(n)$ is absolutely summable over the interval $0 \leq n \leq N$ for any $N > 0$ and the real parameters r_c and K can be chosen such that $\left| r_c^{-n} x_T(n) \right| \leq K$ for $n \geq N$, then the single-sided z-transform sum converges absolutely and uniformly for all z for which $|z| > r_c$.

The smallest value of r_c that we can use to satisfy the existence theorem is called the *radius of absolute convergence for the single-sided z-transform of $x_T(n)$*. This value of r_c defines the circle in the z-plane that is the inside boundary of the region of convergence. Thus, the value that we can use for r in the convergence factor r^{-n} is restricted. When $r > r_c$, then $r^{-n} x_T(n)$ approaches zero sufficiently rapidly so that the infinite sum

$$\sum_{n=0}^{\infty} \left| x_T(n) z^{-n} \right| = \sum_{n=0}^{\infty} \left| x_T(n) \right| \left| r^{-n} e^{-j2\pi n f T} \right| = \sum_{n=0}^{\infty} \left| r^{-n} x_T(n) \right| \quad (14.11)$$

converges (that is, the value of the sum exists).

Since the double-sided z-transform of the signal $x_T(n)$ is the same as the single-sided z-transform of $x_T(n)$ when $x_T(n) = 0$ for $n < 0$, we compute the inverse single-sided z-transform with the same integral we used for the double-sided z-transform. That is, the inverse single-sided z-transform is

$$x_T(n) = \frac{1}{2\pi j} \oint_r X_S(z) z^{n-1} \, dz \equiv \mathcal{Z}_S^{-1}[X_S(z)] \qquad n \geq 0 \qquad (14.12)$$

where the integration circle lies inside the convergence region for $x_T(n)$.

Figure 14.2
Example Regions of Convergence for Single-Sided z-Transforms

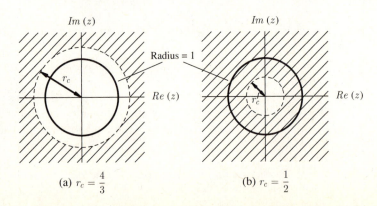

(a) $r_c = \dfrac{4}{3}$ (b) $r_c = \dfrac{1}{2}$

Since the region of convergence for every single-sided z-transform that exists is outside some circle, each single-sided z-transform corresponds to a single convergence region. Therefore, each single-sided z-transform uniquely corresponds to one and only one signal that is zero for $n < 0$, and the region of convergence does not need to be stated to completely characterize the single-sided z-transform. This uniqueness makes it easier for us to use single-sided z-transforms than double-sided z-transforms.

To be consistent with our treatment of Laplace transforms, signals and their z-transforms in Table C.8 of Appendix C are shown in terms of signals of the form $x(n)u_T(n)$, which equals $x_T(n)u_T(n)$, to call attention to the fact that the single-sided z-transform is computed only over $n \geq 0$. Some tables express the signals as $x_T(n)$, $n \geq 0$. This form of expression is equivalent to the form we use. Other tables merely express the signals as $x_T(n)$. In this last case, it must be remembered that $x_T(n)$ is actually assumed to be zero for $n < 0$, even though the signal expression does not so indicate.

Since only single-sided z-transforms are considered in most of the discussion to follow, we will refer to them simply as z-transforms and will drop the S subscript. Therefore, in the double-headed-arrow notation defined for other transforms, a z-transform pair appears as $x_T(n) \leftrightarrow X(z)$. The presence of z as the argument of $X(z)$ indicates that the transform is a z-transform.

z-Plane–s-Plane Correspondence

We can represent a discrete-time signal that is zero for $n < 0$ by the ideally sampled continuous-time signal

$$x_s(t) = \sum_{n=0}^{\infty} x_T(n)\delta(t - nT) \tag{14.13}$$

Using the Laplace transform pair $A\delta(t) \leftrightarrow A$ and the linearity and time-delay theorems for Laplace transforms, we write the Laplace transform of $x_s(t)$ as

$$X_s(s) = \sum_{n=0}^{\infty} x_T(n)e^{-snT} = \sum_{n=0}^{\infty} x_T(n)\left(e^{sT}\right)^{-n} \tag{14.14}$$

Therefore,

$$X_s(s)\big|_{e^{sT}=z} = \sum_{n=0}^{\infty} x_T(n)z^{-n} = X(z) \tag{14.15}$$

Representation of a discrete-time signal as an ideally sampled continuous-time signal points out the correspondence between the s-plane of continuous-time signals and systems and the z-plane of discrete-time signals and systems. This correspondence allows us to parallel discrete-time and continuous-time signal and system concepts in order to clarify characteristics of discrete-time signals and systems and explain their results.

The exponential transformation equation

$$z = e^{sT} \tag{14.16}$$

defines the correspondence between the s-plane and z-plane. Since $s = \sigma + j2\pi f$, then

$$z = e^{\sigma T} e^{j2\pi fT} = re^{j2\pi fT} \tag{14.17}$$

Therefore

$$|z| = r = e^{\sigma T} \tag{14.18}$$

and

$$\underline{/z} = 2\pi fT \tag{14.19}$$

Since $\sigma < 0$ yields $r < 1$, $\sigma = 0$ yields $r = 1$, and $\sigma > 0$ yields $r > 1$, then the left half of the s-plane corresponds to the inside of the circle of radius one centered at the origin in the z-plane, the imaginary axis in the s-plane corresponds to the circle of radius one in the z-plane, and the right half of the s-plane corresponds to the outside of the circle of radius one in the z-plane. We illustrate these correspondences in Figure 14.3. For simplicity, we refer to the circle of radius one as the *unit circle*. Note that many points in the s-plane map into a single point in the z-plane because

$$z = e^{(\sigma + j2\pi f)T} = e^{\sigma T} e^{j2\pi fT}$$

$$= e^{\sigma T} e^{j2\pi fT} e^{\pm j2\pi k} = e^{[\sigma + j2\pi(f \pm k/T)]T} \tag{14.20}$$

for any integer k. The points in the s-plane that map into a single point in the z-plane lie on a vertical line in the s-plane and are separated by $\Delta f = 1/T$ (that is, by $\Delta\omega = 2\pi/T$).

We can use this s-plane-to-z-plane mapping to show that the absolute convergence conditions for the Laplace transform and for the z-transform are parallel. Indeed, the single-sided Laplace transform exists if its defining integral converges for $Re[s] > c$ (see the single-sided Laplace transform existence theorem

Figure 14.3
Corresponding
Regions and Points
in z- and s-Planes

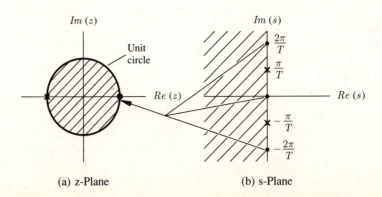

(a) z-Plane   (b) s-Plane

in Chapter 7). Since $\sigma = Re[s]$, then the region in the z-plane corresponding to the region $Re[s] > c$ is determined from

$$|z| = e^{Re[s]T} > e^{cT} = r_c \qquad (14.21)$$

The z-plane region defined by eq. (14.21) is the region outside the circle of radius r_c. This is the region in which the z-transform sum must converge for the z-transform to exist.

14.2 z-Transform Evaluations and Theorems

We will now compute the z-transforms of several simple and useful signals to illustrate z-transform computation and convergence-region determination. We will also present z-transform theorems that can be used with the z-transforms of simple signals to compute the z-transforms of more complicated signals.

z-Transforms of Simple Signals

If a discrete-time signal has nonzero values for only a finite number of values of n, then the z-transform sum is of finite length and can be evaluated easily. This is particularly true if the length is small.

Example 14.1

Find the z-transform of the single-pulse signal $x_T(n) = A\delta_T(n - m)$.

Solution

The single-pulse signal $x_T(n) = A\delta_T(n - m)$ is plotted in Figure 14.4.

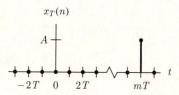

Figure 14.4 Single-Pulse Signal for Example 14.1

The z-transform of $x_T(n)$ is

$$X(z) = \sum_{n=0}^{\infty} x_T(n)z^{-n} = x_T(m)z^{-m} = Az^{-m}$$

and converges for the entire z-plane, except at the point $z = 0$.

Note that if $A = 1$ and $m = 0$, then the signal $x_T(n)$ in Example 14.1 would be the unit pulse signal and its z-transform would be 1.

Example 14.2

Find the z-transform of the finite-length signal shown in Figure 14.5.

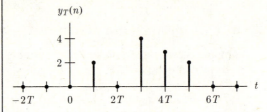

Figure 14.5 Finite-Length Signal for Example 14.2

Solution

$$Y(z) = \sum_{n=0}^{\infty} y_T(n)z^{-n}$$

$$= 0 + 2z^{-1} + 0z^{-2} + 4z^{-3} + 3z^{-4} + 2z^{-5}$$

$$= 2z^{-1} + 4z^{-3} + 3z^{-4} + 2z^{-5}$$

and converges for the entire z-plane, except at the point $z = 0$.

We can write the signal in Example 14.2 as $y_T(n) = 2\delta_T(n-1) + 4\delta_T(n-3) + 3\delta_T(n-4) + 2\delta_T(n-5)$. It is apparent that the z-transform of this signal can be written by inspection from either the signal expression or the signal plot. This is because a single pulse of height A_m that is delayed by m sample spacings from the time origin produces a term in the z-transform of the form $A_m z^{-m}$. Note that the z-transform of any finite-length signal with finite-sequence values converges for any value of z except $z = 0$.

If a discrete-time signal extends with nonzero values from $t = t_1 \geq 0$ to $t = \infty$, then the z-transform sum produces an infinite series. It is desirable to find a closed-form expression for this infinite series. Although generally this is difficult, it is simple for some very important signals.

Example 14.3

Find the z-transform of the exponential signal $x_T(n) = K^n u_T(n)$.

Solution

$$X(z) = \sum_{n=0}^{\infty} x_T(n)z^{-n} = \sum_{n=0}^{\infty} K^n z^{-n} = \sum_{n=0}^{\infty} (Kz^{-1})^n$$

$$= \frac{1}{1 - Kz^{-1}}$$

where the last expression follows because the sum is a geometric infinite series. The closed-form expression is valid (that is, the series converges) if

$$|Kz^{-1}| < 1$$

$$|K|/|z| < 1$$

$$|z| > |K|$$

This shows that the region of convergence in the z-plane for this z-transform is outside a circle of radius equal to $|K|$.

We can verify the geometric-series sum shown in Example 14.3 by performing long division to produce the infinite-series from the closed-form expression for it.

$$
1 - Kz^{-1} \overline{\big)
\begin{array}{l}
1 + Kz^{-1} + K^2z^{-2} + \cdots \\
1 \\
\underline{1 - Kz^{-1}} \\
Kz^{-1} \\
\underline{Kz^{-1} - K^2z^{-2}} \\
K^2z^{-2}
\end{array}}
\tag{14.22}
$$

If we perform the long division with the denominator terms in reverse order, then we obtain

$$
-Kz^{-1} + 1 \overline{\big)
\begin{array}{l}
-K^{-1}z - K^{-2}z^2 - \cdots \\
1 \\
\underline{1 - K^{-1}z} \\
K^{-1}z \\
\underline{K^{-1}z - K^{-2}z^2} \\
K^{-2}z^2
\end{array}}
\tag{14.23}
$$

In this case, the resulting infinite series converges if $|K^{-1}z| < 1$, and thus if $|z| < |K|$; in other words, for z inside a circle of radius $|K|$ in the z-plane. Note that the double-sided z-transform of the signal

$$y_T(n) = -K^n u_T(-n - 1) \tag{14.24}$$

is

$$Y_D(z) = \sum_{n=-\infty}^{\infty} y_T(n)z^{-n}$$

$$= \sum_{n=-\infty}^{-1} -K^n z^{-n}$$

$$= -K^{-1}z - K^{-2}z^{-2} - \cdots$$

$$= \frac{1}{1 - Kz^{-1}} \qquad |z| < |K| \tag{14.25}$$

where the last step follows from eq. (14.23). We refer to the signal $y_T(n)$ defined by eq. (14.24) as a *left-sided signal* because $y_T(n) = 0$ for $n > 0$. Likewise, the signal $x_T(n)$ in Example 14.3 is referred to as a *right-sided signal* because $x_T(n) = 0$ for $n < 0$. The double-sided z-transform of $x_T(n)$ in Example 14.3 is equal to its single-sided z-transform because the signal is equal to zero for $n < 0$. Therefore, $1/(1 - Kz^{-1})$ is the double-sided transform of the right-sided signal $K^n u_T(n)$ if the region of convergence is outside the circle of radius $|K|$; it is the double-sided transform of the left-sided signal $-K^n u_T(-n - 1)$ if the region of convergence is inside the circle of radius $|K|$.

The preceding discussion illustrates the dependence of the double-sided z-transform on the region of convergence; hence, its nonunique character. Since only single-sided z-transforms are considered here, the single-sided z-transform $1/(1 - Kz^{-1})$ uniquely corresponds to the signal $K^n u_T(n)$.

The exponential signal that we considered in Example 14.3 is quite useful in that such signals occur as the solutions of many system difference equations. Two special cases of the exponential signal are of particular importance; therefore we will present them individually. The first is the signal produced by setting $K = e^{-\alpha T}$ to give

$$x_T(n) = e^{-\alpha n T} u_T(n) \tag{14.26}$$

This discrete-time signal consists of samples of the continuous-time exponential signal $x(t) = e^{-\alpha T} u(t)$. From Example 14.3, its z-transform is

$$X(z) = \frac{1}{1 - e^{-\alpha T} z^{-1}} \tag{14.27}$$

and converges for $|z| > e^{-\alpha T}$.

The second special case of the discrete-time exponential signal is the signal produced by setting $K = 1$ to give

$$x_T(n) = (1)^n u_T(n) = u_T(n) \tag{14.28}$$

This is the discrete-time unit step signal, the z-transform of which is found to be

$$U(z) = \frac{1}{1 - z^{-1}} \tag{14.29}$$

by using the result of Example 14.3 with $K = 1$. Note that the z-transform of a unit step function converges in the region outside the unit circle.

Another signal that is often useful is the unit ramp signal

$$r_T(n) = n u_T(n) \tag{14.30}$$

illustrated by Figure 10.6 in Chapter 10. Computing the z-transform of $r_T(n)$ gives

$$
\begin{aligned}
R(z) &= \sum_{n=0}^{\infty} r_T(n) z^{-n} \\
&= z^{-1} + 2z^{-2} + 3z^{-3} + \cdots \\
&= z^{-1}\left(1 + 2z^{-1} + 3z^{-2} + \cdots\right)
\end{aligned}
\tag{14.31}
$$

The infinite series in the last step of eq. (14.31) has the closed-form sum $1/(1 - 2z^{-1} + z^{-2})$, as one can easily verify by performing long division. Therefore,

$$R(z) = \frac{z^{-1}}{1 - 2z^{-1} + z^{-2}} = \frac{z^{-1}}{\left(1 - z^{-1}\right)^2} \tag{14.32}$$

The unit step and unit ramp signals are the first two of a set of signals defined by $(n)^k$ where $k \geq 0$. The z-transforms for signals that are members of this set are given by

$$\mathcal{Z}\left[(n)^k\right] = \lim_{a \to 0} (-1)^k \frac{\partial^k}{\partial a^k}\left[\frac{1}{1 - e^{-a}z^{-1}}\right] \tag{14.33}$$

Other z-transforms of signals can also be computed directly. Table C.8 in Appendix C presents the results for a number of useful signals.

z-Transform Theorems

A number of useful theorems for z-transforms parallel those already presented for Fourier transforms, Laplace transforms, and discrete-time Fourier transforms. We will present and discuss several of the most useful theorems here. These are summarized in Table C.7 of Appendix C, which also includes some additional theorems.

Theorem 1   Linearity
If

$$x_T(n) \leftrightarrow X(z) \quad \text{and} \quad y_T(n) \leftrightarrow Y(z)$$

then

$$a x_T(n) + b y_T(n) \leftrightarrow a X(z) + b Y(z) \tag{14.34}$$

This theorem is easily proved by

$$\mathcal{Z}[ax_T(n) + by_T(n)] = \sum_{n=0}^{\infty} [ax_T(n) + by_T(n)]z^{-n}$$

$$= a \sum_{n=0}^{\infty} x_T(n)z^{-n} + b \sum_{n=0}^{\infty} y_T(n)z^{-n}$$

$$= aX(z) + bY(z)$$

Example 14.4

Find the z-transform of the sinusoidal signal $x_T(n) = \sin(bn)u_T(n)$.

Solution

Since $\sin(bn) = \left(e^{jbn} - e^{-jbn}\right)/2j$, then

$$x_T(n) = \left[\frac{e^{jbn} - e^{-jbn}}{2j}\right]u_T(n)$$

$$= \frac{1}{2j}e^{jbn}u_T(n) - \frac{1}{2j}e^{-jbn}u_T(n)$$

Using the result of Example 14.3 and the linearity theorem, we obtain

$$X(z) = \frac{1}{2j}\left[\frac{1}{1 - e^{jb}z^{-1}}\right] - \frac{1}{2j}\left[\frac{1}{1 - e^{-jb}z^{-1}}\right]$$

$$= \frac{-e^{-jb}z^{-1} + e^{jb}z^{-1}}{2j\left(1 - e^{jb}z^{-1}\right)\left(1 - e^{-jb}z^{-1}\right)}$$

$$= \left[\frac{e^{jb} - e^{-jb}}{2j}\right]\left[\frac{z^{-1}}{1 - \left(e^{jb} + e^{-jb}\right)z^{-1} + z^{-2}}\right]$$

$$= \frac{z^{-1}\sin(b)}{1 - 2\cos(b)z^{-1} + z^{-2}}$$

Theorem 2   Time Delay

If

$$x_T(n) \leftrightarrow X(z)$$

then

$$x_T(n - m) \leftrightarrow z^{-m}X(z) \qquad m \geq 0 \tag{14.35}$$

The proof of this important theorem follows. Since

$$\mathcal{Z}[x_T(n - m)] = \sum_{n=0}^{\infty} [x_T(n - m)]z^{-n} \tag{14.36}$$

then the change of variable $k = n - m$ produces

$$\mathcal{Z}[x_T(n-m)] = \sum_{k=-m}^{\infty} x_T(k)z^{-(k+m)}$$

$$= z^{-m}\sum_{k=0}^{\infty} x_T(k)z^{-k} + \sum_{k=-m}^{-1} x_T(k)z^{-(k+m)}$$

$$= z^{-m}X(z) + \sum_{k=1}^{m} x_T(-k)z^{-(m-k)} \qquad (14.37)$$

Since we are considering single-sided *z*-transforms, $x_T(k) = 0$ for $k < 0$; thus,

$$\mathcal{Z}[x_T(n-m)] = z^{-m}X(z) \qquad (14.38)$$

and the theorem is proved. Note that m cannot be less than zero. This would imply a shift of the signal to the left, and those signal samples shifted to sample locations at $t < 0$ would be cut off by the single-sided *z*-transform. Thus, the signal form would change.

When we use *z*-transforms to solve a difference equation in $y_T(n)$ to obtain $y_T(n)$ for $n \geq 0$ transforms (to be discussed in Section 14.4), we find that the intermediate result given by eq. (14.37) is also useful because the initial conditions are specified as values for $y_T(-1), \ldots, y_T(-M)$, where M is the order of the system. We indicated in Chapter 1 that these values are established by the initial energy stored in the system.

Theorem 3   *z*-Scale
If

$$x_T(n) \leftrightarrow X(z)$$

then

$$a^n x_T(n) \leftrightarrow X(z/a) \qquad (14.39)$$

This theorem parallels the *s*-shift theorem for the Laplace transform when $a = e^{-\alpha T}$. The reason that a shift of s is replaced by a scaling of z becomes apparent if we consider a vertical line in the *s*-plane. The line corresponds to a circle centered at the origin in the *z*-plane, and a shift of the vertical line to the right or left corresponds to a scaling of the radius of the circle. The *z*-scale theorem is easily proved; its proof is requested of the reader in the Problem section.

Example 14.5

Find the *z*-transform of the sinusoidal signal with exponentially changing amplitude $x_T(n) = a^n \sin(bn)u_T(n)$. (When $a < 0$, we refer to $x_T(n)$ as a damped sinusoid because its amplitude decreases with time.)

Solution

From Example 14.4,

$$\mathcal{Z}\left[\sin(bn)u_T(n)\right] = \frac{z^{-1}\sin(b)}{1 - 2\cos(b)z^{-1} + z^{-2}}$$

Therefore, by the z-scale theorem,

$$X(z) = \frac{az^{-1}\sin(b)}{1 - 2a\cos(b)z^{-1} + a^2z^{-2}}$$

Example 14.6

Find the z-transform of the signal $x_T(n) = na^{(n-1)}u_T(n)$.

Solution

We write the signal as

$$x_T(n) = a^{-1}a^n nu_T(n) = a^{-1}a^n r_T(n)$$

From eqs. (14.30) through (14.32),

$$r_T(n) \leftrightarrow \frac{z^{-1}}{\left(1 - z^{-1}\right)^2}$$

Therefore, we obtain

$$X(z) = a^{-1}\frac{az^{-1}}{\left(1 - az^{-1}\right)^2} = \frac{z^{-1}}{\left(1 - az^{-1}\right)^2}$$

by using the z-scale and linearity theorems.

Theorem 4   Convolution

If

$$x_T(n) \leftrightarrow X(z),\ y_T(n) \leftrightarrow Y(z),\ \text{and}\ x_T(n) = y_T(n) = 0,\ \text{for}\ n < 0$$

then

$$x_T(n) * y_T(n) \leftrightarrow X(z)Y(z) \tag{14.40}$$

This theorem is very important because the output signal of a linear time-invariant system with zero initial conditions is the convolution of the input signal and the unit pulse response, as we saw in Chapter 12. We construct the proof of the theorem in the same manner as we did for the convolution theorem for Laplace transforms, in this case with the integrals replaced with summations.

Example 14.7

The unit pulse response of a discrete-time filter that is used in a digital control system is

$$h_T(n) = (0.8)^n u_T(n) - (0.1)(0.8)^{n-1} u_T(n-1)$$

Find the system response to the input signal $x_T(n) = (0.1)^n u_T(n)$ when the initial conditions are zero.

Solution

Using the result of Example 14.3, the time-delay theorem, and the linearity theorem, we find that

$$H(z) = \frac{1}{1 - 0.8z^{-1}} - \frac{0.1z^{-1}}{1 - 0.8z^{-1}} = \frac{1 - 0.1z^{-1}}{1 - 0.8z^{-1}}$$

and

$$X(z) = \frac{1}{1 - 0.1z^{-1}}$$

Since the initial conditions are zero, the output signal (system response) is

$$y_T(n) = x_T(n) * h_T(n)$$

Therefore, the z-transform of the system response is

$$Y(z) = X(z)H(z) = \frac{1}{1 - 0.8z^{-1}}$$

and, by Example 14.3, the system response is

$$y_T(n) = (0.8)^n u_T(n)$$

Theorem 5 Summation

If

$$x_T(n) \leftrightarrow X(z)$$

then

$$\sum_{i=0}^{n} x_T(i) \leftrightarrow \frac{X(z)}{1 - z^{-1}} \qquad (14.41)$$

This theorem parallels the integration theorem because summation of a discrete-time signal parallels integration of a continuous-time signal. The theorem is easily proved, as we will now demonstrate. Consider the signal

$$y_T(n) = x_T(n) * u_T(n) = \sum_{i=-\infty}^{\infty} x_T(i)u_T(n-i)$$

$$= \sum_{i=0}^{n} x_T(i)u_T(n-i) = \sum_{i=0}^{n} x_T(i) \qquad (14.42)$$

where the third equivalence follows because $x_T(i) = 0$ for $i < 0$ and $u_T(n - i) = 0$ for $i > n$. The fourth equivalence follows because $u_T(n - i) = 1$ for $0 \leq i \leq n$. We use the convolution theorem and the z-transform of the unit step function given in eq. (14.29) to write

$$\mathscr{Z}\left[\sum_{i=0}^{n} x_T(i)\right] = Y(z) = X(z)U(z) = \frac{X(z)}{1 - z^{-1}} \tag{14.43}$$

which completes the proof of the theorem.

Theorem 6   Final Value
If (1) $x_T(n) \leftrightarrow X(z)$, and (2) all singularities of $(1 - z^{-1})X(z)$ [points at which $(1 - z^{-1})X(z)$ approaches ∞] are inside the unit circle, then

$$x_T(\infty) = \lim_{z \to 1} \left(1 - z^{-1}\right) X(z) \tag{14.44}$$

The proof of this theorem is beyond the scope of this text. Note that the final value of $x_T(n)$ [that is, $x_T(\infty)$] can be found with the final-value theorem if and only if all the singularities of $X(z)$ are inside the unit circle, except for possibly one at $z = 1$. The resulting finite final value of $x_T(n)$ is nonzero if and only if a singularity of $X(z)$ exists at $z = 1$.

Example 14.8

Find the final value of the signals corresponding to the following z-transforms:

a. $X(z) = \dfrac{1 + z^{-1}}{1 - 0.25z^{-2}}$

b. $Y(z) = \dfrac{2z^{-1}}{1 - 1.8z^{-1} + 0.8z^{-2}}$

c. $W(z) = \dfrac{1}{1 + 2z^{-1} - 3z^{-2}}$

Solution

a. $\left(1 - z^{-1}\right) X(z) = \dfrac{\left(1 - z^{-1}\right)\left(1 + z^{-1}\right)}{\left(1 - 0.5z^{-1}\right)\left(1 + 0.5z^{-1}\right)}$

Note that $(1 - z^{-1})X(z)$ has singularities only inside the unit circle; therefore,

$$x_T(\infty) = \lim_{z \to 1} \frac{\left(1 - z^{-1}\right)\left(1 + z^{-1}\right)}{\left(1 - 0.5z^{-1}\right)\left(1 + 0.5z^{-1}\right)} = \frac{(0)(2)}{(0.5)(1.5)} = 0$$

b. $\left(1 - z^{-1}\right) Y(z) = \dfrac{2z^{-1}\left(1 - z^{-1}\right)}{\left(1 - z^{-1}\right)\left(1 - 0.8z^{-1}\right)} = \dfrac{2z^{-1}}{1 - 0.8z^{-1}}$

Again $(1 - z^{-1})Y(z)$ has singularities only inside the unit circle. The final-value theorem yields

$$y_T(\infty) = \lim_{z \to 1} \frac{2z^{-1}}{1 - 0.8z^{-1}} = \frac{2}{0.2} = 10$$

c. $(1 - z^{-1}) W(z) = \dfrac{(1 - z^{-1})}{(1 - z^{-1})(1 + 3z^{-1})} = \dfrac{1}{1 + 3z^{-1}}$

Therefore, $(1 - z^{-1})W(z)$ has a singularity outside the unit circle and the final-value theorem cannot be used to find $w_T(\infty)$. We can show that $w_T(n) = [1/4 + 3/4(3)^n]u_T(n)$, which can be verified by computing the z-transform of $w_T(n)$. Therefore, $w_T(\infty) = \infty$, whereas the final-value theorem, if applied arbitrarily, would give the erroneous value of 0.25 for $w_T(\infty)$.

Theorem 7   Initial Value

If

$$x_T(n) \leftrightarrow X(z)$$

then

$$x_T(0) = \lim_{z \to \infty} X(z) \qquad (14.45)$$

The proof this theorem is left as an exercise for the reader.

14.3 Evaluation of Inverse z-Transforms

In Section 14.1, we showed that a discrete-time signal can be determined from its corresponding z-transform by performing an inverse z-transformation. This can be accomplished by evaluating the complex integral

$$x_T(n) = \frac{1}{2\pi j} \oint_r X(z)z^{n-1}\, dz \qquad (14.46)$$

around a circle of radius r in the z-plane, where r is greater than the radius of absolute convergence r_c. This integral evaluation requires complex-variable theory with which the reader may not be familiar.[†] Therefore, we will not discuss the direct evaluation of eq. (14.46).

In Example 14.7, we evaluated an inverse z-transform to find $y_T(n)$ by recognizing the form of $Y(z)$ as the z-transform of a known signal type (that is, an exponential signal). We can extend this procedure because linear, time-invariant, lumped-parameter systems with real-valued signals, of the type we usually encounter, result in z-transforms that are rational functions of z^{-1} and have real coefficients. That is, z-transforms of the form

$$X(z) = \frac{b_0 + b_1 z^{-1} + \cdots + b_r z^{-r}}{a_0 + a_1 z^{-1} + \cdots + a_m z^{-m}} \qquad (14.47)$$

[†]R. V. Churchill and J. W. Brown, *Complex Variables and Applications*, 5th ed. (New York: Wiley, 1990), Ch. 4.

where the a_i's and b_i's all have real values. The coefficient a_0 is not zero for a single-sided z-transform because we are only considering z-transforms of right-sided signals. This can be easily shown by performing a power-series expansion.

We can also write the z-transform expression in eq. (14.47) in the form

$$X(z) = \frac{z^m \left(b_0 z^r + b_1 z^{r-1} + \cdots + b_r \right)}{z^r \left(a_0 z^m + a_1 z^{m-1} + \cdots + a_m \right)}$$

$$= \frac{z^{(m-r)} (z - \beta_1) \cdots (z - \beta_r)}{(z - \alpha_1) \cdots (z - \alpha_m)} \tag{14.48}$$

where the values of the β_i's are the zeros of $X(z)$ [that is, the values of z for which $X(z) = 0$] and the values of the α_i's are the poles or singularities of $X(z)$ [that is, the values of z for which $X(z) = \infty$]. Some of the β_i's may be equal (repeated zeros) or equal to zero. The same is true for the α_i's. Also, $X(z)$ has $m - r$ additional zeros at $z = 0$ if $m > r$, or $r - m$ additional poles at $z = 0$ if $m < r$. The number of poles or zeros that occur at a given value of z is referred to as the *order of the pole or zero at the given value of z*.

Some β_i values may be complex. However, if one β_i value is complex, then another β_i value must be its complex conjugate. For example, if $\beta_n = c + jd$, then $X(z)$ must also contain the zero $\beta_l = c - jd$, so that the product of the two factors containing β_n and β_l will have only real coefficients. Otherwise, all the values of the b_i's in $X(z)$ would not be real. The same is true for the α_i values. Therefore, any complex poles or zeros of $X(z)$ must occur as complex-conjugate pairs. This parallels the result indicated in Chapter 7 for rational-function Laplace transforms.

Inverse Transforms of Rational Functions of z^{-1}

Since rational functions of z^{-1} occur as the z-transforms of many useful discrete-time signals, methods for computing the inverse z-transforms (signals) corresponding to rational z-transforms are of interest. The first method we will consider is simple long division, which produces a power-series expansion from which sample values of the signal are easily determined. The second method uses partial-fraction expansion of the rational function to produce a sum of low-degree rational functions, the inverse z-transforms of which are known.

Power-Series Expansion

A z-transform $X(z)$ that is a rational function of z^{-1} is easily expressed in the form of an infinite series in powers of z^{-1} by performing long division. We did this in eq. (14.22) for the z-transform $1/(1 - Kz^{-1})$. Since the single-sided z-transform is unique, then the infinite series obtained by long division equals

$$X(z) = \sum_{n=0}^{\infty} x_T(n)z^{-n} \tag{14.49}$$

Thus, the samples of the signal $x_T(n)$ corresponding to the z-transform $X(z)$ are easily determined by setting them equal to the coefficients of the appropriate power of z^{-1} in the infinite series. Before we perform the long division, however, we must arrange the terms of the numerator and denominator polynomials in z^{-1} in the order of ascending powers of z^{-1} so that a right-sided signal is produced (see eqs. [14.22] and [14.23]).

Example 14.9

Use power-series expansion to find the first four samples of the signals that correspond to the following z-transforms:

a. $X(z) = \dfrac{4 - z^{-1}}{2 - 2z^{-1} + z^{-2}}$

b. $Y(z) = \dfrac{z^{-2} + z^{-1}}{z^{-3} - z^{-2} + 2z + 1}$

Solution

a. Using long division, we obtain

$$
\begin{array}{r}
2 + 1.5x^{-1} + 0.5z^{-2} - 0.25z^{-3} + \cdots \\
2 - 2z^{-1} + z^{-2}\overline{\smash{\big)}\,4 - z^{-1}} \\
4 - 4z^{-1} + 2z^{-2} \\
\hline
3z^{-1} - 2z^{-2} \\
3z^{-1} - 3z^{-2} + 1.5z^{-3} \\
\hline
z^{-2} - 1.5z^{-3} \\
z^{-2} - 1.0z^{-3} + 0.5z^{-4} \\
\hline
-0.5z^{-3} - 0.5z^{-4}
\end{array}
$$

Therefore,

$$X(z) = \sum_{n=0}^{\infty} x_T(n)z^{-n} = 2 + 1.5z^{-1} + 0.5z^{-2} - 0.25z^{-3} + \cdots$$

and

$$x_T(0) = 2, \quad x_T(1) = 1.5, \quad x_T(2) = 0.5, \quad x_T(3) = -0.25$$

The signal can be expressed in the form

$$x_T(n) = 2\delta_T(n) + 1.5\delta_T(n-1) + 0.5\delta_T(n-2) - 0.25\delta_T(n-3) + \cdots$$

b. For $Y(z)$, we must reverse the order of the terms in the numerator and denominator polynomials before we perform long division. The resulting division is

$$
1 + 2z^{-1} - z^{-2} + z^{-3} \overline{\smash{\big)}\ \begin{array}{r} z^{-1} - z^{-2} + 3z^{-3} + \cdots \\ z^{-1} + z^{-2} \phantom{{}+ z^{-3}} \end{array}}
$$

$$
\begin{array}{r}
z^{-1} + 2z^{-2} - z^{-3} + z^{-4} \\
\hline
-z^{-2} + z^{-3} - z^{-4} \\
-z^{-2} - 2z^{-3} + z^{-4} - z^{-5} \\
\hline
3z^{-3} - 2z^{-4} + z^{-5}
\end{array}
$$

Therefore,

$$
Y(z) = \sum_{n=0}^{\infty} y_T(n)z^{-n} = z^{-1} - z^{-2} + 3z^{-3} + \cdots
$$

and

$$
y_T(0) = 0, \quad y_T(1) = 1, \quad y_T(2) = -1, \quad y_T(3) = 3
$$

Again, the signal can be expressed as

$$
y_T(n) = \delta_T(n-1) - \delta_T(n-2) + 3\delta_T(n-3) + \cdots
$$

The power-series expansion of a rational z-transform is only useful if just the first few samples of the corresponding signal are of interest to us.

Partial-Fraction Expansion

In analyzing discrete-time systems, we are usually more interested in finding a general expression for the discrete-time signal corresponding to a rational z-transform rather than in finding just the first few samples by power-series expansion. To find the general expression, we can use partial-fraction expansion and the known inverse z-transforms of low-degree rational functions.

We will now consider the use of partial-fraction expansion to perform the inverse z-transforms of rational z-transforms having no real poles of order higher than two and no complex-conjugate pairs of poles of order higher than one. These low-degree rational functions encompass a large portion of the transforms of interest to us in signal and linear system analysis. The partial-fraction-expansion method can be modified somewhat to be used with poles of higher order; however, in our introductory treatment, we will not discuss these modifications.

We will first present the necessary steps to perform the inverse z-transform with partial-fraction expansion and illustrate them with a general example as they are presented. Discussion of the steps and specific examples then follows. To illustrate the steps, the general example of a rational z-transform that we will use is

$$
X(z) = \frac{b_0 + b_1 z^{-1} + b_2 z^{-2} + b_5 z^{-5}}{a_0 + a_1 z^{-1} + a_2 z^{-2} + a_3 z^{-3} + a_4 z^{-4}} \tag{14.50}
$$

The steps are as follows:

Step 1 Multiply the numerator and denominator of $X(z)$ by the same power of z to eliminate all negative powers of z.

$$X(z) = \frac{b_0 z^5 + b_1 z^4 + b_2 z^3 + b_5}{a_0 z^5 + a_1 z^4 + a_2 z^3 + a_3 z^2 + a_4 z} \tag{14.51}$$

Step 2 Divide each coefficient of the numerator and denominator of $X(z)$ by the coefficient of the highest power of z in the denominator, so that the highest power of z in the denominator has a coefficient of 1.

$$X(z) = \frac{(b_0/a_0) z^5 + (b_1/a_0) z^4 + (b_2/a_0) z^3 + (b_5/a_0)}{z^5 + (a_1/a_0)z^4 + (a_2/a_0)z^3 + (a_3/a_0)z^2 + (a_4/a_0)z} \tag{14.52}$$

Step 3 Factor the denominator to yield

$$X(z) = \frac{(b_0/a_0) z^5 + (b_1/a_0) z^4 + (b_2/a_0) z^3 + (b_5/a_0)}{z(z - \alpha_1)(z - \alpha_2)^2 (z^2 + c_1 z + c_2)} \tag{14.53}$$

The denominator contains one real pole at $z = 0$. For illustration purposes, we have assumed that the denominator also contains a first-order real pole at $z = \alpha_1$, a second order real pole at $z = \alpha_2$, and a pair of complex-conjugate poles that give the quadratic factor $z^2 + c_1 z + c_2$. Thus, all the factors we are considering are included.

Step 4 Multiply $X(z)$ by $1/z$.

$$\frac{X(z)}{z} = \frac{(b_0/a_0) z^5 + (b_1/a_0) z^4 + (b_2/a_0) z^3 + (b_5/a_0)}{z^2(z - \alpha_1)(z - \alpha_2)^2 (z^2 + c_1 z + c_2)} \tag{14.54}$$

Step 5 Find the partial-fraction expansion of $X(z)/z$.

$$\frac{X(z)}{z} =$$

$$\frac{A_{1,1}}{z} + \frac{A_{1,2}}{z^2} + \frac{A_2}{z - \alpha_1} + \frac{A_{3,1}}{z - \alpha_2} + \frac{A_{3,2}}{(z - \alpha_2)^2} + \frac{Ez + F}{z^2 + c_1 z + c_2} \tag{14.55}$$

Note that we use the notation $A_{i,j}$ to indicate the jth coefficient corresponding to the ith pole, except when the pole is of the first order. When the pole is of the first order, there is a single coefficient; hence, we do not need or use the j index.

Step 6 Multiply $X(z)/z$ by z to obtain $X(z)$.

$$X(z) =$$

$$A_{1,1} + \frac{A_{1,2}}{z} + \frac{A_2 z}{z - \alpha_1} + \frac{A_{3,1} z}{z - \alpha_2} + \frac{A_{3,2} z}{(z - \alpha_2)^2} + \frac{Ez^2 + Fz}{z^2 + c_1 z + c_2} \tag{14.56}$$

Step 7 Multiply the numerator and denominator of each term by the negative power of z required to eliminate all positive powers of z.

$$X(z) = A_{1,1} + A_{1,2}z^{-1} + \frac{A_2}{1 - \alpha_1 z^{-1}} + \frac{A_{3,1}}{\left(1 - \alpha_2 z^{-1}\right)} + \frac{A_{3,2}z^{-1}}{\left(1 - \alpha_2 z^{-1}\right)^2}$$

$$+ \frac{E + Fz^{-1}}{1 + c_1 z^{-1} + c_2 z^{-2}} \tag{14.57}$$

Step 8 Use known z-transform pairs corresponding to the rational functions of degree no greater than two (that is, z-transform pairs from Table C.8 of Appendix C) and the linearity theorem to find the signal $x_T(n)$ corresponding to $X(z)$.

In executing step 8, we find the components of $x_T(n)$ corresponding to the first five terms of eq. (14.57) directly from the previously computed transform pairs listed in Table C.8 and the time-delay theorem. These components are

$$A_{1,1} \leftrightarrow A_{1,1}\delta_T(n) \tag{14.58}$$

$$A_{1,2}z^{-1} \leftrightarrow A_{1,2}\delta_T(n - 1) \tag{14.59}$$

$$A_2\left[\frac{1}{1 - \alpha_1 z^{-1}}\right] \leftrightarrow A_2\alpha_1^n u_T(n) \tag{14.60}$$

$$A_{3,1}\left[\frac{1}{1 - \alpha_2 z^{-1}}\right] \leftrightarrow A_{3,1}\alpha_2^n u_T(n) \tag{14.61}$$

and

$$A_{3,2}\left[\frac{z^{-1}}{\left(1 - \alpha_2 z^{-1}\right)^2}\right] \leftrightarrow A_{3,2}n\alpha_2^{(n-1)}u_T(n) \tag{14.62}$$

The partial-fraction-expansion term that corresponds to the quadratic factor produced by the complex-conjugate-pole pair does not appear in Table C.8. However, as we did for the Laplace transform in Chapter 7, we can separate this term into two terms that are in the table and correspond to cosine and sine terms with exponentially changing amplitudes. We will now illustrate this procedure.

From Table C.8,

$$a^n \cos(bn)u_T(n) \leftrightarrow \frac{1 - a\cos(b)z^{-1}}{1 - 2a\cos(b)z^{-1} + a^2 z^{-2}} \tag{14.63}$$

and

$$a^n \sin(bn)u_T(n) \leftrightarrow \frac{a\sin(b)z^{-1}}{1 - 2a\cos(b)z^{-1} + a^2 z^{-2}} \tag{14.64}$$

Therefore,

$$a^n \cos(bn)u_T(n) \leftrightarrow \frac{1 + 0.5c_1 z^{-1}}{1 + c_1 z^{-1} + c_2 z^{-2}} \tag{14.65}$$

and

$$a^n \sin(bn)u_T(n) \leftrightarrow \frac{c_3 z^{-1}}{1 + c_1 z^{-1} + c_2 z^{-2}} \tag{14.66}$$

if

$$c_1 = -2a\cos(b) \tag{14.67}$$

$$c_2 = a^2 \tag{14.68}$$

and

$$c_3 = a\sin(b) \tag{14.69}$$

Separating the sixth term of eq. (14.57) into a weighted sum of terms as in eqs. (14.65) and (14.66) gives

$$\frac{E + Fz^{-1}}{1 + c_1 z^{-1} + c_2 z^{-2}} = E\left[\frac{1 + 0.5c_1 z^{-1}}{1 + c_1 z^{-1} + c_2 z^{-2}}\right]$$

$$+ \frac{(F - 0.5Ec_1)}{c_3}\left[\frac{c_3 z^{-1}}{1 + c_1 z^{-1} + c_2 z^{-2}}\right] \tag{14.70}$$

Therefore,

$$\frac{E + Fz^{-1}}{1 + c_1 z^{-1} + c_2 z^{-2}} \leftrightarrow Ea^n\cos(bn)$$

$$+ \left(\frac{F - 0.5Ec_1}{c_3}\right)a^n\sin(bn) \tag{14.71}$$

where, from eqs. (14.67), (14.68), and (14.69),

$$a = \pm\sqrt{c_2} \tag{14.72}$$

$$b = \cos^{-1}[c_1/(-2a)] = \cos^{-1}\left[c_1/(\mp 2\sqrt{c_2})\right] \tag{14.73}$$

and

$$c_3 = a\sin(b) \tag{14.74}$$

Note that c_2 is positive and $|c_1/(\mp 2\sqrt{c_2})| \leq 1$ since the quadratic-denominator term results from a pair of complex-conjugate poles (see eq. [7.42]); therefore, we can compute the square root and the inverse cosine. Due to the ambiguous nature of samples of a sinusoid, there is more than one solution for the parameters a and b. We chose the parameter b to be the principal angle (that is, $0 \leq b \leq \pi$), with cosine $c_1/(-2a)$. This choice produces sinusoidal terms with normalized frequency r_0 in the interval $0 \leq r_0 \leq 1/2$ because $b = 2\pi r_0$ (see eq. [10.1]). Therefore, the value of b that we obtain with the choice of the principal angle leads to sinusoidal terms having normalized frequencies in the unambiguous interval (that is, $r \leq 1/2$). Also, to enhance visualization of the resulting signal, it is best to choose positive values for the parameter a, so that the damping factor a_n is nonoscillatory.

The first two steps in the preceding set of partial-fraction-expansion steps are included in order that the z-transform $X(z)$ be expressed as a ratio of polynomials in z rather than z^{-1} and have a coefficient of one for the highest power of z in the denominator. This is the polynomial form that is normally used when performing partial-fraction expansion; moreover, it simplifies notation. Alternatively, we could perform the partial-fraction expansion in terms of factors of the form $(1 - \alpha_i z^{-1})$ and $(1 + \gamma_i z^{-1} + \eta_i z^{-2})$.

When we use partial-fraction expansion with polynomials in z rather than z^{-1}, it is necessary to perform the partial-fraction expansion of $X(z)/z$, as indicated in steps 4 and 5, so that the final partial-fraction-expansion terms obtained have forms that match low-degree transforms for which inverse z-transforms are known. Steps 6 and 7 convert the partial-fraction expansion of $X(z)/z$, in terms of z, into the required partial-fraction expansion of $X(z)$, in terms of z^{-1}.

When the rational single-sided z-transform $X(z)$ is converted to a ratio of polynomials in z by step 1, then the degree of the numerator polynomial is always less than or equal to the degree of the denominator polynomial. This is because a_0 in eq. (14.47) is not equal to zero for a single-sided z-transform, as was previously discussed following eq. (14.47). Therefore, the degree of the numerator polynomial of $X(z)/z$ is less than the degree of its denominator polynomial. This means that $X(z)/z$ is a proper rational function, the partial-fraction expansion of which can be found in step 5. (Refer to Chapter 7 to review the most efficient methods for finding the coefficients for the partial-fraction-expansion terms.)

In the three examples that follow, we will illustrate the defined steps for finding the inverse z-transform of a rational function.

Example 14.10

Find the signal corresponding to the z-transform

$$X(z) = \frac{z^{-3}}{2 - 3z^{-1} + z^{-2}}$$

Solution
From steps 1, 2, and 3,

$$X(z) = \frac{0.5}{z^3 - 1.5z^2 + 0.5z} = \frac{0.5}{z(z - 1)(z - 0.5)}$$

From steps 4 and 5,

$$\frac{X(z)}{z} = \frac{0.5}{z^2(z - 1)(z - 0.5)} = \frac{A_{1,1}}{z} + \frac{A_{1,2}}{z^2} + \frac{A_2}{z - 1} + \frac{A_3}{z - 0.5}$$

where

$$A_{1,2} = z^2 [X(z)/z]\big|_{z=0} = \frac{0.5}{(0-1)(0-0.5)} = 1$$

$$A_2 = (z-1)[X(z)/z]\big|_{z=1} = \frac{0.5}{(1)^2(1-0.5)} = 1$$

and

$$A_3 = (z-0.5)[X(z)/z]\big|_{z=0.5} = \frac{0.5}{(0.5)^2(0.5-1)} = -4$$

We find $A_{1,1}$ by setting z equal to -1 ($z = -1$ not being a pole of $X(z)/z$) to produce

$$\frac{0.5}{(-1)^2(-1-1)(-1-0.5)} = \frac{1}{(-1)^2} + \frac{A_{1,1}}{-1} + \frac{1}{(-1-1)} + \frac{-4}{(-1-0.5)}$$

The solution of which is $A_{1,1} = 3$. Now, from steps 6 and 7

$$X(z) = 3 + z^{-1} + \frac{1}{1 - z^{-1}} + (-4)\left[\frac{1}{1 - 0.5z^{-1}}\right]$$

Therefore, we find the signal to be

$$x_T(n) = 3\delta_T(n) + \delta_T(n-1) + u_T(n) - 4(0.5)^n u_T(n)$$

from step 8.

Example 14.11

Find the signal corresponding to the z-transform

$$Y(z) = \frac{1}{(1 + 0.2z^{-1})(1 - 0.2z^{-1})^2}$$

Solution

$$Y(z) = \frac{z^3}{(z+0.2)(z-0.2)^2}$$

$$\frac{Y(z)}{z} = \frac{z^2}{(z+0.2)(z-0.2)^2} = \frac{A_1}{z+0.2} + \frac{A_{2,1}}{(z-0.2)} + \frac{A_{2,2}}{(z-0.2)^2}$$

where

$$A_1 = (z+0.2)[Y(z)/z]\big|_{z=-0.2} = \frac{(-0.2)^2}{(-0.2-0.2)^2} = 0.25$$

and

$$A_{2,2} = (z-0.2)^2 [Y(z)/z]_{z=0.2} = \frac{(0.2)^2}{(0.2+0.2)} = 0.1$$

To find $A_{2,1}$ we set z equal to zero in the partial-fraction-expansion equation to yield

$$\frac{(0)^2}{(0+0.2)(0-0.2)^2} = \frac{0.25}{0+0.2} + \frac{A_{2,1}}{0-0.2} + \frac{0.1}{(0-0.2)^2}$$

The solution of this equation is $A_{2,1} = 0.75$; therefore,

$$Y(z) = \frac{0.25z}{z+0.2} + \frac{0.75z}{z-0.2} + \frac{0.1z}{(z-0.2)^2}$$

$$= 0.25\left[\frac{1}{1+0.2z^{-1}}\right] + 0.75\left[\frac{1}{1-0.2z^{-1}}\right] + 0.1\left[\frac{z^{-1}}{(1-0.2z^{-1})^2}\right]$$

and, using Table C.8, we find the signal $y_T(n)$ to be

$$y_T(n) = 0.25(-0.2)^n u_T(n) + 0.75(0.2)^n u_T(n)$$

$$+ 0.1n(0.2)^{n-1} u_T(n)$$

Example 14.12

Find the signal corresponding to the z-transform

$$X(z) = \frac{2z^3 - 1.5z^2}{z^3 - 2z^2 + 1.5z - 0.5}$$

Solution

In this example, $X(z)$ is already a ratio of polynomials in z with the coefficient of the highest power of z in the denominator equal to one. Therefore steps 1 and 2 are not required. We factor the denominator polynomial to give

$$X(z) = \frac{2z^3 - 1.5z^2}{(z-1)(z^2 - z + 0.5)}$$

where the quadratic term is not factored further, since it has complex-conjugate roots. Then,

$$\frac{X(z)}{z} = \frac{2z^2 - 1.5z}{(z-1)(z^2 - z + 0.5)} = \frac{A}{z-1} + \frac{Ez + F}{z^2 - z + 0.5}$$

where

$$A = (z-1)\left[X(z)/z\right]\big|_{z=1} = \frac{2(1)^2 - 1.5(1)}{(1)^2 - 1 + 0.5} = 1$$

We set z equal to zero in the partial-fraction-expansion equation to solve for F. This gives the equation

$$\frac{2(0) - 1.5(0)}{(0-1)(0^2 - 0 + 0.5)} = \frac{1}{0-1} + \frac{E(0) + F}{0^2 - 0 + 0.5}$$

the solution of which is $F = 0.5$.

To solve for E, we set z equal to -1 in the partial-fraction-expansion equation to produce the equation

$$\frac{2(-1)^2 - 1.5(-1)}{(-1-1)\left[(-1)^2 - (-1) + 0.5\right]} = \frac{1}{-1-1} + \frac{E(-1) + 0.5}{(-1)^2 - (-1) + 0.5}$$

This equation reduces to

$$-\frac{3.5}{5} = -\frac{1}{2} - \frac{E}{2.5} + \frac{0.5}{2.5}$$

for which the solution is $E = 1$. Therefore,

$$X(z) = \frac{z}{z-1} + \frac{z^2 + 0.5z}{z^2 - z + 0.5}$$

$$= \frac{1}{1 - z^{-1}} + \frac{1 + 0.5z^{-1}}{1 - z^{-1} + 0.5z^{-2}}$$

We see from this equation that $c_1 = -1$, $c_2 = 0.5$, $E = 1$, and $F = 0.5$. From eq. (14.72),

$$a = \sqrt{c_2} = \sqrt{0.5} = 1/\sqrt{2}$$

where we have chosen the sign of a to be positive. Using eqs. (14.73) and (14.74), we obtain

$$b = \cos^{-1}[c_1/(-2a)] = \cos^{-1}\left[-1/\left(-2/\sqrt{2}\right)\right]$$

$$= \cos^{-1}\left(1/\sqrt{2}\right) = \pi/4$$

and

$$c_3 = a\sin(b) = \frac{1}{\sqrt{2}}\sin\left(\frac{\pi}{4}\right)$$

$$= \left(\frac{1}{\sqrt{2}}\right)\left(\frac{1}{\sqrt{2}}\right) = 0.5$$

Therefore,

$$\frac{F - 0.5Ec_1}{c_3} = \frac{(0.5) - 0.5(1)(-1)}{0.5} = 2$$

and we use eq. (14.71) and the inverse z-transform of $1/(1 - z^{-1})$ to obtain the signal

$$x_T(n) = u_T(n) + \left(\sqrt{2}\right)^{-n}\cos\left(2\pi n/8\right)$$

$$+ 2\left(\sqrt{2}\right)^{-n}\sin\left(2\pi n/8\right)$$

that corresponds to $X(z)$. Note that the normalized frequency of the sinusoids is $r = 1/8$; thus, it is an unambiguous frequency.

14.4 z-Transform Solutions of Linear Difference Equations

We can use the z-transform to find the solution $y_T(n)$, for $n \geq 0$, of the linear, time-invariant, constant coefficient difference equation that characterizes a causal discrete-time system. In this application, we first compute the z-transforms of the signals represented by each side of the equation. Since these two signals are equal and the z-transform is unique, then the z-transforms of the two signals are equal. Henceforth, we will refer to this z-transform computation as the *z-transform of the difference equation* for simplicity. Next, we solve the transformed difference equation algebraically for the z-transform of the output signal $y_T(n)$ and invert the resulting z-transform to find $y_T(n)$. To find the solution of an Mth-order difference equation, we must know M initial conditions in addition to the input signal $x_T(n)$. These initial conditions could be the first M values of the solution (that is, $y_T(n)$ for $0 \leq n \leq M - 1$). However, we want the solution for $y_T(n)$ to be a function of n for $n \geq 0$. Therefore, it is more convenient to specify the initial conditions as the values $y_T(-1), \ldots,$ $y_T(-M)$ that will give the correct first M values of $y_T(n)$ when substituted in the difference equation. These initial conditions are incorporated directly in the solution obtained with z-transforms when we use the intermediate result in the proof of the time-delay theorem (eq. [14.37]) as the z-transform of the delayed output signals $y_T(n - i)$ in the difference equation. For $i = 1$, 2, and 3, eq. (14.37) gives

$$\mathcal{Z}\left[y_T(n - 1)\right] = z^{-1}Y(z) + y_T(-1) \tag{14.75}$$

$$\mathcal{Z}\left[y_T(n - 2)\right] = z^{-2}Y(z) + z^{-1}y_T(-1) + y_T(-2) \tag{14.76}$$

$$\mathcal{Z}\left[y_T(n - 3)\right] = z^{-3}Y(z) + z^{-2}y_T(-1) + z^{-1}y_T(-2) + y_T(-3) \tag{14.77}$$

Note the simple pattern of decreasing negative powers of z and increasing negative arguments of $y_T(n)$. We do not need the intermediate result of eq. (14.37) to compute the z-transform of the delayed input signals $x_T(n - i)$ in the difference equation because initial conditions are not needed for the input signal. The time-delay theorem applies to the input signal in that, by assumption, the input signal is equal to zero for $n < 0$ when we use single-sided z-transforms.

Example 14.13

A second-order discrete-time system is a part of a flow-control system in a steel mill. It is characterized by the difference equation

$$y_T(n) - 0.1y_T(n - 1) - 0.02y_T(n - 1) = 2x_T(n) - x_T(n - 1)$$

Find $y_T(n)$ for $n \geq 0$ when $x_T(n) = u_T(n)$ and the initial conditions are $y_T(-1) = -10$, $y_T(-2) = 20$.

Solution

Computing the z-transform of the difference equation gives

$$Y(z) - 0.1[z^{-1}Y(z) + y_T(-1)] - 0.02[z^{-2}Y(z) + z^{-1}y_T(-1) + y_T(-2)]$$

$$= 2X(z) - z^{-1}X(z) = (2 - z^{-1})X(z)$$

Note that the time-delay theorem was used in computing the z-transform of $x_T(n-1)$ since $x_T(-1) = 0$. Since $x_T(n) = u_T(n) \leftrightarrow \frac{1}{1-z^{-1}}$, $y_T(-1) = -10$, and $y_T(-2) = 20$, then

$$Y(z) - 0.1[z^{-1}Y(z) - 10] - 0.02[z^{-2}Y(z) - 10z^{-1} + 20]$$

$$= (2 - z^{-1})/(1 - z^{-1})$$

We algebraically solve this equation for $Y(z)$ as follows:

$$(1 - 0.1z^{-1} - 0.02z^{-2})Y(z) = (2 - z^{-1})/(1 - z^{-1}) - 0.2z^{-1} - 0.6$$

$$Y(z) = \frac{1.4 - 0.6z^{-1} + 0.2z^{-2}}{(1 - z^{-1})(1 - 0.1z^{-1} - 0.02z^{-2})}$$

$$= \frac{1.4 - 0.6z^{-1} + 0.2z^{-2}}{(1 - z^{-1})(1 - 0.2z^{-1})(1 + 0.1z^{-1})}$$

$$= \frac{1.4z^3 - 0.6z^2 + 0.2z}{(z - 1)(z - 0.2)(z + 0.1)}$$

Computation of the inverse z-transform of $Y(z)$ to find $y_T(n)$ proceeds as follows:

$$\frac{Y(z)}{z} = \frac{1.4z^2 - 0.6z + 0.2}{(z - 1)(z - 0.2)(z + 0.1)} = \frac{A_1}{z - 1} + \frac{A_2}{z - 0.2} + \frac{A_3}{z + 0.1}$$

where

$$A_1 = (z - 1)[Y(z)/z]\big|_{z=1} = \frac{1.4(1)^2 - 0.6(1) + 0.2}{(1 - 0.2)(1 + 0.1)} = 1.136$$

$$A_2 = (z - 0.2)[Y(z)/z]\big|_{z=0.2} = \frac{1.4(0.2)^2 - 0.6(0.2) + 0.2}{(0.2 - 1)(0.2 + 0.1)} = -0.567$$

and

$$A_3 = (z + 0.1)[Y(z)/z]\big|_{z=-0.1} = \frac{1.4(-0.1)^2 - 0.6(-0.1) + 0.2}{(-0.1 - 1)(-0.1 - 0.2)} = 0.830$$

Therefore,

$$Y(z) = 1.136\left[\frac{1}{1 - z^{-1}}\right] - 0.567\left[\frac{1}{1 - 0.2z^{-1}}\right] + 0.830\left[\frac{1}{1 + 0.1z^{-1}}\right]$$

and the output signal $y_T(n)$ is

$$y_T(n) = 1.136u_T(n) - 0.567(0.2)^n u_T(n) + 0.830(-0.1)^n u_T(n)$$

The input and output signals are shown in Figure 14.6.

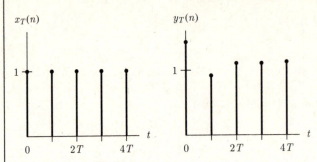

Figure 14.6 Input and Output Signals for the System in Example 14.13 with Initial Conditions $y_T(-1) = -10$ and $y_T(-2) = 20$

Example 14.14

The block diagram in Figure 14.7 represents a discrete-time data-transmission path. Although there is no time delay, the output contains interference from multiple delayed-internal-signal reflections (that is, multipath signals). Find the output signal $y_T(n)$ for the transmission-path system when the input signal is a unit pulse and the initial conditions $w_T(-1) = 0.125$ and $v_T(-1) = 0.25$ characterize the initial energy stored in the two delay components at $n = 0$.

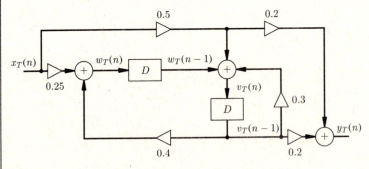

Figure 14.7 System Block Diagram for Example 14.14

Solution
We first write a set of difference equations that characterize the system by writing the expression for the signal at the output of each adder. These expressions are

$$w_T(n) = 0.25x_T(n) + 0.4v_T(n-1)$$

$$v_T(n) = w_T(n-1) + 0.5x_T(n) + 0.3v_T(n-1)$$

$$y_T(n) = 0.2v_T(n-1) + 0.1x_T(n)$$

Computation of the z-transforms of the difference equations in the equation set yields

$$W(z) = 0.25X(z) + 0.4\left[z^{-1}V(z) + v_T(-1)\right]$$

$$V(z) = \left[z^{-1}W(z) + w_T(-1)\right] + 0.5X(z) + 0.3\left[z^{-1}V(z) + v_T(-1)\right]$$

$$Y(z) = 0.2\left[z^{-1}V(z) + v_T(-1)\right] + 0.1X(z)$$

Since $x_T(n) = \delta_T(n)$, then we obtain $X(z) = 1$ from Table C.8 in Appendix C. We substitute $X(z)$ and the initial-condition values to obtain the transformed difference equations

$$W(z) = 0.25 + 0.4z^{-1}V(z) - 0.1$$

$$V(z) = z^{-1}W(z) + 0.125 + 0.5 + 0.3z^{-1}V(z) - 0.075$$

$$Y(z) = 0.2z^{-1}V(z) - 0.05 + 0.1$$

and simplify these equations to give

$$W(z) = 0.4z^{-1}V(z) + 0.15$$

$$V(z) = z^{-1}W(z) + 0.3z^{-1}V(z) + 0.55$$

$$Y(z) = 0.2z^{-1}V(z) + 0.05$$

The simultaneous solution of the first two transformed difference equations for $V(z)$ gives

$$V(z) = \frac{0.55 + 0.15z^{-1}}{1 - 0.3z^{-1} - 0.4z^{-2}}$$

We then substitute $V(z)$ in the third transformed difference equation and simplify to find the z-transform of the output. The result is

$$Y(z) = \frac{0.05 + 0.095z^{-1} + 0.01z^{-2}}{1 - 0.3z^{-1} - 0.4z^{-2}}$$

To compute the inverse z-transform of $Y(z)$ to find the system output signal, we proceed as follows:

$$Y(z) = \frac{0.05z^2 + 0.095z + 0.01}{z^2 - 0.3z - 0.4} = \frac{0.05z^2 + 0.095z + 0.01}{(z - 0.8)(z + 0.5)}$$

$$\frac{Y(z)}{z} = \frac{0.05z^2 + 0.095z + 0.01}{z(z - 0.8)(z + 0.5)} = \frac{A_1}{z} + \frac{A_2}{z - 0.8} + \frac{A_3}{z + 0.5}$$

$$A_1 = z\big[Y(z)/z\big]\big|_{z=0} = \frac{0.05(0)^2 + 0.095(0) + 0.1}{(0 - 0.8)(0 + 0.5)} = -0.025$$

$$A_2 = (z - 0.8)\big[Y(z)/z\big]\big|_{z=0.8} = \frac{0.05(0.8)^2 + 0.095(0.8) + 0.1}{(0.8)(0.8 + 0.5)} = 0.114$$

$$A_3 = (z + 0.5)\big[Y(z)/z\big]\big|_{z=-0.5} = \frac{0.05(-0.5)^2 + 0.095(-0.5) + 0.1}{(-0.5)(-0.5 - 0.8)} = -0.038$$

$$Y(z) = -0.025 + 0.114\left[\frac{1}{1 - 0.8z^{-1}}\right] - 0.038\left[\frac{1}{1 + 0.5z^{-1}}\right]$$

$$y_T(n) = -0.025\delta_T(n) + 0.114(0.8)^n u_T(n) - 0.038(-0.5)^n u_T(n)$$

The system output signal is shown in Figure 14.8. Note that it is not the unit pulse response of the system, even though the input is the unit pulse, the initial conditions being nonzero.

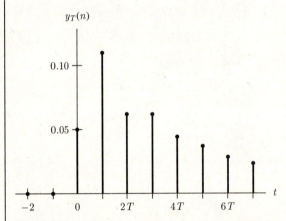

Figure 14.8 System Output Signal for Example 14.14

In Example 14.14, it was more convenient to compute the z-transform of the three difference equations obtained and then solve the transformed equations simultaneously for $Y(z)$ than to combine the three equations into a single equation for $y_T(n)$ in terms of $x_T(n)$ and then compute the z-transform of this equation. If the latter approach had been taken, then the required initial conditions would have been $y_T(-1)$ and $y_T(-2)$, and we would have had to determine them from $w_T(-1)$ and $v_T(-1)$. Note that since $w_T(-1)$ and $v_T(-1)$ define energy stored in each of the delay components at $n = 0$, they are the most convenient initial conditions. We easily obtained the set of three difference equations that characterize the system by writing expressions for the signal at the output of each adder in the system block-diagram representation.

Figure 14.9
Unit Pulse Response
for the System in
Example 14.14

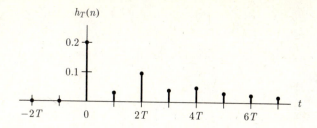

Following the procedure used in Example 14.14 with initial conditions of zero, the unit pulse response for the data-transmission path is found to be

$$h_T(n) = -0.025\delta_T(n) + 0.110(0.8)^n u_T(n) + 0.115(-0.5)^n u_T(n) \quad (14.78)$$

The unit pulse response is plotted in Figure 14.9. Since a single data pulse at $t = 0$ produces some significant output pulses at later times in addition to the desired output data pulse at $t = 0$, considerable multipath interference exists. The additional output pulses are called *intersymbol interference* because they result from one signal pulse (or symbol) and interfere with later signal pulses (or symbols).

Example 14.15

Assume we obtain a loan of C dollars to be repaid in equal installments of P dollars. The interest rate per payment period on the unpaid principal is $100\alpha\%$.

a. Find the difference equation that relates the payments and the unpaid principal after each payment.
b. Use z-transforms to solve for the unpaid principal after each payment.
c. Determine the value of each payment if the loan is to be repaid in N equal installments.
d. Apply the preceding data to determine the monthly payment on an automobile purchased for $5000 using a 36-month loan with an interest rate of 11% per year. Also compute the total cost (principal plus interest) of the automobile.

Solution

a. The difference equation is:

$$y_T(n) = y_T(n - 1) + \alpha y_T(n - 1) - x_T(n)$$

where $y_T(n)$ is the unpaid principal remaining after the nth payment and $x_T(n)$ is the value of the nth payment. Note that $y_T(0) = C$ and $x_T(n) = P u_T(n - 1)$, given that the initial principal value is C and the first payment is made at the end of the first payment period.

b. We compute the z-transform of the preceding difference equation and input signal $x_T(n)$ to give

$$Y(z) = (1 + \alpha)\left[z^{-1}Y(z) + y_T(-1)\right] - X(z)$$

and

$$X(z) = Pz^{-1}/(1 - z^{-1})$$

To find the initial condition $y_T(-1)$, we solve the difference equation at $n = 0$,

$$C = y_T(0) = (1 + \alpha)y_T(-1) + x_T(0) = (1 + \alpha)y_T(-1)$$

to obtain

$$y_T(-1) = C/(1 + \alpha)$$

Therefore,

$$Y(z) = \frac{C}{1 - (1 + \alpha)z^{-1}} - \frac{Pz^{-1}}{\left(1 - z^{-1}\right)\left[1 - (1 + \alpha)z^{-1}\right]}$$

To find $y_T(n)$, we compute $\mathcal{Z}^{-1}[Y(z)]$ as follows:

$$\frac{Y(z)}{z} = \frac{C}{z - (1 + \alpha)} - \frac{P}{(z - 1)[z - (1 + \alpha)]}$$

Partial-fraction expansion of the second term of this equation gives

$$\frac{Y(z)}{z} = \frac{C}{z - (1 + \alpha)} + \frac{P/a}{z - 1} - \frac{P/a}{z - (1 + \alpha)}$$

We then combine terms and multiply both sides of this equation by z to obtain

$$Y(z) = (C - P/a)\left[\frac{1}{1 - (1 + \alpha)z^{-1}}\right] + \frac{P}{a}\left[\frac{1}{1 - z^{-1}}\right]$$

Computation of the inverse z-transform gives

$$y_T(n) = (C - P/a)(1 + \alpha)^n u_T(n) + \frac{P}{a}u_T(n)$$

as the expression for the unpaid principal after the nth payment.
c. For the loan to be paid in full at payment N, $y_T(N) = 0$. Therefore,

$$0 = y_T(N) = (C - P/a)(1 + \alpha)^N u_T(N) + \frac{P}{a}u_T(N)$$

Solving this equation for P gives

$$P = \left[\frac{\alpha(1 + \alpha)^N}{(1 + \alpha)^N - 1}\right]C$$

d. $C = 5000$ $\alpha = (11/100)/12 = 0.00917$ $N = 36$

$$P = \left[\frac{(0.00917)(1.00917)^{36}}{(1.00917)^{36} - 1}\right] 5000$$

$$= \$163.69 \text{ per month}$$

Thus, the total cost of the automobile is $36 \times 163.69 = \$5892.84$

14.5 The System Transfer Function

The output signal for a linear, time-invariant, discrete-time system with zero initial conditions is given by the superposition sum

$$y_T(n) = \sum_{m=-\infty}^{\infty} x_T(m)h_T(n - m) \tag{14.79}$$

(see eq. [12.13] in Chapter 12). In eq. (14.79), $x_T(n)$, $y_T(n)$, and $h_T(n)$ are the input signal, output signal, and unit pulse response, respectively. In addition, if the system is causal [that is, if $h_T(n) = 0$ for $n < 0$] and if $x_T(n) = 0$ for $n < 0$, then

$$y_T(n) = \sum_{m=0}^{n} x_T(m)h_T(n - m) \tag{14.80}$$

If the z-transforms of $h_T(n)$ and $x_T(n)$ [that is, $H(z)$ and $X(z)$] exist, then the convolution theorem for z-transforms gives

$$Y(z) = X(z)H(z) \tag{14.81}$$

as the z-transform of the output signal. Therefore,

$$\mathcal{Z}[h_T(n)] = H(z) = Y(z)/X(z) \tag{14.82}$$

and we see that the z-transform of the system unit pulse response is the ratio of the z-transform of the output signal to the z-transform of the input signal when initial conditions are zero.

Definition _____

The transfer function for a discrete-time system is

$$H(z) = Y(z)/X(z) \tag{14.83}$$

where $Y(z)$ is the z-transform of the zero-initial-condition response that corresponds to the input signal with z-transform $X(z)$.

The discrete-time-system transfer function parallels the system transfer function defined in Chapter 7 for continuous-time systems. It completely characterizes the system except for initial conditions because the unit pulse response completely characterizes the system except for initial conditions and the single-sided z-transform is unique.

When the system initial conditions are zero, we can use the system transfer function to find a causal system's output signal corresponding to any z-transformable input signal that is zero for $n < 0$. The steps are as follows: (1) Compute the z-transform of the input signal, (2) multiply the resulting z-transform by the transfer function, and (3) compute the inverse z-transform of the result of step 2.

We can determine the transfer function for a system from the difference equation that relates the output signal to the input signal by computing the z-transform of the difference equation and solving for $Y(z)/X(z)$. The difference equation for a kth order causal system (see eq. [12.2]) can be written

$$y_T(n) + \sum_{i=1}^{k} a_i y_T(n-i) = \sum_{i=0}^{l} b_i x_T(n-i) \tag{14.84}$$

where the coefficients are all real and constant for practical systems. That is, for time-invariant systems that produce a real output signal when the input signal is real. When initial conditions are zero, we obtain the z-transform of the system difference equation by using the linearity and time delay theorems. The result is

$$Y(z) + \sum_{i=1}^{k} a_i z^{-i} Y(z) = \sum_{i=0}^{l} b_i z^{-i} X(z) \tag{14.85}$$

and gives the transfer function

$$H(z) = \frac{Y(z)}{X(z)} = \frac{\displaystyle\sum_{i=0}^{l} b_i z^{-i}}{1 + \displaystyle\sum_{i=1}^{k} a_i z^{-i}} \tag{14.86}$$

In the following sections, we will refer to this form of the system transfer function as the *standard form*. Note that we can write the standard form of the transfer function from the difference equation by inspection, using the difference-equation coefficients. Likewise, if the system transfer function is known in standard form, then we can write the system difference equation from it, by inspection, using the numerator and denominator polynomial coefficients.

The transfer function that we find from the difference equation is a rational function of z^{-1}; thus, we can use the techniques discussed earlier for computing inverse z-transforms to compute the unit pulse response from it. This method for finding the unit pulse response is a much simpler method than the method of solving the difference equation with a unit pulse input signal.

Example 14.16

A system is characterized by the difference equation

$$y_T(n) - 0.1y_T(n-1) - 0.02y_T(n-2) = 2x_T(n) - x_T(n-1)$$

Find the system transfer function and unit pulse response.

Solution

By inspection of the difference equation, we see that

$$a_1 = -0.1, \quad a_2 = -0.02, \quad b_o = 2, \quad b_1 = -1,$$

and all other a_i's and b_i's are zero. Therefore, the standard form of the transfer function is

$$H(z) = \frac{2 - z^{-1}}{1 - 0.1z^{-1} - 0.02z^{-2}}$$

To find the unit pulse response, we compute the inverse z-transform of $H(z)$ using the partial-fraction-expansion steps. To begin, we construct

$$H(z) = \frac{2z^2 - z}{z^2 - 0.1z - 0.02} = \frac{2z^2 - z}{(z - 0.2)(z + 0.1)}$$

Then,

$$\frac{H(z)}{z} = \frac{2z - 1}{(z - 0.2)(z + 0.1)} = \frac{A_1}{z - 0.2} + \frac{A_2}{z + 0.1}$$

where

$$A_1 = (z - 0.2)\big[H(z)/z\big]\big|_{z=0.2} = \frac{2(0.2) - 1}{0.2 + 0.1} = -2$$

and

$$A_2 = (z + 0.1)\big[H(z)/z\big]\big|_{z=-0.1} = \frac{2(-0.1) - 1}{-0.1 - 0.2} = 4$$

Therefore,

$$H(z) = -2\left[\frac{1}{1 - 0.2z^{-1}}\right] + 4\left[\frac{1}{1 + 0.1z^{-1}}\right]$$

and the unit pulse response is

$$h_T(n) = -2(0.2)^n u_T(n) + 4(-0.1)^n u_T(n)$$

In Chapter 7, we defined the Laplace-transformed block diagram for a continuous-time system. In the same way, we define the z-transformed block diagram of a discrete-time system by replacing signals by their z-transforms and operations in the blocks by their corresponding z-transform operations. Construction of expressions for the outputs of all adders in the transformed block diagram produces a set of equations that are the z-transformed versions of the set of difference equations corresponding to the system block diagram. We solve these equations simultaneously for $Y(z)/X(z)$ to find the system transfer function.

The only operations in a discrete-time-system block diagram, other than additions and multiplications by a constant, are time delays. The linearity theorem shows that multiplication and addition operations are unchanged when transforming the system block diagram. The time-delay theorem indicates that a time-delay operation of m sample-spacings delay transforms to a multiplication by z^{-m}.

We will now illustrate the computation of the system transfer function from the system block-diagram representation and the use of the transfer function to find a system output signal.

Example 14.17

Consider the discrete-time data-transmission path of Example 14.14. Find (a) the transmission-path system transfer function, (b) the system difference equation relating the output and input signals, and (c) the system response when the input signal is a unit step signal and the initial conditions are zero.

Solution

a. The z-transformed block diagram for the system is shown in Figure 14.10.

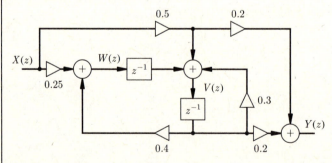

Figure 14.10 Transformed System Block Diagram for Example 14.17

The expressions for the outputs of each adder in the block diagram produce the set of transformed difference equations

$$W(z) = 0.25X(z) + 0.4z^{-1}V(z)$$

$$V(z) = z^{-1}W(z) + 0.5X(z) + 0.3z^{-1}V(z)$$

$$Y(z) = 0.2z^{-1}V(z) + 0.1X(z)$$

that characterizes the system. We substitute the first equation in the second equation and solve for $V(z)$ to obtain

$$V(z) = \frac{\left(0.5 + 0.25z^{-1}\right)X(z)}{1 - 0.3z^{-1} - 0.4z^{-2}}$$

Substituting $V(z)$ in the third equation yields

$$Y(z) = \frac{\left(0.2 - 0.03z^{-1} + 0.01z^{-2}\right)}{1 - 0.3z^{-1} - 0.4z^{-2}}X(z)$$

Therefore, the system transfer function is

$$H(z) = \frac{Y(z)}{X(z)} = \frac{0.2 - 0.03z^{-1} + 0.01z^{-2}}{1 - 0.3z^{-1} - 0.4z^{-2}}$$

and is in standard form.

b. Inspecting $H(z)$, we see that the nonzero coefficients of the system difference equation are $a_1 = -0.3$, $a_2 = -0.4$, $b_0 = 0.2$, $b_1 = -0.03$, and $b_2 = 0.01$. Therefore, the system difference equation is

$$y_T(n) - 0.3y_T(n-1) - 0.4y_T(n-2)$$
$$= 0.2x_T(n) - 0.03x_T(n-1) + 0.01x_T(n-2)$$

c. The z-transform of the system response when the input signal is a unit step signal and the initial conditions are zero is

$$Y(z) = X(z)H(z) = \left(\frac{1}{1 - z^{-1}}\right)\left(\frac{0.2 - 0.03z^{-1} + 0.01z^{-2}}{1 - 0.3z^{-1} - 0.4z^{-2}}\right)$$

$$= \left(\frac{0.2 - 0.03z^{-1} + 0.01z^{-2}}{\left(1 - z^{-1}\right)\left(1 - 0.8z^{-1}\right)\left(1 + 0.5z^{-1}\right)}\right)$$

We compute the inverse z-transform of $Y(z)$ to find the system response. We proceed as follows:

$$Y(z) = \frac{\left(0.2z^3 - 0.03z^2 + 0.01z\right)}{(z-1)(z-0.8)(z+0.5)}$$

$$\frac{Y(z)}{z} = \frac{\left(0.2z^2 - 0.03z + 0.01\right)}{(z-1)(z-0.8)(z+0.5)} = \frac{A_1}{z-1} + \frac{A_2}{z-0.8} + \frac{A_3}{z+0.5}$$

where

$$A_1 = (z-1)\big[Y(z)/z\big]\big|_{z=1} = \frac{0.2(1)^2 - 0.03(1) + 0.01}{(1-0.8)(1+0.5)} = 0.600$$

$$A_2 = (z-0.8)\big[Y(z)/z\big]\big|_{z=0.8} = \frac{0.2(0.8)^2 - 0.03(0.8) + 0.01}{(0.8-1)(0.8+0.5)} = -0.439$$

and

$$A_3 = (z+0.5)\big[Y(z)/z\big]\big|_{z=-0.5} = \frac{0.2(-0.5)^2 - 0.03(-0.5) + 0.01}{(-0.5-1)(-0.5-0.8)} = 0.039$$

Therefore,

$$Y(z) = 0.600 \left[\frac{1}{1 - z^{-1}} \right] - 0.439 \left[\frac{1}{1 - 0.8z^{-1}} \right] + 0.039 \left[\frac{1}{1 + 0.5z^{-1}} \right]$$

and the output signal is

$$y_T(n) = 0.600 u_T(n) - 0.439(0.8)^n u_T(n) + 0.039(-0.5)^n u_T(n)$$

14.6 Use of the Transfer Function for Determining System Stability and Frequency Response

We can perform extensive discrete-time system analysis and design without either applying specific signals and inverting the transform of the system response to determine the output signal or performing a separate spectral analysis. In this section, we will discuss the determination of BIBO system stability and the system frequency response from the system transfer function. The reader will note that the discussion parallels that for continuous-time systems in Section 7.7.

System Stability

We showed in Section 7.7 that a causal continuous-time system, the transfer function of which is a rational function, is BIBO stable if its transfer function satisfies two conditions: (1) The degree of the numerator polynomial must be no larger than the degree of the denominator polynomial and (2) all poles must lie in the left half of the s-plane.

For a causal discrete-time system, the transfer function of which is a rational function in z^{-1}, the relative degree of the numerator and denominator polynomials does not affect the BIBO stability of the system. Additional output-signal terms produced if the degree of the numerator polynomial in z^{-1} were greater than the degree of the denominator polynomial in z^{-1} would represent only delayed versions of the input signal multiplied by constants. This can be demonstrated constructing a transfer function $H(z)$, the numerator-polynomial degree of which is larger than its denominator-polynomial degree, performing a partial-fraction expansion of $H(z)$, and then multiplying this partial-fraction expansion of by $X(z)$.

A causal discrete-time system is BIBO stable only if its transfer function has poles inside the unit circle in the z-plane (that is, $|z| < 1$ at all pole locations). To show that this condition is required, we can use an argument similar to that we used for continuous-time systems in Section 7.7. Also paralleling the results presented for continuous-time systems is the fact that a discrete-time system is called marginally BIBO stable if it has any single-order poles on the unit circle in addition to poles inside the unit circle. The system is unstable in all other cases. Note that we can reasonably expect the preceding results if we keep in mind that the left half of the s-plane maps into the inside of the unit circle in the z-plane.

It is possible to determine whether a transfer function has poles that are not inside the unit circle in the z-plane without factoring the denominator polynomial. A technique used to do this is a modified version of the Routh-Hurwitz test mentioned for continuous-time systems in Chapter 7.[†]

System Frequency Response

We can find the frequency response of a causal, linear, time-invariant, discrete-time system from the system transfer function if the frequency response exists in the nonlimiting sense (that is, if it is the proper discrete-time Fourier transform of a unit pulse response). The technique is similar to the one we used for continuous-time systems in Section 7.7. The frequency response is

$$H_d(f) = \mathcal{F}_d[h_T(n)] = \sum_{n=-\infty}^{\infty} h_T(n)e^{-j2\pi fnT} \qquad (14.87)$$

Since the system is causal, $h_T(n) = 0$ for $t < 0$ and we can write eq. (14.87) as

$$H_d(f) = \sum_{n=0}^{\infty} h_T(n)e^{-j2\pi nfT} = \mathcal{Z}[h_T(n)]\Big|_{z=e^{j2\pi fT}} \qquad (14.88)$$

Therefore, if the frequency response exists in the nonlimiting sense, then it is the transfer function with z replaced by $e^{j2\pi fT}$. That is, the frequency response is the transfer-function values evaluated at $z = e^{j2\pi fT}$. Since $|e^{j2\pi fT}| = 1$ and $\underline{/e^{j2\pi fT}} = 2\pi fT = 2\pi f/f_s$, then $e^{j2\pi fT}$ lies on the unit circle and the frequency response is the transfer function evaluated repeatedly around the unit circle, as shown in Figure 14.11 for a low-pass system with zero phase shift. We could have anticipated this frequency-response-evaluation result by recalling that the unit circle in the z-plane corresponds to the imaginary axis in the s-plane and that the frequency response of a continuous-time system, if it exists, is the system transfer function evaluated along the imaginary axis.

The normalized frequency response is the frequency response with $fT = f/f_s$ replaced by the normalized frequency r. Therefore, we obtain the normalized frequency response of a system by replacing z by $e^{j2\pi r}$ in the system transfer function. That is,

$$H_N(r) = H(z)\big|_{z=e^{j2\pi r}} \qquad (14.89)$$

Note that a system does not have a frequency response if the region of convergence for its transfer function does not include the unit circle. In that case, the frequency-response sum would not converge. Thus, the radius of absolute convergence for the transfer function must be less than unity for a

---

[†]See B. C. Kuo, *Automatic Control Systems*, 2nd ed. (Englewood Cliffs, N.J.: Prentice-Hall, 1967), pp 409–410.

Figure 14.11
Illustration of
Transfer-Function–
Frequency-Response
Relationship

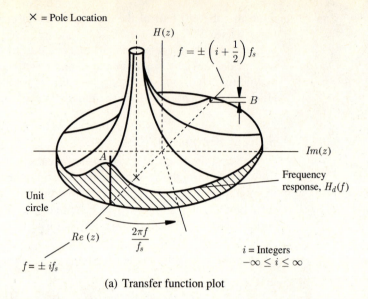

(a) Transfer function plot

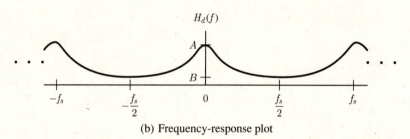

(b) Frequency-response plot

frequency response to exist. This condition means that all system poles must be inside the unit circle. In other words, the system must be stable.

The system frequency response at $f = 0$ (that is, at dc) and at $f = f_s/2$ is particularly easy to determine from the system transfer function for the reason that $z = e^{j2\pi(0)T} = 1$ when $f = 0$ and $z = e^{j2\pi(f_s/2)T} = e^{j\pi} = -1$ when $f = f_s/2$; therefore, $H_d(0) = H(z)|_{z=1}$ and $H_d(f_s/2) = H(z)|_{z=-1}$.

Example 14.18

Find the normalized frequency response of the discrete-time data-transmission path of Example 14.17. Also plot the amplitude and phase responses of the path.

Solution
In Example 14.17 we found that the transfer function of the discrete-time data-transmission path is

$$H(z) = \frac{0.2 - 0.03z^{-1} + 0.01z^{-2}}{1 - 0.3z^{-1} - 0.4z^{-2}}$$

Therefore, the transmission-path normalized frequency response is

$$
\begin{aligned}
H_N(r) &= \frac{0.2 - 0.03e^{-j2\pi r} + 0.01e^{-j4\pi r}}{1 - 0.3e^{-j2\pi r} - 0.4e^{-j4\pi r}} \\
&= \frac{N_R(r) + jN_I(r)}{D_R(r) + jD_I(r)}
\end{aligned}
$$

where

$$N_R(r) = 0.2 - 0.03\cos(2\pi r) + 0.01\cos(4\pi r)$$
$$N_I(r) = 0.03\sin(2\pi r) - 0.01\sin(4\pi r)$$
$$D_R(r) = 1 - 0.3\cos(2\pi r) - 0.4\cos(4\pi r)$$

and

$$D_I(r) = 0.3\sin(2\pi r) + 0.4\sin(4\pi r)$$

We use the procedure described in Chapter 13, following eq. (13.16), to compute values of the amplitude and phase responses. The resulting amplitude and phase responses for the discrete-time data-transmission path are shown in Figure 14.12.

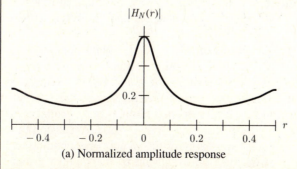

(a) Normalized amplitude response

Figure 14.12 Normalized Amplitude and Phase Responses of the Data-Transmission System in Example 14.18

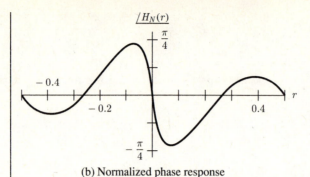

(b) Normalized phase response

Figure 14.12 Normalized Amplitude and Phase Responses of the Data-Transmission System in Example 14.18 *(continued)*

14.7 Summary

The single-sided z-transform is an effective tool for finding the system response of a discrete-time system for the following reasons: (1) It replaces difference equations with algebraic equations, (2) it includes the effects of nonzero initial conditions directly, and (3) it can be used with many signals that do not have a spectrum. The definition of the double-sided z-transform of a signal is conveniently expressed in terms of the discrete-time Fourier transform of the signal multiplied by a convergence factor. A special case of the double-sided z-transform is the single-sided z-transform. The single-sided z-transform can be used to solve problems involving causal systems and signals that are zero for $t < 0$. It is emphasized in this chapter because problems of this nature are of interest to us most often and because the single-sided z-transform provides unique transform pairs.

The z-transform complex variable z is related to the Laplace transform complex variable s through the mapping $z = e^{sT}$. We use this mapping relationship to indicate the parallelism between the Laplace transform for continuous-time signals and systems and the z-transform for discrete-time signals and systems.

The z-transforms of a few simple signals and a set of z-transform theorems can be used to compute other z-transforms. The theorems and a number of z-transform pairs are summarized in Tables C.7 and C.8 of Appendix C.

The z-transforms of many useful signals are rational functions of z^{-1}. We can determine a few of the first samples of the signal corresponding to a rational z-transform by performing a power-series expansion through long division. To find a general expression for the signal corresponding to a rational z-transform, we use partial-fraction expansion of the rational z-transform to express it in terms of a sum of low-degree rational z-transforms with known inverse z-transforms.

We can use the z-transform to solve sets of difference equations that characterize causal, linear, time-invariant systems. Initial conditions are included directly in computing the z-transforms of the difference equations.

The transfer function of a system is defined as the ratio of the z-transform of the output signal to the z-transform of the input signal when initial conditions are zero; it is also the z-transform of the unit pulse response. We can find the standard form of the system transfer function directly from the system difference equation by using the difference-equation coefficients. The transfer function can also be found from a transformed system block diagram without finding the system difference equation. We can determine system stability and the system frequency response directly from the transfer function by techniques that parallel those discussed in Chapter 7 for continuous-time system transfer functions.

Problems

The z-transforms in the following problems are all single-sided z-transforms.

14.1 Find the z-transforms of the signals given in Figure 14.13.

14.2 Find the z-transforms of the signals given in Figure 14.14.

14.3 Use tables and theorems to find the z-transforms of the following signals.

Figure 14.13

a. $x_T(n) = 3\delta_T(n) - 1.5\delta_T(n-2) - \delta_T(n-3) + 4\delta_T(n-5)$

b.

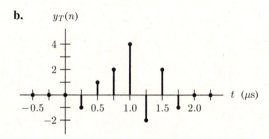

Figure 14.14

a.

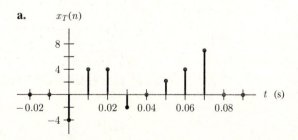

b. $y_T(n) = 1.7\delta_T(n) - 2\delta_T(n-2) + 2\delta_T(n-4) + \delta_T(n-5)$

a. $x_T(n) = 3u_T(n) - 3u_T(n-2)$

b. $y_T(n) = r_T(n) - r_T(n-4) - 4u_T(n-5)$

c. $v_T(n) = (0.2)^n r_T(n)$

d. $w_T(n) = 3nT \cos(0.1nT)u_T(n)$

Simplify each of the transforms found so that it is a ratio of functions of z.

14.4 Use tables and theorems to find the z-transforms of the following signals:

a. $x_T(n) = 2u_T(n-2) - 2u_T(n-4)$

b. $y_T(n) = 3u_T(n) + 2r_T(n-3) - 2r_T(n-5)$

c. $v_T(n) = 0.4ne^{-0.1n}u_T(n-1)$

d. $w_T(n) = (4)^{-n}r_T(n) - 2u_T(n-3)$

Simplify each of the transforms found so that it is a ratio of functions of z.

14.5 Find the z-transforms of the following signals:

a. $x_T(n) = n^2 u_T(n)$

b. $y_T(n) = n^3 u_T(n)$

14.6 Find the z-transforms of the following signals:

a. $x_T(n) = \left\{ \left[\{3, -1, 2, -1\}\, n_0 = 2 \right],\quad 0.1\ \text{s} \right\}$

b. $y_T(n) = \left\{ \left[\{3, -1, 2, -1\}\, n_0 = 2 \right],\quad 10\mu\text{s} \right\}$

c. $w_T(n) = \left[\{0, T, 2T, 3T\},\quad 0.5\ \text{s} \right]$

14.7 Find the z-transforms of the following signals:

a. $x_T(n) = \left[e^{-0.1n},\quad 25\ \text{ms} \right]$

b. $y_T(n) = \left[e^{-0.1n},\quad 0.2\ \text{s} \right]$

c. $w_T(n) = \left[e^{-0.1nT},\quad 0.2\ \text{s} \right]$

14.8 Prove the z-scale theorem (Theorem 3).

14.9 Prove the convolution theorem (Theorem 4).

14.10 Prove the initial-value theorem (Theorem 7).

14.11 Consider the two signals $x_T(n) = u_T(n-1) - u_T(n-4)$ and $y_T(n) = u_T(n) - u_T(n-3)$. Find the z-transform of $w_T(n) = x_T(n) * y_T(n)$ by using the convolution theorem. Verify your result by performing discrete convolution to find $w_T(n)$, computing the z-transform of $w_T(n)$, and comparing this z-transform with the one previously computed.

14.12 Use the initial-value and final-value theorems to find the initial and final values for the signals corresponding to the following z-transforms:

a. $X(z) = \dfrac{2 + z^{-1}}{\left(1 - z^{-1}\right)\left(1 + 0.5z^{-1}\right)}$

b. $Y(z) = \dfrac{1 - 3z^{-1}}{\left(1 - 0.1z^{-1}\right)\left(1 + 0.6z^{-1}\right)}$

c. $V(z) = \dfrac{3 + 0.5z^{-1} + 0.2z^{-2}}{\left(1 - z^{-1}\right)\left(1 + 0.1z^{-1}\right)\left(1 - 2z^{-1}\right)}$

d. $W(z) = \dfrac{0.5 + 0.25z^{-1}}{1 - 1.3z^{-1} + 0.2z^{-2} + 0.1z^{-3}}$

14.13 Find the initial and final values for the signals corresponding to the following z-transforms:

a. $X(z) = \dfrac{z^{-1}}{2\left(1 - 0.2z^{-1}\right)\left(1 + 0.9z^{-1}\right)}$

b. $Y(z) = \dfrac{3 - 6z^{-1}}{\left(1 - z^{-1}\right)^2\left(1 - 0.2z^{-1}\right)}$

c. $V(z) = \dfrac{16 + 4z^{-1} + 2z^{-2}}{1 - 1.5z^{-1} - z^{-2}}$

d. $W(z) = \dfrac{2z^{-1} - 1.5z^{-2}}{1 - z^{-1} - 0.25z^{-2} + 0.25z^{-3}}$

14.14 Use power-series expansion to find the first five samples for the signals corresponding to the following z-transforms:

a. $X(z) = \dfrac{1}{2 - 4z^{-2} + 6z^{-3}}$

b. $Y(z) = \dfrac{3z^{-1} - 33z^{-3}}{1 + 3z^{-1} - 2z^{-2}}$

c. $V(z) = \dfrac{4z^{-1} + 6z^{-2}}{0.5z^{-3} + 0.1z^{-2} + 1.5z^{-1} - 0.2}$

d. $W(z) = \dfrac{3z + 2}{z^2 - 0.5z + 1}$

14.15 Find the first five sequence values for the signals corresponding to the following z-transforms:

a. $X(z) = \dfrac{2 + 4z^{-1}}{2 - 3z^{-1} + z^{-2}}$

b. $Y(z) = \dfrac{2z^{-2} + z^{-1} - 3}{1 - 0.25z^{-2}}$

c. $V(z) = \dfrac{z(z^2 + 3z)}{z^3 - 0.5z^2 + 1.5z - 2}$

d. $W(z) = \dfrac{0.5(1 - 2z^{-1})(1 + 0.5z^{-1})}{(1 - 0.5z^{-1})(1 + 0.2z^{-1})}$

14.16 Find the inverse z-transforms of

a. $X(z) = \dfrac{1}{1 - 0.3z^{-1} + 0.02z^{-2}}$

b. $Y(z) = \dfrac{1 + z^{-1}}{0.5 - 0.1z^{-1}}$

c. $W(z) = \dfrac{z + 1}{z(z - 0.5)^2}$

14.17 Find the expression for the signal corresponding to the z-transform

$$X(z) = \dfrac{z^{-3}}{1 - 1.5z^{-1} + 0.5z^{-2}}$$

Use the expression to compute the signal samples at $n = 0, 1, 2, 3, 4$, and ∞. Verify these values by computing them in another way.

14.18 Repeat Problem 14.17 for the z-transform

$$Y(z) = \dfrac{2 - 3z^{-1}}{(1 - 0.25z^{-1})}$$

14.19 Find the inverse z-transforms of

a. $X(z) = \dfrac{4z^{-3}}{(2 + z^{-1})(1 - 0.1z^{-1} - 0.06z^{-2})}$

b. $Y(z) = \dfrac{z}{(z - 0.2)^2(z + 1)}$

c. $W(z) = \dfrac{z}{(z - 1)(z^2 + 0.1z + 0.01)}$

14.20 Find the inverse z-transforms of

a. $X(z) = \dfrac{1 - 1.5z^{-1} - z^{-2}}{2 - 0.6z^{-1} - 0.1z^{-2}}$

b. $Y(z) = \dfrac{z^{-1} + 2}{(z^{-1} + 0.5)(1 - 0.2z^{-1})^2}$

c. $W(z) = \dfrac{1}{z^2 - 0.5z + 0.25}$

14.21 Find the inverse z-transforms of

a. $X(z) = \dfrac{z}{(z + 0.2)(z^2 - 0.4z + 0.04)}$

b. $Y(z) = \dfrac{1 - 0.1z^{-1} + 0.2z^{-2} + 0.5z^{-3}}{(1 + z^{-1} + 0.24z^{-2})}$

c. $W(z)$

$$= \dfrac{0.6z^{-1} + 0.8z^{-2}}{(1 - 2z^{-1} + z^{-2})(1 + z^{-1} + 0.36z^{-2})}$$

14.22 Use z-transforms to solve the difference equation $y_T(n) + 0.4y_T(n - 1) = 2u_T(n)$ for $n \geq 0$ when the initial condition is $y_T(-1) = 2$. Use the resulting expression to solve for the first four samples of $y_T(n)$. Check these values by solving the difference equation recursively.

14.23 Repeat Problem 14.22 for the difference equation $y_T(n) - 0.2y_T(n - 1) - 0.08y_T(n - 2) = x_T(n)$ when $x_T(n) = 2u_T(n) - 2u_T(n - 2)$, $y_T(-1) = -2$, and $y_T(-2) = 3$.

14.24 Repeat Problem 14.22 for the difference equation $y_T(n) - 0.25y_T(n - 2) = x_T(n)$ when $x_T(n) = 2\delta(n - 1) + 4u_T(n)$, $y_T(-1) = -4$, and $y_T(-2) = 8$.

14.25 An RC low-pass filter for a continuous-time system is characterized by the first-order differential equation

$$\tau\dfrac{dy(t)}{dt} + y(t) = x(t)$$

where $x(t)$ is the input signal, $y(t)$ is the output signal, and $\tau = RC$ (see Example 1.10).

a. Use Laplace transforms to find the system output signal when the input signal is a unit step and the initial condition is zero.

b. A discrete-time system that approximates this RC low-pass filter is one with the difference equation

$$\tau\left[\dfrac{y_T(n) - y_T(n - 1)}{T}\right] + y_T(n) = x_T(n)$$

since

$$\lim_{T \to 0} \tau[\dfrac{y_T(n) - y_T(n - 1)}{T} = \tau\dfrac{dy(t)}{dt}$$

Solve the difference equation for the output signal $y_T(n)$ when the input signal is the sampled unit step signal $u_T(n)$ and the initial condition is zero.

c. Show that $y_T(n)$ of part (b) approaches a sampled version of $y(t)$ of part (a) as T approaches zero.

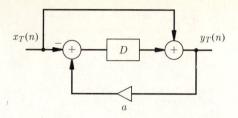

14.26 Find the transfer functions of the systems characterized by the difference equations in Problems 14.22, 14.23, and 14.24.

Figure 14.15

14.28 A flight-control system is implemented by using A/D conversion, discrete-time-signal processing, and D/A conversion. Part of the discrete-time-signal processing system is represented by the block diagram shown in Figure 14.16. Find (a) the system transfer function, (b) the difference equation relating the output and input signals, (c) the system unit pulse response, and (d) the system response to the signal $x_T(n) = 3r_T(n)$.

14.27 The system block diagram shown in Figure 14.15 represents one part of a discrete-time control system used to control a manufacturing process.

a. Write the difference equation that relates the output and input signals.
b. Find the system transfer function.
c. Find the system unit pulse response.
d. Find the system response to a unit step input signal.

14.29 Repeat Problem 14.28 for the system represented by the block diagram shown in Figure 14.17.

14.30 Are the systems corresponding to the following transfer functions stable, unstable, or marginally stable? Explain your answers.

Figure 14.16

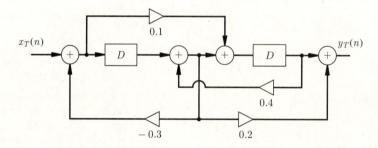

Figure 14.17

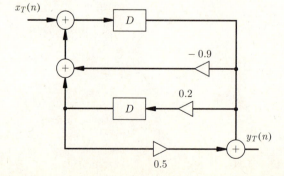

a. $H(z) = \dfrac{z+2}{8z^2 - 2z + 3}$

b. $H(z) = \dfrac{1 + z^{-1}}{1 - z^{-1} + z^{-2}}$

14.31 Repeat Problem 14.30 for the transfer functions

a. $H(z) = \dfrac{z^2 - 3}{z^2 + 2.5z + 1}$

b. $H(z) = \dfrac{1 - 0.5z^{-3}}{1 - 0.5z^{-1} - z^{-2} + 0.5z^{-3}}$

14.32 Find the frequency responses of the systems characterized by the following transfer functions:

a. $H(z) = \dfrac{1 + 0.8z^{-1}}{1 + 0.3z^{-1} - 0.1z^{-2}}$

b. $H(z) = \dfrac{1 - z^{-1}}{1 - 0.5z^{-1} + 0.25z^{-2}}$

Plot the normalized amplitude and phase responses for $|r| \le 0.5$.

14.33 Repeat Problem 14.32 for the systems characterized by the transfer functions

a. $H(z) = \dfrac{1}{z^2 - 0.1z - 0.3}$

b. $H(z) = \dfrac{3 - z^{-1}}{1 + 0.9z^{-1} - 0.9z^{-2}}$

15 Discrete-Time System Realizations and Discrete-Time Filters

We can analyze a wide range of discrete-time signals and systems with the techniques discussed in the preceding chapters. These techniques permit us to characterize a lumped-parameter, linear, time-invariant, discrete-time system with a constant coefficient, linear difference equation, unit pulse response, or system transfer function. In general, we have considered only causal systems, which also will be the case in this chapter. To show how a system can be implemented from these characterizations by hardware or software, we will develop realizations of the system from the characterizations and represent the realizations with block diagrams. The realizations will include several different forms and we will describe the relative merits of each.

In many discrete-time systems of interest, we find subsystems that can be considered discrete-time filters. That is, the subsystem passes signal components at some frequencies and greatly attenuates signal components at other frequencies; or, the subsystem processes the signal in some known way. If signal amplitudes are quantized and discrete-time filters are constructed to process quantized signals (that is, if they are digital systems), then the filters are referred to as *digital filters*. In this chapter we will utilize the spectral and transform tools developed in the preceding chapters to analyze and synthesize discrete-time filters.

In designing a discrete-time filter, we often attempt to approximately match the characteristics of a continuous-time filter in some sense. We can then use the resulting filter with an A/D converter and a D/A converter to approximate the desired filtering of a continuous-time signal with a discrete-time system. Therefore, in this chapter we will consider techniques to produce discrete-time filters that approximate continuous-time filters in some sense. Our treatment of discrete-time signal and system analysis is introductory; therefore, the full range of techniques used to design discrete-time filters will not be explored. Discussion of the more complicated techniques can be found in signal-processing texts and papers at a more advanced level.[†]

15.1 System Types

A discrete-time system can be categorized by the length of its unit pulse response. We refer to a system with a unit pulse response that is infinite in length as an *infinite-impulse-response (IIR) system*. If the system unit pulse response

---

[†]See for example A. V. Oppenheim and R. W. Schafer, *Discrete-Time Signal Processing* (Englewood Cliffs, N.J.: Prentice-Hall, 1989); R. A. Roberts and C. T. Mullis, *Digital Signal Processing*, (Reading, Mass., Addison-Wesley, 1987); T. W. Parks and C. S. Burrus, *Digital Filter Design* (New York: Wiley, 1987).

is of finite length, then we refer to the system as a *finite-impulse-response (FIR)* *system*. Note that impulse response really has no meaning for a discrete-time system; however, the term has been used historically, rather than unit pulse response, in defining these system types.

We can also categorize discrete-time systems as to whether or not the system implementation contains feedback (that is, whether the system transfer function contains poles). A system implementation that contains feedback is called *recursive*; one that contains no feedback is called *nonrecursive*. IIR systems require a recursive implemention. Usually, FIR systems are implemented nonrecursively.

15.2 System Realizations

Before developing the various forms of realizations from system characterizations, we must first define a *system realization*.

Definition _____

A realization of a discrete-time system is a system structure in discrete-time hardware or software that implements the system.

The accuracy with which the output of a realized discrete-time system matches the desired output depends on the precision used in implementing signal and parameter values. In particular, a digital processor has a finite wordlength (number of bits) that it uses to represent numerical values. Therefore, the following quantities are truncated or rounded: (1) the input signal, (2) signals internal to the system, and (3) system parameters (multipliers). The effect of the finite-wordlength restriction becomes more pronounced as the order of the system increases (with more parameters and delays), as the bandwidth of the system becomes narrower, and as the sample rate increases. The different realization forms that we will consider differ in their sensitivity to parameter precision.

Direct-Form Realizations

We generate what is called a *Direct Form I realization* by directly implementing the system difference equation

$$y_T(n) = \sum_{i=1}^{k}(-a_i)\,y_T(n-i) + \sum_{i=0}^{\ell} b_i x(n-i) \tag{15.1}$$

obtained from eq. (12.2) of Chapter 12. The result is shown in Figure 15.1 in a block-diagram representation. We have changed the symbol used for a

Figure 15.1
Block-Diagram
Representation of
Direct Form I
Realization

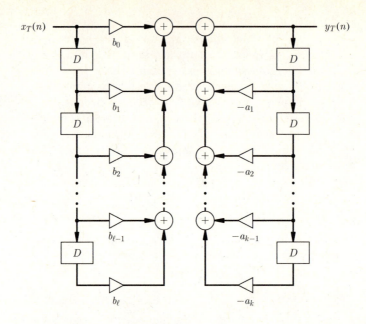

constant-multiplier block to a triangle that points in the direction of signal flow. Thus, a different symbol is used for each type of operation indicated in a block-diagram representation. A separate addition operation is shown for each pair of numbers that must be added. In this way, the number of addition operations that must be performed is readily apparent. The addition operations can be performed sequentially with a single adder or, simultaneously, with multiple adders. Sequential addition requires less hardware but more execution time.

The standard form of the system transfer function that corresponds to eq. (15.1) is

$$H(z) = \frac{\displaystyle\sum_{i=0}^{\ell} b_i z^{-i}}{1 + \displaystyle\sum_{i=0}^{k} a_i z^{-i}} \tag{15.2}$$

We can draw the block-diagram representation for the Direct Form I realization directly, by inspection, from the system's standard-form transfer function because the multipliers in the feedforward paths are the numerator coefficients and the multipliers in the feedback paths are the negatives of the denominator coefficients. Each delay operation corresponds to a storage element required in the implementation. Thus, the Direct Form I realization requires $k + \ell$ storage elements.

We can generate a system realization with a reduced number of delay operations by reorganizing the Direct Form I realization. The resulting realization is called a *Direct Form II realization*, the generation of which we will now describe.

The z-transform of the output signal is

$$Y(z) = H(z)X(z) = \frac{\sum\limits_{i=0}^{\ell} b_i z^{-i}}{1 + \sum\limits_{i=1}^{k} a_i z^{-i}} X(z)$$

$$= \left[\sum_{i=0}^{\ell} b_i z^{-i} \right] W(z) \tag{15.3}$$

where

$$W(z) = \frac{X(z)}{1 + \sum\limits_{i=1}^{k} a_i z^{-i}} \tag{15.4}$$

We can rewrite eq. (15.4) in the form

$$W(z) = X(z) - \sum_{i=1}^{k} a_i z^{-i} W(z) \tag{15.5}$$

Computation of the inverse z-transforms of eqs. (15.5) and (15.3) yields

$$w_T(n) = x_T(n) + \sum_{i=1}^{k} (-a_i) w_T(n - i) \tag{15.6}$$

and

$$y_T(n) = \sum_{i=0}^{\ell} b_i w_T(n - i) \tag{15.7}$$

These two difference equations define the Direct Form II realization, the block-diagram representation of which is shown in Figure 15.2. Note that only the larger of k or ℓ storage elements is required. Thus, fewer storage elements are required for a Direct Form II realization than for a Direct Form I realization. However, if addition is performed sequentially, then a Direct Form II realization requires two adders instead of the single adder needed for a Direct Form I realization. Once again, we can draw the block-diagram representation for the Direct Form II realization directly, by inspection, from the standard form of the system transfer function because the multipliers in the feedforward paths are the numerator coefficients and the multipliers in the feedback paths are the negatives of the denominator coefficients.

Figure 15.2
Block-Diagram
Representation of
Direct Form II
Realization

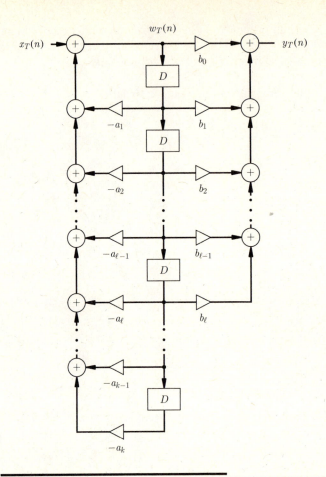

Example 15.1

Draw the block-diagram representations of the Direct Form I and Direct Form II realizations of a system with the following transfer function:

$$H(z) = \frac{6z\left(z^2 - 4\right)}{5z^3 - 4.5z^2 + 1.4z - 0.8}$$

Solution

We obtain the standard form of the transfer function by multiplying the transfer-function numerator and denominator by $0.2z^{-3}$ to obtain

$$H(z) = \frac{1.2 - 4.8z^{-2}}{1 - 0.9z^{-1} + 0.28z^{-2} - 0.16z^{-3}}$$

We then identify the numerator coefficients as the feedforward-path multipliers and the denominator coefficients as the negatives of the feedback multipliers to obtain the Direct Form I and Direct Form II realizations of a system. These realizations are shown in Figure 15.3.

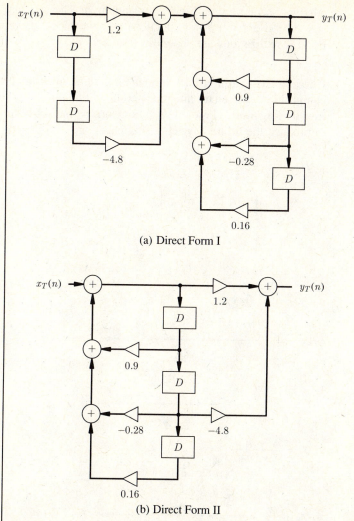

(a) Direct Form I

(b) Direct Form II

Figure 15.3 Block-Diagram Representations of Direct-Form Realizations for the System of Example 15.1

Cascade and Parallel Realizations

Now let us turn to transfer-function-decomposition methods that produce system realizations containing a group of interconnected subsystems of the lowest order. Since the subsystems are of low order, a realization of this type is often less sensitive to reduced parameter and signal precision caused by a finite-wordlength restriction. The lowest order subsystems that we can use are first-order subsystems. However, we also need second-order subsystems if complex-conjugate poles or zeros exist in the system transfer function because, otherwise, complex multipliers are required. The standard form of first-order and second-order transfer functions and the block diagrams for their Direct Form II realizations are shown in Figure 15.4.

Cascade Realizations   To implement the first method of transfer-function decomposition, we factor the transfer function into a product of lowest order transfer functions. These transfer functions define subsystems that are cascaded to produce the desired system. That is,

$$H(z) = \frac{KN(z)}{D(z)} = \frac{K[N_1(z)\cdots N_m(z)]}{D_1(z)\cdots D_r(z)}$$

$$= \begin{cases} K\left[\dfrac{N_1(z)}{D_1(z)}\right]\cdots\left[\dfrac{N_m(z)}{D_m(z)}\right]\left[\dfrac{1}{D_{m+1}(z)}\right]\cdots\left[\dfrac{1}{D_r(z)}\right] & r > m \\[3ex] K\left[\dfrac{N_1(z)}{D_1(z)}\right]\cdots\left[\dfrac{N_r(z)}{D_r(z)}\right]\left[N_{r+1}(z)\right]\cdots\left[N_m(z)\right] & m > r \end{cases}$$

$$\equiv KH_1(z)\cdots H_N(z) \qquad N = \max[m, r] \tag{15.8}$$

Figure 15.4
Block-Diagram
Representations of
Direct Form II
Realizations of First-
and Second-Order
Systems

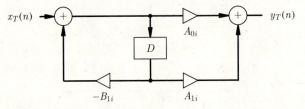

$$H_i(z) = \frac{A_{0i} + A_{1i}z^{-1}}{1 + B_{1i}z^{-1}}$$

(a) First-order system

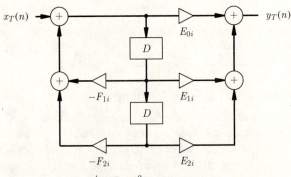

$$H_i(z) = \frac{E_{0i} + E_{1i}z^{-1} + E_{2i}z^{-2}}{1 + F_{1i}z^{-1} + F_{2i}z^{-2}}$$

(b) Second-order system

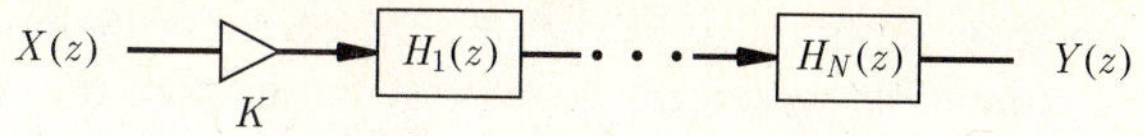

Figure 15.5
Cascade System

$$X(z) \longrightarrow \triangleright \longrightarrow \boxed{H_1(z)} \longrightarrow \cdots \longrightarrow \boxed{H_N(z)} \longrightarrow Y(z)$$
$$K$$

The corresponding transformed block diagram for the *cascade realization* is shown in Figure 15.5. Each of the subsystems in the cascade is realized with a Direct Form II realization. The numerator and denominator polynomial factors in eq. (15.8) are linear factors corresponding to real poles and zeros and quadratic factors corresponding to complex conjugate poles and zeros. The subsystem transfer functions have the form of the first- and second-order factors shown in Figure 15.4. Note that some of the coefficients (A_{0i}, A_{1i}, for example) may be zero.

When pairing numerator and denominator factors to produce the subsystem transfer functions, we should pair as many factors as possible to keep the number of storage elements required at a minimum. This may require construction of one quadratic factor from two linear factors to pair with a quadratic factor. When we use this procedure, the number of storage elements required for a cascade realization is the same as that for a Direct Form II realization. At most, one more multiplier is required.

The factor pairing and subsystem order are not unique; therefore, we can construct several different cascade realizations. With infinite-precision signal and parameter values, all cascade realizations constructed give identical output signals for the same input signal. In a practical system, where finite wordlength is used, the output-signal changes due to parameter truncation or rounding are less for some factor pairings and subsystem orders than for others. A general method for selecting the best pairings and order does not exist; however, sometimes we can use general subsystem characteristics to aid in the selection.[†]

Example 15.2

Find a cascade realization of the system characterized by the transfer function

$$H(z) = \frac{4z^3 + 16z^2 + 4z - 24}{2z^4 + 1.6z^3 + 0.5z^2 + 0.1z}$$

Solution
We factor the numerator and denominator of the transfer function to obtain

$$H(z) = \frac{2(z - 1)(z + 2)(z + 3)}{z(z + 0.5)\left(z^2 + 0.3z + 0.1\right)}$$

where we leave the last factor in the denominator as a quadratic factor since it corresponds to complex-conjugate poles. Also, we have factored out a con-

[†]For comprehensive treatment of factor pairing and subsytem order, see A. V. Oppenheim and R. W. Schafer, *Discrete-Time Signal Processing* (Englewood Cliffs, N.J.: Prentice-Hall, 1989) and L. B. Jackson, *Digital Filters and Signal Processing* (Hingham, Mass.: Kluwer Academic Publishers, 1986).

stant, so that the highest power of z in each factor has a coefficient of one. This factorization produces the fewest number of multiplies in the resulting realization. We multiply the numerator and denominator by z^{-4} to eliminate positive powers of z, combine the last two linear factors in the numerator, and pair the factors to produce the system transfer function in cascade form. The result is

$$H(z) = 2\left(z^{-1}\right)\left(\frac{1-z^{-1}}{1+0.5z^{-1}}\right)\left(\frac{1+5z^{-1}+6z^{-2}}{1+0.3z^{-1}+0.1z^{-2}}\right)$$

The block-diagram representation of the cascade realization that corresponds to this transfer function is shown in Figure 15.6.

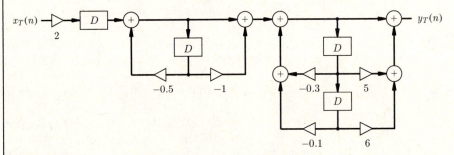

Figure 15.6 Block-Diagram Representation of a Cascade Realization for the System in Example 15.2

The effect of using finite wordlength, rather than infinite precision, for system parameters is to change the system pole and zero locations and the system gain. As a simple illustration of this effect and the improvement that can be achieved by using a cascade realization, the dc gain of a system is considered in the following low-pass filter example.

Example 15.3

Later in the chapter (in Example 15.10) we will design a discrete-time low-pass filter by means of the bilinear-transformation technique (to be discussed). Using infinite precision, this filter's dc gain is 0.5 because bilinear-transformation produces a discrete-time filter, the dc gain of which is equal to the dc gain of the continuous-time filter from which it is derived (0.5 for the filter in Example 15.10). With parameter values rounded to six digits after the decimal, the designed discrete-time filter's transfer function is

$$H(z) = \frac{1.457327 \times 10^{-5}\left(1+3z^{-1}+3z^{-2}+z^{-3}\right)}{1-2.874357z^{-1}+2.756483z^{-2}-0.881893z^{-3}} \qquad (15.9)$$

Analyze the effect of finite wordlength on the dc gain of the designed filter when a Direct Form II realization is used. Repeat the analysis for a cascade realization and compare the results with those obtained for the Direct Form II realization.

Solution

When the Direct Form II realization of the filter is implemented with parameter values rounded to six digits after the decimal, its transfer function is given by eq. (15.9). As for its dc gain,

$$\text{dc gain} = |H(1)| = 0.500370$$

The error of the implemented-filter dc gain with respect to the dc gain of an infinite-precision filter, expressed as a percentage, is

$$\text{Percent error} = \frac{0.500370 - 0.5}{0.5} \times 100 = 0.074\%$$

We use similar computations to compute the dc gain and percent error in dc gain when the filter is implemented with parameter values rounded to fewer digits after the decimal. The results are as follows:

Number of digits after decimal	dc gain	Percent error
5	0.50690	1.38%
4	0.5829	16.5%

When we factor the designed-filter transfer function to produce a cascade realization, we obtain the factored and paired transfer function

$$H(z) =$$

$$1.457327 \times 10^{-5} \left(\frac{1 + z^{-1}}{1 - 0.939062 z^{-1}} \right) \left(\frac{1 + 2z^{-1} + z^{-2}}{1 - 1.935294 z^{-1} + 0.939121 z^{-2}} \right)$$

where we have again rounded the parameter values to six digits after the decimal. Computation of the dc gain and the percent error in dc gain in the manner shown yields the following results:

Number of digits after decimal	dc gain	Percent error
6	0.499920	−0.016%
5	0.49951	−0.098%
4	0.5038	0.760%

The improvement in dc gain when the cascade realization is used is readily apparent.

The error in dc gain can be corrected with appropriate amplification; hence, it is not very important. However, it is indicative of the relative size of changes that can occur in such system properties as pole and zero locations, time response, and frequency response when finite wordlength is used.[†]

Parallel Realizations   The second method of transfer-function decomposition we consider produces a *parallel realization*. In this method, the transfer function is decomposed into lowest order subsections through partial fraction expansion. The subsections are then realized and connected in parallel. In general terms, we start with the transfer function

$$H(z) = \frac{N(z^0, \ldots, z^{-\ell})}{D(z^0, \ldots, z^{-k})} = \frac{z^{-\ell}N(z^\ell, \ldots, z^0)}{z^{-k}D(z^k, \ldots, z^0)}$$

$$= \frac{z^{-\ell}N_s(z^\ell, \ldots, z^0)}{z^{-k}D_s(z^k, \ldots, z^0)} \tag{15.10}$$

where, in the last step, we have multiplied the numerator and denominator of $H(z)$ by the same constant so that the coefficient of z^k in $D_s(z^k, \ldots, z^0)$ is one. We perform this multiplication so that the subsection transfer functions that we obtain are in standard form. We then perform the partial-fraction expansion of $H(z)/z$ to obtain

$$\frac{H(z)}{z} = \frac{N(z^\ell, \ldots, z^0)}{z^{\ell-k+1}D(z^k, \ldots, z^0)} = \frac{N_s(z^\ell, \ldots, z^0)}{z^{\ell-k+1}D_s(z^k, \ldots, z^0)}$$

$$= \begin{cases} \dfrac{G_{\ell-k}}{z^{\ell-k+1}} + \cdots + \dfrac{G_0}{z} + H_{c1}(z) + \cdots + H_{cN}(z) & \ell \geq k \\ H_{c1}(z) + \cdots + H_{cN}(z) & \ell < k \end{cases} \tag{15.11}$$

where

$$H_{ci}(z) = \frac{A_{0i}}{z + B_{1i}} \quad \text{or} \quad H_{ci}(z) = \frac{E_{0i}z + E_{1i}}{z^2 + E_{1i}z + E_{2i}} \tag{15.12}$$

and N is the number of partial-fraction-expansion terms resulting from the factors of $D_s(z^k, \ldots, z^0)$. When we multiply both sides of eq. (15.11) by z and multiply the numerator and denominator terms of each factor by the appropriate negative power of z to eliminate all positive powers of z, then we obtain

$$H(z) = \begin{cases} G_{\ell-k}z^{-(\ell-k)} + G_0 + H_1(z) + \cdots + H_N(z) & \ell \geq k \\ H_1(z) + \cdots + H_N(z) & \ell < k \end{cases} \tag{15.13}$$

where

$$H_i(z) = \frac{A_{0i}}{1 + B_{1i}z^{-i}} \quad \text{or} \quad H_i(z) = \frac{E_{0i} + E_{1i}z^{-1}}{1 + F_{1i}z^{-1} + F_{2i}z^{-2}} \tag{15.14}$$

[†]J. B. Knowles and E. M. Olcayto, "Coefficient Accuracy and Digital Filter Response," *IEEE Transactions on Circuit Theory* CT–15 (March, 1968), pp. 31–41.

Figure 15.7
Parallel System

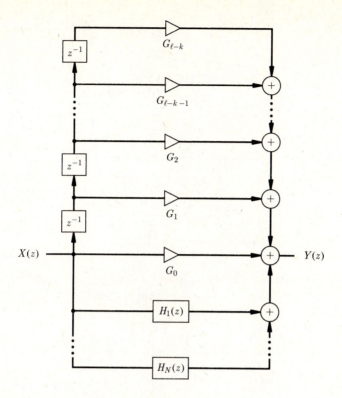

The subsection transfer functions given by eq. (15.14) correspond to the first- and second-order systems, the Direct Form II realizations of which are shown in Figure 15.4. The transformed block-diagram representation of the parallel-form realization defined by eq. (15.13) is shown in Figure 15.7. Just as with a cascade realization, a parallel realization contains no more storage elements than the Direct Form II realization of the same system.

Example 15.4

Find the parallel realization for the system of Example 15.2.

Solution
The transfer function from Example 15.2 is

$$H(z) = \frac{4z^3 + 16z^2 + 4z - 24}{2z^4 + 1.6z^3 + 0.5z^2 + 0.1z}$$

We multiply the numerator and denominator by 0.5 to produce a coefficient of one for the highest power of z in the denominator and then factor the denominator to give

$$H(z) = \frac{2z^3 + 8z^2 + 2z - 12}{z(z + 0.5)\left(z^2 + 0.3z + 0.1\right)}$$

The partial-fraction expansion of $H(z)/z$ is

$$\frac{H(z)}{z} = \frac{2z^3 + 8z^2 + 2z - 12}{z^2(z + 0.5)\left(z^2 + 0.3z + 0.1\right)}$$

$$= \frac{G_1}{z^2} + \frac{G_0}{z} + \frac{A_{01}}{z + 0.5} + \frac{E_{02}z + E_{12}}{z^2 + 0.3z + 0.1}$$

We compute the partial-fraction-expansion coefficients, multiply both sides of the preceding equation by z, and eliminate the positive powers of z to obtain

$$H(z) = -240z^{-1} + 1240 + \frac{(-225)}{1 + 0.5z^{-1}} + \frac{\left(-1015 - 175z^{-1}\right)}{1 + 0.3z^{-1} + 0.1z^{-2}}$$

as the partial-fraction expansion of the system transfer function. The block-diagram representation of the parallel realization of this transfer function is shown in Figure 15.8.

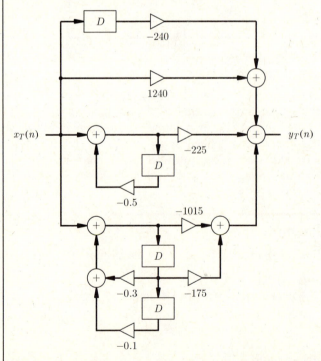

Figure 15.8 Block-Diagram Representation of the Parallel Realization for the System in Example 15.4

15.3 Discrete-Time-Filter Design

In this brief section, we will introduce the subsequent discussion of design techniques for discrete-time filters. Our consideration will be limited to causal filters, signals that are zero for $t < 0$, and systems with zero initial conditions to allow for the determination of filter transfer functions through the use of z-transform techniques.

The design techniques we have selected to discuss produce the transfer function of a discrete-time filter that approximates specified filter characteristics. We indicated in the introduction to this chapter that only a few design techniques will be considered and that they all attempt to approximately match the characteristics of continuous-time filters in some way. We will describe the following approaches to approximate matching:

1. approximately match the continuous-time-filter structure by generating difference equations that approximate the integro-differential equations for the continuous-time filter,
2. approximately match the continuous-time-filter time response, and
3. approximately match the continuous-time-filter frequency response.

We will provide examples of all three approaches for IIR filters. For FIR filters, we emphasize frequency-response matching. This approach is often the most desirable; moreover, the other approaches, excluding approximate impulse-response matching, are generally not appropriate for FIR filter design. Approximate impulse-response matching is discussed only briefly.

15.4 IIR Filter Design

We will now discuss the three approaches listed in the previous section with respect to the design of an IIR discrete-time filter. The design procedures will be illustrated by several simple examples.

Structure Matching—Derivative and Integral Approximation

We can construct difference equations that approximate the integro-differential equations that characterize a continuous-time filter by using discrete-time approximations for the differentiation and integration operations. The z-transform of the resulting difference equations produces a set of equations in z that we can algebraically solve (see Sections 14.4 and 14.5) to find the transfer function of a discrete-time filter that approximately matches the characteristics of the continuous-time filter.

We approximate the derivative

$$x_D(t) = \frac{dx(t)}{dt} = \lim_{\tau \to 0} \frac{x(t) - x(t - \tau)}{\tau} \tag{15.15}$$

of the signal $x(t)$ by the discrete-time operation

$$x_{DT}(n) = \frac{x_T(n) - x_T(n-1)}{T} \qquad (15.16)$$

on the signal $x_T(n)$. The approximation that we use for the integral

$$x_I(t) = \int_0^t x(\alpha)\, d\alpha \qquad (15.17)$$

of the signal $x(t)$ is given by the trapezoidal rule for approximate integration. Thus, the discrete-time operation

$$x_{IT}(n) = \sum_{i=0}^n \frac{T}{2}\,[x_T(i) + x_T(i-1)] \qquad n \geq 0 \qquad (15.18)$$

on the signal $x_T(n)$ is used to approximate an integral operation where the lower limit on the summation and $x_T(-1)$ are zero because the signal is zero for $t < 0$ [that is, $x_T(i) = 0$ for $i < 0$].

Example 15.5

Use discrete-time approximate derivatives and/or approximate integrals to find the transfer function for a discrete-time filter that approximates the continuous-time first-order Butterworth low-pass filter with cutoff frequency f_c Hz implemented by the circuit shown in Figure 15.9 (refer back to Figure 8.9).

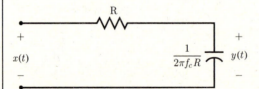

Figure 15.9 First-Order Butterworth Filter Implementation

Solution
The differential equation for the RC circuit shown in Figure 15.9 is

$$\frac{1}{2\pi f_c}\frac{dy(t)}{dt} + y(t) = x(t)$$

(see Example 1.10). For a sample spacing of T, the corresponding approximate difference equation is

$$\frac{y_T(n) - y_T(n-1)}{2\pi f_c T} + y_T(n) = x_T(n)$$

We compute the z-transform of this difference equation and solve it for the transfer function. The steps are

$$\frac{1}{2\pi f_c T}\left[Y(z) - z^{-1}Y(z)\right] + Y(z) = X(z)$$

and

$$H(z) = \frac{Y(z)}{X(z)} = \frac{2\pi f_c T}{1 + 2\pi f_c T}\left[\frac{1}{1 - z^{-1}/(1 + 2\pi f_c T)}\right]$$

The block-diagram representation of the Direct Form II realization of this transfer function is shown in Figure 15.10a.

We can obtain an alternate transfer-function design for the discrete-time filter by first integrating the differential equation to give

$$y(t) = 2\pi f_c \int_0^t x(\alpha)\,d\alpha - 2\pi f_c \int_0^t y(\alpha)\,d\alpha$$

and then using integration approximation to produce the approximate difference equation

$$y_T(n) = \pi f_c T \sum_{i=0}^{n}[x_T(i) + x_T(i-1)] - \pi f_c T \sum_{i=0}^{n}[y_T(i) + y_T(i-1)]$$

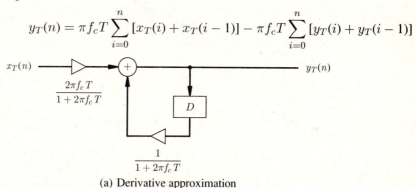

(a) Derivative approximation

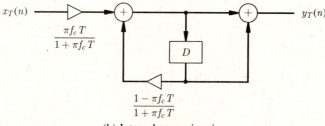

(b) Integral approximation

Figure 15.10 Block-Diagram Representations of Realizations of Discrete-Time Approximations of a First-Order Butterworth Filter

We compute the z-transform of this difference equation by using the time-delay and summation theorems. The result is

$$Y(z) = \pi f_c T\left(\frac{1 + z^{-1}}{1 - z^{-1}}\right)X(z) - \pi f_c T\left(\frac{1 + z^{-1}}{1 - z^{-1}}\right)Y(z)$$

The solution for this equation for $Y(z)/X(z)$ yields the transfer function

$$H(z) = \frac{Y(z)}{X(z)} = \frac{\pi f_c T}{1 + \pi f_c T}\left[\frac{1 + z^{-1}}{1 - (1 - \pi f_c T)z^{-1}/(1 + \pi f_c T)}\right]$$

The block-diagram representation of the Direct Form II realization of this transfer function is shown in Figure 15.10b.

Note that $f_c > 0$ and $T > 0$. The discrete-time filters designed in Example 15.5 are stable for all positive values of f_c and T because $|1/(1+2\pi f_c T)| < 1$ and $|(1 - \pi f_c T)/(1 + \pi f_c T)| < 1$. Thus, the filter poles are inside the unit circle. This being the case, the filter transfer functions converge for values of z on the unit circle. Therefore, the filter frequency responses exist and can be computed and compared with the continuous-time first-order Butterworth filter frequency response to see if a good frequency-response approximation has been achieved. To make this comparison, we first find the frequency response of the continuous-time filter by computing the Laplace transform of the system differential equation, solving for the transfer function $H_a(s)$, and setting $s = j2\pi f$. The result is

$$H_a(j2\pi f) = \frac{f_c}{f_c + jf} \tag{15.19}$$

For the first filter design, the frequency response is

$$H\left(e^{j2\pi fT}\right) = \frac{2\pi f_c T}{(1 + 2\pi f_c T) - e^{-j2\pi fT}}$$

$$= \frac{2\pi f_c T}{[1 + 2\pi f_c T - \cos(2\pi fT)] + j\sin(2\pi fT)} \tag{15.20}$$

For frequencies that are much lower than the sample rate, $f \ll f_s = 1/T$, which implies that $2\pi fT \ll 2\pi$. In this case, $\cos(2\pi fT) \doteq 1$ and $\sin(2\pi fT) \doteq 2\pi fT$, so that

$$H\left(e^{j2\pi fT}\right) \doteq \frac{f_c}{f_c + jf} = H_a(j2\pi f) \qquad f \ll f_s \tag{15.21}$$

For the second filter design, the frequency response is

$$H\left(e^{j2\pi fT}\right) = \frac{\pi f_c T\left(1 + e^{-j2\pi fT}\right)}{\pi f_c T\left(1 + e^{-j2\pi fT}\right) + \left(1 - e^{-j2\pi fT}\right)}$$

$$= \frac{\pi f_c T\left(e^{j\pi fT} + e^{-j\pi fT}\right)}{\pi f_c T\left(e^{j\pi fT} + e^{-j\pi fT}\right) + \left(e^{j\pi fT} - e^{-j\pi fT}\right)}$$

$$= \frac{\pi f_c T \cos(\pi fT)}{\pi f_c T \cos(\pi fT) + j\sin(\pi fT)} \tag{15.22}$$

Again, when $f \ll 4_s$, then $\cos(\pi fT) \doteq 1$ and $\sin(\pi fT) \doteq \pi fT$, so that

$$H\left(e^{j2\pi fT}\right) \doteq \frac{f_c}{f_c + jf} = H_a(j2\pi f) \qquad f \ll f_s \qquad (15.23)$$

Therefore, both discrete-time-filter designs obtained in Example 15.5 approximate the frequency response of the continuous-time first-order Butterworth filter at frequencies that are much lower than the sample rate. This means that the filter cutoff frequency must be much lower than the sample rate if the discrete-time-filter approximations are to respond, essentially, as a first-order Butterworth filter would respond. Since both the derivative and integral approximations become more accurate as T becomes smaller (that is, as the sample rate $f_s = 1/T$ becomes larger), this result is to be expected.

Amplitude and phase responses of the continuous-time first-order Butterworth filter and the two discrete-time approximations derived in Example 15.5 are shown in Figure 15.11 in terms of the normalized frequency $r = f/f_s = fT$. The amplitude and phase responses are shown over the frequency interval $0 \le r \le 0.25$ for a filter with a normalized cutoff frequency of $r_c = 0.05$. Reasonably good amplitude- and phase-response correspondence of the discrete-time filters to the analog filter occurs for frequencies lower than twice the cutoff frequency. This is particularly true for the filter obtained by integral approximation.

In many cases, it is convenient to work with the continuous-time-system block diagram that corresponds to the system differential equation when producing the transfer function of the approximating discrete-time system. In fact, we may define the system by the system block diagram. We generate the discrete-time-system transfer function from the continuous-time-system block diagram by first replacing differentiators and integrators by their discrete-time approximations and replacing signals by their sampled versions. Then we obtain the z-transformed version of the block diagram and determine the transfer function from the transformed block diagram. Actually, we construct the z-transformed block diagram directly from the continuous-time-system block diagram by converting signals to the z-transforms of their sampled versions and replacing differentiators and integrators by transfer functions corresponding to the discrete-time approximate derivative and integral operations.

We will now derive the transfer functions of the discrete-time approximate derivative and integral. For the approximate derivative, we use the time-delay theorem to obtain

$$X_D(z) = \mathcal{Z}[x_{DT}(n)] = \frac{1}{T}\left[X(z) - z^{-1}X(z)\right] \qquad (15.24)$$

from eq. (15.16). We then solve eq. (15.24) for the approximate-derivative transfer function

$$H_D(z) = \frac{X_D(z)}{X(z)} = \frac{1}{T}\left(1 - z^{-1}\right) \qquad (15.25)$$

Figure 15.11
Normalized
Frequency Response
of a First-Order
Butterworth Filter
and Discrete-Time
Approximations

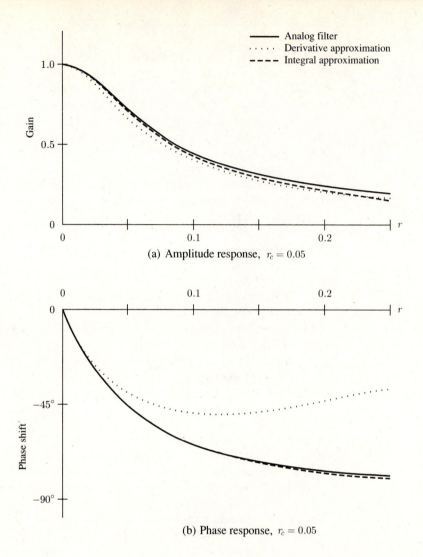

(a) Amplitude response, $r_c = 0.05$

(b) Phase response, $r_c = 0.05$

Similarly, we find the transfer function of the approximate integral by first using the time-delay and summation theorems to compute

$$X_I(z) = \mathcal{Z}[x_{IT}(n)] = \frac{T}{2}\left[\frac{X(z) - z^{-1}X(z)}{1 - z^{-1}}\right] \tag{15.26}$$

from eq. (15.18) and then solving eq. (15.26) to give

$$H_I(z) = \frac{X_I(z)}{X(z)} = \frac{T}{2}\left(\frac{1 + z^{-1}}{1 - z^{-1}}\right) \tag{15.27}$$

Example 15.6

Find the transfer function for the first discrete-time approximation derived in Example 15.5 by using block diagrams.

Solution

The block diagram corresponding to the system differential equation for the continuous-time first-order Butterworth low-pass filter is shown in Figure 15.12.

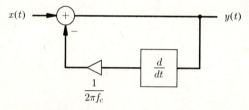

Figure 15.12 Block-Diagram Representation of a Continuous-Time First-Order Butterworth Filter

We obtain the transformed block diagram for the approximating discrete-time filter by (1) substituting the z-transforms of the sampled versions of the signals for the signals and (2) substituting the approximate-derivative transfer function for the derivative operation in the block diagram shown in Figure 15.12. The result is shown in Figure 15.13.

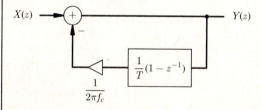

Figure 15.13 Block-Diagram Representation for the Approximating Discrete-Time Filter in Example 15.6

From the block diagram in Figure 15.13, we write the equation

$$Y(z) = X(z) - \frac{1}{2\pi f_c T}(1 - z^{-1})Y(z)$$

The solution of this equation for $Y(z)/X(z)$ is the transfer function of the discrete-time filter:

$$H(z) = \frac{Y(z)}{X(z)} = \frac{2\pi f_c T}{1 + 2\pi f_c T}\left[\frac{1}{1 - z^{-1}/(1 + 2\pi f_c T)}\right]$$

The result is the same as that obtained in Example 15.5, as it must be.

A similar block-diagram-approach derivation of the transfer function for the second filter approximation produced in Example 15.5 is required of the reader in the Problem section.

The approximate-derivative and/or integral-substitution technique for designing discrete-time filters to approximate continuous-time filters is not a performance-related design technique. That is, we make no attempt to match time-response or frequency-response characteristics directly. The resulting filter designs generally require a high sample rate with respect to the highest frequency components in the signals to ensure performance that closely matches the continuous-time-filter performance. We pointed this out in discussing the approximations obtained in Example 15.5. Now, let us turn to techniques that attempt to match the continuous-time-filter time response or frequency response directly.

Time-Response Matching—Time-Invariant Design

Design of a discrete-time filter by time-response matching is frequently referred to as *time-invariant design*. The technique creates a discrete-time-filter design, the input- and output-signal sequence values of which are samples of the corresponding continuous-time-filter input and output signals for a *specified* input signal. That is, if $x(t) = x_1(t)$ yields $y(t) = y_1(t)$ for the continuous-time filter, then $x_T(n) = x_{1T}(n) = x_1(nT)$ yields $y_T(n) = y_{1T}(n) = y_1(nT)$ for the discrete-time filter, where the continuous-time-filter input signal $x_1(t)$ is specified.

The output signal of the discrete-time filter is a sampled version of the continuous-time-filter output signal only if the input signal is a sampled version of $x_1(t)$. For other input signals, the output signal is only an approximation to the sampled version of the continuous-time-filter output signal.

The design procedure consists of the following steps when the continuous-time-filter transfer function $H_a(s)$ and the specified input signal $x_1(t)$ are given.

Step 1 Compute the Laplace transform of the specified input signal

$$X_1(s) = \mathcal{L}[x_1(t)] \tag{15.28}$$

Step 2 Compute the Laplace transform of the continuous-time-filter output signal

$$Y_1(s) = X_1(s)H_a(s) \tag{15.29}$$

Step 3 Compute the continuous-time-filter output signal

$$y_1(t) = \mathcal{L}^{-1}[Y_1(s)] \tag{15.30}$$

Step 4 Write sampled versions of the continuous-time-filter input and output signals

$$x_{1T}(n) = x_1(t)\big|_{t=nT} \quad \text{and} \quad y_{1T}(n) = y_1(t)\big|_{t=nT} \tag{15.31}$$

Step 5   Compute the z-transforms of the sampled input and output signals

$$X_1(z) = Z[x_{1T}(n)] \quad \text{and} \quad Y_1(z) = Z[y_{1T}(n)] \tag{15.32}$$

Step 6   Compute the discrete-time-filter transfer function

$$H(z) = Y_1(z)/X_1(z) \tag{15.33}$$

The design steps are summarized by the overall design equation

$$H(z) = \frac{1}{X_1(z)} \mathcal{Z}\big[\{\mathcal{L}^{-1}[H_a(s)X_1(s)]\}\big|_{t=nT}\big] \tag{15.34}$$

A variation of the time-invariant-design technique sets $y_{1T}(n) = Gy_1(nT)$, where G is an amplitude-scaling constant. The filter design specified by eq. (15.34) is produced if $G = 1$. This variation produces a discrete-time-filter output signal that is an amplitude-scaled version of the sampled continuous-time-filter output signal corresponding to the specified input signal. The overall design equation is

$$H(z) = \frac{G}{X_1(z)} \mathcal{Z}\big[\{\mathcal{L}^{-1}[H_a(s)X_1(s)]\}\big|_{t=nT}\big] \tag{15.35}$$

One reason for using this variation is to match the gain of the two filters at some specified frequency. We will illustrate this use of G following the next example.

The time-invariant-design technique is referred to as *impulse-invariant design* if $x_1(t) = \delta(t)$ and as *step-invariant design* if $x_1(t) = u(t)$. Other types of specified input signals could be considered, but we will discuss only these two.

Impulse-Invariant Design   Strictly speaking, impulse-invariant design is not possible because the sampled version of an impulse function does not exist, as we indicated when we defined the unit pulse signal in Section 10.3. However, we can generate an approximation to impulse-invariant design by matching samples of the impulse response of the continuous-time filter to the discrete-time-filter response corresponding to the input signal $x_{1T}(n) = (1/T)\delta_T(n)$. As we indicated in Section 10.3, $(1/T)\delta_T(n)$ can be thought of as samples of a continuous-time rectangular-pulse signal that approaches a unit impulse signal as T approaches zero. We call it the *approximate sampled-unit-impulse signal*. Therefore, for impulse-invariant design of a discrete-time filter from a continuous-time filter, the transfer function of which is $H_a(s)$, the input signals and corresponding transforms are

$$x_1(t) = \delta(t) \leftrightarrow X_1(s) = 1 \tag{15.36}$$

and

$$x_{1T}(n) = (1/T)\delta_T(n) \leftrightarrow X_1(z) = 1/T \tag{15.37}$$

The overall design equation with $G = 1$ is

$$H(z) = T\mathcal{Z}\big[\{\mathcal{L}^{-1}[H_a(s)]\}\big|_{t=nT}\big] \tag{15.38}$$

The transfer function of the impulse-invariant design of a discrete-time filter is T times the z-transform of the sampled version of the continuous-time-filter impulse response. Therefore, the frequency response of the impulse-invariant design of a discrete-time filter is T times the discrete-time Fourier transform of the samples of the impulse-response corresponding to the continuous-time filter. That is, the designed-filter frequency response is

$$H\left(e^{j2\pi fT}\right) = TH_{ad}(f) = T \sum_{m=-\infty}^{\infty} f_s H_a\left(f - mf_s\right)$$

$$= \sum_{m=-\infty}^{\infty} H_a\left(f - mf_s\right) \qquad (15.39)$$

where $H_a(f)$ is the Fourier transform of the continuous-time-filter impulse response (that is, the filter's frequency response) and $H_{ad}(f)$ is the discrete-time Fourier transform of the sampled version of the continuous-time-filter impulse response. Note that the frequency response $H(e^{j2\pi fT})$ of the discrete-time filter derived by impulse-invariant design matches the continuous-time-filter frequency response for $|f| \leq f_s/2$, except for aliasing. For practical filters, $H_a(f)$ approaches zero as $|f|$ approaches infinity. Therefore, the effect of aliasing becomes smaller as f_s approaches infinity (that is, as T approaches zero), and the discrete-timer-filter frequency response approaches the continuous-time filter frequency response for $|f| \leq f_s/2$.

Example 15.7

Use impulse-invariant design to find the transfer function of a discrete-time filter that corresponds to a first-order Chebyshev low-pass filter with 3 dB of ripple in the passband, a maximum gain of unity, and a cutoff frequency of 100 Hz. The discrete-time filter is to be used in a discrete-time manufacturing-process control system, the sample rate of which is 1000 Hz. Compare the characteristics of the discrete-time and continuous-time filters by plotting their impulse and unit step responses and their amplitude and phase responses. (Use normalized frequency r for the interval $0 \leq r \leq 0.5$.)

Solution
We consider the continuous-time filter first. From Table 8.1, the transfer function of the corresponding normalized low-pass filter is

$$H_{aN}(s_{LN}) = \frac{a_0}{1.0024 + s_{LN}}$$

Since the filter order is odd, $a_0 = G_M b_0 = (1)(1.0024) = 1.0024$. We obtain the transfer function for the continuous-time filter by using eq. (8.44) to scale the cutoff frequency to 100 Hz. The result is

$$H_a(s) = H_{aN}(s_{LN})\big|_{s_{LN}=s/2\pi(100)} = \frac{1.0024}{1.0024 + (s/200\pi)} = \frac{629.83}{629.83 + s}$$

The continuous-time-filter impulse response is

$$y_i(t) = h_a(t) = \mathcal{L}^{-1}[H_a(s)] = 629.83e^{-629.83t}u(t) \qquad (15.40)$$

In order to compare the continuous-time filter with the impulse-invariant design of the discrete-time filter, we also compute the unit step response and normalized frequency response of the continuous-time filter. The unit step response is

$$y_u(t) = \mathcal{L}^{-1}[H_a(s)/s] = \mathcal{L}^{-1}\left[\frac{629.83}{s(s + 629.83)}\right]$$

$$= \mathcal{L}^{-1}\left[\frac{1}{s} - \frac{1}{s + 629.83}\right] = \left[1 - e^{-629.83t}\right]u(t)$$

The frequency response of the continuous-time filter as a function of r is

$$H_a(j2\pi r) = H_a(s)\big|_{s=j2\pi r} = \frac{629.83}{629.83 + j2\pi r}$$

and gives a normalized amplitude response of

$$|H_a(j2\pi r)| = \frac{629.83}{\sqrt{(629.83)^2 + (2\pi r)^2}}$$

and a normalized phase response of

$$\underline{/\,H_a(j2\pi r)} = -\tan^{-1}\left(\frac{2\pi r}{629.83}\right)$$

Now we use impulse-invariant design to generate a discrete-time filter for use in the system with sample rate $f_s = 1000$ Hz. The overall design equation is

$$H(z) = T\mathcal{Z}\left[\left\{\mathcal{L}^{-1}\left[\frac{629.83}{s + 629.83}\right]\right\}\Big|_{t=nT}\right]$$

where $T = 1/f_s = 1/1000 = 0.001$ s. From eq. (15.40),

$$H(z) = 0.001\mathcal{Z}\left[\left\{629.83e^{-629.83t}u(t)\right\}\big|_{t=0.001n}\right]$$

$$= 0.001\mathcal{Z}\left[629.83e^{-0.62983n}u_T(n)\right]$$

$$= \frac{0.62983}{1 - e^{-0.62983}z^{-1}} = \frac{0.62983}{1 - 0.53268z^{-1}}$$

The block-diagram representation of the realization of the discrete-time filter is shown in Figure 15.14.

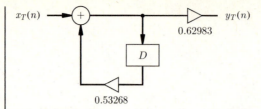

Figure 15.14 Block-Diagram Representation of the Discrete-Time-Filter Realization in Example 15.7

This completes the filter design. Now, using Table C.8, we find that the response of the discrete-time filter to the approximate sampled-unit-impulse signal [that is, $x_T(n) = (1/T)\delta_T(n) = 1000\delta_T(n)$] is

$$y_{iT}(n) = \mathcal{Z}^{-1}[X(z)H(z)] = \mathcal{Z}^{-1}[1000H(z)] = 629.83(0.53268)^n u_T(n)$$

$$= 629.83e^{-0.62983n}u_T(n) = 629.83e^{-629.83nT}u_T(n)$$

As expected, this filter response is a sampled version of the impulse response of the continuous-time filter, the discrete-time filter having been designed to produce this correspondence between the two responses.

The z-transform of the unit step response of the discrete-time filter is

$$Y_u(z) = \mathcal{Z}[u_T(n)]\,H(z) = \left(\frac{1}{1 - z^{-1}}\right)\left(\frac{0.62983}{1 - 0.53268z^{-1}}\right)$$

$$= \frac{0.62983z^2}{(z - 1)(z - 0.53268)}$$

Therefore,

$$\frac{Y_u(z)}{z} = \frac{0.62983z}{(z - 1)(z - 0.53268)} = \frac{A_1}{z - 1} + \frac{A_2}{z - 0.53268}$$

We compute the partial-fraction-expansion coefficients, multiply through by z, and eliminate the positive powers of z to obtain

$$Y_u(z) = 1.34775\left(\frac{1}{1 - z^{-1}}\right) - 0.71792\left(\frac{1}{1 - 0.53268z^{-1}}\right)$$

Using the linearity theorem and the transform pairs in Table C.8, we find that the unit step response is

$$y_{uT}(n) = \left[1.34775 - 0.71792(0.53268)^n\right]u_T(n)$$

$$= \left[1.34775 - 0.71792e^{-629.83nT}\right]u_T(n)$$

The unit step response of the discrete-time filter is only an approximation to a sampled version of the continuous-time-filter unit step response. The output signals of the continuous-time and discrete-time filters are plotted in Figure 15.15, where CT refers to the continuous-time filter and DT refers to the discrete-time filter.

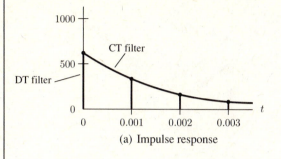

(a) Impulse response

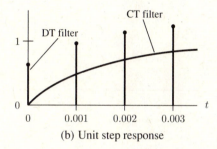

(b) Unit step response

Figure 15.15 Output Signals for the Impulse-Invariant Filter Designed in Example 15.7

The frequency response of the discrete-time filter in terms of normalized frequency is

$$H(e^{j2\pi r}) = \frac{0.62983}{1 - 0.53268e^{-j2\pi r}}$$

and gives a normalized amplitude response of

$$\left|H(e^{j2\pi r})\right| = \frac{0.62983}{\sqrt{[1 - 0.53268\cos(2\pi r)]^2 + [0.53268\sin(2\pi r)]^2}}$$

and a normalized phase response of

$$\underline{/H(e^{j2\pi r})} = -\tan^{-1}\{0.53268\sin(2\pi r)/[1 - 0.53268\cos(2\pi r)]\}$$

The normalized amplitude response and normalized phase response of the discrete-time filter are plotted in Figure 15.16, along with the normalized amplitude and phase responses of the corresponding continuous-time filter. The notation DT and CT is used once again. The DTS filter curves will be discussed immediately following this example.

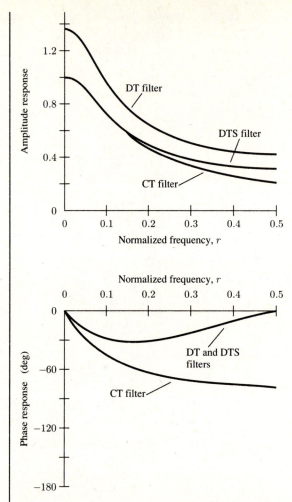

Figure 15.16 Normalized Frequency Responses of the Low-Pass Filter Designed in Example 15.7 and of an Amplitude-Scaled Version of the Filter

Note that the impulse-invariant design of Example 15.7 does not produce a gain in the passband that matches the gain of the continuous-time filter. This gain difference results from extensive aliasing in the passband of the discrete-time-filter frequency response caused by the slow decrease of the continuous-time-filter gain as a function of frequency. We can change the gain difference so that the dc gain of the two filters is the same by using the amplitude-scaling constant $G = 1/(\text{dc gain}) = 1/H(1) = 1/1.34775$ in the impulse-invariant-design variation eq. (15.35). This produces the discrete-time-filter transfer function

$$H_s(z) = \frac{0.46732}{1 - 0.53268z^{-1}}$$

the corresponding amplitude and phase responses of which are also shown in Figure 15.16 (labeled DTS filter). Note that the amplitude response matches the continuous-time-filter amplitude response reasonably well (to within approximately 0.5 dB) for frequencies up to twice the filter cutoff frequency of $r_c = f_c/f_s = 0.1$. The phase response is unchanged from the phase response for the first impulse-invariant design because the only change in the transfer-function is a multiplying constant. It does not match the continuous-time-filter phase response as well as the amplitude responses match.

The response of the amplitude-scaled discrete-time filter to the approximate sampled-unit-impulse signal, $(1/T)\delta_T(n)$, and a sampled unit step signal are the previously computed system responses divided by 1.34775. These system responses are

$$y_{siT}(n) = 467.32e^{-629.83nT}u_T(n)$$

and

$$y_{suT}(n) = \left[1 - 0.53268e^{-629.83nT}\right]u_T(n)$$

The impulse-response matching no longer exists because the discrete-time-filter response to the signal $(1/T)\delta_T(n)$ is matched to an amplitude-scaled version of the continuous-time-filter impulse response. Note that the unit step response of the amplitude-scaled discrete-time filter matches the sampled unit step response of the continuous-time filter for large values of T. At large values of T, the unit step signal essentially appears to be a constant (that is, dc), and thus is affected in the same way by both the amplitude-scaled discrete-time filter and the continuous-time filter (see the frequency responses).

Step-Invariant Design For step-invariant design of a discrete-time filter from a continuous-time filter with transfer function $H_a(s)$, the input signals and corresponding transforms are

$$x_1(t) = u(t) \leftrightarrow X_1(s) = 1/s \tag{15.41}$$

and

$$x_{1T}(t) = u_T(n) \leftrightarrow X_1(z) = 1/\left(1 - z^{-1}\right) \tag{15.42}$$

Therefore, the overall design equation is

$$H(z) = \left(1 - z^{-1}\right)\mathcal{Z}\left[\{\mathcal{L}^{-1}\left[H_a(s)/s\right]\}\,|_{t=nT}\right] \tag{15.43}$$

when $G = 1$.

Example 15.8

Repeat Example 15.7 using step-invariant design.

Solution

The impulse response, unit step response and frequency response for the continuous-time filter were computed in Example 15.7. Now, to generate a discrete-time-filter design for use in the system with sample rate $f_s = 1000$ Hz using step-invariant design, we first substitute $H_a(s)$ from Example 15.7 in eq. (15.43) to obtain

$$H(z) = \left(1 - z^{-1}\right) \mathcal{Z}\left[\left\{\mathcal{L}^{-1}\left[\frac{629.83}{s(s + 629.83)}\right]\right\}\bigg|_{t=nT}\right] \tag{15.44}$$

where $T = 1/f_s = 1/1000 = 0.001$ s. We computed the required inverse Laplace transform in Example 15.7. Substituting that result in eq. (15.44) gives

$$\begin{aligned}
H(z) &= \left(1 - z^{-1}\right) \mathcal{Z}\left[\left\{u(t) - e^{-629.83t}u(t)\right\}\big|_{t=0.001n}\right] \\
&= \left(1 - z^{-1}\right) \mathcal{Z}\left[u_T(n) - e^{-0.62983n}u_T(n)\right] \tag{15.45} \\
&= 1 - \frac{1 - z^{-1}}{1 - e^{-0.62983}z^{-1}} = \frac{0.46732z^{-1}}{1 - 0.53268z^{-1}}
\end{aligned}$$

as the transfer function of the discrete-time filter. The block-diagram representation of the realization of the discrete-time filter is shown in Figure 15.17.

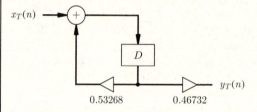

Figure 15.17 Block-Diagram Representation of the Discrete-Time-Filter Realization in Example 15.8

Using Table C.8 and the time-delay theorem, we find that the response of the discrete-time filter to the approximate sampled-unit-impulse signal [that is, $x_T(n) = (1/T)\delta_T(n) = 1000\delta_T(n)$] is

$$\begin{aligned}
y_{iT}(n) &= \mathcal{Z}^{-1}[1000H(z)] = 467.32(0.53258)^{n-1}u_T(n-1) \\
&= 877.29e^{-0.62983n}u_T(n-1) = 877.2e^{-629.83nT}u_T(n-1)
\end{aligned}$$

Note that $y_{iT}(n)$ is only an approximation to a sampled version of the continuous-time-filter impulse response. The z-transform of the unit step response of the discrete-time filter is

$$Y_u(z) = \left(\frac{1}{1 - z^{-1}}\right)H(z) = \mathcal{Z}\left[u_T(n) - e^{-0.62983n}u_T(n)\right]$$

where the second equivalence follows from eq. (15.45). Therefore, the filter unit step response is

$$y_{uT}(n) = \mathcal{Z}^{-1}[Y_u(z)] = u_T(n) - e^{-0.62983n}u_T(n)$$

$$= \left[1 - e^{-629.83nT}\right]u_T(n)$$

Note that $y_{uT}(n)$ is a sampled version of the unit step response of the continuous-time filter. We expect this unit step response correspondence because the discrete-time filter was designed to produce it. The output signals of continuous-time and discrete-time filters are plotted in Figure 15.18. Once again, CT refers to the continuous-time filter and DT refers to the discrete-time filter.

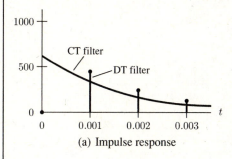

(a) Impulse response

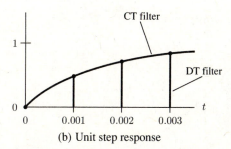

(b) Unit step response

Figure 15.18 Output Signals for the Step-Invariant Filter Designed in Example 15.8

The frequency response of the discrete-time filter in terms of normalized frequency is

$$H\!\left(e^{j2\pi r}\right) = \frac{0.46732e^{-j2\pi r}}{1 - 0.53268e^{-j2\pi r}}$$

and gives a normalized amplitude response of

$$\left|H\!\left(e^{j2\pi r}\right)\right| = \frac{0.46732}{\sqrt{[1 - 0.53268\cos(2\pi r)]^2 + [0.53268\sin(2\pi r)]^2}}$$

and a normalized phase response of

$$\underline{/\,H(e^{j2\pi r})} = -2\pi r - \tan^{-1}\{0.53268\sin(2\pi r)/[1 - 0.53268\cos(2\pi r)]\}$$

The discrete-time-filter normalized amplitude and phase responses are plotted in Figure 15.19, along with the amplitude and phase responses of the corresponding continuous-time filter.

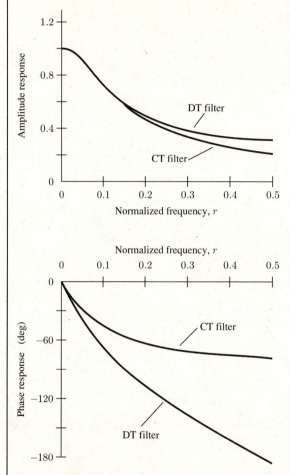

Figure 15.19 Frequency Response of the Step-Invariant Low-Pass Filter Designed in Example 15.8

The transfer function obtained by step-invariant design differs from that for the amplitude-scaled impulse-invariant design only by a z^{-1} factor in the numerator. This result holds only for a first-order filter design, in which case the time responses of the step-invariant-designed filter are the same as those for the amplitude-scaled impulse-invariant design except for a one-sample time delay. The normalized amplitude response is the same for both filter designs because $|e^{-j2\pi r}| = 1$; the normalized phase response differs by $-2\pi r$ rad between the filter designs because $\underline{/e^{-j2\pi r}} = -2\pi r$.

Frequency-Response Matching—Bilinear-Transformation Design

We saw how the time-invariant-design procedure is used to construct a discrete-time filter that gives a specified output-signal response to a specified input signal. In many cases, however, it is preferable to match the frequency response of a discrete-time filter to the frequency response of some continuous-time filter. The discrete-time-filter frequency response is periodic, with period equal to the sample rate f_s; therefore, it cannot match the frequency response of a causal continuous-time filter for all frequencies because the frequency response of a causal continuous-time filter is not periodic and not bandlimited.

Actually, we are interested only in the unambiguous frequency interval $|f| \le f_s/2$ for a discrete-time filter because the frequency response repeats outside this interval. We noted earlier that the impulse-invariant design of a discrete-time filter matches the frequency response in this frequency interval, except for aliasing. However, if the continuous-time-filter frequency response does not decrease rapidly at higher frequencies, then aliasing is pronounced all the way down to very low frequencies. We saw this in Example 15.7.

One method for designing a discrete-time filter with a frequency response similar to that of a continuous-time filter within the unambiguous frequency range is to use a nonlinear transformation of the frequency scale called a *bilinear transformation*. Using this transformation, the entire frequency response of the continuous-time filter is nonlinearly compressed into the frequency interval $-0.5f_s \le f \le 0.5f_s$ and is also repeated in all adjacent intervals of the same size. That is, the frequency response is compressed into the multiple frequency intervals $(n - 0.5)f_s \le f \le (n + 0.5)f_s$ where $-\infty < n < \infty$. This compressing and repeating of the continuous-time-filter frequency response is illustrated in Figure 15.20, where we designate the frequency variable for the continuous-time filter with the subscript a to distinguish it from the frequency variable for the discrete-time filter. Note that both frequency-response plots have the same scale in Figure 15.20.

Since the bilinear transformation compresses the frequency response of the continuous-time filter around the frequency $f = 0$ and does not do any amplitude scaling, then the frequency responses of the discrete-time filter and the continuous-time filter are equal at $f = 0$ (that is, at dc) and the maximum gain of the discrete-time filter equals the maximum gain of the continuous-time filter. We can also match the frequency response of the discrete-time filter at one additional frequency (frequency f_1 in Figure 15.20) to the frequency response of the continuous-time filter at the same or any other specified frequency (frequency f_{a1} in Figure 15.20) because the bilinear transformation, as defined in the following paragraph, contains one arbitrary constant, C.

The discrete-time-filter transfer function $H(z)$ that we obtain by bilinear-transformation design from the continuous-time-filter transfer function $H_a(s)$ is

$$H(z) = H_a(s)\big|_{s=C\frac{1-z^{-1}}{1+z^{-1}}} \tag{15.46}$$

Figure 15.20
Illustration
of Bilinear
Transformation
to Produce a
Discrete-Time-Filter
Frequency Response
from a Continuous-
Time-Filter
Frequency Response

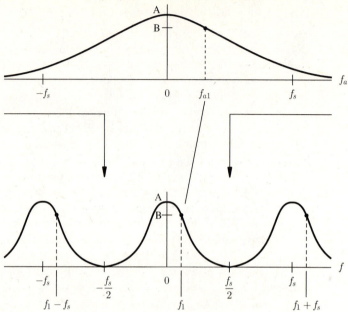

Discrete-Time Filter Frequency Response

The relationship

$$s = C\frac{1 - z^{-1}}{1 + z^{-1}} \tag{15.47}$$

is the bilinear transformation that we use to relate z in the discrete-time-filter transfer function to s in the continuous-time-filter transfer function.

To show the nature of the bilinear transformation, we first obtain the frequency response of the discrete-time filter by setting $z = e^{j2\pi fT}$ in eq. (15.46) to give

$$H\!\left(e^{j2\pi fT}\right) = H_a(s)\Big|_{s=C\frac{1-e^{-j2\pi fT}}{1+e^{-j2\pi fT}}} \tag{15.48}$$

Equation (15.48) shows us that the value of the discrete-time-filter frequency response $H(e^{j2\pi fT})$ at the frequency f equals the value of the continuous-time-filter transfer function $H_a(s)$ at

$$s = C\frac{1 - e^{-j2\pi fT}}{1 + e^{-j2\pi fT}} = C\frac{e^{j\pi fT} - e^{-j\pi fT}}{e^{j\pi fT} + e^{-j\pi fT}}$$

$$= jC\sin(\pi fT)/\cos(\pi fT) \equiv j2\pi f_a \tag{15.49}$$

The values of s given by eq. (15.49) lie on the imaginary axis in the s-plane. Therefore, the frequency response of the continuous-time filter is specified by the values of $H_a(s)$ that correspond to the values $s = j2\pi f_a$ given by eq. (15.49). This means that the discrete-time-filter frequency response at the frequency f equals the continuous-time-filter frequency response at the frequency f_a, where from eq. (15.49), f and fa are related by

$$f_a = \frac{C}{2\pi} \tan(\pi f T) \qquad (15.50)$$

The relationship between the frequencies at which the frequency responses of the continuous-time and discrete-time filters are equal is plotted in Figure 15.21 for three different values of C. Curve 3 corresponds to $C = 5.801 f_s$, which gives the transformation illustrated in Figure 15.20. Curve 2 corresponds to $C = 2f_s$ and is tangent to the line $f_a = f$ at $f = 0$. For the bilinear transformations that correspond to both of these values for C, the frequency-response curve of the continuous-time filter is compressed for all frequencies to produce the discrete-time-filter frequency response for $|f| \leq f_s/2$. Curve 1 corresponds to $C = 0.865 f_s$. In this case, the continuous-time-filter frequency response is expanded for frequencies lower than $0.3927 f_s$ and compressed for frequencies greater than $0.3927 f_s$ to produce the discrete-time-filter frequency response for $|f| \leq f_s/2$ because curve 1 crosses the line $f_a = f$ at $f_a = f = 0.3927 f_s$.

As we mentioned, we can choose the parameter C so that the frequency response of the discrete-time filter at frequency f_m equals the frequency response of the continuous-time filter at f_{am}, where $f_m \leq f_s/2$ and f_{am} are arbitrarily

Figure 15.21
Relationship between
Frequencies at
Which the Frequency
Responses of the
Discrete-Time and
Continuous-Time
Filters Are Equal
Using Bilinear
Transformation

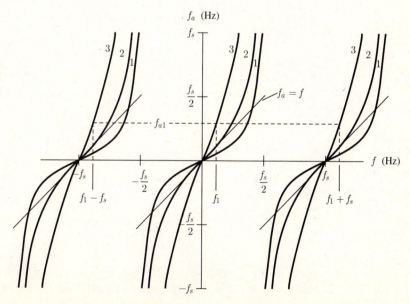

chosen. That is, we can choose C so that eq. (15.50) equals f_{am} when $f = f_m$. With this value of C, the first branch of the tangent curve (center branch) in Figure 15.21 passes, through the point specified by $f_a = f_{am}$, and $f = f_m$. To find the required value for C, we solve eq. (15.50) for C. This solution is

$$C = 2\pi f_{am} \cot(\pi f_m T) \qquad (15.51)$$

We may want to generate a discrete-time low-pass filter with cutoff frequency $f = f_c$ from a normalized continuous-time low-pass filter. In this special case, $\omega_{ac} = 2\pi f_{ac} = 1$ and $C = \cot(\pi f_c T)$.

Low-Pass-Filter Design   Bilinear-transformation design applies directly to the design of discrete-time low-pass filters. We will illustrate this application in the following example in which bilinear transformation is used to obtain the discrete-time filter needed for the manufacturing-process control system in Example 15.7.

Example 15.9

Use bilinear-transformation design to find the transfer function of a discrete-time filter that corresponds to a first-order Chebyshev low-pass filter with 3 dB of ripple in the passband, a maximum gain equal to one, and a cutoff frequency of 100 Hz. The discrete-time filter is to be used in a manufacturing process control system, the sample rate of which 1000 Hz. Compare the frequency responses of the discrete-time and continuous-time filters by plotting their amplitude and phase responses in terms of normalized frequency r for the interval $0 \leq r \leq 0.5$.

Solution
We found in Example 15.7 that the normalized transfer function of the continuous-time Chebyshev low-pass filter is

$$H_{aN}(s) = \frac{1.0024}{1.0024 + s}$$

The continuous-time filter that corresponds to this transfer function has a cutoff frequency of $\omega_{ac} = 1$ rad/s. Therefore, we need a bilinear-transformation parameter value of

$$C = 2\pi f_{ac} \cot(\pi f_c T) = \omega_{ac} \cot(\pi f_c / f_s)$$

$$= (1) \cot(100\pi/1000) = 3.0777$$

to produce a discrete-time filter having a cutoff frequency of 100 Hz. Using this value of C, we find that the discrete-time-filter transfer function is

$$H(z) = \left. H_{aN}(s) \right|_{s=3.0777\frac{1-z^{-1}}{1+z^{-1}}}$$

$$= \frac{1.0024}{1.0024 + 3.0777(1 - z^{-1})/(1 + z^{-1})}$$

$$= \frac{1.0024(1 + z^{-1})}{4.0801 - 2.0753z^{-1}} = \frac{0.2457(1 + z^{-1})}{1 - 0.5086z^{-1}}$$

The block-diagram representation of the realization of the filter that corresponds to $H(z)$ is shown in Figure 15.22.

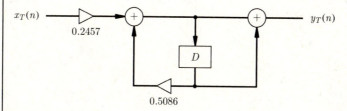

Figure 15.22 Block-Diagram Representation of the Discrete-Time-Filter Realization in Example 15.9

The frequency response of the discrete-time filter, in terms of normalized frequency, is

$$H(e^{j2\pi r}) = \frac{0.2457(1 + e^{-j2\pi r})}{1 - 0.5086e^{-j2\pi r}}$$

and gives the normalized amplitude response

$$\left| H(e^{j2\pi r}) \right| = \frac{0.2457\sqrt{[1 + \cos(2\pi r)]^2 + [-\sin(2\pi r)]^2}}{\sqrt{[1 - 0.5086\cos(2\pi r)]^2 + [0.5086\sin(2\pi r)]^2}}$$

and the normalized phase response

$$\angle H(e^{j2\pi r}) = \tan\{[-\sin(2\pi r)]/[1 + \cos(2\pi r)]\}$$
$$- \tan\{[0.5086\sin(2\pi r)]/[1 - 0.5086\cos(2\pi r)]\}$$

The discrete-time-filter (DT) normalized amplitude and phase responses are plotted in Figure 15.23, along with the normalized amplitude and phase responses of the corresponding continuous-time filter (CT).

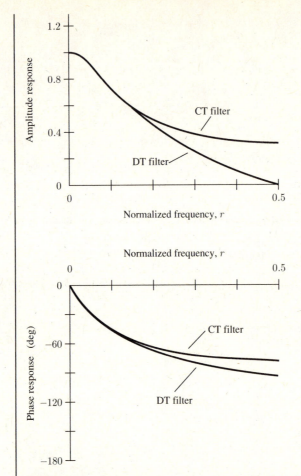

Figure 15.23 Normalized Frequency Response of the Low-Pass Filter Designed with Bilinear Transformation in Example 15.9

For the particular low-pass filter that we considered in Examples 15.7, 15.8, and 15.9, we can see that the bilinear-transformation design, step-invariant design, and amplitude-scaled impulse-invariant design all produce discrete-time filters with amplitude responses (gains) that closely match those of the continuous-time filters in the passband (see Figures 15.23, 15.19, and 15.16). The gain in the stopband for the bilinear-transformation design is less than the stopband gain of the continuous-time filter. This is because the continuous-time-filter frequency response at all frequencies greater than the cutoff frequency ($f_c = 100$ Hz, $r = 0.1$) is compressed into the frequency interval $f_c \leq f \leq f_s/2$ to produce the discrete-time filter. The stopband gain of the time-invariant design, is greater than the stopband gain of the continuous-time filter because of the frequency-response aliasing produced by time-invariant designs. For the low-pass filter considered, the bilinear-transformation design gives a phase response that matches that of the continuous-time filter better than the phase responses of either of the two time-invariant designs.

We will now present a second example of bilinear-transformation design of a low-pass filter in abbreviated form to show how the order of the filter is determined when standard normalized low-pass filters are used to generate the filter design.

Example 15.10

A discrete-time data-transmission system used to transmit slowly varying air-pressure measurements needs a low-pass filter to reduce measurement noise. The measurement sample rate is $1000/\pi$ Hz. Use bilinear-transformation design to develop the transfer function of the needed discrete-time low-pass filter from a continuous-time Butterworth low-pass filter. The filter specifications are as follows: (1) gain variation less than 3 dB for $0 \leq f \leq 10/\pi$ Hz, (2) gain less than or equal to -25 dB with respect to maximum gain for $f \geq 30/\pi$ Hz, and (3) maximum gain of 0.5.

Solution
The required bilinear-transformation parameter value is

$$C = 2\pi f_{ac} \cot\left(\pi f_c T\right) = \omega_{ac} \cot\left(\pi f_c/f_s\right)$$

$$= (1)\cot\left[\pi(10/\pi)/(1000/\pi)\right] = 31.8205$$

The frequency in radians per second at which the frequency-response value of the normalized continuous-time filter equals the discrete-time-filter frequency-response value at $f = f_m = 30/\pi$ Hz is

$$\omega_{am} = 2\pi f_{am} = C \tan\left(\pi f_m T\right) = C \tan\left(\pi f/f_s\right)$$

$$= 31.8205 \tan[\pi(30/\pi)(1000/\pi)] = 3.008 \text{ rad/s}$$

From Figure 8.14, we see that the stopband gain of a third-order normalized Butterworth filter at the frequency $\omega_{am} = 3.008$ is at least -25 dB with respect to the maximum gain. From Table 8.1 and the required maximum gain of 0.5, we find that the transfer function of the normalized continuous-time filter is

$$H_{aN}(s) = \frac{0.5}{1 + 2s + 2s^2 + s^3}$$

Therefore, the required transfer function of the discrete-time filter is

$$H(z) = \left. H_{aN}(s)\right|_{s=31.8205} \frac{1 + z^{-1}}{1 - z^{-1}}$$

$$= \frac{1.457327 \times 10^{-5}\left(1 + 3z^{-1} + 3z^{-2} + z^{-3}\right)}{1 - 2.874357z^{-1} + 2.756483z^{-2} - 0.881893z^{-3}}$$

Bandpass-Filter Design   When we use bilinear transformation to design a bandpass filter, we want to match the frequency response of the discrete-time filter to the frequency response of the continuous-time filter at both the upper and lower cutoff frequencies of the passband. We cannot perform these two frequency-response matches by substituting $s = C(1 - z^{-1})/(1 + z^{-1})$ in the transfer function that corresponds to a continuous-time filter with the desired cutoff frequencies because the single parameter C only permits frequency-response matching at a single frequency. In order to circumvent this problem, we can choose the continuous-time-filter cutoff frequencies and the parameter C so that the frequency response at each of the continuous-time-filter cutoff frequencies matches the frequency response at each of the desired discrete-time-filter cutoff frequencies. Note that we choose three parameters (the two continuous-time-filter cutoff frequencies and the parameter C), whereas only two parameters are needed to produce the desired frequency-response matching at the two cutoff frequencies. Therefore, more than one set of three parameters can be chosen; however, they are interrelated. We can reduce the three nonunique parameters to two unique parameters if we begin our design with a continuous-time normalized low-pass filter.

Consider the continuous-time normalized low-pass filter with a transfer function of $H_{aN}(s)$. This filter has a cutoff frequency of $\omega_c = 1$ rad/s. From eqs. (8.46), (8.47), and (8.51) in Chapter 8, we see that the transfer function of a continuous-time bandpass filter that corresponds in type to the normalized low-pass filter and has upper and lower cutoff frequencies of f_{au} and f_{al} respectively, is

$$H_a(s_1) = H_{aN}(s)\Big|_{s = \frac{s_1^2 + 4\pi f_{au} f_{a\ell}}{2\pi s_1 (f_{au} - f_{a\ell})}} \tag{15.52}$$

We add the subscript 1 to the complex variable in the bandpass-filter transfer function to distinguish this complex variable from the complex variable in the transfer function of the normalized low-pass filter. The transfer function $H(z)$ of the discrete-time filter produced by bilinear-transformation design from the continuous-time bandpass filter with transfer function $H_a(s_1)$ is

$$H(z) = H_a(s_1)\Big|_{s_1 = C\frac{1-z^{-1}}{1+z^{-1}}} = H_{aN}(s)\Big|_{s = \frac{\left[C^2\left(1-z^{-1}\right)^2 / \left(1+z^{-1}\right)^2\right] + 4\pi f_{au} f_{a\ell}}{2\pi C\left(1-z^{-1}\right)(f_{au} - f_{a\ell}) / \left(1+z^{-1}\right)}}$$

$$= H_{aN}(s)\Big|_{s = \frac{C^2\left(1-2z^{-1}+z^{-2}\right)+4\pi f_{au} f_{a\ell}\left(1+2z^{-1}+z^{-2}\right)}{2\pi C\left(1-z^{-2}\right)(f_{au} - f_{a\ell})}} \tag{15.53}$$

We define f_u and f_ℓ to be, respectively, the upper and lower cutoff frequencies of the discrete-time bandpass filter with a transfer function of $H(z)$. Using eq. (15.50), we see that the corresponding upper and lower cutoff frequencies for the continuous-time bandpass filter are

$$f_{au} = \frac{C}{2\pi} \tan\left(\pi f_u T\right) \tag{15.54}$$

and

$$f_{a\ell} = \frac{C}{2\pi} \tan(\pi f_\ell T) \tag{15.55}$$

respectively. We substitute eq. (15.54) and eq. (15.55) in the transformation relationship present in eq. (15.53) to obtain

$$s = \frac{C^2(1 - 2z^{-1} + z^{-2}) + C^2 \tan(\pi f_u T) \tan(\pi f_\ell T)(1 + 2z^{-1} + z^{-2})}{C^2(1 - z^{-2})[\tan(\pi f_u T) - \tan(\pi f_\ell T)]} \tag{15.56}$$

for the relationship between the complex variable s in the transfer function of the continuous-time normalized low-pass filter and the complex variable z in the transfer function of the discrete-time bandpass filter. For simplicity, we define

$$a \equiv \tan(\pi f_u T) \tag{15.57}$$

and

$$b \equiv \tan(\pi f_\ell T) \tag{15.58}$$

When we substitute a and b in eq. (15.56), we obtain the transformation relationship

$$\begin{aligned}
s &= \frac{C^2(1 - 2z^{-1} + z^{-2}) + C^2 ab(1 + 2z^{-1} + z^{-2})}{C^2(1 - z^{-2})(a - b)} \\[2mm]
&= \frac{(1 + ab) - 2(1 - ab)z^{-1} + (1 + ab)z^{-2}}{(a - b)(1 - z^{-2})} \\[2mm]
&= \left[\frac{1 + ab}{a - b}\right]\left[\frac{1 - 2[(1 - ab)/(1 + ab)]z^{-1} + z^{-2}}{(1 - z^{-2})}\right] \\[2mm]
&= D\left[\frac{1 - 2Ez^{-1} + z^{-2}}{(1 - z^{-2})}\right]
\end{aligned} \tag{15.59}$$

where

$$\begin{aligned}
D &= \frac{1 + ab}{a - b} = \frac{1 + \tan(\pi f_u T)\tan(\pi f_\ell T)}{\tan(\pi f_u T) - \tan(\pi f_\ell T)} \\[2mm]
&= \cot[\pi T(f_u - f_\ell)]
\end{aligned} \tag{15.60}$$

and

$$\begin{aligned}
E &= \frac{1 - ab}{1 + ab} = \frac{1 - \tan(\pi f_u T)\tan(\pi f_\ell T)}{1 + \tan(\pi f_u T)\tan(\pi f_\ell T)} \\[2mm]
&= \frac{\cos(\pi f_u T)\cos(\pi f_\ell T) - \sin(\pi f_u T)\sin(\pi f_\ell T)}{\cos(\pi f_u T)\cos(\pi f_\ell T) + \sin(\pi f_u T)\sin(\pi f_\ell T)} \\[2mm]
&= \frac{\cos[\pi T(f_u + f_\ell)]}{\cos[\pi T(f_u - f_\ell)]}
\end{aligned} \tag{15.61}$$

The parameters D and E are the two unique transformation parameters required to produce a discrete-time bandpass filter from a continuous-time normalized low-pass filter.

To use bilinear-transformation design to produce a transfer function $H(z)$ of a discrete-time bandpass filter from a transfer function $H_{aN}(s)$ of a continuous-time normalized low-pass filter, we first determine the parameters D and E by using eqs. (15.60) and (15.61) and the desired discrete-time-filter upper and lower cutoff frequencies. Then we obtain the transfer function of the discrete-time bandpass filter from the transfer function of the continuous-time normalized low-pass filter through the transformation

$$H(z) = H_{aN}(s)\big|_{s=D\left[\frac{1-2Ez^{-1}+z^{-2}}{1-z^{-2}}\right]} \tag{15.62}$$

We obtain the frequency response of the discrete-time bandpass filter by substituting $z = e^{j2\pi fT}$ in eq. (15.62). The result is

$$H\left(e^{j2\pi fT}\right) = H_{aN}(s)\big|_{s=D\left[\frac{1-2Ee^{-j2\pi fT}+e^{-j4\pi fT}}{1-e^{-j4\pi fT}}\right]} \tag{15.63}$$

We see from eq. (15.63) that the value of the frequency response of the discrete-time bandpass filter at the frequency f equals the value of the transfer function $H_{aN}(s)$ of the continuous-time normalized low-pass filter at

$$s = D\left[\frac{1 - 2Ee^{-j2\pi fT} + e^{-j4\pi fT}}{1 - e^{-j4\pi fT}}\right] = D\left[\frac{e^{j2\pi fT} + e^{-j2\pi fT} - 2E}{e^{j2\pi fT} - e^{-j2\pi fT}}\right]$$

$$= D\left[\frac{2\cos(2\pi fT) - 2E}{j2\sin(2\pi fT)}\right] = jD\left[\frac{E - \cos(2\pi fT)}{\sin(2\pi fT)}\right]$$

$$\equiv j2\pi f_a \tag{15.64}$$

The values of s given by eq. (15.64) lie on the imaginary axis in the s-plane. Therefore, the frequency response of the continuous-time normalized low-pass filter is specified by the values of $H_{aN}(s)$ that correspond to the values $s = j2\pi f_a$ given by eq. (15.64). This means that the frequency response of the discrete-time bandpass filter at the frequency f equals the frequency response of the continuous-time normalized low-pass filter at the frequency f_a, where f and f_a are related by

$$f_a = \frac{D}{2\pi}\left[\frac{E - \cos(2\pi fT)}{\sin(2\pi fT)}\right] \tag{15.65}$$

Now, let us illustrate the bilinear-transformation design of a discrete-time bandpass filter with an example.

Example 15.11

We need a bandpass filter in a discrete-time system used for spacecraft attitude control. The bandpass filter is to pass signal components with frequencies varying from 300 to 400 Hz. Find the transfer function, realization, and frequency response for the discrete-time bandpass filter if the system is operating at the rate of 2000 samples per second and the filter specifications are as follows: (1) the filter is derived with bilinear transformation from a normalized low-pass filter of the Butterworth type, (2) the passband, in which gain varies by no more than 3 dB with respect to the maximum gain, is 300 Hz to 400 Hz, (3) the gain is less than or equal to -18 dB with respect to maximum gain for $f \leq 200$ Hz and $f \geq 500$ Hz, and (4) the maximum gain, G_M, is 1.

Solution

The cutoff frequencies of the Butterworth-type filter correspond to gains that are -3 dB with respect to the filter maximum gain. Therefore, from specification 2, the cutoff frequencies of the discrete-time filter are $f_u = 400$ Hz and $f_\ell = 300$ Hz. Also, $T = 1/f_s = 1/2000$ s. Therefore, the bilinear-transformation parameters are

$$D = \cot[\pi T(f_u - f_\ell)] = \cot(0.05\pi) = 6.3138$$

and

$$E = \frac{\cos[\pi T (f_u + f_\ell)]}{\cos[\pi T (f_u - f_\ell)]} = \frac{\cos(0.35\pi)}{\cos(0.05\pi)} = 0.4597$$

In order to use the stopband gain curves of Figure 8.14 to determine the required filter order, we must compute the stopband frequencies of the continuous-time normalized low-pass filter that correspond to the discrete-time-filter stopband frequencies $f_1 = 200$ Hz and $f_2 = 500$ Hz. We use eq. (15.65) to compute these stopband frequencies. The results are

$$\omega_{a1} = D\left[\frac{E - \cos(2\pi f_1 T)}{\sin(2\pi f_1 T)}\right] = D\left[\frac{E - \cos(0.2\pi)}{\sin(0.2\pi)}\right] = -3.753 \text{ rad/s}$$

$$= \omega_{a2} = D\left[\frac{E - \cos(2\pi f_2 T)}{\sin(2\pi f_2 T)}\right] = D\left[\frac{E - \cos(0.5\pi)}{\sin(0.5\pi)}\right] = 2.902 \text{ rad/s}$$

Since $|\omega_{a1}| > |\omega_{a2}|$, then the stopband gain of the normalized low-pass filter at ω_{a2} is greater than the stopband gain at ω_{a1} and we must choose the order of the continuous-time normalized low-pass filter to provide a gain of -18 dB with respect to the filter maximum gain at the normalized frequency $\omega_{a2} = 2.902$ rad/s. From Figure 8.14, we see that a normalized low-pass filter of the second order is required. We obtain the transfer-function denominator polynomial coefficients of the second-order, normalized low-pass filter from Table 8.1. The resulting transfer function is

$$H_{aN}(s) = \frac{a_0}{1 + 1.4142s + s^2}$$

where

$$a_0 = G_M b_0 = (1)(1) = 1$$

since the filter is a Butterworth filter. We use eq. (15.62) to produce the transfer function of discrete-time bandpass filter, with the following result:

$$H(z) = \left. H_{aN}(s) \right|_{s=D\left[\frac{1-2Ez^{-1}+z^{-2}}{1-z^{-2}}\right]}$$

$$= \cfrac{1}{\left\{ 1 + 1.4142\left[6.3138\left(\cfrac{1 - 0.9194z^{-1} + z^{-2}}{1 - z^{-2}}\right)\right] + \left[6.3138\left(\cfrac{1 - 0.9194z^{-1} + z^{-2}}{1 - z^{-2}}\right)\right]^2 \right\}}$$

$$= \frac{0.0201\left(1 - 2z^{-2} + z^{-4}\right)}{1 - 1.6370z^{-1} + 2.2378z^{-2} - 1.3073z^{-3} + 0.6414z^{-4}}$$

The block-diagram representation of the realization of the discrete-time bandpass filter is shown in Figure 15.24.

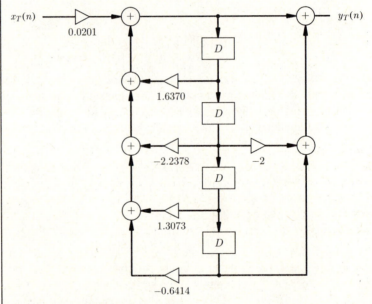

Figure 15.24 Block-Diagram Representation of the Discrete-Time Bandpass-Filter Realization in Example 15.11

Note once again that a fourth-order bandpass filter is produced from a second-order low-pass filter.

The frequency response of the discrete-time bandpass filter in terms of normalized frequency is $H(e^{j2\pi r})$ and is plotted in Figure 15.25 for $0 \leq r \leq 0.5$. Note that (1) the maximum filter gain is unity; (2) the -3-dB cutoff frequencies (frequencies at which the filter gain is 0.707 times the maximum filter gain) are $r = 0.15$ and $r = 0.2$ (that is, $f = 300$ Hz and $f = 400$ Hz); and (3) the filter gains at $r = 0.1$ and $r = 0.25$ (that is, $f = 200$ Hz and $f = 500$ Hz) are less than 0.126 (-18 dB with respect to the maximum filter gain). Thus, the discrete-time bandpass filter designed meets the given specifications.

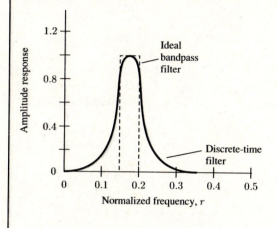

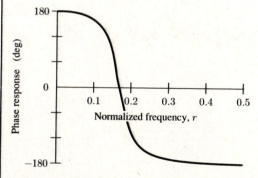

Figure 15.25 Frequency Response of the Bandpass Filter Designed with Bilinear Transformation in Example 15.11

15.5 FIR Filter Design

A finite-impulse-response (FIR) discrete-time filter has a unit pulse response that contains only a finite number of nonzero samples. That is, for a causal FIR filter of length equal to K samples,

$$h_T(n) = \begin{cases} A_n & 0 \leq n \leq K - 1 \\ 0 & \text{elsewhere} \end{cases} \tag{15.66}$$

where $A_0 \neq 0$ and $A_{K-1} \neq 0$. By using discrete convolution, we find that the filter output signal is

$$y_T(n) = \sum_{m=-\infty}^{\infty} h_T(m)x_T(n-m)$$

$$= \sum_{m=0}^{K-1} h_T(m)x_T(n-m) \tag{15.67}$$

when the input signal is $x_T(n)$ and the initial conditions are zero. The second equivalence of eq. (15.67) holds because $h_T(m) = 0$ for $m < 0$ and $m > K-1$. Equation (15.67) shows that the nth (present) output sample from an FIR filter depends on the present and previous $K - 1$ input-signal samples. The input-signal samples are weighted by multiplicative constants that are samples of the filter unit pulse response.

We compute the z-transform of eq. (15.67) to obtain

$$Y(z) = \sum_{m=0}^{K-1} h_T(m)z^{-m}X(z) \tag{15.68}$$

and solve eq. (15.68) for the FIR filter transfer function. The result is

$$H(z) = Y(z)/X(z) = \sum_{m=0}^{K-1} h_T(m)z^{-m} \tag{15.69}$$

The block-diagram representation of the direct-form realization of the FIR filter is shown in Figure 15.26. Note that poles of the FIR filter transfer function

Figure 15.26
Block-Diagram
Representation of a
Direct-Form FIR
Filter Realization

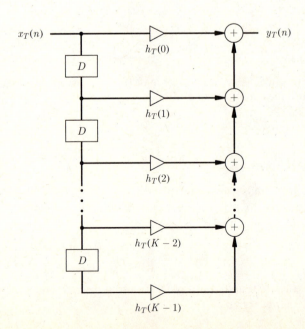

given by eq. (15.69) occur only at $z = 0$. The direct-form realization that corresponds to eq. (15.69) has no feedback, as is apparent in Figure 15.26. Therefore, FIR filters that are realized with direct-form realizations are always stable.

We can see from Figure 15.26 that the direct-form realization consists of a shift register, along which the input-signal data move, and taps along this shift register where input data (present and previous input-signal samples) are extracted, multiplied by the unit-pulse-response values, and added to produce the output. A filter structure of this type is called a *transversal filter*, and the multiplier values are referred to as the *filter coefficients*, *filter weighting coefficients*, or *filter tap coefficients*.

An FIR filter design technique that approximately matches the impulse response of a continuous-time filter uses $h_T(n) = Th_a(nT)$ for the filter coefficients where $h_a(nT)$ are samples of the continuous-time filter impulse response (see impulse-invariant design for IIR filters). The impulse-response match is only approximate because it must be truncated to produce an FIR filter. This design technique is appropriate if the desired filter impulse response is known.

Use of Fourier Series

As we indicated in Section 15.3, we emphasize discrete-time FIR filter design that approximately matches the frequency response of the discrete-time filter to a desired continuous-time-filter frequency response. The technique we develop requires consideration of noncausal IIR filters because we first produce a noncausal IIR filter, the frequency response of which matches a desired discrete-time-filter frequency response. We then convert the noncausal IIR filter to a noncausal FIR filter, the frequency response of which approximately matches the desired frequency response. Finally, we convert the noncausal FIR filter to a causal FIR filter with the same amplitude response as the noncausal FIR filter.

Development of Design   Let us assume that the desired frequency response of the discrete time-filter is

$$H_d(f) = H_R(f) + jH_I(f) \tag{15.70}$$

where $H_R(f)$ and $H_I(f)$ are the real and imaginary parts of the frequency response. Recall from eq. (13.6) that

$$H_d(f) = \sum_{n=-\infty}^{\infty} h_T(n)e^{-j2\pi nfT} \tag{15.71}$$

where $h_T(n)$ is the unit pulse response (possibly noncausal) of the discrete-time filter that produces the desired frequency response. We assume that the desired frequency response is chosen so that the sequence values of $h_T(n)$ are all real, because we are not interested in discrete-time filters that produce complex output-signal samples for real input-signal samples. Therefore, $H_R(f)$ is an even function of frequency and $H_I(f)$ is an odd function of frequency. (Refer to the discussion following eqs. [11.26] and [11.27].)

Recall that the frequency response of a discrete-time filter is periodic, with period equal to the sample rate f_s. Thus, we can represent the frequency response by the complex-exponential Fourier series

$$H_d(f) = \sum_{n=-\infty}^{\infty} k_n e^{-j2\pi n(1/f_s)f} = \sum_{n=-\infty}^{\infty} k_n e^{-j2\pi nfT} \qquad (15.72)$$

where f_s is the period of $H_d(f)$. We have chosen the basis functions to be $e^{-j2\pi nf/f_s}$ rather than the basis functions $e^{j2\pi nf/f_s}$ that we used in Chapter 4. Thus, the Fourier series coefficients are

$$k_n = \frac{1}{f_s} \int_{-f_s/2}^{f_s/2} H_d(f) e^{j2\pi n(1/f_s)f} \, df$$

$$= \frac{1}{f_s} \int_{-f_s/2}^{f_s/2} H_d(f) e^{j2\pi nfT} \, df \qquad (15.73)$$

Note that the same set of series terms is produced with the basis functions $e^{-j2\pi nf/f_s}$ as with the basis functions $e^{j2\pi nf/f_s}$ because n spans both the positive and negative integers in the Fourier series summation.

When we compare eqs. (15.71) and (15.72) and use eq. (15.73), we see that we can compute the unit-pulse-response samples of the filter with

$$h_T(n) = k_n = \frac{1}{f_s} \int_{-f_s/2}^{f_s/2} H_d(f) e^{j2\pi nfT} \, df \qquad (15.74)$$

We use eq. (15.70) and Euler's theorem to rewrite eq. (15.74) as

$$h_T(n) = \frac{1}{f_s} \int_{-f_s/2}^{f_s/2} [H_R(f) + jH_I(f)] e^{j2\pi nfT} \, df$$

$$= \frac{1}{f_s} \int_{-f_s/2}^{f_s/2} H_R(f) \cos(2\pi nfT) \, df - \frac{1}{f_s} \int_{-f_s/2}^{f_s/2} H_I(f) \sin(2\pi nfT) \, df$$

$$+ j\frac{1}{f_s} \int_{-f_s/2}^{f_s/2} H_R(f) \sin(2\pi nfT) \, df$$

$$+ j\frac{1}{f_s} \int_{-f_s/2}^{f_s/2} H_I(f) \cos(2\pi nfT) \, df \qquad (15.75)$$

The last two integrals in eq. (15.75) are equal to zero because they are integrals of odd functions between equal positive and negative limits. Therefore,

$$h_T(n) = \frac{1}{f_s} \int_{-f_s/2}^{f_s/2} H_R(f) \cos(2\pi nfT) \, df - \frac{1}{f_s} \int_{-f_s/2}^{f_s/2} H_I(f) \sin(2\pi nfT) \, df$$

$$\qquad (15.76)$$

The sequence values computed for $h_T(n)$ are generally nonzero for all n (that is, for $-\infty < n < \infty$). Therefore, $h_T(n)$ corresponds to an IIR filter and we cannot use it directly as the unit pulse response of an FIR filter. However, if we truncate $h_T(n)$ to give

$$h_{1T}(n) = \begin{cases} h_T(n) & -M \leq n \leq M \\ 0 & \text{elsewhere} \end{cases} \qquad (15.77)$$

then $h_{1T}(n)$ has a nonzero length of $2M + 1$ samples and can be used as the unit pulse response of an FIR filter. The frequency response of this filter is

$$H_{1d}(f) = \sum_{n=-\infty}^{\infty} h_{1T}(n)e^{-j2\pi nfT}$$

$$= \sum_{n=-M}^{M} h_T(n)e^{-j2\pi nfT} \qquad (15.78)$$

It is an approximation to the desired frequency response in a minimum-integral-square-error sense (see Section 3.6) since the sequence values of $h_T(n)$ are Fourier series coefficients corresponding to the desired frequency response. The approximation becomes better as M becomes larger; however, the length of the finite unit pulse response becomes larger as M becomes larger.

The FIR filter defined by the unit pulse response $h_{1T}(n)$ is still noncausal because $h_{1T}(n)$ is not equal to zero for $-M \leq n < 0$. We can construct the unit pulse response $h_{2T}(n)$ for a causal FIR filter by shifting $h_{1T}(n)$ a distance of M samples to the right. That is, by constructing

$$h_{2T}(n) = h_{1T}(n - M) = \begin{cases} h_T(n - M) & 0 \leq n \leq 2M \\ 0 & \text{elsewhere} \end{cases} \qquad (15.79)$$

We use $h_{2T}(n)$ for the unit pulse response of the causal discrete-time FIR filter. The relation between $h_T(n)$, $h_{1T}(n)$, and $h_{2T}(n)$ is illustrated in Figure 15.27 for $M = 4$.

The frequency response of the causal FIR filter that corresponds to the unit pulse response $h_{2T}(n)$ is

$$H_{2d}(f) = \sum_{n=-\infty}^{\infty} h_{2T}(n)e^{-j2\pi nfT}$$

$$= \sum_{n=-\infty}^{\infty} h_{1T}(n - M)e^{-j2\pi nfT}$$

$$= \sum_{n=0}^{2M} h_T(n - M)e^{-j2\pi nfT} \qquad (15.80)$$

Making the change of variable $m = n - M$ in eq. (15.80), we obtain

Figure 15.27
Steps in Producing
the Unit Pulse
Response for a
Causal FIR Filter
with a Frequency
Response
Approximating a
Desired Frequency
Response

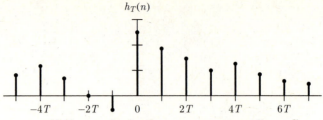

$h_T(n)$

(a) Noncausal unit pulse response corresponding
to desired frequency response

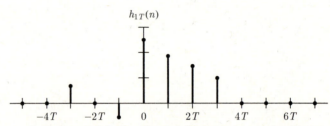

$h_{1T}(n)$

(b) Noncausal finite-length unit pulse response corresponding
to approximation to desired frequency response

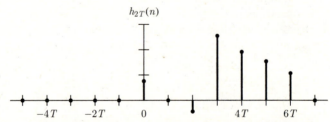

$h_{2T}(n)$

(c) Causal finite-length unit pulse response corresponding
to approximation to desired frequency response

$$H_{2d}(f) = \sum_{m=-M}^{M} h_T(m)e^{-j2\pi(m+M)fT}$$

$$= e^{-j2\pi MT} \sum_{m=-M}^{M} h_T(m)e^{-j2\pi mfT}$$

$$= e^{-j2\pi MT} H_1\left(e^{j2\pi fT}\right) \tag{15.81}$$

Therefore, the amplitude and phase responses for the causal FIR filter design are

$$|H_{2d}(f)| = \left|e^{-j2\pi MT}\right| |H_{1d}(f)|$$

$$= |H_{1d}(f)| \tag{15.82}$$

and

$$\underline{/\,H_{2d}(f)} = \underline{/\,e^{-j2\pi MfT}} + \underline{/\,H_{1d}(f)}$$

$$= \underline{/\,H_{1d}(f)} - 2\pi MfT \qquad (15.83)$$

respectively.

The amplitude response of the designed causal FIR filter is the same minimum-integral-square-error approximation to the desired amplitude response that we obtained with the noncausal FIR filter produced by unit-pulse-response truncation. The phase response is the same as that for the noncausal FIR filter except for the addition of a linear phase term with slope $-2\pi MT$. This linear phase term corresponds to an MT-second time delay (that is, a time delay of M samples). Thus, the causal FIR filter provides the same filtering characteristics as the noncausal FIR filter, except that the causal filter introduces an MT-second delay of the output signal.

The transfer function of the causal FIR filter, the frequency response of which approximates the desired frequency response, is

$$H_2(z) = \mathcal{Z}[h_{2T}(n)] = \sum_{n=0}^{\infty} h_{2T}(n)z^{-n}$$

$$= \sum_{n=0}^{2M} h_T(n - M)z^{-n} \qquad (15.84)$$

and is in the FIR filter form given by eq. (15.69). The block-diagram representation of the realization of this filter is similar to the one shown in Figure 15.26, in this case with $h_T(n)$ replaced by $h_T(n - M)$ and K equal to $2M + 1$, so that the filter coefficients range from $h_T(-M)$ to $h_T(M)$. Note that

$$H_{2d}(f) = H_2\!\left(e^{j2\pi fT}\right) \qquad (15.85)$$

That is, we can obtain the filter frequency response by replacing z by $e^{j2\pi fT}$ in the filter transfer function. This property of causal systems was shown in Section 14.6.

In summary, the steps in the design of causal FIR filters through Fourier series representation of the desired frequency response $H_d(f)$ are as follows:

Step 1   Choose M to produce the desired length $2M + 1$ of the unit pulse response.

Step 2   Compute the $2M + 1$ unit-pulse-response samples for a noncausal filter as

$$h_T(n) = \frac{1}{f_s} \int_{-f_s/2}^{f_s/2} H_d(f)e^{j2\pi nfT}\,df \qquad -M \le n \le M \qquad (15.86)$$

To use normalized frequency, let $r = fT = f/f_s$. Then $dr = df/f_s$, $H_d(f)|_{f=rf_s} = H_d(rf_s) = H_N(r)$, and

$$h_T(n) = \int_{-1/2}^{1/2} H_N(r)e^{j2\pi nr}\, dr \qquad -M \le n \le M \tag{15.87}$$

where $H_N(r)$ is the normalized frequency response that corresponds to the desired frequency response.

Step 3   Find the frequency response of the causal FIR filter as

$$H_{2d}(f) = \sum_{n=0}^{2M} h_T(n - M)e^{-j2\pi nfT} \tag{15.88}$$

or, as the normalized frequency response

$$H_{2N}(r) = \sum_{n=0}^{2M} h_T(n - M)e^{-j2\pi nr} \tag{15.89}$$

Check to see if $H_{2d}(f)$ satisfactorily matches the desired frequency response. For a better match, the gain (amplitude response) can be scaled by multiplying all unit-pulse-response samples by the same value. If the frequency-response match is still not good enough, then increase M and return to step 2 to compute the additional unit-pulse-response samples required for the increased M. Repeat as many times as necessary.

Step 4   Find the transfer function of the causal FIR filter

$$H_2(z) = \sum_{n=0}^{2M} h_T(n - M)z^{-m} \tag{15.90}$$

Special Cases and Examples   Many discrete-time-filter frequency responses that are of interest to us are real and even functions of frequency. For example, ideal and nonideal filters with no phase shift have this characteristic, as illustrated in Figure 15.28. For real and even frequency responses, $H_I(f) = 0$. Therefore, $H_d(f) = H_R(f)$ and, from eq. (15.76),

$$h_T(n) = \frac{1}{f_s} \int_{-f_s/2}^{f_s/2} H_R(f)\cos(2\pi nfT)\, df$$

$$= \frac{2}{f_s} \int_{0}^{f_s/2} H_R(f)\cos(2\pi nfT)\, df \tag{15.91}$$

where the second form of eq. (15.91) follows because $H_R(f)\cos(2\pi nfT)$ is an even function of frequency. Note that $h_T(-n) = h_T(n)$ because $\cos(-2\pi nfT) = \cos(2\pi nfT)$. Therefore, we need only to compute $h_T(n)$ for $n \ge 0$ in this case.

When $H_d(f) = H_R(f)$, then we replace eq. (15.86) in step 2 of the previously listed design procedure by

Figure 15.28
Example of Real and
Even Discrete-Time-
Filter Frequency
Responses

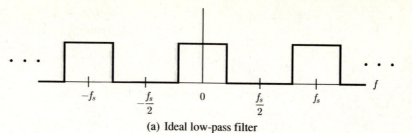

(a) Ideal low-pass filter

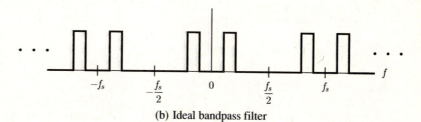

(b) Ideal bandpass filter

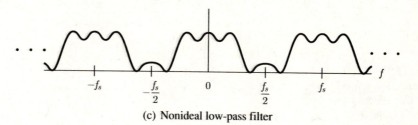

(c) Nonideal low-pass filter

$$h_T(-n) = h_T(n) = \frac{2}{f_s} \int_0^{f_s/2} H_R(f)\cos(2\pi n f T)\,df \qquad 0 \le n \le M$$

(15.92)

or, in terms of normalized frequency, by

$$h_T(-n) = h_T(n) = 2 \int_0^{1/2} H_{RN}(r)\cos(2\pi n r)\,dr \qquad 0 \le n \le M \quad (15.93)$$

where $H_{RN}(r) = H_R(f)\big|_{f=rf_s}$.

Other frequency responses of discrete-time filters of interest to us are imaginary and odd functions of frequency. In this case, $H_R(f) = 0$ because $H_d(f) = jH_I(f)$, and from eq. (15.76),

$$h_T(n) = -\frac{1}{f_s} \int_{-f_s/2}^{f_s/2} H_I(f)\sin(2\pi n f T)\,df$$

$$= -\frac{2}{f_s} \int_0^{f_s/2} H_I(f)\sin(2\pi n f T)\,df$$

(15.94)

where the second form of eq. (15.94) follows because $H_I(f) \sin(2\pi n fT)$ is a product of two odd functions of frequency, and thus is an even function of frequency. Note that $h_T(0) = 0$ and $h_T(-n) = -h_T(n)$ because $\sin(2\pi n fT) = 0$ for $n = 0$ and $\sin(-2\pi n fT) = -\sin(2\pi n fT)$ for $n \neq 0$. Therefore, we need only to compute $h_T(n)$ for $n > 0$ when $H_d(f)$ is imaginary and odd.

Two examples of continuous-time filters, the frequency responses of which are imaginary and odd, are a differentiator and a $-90°$ phase shifter (called a *Hilbert transformer*). From the Fourier transform differentiation theorem of Chapter 4 and the definition of frequency response in Chapter 6, we see that the frequency response of the continuous-time differentiator is $H_a(f) = j2\pi f$. The frequency response of a $-90°$ phase shifter is $H_a(f) = -j \operatorname{sgn}(f)$, since $|-j \operatorname{sgn}(f)| = 1$ and $\underline{/-j \operatorname{sgn}(f)} = -90°$ for $f > 0$. The frequency responses of the differentiator and $-90°$ phase shifter are illustrated in Figure 15.29.

When $H_d(f) = jH_I(f)$, then we replace eq. (15.86) in step 2 of the previously listed design procedure by

$$h_T(0) = 0$$

$$-h_T(-n) = h_T(n) = -\frac{2}{f_s} \int_0^{f_s/2} H_I(f) \sin(2\pi n fT) \, df \qquad 1 \leq n \leq M \qquad (15.95)$$

Figure 15.29
Frequency Responses
of the Continuous-
Time Differentiator
and $-90°$ Phase
Shifter

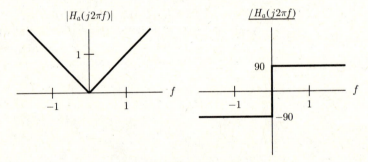

(a) Differentiator

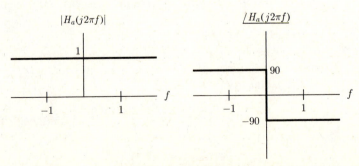

(b) -90 phase shifter

or, in terms of normalized frequency, by

$$h_T(0) = 0$$

$$-h_T(-n) = h_T(n) = -2 \int_0^{1/2} H_{IN}(r) \sin(2\pi nr)\, dr \quad 1 \le n \le M \quad (15.96)$$

where $H_{IN}(r) = H_I(f)\big|_{f=rfs}$.

The imaginary and odd continuous-time filters shown in Figure 15.29 have nonzero amplitude responses for all frequencies. This creates a problem when we use Fourier series to approximate the filters by discrete-time FIR filters because the frequency response of a discrete-time filter is periodic, with period equal to the sample rate f_s. We can construct a discrete-time-filter approximation for a continuous-time filter, the amplitude response of which is nonzero for frequencies greater than $f_s/2$, by using the frequency response

$$H_d(f) = \sum_{n=-\infty}^{\infty} H_{a1}[j2\pi(f - nf_s)] \quad (15.97)$$

where

$$H_{a1}(j2\pi f) = \begin{cases} H_a(j2\pi f) & |f| \le f_s/2 \\ 0 & |f| > f_s/2 \end{cases} \quad (15.98)$$

That is, we truncate the continuous-time frequency response at $-f_s/2$ and $+f_s/2$, and the resulting truncated portion is repeated at frequency locations that are multiples of f_s. This is illustrated in Figure 15.30 for the differentiator and $-90°$ phase shifter.

The two examples that follow serve to illustrate the design procedure outlined for FIR filters.

Figure 15.30
Frequency Responses for Discrete-Time Approximations to a Continuous-Time Differentiator and a $-90°$ Phase Shifter

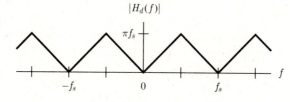

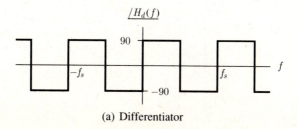

(a) Differentiator

Figure 15.30
Frequency Responses
for Discrete-Time
Approximations to a
Continuous-Time
Differentiator and a
$-90°$ Phase Shifter
(continued)

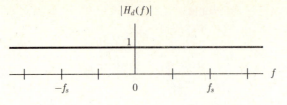

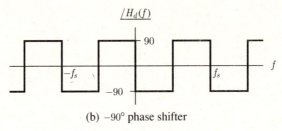

(b) $-90°$ phase shifter

Example 15.12

Design an FIR filter with unit-pulse-response length of seven samples that approximates an ideal low-pass filter with unity gain in the passband and a cutoff frequency of 200 Hz. The filter is for use in a discrete-time data-transmission system, the sample rate of which is 2000 Hz. Its dc gain is to match the dc gain of the desired ideal filter. Plot the amplitude and phase responses that correspond to the desired normalized frequency response and to the normalized frequency response of the designed filter over the interval $0 \leq r \leq 0.5$. Draw the block-diagram representation of a filter realization.

Solution

Step 1   The length of the unit pulse response is specified to be seven samples. Therefore, we choose $M = 3$ so that $2M + 1 = 7$.

Step 2   The desired frequency response is periodic, with period $f_s = 2000$ Hz. It is shown in Figure 15.31, along with the corresponding normalized frequency response.

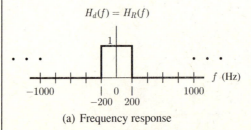

(a) Frequency response

Figure 15.31 Desired Frequency Response for Example 15.12

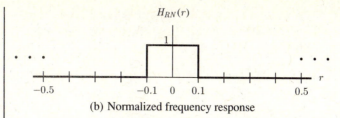

(b) Normalized frequency response

Figure 15.31 Desired Frequency Response for Example 15.12 *(continued)*

We use eq. (15.93) to solve for the unit-pulse-response samples of the non-causal FIR filter. The result is

$$h_T(-n) = h_T(n) = 2\int_0^{1/2} H_{RN}(r)\cos(2\pi nr)\,dr$$

$$= 2\int_0^{0.1} \cos(2\pi nr)\,dr = \frac{1}{\pi n}\sin(2\pi nr)\Big|_0^{0.1}$$

$$= \frac{\sin(0.2\pi n)}{\pi n} = 0.2\ \text{sinc}(0.2n) \qquad 0 \le n \le 3$$

Therefore, the individual samples of the unit pulse response are

$$h_T(0) = 0.2\ \text{sinc}(0) = 0.2$$

$$h_T(-1) = h_T(1) = 0.2\ \text{sinc}(0.2) = 0.18710$$

$$h_T(-2) = h_T(2) = 0.2\ \text{sinc}(0.4) = 0.15137$$

and

$$h_T(-3) = h_T(3) = 0.2\ \text{sinc}(0.6) = 0.10091$$

Step 3 Using eq. (15.89), we obtain the normalized frequency response of the causal FIR filter. The result is

$$H_{2N}(r) = \sum_{n=0}^{6} h_T(n-3)e^{-j2\pi nr}$$

The dc gain of the filter that corresponds to this frequency response is

$$|H_{2N}(0)| = \left|\sum_{n=0}^{6} h_T(n-3)e^{-j2\pi n(0)}\right| = \left|\sum_{n=0}^{6} h_T(n-3)\right|$$

$$= |0.2 + 2(0.18710 + 0.15137 + 0.10091)|$$

$$= 1.07876 \equiv K$$

We want the dc gain of the designed causal FIR filter to be unity. First we define the normalized frequency response of this amplitude-scaled filter as $H_{SN}(r)$. Then,

$$H_{SN}(r) = \frac{1}{K} H_{2N}(r) = \frac{1}{K} \sum_{n=0}^{6} h_T(n-3)e^{-j2\pi nr}$$

$$= \sum_{n=0}^{6} \frac{h_T(n-3)}{K} e^{-j2\pi nr} \equiv \sum_{n=0}^{6} h_{ST}(n-3)e^{-j2\pi nr}$$

where $h_{ST}(n-3)$ are the required samples of the unit pulse response for the amplitude-scaled noncausal FIR filter. The numerical values of these unit-pulse-response samples are

$$h_{ST}(0) = \frac{0.2}{1.07876} = 0.18540$$

$$h_{ST}(-1) = h_{ST}(1) = \frac{0.18710}{1.07876} = 0.17344$$

$$h_{ST}(-2) = h_{ST}(2) = \frac{0.15137}{1.07876} = 0.14032$$

and

$$h_{ST}(-3) = h_{ST}(3) = \frac{0.10091}{1.07876} = 0.09354$$

The normalized frequency response of the resulting causal FIR filter is

$$H_{SN}(r) = 0.09354 + 0.14032e^{-j2\pi r} + 0.17344e^{-j4\pi r}$$

$$+ 0.18540e^{-j6\pi r} + 0.17344e^{-j8\pi r}$$

$$+ 0.14032e^{-j10\pi r} + 0.09354e^{-j12\pi r}$$

$$= [0.09354 + 0.14032\cos(2\pi r) + 0.17344\cos(4\pi r)$$

$$+ 0.18540\cos(6\pi r) + 0.17344\cos(8\pi r)$$

$$+ 0.14032\cos(10\pi r) + 0.09354\cos(12\pi r)]$$

$$- j[0.14032\sin(2\pi r) + 0.17344\sin(4\pi r)$$

$$+ 0.18540\sin(6\pi r) + 0.17344\sin(8\pi r)$$

$$+ 0.14030\sin(10\pi r) + 0.09354\sin(12\pi r)]$$

and the corresponding amplitude and phase responses are plotted in Figure 15.32 for $0 \leq r \leq 0.5$.

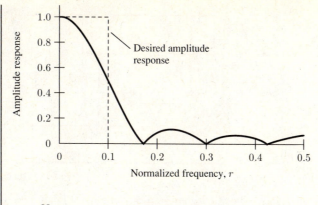

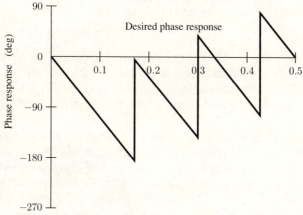

Figure 15.32 Amplitude and Phase Responses for the Discrete-Time Low-Pass FIR Filter Designed in Example 15.12

Step 4 From eq. (15.90), the transfer function of the amplitude-scaled causal FIR filter is

$$H_{S2}(z) = \sum_{n=0}^{6} h_{ST}(n-3)z^{-n}$$

$$= 0.09354 + 0.14032z^{-1} + 0.17344z^{-2}$$

$$+ 0.18540z^{-3} + 0.17344z^{-4} + 0.14032z^{-5}$$

$$+ 0.09354z^{-6}$$

The block-diagram representation of a realization corresponding to this transfer function is shown in Figure 15.33.

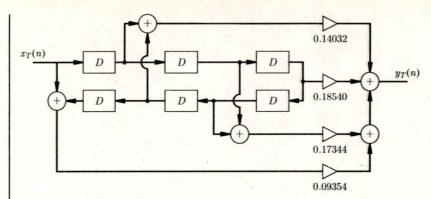

Figure 15.33 Block-Diagram Representation of a Realization of the Discrete-Time Low-Pass Filter Designed in Example 15.12

In Example 15.12, the amplitude response of the designed causal FIR filter deviates greatly from the desired amplitude response, the reason being that we kept too few terms in the Fourier series expansion. The amplitude response can be improved by increasing M. However, increasing M also results in a greater time delay (MT seconds).

Recall that a Fourier series representation of a function with discontinuities approaches the average of the limit from the left and the limit from the right at the discontinuities as we increase the number of terms used in the series. Therefore, the response of the FIR filter approaches 0.5 at the cutoff frequency in Example 15.12. In the filter passband, the phase response of the FIR filter designed in Example 15.12 differs from the desired phase response by the addition of a straight line through the origin with slope $-2\pi MT$. This additional straight line through the origin indicates that the filter introduces no phase distortion, but it does introduce a pure time delay of $MT = 1.5$ ms, or, equivalently, $M = 3$ samples for signals in the passband.

Note that we use a slightly different form of the direct-form realization to take advantage of the fact that multiplier values occur in pairs. This different form requires only $M + 1$ multipliers rather than $2M + 1$ multipliers.

Our second example involves the design of a differentiator. In this case, the solution will be abbreviated.

Example 15.13

Design an FIR filter with unit-pulse-response length of seven samples that approximates a differentiator. The filter is to be used in a discrete-time system, the sample rate of which is 100 Hz. Find and plot the normalized amplitude and phase responses corresponding to the desired normalized frequency response and to the normalized frequency response of the designed filter over the interval $0 \le r \le 0.5$. Draw the block-diagram representation of a filter realization.

Solution

The length of the unit-pulse-response is specified to be seven samples. Therefore, we choose $M = 3$ so that $2M + 1 = 7$. The desired frequency response is

$$H_d(f) = j2\pi f = jH_I(f) \qquad \text{for } -\frac{f_s}{2} \leq f \leq \frac{f_s}{2}$$

and repeats outside this interval. Therefore,

$$H_I(f) = 2\pi f \qquad \text{for } -\frac{f_s}{2} \leq f \leq \frac{f_s}{2}$$

and repeats outside this interval. In terms of normalized frequency,

$$H_{IN}(r) = H_I(rf_s) = 2\pi r f_s = 200\pi r \qquad \text{for } -0.5 \leq r \leq 0.5$$

and repeats outside this interval. $H_{IN}(r)$ is plotted in Figure 15.34.

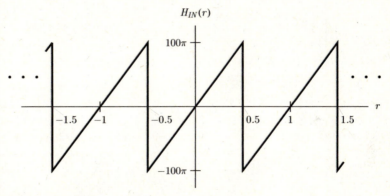

Figure 15.34 Plot of $H_{IN}(r)$ for Example 15.13

The solution for the unit-pulse-response samples for the noncausal FIR filter is

$$h_T(0) = 0$$

$$-h_T(-n) = h_T(n) = -2\int_0^{1/2} H_{IN}(r) \sin(2\pi nr)\, dr$$

$$= -2\int_0^{1/2} (200\pi r) \sin(2\pi nr)\, dr$$

$$= -400\pi \left[\frac{1}{(2\pi n)^2} \sin(2\pi nr) - \frac{r}{2\pi n} \cos(2\pi nr) \right]_0^{1/2}$$

$$= \frac{100}{n} \cos(\pi n) \qquad 1 \leq n \leq 3$$

Therefore, the individual sample values of the unit pulse response are

$$-h_T(-1) = h_T(1) = 100\cos(\pi) = -100$$

$$-h_T(-2) = h_T(2) = 50\cos(2\pi) = 50$$

and

$$-h_T(-3) = h_T(3) = \frac{100}{3}\cos(3\pi) = -(100/3)$$

The normalized frequency response of the causal FIR filter is

$$H_2(r) = \sum_{n=0}^{6} h_T(n-3)e^{-j2\pi nr}$$

$$= (100/3) - 50e^{-j2\pi r} + 100e^{-j4\pi r}$$

$$\quad - 100e^{-j8\pi r} + 50e^{-j10\pi r} - (100/3)e^{-j12\pi r}$$

$$= \big[(100/3) - 50\cos(2\pi r) + 100\cos(4\pi r)$$

$$\quad - 100\cos(8\pi r) + 50\cos(10\pi r) - (100/3)\cos(12\pi r)\big]$$

$$\quad - j\big[-50\sin(2\pi r) + 100\sin(4\pi r) - 100\sin(8\pi r)$$

$$\quad + 50\sin(10\pi r) - (100/3)\sin(12\pi r)\big]$$

and the corresponding amplitude and phase responses are plotted in Figure 15.35 for $0 \le r \le 0.5$.

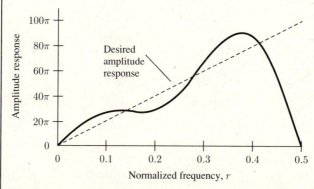

Figure 15.35 Amplitude and Phase Responses for the Discrete-Time FIR Differentiator Designed in Example 15.13

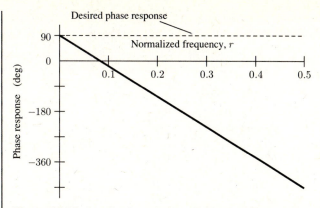

Figure 15.35 Amplitude and Phase Responses for the Discrete-Time FIR Differentiator Designed in Example 15.13 *(continued)*

The transfer function of the causal FIR filter is

$$H_2(z) = \sum_{n=0}^{6} h_T(n-3)z^{-n}$$

$$= (100/3) - 50z^{-1} + 100z^{-2} - 100z^{-4} + 50z^{-5} - (100/3)z^{-6}$$

The block-diagram representation of a realization corresponding to this transfer function is shown in Figure 15.36.

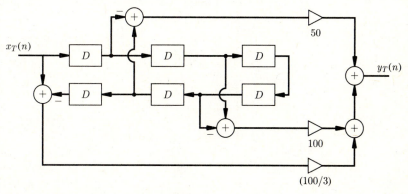

Figure 15.36 Block-Diagram Representation of a Realization of the Discrete-Time Differentiator Designed in Example 15.13

Again, we would need a longer unit pulse response (larger M) than we used in Example 15.13 to produce a designed filter with an amplitude response that more closely approximates the desired amplitude response. The phase response of the designed filter in Example 15.13 differs from the desired phase response by the addition of a straight line through the origin with slope $-2\pi MT$. Therefore, the FIR differentiator design obtained in Example 15.13 introduces no phase distortion, but it does introduce a pure time delay of $MT = 30$ ms, or, equivalently, $M = 3$ samples.

Use of Windows

The frequency response of the FIR filter that we design with the Fourier series technique does not match the desired frequency response because (1) the Fourier series representation of the desired frequency response is truncated and (2) the resulting noncausal finite unit pulse response must be delayed in time to produce a causal FIR filter. The time shift produces only a time delay of the output signal; it does not distort it. This time delay is large if a long FIR unit pulse response is used (that is, if M is large).

As we discussed in Chapter 4, truncated Fourier series representation of a periodic function containing discontinuities (jumps), such as those in Examples 15.12 and 15.13, produces the Gibbs phenomenon. This phenomenon is evidenced by ripples in the representation that do not decrease in size but concentrate near the discontinuity as the length of the truncated series is increased. Thus, increasing the length of the FIR filter unit pulse response does not decrease the maximum stopband gain.

We can decrease the size of the Gibbs phenomenon ripples if we weight (multiply) the Fourier series coefficient [$h_T(n)$ for the FIR filter design] by values, $w_T(n)$, less than or equal to one. They must be an even function of n and approach zero smoothly as the series truncation point is approached. That is, for the Fourier series design technique for the FIR filter, we produce the unit-pulse-response samples of the noncausal FIR filter by

$$h_{1T}(n) = h_{WT}(n) = w_T(n)h_T(n) \tag{15.99}$$

where $w_T(n) = 0$ for $|n| > M$, rather than by eq. (15.77). We refer to $w_T(n)$ as a *window function*, or *window* for short. Note that if

$$w_T(n) = \begin{cases} 1 & -M \leq n \leq M \\ 0 & \text{elsewhere} \end{cases} \tag{15.100}$$

then eq. (15.99) equals eq. (15.77), and direct-series truncation is performed. The window defined by eq. (15.100) is a rectangular window, or a *Fourier window*. Some other useful windows and sketches of their forms are indicated in Figure 15.37. Included in the sketches is the continuous-time function that we can sample to produce the window values. Note that the Hamming window is very similar to the Hanning window. The Hamming window was developed from the Hanning window, with a change of window values to decrease ripple amplitude close to a discontinuity at the expense of ripple amplitude farther from the discontinuity. A number of additional windows have been defined to produce different effects.[†]

[†]F. J. Harris, "On the Use of Windows for Harmonic Analysis with the Discrete Fourier Transform," *Proceedings of the IEEE* 66 (January, 1978): pp. 51–83.

Figure 15.37
Windows for FIR
Filter Design

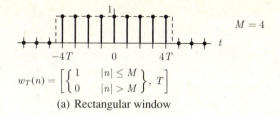

$$w_T(n) = \left[\left\{ \begin{array}{ll} 1 & |n| \le M \\ 0 & |n| > M \end{array} \right\}, T \right]$$

(a) Rectangular window

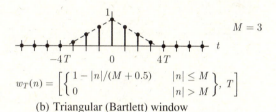

$$w_T(n) = \left[\left\{ \begin{array}{ll} 1 - |n|/(M + 0.5) & |n| \le M \\ 0 & |n| > M \end{array} \right\}, T \right]$$

(b) Triangular (Bartlett) window

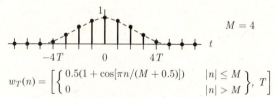

$$w_T(n) = \left[\left\{ \begin{array}{ll} 0.5(1 + \cos[\pi n/(M + 0.5)]) & |n| \le M \\ 0 & |n| > M \end{array} \right\}, T \right]$$

(c) Raised cosine (Hanning) window

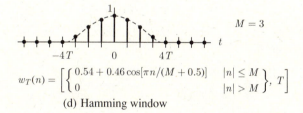

$$w_T(n) = \left[\left\{ \begin{array}{ll} 0.54 + 0.46\cos[\pi n/(M + 0.5)] & |n| \le M \\ 0 & |n| > M \end{array} \right\}, T \right]$$

(d) Hamming window

The terms for larger values of the index n in the Fourier series representation of the desired frequency response change more rapidly with frequency than those for smaller index values. The amplitudes of these terms are reduced by nonrectangular windows. Therefore, all nonrectangular windows have the negative effect of decreasing the rate at which the Fourier series representation can change as a function of frequency. For the FIR filter design, this produces the negative effect of less-sharp filter cutoffs when nonrectangular windows are used.

To use a window in the design of an FIR filter, we must include an additional step, step 2a, after step 2 in the previously stated design procedure:

Step 2a   Compute the $2M + 1$ window values, $w_T(n)$, and compute

$$h_{WT}(n) = w_T(n)h_T(n) \qquad -M \leq n \leq M \qquad (15.101)$$

to produce the $2M + 1$ windowed samples of the unit pulse response for a noncausal filter. When a window function is used, $h_T(k)$ must be replaced by $h_{WT}(k)$ in all design steps following step 2a.

We will now rework the previous filter designs (Examples 15.12 and 15.13) using a nonrectangular window. The solutions will be abbreviated, since they parallel the earlier solutions.

Example 15.14

Rework Example 15.12 using a Hamming window

Solution

From Example 15.12, $M = 3$. The Hamming window sequence values are

$$w_T(n) = \left[\left\{ \begin{matrix} 0.54 + 0.46\cos(\pi n/3.5) & |n| \leq 3 \\ 0 & |n| > 3 \end{matrix} \right\}, \quad T \right]$$

The window sequence values are thus

$$w_T(0) = 0.54 + 0.46\cos(0) = 1.00000$$

$$w_T(-1) = w_T(1) = 0.54 + 0.46\cos(\pi/3.5) = 0.82681$$

$$w_T(-2) = w_T(2) = 0.54 + 0.46\cos(2\pi/3.5) = 0.43764$$

and

$$w_T(-3) = w_T(3) = 0.54 + 0.46\cos(3\pi/3.5) = 0.12555$$

We use the values of $h_T(n)$ from Example 15.12 and the window values computed here to obtain

$$h_{WT}(0) = w_T(0)h_T(0) = (1.0)(0.2) = 0.2$$

$$h_{WT}(1) = w_T(1)h_T(1) = (0.82681)(0.18710) = 0.15470 = h_{WT}(-1)$$

$$h_{WT}(2) = w_T(2)h_T(2) = (0.43764)(0.15137) = 0.06625 = h_{WT}(-2)$$

and

$$h_{WT}(3) = w_T(3)h_T(3) = (0.12555)(0.10091) = 0.01267 = h_{WT}(-3)$$

The normalized frequency response of the causal FIR filter is

$$H_{2N}(r) = \sum_{n=0}^{6} h_{WT}(n-3)e^{-j2\pi nr}$$

The dc gain of the filter corresponding to this frequency response is

$$|H_{2N}(0)| = \sum_{n=0}^{6} h_{WT}(n-3) = 0.66724 \equiv K$$

The unit-pulse-response samples of the amplitude-scaled noncausal filter that produce a causal FIR filter with unity dc gain are

$$h_{SWT}(0) = \frac{0.2}{0.66724} = 0.29974$$

$$h_{SWT}(-1) = h_{SWT}(1) = \frac{0.15470}{0.66724} = 0.23185$$

$$h_{SWT}(-2) = h_{SWT}(2) = \frac{0.06625}{0.66724} = 0.09929$$

and

$$h_{SWT}(-3) = h_{SWT}(3) = \frac{0.01267}{0.66724} = 0.01899$$

The normalized frequency response and transfer function for the resulting causal FIR filter are

$$H_{SN}(r) = 0.01899 + 0.09929e^{-j2\pi r} + 0.23185e^{-j4\pi r}$$
$$+ 0.29974e^{-j6\pi r} + 0.23185e^{-j8\pi r}$$
$$+ 0.09929e^{-j10\pi r} + 0.01899e^{-j12\pi r}$$

and

$$H_{S2}(z) = 0.01899 + 0.09929z^{-1} + 0.23185z^{-2}$$
$$+ 0.29974z^{-3} + 0.23185z^{-4}$$
$$+ 0.09929z^{-5} + 0.01899z^{-6}$$

respectively. The amplitude and phase responses for the filter are plotted in Figure 15.38. The block-diagram representation of a filter realization is the same as that shown in Figure 15.33, except for changed multiplier values.

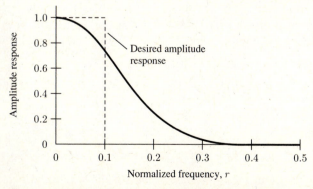

Figure 15.38 Amplitude and Phase Responses for the Discrete-Time Low-Pass FIR Filter Designed in Example 15.14 (Hamming Window)

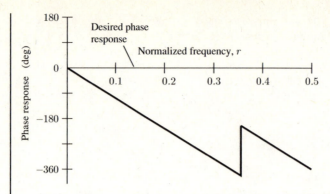

Figure 15.38 Amplitude and Phase Responses for the Discrete-Time Low-Pass FIR Filter Designed in Example 15.14 (Hamming Window) *(continued)*

When we compare Figures 15.38 and 15.32, it is clear that the application of a Hamming window to the FIR low-pass filter design has significantly reduced the unwanted filter-gain sidelobes in the stopband. But, as expected it has also made the filter cutoff less sharp. The gain is not actually zero for frequencies above $r = 0.35$ when the Hamming window is used, as it appears to be in Figure 15.38. The actual gain is simply too small to indicate with the scale used (0.0034 at $r = 0.5$). As for the design with the rectangular window, the phase response differs from the desired phase response in the passband by the addition of a straight line through the origin that has slope $-2\pi MT$. Therefore, the filter introduces a pure time delay of $MT = 1.5$ ms for signal frequencies in the passband.

Example 15.15

Rework Example 15.13 using a Hamming window.

Solution
From Example 15.13, $M = 3$. Using the values of $h_T(n)$ from Example 15.13 and the Hamming window values from Example 15.14, we compute

$$h_{WT}(0) = w_T(0)h_T(0) = (1.0)(0) = 0$$

$$h_{WT}(1) = w_T(1)h_T(1) = (0.82681)(-100) = -82.681 = -h_{WT}(-1)$$

$$h_{WT}(2) = w_T(2)h_T(2) = (0.43764)(50) = 21.882 = -h_W T(-2)$$

and

$$h_{WT}(3) = w_T(3)h_T(3) = (0.12555)(-100/3) = -4.185 = -h_{WT}(-3)$$

for the unit-pulse-response values of the windowed noncausal FIR filter. The normalized frequency response and transfer function for the resulting causal FIR filter are

$$H_{2N}(r) = 4.185 - 21.882e^{-j2\pi r} + 82.681e^{-j4\pi r}$$
$$- 82.681e^{-j8\pi r} + 21.882e^{-j10\pi r} - 4.185e^{-j12\pi r}$$

and

$$H_2(z) = 4.185 - 21.882z^{-1} + 82.681z^{-2}$$
$$- 82.681z^{-4} + 21.882z^{-5} - 4.185z^{-6}$$

respectively. The amplitude and phase responses are plotted in Figure 15.39. The block-diagram representation of a filter realization is the same as that shown in Figure 15.36, except for changed multiplier values.

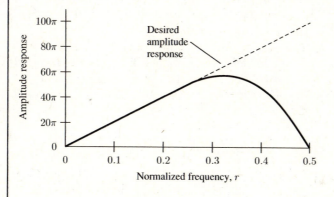

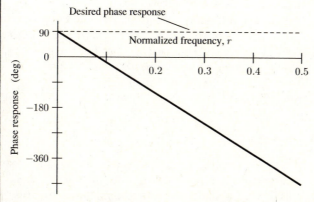

Figure 15.39 Amplitude and Phase Responses for the Discrete-Time FIR Differentiator Designed in Example 15.15 (Hamming Window)

Comparing Figures 15.39 and 15.35, we see that the FIR differentiator design is much improved for low frequencies when the Hamming window is used. The phase response still differs from the desired phase response by the addition of a straight line through the origin, which means that the filter introduces a pure time delay. The amplitude response of the FIR filter design that we obtained when we used the Hamming window does not exhibit any ripples and matches the desired amplitude response very closely for signal-frequency values less than one-fourth of the sample rate ($r < 0.25$).

FIR and IIR Design Comparisons

It is appropriate that we include mention of some of the advantages and disadvantages of FIR discrete-time-filter designs with respect to IIR discrete-time-filter designs. First, let us consider the advantages.

1. Nonrecursive implementations of FIR filter designs, such as those described in this section, have poles only at $z = 0$. Therefore, they have no feedback and are always stable.
2. Errors arising from quantization of signals and parameters (finite wordlength) are usually less critical for FIR filter designs than for IIR filter designs. This is primarily due to the lack of feedback.
3. FIR filters that have ideal linear phase characteristics in the passband and thus no phase distortion in the passband, can be designed. This is not the case with IIR filter designs.

Disadvantages include the following:

1. More delay and multiplier components are usually required in an FIR design to produce the same filter performance (sharpness of cutoff, and so forth), that can be obtained with an IIR filter design.
2. Generating a desirable frequency-response-approximation is more difficult for an FIR filter design than for an IIR filter design because it is more difficult to obtain FIR designs that have a given passband-gain ripple or maximum gain in the stopband. Iterative approaches can be used, as indicated in step 3 of the FIR filter design steps.

15.6 Summary

Discrete-time systems are categorized by the length of their unit pulse response. A system with an infinitely long unit pulse response is called an IIR system. A system with a unit pulse response of finite length is called an FIR system. Systems are also categorized as recursive or nonrecursive with a recursive system having feedback and a nonrecursive system having no feedback.

A realization of a discrete-time system is a system structure in discrete-time hardware or software that implements the system. The Direct Form I realization is obtained by directly implementing the system difference equation. We can reorganize it into a Direct Form II realization that requires fewer storage elements. Transfer-function factorization and partial-fraction expansion produce cascade and parallel realizations, respectively. These realizations contain no more storage elements than the Direct Form II realization. They decompose the system realization into a group of lower order subsystem realizations that are less sensitive to parameter- and signal-precision reduction caused by finite-wordlength restrictions.

In many discrete-time systems, there is a requirement for subsystems that are discrete-time filters. We can use spectral analysis and transform tools to analyze and design discrete-time filters. The product of the filter design is a filter transfer function from which we can construct a filter realization.

In designing discrete-time filters, we often want to achieve a design that approximately matches a continuous-time filter in some sense. Three approaches to approximate matching are the following: (1) approximately match the continuous-time-filter structure, (2) approximately match the continuous-time-filter time response, and (3) approximately match the continuous-time-filter frequency response. In this chapter, we considered all three approaches for IIR filters. We emphasized frequency-response matching for FIR filters because it is usually most desirable to match the frequency response and also because the other techniques are in most cases not appropriate for FIR filter design.

IIR filter design by structure matching involves the substitution of discrete-time approximations for the continuous-time-filter differentiation and integration operations. It is not a performance-related technique and does not supply design flexibility.

The use of time-response matching in IIR filter design is called time-invariant design. In this design procedure, input and output signals of the designed filter are sampled versions of those for the continuous-time filter being matched, for a specified input signal. Matching is imperfect for other input signals.

We use bilinear transformation to perform frequency-response-matching IIR filter design. With this transformation, the entire frequency response of the continuous-time filter is compressed nonlinearly into the frequency interval $-0.5f_s \leq f \leq 0.5f_s$ and is repeated in all adjacent frequency intervals of the same size.

One method for performing frequency-response matching for FIR filter design involves finding the Fourier series of the desired discrete-time-filter frequency response. The Fourier series coefficients are the samples of the unit pulse response for a noncausal IIR filter. We generate a noncausal FIR filter by truncating the noncausal IIR filter unit pulse response symmetrically about the time origin. We then provide a sufficient delay of this truncated unit pulse response to result in unit pulse response values of zero at time less than zero to produce the causal FIR filter design. The filter design often can be improved if a window function is applied to the truncated Fourier series coefficients to produce the noncausal IIR filter unit pulse response. Windowing reduces unwanted ripple and sidelobes in the filter frequency response, but it has the negative effect of reducing the sharpness of the filter cutoff.

The design techniques for discrete-time filters described in this chapter are only a few of the many design techniques available. They provide a basis for study of discrete-time-filter design methodology.

Problems

15.1 Draw the block-diagram representation of the Direct Form I and Direct Form II realizations of the systems with the following transfer functions:

a. $H(z) = \dfrac{1 + z^{-1}}{1 - 0.5z^{-1} + 0.06z^{-2}}$

b. $H(z) = \dfrac{z^{-1} - 3z^{-2}}{\left(10 - z^{-1}\right)\left(1 + 0.5z^{-1} + 0.5z^{-2}\right)}$

15.2 Repeat Problem 15.1 for systems with transfer functions

a. $H(z) = \dfrac{3 - 4z^{-2}}{2 - 2z^{-1} + z^{-2}}$

b. $H(z) = \dfrac{3z^2 + 6z}{(z - 0.2)\left(z^2 + 0.5z - 0.5\right)}$

15.3 Repeat Problem 15.1 for systems with transfer functions

a. $H(z) = \dfrac{2 + 3z^{-2} - 1.5z^{-4}}{1 - 0.2z^{-1} + 0.35z^{-2}}$

b. $H(z) = \dfrac{z(5z + 1)(z - 1)}{2z^3 - 1.2z^2 - 0.14z + 0.12}$

15.4 Repeat Problem 15.1 for systems with transfer functions

a. $H(z) = \dfrac{6\left(1 + 3z^{-1}\right)}{3 + 0.9z^{-1} + 0.06z^{-2}}$

b. $H(z) = \dfrac{3z^2 - 12}{z^2 - 0.4z + 0.03}$

15.5 Find and draw the block-diagram representations for all possible cascade realizations of the systems with the following transfer functions:

a. $H(z) = \dfrac{2z(z + 3)}{z^2 + 0.3z + 0.02}$

b. $H(z) = \dfrac{z^{-1} + 4z^{-2}}{5 - 2z^{-1} + 0.15z^{-2}}$

15.6 Find and draw the block diagram representation for a cascade realization for the systems with transfer functions

a. $H(z) = \dfrac{2 - 10z^{-2}}{1 - 0.8z^{-1} + 0.15z^{-2}}$

b. $H(z) = \dfrac{2(z + 2)}{z(z - 0.1)(z + 0.5)(z + 0.4)}$

15.7 Repeat Problem 15.6 for the systems with the transfer functions given in Problem 15.1.

15.8 Repeat Problem 15.6 for the systems with the transfer functions given in Problem 15.4.

15.9 Repeat Problem 15.6 for the systems with transfer functions

a. $H(z) = \dfrac{z^2 - 1}{z^2 - 0.7z + 0.1}$

b. $H(z) =$

$$\dfrac{1 - z^{-1} - 6z^{-2}}{\left(10 + z^{-1} - 0.2z^{-2}\right)\left(1 - 0.4z^{-1} + 0.13z^{-2}\right)}$$

15.10 Find and draw the block-diagram representations of parallel realizations of the systems with the transfer functions given in Problem 15.5.

15.11 Repeat Problem 15.10 for the systems with the transfer functions given in Problem 15.6.

15.12 Repeat Problem 15.10 for the systems with the transfer functions given in Problem 15.1.

15.13 Repeat Problem 15.10 for the systems with the transfer functions given in Problem 15.4.

15.14 Repeat Problem 15.10 for the systems with the transfer functions given in Problem 15.9.

15.15 A continuous-time filter is characterized by the differential equation

$$\dfrac{dy(t)}{dt} + y(t) = \dfrac{dx(t)}{dt} + 3x(t)$$

Use the discrete-time approximate derivative to find the transfer function for a discrete-time filter, the characteristics of which approximate the continuous-time filter. Plot the amplitude and phase responses for both the continuous-time filter and the discrete-time filter for the frequency interval $0 \le f \le 0.5$. Draw the plots for $T = 1$ s and $T = 0.1$ s. Comment on the results.

15.16 Repeat Problem 15.15 by first integrating the differential equation and then using the approximate integral.

15.17 A continuous-time filter is characterized by the differential equation

$$\frac{d^2y(t)}{dt^2} + 5\frac{dy(t)}{dt} + 6y(t) = 4x(t) + \frac{dx(t)}{dt}$$

Use the discrete-time approximate derivative to find the transfer function for a discrete-time filter, the characteristics of which approximate the continuous-time filter. The discrete-time filter is to be used in a system with a sample rate of 5 Hz. Draw the block-diagram representation for the Direct Form II realization of the resulting discrete-time filter.

15.18 Repeat Problem 15.17 by integrating the differential equation once, and then using the approximate derivative and the approximate integral.

15.19 Use the block-diagram approach to find the discrete-time-filter transfer function for the second filter approximation derived in Example 15.5 (that is, the filter produced by integration approximation).

15.20 A continuous-time filter is represented by the block diagram shown in Figure 15.40. Use the discrete-time approximate derivative and integral to find the transfer function for a discrete-time filter, the characteristics of which approximate the continuous-time filter. Draw the block-diagram representation for the Direct Form II realization of the resulting discrete-time filter.

15.21 The continuous-time, first-order, low-pass filter with transfer function

$$H_a(s) = \frac{1}{s + 1.5}$$

is used as the prototype for a discrete-time low-pass filter.

a. Use impulse-invariant design to find the transfer function for a discrete-time filter that corresponds to the continuous-time filter. The discrete-time system that requires the filter has a sample frequency of 5 Hz. Keep five significant digits in all numbers calculated.

b. Find the dc gain for the resulting discrete-time filter.

c. Modify the discrete-time-filter transfer function obtained in part (a) so that the dc gain of the discrete-time filter is equal to the continuous-time-filter dc gain.

d. Find the responses of the continuous-time filter to input signals of $\delta(t)$, $u(t)$, and $r(t)$. Compare these with the responses of the discrete-time filter to the input signals $\frac{1}{T}\delta(n)$, $u_T(n)$, and $Tr_T(n)$.

15.22 Repeat Problem 15.21 using step-invariant design.

15.23 Repeat Problem 15.21 with the following continuous-time-filter transfer function:

$$H_a(s) = \frac{s + 3}{(s + 1)(s + 2)}$$

15.24 Repeat Problem 15.23 using step-invariant design. Also draw the block-diagram representation of the parallel realization of the filter.

15.25 Use impulse-invariant design to find the transfer function of a discrete-time filter that corresponds to a second-order low-pass Butterworth filter with maximum gain of unity and a cutoff frequency of 10 Hz. The discrete-time filter is to be used in a system with a sample rate of 200 Hz. The filter is to have a maximum gain of unity. Also find and compare the amplitude and phase responses of the

Figure 15.40

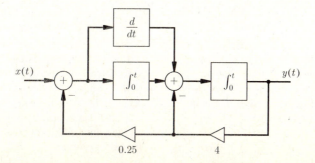

continuous-time and discrete-time filters. (Use normalized frequency over the interval $0 \leq r \leq 0.5$.)

15.26 Repeat Problem 15.25 using step-invariant design.

15.27 Use bilinear-transformation design to find the transfer function for a discrete-time filter that approximately matches the continuous-time filter with the following transfer function:

$$H_a(s) = \frac{s+2}{s^2 + 4.5s + 2}$$

The discrete-time filter is to be used in a system with sample rate of 2 Hz. Its frequency response is to match the continuous-time filter at a frequency of 0.15 Hz. Plot and compare the amplitude and phase responses of the continuous-time and discrete-time filters as a function of normalized frequency over the interval $0 \leq r \leq 0.5$.

15.28 Given the continuous-time filter transfer function

$$H_a(s) = \frac{s^2 + 2s}{(s+1)(s+5)}$$

a. Use bilinear-transformation design to obtain the transfer function for a corresponding discrete-time filter with a frequency response at 30 Hz that matches the continuous-time-filter frequency response at 0.7 Hz. The discrete-time filter is to be used in a discrete-time system with a sample rate of 500 Hz.

b. Draw the block-diagram representation of a cascade realization of the discrete-time filter.

15.29 Use bilinear-transformation design to find the transfer function for a discrete-time low-pass filter that approximately matches a third-order Butterworth filter. Design the discrete-time filter to have a maximum gain of 0.5 and a cutoff frequency of 50 Hz. The filter is part of a system with a sample rate of 250 Hz.

15.30 Find the minimum order of a discrete-time image-processing filter that operates at a sample rate of 2 kHz. Specifications for the filter are as follows: (1) a bilinear-transformation design based on a Chebyshev continuous-time filter, (2) a gain variation less than or equal to 1 dB for $0 \leq f \leq$ 250 Hz, (3) a gain less than or equal to -55 dB, with respect to maximum gain, for $f \geq 700$ Hz and

(4) a maximum gain of 0.8. Also find the approximate gain of the resulting filter at a frequency of 400 Hz.

15.31 A discrete-time bandpass filter is to be designed to reject noise in a discrete-time data-transmission system with a sample rate of 5 kHz. The filter is to satisfy the following specifications: (1) a bilinear-transformation design based on a Chebyshev continuous-time filter, (2) a center frequency of 1 kHz, (3) a bandwidth of 200 Hz, (4) a gain variation in the passband less than or equal to 1 dB, (5) a gain with respect to maximum gain less than or equal to -20 dB for $f \leq 600$ Hz and $f \geq$ 1.4 kHz, and (6) a maximum gain of 1. Find the minimum order required for the discrete-time filter, and the transfer function for the discrete-time filter. Draw the block-diagram representation of the Direct Form II realization of the filter.

15.32 A discrete-time bandpass filter is to be used in an image processor for image enhancement. The processor operates at a sample rate of 2 kHz. The filter is to satisfy the following specifications: (1) a bilinear-transformation design based on a Chebyshev continuous-time filter, (2) a passband from 790 Hz to 810 Hz with a gain variation less than or equal to 3 dB in the passband, (3) a gain with respect to maximum gain less than or equal to -16 dB for $f \leq 780$ Hz and $f \geq 820$ Hz, and (4) a maximum gain of 1. Find the minimum order required for the discrete-time filter and the transfer function for the discrete-time filter. Draw the block-diagram representation of the Direct Form II realization of the filter.

15.33 We want to design an FIR filter with a unit-pulse-response length of 11 samples to suppress noise in a digital temperature-sensing unit that operates at a sample rate of 25 Hz. The filter is to approximate an ideal low-pass filter with a gain of 0.8 in the passband and a cutoff frequency of 2 Hz. A Hanning window is to be used. The filter is to have a dc gain that matches the dc gain of the desired filter. Find the filter transfer function and the filter frequency response. Draw the block-diagram representation of the filter realization and plot the amplitude and phase responses of the desired filter and the designed filter as a function of normalized frequency over the interval $0 \leq r \leq 0.5$.

15.34 A discrete-time system has a sample rate of 5 kHz. You are to design an FIR discrete-time filter, the unit pulse response of which is nine samples long for this system. The filter's frequency response is to approximate an ideal bandpass filter with a center frequency of 1.5 kHz, a bandwidth of 500 Hz, and a gain of unity. The gain of the discrete-time filter is to match the gain of the desired filter at its center frequency.

a. Using a rectangular window, find the discrete-time-filter transfer function and frequency response. Draw the block-diagram representation of the filter realization.
b. Repeat part (a) using a Hanning window.
c. Plot and compare the amplitude and phase responses of the desired filter and the two filter designs as a function of normalized frequency over the interval $0 \leq r \leq 0.5$.

15.35 A discrete-time system has a sample rate of 5 kHz. Develop a design for an FIR discrete-time high-pass filter for use with this system. The filter's frequency response is to approximate an ideal high-pass filter with a cutoff frequency of 0.5 kHz. Its unit-pulse-response length is to be seven samples. The gain of the discrete-time filter is to match the gain of the desired filter at $f = 2.5$ kHz.

a. Using a Bartlett window, find the discrete-time-filter transfer function, and frequency response. Draw the block-diagram representation of the filter realization.

b. Repeat part (a) using a Hanning window.
c. Plot and compare the amplitude and phase responses for the desired filter and the two filter designs as a function of normalized frequency over the interval $0 \leq r \leq 0.5$.

15.36 A discrete-time FIR filter that approximates an ideal $-90°$ phase shifter (Hilbert transformer) is to be designed for a system with a sample rate of 100 Hz. The length of the FIR filter unit pulse response is to be 11 samples and the filter's gain is to match the ideal $-90°$ phase-shifter gain at $f = 0$.

a. Using a rectangular window, find the discrete-time-filter transfer function and frequency response. Draw the block-diagram representation of the filter realization.
b. Repeat part (a) using Hamming window.
c. Plot and compare the amplitude and phase responses of the desired filter and the two filter designs as a function of normalized frequency over the interval $0 \leq r \leq 0.5$.

15.37 Rework Examples 15.12 and 15.14 using a unit pulse response of length 15. Compare the resulting frequency responses with those shown in the examples.

15.38 Rework Examples 15.13 and 15.15 using a unit pulse response of length 15. Compare the resulting frequency response with those shown in the examples.

16 State-Variable Concepts for Discrete-Time Linear Systems

In Chapter 9, we briefly introduced the concepts of continuous-time state variables and illustrated how these state variables are used in the analysis of linear continuous-time systems. We showed that state variables provide an organized approach to the analysis of multiple-input, multiple-output systems (see Figure 9.1) or very complicated single-input, single-output systems. Also, solutions obtained with the state-variable approach provide information about internal energy storage in the system in addition to output signals.

This chapter parallels Chapter 9 in that it presents the same type of brief introduction to state variables, state-variable equations, output equations, and the solution of these equations for discrete-time signals and systems as Chapter 9 did for continuous-time signals and systems. Hence, the basic concepts of system state, zero state, zero-input response, zero-state response, and system linearity will not be defined here. Their definitions are the same as those given in Section 9.2, only with "continuous-time signals and systems" in those definitions replaced by "discrete-time signals and systems" here. Once again, the state-variable concepts and techniques for discrete-time systems can be studied in greater depth and breadth using texts devoted entirely to state variables.[†]

16.1 State Variables and Equations

To illustrate the concept of discrete-time state and output equations, let us first consider a kth-order, lumped-parameter, linear, time-varying, causal, single-input–single-output, discrete-time system. We will then extend the concept to systems with multiple inputs and outputs. Initially, we will permit the system to be time-varying for generality. However, for most of the chapter, we will consider only time-invariant systems for which we can use the convolution and transform techniques previously discussed in Chapters 12 and 14.

The difference equation that models a kth-order, lumped-parameter, linear, time-varying, causal, discrete-time system is

$$y_T(n) + \sum_{i=1}^{k} a_{iT}(n)y_T(n-i) = \sum_{i=0}^{\ell} b_{iT}(n)x_T(n-i) \qquad (16.1)$$

This difference equation for a time-varying system resembles the difference equation for a time-invariant system in Chapter 12 (eq. [12.2]), except that the coefficients a_i and b_i are now time-varying; that is, they are a function of the sample number n. The block-diagram representation of the Direct Form II realization of the system corresponding to eq. (16.1) is shown in Figure 16.1 for the case in which $k > \ell$.

[†]See for example DeRusso, Roy, and Close, *State Variables for Engineers*; Timothy and Bona, *State Space Analysis*.

Figure 16.1
Block-Diagram Representation of the Direct Form II Realization of a Time-Varying System

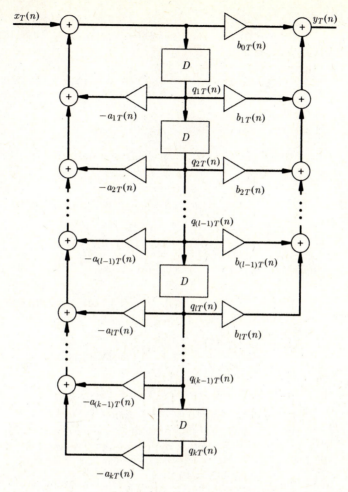

The signals $q_{iT}(n)$ shown in Figure 16.1 are the outputs of each of the delay components at the nth sample time (that is, at time $t = nT$). They characterize the energies stored in the delay components at time $t = nT$; thus, just as for continuous-time systems, we can use them as a set of state variables for the system. From Figure 16.1, we see that difference equations that can be used to solve for the state variables are

$$q_{1T}(n+1) = \left\{ \sum_{j=1}^{k} \left[-a_{jT}(n) \right] q_{jT}(n) \right\} + x_T(n)$$

$$q_{iT}(n+1) = q_{(i-1)T}(n) \qquad 2 \le i \le N$$

(16.2)

These equations are called the *state equations* for the system. The output equation for the system is easily obtained from Figure 16.1. It is

$$y_T(n) = b_{0T}(n)q_{1T}(n+1) + \sum_{i=1}^{\ell} b_{iT}(n)q_{iT}(n) \tag{16.3}$$

Since

$$q_{1T}(n+1) = x_T(n) + \sum_{i=1}^{k} [-a_{iT}(n)]\, q_{iT} \tag{16.4}$$

then eq. (16.3) becomes

$$y_T(n) = \sum_{i=1}^{\ell} [b_{iT}(n) - a_{iT}(n)b_{0T}(n)]\, q_{iT}(n)$$

$$+ \sum_{i=\ell+1}^{k} [-a_{iT}(n)b_{0T}(n)]\, q_{iT}(n)$$

$$+ b_{0T}(n)x_T(n) \tag{16.5}$$

if $k \geq \ell$, and eq. (16.3) becomes

$$y_T(n) = \sum_{i=1}^{k} [b_{iT}(n) - a_{iT}(n)b_{0T}(n)]\, q_{iT}(n)$$

$$+ \sum_{i=k+1}^{\ell} b_{iT}(n)q_{iT}(n)$$

$$+ b_{0T}(n)x_T(n) \tag{16.6}$$

if $k < \ell$. Using eqs. (16.2), (16.5), and (16.6), we can define state equations and output equations for a discrete-time system. These definitions are generalized to include multiple-input, multiple-output systems.

Definition _____

The state equations for a kth-order, lumped-parameter, linear, discrete-time system are N first-order, linear, ordinary difference equations that express the N state variables at sample number $n+1$ in terms of the N state variables and the system input signals at sample number n, where $N = \max[k, \ell]$ and ℓ is the largest delay of an input signal in the system difference equations.

Definition _____

The output equations for a kth-order, lumped-parameter, linear, discrete-time system are linear algebraic equations that express the system output signals as weighted sums of the N system state variables and the system input signals, where $N = \max[k, \ell]$ and ℓ is the largest delay of an input signal in the system difference equations.

For many IIR systems of interest, $N = k$. Henceforth, we will assume that $N = k$ for simplicity in this introductory treatment.

Using the preceding definitions and $N = k$, we write the state equations and output equations for a kth-order, lumped-parameter, linear, time-varying, causal, discrete-time system with m inputs and p outputs. These equations are

$$q_{iT}(n+1) = \sum_{r=1}^{k} a_{irT}(n)q_{rT}(n) + \sum_{s=1}^{m} b_{isT}(n)x_{sT}(n) \qquad 1 \le i \le k \quad (16.7)$$

and

$$y_{jT}(n) = \sum_{r=1}^{k} c_{jrT}(n)q_{rT}(n) + \sum_{s=1}^{m} d_{jsT}(n)x_{sT}(n) \qquad 1 \le j \le p \quad (16.8)$$

The system block-diagram representation corresponding to these equations for a single state variable and single output signal is shown in Figure 16.2. This

Figure 16.2
Portion of a
Multiple-Input,
Multiple-Output
System Block
Diagram

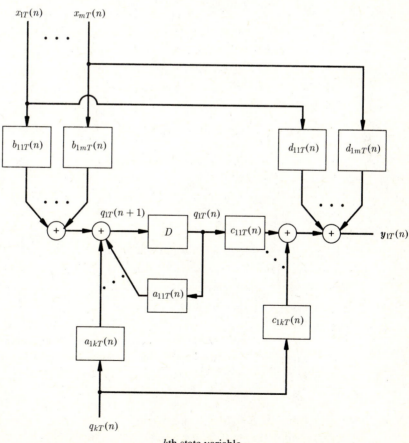

kth state variable
(internal signal)

block diagram is a portion of the complete system block diagram. The complete system block diagram contains k delay components. The inputs to the delay components are the weighted sums of the k state variables and the m input signals. The outputs of the delay components are the state variables. The complete system block diagram also contains p output summations of k weighted state variables and m weighted input signals.

We found for continuous-time systems that it is more convenient to express state equations and output equations in matrix notation, given its compactness and the ease with which matrix equations are manipulated. The matrix state equation and matrix output equation for a discrete-time system express eqs. (16.7) and (16.8) in matrix notation. They are

$$\mathbf{q}_T(n+1) = \mathbf{A}_T(n)\mathbf{q}_T(n) + \mathbf{B}_T(n)\mathbf{x}_T(n) \tag{16.9}$$

and

$$\mathbf{y}_T(n) = \mathbf{C}_T(n)\mathbf{q}_T(n) + \mathbf{D}_T(n)\mathbf{x}_T(n) \tag{16.10}$$

In these equations,

$$\mathbf{x}_T(n) = \begin{bmatrix} x_{1T}(n) \\ \vdots \\ x_{mT}(n) \end{bmatrix} \qquad \mathbf{q}_T(n) = \begin{bmatrix} q_{1T}(n) \\ \vdots \\ q_{kT}(n) \end{bmatrix} \qquad \mathbf{y}_T(n) = \begin{bmatrix} y_{1T}(n) \\ \vdots \\ y_{pT}(n) \end{bmatrix} \tag{16.11}$$

are the $m \times 1$ input-signal matrix, the $k \times 1$ state-variable matrix (called simply the *state matrix*), and the $p \times 1$ output-signal matrix, respectively. These matrices, because they consist of only one column, are also called *vectors*. The matrices

$$\mathbf{A}_T(n) = \begin{bmatrix} a_{11T}(n) & \cdots & a_{1kT}(n) \\ \vdots & & \vdots \\ a_{k1T}(n) & \cdots & a_{kkT}(n) \end{bmatrix} \qquad \mathbf{B}_T(n) = \begin{bmatrix} b_{11T}(n) & \cdots & b_{1mT}(n) \\ \vdots & & \vdots \\ b_{k1T}(n) & \cdots & b_{kmT}(n) \end{bmatrix} \tag{16.12}$$

$$\mathbf{C}_T(n) = \begin{bmatrix} c_{11T}(n) & \cdots & c_{1kT}(n) \\ \vdots & & \vdots \\ c_{p1T}(n) & \cdots & c_{pkT}(n) \end{bmatrix} \qquad \mathbf{D}_T(n) = \begin{bmatrix} d_{11T}(n) & \cdots & d_{1mT}(n) \\ \vdots & & \vdots \\ d_{p1T}(n) & \cdots & d_{pmT}(n) \end{bmatrix}$$

are the parameter matrices of the system. Matrices $\mathbf{A}_T(n)$, $\mathbf{B}_T(n)$, $\mathbf{C}_T(n)$, and $\mathbf{D}_T(n)$ have dimensions of $k \times k$, $k \times m$, $p \times k$, and $p \times m$, respectively. The block-diagram representation corresponding to the matrix-equation representation of the system parallels that shown in Figure 9.5, with $\mathbf{A}(t)$, $\mathbf{B}(t)$, $\mathbf{C}(t)$, and $\mathbf{D}(t)$ of Figure 9.5 replaced by $\mathbf{A}_T(n)$, $\mathbf{B}_T(n)$, $\mathbf{C}_T(n)$, and $\mathbf{D}_T(n)$, respectively; $d\mathbf{q}(t)/dt$, $\mathbf{q}(t)$, $\mathbf{x}(t)$, and $\mathbf{y}(t)$ replaced by $\mathbf{q}_T(n+1)$, $\mathbf{q}_T(n)$, $\mathbf{x}_T(n)$, and $\mathbf{y}_T(n)$, respectively; and the integration component replaced with a delay component.

The values of the state variables $\mathbf{q}_T(n)$ and output signals $\mathbf{y}_T(n)$ are easily determined for a time-varying system for $n \geq n_0$ when the input signals $\mathbf{x}_T(n)$ and the parameter matrices $\mathbf{A}_T(n)$, $\mathbf{B}_T(n)$, $\mathbf{C}_T(n)$, and $\mathbf{D}_T(n)$ are known for

$n \geq n_0$ and the initial system state vector $\mathbf{q}_T(n_0)$ is known. We accomplish this signal determination by recursively solving the state equations for the samples corresponding to $\mathbf{q}_T(n)$ for $n \geq n_0$ and substituting these samples and $\mathbf{x}_T(n)$ in the output equations for $n \geq n_0$. We will illustrate this procedure by using it to check the first few output-signal samples obtained in Example 16.1, presented in the following section.

As previously mentioned, the linear systems that we will consider will be limited to time-invariant systems, consistent with our coverage in the preceding chapters. For these systems, we can develop analytic solutions simply. For linear time-invariant systems, the matrix state equation and output equation are

$$\mathbf{q}_T(n+1) = \mathbf{A}\mathbf{q}_T(n) + \mathbf{B}\mathbf{x}_T(n) \tag{16.13}$$

and

$$\mathbf{y}_T(n) = \mathbf{C}\mathbf{q}_T(n) + \mathbf{D}\mathbf{x}_T(n) \tag{16.14}$$

respectively, where $\mathbf{A}$, $\mathbf{B}$, $\mathbf{C}$, and $\mathbf{D}$ are matrices with components that are constants.

16.2 Time-Domain Analysis of Systems

Now let us turn to the solution of the state and output equations to obtain the state vector and output-signal vector. Since we are dealing with only linear and time-invariant systems, the state and output equations are those with constant-parameter matrices given by eqs. (16.13) and (16.14). In addition, for simplicity, the zero state $\mathbf{q}_0(n) = \mathbf{0}$ will be the only zero state that we will consider. Recall that $\mathbf{0}$ is always a zero state for a linear system and is the only zero state for most linear systems of interest to us.

Solutions for State and Output Signals

We use the state and output equations (eqs. [6.13] and [6.14]) to determine the state vector $\mathbf{q}_T(n)$ for $n \geq n_0$ and the output-signal vector $\mathbf{y}_T(n)$ for $n \geq n_0$, given the initial state vector $\mathbf{q}_T(n_0)$ and the input-signal vector $\mathbf{x}_T(n)$ for $n \geq n_0$. Since $\mathbf{x}_T(n)$ is needed only for $n \geq n_0$, we assume that it is the zero vector for $n < n_0$. This assumption simplifies discussion of the solution and permits us to use single-sided z-transforms later in the chapter.

We prefer to obtain solutions for $n \geq n_0$ for both the state vector $\mathbf{q}_T(n)$ and output-signal vector $\mathbf{y}_T(n)$. The solution for $\mathbf{q}_T(n)$ for $n \geq 0$ is accomplished by rewriting eq. (16.13) as

$$\mathbf{q}_T(n) = \mathbf{A}\mathbf{q}_T(n-1) + \mathbf{B}\mathbf{x}_T(n-1) \tag{16.15}$$

and using the values of $\mathbf{q}_T(n_0 - 1)$ that produce the desired initial state vector $\mathbf{q}_T(n_0)$ as the initial conditions. Note that $\mathbf{x}_T(n_0 - 1) = \mathbf{0}$ in accordance with our previous assumption. This difference-equation-solution procedure parallels the procedure we used in Chapter 14 when we solved difference equations with z-transforms. It permits direct consideration of the use of z-transforms in solving state and output equations later in this chapter.

To find $\mathbf{q}_T(n)$ for $n \geq n_0$, we start with the first two terms of the solution $\mathbf{q}_T(n)$ obtained by using eq. (16.15) and the initial conditions. These first two terms are

$$\mathbf{q}_T(n_0) = \mathbf{A}\mathbf{q}_T(n_0 - 1) + \mathbf{B}\mathbf{x}_T(n_0 - 1) \tag{16.16}$$

and

$$\mathbf{q}_T(n_0 + 1) = \mathbf{A}\mathbf{q}_T(n_0) + \mathbf{B}\mathbf{x}_T(n_0)$$
$$= \mathbf{A}^2\mathbf{q}_T(n_0 - 1) + \mathbf{A}\mathbf{B}\mathbf{x}_T(n_0 - 1) + \mathbf{B}\mathbf{x}_T(n_0) \tag{16.17}$$

where we obtain the second equivalence in eq. (16.17) by substitution of eq. (16.16). Continued substitution of the evaluated state vector $\mathbf{q}_T(n - 1)$ in the state equation for $\mathbf{q}_T(n)$ gives

$$\mathbf{q}_T(n_0 + \ell) = \mathbf{A}^{\ell+1}\mathbf{q}_T(n_0 - 1) + \sum_{k=0}^{\ell} \mathbf{A}^{\ell-k}\mathbf{B}\mathbf{x}_T(k + n_0 - 1) \quad \ell \geq 0 \tag{16.18}$$

as the solution for the state vector at sample number $n_0 + \ell$. Note that $\mathbf{A}^0 = \mathbf{I}$ since

$$\mathbf{A}^0\mathbf{A} = \mathbf{A}^{(0+1)} = \mathbf{A} \tag{16.19}$$

We make the variable changes $\ell = n - n_0$ and $k = i - n_0 + 1$ to obtain

$$\mathbf{q}_T(n) = \mathbf{A}^{n-n_0+1}\mathbf{q}_T(n_0 - 1) + \sum_{i=n_0-1}^{n-1} \mathbf{A}^{n-1-i}\mathbf{B}\mathbf{x}_T(i) \quad n \geq n_0 \tag{16.20}$$

as the solution for the state vector for $n \geq n_0$. Since $\mathbf{x}_T(n_0 - 1) = \mathbf{0}$, then, from eq. (16.16), $\mathbf{q}_T(n_0) = \mathbf{A}\mathbf{q}_T(n_0 - 1)$, and we can write eq. (16.20) in terms of the initial state vector as

$$\mathbf{q}_T(n) = \mathbf{A}^{n-n_0}\mathbf{q}_T(n_0) + \sum_{i=n_0-1}^{n-1} \mathbf{A}^{n-1-i}\mathbf{B}\mathbf{x}_T(i) \quad n \geq n_0 \tag{16.21}$$

The assumption that $\mathbf{x}_T(n_0 - 1) = \mathbf{0}$ also permits changing the lower limit on the summation to n_0. However, to produce equation similarity for all $n \geq n_0$, we find that it is more convenient to leave the lower limit as is so that the summation will have at least one term (even though it may be zero) for all $n \geq n_0$.

The zero-input state vector for $n \geq n_0$ is defined to be $\mathbf{q}_{ziT}(n)$. That is, $\mathbf{q}_{ziT}(n)$ is the state vector for $n \geq n_0$ when $\mathbf{x}_T(n) = \mathbf{0}$ for $n \geq n_0$. Equation (16.21) then gives

$$\mathbf{q}_{ziT}(n) = \mathbf{A}^{n-n_0}\mathbf{q}_T(n_0) \quad n \geq n_0 \tag{16.22}$$

Note that the matrix $\mathbf{A}^{n-n_0}$ is a $k \times k$ matrix that produces the zero-input state vector at sample number n, where $n \geq n_0$, when it is multiplied by the zero-input state vector at sample number n_0. Thus, we call $\mathbf{A}^{n-n_0}$ the *state transition matrix*. It is a function of n and thus corresponds to a sequence of matrices for $n \geq n_0$. When used in the preceding equations, the sample spacing between successive matrices is T. Therefore, we give the state transition matrix the discrete-time matrix signal notation

$$\mathbf{r}_T(n - n_0) \equiv \mathbf{A}^{n-n_0} \qquad n \geq n_0 \tag{16.23}$$

in keeping with the continuous-time state-variable analysis discussed in Chapter 9. Using the notation defined by eq. (16.23) in eq. (16.21), the solution for the state vector is

$$\mathbf{q}_T(n) = \mathbf{r}_T(n - n_0)\,\mathbf{q}_T(n_0)$$

$$+ \sum_{i=n_0-1}^{n-1} \mathbf{r}_T(n - 1 - i)\,\mathbf{Bx}_T(i) \qquad (n \geq n_0) \tag{16.24}$$

We find the output-signal vector (system response vector) by substituting eq. (16.24) in eq. (16.14) to yield

$$\mathbf{y}_T(n) = \mathbf{Cr}_T(n - n_0)\,\mathbf{q}_T(n_0)$$

$$+ \sum_{i=n_0-1}^{n-1} \mathbf{Cr}_T(n - 1 - i)\mathbf{Bx}_T(i)$$

$$+ \mathbf{Dx}_T(n) \qquad (n \geq n_0) \tag{16.25}$$

We see that the first term of eq. (16.25) is the zero-input response of the system and that the second and third terms are the zero-state response of the system.

When we considered single-input, single-output systems, we found the unit pulse response of the system to be important. By computing the convolution of the unit pulse response and the input signal, we could find the zero-initial-condition response to any input signal when the system was linear and time-invariant. We can obtain a similar result for a linear, time-invariant system described by state and output equations if the system initial state is a zero state, as we will now demonstrate.

With $\mathbf{q}_T(n_0) = \mathbf{q}_0(n_0) = \mathbf{0}$, the system response is

$$\mathbf{y}_T(n) = \sum_{i=n_0-1}^{n-1} [\mathbf{Cr}_T(n - 1 - i)\mathbf{Bx}_T(i)] + \mathbf{Dx}_T(n) \qquad n \geq n_0 \tag{16.26}$$

which can be expressed as

$$\mathbf{y}_T(n) = \sum_{i=n_0-1}^{n} [u_T(n - 1 - i)\mathbf{Cr}_T(n - 1 - i)\mathbf{B} + \delta_T(n - i)\mathbf{D}]\,\mathbf{x}_T(i)$$

$$n \geq n_0 \tag{16.27}$$

because $u_T(n - i - 1) = 0$ for $i > n - 1$ and $\delta_T(n - i)$ equals one for $i = n$ and zero for $i \neq n$. Equation (16.27) can be written as

$$\mathbf{y}_T(n) = \sum_{i=n_0-1}^{n} \mathbf{h}_T(n - i)\mathbf{x}_T(i) \qquad n \geq n_0 \qquad (16.28)$$

if

$$\begin{bmatrix} h_{11T}(n) & \cdots & h_{1mT}(n) \\ \vdots & & \vdots \\ h_{p1T}(n) & & h_{pmT}(n) \end{bmatrix} = \mathbf{h}_T(n) \equiv u_T(n - 1)\mathbf{Cr}_T(n - 1)\mathbf{B} + \delta_T(n)\mathbf{D} \qquad (16.29)$$

Since $\mathbf{x}_T(i) = 0$ for $i < n_0$ and $\mathbf{h}_T(n - i) = u_T(n - 1 - i)\mathbf{Cr}_T(n - 1 - i)\mathbf{B} + \delta_T(n - i)\mathbf{D} = 0$ for $i > n$, then we can write eq. (16.28) in the form

$$\mathbf{y}_T(n) = \sum_{i=-\infty}^{\infty} \mathbf{h}_T(n - i)\mathbf{x}_T(i) \qquad n \geq n_0 \qquad (16.30)$$

Equation (16.30) is defined as the *discrete convolution of the matrices* $\mathbf{h}_T(n)$ and $\mathbf{x}_T(n)$; thus, we can write it in simple notation as

$$\mathbf{y}_T(n) = \mathbf{h}_T(n) * \mathbf{x}_T(n) \qquad n \geq n_0 \qquad (16.31)$$

Therefore, when the initial state is the zero state, we can obtain the system response vector $\mathbf{y}_T(n)$ for any input-signal vector by computing the convolution of the matrix $\mathbf{h}_T(n)$ and the input-signal vector $\mathbf{x}_T(n)$. As we indicated in Chapter 9, the convolution of two matrices is performed in the manner of matrix multiplication defined by eqs. (A.11) and (A.12) in Appendix A, except that the individual multiplications in eq. (A.12) are replaced with convolutions.

The $p \times m$ matrix $\mathbf{h}_T(n)$ is called the *unit-pulse-response matrix* of the system because its components are the responses at an individual output to a unit pulse input applied to a single-input terminal at $n = 0$. We illustrate this statement by considering the jth component of the output-signal vector in eq. (16.30). When the ith component $x_{iT}(n)$ of the input-signal vector is $\delta_T(n)$ and all other input-signal vector components are zero, then

$$y_{jT}(n) = \sum_{\ell=-\infty}^{\infty} h_{jiT}(n - \ell)\delta_T(\ell) = h_{jiT}(n) \qquad n \geq 0 \qquad (16.32)$$

As we saw for continuous-time systems in Chapter 9, all system parameter matrices enter into the computation of the unit-pulse-response matrix $\mathbf{h}_T(n)$. The matrices $\mathbf{B}$, $\mathbf{C}$, and $\mathbf{D}$ enter directly and the matrix $\mathbf{A}$ enters through the state transition matrix $\mathbf{r}_T(n - n_0)$.

Computation of the State Transition Matrix

We must find the state transition matrix that corresponds to a system before we can compute the solutions for the state and output-signal vectors for $n \geq n_0$.

The state transition matrix for a system with k states is the $k \times k$ matrix

$$\mathbf{r}_T(\ell) = \mathbf{A}^\ell \qquad \ell \geq 0 \tag{16.33}$$

This matrix is the matrix defined by eq. (16.23), with $n - n_0$ replaced by ℓ for notational simplicity.

Note that the computation of the state transition matrix for a given value of ℓ requires the computation of the ℓth power of the parameter matrix $\mathbf{A}$. We will now describe a method for finding a matrix equal to $\mathbf{A}^\ell$, with terms that are closed-form functions of ℓ, for the case when the eigenvalues of $\mathbf{A}$ are distinct (that is, when they all have different values). Should the eigenvalues not be distinct, the method can be modified, but we do not consider these modifications in this introductory treatment.

Equation (9.51) in Chapter 9 shows that

$$\mathbf{A}^\ell = \sum_{i=0}^{k-1} w_{i\ell} \mathbf{A}^i \tag{16.34}$$

where $\mathbf{A}$ is a square $k \times k$ matrix and the $w_{i\ell}$'s are constants. We developed eq. (16.34) in the discussion just prior to eq. (9.51) in Chapter 9 by using the characteristic equation for the matrix $\mathbf{A}$ and the Cayley-Hamilton theorem. From eq. (9.55),

$$\mathbf{P}\mathbf{A}^\ell\mathbf{P}^{-1} = \mathbf{A}_D^\ell = \begin{bmatrix} \lambda_1^\ell & 0 & \cdots & 0 \\ 0 & \lambda_2^\ell & & \vdots \\ \vdots & & \ddots & \\ 0 & \cdots & & \lambda_k^\ell \end{bmatrix} \tag{16.35}$$

where the λ_i's are the eigenvalues of $\mathbf{A}$ and $\mathbf{P}\mathbf{A}^\ell\mathbf{P}^{-1}$ is the similarity transformation that we can apply to $\mathbf{A}^\ell$ to produce the diagonal matrix $\mathbf{A}_D^\ell$. Substituting eq. (16.34) in eq. (16.35), we obtain

$$\mathbf{A}_D^\ell = \mathbf{P}\mathbf{A}^\ell\mathbf{P}^{-1} = \mathbf{P}\left[\sum_{i=0}^{k-1} w_{i\ell}\mathbf{A}^i\right]\mathbf{P}^{-1}$$

$$= \sum_{i=0}^{k-1} w_{i\ell}\mathbf{P}\mathbf{A}^i\mathbf{P}^{-1}$$

$$= \sum_{i=0}^{k-1} w_{i\ell}\mathbf{A}_D^i \qquad \ell \geq 0 \tag{16.36}$$

which, in turn, gives the k scalar equations

$$\lambda_j^\ell = \sum_{i=0}^{k-1} w_{i\ell}\lambda_j^i \qquad 1 \leq j \leq k \tag{16.37}$$

that correspond to the diagonal terms of the matrices in eq. (16.36). All other terms in the matrices in eq. (16.36) are zero.

We can simultaneously solve the k equations in eq. (16.37) to find the k coefficients $w_{i\ell}$, $0 \le i \le k - 1$, as functions of ℓ after first determining the eigenvalues from the characteristic equation of $\mathbf{A}$. Note that this simultaneous solution would not be possible if the eigenvalues were not distinct; then, fewer than k different equations would be specified by eq. (16.37). (As previously mentioned, a modified method can be used to obtain a solution in such a case.) After finding the functions for the $w_{i\ell}$'s and computing the constant matrices $\mathbf{A}^i$ for $0 \le i \le k - 1$, then we can compute the state transition matrix $\mathbf{r}_T(\ell)$ by using eq. (16.34).

System Analysis Example

We will now illustrate the techniques described for time-domain system analysis. For convenience, the time reference in our example is selected so that $n_0 = 0$.

Example 16.1

The second-order system shown in Figure 16.3 is the discrete-time portion of a system used to generate three control signals for a manufacturing process from two signals generated by process sensors. Find the output signals for $n \ge 0$ when the input signals are $x_{1T}(n) = \delta_T(n)$ and $x_{2T}(n) = u_T(n)$ and the initial state values are (a) $q_1(0) = 1$, $q_2(0) = -2$, and (b) $q_1(0) = q_2(0) = 0$. In part (a), check the results for $0 \le n \le 2$ by recursively solving the state and output equations.

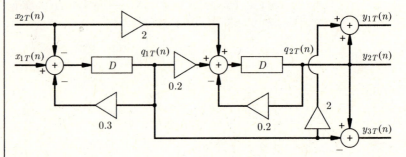

Figure 16.3 Block-Diagram Representation of the Second-Order System in Example 16.1

Solution

From the block diagram, we write the state and output equations as follows:

$$q_{1T}(n+1) = (-0.3)q_{1T}(n) + (0)q_{2T}(n) + (1)x_{1T}(n) + (-1)x_{2T}(n)$$

$$q_{2T}(n+1) = (0.2)q_{1T}(n) + (-0.2)q_{2T}(n) + (0)x_{1T}(n) + (2)x_{2T}(n)$$

$$y_{1T}(n) = (2)q_{1T}(n) + (1)q_{2T}(n)$$

$$y_{2T}(n) = (0)q_{1T}(n) + (1)q_{2T}(n)$$

$$y_{3T}(n) = (-1)q_{1T}(n) + (1)q_{2T}(n)$$

In matrix form, these equations are

$$\mathbf{q}_T(n+1) = \mathbf{A}\mathbf{q}_T(n) + \mathbf{B}\mathbf{x}_T(n)$$

and

$$\mathbf{y}_T(n) = \mathbf{C}\mathbf{q}_T(n) + \mathbf{D}\mathbf{x}_T(n)$$

where

$$\mathbf{A} = \begin{bmatrix} -0.3 & 0 \\ 0.2 & -0.2 \end{bmatrix} \qquad \mathbf{B} = \begin{bmatrix} 1 & -1 \\ 0 & 2 \end{bmatrix} \qquad \mathbf{C} = \begin{bmatrix} 2 & 1 \\ 0 & 1 \\ -1 & 1 \end{bmatrix} \qquad \mathbf{D} = 0$$

The characteristic equation for the system is

$$|\mathbf{A} - \lambda\mathbf{I}| = \begin{vmatrix} -0.3 - \lambda & 0 \\ 0.2 & -0.2 - \lambda \end{vmatrix} = (-0.3 - \lambda)(-0.2 - \lambda) = 0$$

(see eq. [9.45]). The roots of the characteristic equation are the eigenvalues

$$\lambda_1 = -0.2 \quad \text{and} \quad \lambda_2 = -0.3$$

Since $k = 2$ and $n_0 = 0$, then

$$\mathbf{r}_T(n) = \mathbf{A}^n = w_{0n}\mathbf{I} + w_{1n}\mathbf{A} \qquad n \geq 0$$

where w_{0n} and w_{1n} can be determined from the two equations given by eq. (16.37); that is, from the equations

$$(-0.2)^n = w_{0n} + w_{1n}(-0.2)$$

and

$$(-0.3)^n = w_{0n} + w_{1n}(-0.3)$$

Simultaneous solution of these equations gives

$$w_{0n} = 3(-0.2)^n - 2(-0.3)^n$$

and

$$w_{1n} = 10(-0.2)^n - 10(-0.3)^n$$

Substituting w_{0n} and w_{1n} in the equation for $\mathbf{r}_T(n)$ yields

$$\mathbf{r}_T(n) = \begin{bmatrix} 3(-0.2)^n - 2(-0.3)^n \end{bmatrix} \begin{bmatrix} 1 & 0 \\ 0 & 1 \end{bmatrix}$$

$$+ \begin{bmatrix} 10(-0.2)^n - 10(-0.3)^n \end{bmatrix} \begin{bmatrix} -0.3 & 0 \\ 0.2 & -0.2 \end{bmatrix}$$

$$= \begin{bmatrix} (-0.3)^n & 0 \\ 2(-0.2)^n - 2(-0.3)^n & (-0.2)^n \end{bmatrix} \qquad n \geq 0$$

From eq. (16.25), the matrix output equation is

$$\mathbf{y}_T(n) = \mathbf{Cr}_T(n)\mathbf{q}_T(0) + \sum_{i=-1}^{n-1} \mathbf{Cr}_T(n-1-i)\mathbf{Bx}_T(i) + \mathbf{Dx}_T(n) \qquad n \geq 0$$

Since

$$\mathbf{Cr}_T(n) = \begin{bmatrix} 2 & 1 \\ 0 & 1 \\ -1 & 1 \end{bmatrix} \begin{bmatrix} (-0.3)^n & 0 \\ 2(-0.2)^n - 2(-0.3)^n & (-0.2)^n \end{bmatrix}$$

$$= \begin{bmatrix} 2(-0.2)^n & (-0.2)^n \\ 2(-0.2)^n - 2(-0.3)^n & (-0.2)^n \\ 2(-0.2)^n - 3(-0.3)^n & (-0.2)^n \end{bmatrix} \qquad n \geq 0$$

and

$$\mathbf{Cr}_T(n)\mathbf{B} = \begin{bmatrix} 2(-0.2)^n & (-0.2)^n \\ 2(-0.2)^n - 2(-0.3)^n & (-0.2)^n \\ 2(-0.2)^n - 3(-0.3)^n & (-0.2)^n \end{bmatrix} \begin{bmatrix} 1 & -1 \\ 0 & 2 \end{bmatrix}$$

$$= \begin{bmatrix} 2(-0.2)^n & 0 \\ 2(-0.2)^n - 2(-0.3)^n & 2(-0.3)^n \\ 2(-0.2)^n - 3(-0.3)^n & 3(-0.3)^n \end{bmatrix} \qquad n \geq 0$$

then, for part (a),

$$\mathbf{y}_T(n) = \begin{bmatrix} 2(-0.2)^n & (-0.2)^n \\ 2(-0.2)^n - 2(-0.3)^n & (-0.2)^n \\ 2(-0.2)^n - 3(-0.3)^n & (-0.2)^n \end{bmatrix} \begin{bmatrix} 1 \\ -2 \end{bmatrix}$$

$$+ \sum_{i=-1}^{n-1} \begin{bmatrix} 2(-0.2)^{n-1-i} & 0 \\ 2(-0.2)^{n-1-i} - 2(-0.3)^{n-1-i} & 2(-0.3)^{n-1-i} \\ 2(-0.2)^{n-1-i} - 3(-0.3)^{n-1-i} & 3(-0.3)^{n-1-i} \end{bmatrix} \begin{bmatrix} \delta_T(i) \\ u_T(i) \end{bmatrix}$$

for $n \geq 0$.

Performing the indicated matrix multiplications, we find that the samples of the first output signal, $y_{1T}(n)$, are

$$y_{1T}(0) = 2(-0.2)^0 - 2(-0.2)^0 = 0$$

$$y_{1T}(n) = 2(-0.2)^n - 2(-0.2)^n + \sum_{i=-1}^{n-1} 2(-0.2)^{n-1-i}\delta_T(i)$$

$$= 2(-0.2)^{n-1} = -10(-0.2)^n \qquad n > 0$$

By using the unit pulse and unit step signals, we can write the first output signal as

$$y_{1T}(n) = 10\delta_T(n) - 10(-0.2)^n u_T(n)$$

The samples of the second output signal, $y_{2T}(n)$, are

$$y_{2T}(0) = 2(-0.2)^0 - 2(-0.3)^0 - 2(-0.2)^0 = -2$$

$$y_{2T}(n) = 2(-0.2)^n - 2(-0.3)^n - 2(-0.2)^n$$

$$+ \sum_{i=-1}^{n-1} \left[2(-0.2)^{n-1-i} - 2(-0.3)^{n-1-i} \right]\delta_T(i)$$

$$+ \sum_{i=-1}^{n-1} 2(-0.3)^{n-1-i} u_T(i)$$

$$= -2(-0.3)^n + 2(-0.2)^{n-1} - 2(-0.3)^{n-1} + \sum_{i=0}^{n-1} 2(-0.3)^{n-1-i}$$

$$= -2(-0.3)^n - 10(-0.2)^n + \frac{20}{3}(-0.3)^n + 2\sum_{j=0}^{n-1} (-0.3)^j$$

$$= -10(-0.2)^n - \frac{14}{3}(-0.3)^n + 2\left[\frac{1-(-0.3)^n}{1-(-0.3)} \right]$$

$$= \frac{20}{13} - 10(-0.2)^n + \frac{122}{39}(-0.3)^n \qquad n > 0$$

where we have used the change of variable $j = n - 1 - i$ to produce the third equivalence and the finite-geometric-sum relation

$$\sum_{j=0}^{m} a^j = \frac{1 - a^{m+1}}{1 - a}$$

to produce the fourth equivalence. By using the unit pulse and unit step signals, we write the second output signal as

$$y_{2T}(n) = \frac{10}{3}\delta_T(n) + \left[\frac{20}{13} - 10(-0.2)^n + \frac{122}{39}(-0.3)^n \right] u_T(n)$$

The matrix multiplications in the matrix equation for $\mathbf{y}_T(n)$ produce the third output signal sample values

$$y_{3T}(0) = 2(-0.2)^0 - 3(-0.3)^0 - 2(-0.2)^0 = -3$$

$$y_{3T}(n) = 2(-0.2)^n - 3(-0.3)^n - 2(-0.2)^n$$

$$+ \sum_{i=-1}^{n-1} \left[2(-0.2)^{n-1-i} - 3(-0.3)^{n-1-i} \right] \delta_T(i)$$

$$+ \sum_{i=-1}^{n-1} 3(-0.3)^{n-1-i} u_T(i)$$

$$= -3(-0.3)^n + 2(-0.2)^{n-1} - 3(-0.3)^{n-1} + \sum_{i=0}^{n-1} 3(-0.3)^{n-1-i}$$

$$= -3(-0.3)^n - 10(-0.2)^n + 10(-0.3)^n + 3 \sum_{j=0}^{n-1} (-0.3)^j$$

$$= -10(-0.2)^n + 7(-0.3)^n + 3 \left[\frac{1 - (-0.3)^n}{1 - (-0.3)} \right]$$

$$= \frac{30}{13} - 10(-0.2)^n + \frac{61}{13}(-0.3)^n \qquad n > 0$$

The expression for $y_{3T}(n)$ for $n > 0$ also equals $y_T(n)$ when $n = 0$. Therefore, the third output signal is

$$y_{3T}(n) = \left[\frac{30}{13} - 10(-0.2)^n + \frac{61}{13}(-0.3)^n \right] u_T(n)$$

The expressions we have found for $y_{1T}(n)$, $y_{2T}(n)$, and $y_{3T}(n)$ produce the following values for the output-signal vector $\mathbf{y}_T(n)$ for $0 \le n \le 2$:

$$\mathbf{y}_T(0) = \begin{bmatrix} 0 \\ -2 \\ -3 \end{bmatrix} \qquad \mathbf{y}_T(1) = \begin{bmatrix} 2 \\ 2.6 \\ 2.9 \end{bmatrix} \qquad \mathbf{y}_T(2) = \begin{bmatrix} -0.4 \\ 1.42 \\ 2.33 \end{bmatrix}$$

A recursive solution of the output and state equations for $0 \le n \le 2$ gives

$$\mathbf{y}_T(0) = \mathbf{C}\mathbf{q}_T(0) + \mathbf{D}\mathbf{x}_T(0)$$

$$= \begin{bmatrix} 2 & 1 \\ 0 & 1 \\ -1 & 1 \end{bmatrix} \begin{bmatrix} 1 \\ -2 \end{bmatrix} + \mathbf{0}\mathbf{x}_T(0) = \begin{bmatrix} 0 \\ -2 \\ -3 \end{bmatrix}$$

$$\mathbf{q}_T(1) = \mathbf{A}\mathbf{q}_T(0) + \mathbf{B}\mathbf{x}_T(0)$$

$$= \begin{bmatrix} -0.3 & 0 \\ 0.2 & -0.2 \end{bmatrix} \begin{bmatrix} 1 \\ -2 \end{bmatrix} + \begin{bmatrix} 1 & -1 \\ 0 & 2 \end{bmatrix} \begin{bmatrix} 1 \\ 1 \end{bmatrix} = \begin{bmatrix} -0.3 \\ 2.6 \end{bmatrix}$$

$$\mathbf{y}_T(1) = \mathbf{C}\mathbf{q}_T(1) + \mathbf{D}\mathbf{x}_T(1)$$

$$= \begin{bmatrix} 2 & 1 \\ 0 & 1 \\ -1 & 1 \end{bmatrix} \begin{bmatrix} -0.3 \\ 2.6 \end{bmatrix} + \mathbf{0}\mathbf{x}_T(1) = \begin{bmatrix} 2 \\ 2.6 \\ 2.9 \end{bmatrix}$$

$$\mathbf{q}_T(2) = \mathbf{A}\mathbf{q}_T(1) + \mathbf{B}\mathbf{x}_T(1)$$

$$= \begin{bmatrix} -0.3 & 0 \\ 0.2 & -0.2 \end{bmatrix} \begin{bmatrix} -0.3 \\ 2.6 \end{bmatrix} + \begin{bmatrix} 1 & -1 \\ 0 & 2 \end{bmatrix} \begin{bmatrix} 0 \\ 1 \end{bmatrix} = \begin{bmatrix} -0.91 \\ 1.42 \end{bmatrix}$$

$$\mathbf{y}_T(2) = \mathbf{C}\mathbf{q}_T(2) + \mathbf{D}\mathbf{x}_T(2)$$

$$= \begin{bmatrix} 2 & 1 \\ 0 & 1 \\ -1 & 1 \end{bmatrix} \begin{bmatrix} -0.91 \\ 1.42 \end{bmatrix} + \mathbf{0}\mathbf{x}_T(2) = \begin{bmatrix} -0.4 \\ 1.42 \\ 2.33 \end{bmatrix}$$

These results confirm the results we obtained previously.

Since $\mathbf{q}_T(0) = \mathbf{0}$ for part (b), then we will solve part (b) by using the unit-pulse-response matrix $\mathbf{h}_T(n)$ to illustrate its use. We find the unit-pulse-response matrix from eq. (16.29), with the following result:

$$\mathbf{h}_T(n) = u_T(n-1)\mathbf{C}\mathbf{r}_T(n-1)\mathbf{B} + \delta_T(n)\mathbf{D}$$

$$= u_T(n-1) \begin{bmatrix} 2(-0.2)^{n-1} & 0 \\ 2(-0.2)^{n-1} - 2(-0.3)^{n-1} & 2(-0.3)^{n-1} \\ 2(-0.2)^{n-1} - 3(-0.3)^{n-1} & 3(-0.3)^{n-1} \end{bmatrix}$$

We then compute the output-signal vector by using eq. (16.28). The result is

$$\mathbf{y}_T(n) = \sum_{i=-1}^{n} \mathbf{h}_T(n-i)\mathbf{x}_T(i)$$

$$= \sum_{i=-1}^{n} u_T(n-1-i) \begin{bmatrix} 2(-0.2)^{n-1-i} & 0 \\ 2(-0.2)^{n-1-i} - 2(-0.3)^{n-1-i} & 2(-0.3)^{n-1-i} \\ 2(-0.2)^{n-1-i} - 3(-0.3)^{n-1-i} & 3(-0.3)^{n-1-i} \end{bmatrix}$$

$$\times \begin{bmatrix} \delta_T(i) \\ u_T(i) \end{bmatrix}$$

$$= \sum_{i=-1}^{n-1} \begin{bmatrix} 2(-0.2)^{n-1-i} & 0 \\ 2(-0.2)^{n-1-i} - 2(-0.3)^{n-1-i} & 2(-0.3)^{n-1-i} \\ 2(-0.2)^{n-1-i} - 3(-0.3)^{n-1-i} & 3(-0.3)^{n-1-i} \end{bmatrix} \begin{bmatrix} \delta_T(i) \\ u_T(i) \end{bmatrix} \quad n \geq 0$$

where the last step follows because $u_T(n-1-i) = 0$ for $i > n-1$. Performing the indicated matrix multiplication, we find the samples of the first output signal, $y_{1T}(n)$, to be

$$y_{1T}(0) = 0$$

$$y_{1T}(n) = \sum_{i=-1}^{n-1} 2(-0.2)^{n-1-i}\delta_T(i) = 2(-0.2)^{n-1} = -10(-0.2)^n \qquad n > 0$$

We use the unit pulse and unit step signals to write the first output signal as

$$y_{1T}(n) = 10\delta_T(n) - \big[10(-0.2)^n\big]u_T(n)$$

The matrix multiplication in the matrix equation for $\mathbf{y}_T(n)$ produces the samples for the second output signal

$$y_{2T}(0) = 0$$

$$y_{2T}(n) = \sum_{i=-1}^{n-1} \big[2(-0.2)^{n-1-i} - 2(-0.3)^{n-1-i}\big]\delta_T(i)$$

$$+ \sum_{i=-1}^{n-1} 2(-0.3)^{n-1-i}u_T(i)$$

$$= 2(-0.2)^{n-1} - 2(-0.3)^{n-1} + 2\sum_{j=0}^{n-1}(-0.3)^j$$

$$= -10(-0.2)^n + \frac{20}{3}(-0.3)^n + 2\left[\frac{1-(-0.3)^n}{1-(-0.3)}\right]$$

$$= \frac{20}{13} - 10(-0.2)^n + \frac{200}{39}(-0.3)^n \qquad n > 0$$

The corresponding output signal can be written as

$$y_{2T}(n) = \frac{10}{3}\delta_T(n) + \left[\frac{20}{13} - 10(-0.2)^n + \frac{200}{39}(-0.3)^n\right]u_T(n)$$

by using the unit pulse and unit step signals.

The samples for the third output signal, $y_{3T}(n)$, are

$$y_{3T}(0) = 0$$

$$y_{3T}(n) = \sum_{i=-1}^{n-1} \big[2(-0.2)^{n-1-i} - 3(-0.3)^{n-1-i}\big]\delta_T(i)$$

$$+ \sum_{i=-1}^{n-1} 3(-0.3)^{n-1-i}u_T(i)$$

$$= 2(-0.2)^{n-1} - 3(-0.3)^{n-1} + 3\sum_{j=0}^{n-1}(-0.3)^j$$

$$= -10(-0.2)^n + 10(-0.3)^n + 3\left[\frac{1 - (-0.3)^n}{1 - (-0.3)}\right]$$

$$= \frac{30}{13} - 10(-0.2)^n + \frac{100}{13}(-0.3)^n \qquad n > 0$$

The expression for $y_{3T}(n)$ for $n > 0$ also equals $y_{3T}(n)$ when $n = 0$. Therefore, the third output signal is

$$y_{3T}(n) = \left[\frac{30}{13} - 10(-0.2)^n + \frac{100}{13}(-0.3)^n\right]u_T(n)$$

The summations in Example 16.1 were easily evaluated, because the input signals consisted of a unit pulse signal and a unit step signal. In general, however, the evaluation of these summations in closed form is not easy. We can use z-transforms and the convolution theorem to make this summation evaluation easier if, as in Example 16.1, $n_0 \geq 0$. But if z-transforms are used, it is preferable to use them throughout the solution. This approach allows us to find the total solution for the state vector and the output-signal vector by using z-transforms because the initial state values are included directly.

16.3 System Analysis Using z-Transforms

If the time reference is chosen to be $n_0 = 0$, then we can assume that all signals are zero for $n < 0$ because earlier signal contributions are incorporated through the state values at $n = 0$. In this case, we can use the single-sided z-transform to find solutions for the state variables and the output signals for $n \geq 0$, given that the system is linear and time-invariant. We will now present the analysis techniques and illustrate them by an example in which we use the same system and signals as we did in Example 16.1.

To begin the discussion, recall that the state and output equations can be expressed as

$$\mathbf{q}_T(n) = \mathbf{A}\mathbf{q}_T(n - 1) + \mathbf{B}\mathbf{x}_T(n - 1) \tag{16.38}$$

and

$$\mathbf{y}_T(n) = \mathbf{C}\mathbf{q}_T(n) + \mathbf{D}\mathbf{x}_T(n) \tag{16.39}$$

where $\mathbf{x}_T(-1) = \mathbf{0}$ and $\mathbf{q}_T(-1)$ is the initial condition that produces the desired initial system state $\mathbf{q}_T(0)$. Computing the z-transform of these equations, we obtain

$$\mathbf{Q}(z) = \mathbf{A}\left[z^{-1}\mathbf{Q}(z) + \mathbf{q}_T(-1)\right] + \mathbf{B}\left[z^{-1}\mathbf{X}(z)\right] \tag{16.40}$$

and

$$\mathbf{Y}(z) = \mathbf{C}\mathbf{Q}(z) + \mathbf{D}\mathbf{X}(z) \tag{16.41}$$

where the z-transform of a matrix or vector is computed as the z-transform of each component (see A.20 in Appendix A). Rearranging eq. (16.40), we obtain

$$\left[\mathbf{I} - z^{-1}\mathbf{A}\right]\mathbf{Q}(z) = \mathbf{A}\mathbf{q}_T(-1) + z^{-1}\mathbf{B}\mathbf{X}(z)$$

$$= \mathbf{q}_T(0) + z^{-1}\mathbf{B}\mathbf{X}(z) \tag{16.42}$$

where the last step follows from eq. (16.38), since $\mathbf{x}_T(-1) = \mathbf{0}$. The matrix $\mathbf{I} - z^{-1}\mathbf{A}$ is a $k \times k$ matrix and has the inverse $(\mathbf{I} - z^{-1}\mathbf{A})^{-1}$. Therefore,

$$\mathbf{Q}(z) = \left(\mathbf{I} - z^{-1}\mathbf{A}\right)^{-1}\mathbf{q}_T(0) + z^{-1}\left(\mathbf{I} - z^{-1}\mathbf{A}\right)^{-1}\mathbf{B}\mathbf{X}(z) \tag{16.43}$$

Substituting eq. (16.43) in eq. (16.41), we obtain

$$\mathbf{Y}(z) = \mathbf{C}\left(\mathbf{I} - z^{-1}\mathbf{A}\right)^{-1}\mathbf{q}_T(0) + z^{-1}\mathbf{C}\left(\mathbf{I} - z^{-1}\mathbf{A}\right)^{-1}\mathbf{B}\mathbf{X}(z) + \mathbf{D}\mathbf{X}(z) \tag{16.44}$$

We now want to identify the z-transform of the state transition matrix and use it to simplify eqs. (16.43) and (16.44). To do so, we begin with eq. (16.24). Since $n_0 = 0$, eq. (16.24) yields

$$\mathbf{q}_T(n) = \mathbf{r}_T(n)\mathbf{q}_T(0) + \sum_{i=-1}^{n-1} \mathbf{r}_T(n-1-i)\mathbf{B}\mathbf{x}_T(i) \qquad n \geq 0 \tag{16.45}$$

We have defined the state transition matrix $\mathbf{r}_T(n)$ only for $n \geq 0$ (see eq. [16.23]). Therefore, we can set it equal to the zero matrix, $\mathbf{0}$, for $n < 0$, which means that $\mathbf{r}_T(n-1-i) = \mathbf{0}$ for $i > n - 1$. Also, we have previously assumed that $\mathbf{x}_T(i) = \mathbf{0}$ for $i < 0$. Therefore, we can write eq. (16.45) as

$$\mathbf{q}_T(n) = \mathbf{r}_T(n)\mathbf{q}_T(0) + \sum_{i=-\infty}^{\infty} \mathbf{r}_T(n-1-i)\mathbf{B}\mathbf{x}_T(i)$$

$$= \mathbf{r}_T(n)\mathbf{q}_T(0) + \mathbf{r}_T(n-1) * \mathbf{B}\mathbf{x}_T(n) \qquad n \geq 0 \tag{16.46}$$

We use the time-delay, convolution, and linearity theorems to compute the z-transform of eq. (16.46), with the following result:

$$\mathbf{Q}(z) = \mathbf{R}(z)\mathbf{q}_T(0) + z^{-1}\mathbf{R}(z)\mathbf{B}\mathbf{X}(z) \tag{16.47}$$

Comparing eq. (16.47) with eq. (16.43), we see that

$$\mathbf{R}(z) = \left(\mathbf{I} - z^{-1}\mathbf{A}\right)^{-1} \tag{16.48}$$

Therefore, $(\mathbf{I} - z^{-1}\mathbf{A})^{-1}$ is the z-transform of the state transition matrix. Using eq. (16.48) we write the z-transform of the output equation (eq. [6.44]) as

$$\mathbf{Y}(z) = \mathbf{C}\mathbf{R}(z)\mathbf{q}_T(0) + z^{-1}\mathbf{C}\mathbf{R}(z)\mathbf{B}\mathbf{X}(z) + \mathbf{D}\mathbf{X}(z) \tag{16.49}$$

If the initial state is the zero state (that is, $\mathbf{q}_T(0) = \mathbf{0}$), then

$$\mathbf{Y}(z) = \left[z^{-1}\mathbf{C}\mathbf{R}(z)\mathbf{B} + \mathbf{D}\right]\mathbf{X}(z) \equiv \mathbf{H}(z)\mathbf{X}(z) \tag{16.50}$$

where we define

$$\mathbf{H}(z) = z^{-1}\mathbf{C}\mathbf{R}(z)\mathbf{B} + \mathbf{D} \tag{16.51}$$

to be the transfer-function matrix of the system. Now

$$\mathcal{Z}^{-1}[\mathbf{H}(z)] = \mathbf{Cr}_T(n-1)\mathbf{B} + \delta_T(n)\mathbf{D}$$

$$= u_T(n-1)\mathbf{Cr}_T(n-1)\mathbf{B} + \delta_T(n)\mathbf{D} \qquad (16.52)$$

where the last step follows because we have defined $\mathbf{r}_T(n)$ to be equal to $\mathbf{0}$ for $n < 0$. Thus, the transfer-function matrix is the z-transform of the unit-pulse-response matrix.

We will now reconsider the previous example, this time using z-transform techniques to find the solution.

Example 16.2

Use z-transforms to perform the analysis requested in Example 16.1.

Solution
Recall from Example 16.1 that

$$\mathbf{A} = \begin{bmatrix} -0.3 & 0 \\ 0.2 & -0.2 \end{bmatrix} \qquad \mathbf{B} = \begin{bmatrix} 1 & -1 \\ 0 & 2 \end{bmatrix} \qquad \mathbf{C} = \begin{bmatrix} 2 & 1 \\ 0 & 1 \\ -1 & 1 \end{bmatrix} \qquad \mathbf{D} = \mathbf{0}$$

We next find the z-transform of the state transition matrix. To do so, we first compute

$$\mathbf{I} - z^{-1}\mathbf{A} = \begin{bmatrix} 1 & 0 \\ 0 & 1 \end{bmatrix} - z^{-1}\begin{bmatrix} -0.3 & 0 \\ 0.2 & -0.2 \end{bmatrix} = \begin{bmatrix} 1 + 0.3z^{-1} & 0 \\ -0.2z^{-1} & 1 + 0.2z^{-1} \end{bmatrix}$$

The z-transform of the state transition matrix is the inverse of this 2×2 matrix, which is

$$\mathbf{R}(z) = (\mathbf{I} - z^{-1}\mathbf{A})^{-1} = \frac{\begin{bmatrix} 1 + 0.2z^{-1} & 0 \\ 0.2z^{-1} & 1 + 0.3z^{-1} \end{bmatrix}}{\left(1 + 0.3z^{-1}\right)\left(1 + 0.2z^{-1}\right) - (0)\left(-0.2z^{-1}\right)}$$

$$= \begin{bmatrix} \dfrac{1}{1 + 0.3z^{-1}} & 0 \\ \dfrac{2z^{-1}}{\left(1 + 0.2z^{-1}\right)\left(1 + 0.3z^{-1}\right)} & \dfrac{1}{1 + 0.2z^{-1}} \end{bmatrix}$$

$$= \begin{bmatrix} \dfrac{1}{1 + 0.3z^{-1}} & 0 \\ \dfrac{2}{\left(1 + 0.2z^{-1}\right)} - \dfrac{2}{\left(1 + 0.3z^{-1}\right)} & \dfrac{1}{1 + 0.2z^{-1}} \end{bmatrix}$$

(see Example 9.4). Note that

$$\mathbf{r}_T(n) = \mathcal{Z}^{-1}[\mathbf{R}(z)] = \begin{bmatrix} (-0.3)^n & 0 \\ 2(-0.2)^n - 2(-0.3)^n & (-0.2)^n \end{bmatrix} \qquad n \geq 0$$

This result is consistent with the state transition matrix that we found in Example 16.1.

From eqs. (16.49) and (16.51), the z-transform of the output-signal vector is

$$\mathbf{Y}(z) = \mathbf{CR}(z)\mathbf{q}_T(0) + \left[z^{-1}\mathbf{CR}(z)\mathbf{B} + \mathbf{D}\right]\mathbf{X}(z)$$
$$= \mathbf{CR}(z)\mathbf{q}_T(0) + \mathbf{H}(z)\mathbf{X}(z)$$

Also, the z-transform of the input-signal vector is

$$\mathbf{X}(z) = \begin{bmatrix} 1 \\ \dfrac{1}{1 - z^{-1}} \end{bmatrix}$$

since $x_{1T}(n) = \delta_T(n)$ and $x_{2T}(n) = u_T(n)$. Now,

$$\mathbf{CR}(z) = \begin{bmatrix} 2 & 1 \\ 0 & 1 \\ -1 & 1 \end{bmatrix} \begin{bmatrix} \dfrac{1}{1 + 0.3z^{-1}} & 0 \\ \dfrac{2}{1 + 0.2z^{-1}} - \dfrac{2}{1 + 0.3z^{-1}} & \dfrac{1}{1 + 0.2z^{-1}} \end{bmatrix}$$

$$= \begin{bmatrix} \dfrac{2}{1 + 0.2z^{-1}} & \dfrac{1}{1 + 0.2z^{-1}} \\ \dfrac{2}{1 + 0.2z^{-1}} - \dfrac{2}{1 + 0.3z^{-1}} & \dfrac{1}{1 + 0.2z^{-1}} \\ \dfrac{2}{1 + 0.2z^{-1}} - \dfrac{3}{1 + 0.3z^{-1}} & \dfrac{1}{1 + 0.2z^{-1}} \end{bmatrix}$$

and, since $\mathbf{D} = \mathbf{0}$, then the transfer-function matrix is

$$\mathbf{H}(z) = z^{-1}\mathbf{CR}(z)\mathbf{B}$$

$$= z^{-1} \begin{bmatrix} \dfrac{2}{1 + 0.2z^{-1}} & \dfrac{1}{1 + 0.2z^{-1}} \\ \dfrac{2}{1 + 0.2z^{-1}} - \dfrac{2}{1 + 0.3z^{-1}} & \dfrac{1}{1 + 0.2z^{-1}} \\ \dfrac{2}{1 + 0.2z^{-1}} - \dfrac{3}{1 + 0.3z^{-1}} & \dfrac{1}{1 + 0.2z^{-1}} \end{bmatrix} \begin{bmatrix} 1 & -1 \\ 0 & 2 \end{bmatrix}$$

$$= \begin{bmatrix} \dfrac{2z^{-1}}{1 + 0.2z^{-1}} & 0 \\ \dfrac{2z^{-1}}{1 + 0.2z^{-1}} - \dfrac{2z^{-1}}{1 + 0.3z^{-1}} & \dfrac{2z^{-1}}{1 + 0.3z^{-1}} \\ \dfrac{2z^{-1}}{1 + 0.2z^{-1}} - \dfrac{3z^{-1}}{1 + 0.3z^{-1}} & \dfrac{3z^{-1}}{1 + 0.3z^{-1}} \end{bmatrix}$$

We use Table C.8 and the time-delay theorem for z-transforms to obtain

$\mathbf{h}_T(n)$

$$= \begin{bmatrix} 2(-0.2)^{n-1}u_T(n-1) & 0 \\ 2(-0.2)^{n-1}u_T(n-1) - 2(-0.3)^{n-1}u_T(n-1) & 2(-0.3)^{n-1}u_T(n-1) \\ 2(-0.2)^{n-1}u_T(n-1) - 3(-0.3)^{n-1}u_T(n-1) & 3(-0.3)^{n-1}u_T(n-1) \end{bmatrix}$$

as the system unit-pulse-response matrix. The matrix $\mathbf{h}_T(n)$ is consistent with the result obtained in Example 16.1.

For part (a), $q_{1T}(0) = 1$ and $q_{2T}(0) = -2$; therefore,

$$\mathbf{Y}(z) = \begin{bmatrix} \dfrac{2}{1+0.2z^{-1}} & \dfrac{1}{1+0.2z^{-1}} \\ \dfrac{2}{1+0.2z^{-1}} - \dfrac{2}{1+0.3z^{-1}} & \dfrac{1}{1+0.2z^{-1}} \\ \dfrac{2}{1+0.2z^{-1}} - \dfrac{3}{1+0.3z^{-1}} & \dfrac{1}{1+0.2z^{-1}} \end{bmatrix} \begin{bmatrix} 1 \\ -2 \end{bmatrix}$$

$$+ \begin{bmatrix} \dfrac{2z^{-1}}{1+0.2z^{-1}} & 0 \\ \dfrac{2z^{-1}}{1+0.2z^{-1}} - \dfrac{2z^{-1}}{1+0.3z^{-1}} & \dfrac{2z^{-1}}{1+0.3z^{-1}} \\ \dfrac{2z^{-1}}{1+0.2z^{-1}} - \dfrac{3z^{-1}}{1+0.3z^{-1}} & \dfrac{3z^{-1}}{1+0.3z^{-1}} \end{bmatrix} \begin{bmatrix} 1 \\ \dfrac{1}{1-z^{-1}} \end{bmatrix}$$

Performing the matrix multiplications, adding, and simplifying yields

$\mathbf{Y}(z)$

$$= \begin{bmatrix} \dfrac{2z^{-1}}{1+0.2z^{-1}} \\ -\dfrac{2}{1+0.3z^{-1}} + \dfrac{2z^{-1}}{1+0.2z^{-1}} - \dfrac{2z^{-1}}{1+0.3z^{-1}} + \dfrac{2z^{-1}}{(1+0.3z^{-1})(1-z^{-1})} \\ -\dfrac{3}{1+0.3z^{-1}} + \dfrac{2z^{-1}}{1+0.2z^{-1}} - \dfrac{3z^{-1}}{1+0.3z^{-1}} + \dfrac{3z^{-1}}{(1+0.3z^{-1})(1-z^{-1})} \end{bmatrix}$$

$$= \begin{bmatrix} \dfrac{2}{z+0.2} \\ \dfrac{-2z^3 + 3.6z^2 + z - 0.2}{(z-1)(z+0.2)(z+0.3)} \\ \dfrac{-3z^3 + 4.4z^2 + 2.2z}{(z-1)(z+0.2)(z+0.3)} \end{bmatrix}$$

The z-transforms of the output signals are the three components in the $\mathbf{Y}(z)$ vector. They are

$$Y_1(z) = \frac{2}{z + 0.2}$$

$$Y_2(z) = \frac{-2z^3 + 3.6z^2 + z - 0.2}{(z - 1)(z + 0.2)(z + 0.3)}$$

$$Y_3(z) = \frac{-3z^3 + 4.4z^2 + 2.2z}{(z - 1)(z + 0.2)(z + 0.3)}$$

Partial-fraction expansion of $Y_1(z)/z$, $Y_2(z)/z$, and $Y_3(z)/z$ gives

$$\frac{Y_1(z)}{z} = \frac{2}{z(z + 0.2)} = \frac{10}{z} - \frac{10}{z + 0.2}$$

$$\frac{Y_2(z)}{z} = \frac{-2z^3 + 3.6z^2 + z - 0.2}{z(z - 1)(z + 0.2)(z + 0.3)}$$

$$= \frac{10}{3z} + \frac{20}{13(z - 1)} - \frac{10}{z + 0.2} + \frac{122}{39(z + 0.3)}$$

and

$$\frac{Y_3(z)}{z} = \frac{-3z^2 + 4.4z + 2.2}{(z - 1)(z + 0.2)(z + 0.3)}$$

$$= \frac{30}{13(z - 1)} - \frac{10}{z + 0.2} + \frac{61}{13(z + 0.3)}$$

Multiplication of the denominator of both sides of the equations by z^{-1} yields

$$Y_1(z) = 10 - \frac{10}{1 + 0.2z^{-1}}$$

$$Y_2(z) = \frac{10}{3} + \frac{20}{13}\left[\frac{1}{1 - z^{-1}}\right] - 10\left[\frac{1}{1 + 0.2z^{-1}}\right] + \frac{122}{39}\left[\frac{1}{1 + 0.3z^{-1}}\right]$$

and

$$Y_3(z) = \frac{30}{13}\left[\frac{1}{1 - z^{-1}}\right] - 10\left[\frac{1}{1 + 0.2z^{-1}}\right] + \frac{61}{13}\left[\frac{1}{1 + 0.3z^{-1}}\right]$$

Computing the inverse z-transforms of these equations, we obtain

$$y_{1T}(n) = 10\delta_T(n) - 10(-0.2)^n u_T(n)$$

$$y_{2T}(n) = \frac{10}{3}\delta_T(n) + \left[\frac{20}{13} - 10(-0.2)^n + \frac{122}{39}(-0.3)^n\right]u_T(n)$$

and

$$y_{3T}(n) = \left[\frac{30}{13} - 10(-0.2)^n + \frac{61}{13}(-0.3)^n\right]u_T(n)$$

as the output signals. Note that these output signals match those obtained in part (a) of Example 16.1, as they should.

For part (b), $q_{1T}(0) = q_{2T}(0) = 0$; therefore,

$$\mathbf{Y}(z) = \begin{bmatrix} \dfrac{2z^{-1}}{1+0.2z^{-1}} & 0 \\[4mm] \dfrac{2z^{-1}}{1+0.2z^{-1}} - \dfrac{2z^{-1}}{1+0.3z^{-1}} & \dfrac{2z^{-1}}{1+0.3z^{-1}} \\[4mm] \dfrac{2z^{-1}}{1+0.2z^{-1}} - \dfrac{3z^{-1}}{1+0.3z^{-1}} & \dfrac{3z^{-1}}{1+0.3z^{-1}} \end{bmatrix} \begin{bmatrix} 1 \\[1mm] 1 \\[1mm] 1-z^{-1} \end{bmatrix}$$

$$= \begin{bmatrix} \dfrac{2z^{-1}}{1+0.2z^{-1}} \\[4mm] \dfrac{2z^{-1}}{1+0.2z^{-1}} - \dfrac{2z^{-1}}{1+0.3z^{-1}} + \dfrac{2z^{-1}}{\left(1+0.3z^{-1}\right)\left(1-z^{-1}\right)} \\[4mm] \dfrac{2z^{-1}}{1+0.2z^{-1}} - \dfrac{3z^{-1}}{1+0.3z^{-1}} + \dfrac{3z^{-1}}{\left(1+0.3z^{-1}\right)\left(1-z^{-1}\right)} \end{bmatrix}$$

$$= \begin{bmatrix} \dfrac{2}{z+0.2} \\[4mm] \dfrac{2z^2+0.6z-0.2}{(z-1)(z+0.2)(z+0.3)} \\[4mm] \dfrac{2z^2+1.6z}{(z-1)(z+0.2)(z+0.3)} \end{bmatrix}$$

The z-transforms of the output signals are

$$Y_1(z) = \frac{2}{z+0.2}$$

$$Y_2(z) = \frac{2z^2+0.6z-0.2}{(z-1)(z+0.2)(z+0.3)}$$

and

$$Y_3(z) = \frac{2z^2+1.6z}{(z-1)(z+0.2)(z+0.3)}$$

Partial-fraction expansion of $Y_1(z)/z$, $Y_2(z)/z$, and $Y_3(z)/z$, and multiplication of the denominator of both sides of the equations by z^{-1} yields

$$Y_1(z) = 10 - \left[\frac{10}{1+0.2z^{-1}}\right]$$

$$Y_2(z) = \frac{10}{3} + \frac{20}{3}\left[\frac{1}{1-z^{-1}}\right] - 10\left[\frac{1}{1+0.2z^{-1}}\right] + \frac{200}{39}\left[\frac{1}{1+0.3z^{-1}}\right]$$

and

$$Y_3(z) = \frac{30}{13}\left[\frac{1}{1-z^{-1}}\right] - 10\left[\frac{1}{1+0.2z^{-1}}\right] + \frac{100}{39}\left[\frac{1}{1+0.3z^{-1}}\right]$$

Computing the inverse z-transforms, we obtain

$$y_{1T}(n) = 10\delta_T(n) - 10(-0.2)^n u_T(n)$$

$$y_{2T}(n) = \frac{10}{3}\delta_T(n) + \left[\frac{20}{13} - 10(-0.2)^n + \frac{200}{39}(-0.3)^n\right] u_T(n)$$

and

$$y_{3T}(n) = \left[\frac{30}{13} - 10(-0.2)^n + \frac{100}{13}(-0.3)^n\right] u_T(n)$$

as the output signals. Note that they match those obtained in part (b) of Example 16.1, as they should.

We see from Examples 16.1 and 16.2 that the use of z-transforms simplifies the solutions for the signals. In particular, it eliminates the need to perform the convolution sums, which may prove difficult if input signals are not unit pulses or steps.

16.4 Summary

State variables of a discrete-time system are internal signals and, possibly, output signals. The number of state variables needed to characterize a system is equal to the order of the system. The choice of state variables is not unique. In general, the most convenient set of state variables are the signals at the output of the delay operations in the system. This set of state variables characterizes the energy stored in the energy-storage elements of the system.

The $(n+1)$th sample of each system state variable is expressible as a weighted sum of the nth sample of the state variables and input signals. The resulting expressions are called the system state equations. The output signals are computed with output equations that express the output signals at sample number n in terms of a weighted sum of the state variables and input signals at sample number n.

Analysis techniques using state variables, state equations, and output equations constitute an organized approach for finding the output signals of multiple-input, multiple-output systems or of complicated single-input, single-output systems. They also provide information about internal system characteristics through the solutions found for the state variables.

The state equations and output equations used in the state-variable method of system analysis are most easily expressed and solved in matrix format. We can use either time-domain or z-transform techniques to solve these matrix equations to find the state variables and output signals for $n \geq n_0$, where n_0 is the initial time of interest. Use of these techniques produces the unit-pulse-response matrix $\mathbf{h}_T(n)$ and the transfer-function matrix $\mathbf{H}(z)$ of the system. The two matrices are a z-transform pair [that is, $\mathbf{h}_T(n) \leftrightarrow \mathbf{H}(z)$].

In the introductory treatment of the basic concepts associated with state variables and state-variable analysis of linear discrete-time systems in this chapter, our discussion built on concepts defined and discussed in Chapter 9 for continuous-time signals and systems. The full power of state-variable techniques for characterizing and analyzing discrete-time systems is not evident in an introductory treatment. Texts devoted to state variables can be consulted for further study.

Problems

16.1 Write a set of state equations corresponding to the difference equation

$$y_T(n) + 0.5y_T(n-1) + 0.04y_T(n-2) = x_T(n)$$

Also express them as a matrix state equation.

16.2 Repeat Problem 16.1 for the difference equation

$$y_T(n) + 0.5y_T(n-1) - 0.24y_T(n-2)$$
$$- 0.13y_T(n-3) = x_T(n) - x_T(n-1)$$

16.3 Repeat Problem 16.1 for the difference equation

$$y_T(n) - 0.4y_T(n-1) - 0.6y_T(n-2)$$
$$= 3x_T(n) + 2x_T(n-1)$$

16.4 Repeat Problem 16.1 for the difference equation

$$y_T(n) + 0.75y_T(n-1) - 0.11y_T(n-2)$$
$$= x_T(n) - 2x_T(n-1) + 0.5x_T(n-2)$$

16.5 The system shown in Figure 16.4 is a discrete-time filter used to enhance some of the features in a digital image received from a satellite. Identify the system state variables and write the matrix state and output equations for the system.

16.6 The block diagram shown in Figure 16.5 represents a system for processing two discrete-time control signals in a machine-tool control system. Identify the system state variables and write the matrix state and output equations for the system.

16.7 The block diagram shown in Figure 16.6 represents the discrete-time portion of a system for controlling two valve openings in a refinery piping system from two sensor signals. Identify the system state variables and write the matrix state and output equations for the system.

16.8 Consider the system represented by the block diagram shown in Figure 16.7. Identify the system

Figure 16.4

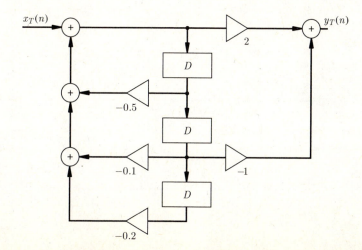

state variables and write the matrix state and output equations for the system.

16.9 Use the eigenvalue method to find the state transition matrix and impulse-response matrix for the

Figure 16.5

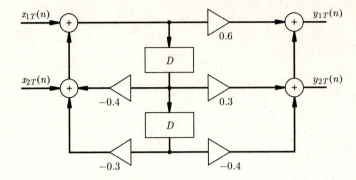

Figure 16.6

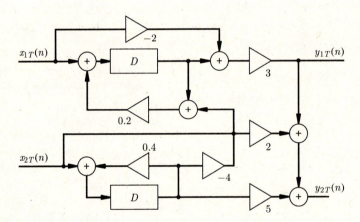

Figure 16.7

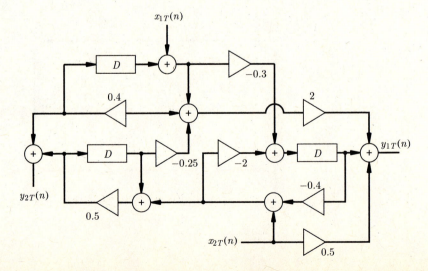

system having the following parameter matrices:

$$\mathbf{A} = \begin{bmatrix} 0.4 & -0.1 \\ 0.2 & 0.1 \end{bmatrix} \quad \mathbf{B} = \begin{bmatrix} 1 & 2 & 0 \\ 0 & -1 & 1 \end{bmatrix}$$

$$\mathbf{C} = [-3 \quad 1] \quad \mathbf{D} = [-1 \quad 0 \quad 1]$$

How many inputs, outputs, and states does the system have?

16.10 Repeat Problem 16.9 for the system with parameter matrices

$$\mathbf{A} = \begin{bmatrix} -0.3 & 0.2 \\ 0.1 & -0.4 \end{bmatrix} \quad \mathbf{B} = \begin{bmatrix} 2 \\ 1 \end{bmatrix}$$

$$\mathbf{C} = \begin{bmatrix} 1 & 0 \\ 0 & 1 \\ 1 & 0 \end{bmatrix} \quad \mathbf{D} = \begin{bmatrix} 0 \\ 0 \\ 1 \end{bmatrix}$$

16.11 Repeat Problem 16.9 using z-transforms.

16.12 Repeat Problem 16.10 using z-transforms.

16.13 Use time-domain analysis to find the output signals for $n \geq 0$ for the system of Problem 16.9 if the input signals are all $u_T(n)$ and all initial state values equal 3.

16.14 Use time-domain analysis to find the output signals for $n \geq 0$ for the system of Problem 16.10 if the input signals are all unit pulse functions and the initial state values equal -2.

16.15 Repeat Problem 16.13 using z-transform analysis techniques.

16.16 Repeat Problem 16.14 using z-transform analysis techniques.

16.17 Use z-transforms to find the output signals for the system of Problem 16.7 if $x_{1T}(n) = r_T(n)$, $x_{2T}(n) = u_T(n)$, and all initial state values equal zero.

16.18 Consider the system represented by the block diagram shown in Figure 16.8, where the state values at $n = 0$ are equal to zero. Use time-domain analysis to find $y_{1T}(n)$ and $y_{2T}(n)$ for $n \geq 0$ when $x_{1T}(n) = 4\delta(n - 2)$ and $x_{2T}(n) = -2u_T(n - 1)$.

16.19 Repeat Problem 16.18 using z-transform analysis techniques.

16.20 Repeat Problem 16.19 using z-transform analysis techniques and input signals of $x_{1T}(n) = (0.5)^n u_T(n)$ and $x_{2T}(n) = 3\delta_T(n)$.

16.21 Use state-variable techniques to find the output signals for $n \geq 0$ for the system of Problem 16.8 when $x_{1T}(n) = 2\delta_T(n)$, $x_{2T}(n) = -u_T(n)$, and all state values are equal to -2 at $n = 0$.

Figure 16.8

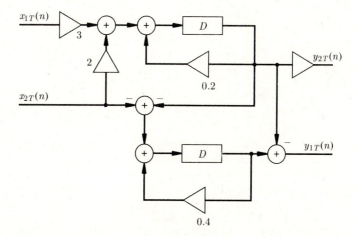

17 The Discrete Fourier Transform

The frequency content of a signal is indicated by its spectrum. We can find the spectrum of a continuous-time signal by computing the Fourier transform of the signal. Likewise, we can compute the discrete-time Fourier transform of a discrete-time signal to find the spectrum of the signal. Both of these spectrum computations produce continuous-frequency functions.

If we want to use a digital computer to compute signal spectral information, which is frequently the case, we can use only a finite number of signal samples. Also, we can compute only a finite number of samples of the continuous-frequency spectrum that correspond to either a continuous- or discrete-time signal. In this chapter, we will discuss the characteristics of the spectrum samples and their computation. We will start with a continuous-time signal and its spectra so that we can relate the spectrum samples obtained to the spectrum of the continuous-time signal. As we saw in Chapter 10, the results are directly applicable to discrete-time signals and their spectra because discrete-time signals can be represented as ideally sampled continuous-time signals. The computation of a finite-length sequence of spectrum samples $X(m)$ directly from a finite-length sequence of signal samples $x(n)$ is accomplished with the *discrete Fourier transform* (DFT) given by

$$X(m) = \sum_{n=0}^{N-1} x(n) W_N^{mn} \qquad 0 \le m \le N - 1 \qquad (17.1)$$

where $W_N = e^{-j2\pi/N}$. We derive this equation in Section 17.1.

Another use for the DFT is in the implementation of filters. The general block diagram of a continuous-time filter in which the DFT is used is shown in Figure 17.1. The signal-filtering steps are as follows:

1. Samples of the input signal, with spacing of T seconds, are produced by an A/D converter.
2. Samples, spaced by F Hz, of the spectrum corresponding to the input-signal samples are computed with the DFT.
3. The input-signal spectrum samples are multiplied by the desired frequency-response values, $H(m)$, to produce output-signal spectrum samples.
4. The output-signal samples are computed by using the inverse discrete Fourier transform (IDFT).
5. D/A conversion is used to produce the continuous-time signal.

It may appear that the required computation to implement this filter would be so extensive as to make the filter impractical. However, very efficient and rapid techniques have been developed for computing DFTs and IDFTs.

We can design the filter shown in Figure 17.1 to give a prescribed frequency response at some given sample frequencies. However, various characteristics

Figure 17.1
Continuous-Time
Filter Using
Spectrum-Sample
Processing

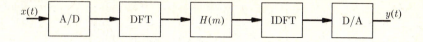

of the DFT produce spectrum-sample effects that must be understood before using the DFT. These effects are highlighted in our derivation and discussion of the DFT.

We will conclude Chapter 17 with an introduction to the *fast Fourier transform* (FFT). The FFT is a technique that permits rapid computation of the discrete Fourier transform. The time-saving approach is enabled by the symmetry of many of the computations required in the DFT.

17.1 The Discrete Fourier Transform and Inverse Discrete Fourier Transform

In our derivation of the discrete Fourier transform (DFT), we will use illustrations as an aid in presenting the mathematical concepts. The presentation is quite detailed in order to explain how various characteristics of the DFT arise. These characteristics are discussed further in Section 17.2, where they are also summarized.

Derivation of the DFT

We will first outline the DFT derivation and then examine the various steps in detail. Let us begin by considering a continuous-time signal and its spectrum. We use ideal sampling of the continuous-time signal with sample spacing T to produce signal samples that can be used in the DFT. These samples could also represent a discrete-time signal. The spectrum of the sampled signal is a periodic continuous-frequency function with period $f_s = 1/T$.

We then use ideal sampling to produce samples of the spectrum of the sampled signal. We choose an integer number, N, of samples per period of the spectrum, so that the spectrum samples are periodic with period N and thus we need only to compute N samples to characterize the spectrum. These spectrum samples have sample spacing $F = f_s/N = 1/NT$ and are the spectrum samples computed by the DFT.

We next find the signal that corresponds to the spectrum samples. It is $1/F$ times the original sampled signal plus $1/F$ times the original sampled signal shifted in time by all multiples of $t_p = 1/F = NT$. F times this signal is the periodic extension with period t_p, of the original sampled signal. If the original sampled signal is longer than t_p, then it must be truncated so that no sample overlap (time aliasing) will occur in the periodic extension. That is, no more than N signal samples can be used.

Finally, we simplify the notation to obtain the DFT equation (eq. 17.1) that we use to compute the N samples $X(m)$ of the spectrum of a signal from the first N samples of the periodic extension, $x(n)$, of the signal. We now examine the indicated steps in detail.

The Continuous-Time Signal and Its Spectrum   We begin by considering a continuous-time signal, $x(t)$, and its continuous-frequency spectrum, $X(f)$, so that we can relate the spectrum samples generated by the DFT to the spectrum of a corresponding continuous-time signal. For illustration purposes, the continuous-time signal and corresponding spectrum sketches shown in Figure 17.2 are used throughout this derivation. As shown, the signal width is t_s seconds. We assume that the amplitudes of signal components at frequencies greater than B Hz are negligible, so that the non-negligible-component bandwidth of the signal is B Hz.

The Sampled Signal and Its Spectrum   We ideally sample the continuous-time signal $x(t)$ to obtain the ideally sampled signal

$$x_s(t) = x(t)\delta_s(t) = x(t)\left[\sum_{n=-\infty}^{\infty} \delta(t - nT)\right]$$

$$= \sum_{n=-\infty}^{\infty} x(nT)\delta(t - nT) \tag{17.2}$$

Figure 17.2
Continuous-Time
Signal and
Corresponding
Spectrum to Illustrate
the DFT Derivation

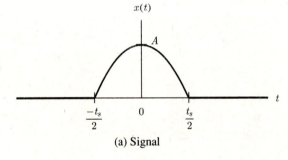

(a) Signal

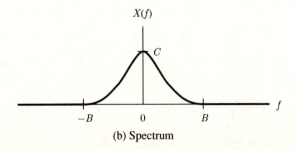

(b) Spectrum

where T is the sample spacing produced by the sampling rate $f_s = 1/T$ and $x(nT)$ are the sample values of the signal. Note that eq. (17.2) is also the ideally sampled continuous-time-signal representation of the discrete-time signal $x_T(n) = \{x(nT), \quad T\}$. Thus, all of the following derivation also applies to a discrete-time signal specified by the sequence of values $x(nT)$ and sample spacing T. The signal defined by eq. (17.2) is illustrated in Figure 17.3a for the case in which $x(t)$ is the signal of Figure 17.2. Note that there are M nonzero samples in the ideally sampled signal, where $(M-1)T \leq t_s < MT$, and that the first nonzero sample occurs at $t = aT$. Thus, in general, eq. (17.2) can be written as

$$x_s(t) = \sum_{n=a}^{a+M-1} x(nT)\delta(t - nT) \tag{17.3}$$

Note that, for illustrative purposes, we have chosen $a = -3$ and $M = 7$.

The spectrum of the ideally sampled, continuous-time signal is

$$X_s(f) = \mathcal{F}[x_s(t)] = \mathcal{F}[x(t)\delta_s(t)]$$

$$= X(f) * \left[f_s \sum_{k=-\infty}^{\infty} \delta(f - kf_s) \right]$$

$$= \sum_{k=-\infty}^{\infty} \frac{1}{T} X(f - kf_s) \tag{17.4}$$

Figure 17.3
Ideally Sampled
Version of the Signal
of Figure 17.2 and
Its Spectrum

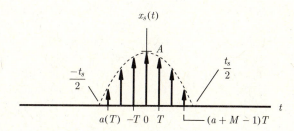

(a) Signal

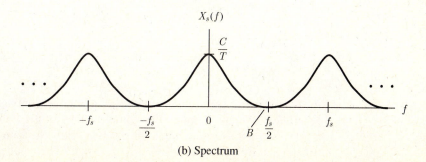

(b) Spectrum

where $f_s = 1/T$. The spectrum is shown in Figure 17.3b for the signal $x(t)$ of Figure 17.2. As previously noted (see Chapter 11), the spectrum of an ideally sampled continuous-time signal is periodic with period f_s. The ideally sampled continuous-time signal possesses negligible frequency-component aliasing when $f_s > 2B$ if, as we have assumed, B is the nonnegligible-component bandwidth of the continuous-time signal. Note that, with negligible aliasing,

$$X_s(f) \doteq \frac{1}{T} X(f) \qquad |f| < f_s/2 \tag{17.5}$$

Direct evaluation of the Fourier transform of $x_s(t)$ yields the following alternate expression for the spectrum of the ideally sampled continuous-time signal:

$$
\begin{aligned}
X_s(f) &= \int_{-\infty}^{\infty} x_s(t) e^{-j2\pi ft}\, dt \\
&= \int_{-\infty}^{\infty} \left[\sum_{n=a}^{a+M-1} x(nT)\delta(t - nT) \right] e^{-j2\pi ft}\, dt \\
&= \sum_{n=a}^{a+M-1} x(nT) \int_{-\infty}^{\infty} \delta(t - nT) e^{-j2\pi ft}\, dt \\
&= \sum_{n=a}^{a+M-1} x(nT) e^{-j2\pi nfT} \tag{17.6}
\end{aligned}
$$

Samples of the Sampled-Signal Spectrum We can produce samples of the spectrum $X_s(f)$ of the ideally sampled continuous-time signal by ideally sampling $X_s(f)$. Since $X_s(f)$ is periodic with period f_s, then it is convenient to sample $X_s(f)$ at a rate that produces N samples per period, where N is an integer. The resulting spectrum samples are then periodic, and we need only to compute N of them to completely specify the spectrum samples of $X_s(f)$. With N samples per period, the spacing between the samples is

$$F = f_s/N = 1/NT \tag{17.7}$$

and the ideally sampled spectrum of the ideally sampled continuous-time signal is

$$X_{ss}(f) = X_s(f) \left[\sum_{m=-\infty}^{\infty} \delta(f - mF) \right]$$

$$= \sum_{m=-\infty}^{\infty} X_s(mF)\delta(f - mF) \tag{17.8}$$

We illustrate this spectrum in Figure 17.4 for the continuous-time signal of Figure 17.2. We can compute the sample values $X_s(mF)$ of the spectrum of the ideally sampled continuous-time signal from the sample values, $x(nT)$, of the signal by using eq. (17.6) with $f = mF$. The result is

$$X_s(mF) = \sum_{n=a}^{a+M-1} x(nT)e^{-j2\pi nmFT} \tag{17.9}$$

The Signal Corresponding to the Sampled Spectrum   The next step in the derivation of the DFT is the determination of the signal that corresponds to the ideally sampled spectrum of the ideally sampled continuous-time signal. This signal is

$$x_{ss}(t) = \mathcal{F}^{-1}[X_{ss}(f)] \tag{17.10}$$

From eq. (17.8),

$$X_{ss}(f) = X_s(f)\left[\sum_{m=-\infty}^{\infty} \delta(f - mF)\right] \tag{17.11}$$

Therefore, we use Table C.2, the convolution and duality theorems for the Fourier transform, and eq. (17.7) to obtain

$$x_{ss}(t) = x_s(t) * \mathcal{F}^{-1}\left[\sum_{m=-\infty}^{\infty} \delta(f - mF)\right]$$

$$= x_s(t) * \left[\frac{1}{F}\sum_{k=-\infty}^{\infty} \delta\left(t - \frac{k}{F}\right)\right]$$

$$= x_s(t) * \sum_{k=-\infty}^{\infty} \frac{1}{F}\delta(t - kNT)$$

$$= \sum_{k=-\infty}^{\infty} \frac{1}{F}x_s(t - kNT) \tag{17.12}$$

This signal is illustrated in Figure 17.5 for the continuous-time signal of Figure 17.2. It is apparent that $x_{ss}(t)$ is $1/F$ times a sum of time-shifted versions of the ideally sampled signal $x_s(t)$. It is periodic with period $t_p = NT$.

Figure 17.4
Ideally-Sampled
Spectrum of the
Ideally Sampled
Version of the
Continuous-Time
Signal of Figure 17.2

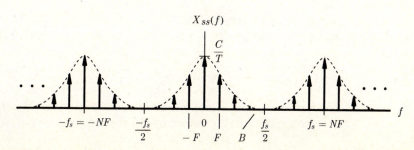

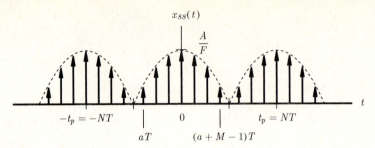

The Periodic Extension of the Sampled Signal   We define the periodic extension of $x_s(t)$ as

$$x_{sp}(t) = Fx_{ss}(t) = \sum_{k=-\infty}^{\infty} x_s(t - kNT) \tag{17.13}$$

Note that $x_{sp}(t)$ is a sum of time-shifted versions of the ideally sampled signal $x_s(t)$, and is periodic with period $t_p = NT$. Note, too, that no overlap of the time-shifted versions (time aliasing) occurs if $(a + M - 1)T - aT < NT$; that is, if $M - 1 < N$ or, equivalently, if $M \leq N$. When no overlap occurs, we can recover the original signal samples from the spectrum samples as the N samples of $x_{sp}(t)$ in one period. Therefore, we can use no more than N signal samples in computing N spectrum samples if the original signal samples are to be recoverable from the spectrum samples. Otherwise, time aliasing would occur in the periodic extension $x_{sp}(t)$ corresponding to the samples of the spectrum of $x_s(t)$.

We can also write the periodic extension of $x_s(t)$ as

$$x_{sp}(t) = \sum_{n=-\infty}^{\infty} x_p(nT)\delta(t - nT) \tag{17.14}$$

where the values $x_p(nT)$ are the strengths of the impulses and are periodic with period N. If $M \leq N$, then

$$x_p(nT) = x(nT) \qquad a \leq n \leq a + N - 1 \tag{17.15}$$

and repeats outside the interval $a \leq n \leq a + N - 1$.

Signal Truncation   We always use N signal samples when computing N spectrum samples to avoid the time-aliasing problem. If $t_s \geq t_p = NT$, then we truncate the original sampled signal to give the samples

$$x_t(nT) = \begin{cases} x(nT) & a \leq n \leq a + N - 1 \\ 0 & \text{elsewhere} \end{cases} \tag{17.16}$$

before computing the spectrum samples. Therefore, the spectrum samples that we compute actually correspond to the spectrum of a truncated version of the ideally sampled signal. The truncation produces no change in the spectrum samples if $t_s < t_p = NT$ because $x_t(nT) = x(nT)$ for all n in this case. The subscript t is added to the previously derived equations when they are used in the subsequent development to highlight the fact that, in general, the signal being sampled is truncated.

The Discrete Fourier Transform From eq. (17.9), the spectrum samples computed for a truncated signal that contains N samples are

$$X_{ts}(mF) = \sum_{n=a}^{a+N-1} x_t(nT)e^{-j2\pi mnFT}$$

$$= \sum_{n=a}^{a+N-1} x_{tp}(nT)e^{-j2\pi mn/N} \qquad (17.17)$$

where we obtain the last expression in eq. (17.17) by using eqs. (17.7) and (17.15). The periodic-extension sample values $x_{tp}(nT)$ and the values of $e^{-j2\pi mn/N}$ are both periodic in n, with period N. Therefore, we can use any set of N contiguous samples of the periodic extension of the truncated, ideally sampled continuous-time signal. It is convenient to use the N samples starting with $n = 0$. Also, for convenience, we define

$$e^{-j2\pi/N} \equiv W_N \qquad (17.18)$$

Therefore,

$$X_{ts}(mF) \equiv \sum_{n=0}^{N-1} x_{tp}(nT)W_N^{mn} \qquad (17.19)$$

The spectrum samples that we compute with eq. (17.19) are obtained from the first N samples of the periodic extension of the samples of the truncated continuous-time signal. Note from eq. (17.5), eq. (17.8), and Figure 17.4 that these spectrum samples are approximately $1/T$ times the samples of the spectrum of the truncated continuous-time signal over the interval $|f| \leq f_s/2$ if negligible spectrum aliasing occurs. That is,

$$X_{ts}(mF) \doteq X_t(mF)/T \qquad |m| \leq f_s/2F = 1/2FT = N/2 \qquad (17.20)$$

where the values of $X_t(mF)$ are samples of the spectrum of the truncated continuous-time signal taken at a spacing of F Hz. From eq. (11.9) in Chapter 11 and eq. (17.17), we see that the spectrum samples that we compute are equal to spectrum samples of a truncated discrete-time signal having the samples $x_{tT}(n) = x_t(nT)$. In addition, if $x_{tT}(n)$ are samples that span one period of a periodic discrete-time signal, then the spectrum samples computed are equal to the strength of the impulses in the spectrum of the periodic discrete-time signal [see eq. (11.62)] because the periodic extension of $x_{tT}(n)$ is equal to the periodic discrete-time signal.

Since $X_{ts}(mF)$ is periodic with period N, then only N samples of $X_{ts}(mF)$ must be computed. The samples that we compute are the samples in the range

$0 \leq m \leq N - 1$. All other spectrum samples can be obtained from these samples because of the spectrum periodicity. We suppress the subscripts on the signal and spectrum samples in eq. (17.19) and replace the arguments mF and nT by m and n, respectively, for simplicity. Equation (17.19) then becomes

$$X(m) = \sum_{n=0}^{N-1} x(n)W_N^{mn} \qquad 0 \leq m \leq N - 1 \qquad (17.21)$$

This equation is called the *discrete Fourier transform of the N sample sequence x(n)*. It is frequently called an *N-point discrete Fourier transform*.

Since the explicit notation has been suppressed in eq. (17.21), it is important to remember that $x(n)$ is the periodic extension, with period N, of the truncated signal samples, and that the first N samples of the periodic extension are used in the DFT computation. Also, it must be remembered that the computed values of $X(m)$ consist of the first N samples of the periodic spectrum corresponding to samples of the truncated signal. The spectrum samples computed encompass one period of the periodic spectrum.

Now, let us assume that $x(n)$ is the periodic extension of samples of a truncated continuous-time signal, and that the samples are obtained at a high enough sampling rate so that spectrum aliasing is negligible. In this case, the spectrum samples obtained with the DFT for $|m| \leq N/2$ are approximately equal to $1/T$ times the samples of the spectrum of the truncated continuous-time signal. Note that spectrum aliasing always occurs when samples of a continuous-time signal are considered because the sequence $x(n)$ has finite extent. This means that $x(n)$ corresponds to samples of a continuous-time signal that is time-limited, and thus cannot be bandlimited.

Qualitative Illustration of the DFT   We will now illustrate the computation of a DFT by considering the continuous-time signal and spectrum shown in Figure 17.6. Note that the spectrum is not the actual spectrum of the signal shown, but rather a simple general shape that can be easily drawn. It is assumed that the spectrum amplitude is negligible for $|f| > 4/\tau$.

We first consider the case where a 4-point DFT is used to compute spectrum samples. To avoid truncation of the signal, we choose $t_p = NT = \tau$. The resulting signal-sample spacing is $T = t_p/N = \tau/4$, and the spectrum-sample spacing is $F = 1/NT = 1/t_p = 1/\tau$. The signal samples that we take, $x(nT)$, the periodic extension of these signal samples, $x(n)$, and the samples that we use in the DFT computation are shown in Figure 17.7a. Note that the signal samples used consist of the first N samples of the periodic extension of $x(nT)$. The spectrum samples are computed with

$$X(m) = \sum_{n=0}^{3} x(n)W_4^{mn} \qquad 0 \leq m \leq 3 \qquad (17.22)$$

The computed spectrum samples are shown in Figure 17.7b. Since these samples are for one period of the sampled-signal spectrum, they are repeated to

Figure 17.6
Continuous-Time
Signal and Spectrum
for Qualitative
Illustration of the
DFT

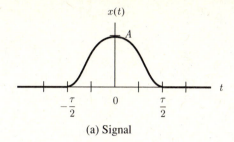

(a) Signal

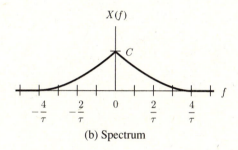

(b) Spectrum

show samples of the sampled-signal spectrum. The spectrum of the sampled signal is

$$X_s(f) = \sum_{k=-\infty}^{\infty} \frac{1}{T} X(f - kf_s) \qquad (17.23)$$

and is also shown in Figure 17.7b along with the computed spectrum samples corresponding to samples of $X_s(f)$ over the frequency range $|f| \leq f_s/2$. Equation (17.20) shows that multiplication of these samples by T produces samples of the continuous-time-signal spectrum $X(f)$ if there is negligible spectrum aliasing. Note that there is significant spectrum aliasing in Figure 17.7b.

To reduce the spectrum aliasing, we need to increase the signal sampling rate. An increased sampling rate is illustrated by signal- and spectrum-sample plots in Figure 17.8. In Figure 17.8, we increase N to 8 and maintain t_p equal to τ so that the sampling rate is doubled and there is no signal truncation. In this case, $T = t_p/N = \tau/8$, $f_s = 1/T = 8/\tau$, and $F = 1/t_p = 1/\tau$. Note that the spectrum-sample spacing is unchanged because the length of signal sampled (that is, $NT = t_p$) is unchanged. Only negligible spectrum aliasing exists. Therefore, the spectrum samples obtained for $|f| \leq f_s/2$ are approximately equal samples of $X(f)/T$ and can be multiplied by T to produce approximate samples of the continuous-time-signal spectrum.

Derivation of the IDFT

We must be able to compute the signal-sample values from the spectrum-sample values if we want to use the DFT in signal-analysis and signal-processing

Figure 17.7
Signal Samples Used
and Spectrum
Samples Computed
for the DFT of the
Signal of Figure 17.6
with $N = 4$ and
$NT = \tau$

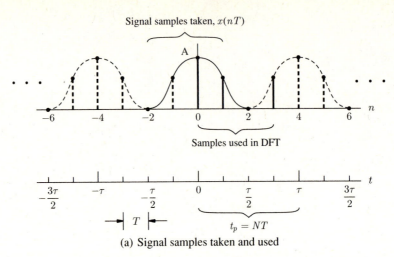

(a) Signal samples taken and used

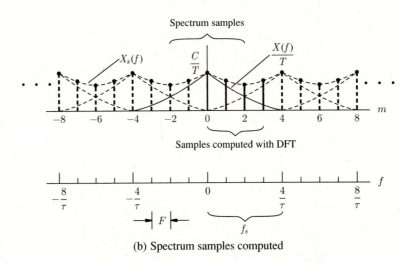

(b) Spectrum samples computed

applications. Therefore, we need an *inverse discrete Fourier transform* (IDFT) to compute signal-sample values from spectrum-sample values.

We begin the derivation of the IDFT by first noting from eq. (17.10) that the ideally sampled spectrum $X_{tss}(f)$ of the ideally sampled truncated signal $x_{ts}(t)$ corresponds to the ideally sampled signal $x_{tss}(t)$. Therefore, we can write the ideally sampled spectrum as

$$X_{tss}(f) = \mathcal{F}[x_{tss}(t)]$$

$$= \int_{-\infty}^{\infty} x_{tss}(t)e^{-j2\pi ft}\,dt \qquad (17.24)$$

Figure 17.8
Signal Samples Used
and Spectrum
Samples Computed
for the DFT of the
Signal of Figure 17.6
with $N = 8$ and
$NT = \tau$

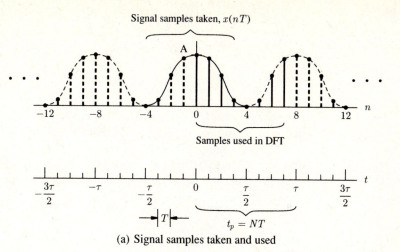

(a) Signal samples taken and used

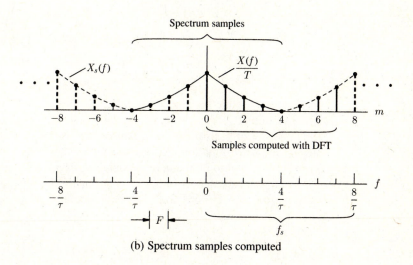

(b) Spectrum samples computed

From eqs. (17.13) and (17.14),

$$x_{tss}(t) = \frac{1}{F}x_{tsp}(t)$$

$$= \frac{1}{F}\sum_{n=-\infty}^{\infty} x_{tp}(nT)\delta(t - nT) \qquad (17.25)$$

Substituting eq. (17.25) in eq. (17.24), we obtain

$$X_{tss}(f) = \int_{-\infty}^{\infty} \left[\frac{1}{F}\sum_{n=-\infty}^{\infty} x_{tp}(nT)\delta(t - nT)\right]e^{-j2\pi ft}\,dt$$

$$= \frac{1}{F} \sum_{n=-\infty}^{\infty} x_{tp}(nT) \int_{-\infty}^{\infty} \delta(t - nT)e^{-j2\pi ft}\, dt$$

$$= \sum_{n=-\infty}^{\infty} \frac{1}{F} x_{tp}(nT)e^{-j2\pi nfT} \tag{17.26}$$

The ideally sampled spectrum $X_{tss}(f)$ consists of impulses and is a periodic function of f, with period $f_s = 1/T$. Even though $X_{tss}(f)$ does not contain finite energy in a one-period interval, we can express it as a complex-exponential Fourier series in the limit (a concept similar to the Fourier transform in the limit), where the independent variable is f. This Fourier series in the limit is

$$X_{tss}(f) = \sum_{n=-\infty}^{\infty} k_n e^{-j2\pi n(1/f_s)f} = \sum_{n=-\infty}^{\infty} k_n e^{-j2\pi nfT} \tag{17.27}$$

where

$$k_n = \frac{1}{f_s} \int_{-F/2}^{f_s-F/2} X_{tss}(f)e^{j2\pi n(1/f_s)f}\, df$$

$$= T \int_{-F/2}^{f_s-F/2} X_{tss}(f)e^{j2\pi nfT}\, df \tag{17.28}$$

Note that we use the Fourier series basis functions $e^{-j2\pi n(1/f_s)f}$ rather than the basis functions $e^{j2\pi n(1/f_s)f}$ that we used in Chapter 4. The same series results with either set of basis functions because n spans both the positive and negative integers in the summation. Also note that we have chosen the Fourier series expansion interval to be the period extending from $f = -F/2$ to $f = f_s - F/2$ to avoid the occurrence of impulse functions on the integration boundaries.

Comparing eqs. (17.26) and (17.27), we note that $k_n = x_{tp}(nT)/F$. Therefore,

$$x_{tp}(nT) = FT \int_{-F/2}^{f_s-F/2} X_{tss}(f)e^{j2\pi nfT}\, df \tag{17.29}$$

Equation (17.8) gives the expression

$$X_{tss}(f) = \sum_{m=-\infty}^{\infty} X_{ts}(mF)\delta(f - mF) \tag{17.30}$$

for the ideally sampled spectrum of the ideally sampled truncated signal. Therefore,

$$x_{tp}(nT) = \frac{1}{N} \int_{-F/2}^{f_s-F/2} \left[\sum_{m=-\infty}^{\infty} X_{ts}(mF)\delta(f - mF) \right] e^{j2\pi nfT}\, df$$

$$= \frac{1}{N} \sum_{m=-\infty}^{\infty} X_{ts}(mF) \int_{-F/2}^{f_s-F/2} \delta(f - mF)e^{j2\pi nfT}\, df \tag{17.31}$$

since $FT = 1/N$. The impulse is within the interval of integration only if $0 \le mF < f_s$. Therefore, the only terms of the summation that can be nonzero are those corresponding to $0 \le m < 1/FT = N$. Therefore,

$$x_{tp}(nT) = \frac{1}{N} \sum_{m=0}^{N-1} X_{ts}(mF) e^{j2\pi mnFT}$$

$$= \frac{1}{N} \sum_{m=0}^{N-1} X_{ts}(mF) e^{j2\pi mn/N}$$

$$= \frac{1}{N} \sum_{m=0}^{N-1} X_{ts}(mF) W_N^{-mn} \qquad (17.32)$$

where we obtain the last equivalence by using the definition for W_N given by eq. (17.18).

The sequence $x_{tp}(nT)$ consists of samples of the periodic extension of the truncated signal. The periodic extension has a period of $t_p = NT$, which means that $x_{tp}(nT)$ has a period of N. Therefore, only N samples of $x_{tp}(nT)$ must be computed. The samples that we compute are for $0 \le n \le N - 1$. Again, we suppress the subscripts and replace the arguments nT and mF by n and m, respectively, to produce

$$x(n) = \frac{1}{N} \sum_{m=0}^{N-1} X(m) W_N^{-mn} \qquad 0 \le n \le N - 1 \qquad (17.33)$$

as the *inverse discrete Fourier transform.*

As we noted for the discrete Fourier transform, we must remember that $x(n)$ is the periodic extension, with period N, that corresponds to the truncated signal samples. We compute the first N samples of this periodic extension. Also, the values of $X(m)$ that we use in eq. (17.33) consist of the first N samples of the periodic spectrum that corresponds to samples of the truncated signal. The spectrum samples that we use encompass one period of the periodic spectrum.

Since the IDFT produces the first N samples of the periodic extension of the sampled signal that corresponds to the spectrum samples, then it is most convenient to choose the time origin so that this set of samples encompasses the signal samples in order. That is, if possible, the time origin should be chosen so that the initial signal sample occurs at $t \ge 0$ and the final signal sample occurs at $t < NT$. With this choice of time origin, the IDFT produces the corresponding signal samples directly and is easily interpreted. In many cases, we can select this time origin because the time origin is not critical.

The Discrete Fourier Transform Pair

In summary, the discrete Fourier transform pair is

$$X(m) = \sum_{n=0}^{N-1} x(n) W_N^{mn} \qquad 0 \le m \le N - 1 \qquad (17.34)$$

$$x(n) = \frac{1}{N} \sum_{n=0}^{N-1} X(m) W_N^{-mn} \qquad 0 \le n \le N - 1 \qquad (17.35)$$

where

$$W_N = e^{-j2\pi/N} \qquad (17.36)$$

The sequence $x(n)$ is the periodic extension, with period N, that corresponds to the truncated-signal samples. We use, or compute, the first N samples of this periodic extension. The sequence $X(m)$ is periodic with period N and consists of the samples of the periodic spectrum that corresponds to the truncated-signal samples. We compute, or use, the first N samples of $X(m)$. Note that W_N^{mn} is periodic in m with period N and periodic in n with period N.

Many times the subscript on W is dropped. It is redundant in the preceding DFT and IDFT equations because the summation limit indicates the number of samples being considered. However, in some of the subsequent discussion, we will consider DFTs of different lengths in the same derivation or example. In this case, the explicit indication of the subscript is necessary.

The $1/N$ factor associated with the IDFT could be associated with the DFT instead. In this case, the spectrum samples are $1/NT = F$ times the continuous-time-signal spectrum values for $|m| \le N/2$, rather than $1/T$ times these values. This form appears sometimes in the literature.

In addition, W_N is sometimes defined as $e^{j2\pi/N}$ rather than as $e^{-j2\pi/N}$. This definition is used if the Fourier transform is defined using $e^{j2\pi ft}$ rather than $e^{-j2\pi ft}$. It results in a continuous-time-signal spectrum and spectrum samples that are reversed on the frequency axis. If software is available to compute the DFT or IDFT, then the user must be aware of the particular form being used, so that the results can be properly interpreted.

To indicate the DFT, we will use simplified notations similar to those we used for other transforms. These notations include

$$X(m) = \mathcal{D}[x(n)] \quad \text{and} \quad x(n) = \mathcal{D}^{-1}[X(m)] \qquad (17.37)$$

and

$$x(n) \leftrightarrow X(m) \qquad (17.38)$$

Example Transform Computation

We will now consider a simple continuous-time signal to illustrate the computation of spectrum samples with the DFT. We will also illustrate how the IDFT is used in computing the signal samples from the spectrum samples.

Example 17.1

Consider the continuous-time rectangular-pulse signal $x(t) = (t - 0.1)/0.8$. Sample $x(t)$ at a rate of five samples per second and (a) use a 6-point DFT to compute samples, $X(m)$, of the spectrum of the sampled signal; (b) plot the amplitude and phase of $X(f)$ and of $TX(m)$ for $0 \le f < f_s$; and (c) compute the IDFT of $X(m)$.

Solution
The signal samples are shown in Figure 17.9a.

a. To compute a 6-point DFT, the six signal samples starting at $t = -0.4$ are taken so as to preclude signal truncation. The first N samples (that is, six samples) of the periodic extension of the signal samples taken are shown in Figure 17.9b. We use these samples to compute the DFT.

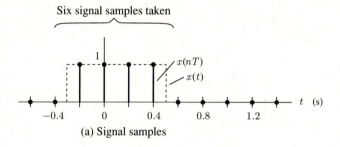

(a) Signal samples

(b) Periodic-extension samples used to compute DFT

Figure 17.9 Signal Samples Taken and Used to Compute the DFT in Example 17.1

Since $N = 6$, $W_N = W_6 = e^{-j2\pi/6} = e^{-j\pi/3}$. Therefore,

$$W_6^0 = 1 \qquad W_6^1 = \frac{1}{2} - j\frac{\sqrt{3}}{2} \qquad W_6^2 = -\frac{1}{2} - j\frac{\sqrt{3}}{2} \qquad W_6^3 = -1$$

$$W_6^4 = -\frac{1}{2} + j\frac{\sqrt{3}}{2} \qquad W_6^5 = \frac{1}{2} + j\frac{\sqrt{3}}{2}$$

Since W_6^i is periodic with period 6, it repeats after W_6^5. The samples of the spectrum of the sampled signal are

$$X(m) = \sum_{n=0}^{5} x(n)W_6^{mn}$$

Therefore,

$$X(0) = \sum_{n=0}^{5} x(n)W_6^0 = \sum_{n=0}^{5} x(n) = 4$$

$$X(1) = \sum_{n=0}^{5} w(n)W_6^n$$

$$= x(0)W_6^0 + x(1)W_6^1 + x(2)W_6^2 + x(3)W_6^3 + x(4)W_6^4 + x(5)W_6^5$$

$$= (1)(1) + (1)\left(\frac{1}{2} - j\frac{\sqrt{3}}{2}\right) + (1)\left(-\frac{1}{2} - j\frac{\sqrt{3}}{2}\right) + (0)(-1)$$

$$+ (0)\left(-\frac{1}{2} + j\frac{\sqrt{3}}{2}\right) + (1)\left(\frac{1}{2} + j\frac{\sqrt{3}}{2}\right) = \frac{3}{2} - j\frac{\sqrt{3}}{2} = \sqrt{3}e^{-j\pi/6}$$

$$X(2) = \sum_{n=0}^{5} w(n)W_6^{2n}$$

$$= x(0)W_6^0 + x(1)W_6^2 + x(2)W_6^4 + x(3)W_6^6 + x(4)W_6^8 + x(5)W_6^{10}$$

$$= x(0)W_6^0 + x(1)W_6^2 + x(2)W_6^4 + x(3)W_6^0 + x(4)W_6^2 + x(5)W_6^4$$

$$= -\frac{1}{2} + j\frac{\sqrt{3}}{2} = e^{j2\pi/3}$$

$$X(3) = \sum_{n=0}^{5} x(n)W_6^{3n} = \sum_{n=0}^{5} x(n)e^{-j6\pi n/6} = \sum_{n=0}^{5} x(n)(-1)^n$$

$$= x(0) - x(1) + x(2) - x(3) + x(4) - x(5) = 0$$

$$X(4) = \sum_{n=0}^{5} w(n)W_6^{4n}$$

$$= x(0)W_6^0 + x(1)W_6^4 + x(2)W_6^2 + x(3)W_6^0 + x(4)W_6^4 + x(5)W_6^2$$

$$= -\frac{1}{2} - j\frac{\sqrt{3}}{2} = e^{-j2\pi/3}$$

$$X(5) = \sum_{n=0}^{5} w(n)W_6^{5n}$$

$$= x(0)W_6^0 + x(1)W_6^5 + x(2)W_6^4 + x(3)W_6^3 + x(4)W_6^2 + x(5)W_6^1$$

$$= \frac{3}{2} + j\frac{\sqrt{3}}{2} = \sqrt{3}e^{j\pi/6}$$

b. The signal-sample spacing is

$$T = 1/f_s = 1/5 = 0.2 \text{ s}$$

and the spectrum-sample spacing is

$$F = 1/NT = f_s/N = 5/6 \text{ Hz}$$

Using Tables C.1 and C.2, the continuous-time-signal spectrum is

$$X(f) = 0.8 \text{ sinc}(0.8f)e^{-j2\pi f(0.1)}$$

Therefore,

f	m	$X(mF)$	$TX(m)$
0	0	0.8	0.8
5/6	1	$0.3308e^{-j\pi/6}$	$0.3464e^{-j\pi/6}$
10/6	2	$0.1654e^{j2\pi/3}$	$0.2e^{j2\pi/3}$
15/6	3	0	0
20/6	4	$0.1372e^{-j2\pi/3}$	$0.2e^{-j2\pi/3}$
25/6	5	$0.0827e^{j\pi/6}$	$0.3464e^{j\pi/6}$

The amplitude of $X(f)$ and $TX(m)$ are plotted for $0 \le f < f_s = 5$ Hz in Figure 17.10a. Likewise, the phase of $X(f)$ and $TX(m)$ are plotted in Figure 17.10b. The values of the samples $|TX(m)|$ are not equal to the values of $|X(f)|$ for all frequencies lower than $f_s/2 = 2.5$ Hz because the signal sample rate is too slow, resulting in spectrum aliasing. The values of $\angle TX(m)$ do match the values of $\angle X(f)$ in the frequency range $f \le f_s/2$.

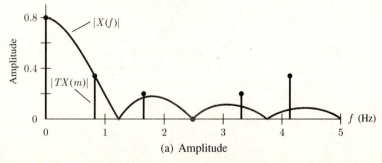

(a) Amplitude

Figure 17.10 Amplitude and Phase of the Spectrum and Spectrum Samples for Example 17.1

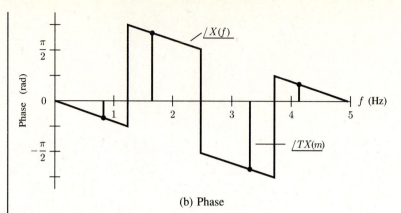

(b) Phase

Figure 17.10 Amplitude and Phase of the Spectrum and Spectrum Samples for Example 17.1 *(continued)*

c. The IDFT is computed with

$$x(n) = \frac{1}{6} \sum_{m=0}^{5} X(m) W_6^{-mn}$$

where

$$W_6^0 = 1 \qquad W_6^{-1} = \frac{1}{2} + j\frac{\sqrt{3}}{2} \qquad W_6^{-2} = -\frac{1}{2} + j\frac{\sqrt{3}}{2}$$

$$W_6^{-3} = -1 \qquad W_6^{-4} = -\frac{1}{2} - j\frac{\sqrt{3}}{2} \qquad W_6^{-5} = \frac{1}{2} - j\frac{\sqrt{3}}{2}$$

Since W_6^{-i} is periodic with period 6, it repeats after W_6^{-5}.

The signal samples computed are

$$x(0) = \frac{1}{6} \sum_{m=0}^{5} X(m) W_6^0 = \frac{1}{6} \sum_{m=0}^{5} X(m)$$

$$= \frac{1}{6} \left[4 + \left(\frac{3}{2} - j\frac{\sqrt{3}}{2} \right) + \left(-\frac{1}{2} + j\frac{\sqrt{3}}{2} \right) + (0) + \left(-\frac{1}{2} - j\frac{\sqrt{3}}{2} \right) \right.$$

$$\left. + \left(\frac{3}{2} + j\frac{\sqrt{3}}{2} \right) \right] = 1$$

$$x(1) = \frac{1}{6} \sum_{m=0}^{5} X(m) W_6^{-m}$$

$$= \frac{1}{6} \left[X(0) W_6^0 + X(1) W_6^{-1} + X(2) W_6^{-2} + X(3) W_6^{-3} \right.$$

$$\left. + X(4) W_6^{-4} + X(5) W_6^{-5} \right]$$

$$= \frac{1}{6}\left[(4)(1) + \left(\frac{3}{2} - j\frac{\sqrt{3}}{2}\right)\left(\frac{1}{2} + j\frac{\sqrt{3}}{2}\right)\right.$$

$$+ \left(-\frac{1}{2} + j\frac{\sqrt{3}}{2}\right)\left(-\frac{1}{2} + j\frac{\sqrt{3}}{2}\right) + (0)(-1)$$

$$\left. + \left(-\frac{1}{2} - j\frac{\sqrt{3}}{2}\right)\left(-\frac{1}{2} - j\frac{\sqrt{3}}{2}\right) + \left(\frac{3}{2} + j\frac{\sqrt{3}}{2}\right)\left(\frac{1}{2} - j\frac{\sqrt{3}}{2}\right)\right]$$

$$= 1$$

$$x(2) = \frac{1}{6}\sum_{m=0}^{5} X(m)W_6^{-2m}$$

$$= \frac{1}{6}\left[X(0)W_6^0 + X(1)W_6^{-2} + X(2)W_6^{-4} + X(3)W_6^{-6}\right.$$

$$\left. + X(4)W_6^{-8} + X(5)W_6^{-10}\right]$$

$$= \frac{1}{6}\left[X(0)W_6^0 + X(1)W_6^{-2} + X(2)W_6^{-4} + X(3)W_6^0\right.$$

$$\left. + X(4)W_6^{-2} + X(5)W_6^{-4}\right] = 1$$

$$x(3) = \frac{1}{6}\sum_{m=0}^{5} X(m)W_6^{-3m} = \frac{1}{6}\sum_{m=0}^{5} X(m)e^{-j\frac{6\pi m}{6}} = \frac{1}{6}\sum_{m=0}^{5} X(m)(-1)^m$$

$$= \frac{1}{6}[X(0) - X(1) + X(2) - X(3) + X(4) - X(5)] = 0$$

$$x(4) = \frac{1}{6}\sum_{m=0}^{5} X(m)W_6^{-4m}$$

$$= \frac{1}{6}\left[X(0)W_6^0 + X(1)W_6^{-4} + X(2)W_6^{-2} + X(3)W_6^0\right.$$

$$\left. + X(4)W_6^{-4} + X(5)W_6^{-2}\right] = 0$$

$$x(5) = \frac{1}{6}\sum_{m=0}^{5} X(m)W_6^{-5m}$$

$$= \frac{1}{6}\left[X(0)W_6^0 + X(1)W_6^{-5} + X(2)W_6^{-4} + X(3)W_6^{-3}\right.$$

$$\left. + X(4)W_6^{-2} + X(5)W_6^{-1}\right]$$

$$= 1$$

As expected, our computation of the IDFT has produced the first $N = 6$ samples of the periodic extension of the signal samples taken (see Figure 17.9b).

17.2 Discrete Fourier Transform Characteristics and Properties

If we are to use the DFT effectively, we must be aware of its characteristics and properties. In this section we will first summarize some of the general characteristics and then discuss symmetry properties that apply to the DFT.

General Characteristics

There are a number of general characteristics associated with the values of the signal samples used and the values of the spectrum samples computed with the DFT, some of which were pointed out in Section 17.1. We will summarize these characteristics here.

Signal Samples Used   If we want to compute an N-point DFT of samples of the continuous-time signal $x(t)$, then N signal samples are required with a sample spacing of T seconds [that is, $x(nT)$, $n_1 \leq n \leq N + n_1 - 1$]. If the length of the signal is greater than $t_p = NT$ seconds, then we use the samples $x_t(nT)$ of a truncated version of the signal. We define the periodic extension of the signal samples to be the sequence $x(n)$ and use the first N samples of this sequence to compute the values of the spectrum samples. These concepts are illustrated in Figure 17.11.

If the signal samples were previously taken and there are fewer than N of them, then we add samples of zero value to the end of the set of signal samples (zero padding) to give a total of N samples. The set of samples used to compute the DFT are the first N samples of the periodic extension of the zero-padded signal samples. These concepts are illustrated in Figure 17.12. Note that the length of the truncated signal is MT seconds and the length of the zero-padded truncated signal is NT seconds.

Spectrum Samples Computed   The spectrum samples that we compute with the DFT are the first N samples of the spectrum $X_{ts}(f)$ that corresponds to the sampled truncated signal $x_{ts}(t)$. They span one period of the spectrum, as shown in Figure 17.13. The spectrum samples for $0 \leq m \leq INT(N/2)$ correspond to the frequency interval $0 \leq f \leq f_s/2$, where $INT(N/2)$ is the integer part of $N/2$. These are samples of the spectrum of the sampled signal, or corresponding discrete-time signal, and are an aliased version of the samples of $X_t(f)/T$, where $X_t(f)$ is the spectrum of the truncated continuous-time signal corresponding to the signal samples used. We can see these spectrum-sample characteristics in the first five spectrum samples shown in Figure 17.13.

Figure 17.11
Examples of Signal
Samples Taken and
Used for DFT
Computation

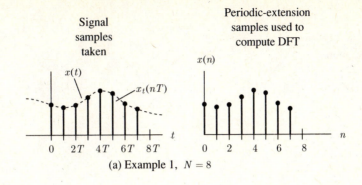

Signal
samples
taken

Periodic-extension
samples used to
compute DFT

(a) Example 1, $N = 8$

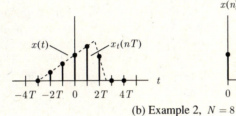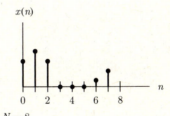

(b) Example 2, $N = 8$

Figure 17.12
Examples of
Signal-Sample Zero
Padding

M signal
samples
taken

N periodic-extension
samples used to
compute N-point DFT

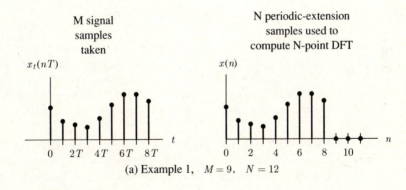

(a) Example 1, $M = 9$, $N = 12$

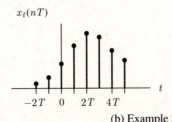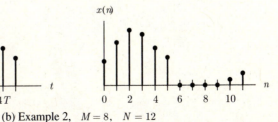

(b) Example 2, $M = 8$, $N = 12$

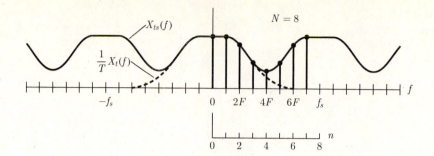

The spectrum samples for $INT(N/2) < m \leq N - 1$ when N is odd, or for $INT(N/2) \leq m \leq N - 1$ when N is even, correspond to the frequency interval $-f_s/2 \leq f < 0$ because $X(m)$ is periodic with period N, and thus $X(N - m) = X(-m)$. These samples are an aliased version of the samples of $X_t(f)/T$ for the frequency interval $-f_s/2 \leq f < 0$, as is apparent from the last four spectrum samples shown in Figure 17.13 and the periodic nature of $X_{ts}(f)$.

Spectrum-Sample Spacing The spacing of the spectrum samples is $F = 1/NT = 1/t_p$ Hz. Thus, it is the reciprocal of the length (including any zero samples added) of the signal being transformed.

Aliasing If the spectrum samples are to be computed for a continuous-time signal, then the chosen continuous-time-signal sampling rate, $f_s = 1/T$, must be large enough to minimize distortion due to aliasing in the spectrum-sample values computed. The required sampling rate is specified by the sampling theorem. Aliasing does not exist for a discrete-time signal.

Spectrum Leakage Distortion of the spectrum-sample values is obtained because a truncated signal is used. We refer to this distortion as spectrum-leakage distortion because it causes a spectrum value at one frequency to appear at other frequencies as well.

To illustrate spectrum leakage, consider the single-frequency cosine signal

$$x(t) = \cos(2\pi f_1 t) \tag{17.39}$$

We truncate the signal to 2-1/2 cycles, as illustrated by Figure 17.14a and expressed by

$$x_t(t) = \cos(2\pi f_1 t) \qquad -\frac{5}{4f_1} \leq t \leq \frac{5}{4f_1}$$

$$= \cos(2\pi f_1 t) \, \Pi\!\left(\frac{2f_1 t}{5}\right) \tag{17.40}$$

Figure 17.14
Truncated Cosine
Signal and Its
Spectrum

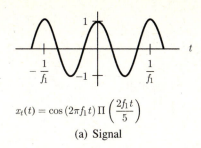

$$x_t(t) = \cos\left(2\pi f_1 t\right) \Pi\left(\frac{2f_1 t}{5}\right)$$

(a) Signal

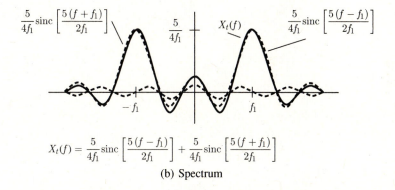

$$X_t(f) = \frac{5}{4f_1}\operatorname{sinc}\left[\frac{5\left(f - f_1\right)}{2f_1}\right] + \frac{5}{4f_1}\operatorname{sinc}\left[\frac{5\left(f + f_1\right)}{2f_1}\right]$$

(b) Spectrum

Using Tables C.1 and C.2, we find that the spectrum of the truncated signal is

$$X_t(f) = \frac{5}{4f_1}\operatorname{sinc}\left[\frac{5\left(f - f_1\right)}{2f_1}\right] + \frac{5}{4f_1}\operatorname{sinc}\left[\frac{5\left(f + f_1\right)}{2f_1}\right] \tag{17.41}$$

This spectrum is illustrated in Figure 17.14b.

The spectrum of the nontruncated cosine signal is concentrated in impulses at $f = f_1$ and $f = -f_1$ (refer back to Figure 4.42). Figure 17.14b shows us that truncation of the cosine signal causes its spectrum to spread over all frequencies. This spectrum spreading is spectrum leakage.

The spectrum leakage into frequencies in the vicinity of $f = f_1$ and $f = -f_1$ (that is, the large lobes in the spectrum) makes it difficult to determine if two closely adjacent frequencies are present in a signal. Thus, the width of these large lobes surrounding $f = f_1$ and $f = -f_1$ define the *spectrum resolution*. Note that better (narrower) spectrum resolution is achieved if a longer truncated portion of the signal is used because the large lobe width is inversely proportional to the truncated signal length.

The smaller lobes in the spectrum leakage are called *sidelobe leakage*; they show us that part of the truncated cosine-signal energy appears at frequencies that are significantly removed from the cosine-signal frequency. If a truncated signal contains components at several frequencies, then the signal contribution from a small signal component may be obscured in the signal spectrum by a large signal-component sidelobe.

Now let us assume that $x_t(t)$ is sampled at a rate $f_s = 1/T$ such that N samples occur over the truncation interval of the signal; that is,

$$NT = 5/2f_1 \qquad (17.42)$$

The spectrum of the sampled signal is

$$X_{ts}(f) = \sum_{i=-\infty}^{\infty} \frac{1}{T} X_t(f - if_s) = \sum_{i=-\infty}^{\infty} \frac{2Nf_1}{5} X_t(f - if_s) \qquad (17.43)$$

We illustrate this spectrum in Figure 17.15 for $|f| \leq 2f_1$, with the assumption that $f_s/2 \gg 2f_1$ so that aliasing is negligible. The spacing of the spectrum samples obtained with the DFT is

$$F = 1/NT = 2f_1/5 \qquad (17.44)$$

The spectrum samples are also shown in Figure 17.15. The signal with frequency f_1 shows up at a number of spectrum-sample locations, indicating spectrum leakage. In fact, it shows up only through leakage because no spectrum sample exists at $f = f_1$.

Picket-Fence Effect   Only samples of a continuous-frequency spectrum are produced by the DFT. Because looking at samples of a spectrum is like viewing the spectrum through a picket fence, we call this DFT characteristic the picket-fence effect. It can cause a problem if the spectrum-sample spacing is too wide to adequately indicate spectrum variations. The effect is apparent in the spectrum samples shown in Figure 17.15 for the truncated cosine signal. We can reduce the problem by decreasing the spectrum-sample spacing F. Since $F = 1/NT$, this requires increasing the sequence length, NT, of the signal samples used in computing the DFT. We can increase this length by adding zero samples to the end of the signal samples to increase N (a desirable approach when signal samples already exist), or taking additional samples of the original signal at the

Figure 17.15
Spectrum $X_{ts}(f)$ of a Sampled Truncated Cosine and Samples of the Spectrum Computed by an N-Point DFT

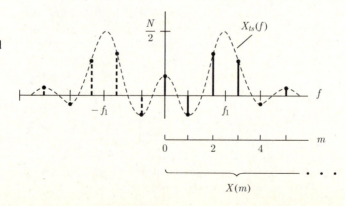

same sample spacing to increase N. In the case of a sampled continuous-time signal, we can also decrease the signal sampling rate to increase T (which also has the effect of increasing aliasing and which requires sampling of the signal at the new rate).

Figure 17.16 shows the spectrum samples obtained for the truncated cosine signal of Figure 17.14a when the three techniques are applied to decrease F to one-third of its previous value. In Figure 17.16a we show the spectrum samples obtained with technique 1 (that is, using zero padding to increase N). Note that the spectrum samples are closer together than those of Figure 17.15. However, the spectrum resolution and sidelobe leakage remains the same because the length of the truncated signal is still the same. With the second and third techniques, we increase the truncation length of the signal by a factor of three. The spectrum samples obtained with technique 2 (that is, taking additional signal

Figure 17.16
Results of Applying Picket-Fence-Reduction Techniques in Computing Spectrum Samples for a Sampled Truncated Cosine Signal

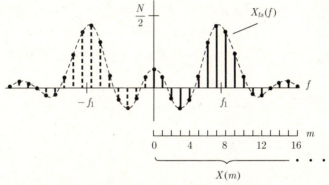

(a) Same signal truncation width, zero padding to three times truncation width

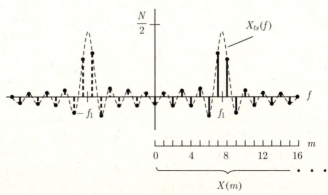

(b) Three times signal truncation width

samples with the same spacing to increase N) are shown in Figure 17.16b. Note that the spectrum resolution and leakage is improved in this case because the signal truncation length is longer, and thus there is reduced spectrum-leakage distortion. Figure 17.16b also applies to technique 3 (that is, decreasing the signal sampling rate to increase T) if the sampling rate remains high enough so that aliasing is negligible over the frequency range shown.

Symmetry Properties

In our discussion of the Fourier transform in Chapter 4, we showed that the Fourier transform of a real continuous-time function has a real part that is an even function of frequency and an imaginary part that is an odd function of frequency. We also showed that these even and odd characteristics imply that the amplitude spectrum is an even function of frequency and that the phase spectrum is an odd function of frequency for a real signal. Therefore, $|X(-f)| = |X(f)|$, $\underline{/X(-f)} = -\underline{/X(f)}$ or, equivalently, $X(-f) = X * (f)$.

Since the DFT is a sampled version of a Fourier transform, these same even and odd symmetry properties hold for it. That is, $|X(-m)| = |X(m)|$, $\underline{/X(-m)} = -\underline{/X(m)}$, or, equivalently, $X(-m) = X * (m)$ for a real signal. However, with the DFT, we compute only the first N samples of the spectrum from the first N samples of the periodic extension of the signal. Therefore, it is convenient to express the symmetry conditions in terms of these samples. This is easily achieved because both $x(n)$ and $X(m)$ are periodic with period N; thus,

$$x(N + n) = x(n) \qquad (17.45)$$

and

$$X(N + m) = X(m) \qquad (17.46)$$

Using eqs. (17.45) and (17.46) we can express the even and odd symmetry conditions for the DFT of real-signal samples as

$$|X(N - m)| = |X(-m)| = |X(m)| \qquad (17.47)$$

and

$$\underline{/X(N - m)} = \underline{/X(-m)} = -\underline{/X(m)} \qquad (17.48)$$

or, equivalently, as

$$X(N - m) = X * (m) \qquad (17.49)$$

This symmetry is apparent in Figure 17.10, where a real signal was being considered. Note that we need to compute the spectrum samples only through $m = INT(N/2)$ if the signal is real; we can find the remaining spectrum samples from these samples by complex conjugation.

Other symmetry properties of the Fourier transform that also apply to the DFT are (1) that real and even signals yield real and even transforms and (2) that real and odd signals yield imaginary and odd transforms. The proof of these properties remains for the reader in the Problem section.

17.3 Use of Windows in DFT Computation

The direct truncation of a signal is equivalent to multiplying the signal by a rectangular-pulse function (see eq. [17.40]). Therefore, we can use the time-multiplication theorem for the Fourier transform to compute the spectrum of the truncated continuous-time signal $x_t(t)$, with the following result:

$$X_t(f) = X(f) * \mathcal{F}\left[\Pi\left(\frac{t - t_1}{\tau}\right)\right] \tag{17.50}$$

where $X(f)$ is the spectrum of the nontruncated signal $x(t)$, τ is the width of the truncated signal, and t_1 is the location of the center of the truncated signal.

Equation (17.50) shows us that the spectrum-leakage characteristics are determined by the Fourier transform of the rectangular-pulse function. Therefore, we can change the spectrum-leakage characteristics if we replace the rectangular-pulse function by some other function that is also nonzero only over the signal-truncation time interval. A function of this type is called a *window function*. If we choose a window function that approaches zero smoothly at its ends, rather than changing abruptly, then the Fourier transform of the window function will have smaller sidelobes but a broader mainlobe than the sinc-function transform of the rectangular-pulse function. The smaller sidelobes reduce leakage of a signal component to frequencies in the spectrum that are sufficiently separated from the signal-component frequency. However, leakage of a signal component to frequencies close to the signal-component frequency is increased (that is, spectrum resolution is not as good). We illustrate the change in spectrum-leakage characteristics in Figure 17.17 by using the triangular window $\Lambda(4f_1t/5)$ to perform windowed truncation of the cosine signal. Comparison of the spectrum in Figure 17.17 with the spectrum in Figure 17.14 shows the decreased sidelobes and broadened resolution.

Note that the application of the window function also produces an overall signal-spectrum amplitude change. Since we are usually interested in the relative amplitude of various frequency components in a signal rather than in the absolute amplitude of the spectrum, this amplitude change is not a serious problem.

When we compute the DFT of a truncated signal, it is more convenient to apply a sampled window function directly to the signal samples taken than to use a continuous-time window function. Sampled window functions that we use are even discrete-time functions about their center. We must center the window function over the signal samples taken. If we use zero padding, then the sampled window function must remain centered over the signal samples taken and not be extended in width to include the zero samples added. The location of the window samples that we use is illustrated in Figure 17.18 for three different sets of eight signal samples. Note that no leakage occurs, and thus no window is required, if the signal length is less than or equal to the time interval over which we take signal samples (that is, $t_s \le MT$). The sequences

Figure 17.17
Truncated Cosine
Signal and Its
Spectrum (Triangular
Window)

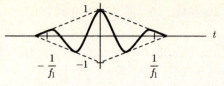

$$x_w(t) = \cos\left(2\pi f_1 t\right)\Lambda\left(\frac{4 f_1 t}{5}\right)$$

(a) Signal

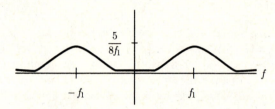

$$X_w(f) = \frac{5}{8 f_1}\operatorname{sinc}^2\left[\frac{5\left(f + f_1\right)}{4 f_1}\right] + \frac{5}{8 f_1}\operatorname{sinc}^2\left[\frac{5\left(f - f_1\right)}{4 f_1}\right]$$

(b) Spectrum

Figure 17.18
Window-Sample
Locations

Signal samples taken   Window samples

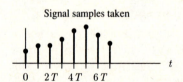 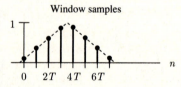

(a) Example 1, $M = 8$, $N = 8$

Signal samples taken with zero samples added   Window samples

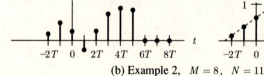 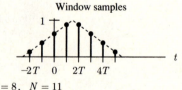

(b) Example 2, $M = 8$, $N = 11$

No signal truncation, thus no window required

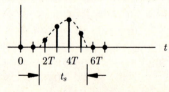

(c) Example 3, $M = 8$, $N = 8$, $t_s \le MT$

for several sampled window functions that are typically used are defined in Figure 17.19. These are the same as the sampled window functions we used in Chapter 15 for FIR filter design, except for a change in notation in this case to produce windows centered on the signal samples. The signal samples begin at index $n = n_1$ and there are M of them. The DFTs of the different windows produce different mainlobe and sidelobe characteristics. Other windows have been defined and are described in the literature.[†]

Figure 17.19
Window-Function Sequences for DFT Computation

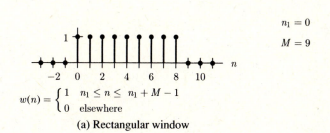

$$w(n) = \begin{cases} 1 & n_1 \leq n \leq n_1 + M - 1 \\ 0 & \text{elsewhere} \end{cases}$$

(a) Rectangular window

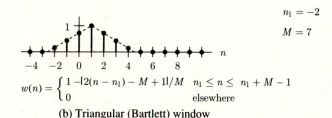

$$w(n) = \begin{cases} 1 - |2(n - n_1) - M + 1|/M & n_1 \leq n \leq n_1 + M - 1 \\ 0 & \text{elsewhere} \end{cases}$$

(b) Triangular (Bartlett) window

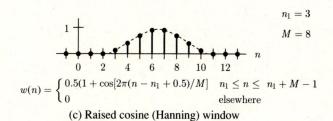

$$w(n) = \begin{cases} 0.5(1 + \cos[2\pi(n - n_1 + 0.5)/M] & n_1 \leq n \leq n_1 + M - 1 \\ 0 & \text{elsewhere} \end{cases}$$

(c) Raised cosine (Hanning) window

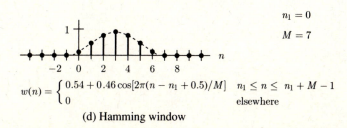

$$w(n) = \begin{cases} 0.54 + 0.46\cos[2\pi(n - n_1 + 0.5)/M] & n_1 \leq n \leq n_1 + M - 1 \\ 0 & \text{elsewhere} \end{cases}$$

(d) Hamming window

[†]Harris, "On the Use of Windows," pp. 51–83.

Example 17.2

The signal samples shown in Figure 17.20 are received from a sensor that measures pressure in a water tank (the units are normalized). To determine the dominant rate of pressure variation, the DFT will be used to compute the signal spectrum. Apply a Bartlett window to the data to reduce spectrum leakage and add four zero samples to decrease the spectrum-sample spacing. Find and sketch the samples used to compute the DFT.

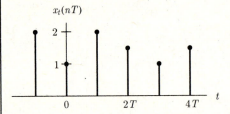

Figure 17.20 Signal Samples Taken for Example 17.2

Solution

The number of signal samples is $M = 6$ and the number of samples that we use to compute the DFT is $N = 6 + 4 = 10$. The signal and window-function samples are shown in Figure 17.21.

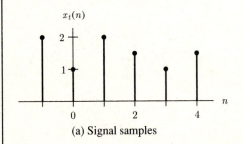

(a) Signal samples

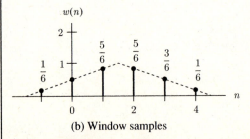

(b) Window samples

Figure 17.21 Signal and Window-Function Samples for Example 17.2

The samples that we use to compute the DFT are

$$x(0) = x_t(0)w(0) = 1/2 \qquad x(1) = x_t(1)w(1) = 5/3$$

$$x(2) = x_t(2)w(2) = 5/4 \qquad x(3) = x_t(3)w(3) = 1/2$$

$$x(4) = x_t(4)w(4) = 1/4 \qquad x(5) = x(6) = x(7) = x(8) = 0$$

$$x(9) = x_t(-1)w(-1) = 1/3$$

These samples are shown in Figure 17.22.

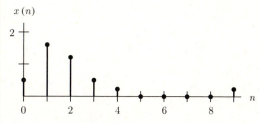

Figure 17.22 Samples Used in DFT Computation for Example 17.2

17.4 Discrete Fourier Transform Parameter Selection

If a continuous-time signal has a finite length of t_s seconds and significant bandwidth of B Hz, then we can select the signal-sample spacing T, the spectrum-sample spacing F, and the DFT length N so that aliasing will be negligible in the spectrum samples produced and there will be no truncation distortion. The sampling theorem states that negligible aliasing results if we select the signal-sample spacing

$$T = \frac{1}{f_s} < \frac{1}{2B} \tag{17.51}$$

No truncation distortion (spectrum leakage) is produced if the spectrum-sample spacing is

$$F = \frac{1}{t_p} \le \frac{1}{t_s} \tag{17.52}$$

since, in that case $t_p \ge t_s$. If the two conditions specified by eqs. (17.51) and (17.52) are satisfied, then we must select a DFT of length greater than twice the time-bandwidth product of the signal because

$$N = \frac{1}{FT} > 2t_s B \tag{17.53}$$

Note that the condition given by eq. (17.53) is not sufficient to guarantee negligible aliasing and no truncation distortion because eqs. (17.51) and (17.52) must also be satisfied.

Example 17.3

The signal to an attitude-control thruster on a satellite has a time duration of 0.123 s and a significant bandwidth of 100 Hz. Find the DFT length, signal-sample spacing, and spectrum-sample spacing that will produce spectrum samples with negligible aliasing and no truncation distortion.

Solution

$$N > 2t_s B = 2(0.123)(100) = 24.6 \qquad \text{Select } N = 25$$

If we select T to be the maximum value that produces negligible aliasing, then

$$T = \frac{1}{2B} = \frac{1}{2(100)} = 0.005 \text{ s} = 5 \text{ ms}$$

The resulting spectrum-sample spacing is

$$F = \frac{1}{NT} = \frac{1}{(25)(0.0005)} = 8 \text{ Hz}$$

If we select F to be the maximum value that produces no truncation distortion, then

$$F = \frac{1}{t_s} = \frac{1}{0.123} = 8.13 \text{ Hz}$$

and the required signal-sample spacing is

$$T = \frac{1}{NF} = \frac{1}{(25)(8.13)} = 0.0049 \text{ s} = 4.9 \text{ ms}$$

Suppose, now, that we want to make the spectrum-sample spacing less than or equal to 3.5 Hz in Example 17.3. If we select the maximum signal-sample spacing, then

$$T = 0.005 \text{ s}$$

and

$$N = \frac{1}{FT} = \frac{1}{(3.5)(0.005)} = 57.14$$

Since N must be an integer, then we select $N = 58$. These choices of T and N produce a spectrum-sample spacing of

$$F = \frac{1}{NT} = \frac{1}{(58)(0.005)} = 3.448 \text{ Hz}$$

If the signal samples have already been taken, then we must add $58 - 25 = 33$ zero samples to the nonzero signal samples to produce the narrower spectrum-sample spacing. Recall that zero padding does not improve spectrum resolution. It merely decreases the spacing between the computed spectrum samples.

Example 17.4

Repeat Example 17.3 for a signal with a time duration of 1 s and a significant bandwidth of 10 kHz.

Solution

For no spectrum aliasing or truncation distortion,

$$N > 2t_s B = 2(1)(10000) = 20000$$

We select the minimum value for N; that is, $N = 20{,}000$. If we select T to be the maximum value that produces negligible aliasing, then

$$T = \frac{1}{2B} = \frac{1}{2(10{,}000)} = 0.00005 \text{ s} = 0.05 \text{ ms}$$

and the spectrum-sample spacing is

$$F = \frac{1}{NT} = \frac{1}{(20{,}000)(0.00005)} = 1 \text{ Hz}$$

In Example 17.4, the value for N is larger than practical for most DFT processors. Assume that a DFT processor is available that has a length of $N = 2^{12} = 4096$ (powers of 2 produce efficient DFT processors). We must increase T, or F, or both, to permit the use of the smaller value of N. Assume first that we select the same value for T, so that no further aliasing is produced. The spectrum-sample spacing that we achieve is

$$F = \frac{1}{NT} = \frac{1}{(4096)(0.00005)} = 4.883 \text{ Hz}$$

and the truncated signal length is

$$t_p = NT = (4096)(0.00005) = 0.205 \text{ s}$$

Thus, truncation distortion of the spectrum-samples occurs. This truncation distortion must not be severe if the DFT computations are to produce values that closely approximate samples of the signal spectrum.

If we want spectrum-sample spacing that is equal to 2.5 Hz for the signal of Example 17.4 and the shorter DFT, then we must select

$$T = \frac{1}{NF} = \frac{1}{(4096)(2.5)} = 0.00009766 \text{ s} = 0.09766 \text{ ms}$$

for the signal-sample spacing. The resulting truncated signal length is

$$t_p = NT = (4096)(0.00009766) = 0.40 \text{ s}$$

With this selection of parameters, less truncation distortion results but additional aliasing is introduced. Whether the computed samples approximate the signal-spectrum samples more closely depends on the relative effect of the aliasing and truncation distortion.

17.5 Discrete Fourier Transform Theorems

Several theorems are helpful in the computation and use of DFTs. These theorems parallel those already discussed for Fourier transforms and discrete-time Fourier transforms in Chapters 4 and 11, respectively. A few of the most important theorems are listed in Table C.9 in Appendix C. These include the linearity theorem, the time-shift theorem, the frequency-shift theorem, the duality theorem, the circular-convolution theorem, and the signal-multiplication theorem. Circular convolution is a new term here; we will consider it in more detail in the next section. Proofs of most of the theorems parallel or nearly parallel those of the theorems for discrete-time Fourier transforms; hence, they are left for the reader in the Problem section.

In using the theorems, it must be remembered that $x(n)$ is the periodic extension, with period N, of the truncated signal samples, and that the first N samples of this periodic extension are used in DFT computation. Also, $X(m)$ contains the first N samples of the periodic spectrum associated with the sampled signal. The samples span one period of the spectrum.

To illustrate the procedure for proving DFT theorems, let us consider the time-shift theorem; that is, if

$$x(n) \leftrightarrow X(m)$$

then

$$x(n-k) \leftrightarrow W_N^{mk} X(m) \tag{17.54}$$

We construct the proof of this theorem as follows:

$$\mathcal{D}[x(n-k)] = \sum_{n=0}^{N-1} x(n-k) W_N^{mn} \tag{17.55}$$

Since $x(n-k)$ and W_N^{mn} are both the periodic in n with period N, then $x(n-k)W_N^{mn}$ is periodic in n with period N. Therefore, we can perform the summation over any period of N samples. Choosing the period $k \leq n \leq k+N-1$, we obtain

$$\mathcal{D}[x(n-k)] = \sum_{n=k}^{k+N-1} x(n-k) W_N^{mn} \tag{17.56}$$

which can be expressed as

$$\mathcal{D}[x(n-k)] = \sum_{i=0}^{N-1} x(i) W_N^{m(k+i)} \tag{17.57}$$

by using the change of variable $i = n - k$. Simplifying eq. (17.57), we obtain

$$\mathcal{D}[x(n-k)] = W_N^{km} \sum_{i=0}^{N-1} x(i)W_N^{mi}$$

$$= W_N^{km} X(m) \tag{17.58}$$

Recall that $x(n)$ is a periodic sequence that is the periodic extension of the truncated signal-sample sequence; therefore, $x(n-k)$ is a time-shifted version of this periodic sequence. This property is illustrated in the example that follows.

Example 17.5

Consider the truncated signal samples shown in Figure 17.23. Find and sketch the sample sequences $y(n)$ and $z(n)$ when $Y(m) = W_4^{2m} X(m)$ and $Z(m) = W_4^{-m} X(m)$.

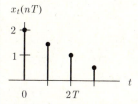

Figure 17.23 Signal Samples for Example 17.5

Solution
The periodic extension of the signal samples is shown in Figure 17.24.

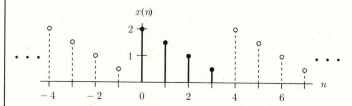

Figure 17.24 Periodic Extension of the Signal Samples in Example 17.5

We use the first $N = 4$ samples of the periodic extension to compute the DFT. These samples are highlighted in Figure 17.24. Since

$$X(m) \leftrightarrow x(n)$$

then the time-shift theorem gives

$$y(n) = x(n-2) \quad \text{and} \quad z(n) = x(n+1)$$

These sequences are illustrated in Figure 17.25.

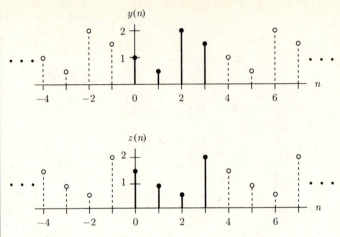

Figure 17.25 Time-Shift-Theorem Results for Example 17.5

If we compute $X(m) = \mathcal{D}[x(n)]$, followed by $Y(m) = W_4^{2m} X(m)$, followed by $y(n) = \mathcal{D}^{-1}[Y(m)]$, we then obtain the first $N = 4$ samples of $y(n)$. These samples are highlighted in Figure 17.25. The same type of computations produce the highlighted samples of $z(n)$ shown in Figure 17.25.

The time-shift theorem is sometimes called the *circular time-shift theorem* because the signal-sequence samples used by the DFT and those produced by the IDFT (that is, the first N samples) shift in a circular fashion. In other words, the last sample replaces the first sample on a right shift and the first sample replaces the last sample on a left shift. Since the spectrum samples produced by the DFT are the first N samples of a periodic spectrum that has a period of N samples, then the frequency-shift theorem possesses the same circular-shift property.

17.6 Discrete-Convolution Computation Using DFTs

The convolution theorem for the discrete-time Fourier transform (see Chapter 11) allows us to use discrete-time Fourier transforms in computing discrete convolutions. According to this procedure,

$$x_T(n) * y_T(n) = \mathcal{F}_d^{-1}[X_d(f)Y_d(f)] \tag{17.59}$$

where $X_d(f)$ and $Y_d(f)$ are the discrete-time Fourier transforms that we compute for $x_T(n)$ and $y_T(n)$. Our question now is whether we can use DFTs in a similar fashion to compute discrete convolutions.

Circular Convolution

Let us consider the IDFT of the product of the N-point DFTs corresponding to two sample sequences. This IDFT is

$$\mathcal{D}^{-1}[X(m)Y(m)] = \sum_{m=0}^{N-1} X(m)Y(m)W_N^{-mn}$$

$$= \sum_{m=0}^{N-1} \left[\sum_{k=0}^{N-1} x(k)W_N^{mk}\right]Y(m)W_N^{-mn}$$

$$= \sum_{k=0}^{N-1} x(k)\left\{\sum_{m=0}^{N-1}\left[W_N^{mk}Y(m)\right]W_N^{-mn}\right\}$$

$$= \sum_{k=0}^{N-1} x(k)y(n-k) \tag{17.60}$$

We used the time-shift theorem to produce the last step in the preceding equation. The expression

$$\sum_{k=0}^{N-1} x(k)y(n-k) \tag{17.61}$$

has the form of a discrete convolution, except that the summation is only over N sample products. Also the sample sequences $x(n)$ and $y(n)$ are periodic sequences with period equal to N because they are samples of periodic extensions of truncated signals. Since $y(n-k)$ is a time-shifted version (circular time shift) of $y(n)$, then eq. (17.61) is called the *circular convolution of x(n) and y(n)*. We use the notation

$$x(n)\textcircled{N}y(n) = \sum_{k=0}^{N-1} x(k)y(n-k) \tag{17.62}$$

for an N-point circular convolution. Note that if

$$x(n) \leftrightarrow X(m) \quad \text{and} \quad y(n) \leftrightarrow Y(m)$$

then

$$x(n)\textcircled{N}y(n) \leftrightarrow X(m)Y(m) \tag{17.63}$$

This theorem is called the *circular convolution theorem*. Note that $X(m)$ and $Y(m)$ must be DFTs of the same length, N. Also note that the values computed for $x(n)\textcircled{N}y(n)$ are the first N samples of the periodic extension of the circular convolution.

To illustrate the results of circular convolution, we will consider the two discrete time signals shown in Figure 17.26a. We select $N = 3$, so that no signal truncation exists. The resulting periodic-extension sequences corresponding

Figure 17.26
Discrete-Time
Signals and Their
Periodic Extensions
to Illustrate Circular
Convolution

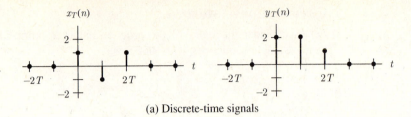

(a) Discrete-time signals

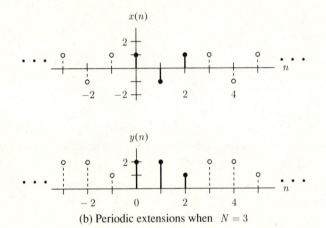

(b) Periodic extensions when $N = 3$

to the signals in 17.26a are shown in Figure 17.26b. The first three sequence values of the periodic extensions are highlighted; they are the values we use in computing the DFT. The sequence $x(k)$ and the sequences $y(0 - k)$, $y(1 - k)$, and $y(2 - k)$ are shown in Figure 17.27, along with the circular convolution $x(n)\,\text{③}\,y(n)$ that results when we use these sequences in eq. (17.62). The first three sequence values are highlighted in all sequences shown in Figure (17.27). These sequence values are the ones we use in eq. (17.62) and the values computed for $x(n)\,\text{Ⓝ}\,y(n)$ by $\mathcal{D}^{-1}[X(m)Y(m)]$ when we use DFTs.

Discrete Convolution

In Figure 17.28a, we show the discrete convolution of the two signals defined in Figure 17.26a [that is, $x_T(n) * y_T(n)$]. The circular convolution of these two signals does not produce the same sequence of values (compare Figures 17.27 and 17.28a). The problem is that $x(n)\,\text{Ⓝ}\,y(n)$ is the IDFT of an N-point DFT, while $x_T(n) * y_T(n)$ has a length of $N + N - 1 = 2N - 1$ sequence values. Therefore, the signal produced by $x_T(n) * y_T(n)$ has a sequence length, $M = 2N - 1$, that is longer than the DFT length, N. This length difference produces time aliasing when we compute the IDFT of $X(m)Y(m)$ to obtain $x(n)\,\text{Ⓝ}\,y(n)$. Therefore, $x(n)\,\text{Ⓝ}\,y(n)$ is a time-aliased version of the sequence corresponding to $x_T(n) * y_T(n)$. We show this time aliasing in Figure 17.28b

Figure 17.27
Sequences Used in
Circular Convolution
and the Circular-
Convolution Result
for the Signals of
Figure 17.26

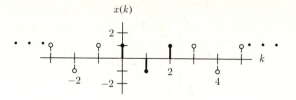

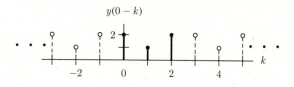

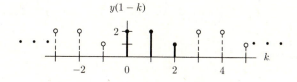

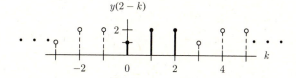

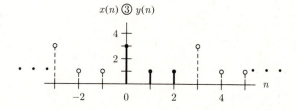

and 17.28c by repeating the sequence corresponding to $x_T(n) * y_T(n)$ with shifts that are multiples of N and adding these repeated sequences. Recall that $N = 3$ for the illustration case.

We can compute the discrete convolution of two signals by using the DFT if we use zero padding to extend the signal lengths to equal the length of the corresponding discrete convolution. That is, if $x_T(n)$ has a length of N_x samples and $y_T(n)$ has a length of N_y samples, then we pad $x_T(n)$ with $N_y - 1$ zeros and $y_T(n)$ with $N_x - 1$ zeros so the zero-padded sequences are both of length $N_x + N_y - 1$. We use an $(N_x + N_y - 1)$-point DFT to compute $X(m)$ and $Y(m)$. The length of the sequence that we obtain from the IDFT $\mathcal{D}^{-1}[X(m)Y(m)]$ is

Figure 17.28
Circular-
Convolution–
Discrete-Convolution
Relationship for the
Signals of
Figure 17.26

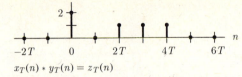

$x_T(n) * y_T(n) = z_T(n)$

(a) Discrete convolution of signals in Figure 17.26a

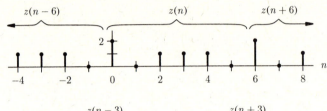

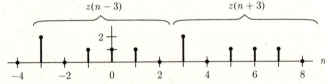

(b) Time-shifted versions of the sequence
corresponding to $x_T(n) * y_T(n)$

$$\sum_{m=-\infty}^{\infty} z(n - 3m) = x(n) \; ③ \; y(n)$$

(c) Sum of time-shifted sequences

sufficiently long ($N_x + N_y - 1$) so that the discrete convolution is produced by the IDFT of the signal-transform product.

Example 17.6

Use the DFT to compute the discrete convolution of the signals shown in Figure 17.26a.

Solution
The signal lengths are $N_x = N_y = 3$; therefore, we add $N_y - 1 = 2$ zero samples to the nonzero $x_T(n)$ signal samples and $N_x - 1 = 2$ zero samples

to the nonzero $y_T(n)$ signal samples so that the zero-padded signals are both of length $N_x + N_y - 1 = 5$. The length of the DFT and IDFT that we use is $N = 5$. The first five samples of the periodic extensions of the zero-padded signals are

$$x(0) = 1 \qquad x(1) = -1 \qquad x(2) = 1 \qquad x(3) = 0 \qquad x(4) = 0$$

and

$$y(0) = 2 \qquad y(1) = 2 \qquad y(2) = 1 \qquad y(3) = 0 \qquad y(4) = 0$$

The 5-point DFTs of the zero-padded-signal sequences give

$$X(m) = \sum_{n=0}^{4} x(n) W_5^{mn} \qquad 0 \le m \le 4$$

and

$$Y(m) = \sum_{n=0}^{4} y(n) W_5^{mn} \qquad 0 \le m \le 4$$

When we perform the computations indicated by these equations, we obtain

$$X(0) = 1 \qquad X(1) = 0.3820 e^{j0.6\pi} \qquad X(2) = 2.6180 e^{j0.2\pi}$$
$$X(3) = 2.6180 e^{-j0.2\pi} \qquad X(4) = 0.3820 e^{-j0.6\pi}$$

and

$$Y(0) = 5 \qquad Y(1) = 3.0777 e^{-j0.3\pi} \qquad Y(2) = 0.7265 e^{-j0.1\pi}$$
$$Y(3) = 0.7265 e^{j0.1\pi} \qquad Y(4) = 3.0777 e^{j0.3\pi}$$

The product of these two sets of spectrum samples is

$$Y(0)X(0) = 5 \qquad X(1)Y(1) = 1.1757 e^{j0.3\pi} \qquad X(2)Y(2) = 1.9020 e^{j0.1\pi}$$
$$X(3)Y(3) = 1.9020 e^{-j0.1\pi} \qquad X(4)Y(4) = 1.1757 e^{-j0.3\pi}$$

When we compute the IDFT with

$$\mathcal{D}^{-1}[X(m)Y(m)] = \frac{1}{5} \sum_{m=0}^{4} X(m)Y(m) W_5^{-mn} \equiv z(n)$$

we obtain

$$z(0) = 2 \qquad z(1) = 0 \qquad z(2) = 1 \qquad z(3) = 1 \qquad z(4) = 1$$

These sequence values are the sequence values of the discrete convolution of the two signals $x_T(n)$ and $y_T(n)$. (See Figure 17.28a.)

17.7 Discrete-Time-Signal Filtering Using The DFT

In the introduction to this chapter, we briefly discussed the use of the DFT in filter construction. The filter implementation using DFTs, as shown in Figure 17.1, provides considerable flexibility in filter design.

Implementation Considerations

In the filter of Figure 17.1, the discrete Fourier transform is used to compute sample values of the input-signal spectrum. As previously mentioned, we must choose the signal-sample spacing T and spectrum-sample spacing F that allows for a good compromise between aliasing and signal-truncation effects, at the same time maintaining a reasonable DFT length, N.

The discrete-time portion of the filter actually performs a circular convolution. Therefore, the sampled unit pulse response corresponding to the spectrum weighting coefficients $H(m)$ (that is, the frequency-response samples) must be nonnegligible only for $n \ll N$ if the time aliasing of the output signal is to remain small.

The spectrum-sample weighting coefficients $H(m)$ that we use must possess complex-conjugate symmetry $[H(N-m) = H*(m)]$ in order for the output-signal samples to be real when the input-signal samples are real. Actually, round-off effects will probably cause complex outputs, even with the correct symmetry. We must design the filter with sufficient numerical precision so as to keep the imaginary part of the output signal caused by round-off effects small enough to be neglected. We can then use the real part of the output signal as the output signal.

It may appear that completely arbitrary filter frequency responses can be constructed with the appropriate choice of spectrum-sample weighting coefficients $H(m)$. For example, we can choose the values of $H(m)$ to be the values shown in Figure 17.29. These are the periodic-extension samples for the frequency response of an ideal low-pass filter with cutoff frequency $f_c = 3.5F$ when we use a 15-point DFT. If we use this set of spectrum-sample weighting coefficients, then, in the unambiguous frequency interval $0 \le f \le f_s/2$, the filter will pass signal components with frequencies of 0, F, $2F$, and $3F$ without change and will eliminate signal components with frequencies of $4F$, $5F$, $6F$, and $7F$. However, signal components with frequencies between frequencies that are multiples of F show up in the spectrum samples computed at these frequencies only through spectrum leakage. Thus, these signal com-

Figure 17.29
Periodic-Extension Samples of the Frequency Response of an Ideal Low-Pass Filter with Cutoff Frequency of 3.5F

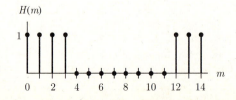

ponents appear in several spectrum samples; they are not simply multiplied by either one or zero in passing through the filter. In Problem 17.31, the reader is asked to find the frequency response of the filter using the spectrum weighting coefficients shown in Figure 17.29.

Selective-Save Implementation

To use the filter implementation of Figure 17.1 directly, we must first read in all N samples of the signal and then perform the DFT. This would not be practical for very long input signals, however, because the length of the DFT required would be very large. Also, the time delay through the filter would be exceedingly large because no output signal samples could be obtained until all input signal samples were read in. An alternate filter implementation that effectively resolves this problem uses the *selective-save method* (sometimes called the *overlap-save method*) to perform a discrete convolution by using DFTs. Actually, the use of this method for performing a discrete convolution is not limited to filtering applications. It can be used any time we want to convolve discrete-time signals of greatly differing lengths.

If we want to use the selective-save method, then the desired unit pulse response, $h_T(n)$, of the discrete-time portion of the filter must have negligibly small sample values after a reasonable length of time so that it can be truncated with little effect. Let us assume that the unit pulse response is truncated to a length of N samples. We implement the selective-save method by using two processing channels that each read and process successive sets of $2N$ samples of the input signal $x_T(n)$. The first channel begins reading input-signal samples N samples before the time corresponding to the first output-signal sample that we want. The second channel begins reading input-signal samples N samples later than the first channel. Thus, overlapping sets of $2N$ signal samples are read and processed by each of the two channels. The processing performed by the channels uses DFTs of length $2N$ to obtain a circular convolution of each signal-sample set of length $2N$ and $\hat{h}_T(n)$, where $\hat{h}_T(n)$ is the N-sample truncated portion of $h_T(n)$ augmented with N zero samples. The last N samples of each of these circular convolutions are samples that correspond to the discrete convolution $x_T(n) * h_T(n)$. They are selected, saved, and put together successively to construct the sample sequence corresponding to the desired output discrete-time signal $y_T(n) = x_T(n) * h_T(n)$. Note that an indefinitely long input signal can be processed with a DFT of length $2N$. We will show later that the output signal is delayed in time by $3N$ samples.

We will now demonstrate the selective-save method for computing a discrete convolution with an illustrative case. This demonstration is performed by considering the sample sequences being convolved. Actually, the processing would be performed with DFTs, and we will show the required implementation following the demonstration of the method.

Our illustrative case is for $N = 2$, in order to keep it simple. The input-signal samples, $x_T(n)$; unit-pulse-response samples, $h_T(n)$; and desired output signal samples, $y_T(n) = x_T(n) * h_T(n)$, are shown in Figure 17.30. The zero time reference chosen is the time at which we begin observation of the output signal. Therefore, $y_T(n)$ is only shown for $n \geq 0$. Only a segment of the input signal is shown; the assumption is that it extends further into the past and future. The first two sets of input-signal samples read and processed by channel 1, $x_{11T}(n)$ and $x_{12T}(n)$, and by channel 2, $x_{21T}(n)$ and $x_{22T}(n)$, are also defined in Figure 17.30. They are of length $2N = 4$ and overlap between channels, as previously indicated.

Figures 17.31 and 17.32 show the periodic extensions, with period $2N$, of input-signal and unit-pulse-response sample sequences that correspond to the

Figure 17.30
Signal-Sample Sequences Used to Illustrate the Selective-Save Convolution

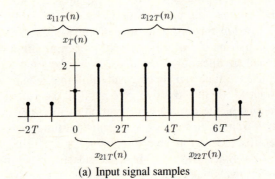

(a) Input signal samples

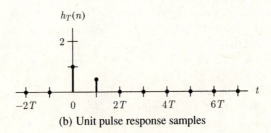

(b) Unit pulse response samples

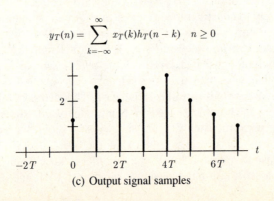

(c) Output signal samples

Figure 17.31
Results of Circular
Convolution for the
Selective-Save-
Method
Illustration

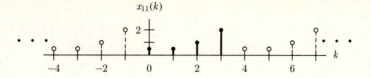

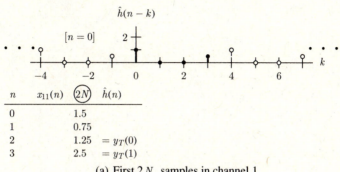

n	$x_{11}(n)$	$(2N)$ $\hat{h}(n)$	
0	1.5		
1	0.75		
2	1.25	$= y_T(0)$	
3	2.5	$= y_T(1)$	

(a) First $2N$ samples in channel 1

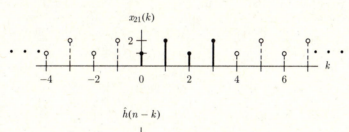

n	$x_{21}(n)$	$(2N)$ $\hat{h}(n)$	
0	2.0		
1	2.5		
2	2.0	$= y_T(2)$	
3	2.5	$= y_T(3)$	

(b) First $2N$ samples in channel 2

circular convolutions (see eq. [17.62)] performed by the DFT processor on the first two sets of samples read and processed by each channel. The results of the circular convolutions of length $2N$ are also indicated in the figures. Note that the last N samples produced by each circular convolution produce the output-signal samples when they are saved and put together successively by alternating between the two channels.

To promote understanding of the sequence of events in the computation of output-signal samples using the selective-save method, a time-line of the

Figure 17.32
Results of Circular
Convolution for the
Selective-Save-
Method
Illustration

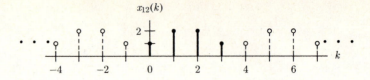

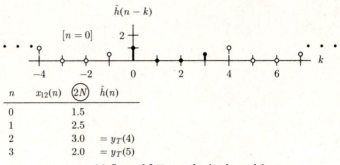

n	$x_{12}(n)$	$2N$	$\hat{h}(n)$
0		1.5	
1		2.5	
2		3.0	$= y_T(4)$
3		2.0	$= y_T(5)$

(a) Second $2N$ samples in channel 1

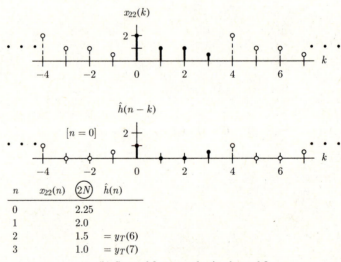

n	$x_{22}(n)$	$2N$	$\hat{h}(n)$
0		2.25	
1		2.0	
2		1.5	$= y_T(6)$
3		1.0	$= y_T(7)$

(b) Second $2N$ samples in channel 2

occurrence of events for the illustrative case is shown in Figure 17.33. Note
the six-sample delay of the output signal. In general, there is a delay of $3N$
samples, as we indicated earlier.

Figure 17.34 shows a block diagram of a discrete-time filter implemented
with the DFT and the selective-save method. The switches shown are synchro-
nized. Those routing the signal samples in and out of storage switch every $2N$
samples and the switch routing signal samples to the output switches every N
samples, as shown by the time-line of Figure 17.33.

Figure 17.33
Time-Line of
Events for DFT
Discrete-Time-Filter
Implementation
Using the
Selective-Save
Method

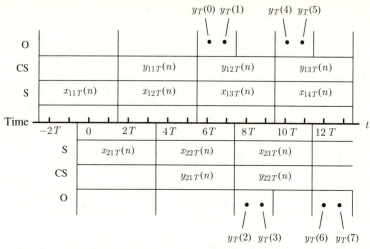

S = Read and Store

CS = Compute and Save

O = Output Samples

Figure 17.34
Block-Diagram
Representation
of the DFT
Discrete-Time-Filter
Implementation
Using the
Selective-Save
Method

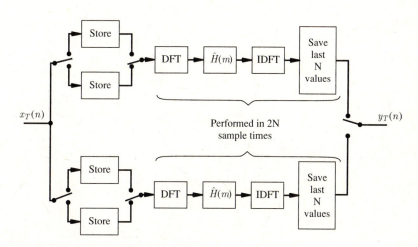

17.8 The Fast Fourier Transform

Considerable symmetry is evident in the operations and coefficients required to compute a DFT. This symmetry is apparent in Example 17.1, where the DFT computations are shown in detail for an example signal. The symmetry can be exploited to reduce the number of operations required, thus reducing the time required for the DFT computation. The resulting computation algorithm is known as the *fast Fourier transform*, or FFT for short. The FFT is most efficient in terms of time utilization when the number of sample values used and computed is a power of 2 (that is, $N = 2^L$). Consequently, we will consider only these types of FFT algorithms here. They are known as *FFTs of radix 2*.

Derivation of the FFT

There are several FFT algorithms; however, in this introductory treatment we will derive and use only one. It is obtained by a method called *decimation in time*—a method that works when $N = 2^L$. The name of the method stems from the fact that the set of input samples is successively split into smaller and smaller subsets of even-numbered and odd-numbered samples.

To illustrate our derivation of an N-point decimation-in-time FFT algorithm, we will use an 8-point FFT. To begin the derivation, we consider the input sequence $x(n)$ with length equal to N samples. We define two $N/2$-point sequences as

$$x_e(n) = x(2k) \qquad 0 \le k \le \frac{N}{2} - 1 \tag{17.64}$$

and

$$x_o(n) = x(2k + 1) \qquad 0 \le k \le \frac{N}{2} - 1 \tag{17.65}$$

One can see that $x_e(n)$ and $x_o(n)$ are, respectively, sequences of the even-numbered and odd-numbered samples of $x(n)$. The DFT transforms of these sequences are

$$X_e(m) = \sum_{k=0}^{(N/2)-1} x_e(k) W_{N/2}^{mk} \qquad 0 \le m \le \frac{N}{2} - 1 \tag{17.66}$$

and

$$X_o(m) = \sum_{k=0}^{(N/2)-1} x_o(k) W_{N/2}^{mk} \qquad 0 \le m \le \frac{N}{2} - 1 \tag{17.67}$$

We now define

$$x_1(n) = \begin{cases} x(n) & \text{n even} \\ 0 & \text{n odd} \end{cases} \qquad 0 \le n \le N - 1 \tag{17.68}$$

and

$$x_2(n) = \begin{cases} 0 & \text{n even} \\ x(n) & \text{n odd} \end{cases} \qquad 0 \le n \le N - 1 \tag{17.69}$$

so that $x(n) = x_1(n) + x_2(n)$ for $0 \le n \le N - 1$. The DFT of $x_1(n)$ is

$$X_1(m) = \sum_{n=0}^{N-1} x_1(n) W_N^{mn} \qquad 0 \le m \le N - 1 \tag{17.70}$$

Since $x_1(n) = 0$ for n odd, then we can write eq. (17.70) as

$$X_1(m) = \sum_{k=0}^{(N/2)-1} x_1(2k)W_N^{2mk} \qquad 0 \le m \le N-1 \qquad (17.71)$$

However, $x_1(2k) = x(2k) = x_e(k)$ for $0 \le k \le (N/2) - 1$ and

$$W_N^{2mk} = e^{-j[2\pi(2mk)/N]} = e^{-j[2\pi mk/(N/2)]} = W_{N/2}^{mk} \qquad (17.72)$$

Therefore,

$$X_1(m) = \sum_{k=0}^{(N/2)-1} x_e(k)W_{N/2}^{mk} = X_e(m) \qquad 0 \le m \le N-1 \qquad (17.73)$$

Since $X_e(m)$ is periodic with period $N/2$, then $X_e(\frac{N}{2} + m) = X_e(m)$ and $X_e(m)$ cycles through two periods as we compute eq. (17.73) for $0 \le m \le N - 1$.

The DFT of $x_2(n)$ is

$$X_2(m) = \sum_{n=0}^{N-1} x_2(n)W_N^{mn} \qquad 0 \le m \le N-1 \qquad (17.74)$$

Since $x_2(n) = 0$ for n even, then we can write eq. (17.74) as

$$X_2(m) = \sum_{k=0}^{N/2-1} x_2(2k+1)W_N^{m(2k+1)} \qquad 0 \le m \le N-1 \qquad (17.75)$$

However, $x_2(2k + 1) = x(2k + 1) = x_o(k)$ for $0 \le k \le \frac{N}{2} - 1$ and

$$W_N^{m(2k+1)} = W_N^m W_N^{2mK} = W_N^m W_{N/2}^{mk}$$

Therefore,

$$X_2(m) = W_N^m \sum_{k=0}^{(N/2)-1} x_o(k)W_{N/2}^{mk} = W_N^m X_o(m) \qquad 0 \le m \le N-1 \qquad (17.76)$$

Since $X_o(m)$ is periodic with period $N/2$, then $X_o(\frac{N}{2} + m) = X_o(m)$ and $X_o(m)$ cycles through two periods as we compute eq. (17.76) for $0 \le m \le N - 1$.

Now, $x(n) = x_1(n) + x_2(n)$ for $0 \le n \le N - 1$; therefore, we obtain

$$X(m) = X_1(m) + X_2(m)$$
$$= X_e(m) + W_N^m X_o(m) \qquad 0 \le m \le N-1 \qquad (17.77)$$

by using the linearity theorem. We can split eq. (17.77) into two parts that compute the first $N/2$ and last $N/2$ spectrum samples, respectively. These are

$$X(k) = X_e(k) + W_N^k X_o(k) \qquad 0 \le k \le \frac{N}{2} - 1 \qquad (17.78)$$

and

$$X\left(k + \frac{N}{2}\right) = X_e\left(k + \frac{N}{2}\right) + W_N^{k+N/2}X_o\left(k + \frac{N}{2}\right)$$

$$= X_e(k) + W_N^{k+N/2}X_o(k) \qquad 0 \le k \le \frac{N}{2} - 1 \qquad (17.79)$$

We call eqs. (17.78) and (17.79) *connection equations*. They show that the spectrum samples $X(k)$ and $X(k + N/2)$ can both be computed from $X_e(k)$ and $X_o(k)$. These computations are illustrated in Figure 17.35 using a signal-flow-graph representation, rather than a block diagram, for simplicity. In a *signal-flow graph*, a node is a summing node if more than one signal enters the node. Signal multipliers are written beside the signal-flow paths in which multiplication occurs. The arrow indicates the direction of signal flow. Since all signal flow is from left to right in the signal-flow graphs representing FFT algorithms, we do not need the arrows on FFT signal-flow graphs. In the remainder of this chapter, they are not shown. The signal-flow-graph structure of Figure 17.35 is called a *butterfly connection* because of its characteristic shape.

Figure 17.36 consists of a hybrid (block-diagram–signal-flow-graph) representation of DFT computation using the results obtained thus far as applied to an 8-point DFT computation.

The decimation-in-time technique that we have used can be repeated to compute the two $N/2$-point DFTs $X_e(m)$ and $X_o(m)$. The four subsets of input samples are

$$x_{ee}(j) = x_e(2j) = x(4j) \qquad 0 \le j \le \frac{N}{4} - 1 \qquad (17.80)$$

$$x_{eo}(j) = x_e(2j + 1) = x(4j + 2) \qquad 0 \le j \le \frac{N}{4} - 1 \qquad (17.81)$$

$$x_{oe}(j) = x_o(2j) = x(4j + 1) \qquad 0 \le j \le \frac{N}{4} - 1 \qquad (17.82)$$

and

$$x_{oo}(j) = x_o(2j + 1) = x(4j + 3) \qquad 0 \le j \le \frac{N}{4} - 1 \qquad (17.83)$$

Figure 17.35
Signal-Flow-Graph
Representation of
Eqs. (17.78) and
(17.79) (Butterfly
Connection)

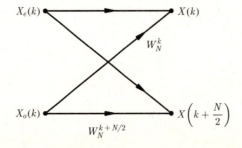

Figure 17.36
Hybrid
Representation of
DFT Computation
after the First Signal
Split in the
Decimation-in-Time
Algorithm

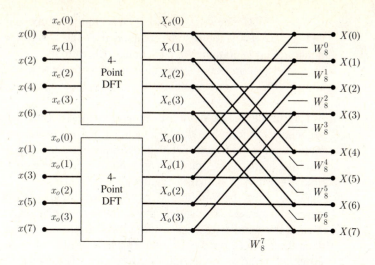

The resulting connection equations are

$$X_e(k) = X_{ee}(k) + W_{N/2}^k X_{eo}(k) \qquad 0 \le k \le \frac{N}{4} - 1 \qquad (17.84)$$

$$X_e\left(k + \frac{N}{4}\right) = X_{ee}(k) + W_{N/2}^{k+N/4} X_{eo}(k) \qquad 0 \le k \le \frac{N}{4} - 1 \qquad (17.85)$$

$$X_o(k) = X_{oe}(k) + W_{N/2}^k X_{oo}(k) \qquad 0 \le k \le \frac{N}{4} - 1 \qquad (17.86)$$

and

$$X_o\left(k + \frac{N}{4}\right) = X_{oe}(k) + W_{N/2}^{k+N/4} X_{oo}(k) \qquad 0 \le k \le \frac{N}{4} - 1 \qquad (17.87)$$

It is more convenient for us to express these equations in terms of powers of W_N rather than powers of $W_{N/2}$, so that only W_N appears in the final FFT signal-flow graph. This change is possible because $W_{N/2}^i = W_N^{2i}$, as shown by eq. (17.72). With this change, the connection equations become

$$X_e(k) = X_{ee}(k) + W_N^{2k} X_{eo}(k) \qquad 0 \le k \le \frac{N}{4} - 1 \qquad (17.88)$$

$$X_e\left(k + \frac{N}{4}\right) = X_{ee}(k) + W_N^{2k+N/2} X_{eo}(k) \qquad 0 \le k \le \frac{N}{4} - 1 \qquad (17.89)$$

$$X_o(k) = X_{oe}(k) + W_N^{2k} X_{oo}(k) \qquad 0 \le k \le \frac{N}{4} - 1 \qquad (17.90)$$

and

$$X_o\left(k + \frac{N}{4}\right) = X_{oe}(k) + W_N^{2k+N/2} X_{oo}(k) \qquad 0 \le k \le \frac{N}{4} - 1 \qquad (17.91)$$

Figure 17.37 shows a hybrid representation of the computation of $X_e(m)$ and $X_o(m)$ using the results obtained thus far as applied to an 8-point DFT transform computation. We can connect the outputs of this representation using the butterfly connections specified by the connection equations (eqs. [17.78] and [17.79]) to complete the computation of the spectrum samples $X(m)$ (see Figure 17.36).

The preceding decimation-in-time technique can be repeated until the input sequences to the individual DFTs can be divided no further. This occurs in one more step for the 8-point DFT computation we are using as an illustration. The resulting signal-flow graph for an 8-point FFT produced by the decimation-in-time algorithm is shown in Figure 17.38.

We see from the 8-point FFT example that the number of input-sample splits required is three. In general, the number of splits required is $L = \log_2 N$. This is the number of steps required in the decimation-in-time FFT algorithm derivation. These steps each produce a different set of connection equations and different butterfly-connection characteristics. These characteristics are apparent in Figure 17.38. The set of connections produced by each step of the derivation is called a *stage* of the FFT algorithm. Thus, there are $L = \log_2 N$ stages in a radix-2 decimation-in-time FFT algorithm (three stages for the 8-point FFT example).

At each stage of the FFT algorithm, output nodes of the previous stage are connected to input nodes of the succeeding stage by butterfly connections. Therefore, we compute a pair of inputs for the succeeding stage from a pair of outputs of the previous stage, and that pair of outputs of the previous stage is not needed to compute any other inputs of the succeeding stage. Therefore, we can

Figure 17.37
Hybrid
Representation of
DFT Computation of
$X_e(m)$ and $X_o(m)$
after the Second
Signal Split in the
Decimation-in-Time
Algorithm

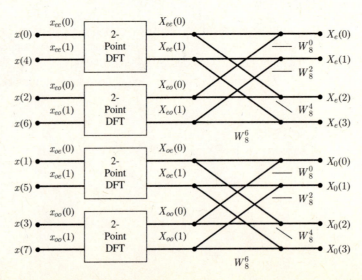

Figure 17.38
Signal-Flow-Graph
Representation of an
8-Point Decimation-
in-Time FFT
Algorithm

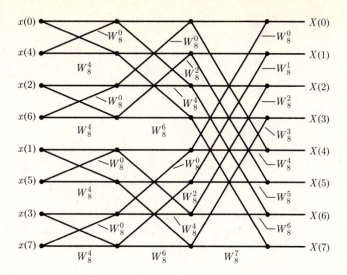

store the succeeding-stage inputs in the locations that the previous-stage outputs had occupied immediately before computation. This characteristic results in a requirement for only N storage locations, rather than $2N$. An algorithm with this characteristic is called an *in-place algorithm*; it has the advantage of a reduced storage requirement.

Figure 17.38 shows that the upper signal-flow path entering each node does not provide any multiplication. Since the output of one-half of the nodes feeds only upper signal-flow paths to succeeding nodes, then these outputs are not multiplied in passing through a stage. It is also apparent from Figure 17.38 that the output of each remaining node feeds two paths that provide multiplications of W_N^i and $W_N^{i+N/2}$. However,

$$W_N^{i+N/2} = W_N^{N/2} W_N^i = e^{-j[2\pi(N/2)/N]} W_N^i$$
$$= e^{-j\pi} W_N^i = -W_N^i \tag{17.92}$$

Therefore, we can multiply the node output by W_N^i before splitting into the two paths and then subtract at one of the succeeding nodes instead of adding. Using this technique, we reduce the number of multiplications required, which increases the speed of computation. The signal-flow graph that results when this technique is used is illustrated in Figure 17.39 for the 8-point FFT example, where the minus sign beside a signal-flow path entering a node indicates subtraction of the signal arriving by that path instead of addition. Since $W_N^0 = 1$, then W_N^0 is not a multiply; thus, the number of multiplications required to implement the 8-point FFT using the implementation of Figure 17.39 is only five.

Figure 17.39
Signal-Flow-Graph
Representation of an
8-Point Decimation-
in-Time FFT
Algorithm with
Minimum Number of
Multiplications

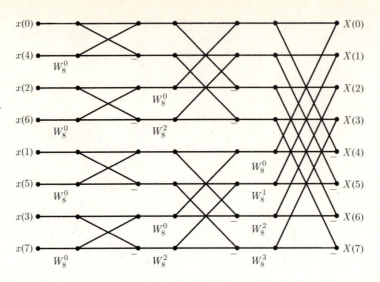

For the decimation-in-time FFT algorithm derived, the output-spectrum samples are in order but the input samples are out of order. This reordering of the input samples is a result of successively splitting the input sequence into even-numbered and odd-numbered samples. We can determine the order required for the input by keeping track of the resulting order from the successive splits of the input sequence. However, there is an easier method. To use it, we first write the binary-number representation for the index of each output-spectrum sample. Reversing the binary digits, we obtain the binary-number representation of the corresponding input-sample index. This method is illustrated in Figure 17.40

Figure 17.40
Bit-Reversal
Determination of the
Input-Sample Order
for the Decimation-
in-Time FFT
Algorithm

000	$x(0)$ ▪	▪	$X(0)$	000
100	$x(4)$ ▪	▪	$X(1)$	001
010	$x(2)$ ▪	▪	$X(2)$	010
110	$x(6)$ ▪	▪	$X(3)$	011
001	$x(1)$ ▪	▪	$X(4)$	100
101	$x(5)$ ▪	▪	$X(5)$	101
011	$x(3)$ ▪	▪	$X(6)$	110
111	$x(7)$ ▪	▪	$X(7)$	111

for an 8-point FFT, where the arrows indicate the steps required to determine the required input-sample order. The required input-sample order is frequently called the *bit-reversed order*; it is easily obtained in a discrete-time system by comparing the outputs of direct and bit-reversed counters. Digital-signal processing chips that contain bit-reversed addressing hardware are available.

When N is a power of 2, we can easily construct the signal-flow-graph representation of the FFT algorithm shown in Figure 17.38 (and thus define the implementation for an N-point FFT), without going through the mathematical derivation. A description of the steps in the signal-flow-graph construction follows. The reader is advised to use Figure 17.38 when studying these steps to better understand them.

Step 1   There are $L = \log_2 N$ stages. Therefore, draw $L + 1$ columns of N nodes.

Step 2   Split the last two columns of nodes into two equal subsets of nodes and connect corresponding node pairs in the two subsets with butterfly connections.

Step 3   Split each of the subsets of nodes of the next-to-last column of nodes and the corresponding subsets of nodes of the preceding column of nodes into two equal sub-subsets of nodes. With butterfly connections, connect corresponding pairs of nodes in the two sub-subsets of each subset.

Step 4   Continue the splitting and connecting until the first set of nodes is reached. Adjacent pairs of nodes in the first and second columns will be connected by butterfly connections if the procedure has been followed correctly.

Step 5   Add signals entering nodes along upper paths without multiplication. Multipliers for signals entering nodes along lower paths are W_N^i for the last stage, where $0 \leq i \leq N - 1$ is the node index counted from the top. For the next-to-last stage, the multipliers are W_N^{2i}, where $0 \leq i \leq \frac{N}{2} - 1$, for the first $N/2$ nodes; they repeat for the last $N/2$ nodes. For the second-from-the-last stage, the multipliers are W_N^{4i}, where $0 \leq i \leq \frac{N}{4} - 1$, for the first $N/4$ nodes; they repeat successively for the remaining nodes. Continue this procedure with repeating sets of multipliers of W_N^{ki}, $0 \leq i \leq [N/k] - 1$, where $k = (2)^{L-j}$, for the jth stage, until multipliers have been identified for all lower paths.

Step 6   Write spectrum-sample identifiers for each output node. These start with $X(0)$ at the top node and continue in order to $X(N - 1)$ at the bottom node.

Step 7   Write input-sample identifiers, $x(n)$, for each input node. The particular value of n for each node is obtained with the bit-reversal rule.

We can obtain a signal-flow graph with a minimum number of multipliers, such as the one shown in Figure 17.39, from the signal-flow graph constructed using the preceding steps. First, we multiply the output of each node that has succeeding paths with multipliers before splitting into the succeeding paths.

The multiplier used is that on the upper succeeding path. We then eliminate the multipliers on the succeeding paths, add the upper succeeding path at the succeeding node, and subtract the lower succeeding path at the succeeding node. Actually, this signal-flow graph can be constructed directly by modifying the preceding steps.

In our introductory treatment, we emphasized the decimation-in-time FFT algorithm; however, as we mentioned, there are several other FFT algorithms.[†] One of them is called the *decimation-in-frequency* algorithm. Its derivation is similar to that of the decimation-in-time algorithm, except that in this case the spectrum samples are successively split into even- and odd-numbered subsets. It has naturally ordered input samples and bit-reversed spectrum samples and is an in-place algorithm.

Algorithms have been constructed that have both the input samples and the output-spectrum samples in order. They are not in-place algorithms, however, and thus require twice as much storage.

Still another algorithm is neither in-place nor naturally ordered at the input. However, because the structure of each stage is identical, software generation is simplified. The algorithm also permits sequential data access; hence, random data access is not required.

Inverse FFT

The equations for the DFT and IDFT are

$$X(m) = \sum_{n=0}^{N-1} x(n)W_N^{mn} = \sum_{n=0}^{N-1} x(n)e^{-j(2\pi mn/N)} \qquad 0 \le m \le N-1 \quad (17.93)$$

and

$$x(n) = \frac{1}{N}\sum_{n=0}^{N-1} X(m)W_N^{-mn} = \frac{1}{N}\sum_{n=0}^{N-1} X(m)e^{j(2\pi mn/N)} \qquad 0 \le n \le N-1$$
$$(17.94)$$

We see from these equations that the FFT algorithm signal flow graph we developed for computing the DFT will work equally well for computing the IDFT. All we must do is replace all the multiplier values by their complex conjugates and multiply the output quantities by $1/N$. With these changes, the signal flow graph corresponds to the inverse FFT (IFFT) algorithm for computing the IDFT. Since the basic signal flow graph is unchanged, we need only one discrete-time-system structure to compute either the FFT or the IFFT.

[†]A. V. Oppenheim and R. W. Schafer, *Discrete-Time Signal Processing* (Englewood Cliffs, N.J.: Prentice-Hall, 1989), pp. 596–605.

We can construct an IFFT algorithm in another way as follows. Since

$$x(n) = \frac{1}{N} \sum_{n=0}^{N-1} X(m)e^{j(2\pi mn/N)} \qquad 0 \le n \le N-1 \qquad (17.95)$$

then

$$x*(n) = \frac{1}{N} \sum_{m=0}^{N-1} X*(m)e^{-j(2\pi mn/N)} \qquad 0 \le n \le N-1 \qquad (17.96)$$

because the complex conjugate of a sum is the sum of the complex conjugates and the complex conjugate of a product is the product of the complex conjugates. Therefore,

$$x*(n) = \frac{1}{N}\mathcal{D}[X*(m)] \qquad 0 \le n \le N-1 \qquad (17.97)$$

We obtain

$$x(n) = \left\{\frac{1}{N}\mathcal{D}[X*(m)]\right\}^* \qquad 0 \le n \le N-1 \qquad (17.98)$$

by taking the complex conjugate of both sides of eq. (17.97). Equation (17.98) shows that we can compute the IDFT by (1) forming the complex conjugate of the spectrum samples, $X*(m)$; (2) computing the DFT of $X*(m)$; (3) multiplying the DFT outputs by $1/N$; and (4) forming the complex conjugate of the resulting sequence values. When the FFT is used for the DFT computation, the steps indicated correspond to the IFFT computation of the IDFT.

Example FFT Computation

To demonstrate the computations involved in using the FFT algorithm, we will now compute the DFT of samples of a single-rectangular-pulse signal.

Example 17.7

Consider the single-rectangular-pulse continuous-time signal $x(t) = \Pi(t)$. Take eight samples of $x(t)$ at a rate of five samples per second, with the first sample at $t = -0.6$. No signal truncation occurs when these samples are used. Draw the signal-flow graph for an 8-point FFT, labeling all inputs, outputs, and multipliers. Use this signal-flow graph to compute the spectrum samples $X(m)$ for the sampled signal.

Solution

The signal samples taken are shown in Figure 17.41a. The first $N = 8$ samples of the periodic extension of these signal samples are shown in Figure 17.41b. We use these periodic-extension samples as input to the FFT when we compute the DFT. Multipliers that are required for the FFT computations are powers of $W_8 = e^{-j2\pi/8} = e^{-j\pi/4}$. They are

$$W_8^0 = 1 \qquad W_8^1 = \frac{1}{\sqrt{2}} - j\frac{1}{\sqrt{2}}$$

$$W_8^2 = -j \qquad W_8^3 = -\frac{1}{\sqrt{2}} - j\frac{1}{\sqrt{2}}$$

$$W_8^4 = -1 \qquad W_8^5 = -\frac{1}{\sqrt{2}} + j\frac{1}{\sqrt{2}}$$

$$W_8^6 = j \qquad W_8^7 = \frac{1}{\sqrt{2}} + j\frac{1}{\sqrt{2}}$$

Using the construction steps previously defined and labeling all inputs, outputs, and multipliers, we obtain the signal-flow graph shown in Figure 17.42. Note that this is the same signal-flow graph as the one in Figure 17.38, except that in this case numerical values are shown for the multipliers. We can most easily evaluate the DFT using this signal-flow graph if we construct the values at each node in tabular form, as illustrated by Table 17.1. We label each vertical set of nodes of the signal-flow graph in Figure 17.42 and use these labels to identify the columns of sample values in Table 17.1 that correspond to these vertical sets of nodes.

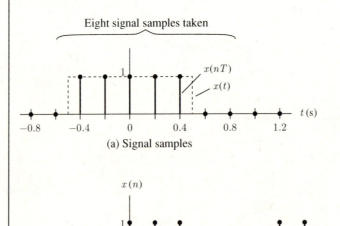

(a) Signal samples

(b) Periodic-extension samples used to compute DFT

Figure 17.41 Signal Samples Taken and Used to Compute the DFT in Example 17.7

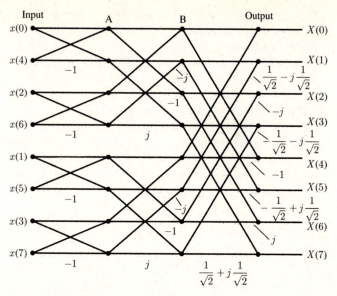

Figure 17.42 Signal-Flow-Graph Representation of the 8-Point FFT Algorithm Used in Example 17.7

Table 17.1 Sample Values at the Nodes of the FFT Signal-Flow Graph of Figure 17.42 for Example 17.7.

Input	**A**	**B**	*Output*	
1	$1 + 0 = 1$	$1 + 2 = 3$	$3 + 2$	$= 5$
0	$1 - 0 = 1$	$1 - j(0) = 1$	$1 + \left(\sqrt{2}\underline{/45°}\right)\left(1\underline{/-45°}\right)$	$= 2.414$
1	$1 + 1 = 2$	$1 - 2 = -1$	$-1 - j(0)$	$= -1$
1	$1 - 1 = 0$	$1 + j(0) = 1$	$1 + \left(\sqrt{2}\underline{/-45°}\right)\left(1\underline{/-135°}\right)$	$= -0.414$
1	$1 + 0 = 1$	$1 + 1 = 2$	$3 - 2$	$= 1$
0	$1 - 0 = 1$	$1 - j(-1) = 1 + j$	$1 + \left(\sqrt{2}\underline{/45°}\right)\left(1\underline{/135°}\right)$	$= -0.414$
0	$0 + 1 = 1$	$1 - 1 = 0$	$-1 + j(0)$	$= -1$
1	$0 - 1 = -1$	$1 + j(-1) = 1 - j$	$1 + \left(\sqrt{2}\underline{/-45°}\right)\left(1\underline{/45°}\right)$	$= 2.414$

Using the output-node correspondences, we see that the samples of the spectrum of the sampled signal are

$$X(0) = 5 \qquad X(1) = 2.414 \qquad X(2) = -1 \qquad X(3) = -0.414$$

$$X(4) = 1 \qquad X(5) = -0.414 \qquad X(6) = -1 \qquad \text{and} \qquad X(7) = 2.414$$

We can also use the FFT signal-flow graph of Figure 17.39 to indicate the required FFT computations for Example 17.7 if we substitute numerical values for the multipliers. If we label the vertical sets of nodes from the left as Input, A, B, C, D, and Output, then we can construct the table of values at each node, and thus compute the spectrum samples, as shown in Table 17.2, where $\alpha = 1\underline{/-45^{\circ}}$ and $\beta = 1\underline{/-135^{\circ}}$.

In the Problem section, the reader is asked to perform manual computations of FFTs, such as those in Table 17.2 and Table 17.1, so as to gain familiarity with the concept and computation procedure. In actual practice, we compute the FFT with discrete-time-system software or hardware. A FORTRAN subroutine that can be used to perform either the FFT or IFFT computations is shown in Appendix D.

Computation-Time Comparison

In this section, we will compare the speed of DFT computation when using direct evaluation of the DFT to the computation speed when using the FFT algorithm. For the FFT evaluation, we will use the decimation-in-time FFT algorithm derived and configured with the minimum number of multiplies.

Since complex coefficients are used and complex values result, we must compute DFTs with complex arithmetic. Comparisons are based on the number of real additions and real multiplications required because these are the operations actually performed. Recall that each complex addition requires two real additions, since $(a + jb) + (c + jd) = (a + c) + j(b + d)$; also, each complex multiplication requires four real multiplications and two real additions, since $(a + jb)(c + jd) = (ac - bd) + j(ad + bc)$. We assume that a real addition requires t_a seconds, that a real multiplication requires $10t_a$ seconds, and that the amount of time required to perform other operations is small with respect to the amount of time required for the additions and multiplications. These are reasonable assumptions for typical discrete-time systems.

Table 17.2 Sample Values for Example 17.7 Corresponding to the FFT Signal-Flow Graph of Figure 17.39.

Input	A	B	C	D	Output
1	$1 + 0 = 1$	1	$1 + 2 = 3$	3	$3 + 2 = 5$
0	$1 - 0 = 1$	1	$1 + 0 = 1$	1	$1 + \sqrt{2} = 2.414$
1	$1 + 1 = 2$	2	$1 - 2 = -1$	-1	$-1 + 0 = -1$
1	$1 - 1 = 0$	$(-j)(0) = 0$	$1 - 0 = 1$	1	$1 - \sqrt{2} = -0.414$
1	$1 + 0 = 1$	1	$1 + 1 = 2$	2	$3 - 2 = 1$
0	$1 - 0 = 1$	1	$1 + j$	$\alpha(1 + j) = \sqrt{2}$	$1 - \sqrt{2} = -0.414$
0	$0 + 1 = 1$	1	$1 - 1 = 0$	$(-1)(0) = 0$	$-1 - 0 = -1$
1	$0 - 1 = -1$	$(-j)(-1) = j$	$1 - j$	$\beta(1 - j) = -\sqrt{2}$	$1 - (-\sqrt{2}) = 2.414$

In eq. (17.21) we defined the direct evaluation of the DFT. We repeat it here as Eq. (17.99) for the reader's convenience:

$$X(m) = \sum_{n=0}^{N-1} x(n) W_N^{mn} \qquad 0 \leq m \leq N-1 \tag{17.99}$$

Since $W_N^0 = 1$ and $W_N^{mN/2} = e^{-j2\pi mN/2N} = e^{-jm\pi} = \pm 1$, then $x(0)W_N^0$ and $x(N/2)W_N^{mN/2}$ do not require multiplications. Therefore, $N-2$ complex multiplications and $N-1$ complex additions are required to compute each spectrum sample. Since there are N spectrum samples computed, then

$$2N(N-2) + 2N(N-1) = 4N^2 - 6N \tag{17.100}$$

real additions and

$$4N(N-2) = 4N^2 - 8N \tag{17.101}$$

real multiplications are required to perform a direct computation of a discrete Fourier transform.

The decimation-in-time FFT algorithm configured with the minimum number of complex multiplications, as shown in Figure 17.39 for $N = 8$, contains N complex additions and $N/2$ complex multiplications in each of its $L = \log_2 N$ stages. However, there are $N/2^i$ multiplication coefficients that are W_N^0 in the ith stage, where $1 \leq i \leq L$. These are not multiplications, and thus must be subtracted from the total number of complex multiplications. Therefore,

$$NL = N \log_2 N \tag{17.102}$$

complex additions and

$$\frac{N}{2}L - \sum_{i=1}^{L} \frac{N}{2^i} = \frac{N}{2} \log_2 N - N \left[\sum_{i=0}^{\log_2 N} \left(\frac{1}{2} \right)^i - 1 \right]$$

$$= \frac{N}{2} \log_2 N - N + 1 \tag{17.103}$$

complex multiplications are required to compute the discrete Fourier transform with the FFT. To compute these numbers of complex additions and multiplications requires

$$2 \left[\frac{N}{2} \log_2 N - N + 1 \right] + 2N \log_2 N = 3N \log_2 N - 2N + 2 \tag{17.104}$$

real additions and

$$4 \left[\frac{N}{2} \log_2 N - N + 1 \right] = 2N \log_2 N - 4N + 4 \tag{17.105}$$

real multiplications.

The computation time required to perform the multiplications and additions for the direct evaluation and FFT evaluation of the discrete Fourier transform are

$$t_D = 10t_a(4N^2 - 8N) + t_a(4N^2 - 6N)$$

$$= (44N^2 - 86N)t_a \qquad (17.106)$$

and

$$t_F = 10t_a(2N \log_2 N - 4N + 4) + t_a(3N \log_2 N - 2N + 2)$$

$$= (23N \log_2 N - 42N + 42)t_a \qquad (17.107)$$

respectively. The FFT evaluation is $r_s = t_D/t_F$ times faster than the direct evaluation. Using eqs. (17.106) and (17.107), we find that $r_s = 6.21, 46.48$, and 769.79 for $N = 4, 128$, and 4096, respectively. Thus, the FFT provides a computation-speed increase by a factor of nearly 800 when a 4096-point DFT is computed. For the very short 4-point transform, the speed increase is approximately a factor of 6, which is still significant. However, the speed factor is somewhat optimistic in this case because the number of multiplication and addition operations is small enough for either method; hence, other computation overhead (for example, data retrieval and storage and logical operations) must be considered.

17.9 Summary

The discrete Fourier transform (DFT) is a method for computing samples of the spectrum of a continuous-time signal or a discrete-time signal from samples of the signal. We can use the DFT to produce signal-spectrum information with a digital computer because only samples of signals and their spectra are used and computed. The inverse discrete Fourier transform (IDFT) is used to compute signal samples from spectrum samples.

The DFT has a number of important characteristics that must be understood if we are to properly interpret the spectrum samples we compute. The following are characteristics of an N-point DFT:

1. The signal samples used are the first N samples of the periodic extension of the set of N samples consisting of the signal samples plus any zero samples added.
2. The spectrum samples computed are the first N samples of the periodic spectrum of the sampled signal.
3. The spectrum-sample spacing, in hertz, is the inverse of the length, in seconds, of the discrete-time signal consisting of the signal samples plus any zero samples added.
4. Spectrum aliasing occurs for sampled continuous-time signals if signal-sample spacing exceeds the spacing specified by the sampling theorem.
5. The picket-fence effect occurs from the fact that only samples of a continuous-frequency spectrum are computed.

6. If the signal must be truncated to produce no more than N samples, then spectrum leakage (that is, truncation distortion of the computed spectrum samples) occurs.

The three basic parameters of the DFT are the signal-sample spacing T, the spectrum-sample spacing F, and the DFT length N. If there is to be no spectrum leakage and negligible spectrum aliasing, then N must be greater than twice the time-bandwidth product of the signal.

A number of useful DFT theorems parallel those previously discussed for Fourier transforms and discrete-time Fourier transforms. The circular-convolution theorem states that the IDFT of the product of the DFTs of two signals is a signal that we call the circular convolution of the two signals.

We can use the DFT to compute the discrete convolution of two discrete-time signals. In this application, each of the two signals must be padded with sufficient zeros so that the DFT we use will contain enough samples to encompass the discrete-convolution result.

In the selective-save method, the DFT is used to compute a discrete convolution of two signals. The method permits us to reduce the length of the DFT should one of the signals be significantly shorter than the other signal. The method is particularly useful when one of the signals is of indefinite length; for example, in the implementation of an FIR filter.

The fast Fourier transform (FFT) is an algorithm used to compute the DFT. It makes use of the symmetry in the operations and coefficients of the DFT to effectively reduce the DFT computation time. The FFT algorithm provides speed-increase factors, when compared with direct computation of the DFT, of approximately 46 and 770 for 128-point and 4096-point transforms, respectively.

Problems

In the following problems, time factors are in seconds unless otherwise indicated.

17.1 For the continuous-time rectangular-pulse signal

$$x(t) = 2\Pi\left(\frac{t-3}{5}\right)$$

sample the signal using a sample spacing of 2 s, with the first sample at $t = 0$.

a. Use the DFT defining equation with $N = 4$ to compute samples, $X(m)$, of the spectrum of the sampled signal.

b. Plot the amplitude and phase of T times the spectrum samples computed and the amplitude and phase of the spectrum of $x(t)$. Plot both amplitudes on the same plot and both phases on the same plot over the frequency range $|f| \le f_s/2$ and comment on the results.

17.2 Repeat Problem 17.1 using a DFT with $N = 8$.

17.3 Repeat Problem 17.1 using a signal sampling rate of 1 Hz and a DFT with $N = 8$.

17.4 The discrete-time signal shown in Figure 17.43 is received over a digital data link that transmits digitized machine-tool control commands in a factory. It causes a stepping motor to reposition a milling head. We want to determine samples of the signal spectrum to assess the characteristics of the rate of position change commanded.

a. Use the DFT defining equation with $N = 4$

to compute samples, $X(m)$, of the spectrum of $x_T(n)$.

b. Plot the amplitude and phase of the spectrum samples computed and the amplitude and phase of the spectrum of $x_T(n)$. Plot both amplitudes on the same plot and both phases on the same plot over the frequency range $|f| \leq f_s/2$ and comment on the results.

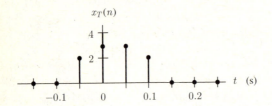

Figure 17.43

17.5 Repeat Problem 17.4 using $N = 8$.

17.6 Repeat Problem 17.4 using a value of N that results in a spectrum sample spacing of 2 Hz or less.

17.7 The continuous-time signal $x(t) = \pi e^{-\pi|t+0.25|}$ represents the liquid level in a temporary storage tank in a refinery.

a. Take eight samples of the signal at a rate of four samples per second with the first sample at $t = -0.75$ s and use the 8-point DFT defining equation to solve for the samples, $X(m)$, of the spectrum of the sampled signal.

b. Plot the amplitude and phase of T times the spectrum samples and the amplitude and phase of the spectrum of $x(t)$ for $|f| \leq f_s/2$. Draw both amplitude plots on one set of axes and both phase plots on one set of axes.

c. Repeat parts (a) and (b) with the first sample taken at $t = 0$. Comment on the difference in results.

17.8 Repeat Problem 17.7 using the signal $x(t) = 4e^{-0.6(t+0.75)}u(t+0.75)$, a sampling rate of two samples per second, and a first-sample location at $t = -1$ s.

17.9 An aircraft control-surface-position sensor produces a continuous-time signal that is sampled at a rate of 10 samples per second. The first sample is taken at $t = 0$ and the sequence of sample values obtained is $\{0.5, -0.2, -0.4, 0.1, 0.5, 1.0, 1.2, 1.0, 0.8\}$. Use the DFT to compute samples of the spectrum of the continuous-time signal. Plot the amplitude and phase of the spectrum samples for $|f| \leq f_s/2$. What range of frequency components dominate the control-surface-position variation?

17.10 Repeat Problem 17.9 with an upper limit on the spacing of the spectrum samples of 0.9 Hz.

17.11 Consider the spectrum samples $X(0) = 4$, $X(1) = 1 - j$, $X(2) = 2 + j$, $X(3) = 2 - j$, and $X(4) = 1 + j$, spaced in frequency by 10 Hz. Find the discrete-time-signal samples corresponding to these spectrum samples. Is the signal real or complex? Could you have answered the question without computing the IDFT? Explain your answers.

17.12 Repeat Problem 17.11 for the spectrum samples $X(0) = 2$, $X(1) = 1.5 / -30°$, $X(2) = 1 / -10°$, $X(3) = 1 / 20°$, and $X(4) = 1.5 / 30°$.

17.13 The signal $x(t) = 4\cos(16\pi t) + 3\cos(66\pi t)$ is sampled with ten samples taken at a rate of 100 samples per second starting at $t = 0$. Use the DFT to compute the samples of the spectrum of the resulting truncated sampled signal. Plot the amplitude of these spectrum samples for $|f| \leq f_s/2$ and comment on the result.

17.14 The signal $x(t) = 2\cos(40\pi t) + 2\cos(70\pi t)$ is sampled with eight samples taken at a rate of 160 samples per second starting at $t = 0$. Use the DFT to compute the samples of the spectrum of the resulting truncated sampled signal. Plot the amplitude of these spectrum samples for $|f| \leq f_s/2$ and comment on the result.

17.15 Consider the continuous-time signal $x(t) = e^{-\pi|t|}$. Sample the signal at a rate of five samples per second with the first sample at $t = t_1$.

a. Find $X(m)$ when $t_1 = -0.2$ and $N = 3$ and plot $|TX(m)|$ and $|X(f)|$ on one set of axes and $/ TX(m)$ and $/ X(f)$ on another set of axes for $|f| \leq f_s/2$.

b. Repeat part (a) with $t_1 = -0.4$ and $N = 5$.

c. Repeat part (a) with $t_1 = -0.6$ and $N = 7$.

d. Repeat part (a) with $t_1 = -0.8$ and $N = 9$.

Comment on the relative characteristics of the spectrum samples produced in each part of this problem.

17.16 Repeat Problem 17.15 using a Hamming window.

17.17 Repeat Problem 17.7 using a Bartlett window.

17.18 Samples of a continuous-time noise burst in a communications channel are shown in Figure 17.44. We want to determine the frequency components present in this noice burst so that we can design a filter to reduce the noise burst. An 8-point DFT is available to compute eight spectrum samples for the signal. (a) Indicate the eight signal samples that you will use to compute the spectrum samples and (b) sketch the sequence of values $x(n)$ that you will use to compute the DFT.

17.19 Show that $X(N - m) = X*(m)$ for the DFT of a real sequence.

17.20 Show that the DFT of a real and even sequence is real and even.

17.21 Show that the DFT of a real and odd sequence is imaginary and odd.

17.22 Consider the four signal samples shown in Figure 17.45. These samples have been taken from a longer signal. Spectrum samples for the signal are to be computed by direct computation of the DFT. A Bartlett window is to be used, and the spectrum samples are to be no further apart than 1 Hz. (a) Determine the order of the required DFT and (b) find and plot the sequence of values $x(n)$ to use in the DFT equation for the spectrum-sample computation.

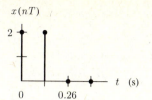

Figure 17.45

17.23 A continuous-time signal in a chemical-process control system has a significant bandwidth of 300 Hz.

a. Find the minimum sampling rate and DFT length required so as to obtain samples of the spectrum of this signal in which aliasing is negligible and samples are spaced no further than 5 Hz apart. The DFT length is to be a power of 2.

b. Determine the resulting spectrum-sample spacing and the required signal-portion length.

17.24 A continuous-time signal has a length of 128 ms and no significant content at frequencies above 400 Hz.

a. Find the minimum sampling frequency and DFT length required so that samples of the spectrum of this signal will contain negligible aliasing and be spaced no further apart than 7 Hz. The DFT length is to be a power of 2.

b. State whether or not it is necessary to truncate the signal. Explain your response.

17.25 In two sets of signal samples, $x(nT)$ and $y(nT)$, samples are all zero except for $x(0) = 1$, $x(T) = 2$, $x(2T) = 1$, $x(3T) = 2$, and $y(0) = 1$, $y(T) = 1$. The 4-point DFTs of these two sets of signal samples are $X(m) = [6, 0, -2, 0]$ and $Y(m) = [2, 1 - j, 0, 1 + j]$.

Figure 17.44

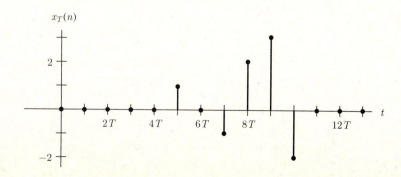

a. Use the DFT theorems to compute the spectrum samples $Z(m)$ that correspond to $z(n) = x(n) - 3y(n - 2)$, where $x(n)$ and $y(n)$ are the periodic extensions of the first four samples of each signal.

b. Use the IDFT to compute $z(n)$.

c. Is $z(n)$ equal to the signal samples $z(nT) = x(nT) - 3y(nT - 2T)$ for $0 \le n \le 4$?

d. Repeat parts (a), (b), and (c) with $y(nT)$ changed to $y(0) = 1, y(T) = 1, y(2T) = 1$, and $y(nT) = 0$ elsewhere. The DFT of this set of signal samples is $Y(m) = [3, -j, 1, j]$. Comment on any differences noted.

17.26 Prove the superposition theorem for DFTs.

17.27 Prove the frequency-shift theorem for DFTs.

17.28 Prove the duality theorem for DFTs.

17.29 A discrete-time system has the unit pulse response and input signal shown in Figure 17.46. (a) Use discrete convolution to find the output signal $y_T(n)$ and (b) use DFTs to find the output signal $y_T(n)$.

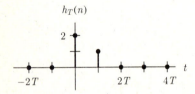

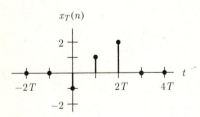

Figure 17.46

17.30 Repeat Problem 17.29 with the unit pulse response and input signal shown in Figure 17.47.

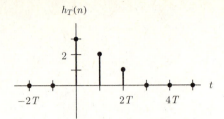

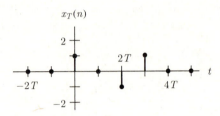

Figure 17.47

17.31 Consider the frequency-response samples, $H(m)$, of Figure 17.29 with $F = 10$. Use the IDFT to compute the first $N = 15$ samples of the periodic extension, $h(n)$, of the unit pulse response corresponding to these frequency-response samples. Since the spectrum samples are real, then $h(n)$ is real and even. Thus, $h_T(n) = h(n)$ for $|n| < N/2 = 7.5$ and $h_T(n) = 0$ elsewhere, where $h_T(n)$ is the system unit pulse response. Compute the discrete-time Fourier transform of $h_T(n)$ to find the frequency response of the system. Plot it for $|f| \le f_s/2 = 75$. How does it compare with the frequency response of the discrete-time ideal low-pass filter?

17.32 A discrete-time filter with the unit pulse response $h_T(n)$ is used to process the data signal $x_T(n)$ received from a radiation detector on a deep-space probe. The unit pulse response and data signal are shown in Figure 17.48. The filter is constructed with DFTs and the selective-save method, using the implementation of Figure 17.34. Compute eight of the output-signal samples starting at $t = 0$ by performing the operations indicated in Figure 17.34. What effect does the filter have on the signal?

17.33 Use the FFT to solve Problem 17.1 by drawing the 4-point FFT signal-flow graph and using it to manually compute the 4-point FFT required.

Figure 17.48

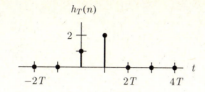

17.34 Use the FFT to solve Problem 17.3 by drawing the 8-point FFT signal-flow graph and using it to manually compute the required 8-point FFT.

17.35 Consider the noise-burst samples shown in Figure 17.44. An 8-point FFT is available to compute eight spectrum samples for the signal. (a) Indicate the eight samples to use in computing the spectrum samples and (b) indicate the input value at each input node of the FFT signal-flow graph.

17.36 Use the FFT to solve Problem 17.4 by drawing the 4-point FFT signal-flow graph and using it to manually compute the required 4-point FFT.

17.37 Use the FFT to solve Problem 17.7 by drawing the 8-point FFT signal-flow graph and using it to manually compute the required 8-point FFT.

17.38 Use the steps of Section 17.8 to construct the 16-point FFT signal-flow graph. Label all inputs and outputs and all multiplier values.

17.39 Using an FFT computer program, repeat Problem 17.1 for the following sampled rates and DFT lengths.

  a. $f_s = 2$ Hz $N = 16$

  b. $f_s = 2$ Hz $N = 32$

  c. $f_s = 4$ Hz $N = 32$

  d. $f_s = 4$ Hz $N = 64$

  e. $f_s = 8$ Hz $N = 64$

  f. $f_s = 8$ Hz $N = 128$

17.40 Using an FFT computer program, repeat parts (a) and (b) of Problem 17.7 for the following first-sample times, rates, and DFT lengths.

  a. $t_1 = -0.75$ s $fs = 4$ Hz $N = 4$

  b. $t_1 = -1.25$ s $fs = 4$ Hz $N = 8$

  c. $t_1 = -2.25$ s $fs = 4$ Hz $N = 16$

  d. $t_1 = -0.75$ s $fs = 8$ Hz $N = 8$

  e. $t_1 = -1.25$ s $fs = 8$ Hz $N = 16$

  f. $t_1 = -2.25$ s $fs = 8$ Hz $N = 32$

  g. $t_1 = -0.75$ s $fs = 16$ Hz $N = 16$

  h. $t_1 = -1.25$ s $fs = 16$ Hz $N = 32$

  i. $t_1 = -2.25$ s $fs = 16$ Hz $N = 64$

17.41 Repeat Problem 17.40 using a Hamming window.

17.42 Repeat Problem 17.40 using a Bartlett window.

A Matrix Properties and Operations

In the discussions of state-variable concepts for continuous-time and discrete-time linear systems in Chapters 9 and 16, respectively, we indicated that it is convenient to express state equations and output equations in matrix notation for analysis purposes. The properties of matrices and operations associated with matrices are briefly summarized in this appendix.

1. An $n \times m$ *matrix* is a rectangular array of elements arranged in n horizontal rows and m vertical columns. The notation used for the matrix $\mathbf{A}$ is

$$\mathbf{A} = \begin{bmatrix} a_{11} & a_{12} & \cdots & a_{1m} \\ a_{21} & a_{22} & \cdots & a_{2m} \\ \vdots & \vdots & & \vdots \\ a_{n1} & a_{n2} & \cdots & a_{nm} \end{bmatrix} \quad (A.1)$$

where a_{ij} indicates the element in the ith row and jth column. Elements can be either real or complex numbers or real or complex functions of a variable such as time.

2. A matrix with a single column (that is, $m = 1$) is referred to as a *column matrix* or a *column vector*. A matrix with a single row (that is, $n = 1$) is referred to as a *row matrix* or *row vector*. Example column and row vectors are

$$\mathbf{a} = \begin{bmatrix} a_{11} \\ a_{21} \\ \vdots \\ a_{n1} \end{bmatrix} \equiv \begin{bmatrix} a_1 \\ a_2 \\ \vdots \\ a_n \end{bmatrix} \quad (A.2)$$

and

$$\mathbf{b} = \begin{bmatrix} b_{11} & b_{12} & \cdots & b_{1m} \end{bmatrix}$$
$$\equiv \begin{bmatrix} b_1 & b_2 & \cdots & b_m \end{bmatrix} \quad (A.3)$$

3. Matrix $\mathbf{A}'$ formed by interchanging the rows and columns of the matrix $\mathbf{A}$ is called *the transpose of* $\mathbf{A}$. Note that the transpose of a column vector is a row vector and vice versa.

4. An $n \times n$ matrix, or *square matrix*, contains an equal number of rows and columns; that is, $m = n$.

5. The *main diagonal* of an $n \times n$ matrix consists of the elements a_{ii}, where $1 \le i \le n$.

6. A square matrix is called a *diagonal matrix* if its elements that are not on the main diagonal are all equal to zero; that is, if $a_{ij} = 0$ for $i \ne j$.

7. The *identity matrix* is the diagonal matrix with all main diagonal elements equal to unity. It is denoted by the symbol $\mathbf{I}$.

8. The *determinant* of a square matrix of size $n \times n$ is denoted as

$$|\mathbf{A}| = \begin{vmatrix} a_{11} & a_{12} & \cdots & a_{1n} \\ a_{21} & a_{22} & \cdots & a_{2n} \\ \vdots & \vdots & & \\ a_{n1} & a_{n2} & \cdots & a_{nn} \end{vmatrix} \quad (A.4)$$

9. The minor m_{ij} corresponding to the element a_{ij} in the square matrix $\mathbf{A}$ is the determinant of the matrix found by deleting the ith row and jth column from matrix $\mathbf{A}$. For example, m_{12} corresponding to the matrix $\mathbf{A}$ in eq. (A.4) is

$$m_{12} = \begin{vmatrix} a_{21} & a_{23} & \cdots & a_{2n} \\ a_{31} & a_{33} & \cdots & a_{3n} \\ \vdots & \vdots & & \\ a_{n1} & a_{n3} & \cdots & a_{nn} \end{vmatrix} \quad (A.5)$$

10. The *cofactor* c_{fij} corresponding to the element a_{ij} in the square matrix $\mathbf{A}$ is defined as $c_{fij} = (-1)^{i+j} m_{ij}$.

11. The determinant of the $n \times n$ matrix $\mathbf{A}$ is computed with

$$|\mathbf{A}| = \sum_{i=1}^{n} a_{ij} c_{fij} \quad (A.6)$$

or

$$|\mathbf{A}| = \sum_{j=1}^{m} a_{ij} c_{fij} \quad (A.7)$$

where j can be any value from to 1 to n in eq. (A.6) and i can be any value from 1 to n in eq. (A.7). As examples, the determinants of 2×2 and 3×3 matrices are

$$\begin{vmatrix} a_{11} & a_{12} \\ a_{21} & a_{22} \end{vmatrix} = a_{11}a_{22} + a_{12}(-1)a_{21}$$

$$= a_{11}a_{22} - a_{12}a_{21} \qquad (A.8)$$

and

$$\begin{vmatrix} a_{11} & a_{12} & a_{13} \\ a_{21} & a_{22} & a_{23} \\ a_{31} & a_{32} & a_{33} \end{vmatrix} = a_{11}\begin{vmatrix} a_{22} & a_{23} \\ a_{32} & a_{33} \end{vmatrix}$$

$$+ a_{21}(-1)\begin{vmatrix} a_{12} & a_{13} \\ a_{32} & a_{33} \end{vmatrix}$$

$$+ a_{31}\begin{vmatrix} a_{12} & a_{13} \\ a_{22} & a_{23} \end{vmatrix}$$

$$= a_{11}[a_{22}a_{33} + a_{32}(-1)a_{23}]$$

$$+ a_{21}(-1)$$

$$\times [a_{12}a_{33} + a_{32}(-1)a_{13}]$$

$$+ a_{31}[a_{12}a_{23} + a_{22}(-1)a_{13}]$$

$$= a_{11}a_{22}a_{33} - a_{11}a_{32}a_{23}$$

$$- a_{21}a_{12}a_{33} + a_{21}a_{32}a_{13}$$

$$+ a_{31}a_{12}a_{23}$$

$$- a_{31}a_{22}a_{13} \qquad (A.9)$$

where eq. (A.7) was used to compute eq. (A.8) and eq. (A.6) was used to compute eq. (A.9) to illustrate the use of both eqs. (A.6) and (A.7).

12. If the determinant of a square matrix equals zero, then the matrix is said to be *singular*.

13. The sum of two matrices (*matrix addition*) is denoted as

$$C = A + B \qquad (A.10)$$

and is defined such that the elements of **C** are the sum of the corresponding elements of **A** and **B**; that is, $c_{ij} = a_{ij} + b_{ij}$. Note that the sum can be performed only if the matrices **A** and **B** are of the same size. If the plus sign is replaced by a minus sign, then the two matrices are *subtracted* rather than added.

14. The multiplication of a matrix by a scalar constant produces a matrix of the same size, the elements of which are all multiplied by the scalar constant; that is, the elements of $B = kA$ are $b_{ij} = ka_{ij}$.

15. The product of two matrices (*matrix multiplication*) is denoted as

$$C = AB \qquad (A.11)$$

where the elements of **C** are defined as

$$c_{ij} = \sum_{\ell=1}^{k} a_{i\ell}b_{\ell j} \qquad (A.12)$$

where k is the number of columns in **A** and rows in **B**. Note that **A** and **B** can be multiplied only if the number of columns in **A** equals the number of rows in **B**. If **A** is $n \times k$ and **B** is $k \times m$, then **C** is $n \times m$. Generally, matrix multiplication is not commutative; that is

$$AB \neq BA \qquad (A.13)$$

in general. However, it is associative and distributive; thus,

$$A(BC) = (AB)C \quad \text{and} \quad A(B + C) = AB + AC \qquad (A.14)$$

Multiplication by the identity matrix is commutative and gives

$$AI = IA = A \qquad (A.15)$$

16. A matrix A^{-1} can be found for any nonsingular square matrix **A** such that

$$AA^{-1} = I = A^{-1}A \qquad (A.16)$$

This matrix is referred to as the inverse matrix corresponding to the matrix **A**. It can be computed by

$$A^{-1} = C'_f/|A|$$

where C_f is the matrix with elements that are the cofactors of matrix **A**.

17. Any linear operation performed on a matrix **A** produces a matrix, the elements of which are obtained by performing the linear operation on the elements of **A**. Some commonly used operations and the resulting matrix elements are

$$B = \frac{dA}{dt} : \qquad b_{ij} = \frac{da_{ij}}{dt} \qquad (A.17)$$

$$B = \int_{\alpha}^{\beta} A \, dt : \qquad b_{ij} = \int_{\alpha}^{\beta} a_{ij} \, dt \qquad (A.18)$$

$$B = \mathcal{L}[A] : \qquad b_{ij} = \mathcal{L}[a_{ij}] \qquad (A.19)$$

$$B = \mathcal{Z}[A] : \qquad b_{ij} = \mathcal{Z}[a_{ij}] \qquad (A.20)$$

B Mathematical Tables

Table B.1 Trigonometric Identities.

1. $e^{\pm jx} = \cos(x) \pm j\sin(x)$

2. $\left| e^{\pm jx} \right| = 1, \; \underline{/e^{\pm jx}} = \pm x$

3. $\cos(x) = \left(e^{jx} + e^{-jx} \right) \big/ 2$

4. $\sin(x) = \left(e^{jx} - e^{-jx} \right) \big/ 2j$

5. $\cos(x \pm y) = \cos(x)\cos(y) \mp \sin(x)\sin(y)$

6. $\sin(x \pm y) = \sin(x)\cos(y) \pm \cos(x)\sin(y)$

7. $\cos(x)\cos(y) = [\cos(x - y) + \cos(x + y)] \big/ 2$

8. $\sin(x)\sin(y) = [\cos(x - y) - \cos(x + y)] \big/ 2$

9. $\sin(x)\cos(y) = [\sin(x - y) + \sin(x + y)] \big/ 2$

10. $\cos^2(x) = [1 + \cos(2x)] \big/ 2$

11. $\sin^2(x) = [1 - \cos(2x)] \big/ 2$

12. $\cos^2(x) + \sin^2(x) = 1$

Table B.2 Derivatives.

1. $\dfrac{d}{dx}\left[x^n\right] = nx^{(n-1)}$

2. $\dfrac{d}{dx}[\ln(x)] = x^{-1}$

3. $\dfrac{d}{dx}\left[e^{ax}\right] = ae^{ax}$

4. $\dfrac{d}{dx}[\sin(ax)] = a\cos(ax)$

5. $\dfrac{d}{dx}[\cos(ax)] = -a\sin(ax)$

6. $\dfrac{d}{dx}[\tan(ax)] = a\sec^2(ax)$

7. $\dfrac{d}{dx}\left[\sin^{-1}(ax)\right] = a\left/\sqrt{1-(ax)^2}\right.$

8. $\dfrac{d}{dx}\left[\cos^{-1}(ax)\right] = -a\left/\sqrt{1-(ax)^2}\right.$

9. $\dfrac{d}{dx}\left[\tan^{-1}(ax)\right] = a\left/\left[1+(ax)^2\right]\right.$

10. $\dfrac{d}{dx}[f(x)g(x)] = f(x)\dfrac{d}{dx}[g(x)] + g(x)\dfrac{d}{dx}[f(x)]$

11. $\dfrac{d}{dx}\left[f(x)/g(x)\right] = \left\{g(x)\dfrac{d}{dx}[f(x)] - f(x)\dfrac{d}{dx}[g(x)]\right\}\left/g^2(x)\right.$

Table B.3 Indefinite Integrals.

1. $\displaystyle\int x^n\,dx = x^{n+1}\big/(n+1) \qquad n \neq -1$

2. $\displaystyle\int x^{-1}\,dx = \ln(x)$

3. $\displaystyle\int e^{ax}\,dx = e^{ax}\big/a \qquad a \neq 0$

4. $\displaystyle\int x^n e^{ax}\,dx = \left[x^n e^{ax} - n\int x^{(n-1)}e^{ax}\,dx\right]\Big/a$

5. $\displaystyle\int \sin(ax)\,dx = -[\cos(ax)]/a$

6. $\displaystyle\int \cos(ax)\,dx = [\sin(ax)]/a$

7. $\displaystyle\int \sin^2(ax)\,dx = [2ax - \sin(2ax)]/4a$

8. $\displaystyle\int \cos^2(ax)\,dx = [2ax + \sin(2ax)]/4a$

9. $\displaystyle\int x^n \sin(ax)\,dx = \left[-x^n\cos(ax) + n\int x^{n-1}\cos(ax)\,dx\right]\Big/a$

10. $\displaystyle\int x^n \cos(ax)\,dx = \left[x^n\sin(ax) - n\int x^{n-1}\sin(ax)\,dx\right]\Big/a$

11. $\displaystyle\int \cos(ax)\cos(bx)\,dx = \frac{\sin[(a-b)x]}{2(a-b)} + \frac{\sin[(a+b)x]}{2(a+b)} \qquad a^2 \neq b^2$

12. $\displaystyle\int \sin(ax)\sin(bx)\,dx = \frac{\sin[(a-b)x]}{2(a-b)} - \frac{\sin[(a+b)x]}{2(a+b)} \qquad a^2 \neq b^2$

13. $\displaystyle\int \sin(ax)\cos(bx)\,dx = -\frac{\cos[(a-b)x]}{2(a-b)} - \frac{\cos[(a+b)x]}{2(a+b)} \qquad a^2 \neq b^2$

14. $\displaystyle\int e^{ax}\sin(bx)\,dx = e^{ax}[a\sin(bx) - b\cos(bx)]\big/\!\left(a^2+b^2\right)$

15. $\displaystyle\int e^{ax}\cos(bx)\,dx = e^{ax}[a\cos(bx) + b\sin(bx)]\big/\!\left(a^2+b^2\right)$

16. $\displaystyle\int f(x)\left[\frac{dg(x)}{dx}\right]dx = f(x)g(x) - \int g(x)\left[\frac{df(x)}{dx}\right]dx$

Table B.4 Definite Integrals.

1. $\displaystyle\int_0^\infty \left[\frac{a}{a^2+x^2}\right]\,dx = \frac{\pi}{2} \qquad a > 0$

2. $\displaystyle\int_0^\infty x^n e^{-ax}\,dx = \frac{n!}{a^{n+1}} \qquad a > 0,\ n \text{ is a positive integer.}$

3. $\displaystyle\int_0^\infty e^{-a^2 x^2}\,dx = \sqrt{\pi}/2a \qquad a > 0$

4. $\displaystyle\int_0^\pi \sin^2(nx)\,dx = \int_0^\pi \cos^2(nx)\,dx = \pi/2 \qquad n \text{ is an integer.}$

5. $\displaystyle\int_0^{\pi/2} \sin^n(x)\,dx = \int_0^{\pi/2} \cos^n(x)\,dx = \begin{cases} \dfrac{(1)(3)\cdots(n-1)(\pi)}{(2)(4)\cdots(n)(2)} & \begin{array}{l} n \text{ is an even integer.} \\ n \neq 0 \end{array} \\[2ex] \dfrac{(2)(4)\cdots(n-1)}{(1)(3)\cdots(n)} & \begin{array}{l} n \text{ is an odd integer.} \\ n \neq 1 \end{array} \end{cases}$

6. $\displaystyle\int_0^\pi \sin(nx)\sin(mx)\,dx = \int_0^\pi \cos(nx)\cos(mx)\,dx = 0 \qquad n \text{ and } m \text{ are unequal integers.}$

7. $\displaystyle\int_0^\pi \sin(nx)\cos(mx)\,dx = \begin{cases} 2n/(n^2-m^2) & n-m \text{ is an odd integer.} \\ 0 & n-m \text{ is an even integer.} \end{cases}$

8. $\displaystyle\int_0^\infty \operatorname{sinc}(ax) = 1/2a \qquad a > 0$

9. $\displaystyle\int_0^\infty \operatorname{sinc}^2(ax) = 1/2a \qquad a > 0$

10. $\displaystyle\int_0^\infty e^{-ax}\cos(bx)\,dx = \frac{a}{a^2+b^2} \qquad a > 0$

11. $\displaystyle\int_0^\infty e^{-ax}\sin(bx)\,dx = \frac{b}{a^2+b^2} \qquad a > 0$

12. $\displaystyle\int_a^b f(x)\left[\frac{dg(x)}{dx}\right]\,dx = f(x)g(x)\Big|_a^b - \int_a^b g(x)\left[\frac{df(x)}{dx}\right]\,dx$

Table B.5 Series.

1. $f(x + a) = f(x) + a\dfrac{df(x)}{dx} + \dfrac{a^2}{2!}\dfrac{d^2 f(x)}{dx^2} + \dfrac{a^3}{3!}\dfrac{d^3 f(x)}{dx^3} + \cdots$

2. $(x + y)^n = x^n + nx^{n-1}y + \dfrac{n(n-1)}{2!}x^{n-2}y^2 + \dfrac{n(n-1)(n-2)}{3!}x^{n-3}y^3 + \cdots \qquad y^2 < x^2$

3. $e^x = 1 + x + \dfrac{x^2}{2!} + \dfrac{x^3}{3!} + \cdots$

4. $a^x = 1 + x\ln a + \dfrac{(x\ln a)^2}{2!} + \dfrac{(x\ln a)^3}{3!} + \cdots$

5. $\ln(x) = 2\left(\dfrac{x-1}{x+1}\right) + \dfrac{2}{3}\left(\dfrac{x-1}{x+1}\right)^3 + \dfrac{2}{5}\left(\dfrac{x-1}{x+1}\right)^5 + \cdots \qquad x > 0$

6. $\ln(1 + x) = x - \dfrac{x^2}{2} + \dfrac{x^3}{3} - \dfrac{x^4}{4} + \cdots \qquad |x| < 1$

7. $\sin x = x - \dfrac{x^3}{3!} + \dfrac{x^5}{5!} - \dfrac{x^7}{7!} + \cdots$

8. $\cos x = 1 - \dfrac{x^2}{2!} + \dfrac{x^4}{4!} - \dfrac{x^6}{6!} + \cdots$

9. $\dfrac{1}{1-x} = \displaystyle\sum_{n=0}^{\infty} x^n \qquad |x| < 1 \qquad$ (Geometric series)

10. $\dfrac{1-x^n}{1-x} = 1 + x + x^2 + \cdots + x^{n-1}$

C Transform Tables

Table C.1 Fourier Transform Theorems.

Name	Transform Pair
1. Linearity	$ax(t) + by(t) \leftrightarrow aX(f) + bY(f)$
2. Scale Change	$x(at) \leftrightarrow \dfrac{1}{\|a\|}X\left(f/a\right)$
3. Time Reversal	$x(-t) \leftrightarrow X(-f)$
4. Complex Conjugation	$x*(t) \leftrightarrow X*(-f)$
5. Duality	$X(t) \leftrightarrow x(-f)$
6. Time Shift	$X(t - t_0) \leftrightarrow X(f)e^{-j2\pi f t_0}$
7. Frequency Translation	$x(t)e^{j2\pi f_0 t} \leftrightarrow X(f - f_0)$
8. Modulation	$x(t)\cos 2\pi f_0 t \leftrightarrow \dfrac{1}{2}X(f - f_0) + \dfrac{1}{2}X(f + f_0)$
9. Time Differentiation	$\dfrac{d^n x(t)}{dt^n} \leftrightarrow (j2\pi f)^n X(f)$
10. Time Integration	$\displaystyle\int_{-\infty}^{t} x(\lambda)\,d\lambda \leftrightarrow (j2\pi f)^{-1}X(f) + \dfrac{1}{2}X(0)\delta(f)$
11. Convolution	$x(t) * y(t) \leftrightarrow X(f)Y(f)$
12. Multiplication	$x(t)y(t) \leftrightarrow X(f) * Y(f)$

Table C.2 Fourier Transform Pairs.

	$x(t)$	$X(f)$
1.	$\Pi\left(t/\tau\right)$	$\tau\,\mathrm{sinc}(\tau f)$
2.	$\Lambda\left(t/\tau\right)$	$\tau\,\mathrm{sinc}^2(\tau f)$
3.	$\mathrm{sinc}(at)$	$\dfrac{1}{|a|}\Pi\left(f/a\right)$
4.	$e^{-at}u(t) \qquad a>0$	$\dfrac{1}{a+j2\pi f}$
5.	$te^{-at}u(t) \qquad a>0$	$\dfrac{1}{(a+j2\pi f)^2}$
6.	$e^{-a|t|} \qquad a>0$	$\dfrac{2a}{a^2+(2\pi f)^2}$
7.	$e^{-a^2t^2}$	$\dfrac{\sqrt{\pi}}{a}e^{-(\pi f/a)^2}$
8.	1	$\delta(f)$
9.	$\delta(t)$	1
10.	$\mathrm{sgn}(t)$	$1/j\pi f$
11.	$u(t)$	$\dfrac{1}{2}\delta(f)+\dfrac{1}{j2\pi f}$
12.	$\cos\left(2\pi f_0 t+\theta\right)$	$\dfrac{1}{2}e^{j\theta}\delta\left(f-f_0\right)+\dfrac{1}{2}e^{-j\theta}\delta\left(f+f_0\right)$
13.	$\displaystyle\sum_{n=-\infty}^{\infty}\delta\left(t-nT_s\right)$	$\displaystyle f_s\sum_{m=-\infty}^{\infty}\delta\left(f-mf_s\right) \qquad f_s=\dfrac{1}{T_s}$

Table C.3 Laplace Transform Theorems.

Name	Transform Pair
1. Linearity	$ax(t) + by(t) \leftrightarrow aX(s) + bY(s)$
2. Scale Change	$x(at) \leftrightarrow \dfrac{1}{a} X\left(\dfrac{s}{a}\right) \qquad a > 0$
3. Time Delay	$x(t - t_0) \leftrightarrow X(s)e^{-st_0} \qquad t_0 > 0$
4. s-Shift	$e^{-at}x(t) \leftrightarrow X(s + a)$
5. Multiplication by t^n	$t^n x(t) \leftrightarrow (-1)^n \dfrac{d^n X(s)}{ds^n} \qquad n = 1, 2, \ldots$
6. Time Differentiation	$\dfrac{d^n x(t)}{dt^n} \leftrightarrow s^n X(s) - \displaystyle\sum_{i=0}^{n-1} s^{n-1-i} x^{(i)}(0^-)$
	where $\quad x^{(i)}(t) = \dfrac{d^i x(t)}{dt^i}$
7. Time Integration	$y(t) = \displaystyle\int_{0^-}^{t} x(\lambda)\, d\lambda + y(0^-) \leftrightarrow \dfrac{X(s)}{s} + \dfrac{y(0^-)}{s}$
8. Convolution	$x(t) * y(t) \leftrightarrow X(s)Y(s)$

In Theorems 9 and 10 equal expressions are used, rather than transform pairs. Also, these two theorems are valid only if the conditions stated in Chapter 7 are satisfied.

9. Final Value	$\displaystyle\lim_{t\to\infty} x(t) = \lim_{s\to 0} sX(s)$
10. Initial Value	$\displaystyle\lim_{t\to 0^+} x(t) = \lim_{s\to\infty} sX(s)$

Table C.4 Laplace Transform Pairs.

$x(t)$	$X(s)$
1. $\delta(t)$	1
2. $u(t)$	$\dfrac{1}{s}$
3. $\dfrac{t^n}{n!}u(t)$	$\dfrac{1}{s^{n+1}}$
4. $e^{-at}u(t)$	$\dfrac{1}{s+a}$
5. $\dfrac{t^n e^{-at}}{n!}u(t)$	$\dfrac{1}{(s+a)^{n+1}}$
6. $\sin(\omega_0 t)u(t)$	$\dfrac{\omega_0}{s^2+\omega_0^2}$
7. $\cos(\omega_0 t)u(t)$	$\dfrac{s}{s^2+\omega_0^2}$
8. $t\sin(\omega_0 t)u(t)$	$\dfrac{2\omega_0 s}{\left(s^2+\omega_0^2\right)^2}$
9. $t\cos(\omega_0 t)u(t)$	$\dfrac{\omega_0}{\left(s^2+\omega_0^2\right)^2}$
10. $e^{-at}\sin(\omega_0 t)u(t)$	$\dfrac{\omega_0\left(s^2-\omega_0^2\right)}{(s+a)^2+\omega_0^2}$
11. $e^{-at}\cos(\omega_0 t)u(t)$	$\dfrac{s+a}{(s+a)^2+\omega_0^2}$

Table C.5 Discrete-Time Fourier Transform Theorems.

Name	Transform Pair
1. Linearity	$ax_T(n) + by_T(n) \leftrightarrow aX_d(f) + bY_d(f)$
2. Scale Change	$x_{T/a}(n) \leftrightarrow X_d(f/a)$
3. Time Reversal	$x_T(-n) \leftrightarrow X_d(-f)$
4. Complex Conjugation	$x_T^*(n) \leftrightarrow X_d^*(-f)$
5. Time Shift	$x_T(n - n_1) \leftrightarrow X_d(f)e^{-j2\pi n_1 fT}$
6. Frequency Translation	$x_T(n)e^{j2\pi n f_0 T} \leftrightarrow X_d(f - f_0)$
7. Modulation	$x_T(n)\cos(2\pi n f_0 T) \leftrightarrow \frac{1}{2}X_d(f - f_0) + \frac{1}{2}X_d(f + f_0)$
8. Time Differencing	$x_T(n) - x_T(n - 1) \leftrightarrow \left(1 - e^{-j2\pi fT}\right)X_d(f)$
9. Summation	$\displaystyle\sum_{i=-\infty}^{n} x_T(i) \leftrightarrow \frac{X_d(f)}{1 - e^{-j2\pi fT}} + \frac{f_s X_d(0)}{2} \sum_{m=-\infty}^{\infty} \delta(f - mf_s)$
10. Convolution	$x_T(n) * y_T(n) \leftrightarrow X_d(f)Y_d(f)$
11. Multiplication	$x_T(n)y_T(n) \leftrightarrow \dfrac{1}{f_s}\displaystyle\int_{-f_s/2}^{f_s/2} X_d(\alpha)Y_d(f - \alpha)\, d\alpha$

Table C.6 Discrete-Time Fourier Transform Pairs.

$x_T(n)$	$X_d(f)$
1. $\left[\left\{\begin{array}{ll} 1 & \|n\| \leq n_1 \\ 0 & \|n\| > n_1 \end{array}\right\}, \quad T\right]$	$\dfrac{\sin[(2n_1 + 1)\pi fT]}{\sin \pi fT}$
2. $\left[\left\{\begin{array}{ll} (1 - \|n/n_1\|) & \|n\| \leq n_1 \\ 0 & \|n\| > n_1 \end{array}\right\}, \quad T\right]$	$1 + \displaystyle\sum_{k=1}^{n_1-1} 2\left(1 - \left\|\dfrac{k}{n_1}\right\|\right)\cos 2\pi k fT$
3. $\left[a^n u(n), \quad T\right] \qquad \|a\| < 1$	$\dfrac{1}{1 - ae^{-j2\pi fT}}$
4. $\left[a^{\|n\|}, \quad T\right] \qquad \|a\| < 1$	$\dfrac{1 - a^2}{1 - 2a\cos 2\pi fT + a^2}$
5. $\left[e^{-an} u(n), \quad T\right] \qquad a > 0$	$\dfrac{1}{1 - \exp(-a - j2\pi fT)}$
6. $\left[e^{-a\|n\|}, \quad T\right] \qquad a > 0$	$\dfrac{1 - e^{-2a}}{1 - 2e^{-a}\cos 2\pi fT + e^{-2a}}$
7. $\delta_T(n)$	1
8. 1	$f_s \displaystyle\sum_{k=-\infty}^{\infty} \delta(f - kf_s)$
9. $\mathrm{sgn}_T(n) = \left[\left\{\begin{array}{ll} 1 & n \geq 0 \\ -1 & n < 0 \end{array}\right\}, \quad T\right]$	$\dfrac{2}{1 - e^{-j2\pi fT}}$
10. $u_T(n)$	$\dfrac{1}{1 - e^{-j2\pi fT}} + \dfrac{f_s}{2}\displaystyle\sum_{k=-\infty}^{\infty} \delta(f - kf_s)$
11. $\left[\cos(2\pi r_0 n + \theta), \quad T\right]$	$\dfrac{f_s}{2}\displaystyle\sum_{k=-\infty}^{\infty}\left[e^{j\theta}\delta(f - r_0 f_s - kf_s) + e^{-j\theta}\delta(f + r_0 f_s - kf_s)\right]$

Table C.7 z-Transform Theorems.

Name	Transform Pair
1. Linearity	$ax_T(n) + by_T(n) \leftrightarrow aX(z) + bY(z)$
2. Time Delay	$x_T(n - m) \leftrightarrow z^{-m}X(z) \qquad m \geq 0$
3. z-Scale	$a^n x_T(n) \leftrightarrow X(z/a)$
4. Multiplication by $(nT)^k \qquad k > 0$	$(nT)^k x_T(n) \leftrightarrow -zT\dfrac{d\left\{\mathcal{Z}\left[(nT)^{k-1}x_T(n)\right]\right\}}{dz}$
5. Differencing	$x_T(n) - x_T(n-1) \leftrightarrow \left(1 - z^{-1}\right)X(z)$
6. Summation	$\displaystyle\sum_{i=0}^{n} x_T(i) \leftrightarrow X(z)\Big/\left(1 - z^{-1}\right)$
7. Convolution	$x_T(n) * y_T(n) \leftrightarrow X(z)Y(z)$

In theorems 8 and 9 equal expressions are used, rather than transform pairs.

8. Initial Value	$x_T(0) = \lim\limits_{z \to \infty} X(z)$
9. Final Value	$\lim\limits_{n \to \infty} x_T(n) = \lim\limits_{z \to 1}\left(1 - z^{-1}\right)X(z)$
	(Singularities of $\left(1 - z^{-1}\right)X(z)$ must lie inside the unit circle.)

Table C.8 z-Transform Pairs.[†]

$x_T(n)$	$X(z)$
1. $\delta_T(n)$	1
2. $u_T(n)$	$\dfrac{1}{1 - z^{-1}}$
3. $a^n u_T(n)$	$\dfrac{1}{1 - az^{-1}}$
4. $nTu_T(n)$	$\dfrac{Tz^{-1}}{\left(1 - z^{-1}\right)^2}$
5. $(nT)^2 u_T(n)$	$\dfrac{Tz^{-1}\left(1 + z^{-1}\right)}{\left(1 - z^{-1}\right)^3}$
6. $na^{(n-1)} u_T(n)$	$\dfrac{z^{-1}}{\left(1 - az^{-1}\right)^2}$
7. $n^2 a^{(n-1)} u_T(n)$	$\dfrac{z^{-1}\left(1 + az^{-1}\right)}{\left(1 - az^{-1}\right)^3}$
8. $\cos(bn)u_T(n)$	$\dfrac{1 - z^{-1}\cos(b)}{1 - 2z^{-1}\cos(b) + z^{-2}}$
9. $\sin(bn)u_T(n)$	$\dfrac{z^{-1}\sin(b)}{1 - 2z^{-1}\cos(b) + z^{-2}}$
10. $a^n \cos(bn)u_T(n)$	$\dfrac{1 - az^{-1}\cos(b)}{1 - 2az^{-1}\cos(b) + a^2 z^{-2}}$
11. $a^n \sin(bn)u_T(n)$	$\dfrac{az^{-1}\sin(b)}{1 - 2az^{-1}\cos(b) + a^2 z^{-2}}$

[†] Replacing a by $e^{-\alpha T}$ produces a number of additional useful z-transform pairs.

Table C.9 Discrete Fourier Transform (DFT) Theorems.

Name	Transform Pair
1. Linearity	$ax(n) + by(n) \leftrightarrow aX(m) + bY(m)$
2. Time Shift	$x(n-k) \leftrightarrow W_N^{km} X(m)$
3. Frequency Shift	$x(n)e^{j2\pi nk/N} \leftrightarrow X(m-k)$
4. Duality	$N^{-1}X(n) \leftrightarrow x(-m)$
5. Circular Convolution[†]	$x(n)\;\textcircled{N}\;y(n) \leftrightarrow X(m)Y(m)$
6. Multiplication	$x(n)y(n) \leftrightarrow N^{-1}X(m)\;\textcircled{N}\;Y(m)$

$$^{†}\; x(n)\,\textcircled{N}\,y(n) = \sum_{k=0}^{N-1} x(k)y(n-k)$$

D Computer Program for FFT and IFFT Computation

A FORTRAN computer subroutine that implements the fast Fourier transform and inverse fast Fourier transform presented in Chapter 17 is shown in this appendix. The program is an adaptation of the subroutine originally given by J. W. Cooley, and others.[†] It is not optimized to produce the greatest possible computation time efficiency. However, it does provide rapid computation of the discrete Fourier transform and inverse discrete Fourier transform. The DO loop that ends on statement number 30 performs the needed sorting of the input data samples into bit-reversed order. The DO loop that ends on statement number 60 implements the FFT in the form that is characterized by the signal-flow graph shown in Figure 17.39 on page 700.

```
C       SUBROUTINE FFT.  A FORTRAN SUBROUTINE THAT COMPUTES THE
C       DFT OR IDFT OF N = 2**L DATA POINTS, X(J), BY USING AN
C       N-POINT FFT.  A DFT IS PRODUCED IF INV = F.  AN IDFT IS
C       PRODUCED IF INV = T.  XR(J) AND XI(J) MUST CONTAIN THE
C       REAL AND IMAGINARY COMPONENTS OF THE INPUT DATA AT THE
C       BEGINNING OF EXECUTION.  THEY CONTAIN THE REAL AND
C       IMAGINARY COMPONENTS OF THE TRANSFORM OR INVERSE
C       TRANSFORM AT EXECUTION TERMINATION.
C
        SUBROUTINE FFT(XR, XI, L, INV)
C
        DIMENSION XR(512), XI(512)
        LOGICAL INV
        N = 2**L
        N2 = N/2
        N1 = N - 1
        PI = 3.14159265358979
        IF(INV) PI = -PI
        J = 1
        DO 30 I=1,N1
          IF(I.GE.J) GO TO 10
          TR = XR(J)
          TI = XI(J)
          XR(J) = XR(I)
          XI(J) = XI(I)
          XR(I) = TR
          XI(I) = TI
10        K = N2
20        IF(K.GE.J) GO TO 30
          J = J - K
```

[†]Cooley, J. W., Lewis, P. A. W. and Welch, P. D. "The Fast Fourier Transform and Its Applications," *IEEE Trans. on Education*, Vol 12, March, 1969.

```
              K = K/2
              GO TO 20
30            J = J + K
           DO 60 L1=1,L
              L2 = 2**L1
              L3 = L2/2
              UR = 1.0
              UI = 0.0
              AG = -PI/FLOAT(L3)
              WR = COS(AG)
              WI = SIN(AG)
              DO 50 J=1,L3
                 DO 40 I=J,N,L2
                    II = I + L3
                    TR = XR(II)*UR - XI(II)*UI
                    TI = XR(II)*UI + XI(II)*UR
                    XR(II) = XR(I) - TR
                    XI(II) = XI(I) - TI
                    XR(I) = XR(I) + TR
40                  XI(I) = XI(I) + TI
                 UA = UR
                 UR = UR*WR - UI*WI
50               UI = UA*WI + UI*WR
60         CONTINUE
           IF(.NOT.INV) GO TO 80
           DO 70 I=1,N
              XR(I) = XR(I)/FLOAT(N)
70            XI(I) = XI(I)/FLOAT(N)
80         RETURN
           END
```

Index